HANDBOOK OF NANOSCALE OPTICS AND ELECTRONICS

HANDBOOK OF NANOSCALE OPTICS AND ELECTRONICS

Editor
Gary P. Wiederrecht

ELSEVIER

AMSTERDAM • BOSTON • HEIDELBERG • LONDON • NEW YORK • OXFORD
PARIS • SAN DIEGO • SAN FRANCISCO • SINGAPORE • SYDNEY • TOKYO

Elsevier B.V.
Radarweg 29, 1043 NX Amsterdam, the Netherlands

First edition 2010

British Library Cataloguing in Publication Data
A catalogue record for this book is available from the British Library

Library of Congress Control Number: 2009937757

ISBN: 978-0-12-375178-2

For information on all Elsevier publications visit our website at elsevierdirect.com

Printed and bound in Spain

09 10 11 12 10 9 8 7 6 5 4 3 2 1

CONTENTS

Preface

The driving forces for nanoscale optics and electronics come from both ends of the research spectrum, from applied to basic research. On the applied side, the trend towards smaller, faster, and more efficient optical and electronic devices necessitates the miniaturization of the optical and electronic components that make up these devices. Examples of such products can be found all around us in modern life: in our computers, in optical data storage media, in telecommunications devices, in sensors, in analytical healthcare and treatment tools, to name only a few. Even products not traditionally viewed as benefiting from nanoscale optics and electronics, such as paints and fabrics, now have been created with just such nanoscale structures to incorporate new capabilities ranging from opto-electronic detectors and even solar energy conversion. From the other end of the research spectrum, in basic science, extraordinary discoveries of the novel optical and electronic behavior of nanoscale materials has motivated researchers to parse materials to ever smaller sizes. One example of new phenomena is the confinement of electronic states in semiconductor materials to produce quantized electronic transitions rather than band structure behavior that typifies bulk semiconductors. Entirely new optical and electronic behavior is produced in such materials compared to their bulk counterparts, such as narrow linewidths for emission, multiexciton generation for solar energy conversion, etc. Furthermore, these discoveries and others underscore a maxim of nanoscale structures: that the new properties can be dramatically tuned with size and shape of the nanostructures. This leads to an enormous new landscape over which to manipulate optical and electronic properties, previously unavailable to engineers and scientists only a decade or two ago. These materials and applications have also spawned an equally strong effort to develop the microscopy and spectroscopy tools that allow the researcher to visualize and manipulate optical and electronic properties with nanoscale resolution. It is just these sort of characterization tools that will advance nanoscience, just as the telescope enabled Galileo to provide a new knowledge of the stars.

A definition of "nanooptics" and "nanoelectronics" is useful to define the scope of this volume. For the purposes of this work, "nanooptics" is defined as the optical phenomena that result at length scales below the diffraction limit of light. In dielectric materials, this can be approximated as equal to "$\lambda/2n$", where λ is the wavelength for light incident on a medium and n is the refractive index of the medium. For visible light, this length scale is just under 200 nm. Thus, we seek in this volume to discuss the most recent efforts to understand, control, and visualize the behavior of light and nanooptical materials with 100 nm spatial features or less. For "nanoelectronics," we refer to behavior in structures small enough to produce quantized effects in electronic behavior. This generally occurs for electronic nanostructures with at least one dimension smaller than 100 nm. Thus, both nanooptical and nanoelectronic materials produce decidedly different physical behavior than bulk materials at length scales on the order of 100 nm and below, albeit for completely different reasons.

The chapters herein describe many of the most cutting edge topics in these fields. The challenges are great at this length scale. In many cases, it is evident that nature will not cooperate with a business as usual approach. For example, entirely new architectures in electronic integrated circuitry will likely be needed soon, simply because silicon transistors are approaching sizes that will rely on the transport of a single electron. Furthermore, enormous engineering challenges are present at this length scale, as the high density of transistors create tremendous heat loads and the lithography requires ever more complex and costly fabrication tools. As transistors scale down to the 10 nm length scale, it is likely that new nanostructures will be needed that mitigate these concerns. Possible examples of the future of integrated circuits include the evolution of carbon nanotubes and graphene transistors, or even single molecule transistors. Similarly, in optics, the optical waveguides the makeup optical interconnects face fundamental limits of width because light will not propagate, or becomes evanescent, at length scales below the diffraction limit of fiber-based systems. Thus, entirely new platforms for propagating optical energy has been proposed, such as using propagating coupled light and electron waves in metals (known as surface plasmon polaritons, or SPPs). In these plasmonic materials, the diffraction limit as compared to conventional dielectrics is reduced to the nanometer length scale, thus allowing in principle the creation of "optical" circuitry far smaller than today's technology.

This book presents the work of true leaders in the fields of nanoscale optics and electronics who are tackling the challenges head-on. These researchers are unlocking the secrets of the nanoscale, with new phenomena that may be tomorrow's imaging or device technology. In nano-optics the chapters focus on recent advances towards (a) manipulating photons and energy flow at the nanoscale and (b) imaging and sensing at ultra-small length scales. These are needed to understand nano-optical phenomena and to realize optical and opto-electronic devices based on those phenomena. For (a), important contributions in experiment and theory are described herein that focus on plasmonics, particularly in the area of producing strongly coupled plasmonic nanostructures for the control of energy flow. These chapters include groundbreaking computational work by Schatz et al. and novel experimental structures by Wurtz et al. Nonlinear processes, so important as a means to manipulate the optical properties of nanomaterials, are discussed by Brevet. For (b) new developments for ultra-sensitive detection of molecules via coupling to nanostructures are described by Van Duyne et al. Furthermore, photosensitive materials are explored by Dhar et al. as a route to advanced optical data storage methods at ultrasmall length scales. These chapters present a cohesive and visionary picture of nanophotonics research.

At the nanoscale, optical and electronic properties are frequently modulated hand-in-hand. This book therefore also covers electronic and opto-electronic properties of nanostructured materials (Kafafi et al.). This further connects to the nanoelectronic thematic chapters that describe new means to propagate energy and information though nanoscale structures. These include electron transport in molecular systems (Lee and Reed), advanced inorganic nanoscale transistors (Li), and organic nanostructured electronic structures (Loo). Equally important, new routes towards information storage in nanoscale structures is presented by Ozatay et al.

This volume as a whole illustrates that nanoscience, best known for its goal of creating miniaturized devices, continues to evolve with a revolutionary impact on virtually all scientific and engineering disciplines. This impact is expected to only grow largely as a result of emerging abilities to design the structure and composition of materials on the nanometer scale in many types of environments with ever increasing precision and complexity, as well as to reliably image, manipulate, and spectroscopically probe materials over smaller length scales and with greater sensitivity. This book presents a snapshot of the amazing engineering and scientific discoveries of recent years, and provides a vision of the future for nanooptics and nanoelectronics.

CONTRIBUTORS

J M Bingham
Northwestern University, Evanston, IL, USA

P M Braganca
Hitachi Global Storage Technologies, San Jose Research Center, San Jose, CA, USA

P-F Brevet
Université Claude Bernard Lyon 1, Villeurbanne, France

K R Choudhury
University of Florida, Gainesville, FL, USA

K Curtis
InPhase Technologies, Longmont, CO, USA

L Dhar
InPhase Technologies, Longmont, CO, USA

S K Gray
Argonne National Laboratory, Argonne, IL, USA

T Hauet
Hitachi Global Storage Technologies, San Jose Research Center, San Jose, CA, USA

C-H Hwang
National Chiao Tung University, Hsinchu 300, Taiwan

Z Kafafi
National Science Foundation, Arlington, VA, USA

T-W Kim
Gwangju Institute of Science and Technology, Gwangju, Korea

K M Kosuda
Northwestern University, Evanston, IL, USA

T Lee
Gwangju Institute of Science and Technology, Gwangju, Korea

Y Li
National Chiao Tung University, Hsinchu 300, Taiwan

Y-L Loo
Princeton University, Princeton, NJ, USA

P G Mather
Everspin Technologies, Chandler, AZ, USA

J M McMahon
Northwestern University, Evanston, IL, USA

O Ozatay
San Jose Research Center, San Jose, CA, USA

R J Pollard
The Queen's University of Belfast, Belfast, Northern Ireland, UK

M A Reed
Yale University, New Haven, CT, USA

G C Schatz
Northwestern University, Evanston, IL, USA

F So
University of Florida, Gainesville, FL, USA

J Tarver
Princeton University, Princeton, NJ, USA

J-U Thiele
Seagate Recording Media Operations, Fremont, CA, USA

R P Van Duyne
Northwestern University, Evanston, IL, USA

G Wang
Gwangju Institute of Science and Technology, Gwangju, Korea

W Wang
University of Wyoming, Laramie, WY, USA

W L Wilson
InPhase Technologies, Longmont, CO, USA

G A Wurtz
University of North Florida, Jacksonville, FL, USA

K L Wustholz
Northwestern University, Evanston, IL, USA

J E Yoo
University of Texas at Austin, Austin, TX, USA

A V Zayats
The Queen's University of Belfast, Belfast, Northern Ireland, UK

EDITORIAL ADVISORY BOARD

1 Optics of Metallic Nanostructures

G A Wurtz, University of North Florida, Jacksonville, FL, USA

R J Pollard and A V Zayats, The Queen's University of Belfast, Belfast, Northern Ireland, UK

1.1 Introduction

Nanostructured materials are witnessing an expanding development and utilization in a variety of applications. They demonstrate a breadth of fundamental properties that can be tailored extensively to design new and adapted functionalities. In particular, nanostructured optical media are regarded as strong contenders for the development of the next generation of ultra-integrated optical and optoelectronic applications.

A class of metallo-dielectric nanostructures, considered especially suited for these nanophotonic applications, have optical properties determined by on the collective excitation of the free-electron density in a metal [1–5]. These electromagnetic modes, known as surface plasmons, offer unique electromagnetic field properties. They demonstrate strong spatial field confinement and enhancement, and prove extremely sensitive to the geometry of the supporting structure, thus allowing an extensive tuning of both their spectral properties and associated electromagnetic field distribution by appropriate structural design [6–9]. This allows triggering nonlinear optical effects at very low excitation powers and enables single-molecule sensitivity in optical spectroscopies [10–12]. Plasmonic nanostructures, often referred to as metamaterials in this context, are already used for negative-refractive-index engineering, superlensing, optical cloaking, and slow light applications [13–16]. The unique electromagnetic properties of plasmonic modes also represent an opportunity to develop dynamic functionalities based on metallo-dielectric materials enabling an active control over optical signals with electric, magnetic, or optical stimuli [10,17,18].

The flexibility in the design of plasmonic modes already emerges at the single nanoparticle level-metallic particle or dielectric void in a metal film. Simple geometrical modifications in the particle shape, size, and environment may alter the extinction of the supported plasmonic modes quite significantly with the possibility of generating multipolar resonances, associated with characteristic spatial field distributions, and tuning the resonant wavelengths from the UV, through the visible, to the infrared part of the spectrum [19,20]. An additional level of sophistication in the design of single-particle extinction comes with the possibility of creating hybrid plasmonic modes [21–26]. These modes are generated from a strong electromagnetic coupling between interacting nanoparticles. In such a geometry, the spectral and spatial overlap between plasmonic modes supported by two or more nanoparticles will coherently interfere to produce new mixed modes representing the new eigenmodes of the coupled system. The system of nanoparticles contributing to the hybrid structure can then no longer be described as a mere sum of individual nanoparticle resonances but must be viewed as a new singular entity. In fact, in strong-coupling situations the optical properties associated with the hybrid system are unique relative to those of the constituents providing a powerful means to extensively tailor the optical properties of plasmonic systems in order to design new and enhanced optical functionalities [21]. Based on this principle, geometries involving the mixing of smooth interface surface plasmons [27], localized modes [28–32], localized and delocalized plasmonic modes [33, 34], and even heterogeneous states resulting from mixed plasmons and excitons modes were created with impact on the spectral, spatial, and dynamical properties of the mixed eigenmodes [35–40].

1.1.1 Scope of the Article

This article focuses on the optical properties of two complementary families of metallic nanostructured systems supporting plasmonic excitations. These are surface plasmon polaritonic crystals and arrays of aligned metallic nanorods. These structures can be thought of as originating from two generic geometries: surface plasmonic crystals are made from a continuous metal film, nanostructured with a dielectric material while preserving the film's continuity. The complementary nanorod structure is created from an assembly of metal nanoparticles embedded in a dielectric host. As a result, the optical properties of plasmonic crystals stem from propagating SPP modes, while the eigenmodes supported by arrays of nanorods originate from localized surface plasmons. Both types of structures demonstrate unique optical properties governed by these fundamental resonances, which prove remarkable in the development of nano-photonic devices including active functionalities where optical signals are controlled by light or other external electric or magnetic field stimuli. A common advantage of these two types of systems rests on the possibility to finely tune their geometry in order to design their optical response with tailored dispersion and electromagnetic field distribution. In this context, this article describes design principles and applications of metallic nanostructures for achieving a desired and dynamically controllable photonic response. This optical response is analyzed in the far-field (Fraunhofer) region but originates from the near-field optical interactions between nanostructures.

The article is split into two sections discussing surface plasmon polaritonic crystals (Section 1.2) and plasmonic nanorod arrays (Section 1.3). In Section 1.2, after a brief summary on SPP properties on a smooth surface, plasmonic excitations in a periodic lattice with a period comparable to the wavelength of light are discussed. Effects related to both the lattice, such as band-gap effects and coupling to photons, and the lattice basis are elucidated. The polarization properties of plasmonic crystals, emanating from the interaction of light with the crystals' Bloch modes, are also described. Finally, this section presents several approaches to active (dynamically tunable) plasmonic crystals. They include the use of external optical, electric, and magnetic fields to control the transmission of light through plasmonic crystals. These examples illustrate the sensitivity of the crystals' eigenmodes to local changes in the dielectric environment through both linear and nonlinear optical effects.

Section 1.3 describes the optical properties of arrays of plasmonic nanorods. These are novel materials, not yet widely studied by the scientific community, but that demonstrate great potential for the development of nanophotonic devices for light manipulation at the subwavelength scales. These structures consist of aligned, closely spaced (~30 nm), high-aspect-ratio (>10) metallic nanorods standing on a transparent substrate. The fabrication procedure of these plasmonic metamaterials is described briefly. The focus of the section is on their optical properties, which are discussed in detail with regard to their geometry. Key applications in refractive index sensing, electrical control of polarization and optical extinction, as well as nonlinear effects of hybrid nanorod arrays will be presented.

1.2 Surface Plasmon Polaritonic Crystals

1.2.1 Optical Properties

1.2.1.1 Smooth film surface plasmon polaritons

SPPs are surface electromagnetic waves associated with the coherent oscillations of the free-electron density at the interface between a metal and a dielectric material. **Figure 1(a)** illustrates the simplest geometry for which SPP may exist. It consists of a smooth interface separating two semi-infinite regions made of a metal with a complex permittivity $\varepsilon_m = \varepsilon'_m + \varepsilon''_m$ and a dielectric with a real permittivity ε_d. Maxwell's equations show that this geometry supports transverse magnetic (TM) electromagnetic (EM) solutions of the form of **equation 1**:

In the dielectric:

$$\begin{cases} H_d(x,t) = (0, A, 0)e^{k_z^d z} e^{i(k_x x - \omega t)} \\ E_d(x,t) = -A\dfrac{c}{i\omega\varepsilon_d}(k_z^d, 0, ik_x)e^{k_z z} e^{i(k_x x - \omega t)} \end{cases} \tag{1}$$

In the metal:

$$\begin{cases} H_m(x,t) = (0, A, 0)e^{k_z^m z} e^{i(k_x x - \omega t)} \\ E_m(x,t) = -A\dfrac{c}{i\omega\varepsilon_m}(-k_z^m, 0, ik_x)e^{k_z^m z} e^{i(k_x x - \omega t)} \end{cases}$$

with $k_z^d = \sqrt{k_x^2 - \varepsilon_d(\omega/c)^2}$ and $k_z^m = \sqrt{k_x^2 - \varepsilon_m(\omega/c)^2}$ if their momentum k_x obeys **equation 2**:

$$k_x = k_{spp} = \frac{2\pi}{\lambda_0}\sqrt{\frac{\varepsilon_m(\omega)\varepsilon_d}{\varepsilon_m(\omega) + \varepsilon_d}} \tag{2}$$

where k_{SPP} represents the complex SPP wave vector, $\lambda_0 = 2\pi c/\omega$ is the wavelength of photons in vacuum, H and E are magnetic and electric fields, respectively, and A is a constant. For frequencies ω smaller than ω_p, the plasma frequency of the metal, **equation 1** represents waves propagating along the metal/dielectric interface and bound to this interface. The mode wavelength and propagation length are then determined as

$$\lambda_{spp} = \frac{2\pi}{\mathrm{Re}(k_{spp})} \text{ and } L_{spp} = \frac{2\pi}{2\mathrm{Im}(k_{spp})}$$

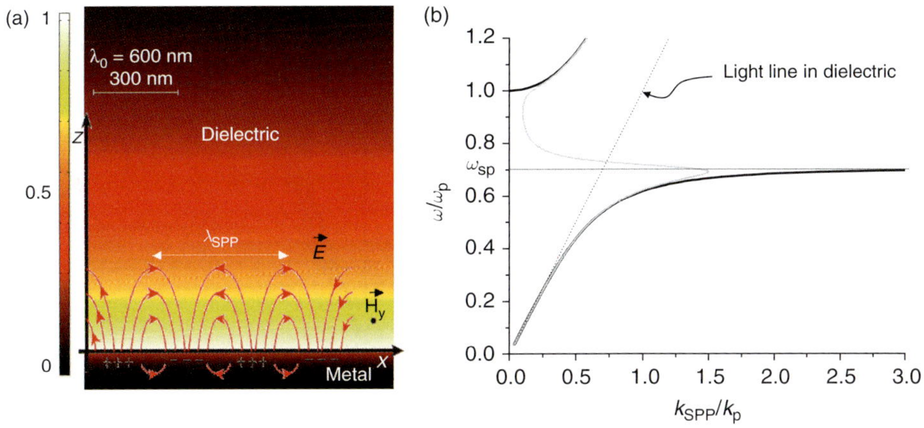

Figure 1 SPP dispersion on a smooth interface. (a) Electric-field amplitude distribution of the SPP mode at a free-space wavelength of 600 nm. The geometry is determined by a smooth Au semi-space in contact with a dielectric with $n_d = 1$ as a superstrate. A schematic of the isostrength field lines is shown as well. (b) SPP dispersion for the geometry shown in (a) for a Drude metal (gray curve) and a lossless Drude metal (black curve).

respectively. As the wave is bound to the metal/dielectric interface, the SPP wave amplitude is decaying exponentially in both the metal and the dielectric. Assuming $|\varepsilon'_{\mathrm{m}}| >> |\varepsilon''_{m}|$, the decay rates are expressed as

$$\delta_{\mathrm{d}} = k_0 \left| \frac{\varepsilon_{\mathrm{d}}^2}{\varepsilon_{\mathrm{d}} + \varepsilon'_{\mathrm{m}}} \right|^{1/2} \text{ and } \delta_{\mathrm{m}} = k_0 \left| \frac{\varepsilon'_{\mathrm{m}}}{\varepsilon_{\mathrm{d}} + \varepsilon'_{\mathrm{m}}} \right|^{1/2}$$

in the dielectric and the metal, respectively [41]. While δ_{d} is an increasing function of frequency, it can be shown that the decay rate δ_{m} in the metal does not vary significantly within the spectral range of the visible frequencies and is determined by the metal's skin depth. The dispersion relation $\omega = f(k)$ for SPPs is plotted in **Figure 1b** using **equation 2** for a Drude-like metal of plasma frequency ω_{p} in air. The dispersion $\omega = n_{\mathrm{d}} k_0 c$ corresponding to the maximum possible momentum for EM radiation in the dielectric of refractive index $n_{\mathrm{d}} = 1$, that is, the light line, is plotted alongside. For a lossless metal, the dispersion for surface modes is bounded by an asymptotical limit corresponding to the surface plasmon frequency $\omega_{\mathrm{SP}} = \omega_{\mathrm{p}}/\sqrt{1+\varepsilon_{\mathrm{d}}}$. When losses are included in ε_{m}, the SPP dispersion bends back into the radiation region. This situation is also illustrated in **Figure 1(a)**. The spatial distribution of the SPP field is plotted in **Figure 1(a)** for a frequency $\omega = 0.225\omega_{\mathrm{p}}$. The bound nature of the field at that frequency is evident from the cross-section taken from **Figure 1(a)**. This is shown in **Figure 2(a)** and reveals the exponentially decaying EM field of the mode in both the metal and the dielectric media.

As shown in **Figure 1(b)**, the momentum of SPP modes in air always exceeds the momentum of free radiation in this medium at the same frequency. For example, the SPP wavelength of the mode of **Figure 1(b)** is $\lambda_{\mathrm{spp}} = 566$ nm, while the free-space wavelength at this frequency is $\lambda_0 = 600$ nm. Consequently, photons in the dielectric do not simultaneously satisfy energy and momentum conservation for SPP excitation at the interface and SPP excitation with photons propagating in the adjacent dielectric is not possible in the geometry shown in **Figure 1(a)**.

The situation is different for a metal film of finite thickness in an asymmetric environment. A common geometry consists of a 50 nm Au film supported by a glass ($n_{\mathrm{d}} = 1.5$) substrate in air ($n_{\mathrm{d}} = 1$). In this geometry, both interfaces support SPP modes with different dispersion. These are plotted in **Figure 2(b)** for both the air/metal (black curve) and the glass/metal (gray curve), respectively. As observed in **Figure 2(b)**, the higher refractive index of the substrate shifts the SPP dispersion at the glass interface to lower frequencies compared to that at the air interface. Here again, momentum and energy conservation prevent bulk EM modes in the glass substrate to couple to SPP mode at the glass/metal interface. However, the finite thickness of the film enables an overlap between EM modes in the glass with certain SPP modes at the air/metal interface: these SPP states are situated between the two light lines in **Figure 2(b)**. This is clearly seen from the dispersion plot in **Figure 2(b)** where SPP modes at the air/metal interface with frequency ω

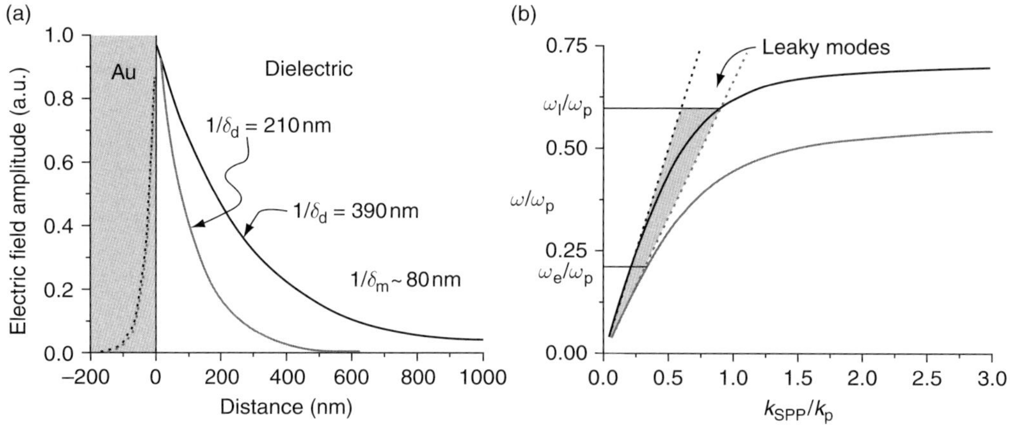

Figure 2 (a) SPP electric-field amplitude at the interface between a semi-infinite metal (Au) and a dielectric with $n_d = 1$ (black curve) and $n_d = 1.5$ (gray curve). (b) SPP dispersion for the geometry of **Figure 1(a)** with $n_d = 1$ (black curve) and $n_d = 1.5$ (gray curve). The light lines in both dielectrics are plotted in dotted lines. For a metal film of finite thickness in air ($n_d = 1$) and supported by glass ($n_d = 1.5$), SPP modes at the air interface present in the gray region comprised between the two light lines leak into the glass.

lower then ω_l can phase match with free radiation in the glass substrate. These SPP modes are referred to as leaky modes and their coupling efficiency to bulk EM radiation is conditioned by the thickness of the metal film. The optimal film thickness for Au and Ag films is on the order of 50 nm to achieve a compromise between tunnel-coupling strength and light re-radiation back into the dielectric [8,9,41].

One important consequence of increasing the index of refraction n_d of the dielectric adjacent to the metal is to decrease the penetration depth $1/\gamma_d$ of the SPP field in this dielectric at the frequency considered. An example is plotted in **Figure 2(a)** where the electric field distribution was plotted as a function of spatial coordinate along a direction perpendicular the metal/dielectric interface for the geometry defined in **Figure 1(a)**. The field distributions correspond to an SPP mode of frequency ω_e at either the air/metal (black curve) or the glass/metal (gray curve) interface. The respective penetration depths in the dielectric can be estimated from **Figure 2(a)** to decrease from 400 nm in air to 200 nm in glass, corresponding to a fourfold increase in the spatial confinement of the SPP field at the considered frequency.

This confinement is concomitant with a modification of the SPP momentum which can be written as $k_{spp} = n_{eff} k_O$, where

$$n_{eff} = \sqrt{\frac{\varepsilon_m(\omega)\varepsilon_d}{\varepsilon_m(\omega) + \varepsilon_d}}$$

takes a form equivalent to the effective index of refraction of the mode. The bound nature of the SPP implies $\mathrm{Re}(n_{eff}) > n_d$, that is, the norm of k_{spp} exceeds the wave number of the EM wave in the adjacent dielectric at the same frequency. In the example of **Figure 2(a)** corresponding to the geometry of **Figure 1**, the effective index of the SPP mode at ω_e is $n_{eff} = 1.06 + i0.0072$ at the air interface and $n_{eff} = 1.73 + i0.031$ at the glass interface. Both SPP modes are then true surface modes at the air/metal and glass/metal interfaces, respectively. However, in a geometry consisting of a metal film in contact with a glass substrate and an air superstrate, the SPP mode supported by the air/metal interface is leaking into the superstrate since $\mathrm{Re}(n_{eff}) = 1.06 < n_d = 1.5$. This corresponds to the situation of **Figure 2(b)**. The modulation of the dielectric's permittivity represents, therefore, an effective avenue to tailor SPP mode dispersion, including group velocity and field confinement, which are of paramount importance to design both the linear and nonlinear optical properties of surface plasmonic materials.

1.2.1.2 Surface plasmon polaritonic crystals

Surface plasmon polaritonic crystals (SPCs) are created from the periodic modulation of the permittivity of a metal/dielectric interface. This periodic modulation generates Bragg reflected SPP modes and folds the smooth interface SPP dispersion curve by integral number of the Bragg momentum $k_{Bragg} = \frac{2\pi}{d}$, where d is the modulation period or lattice period. In this approximation not taking into account band-gaps' formation the momentum for SPP Bloch modes (SPP-BMs) can be expressed as

$$\boldsymbol{k}_{\mathrm{SPP-BM}} = \boldsymbol{k}_{\mathrm{SPP}} + \boldsymbol{k}_{\mathrm{Bragg}} = \frac{2\pi}{\lambda_0}\sqrt{\frac{\varepsilon_{\mathrm{m}}(\omega)\varepsilon_{\mathrm{d}}}{\varepsilon_{\mathrm{m}}(\omega)+\varepsilon_{\mathrm{d}}}}\boldsymbol{u} + l\frac{2\pi}{d_x}\boldsymbol{i} + m\frac{2\pi}{d_y}\boldsymbol{j} \tag{3}$$

where $\boldsymbol{k}_{\mathbf{SPP}}$ and $\boldsymbol{k}_{\mathbf{SPP\text{-}BM}}$ are the SPP wave vectors before and after Bragg scattering in the periodic lattice, and $\boldsymbol{k}_{\mathbf{Bragg}}$ is the momentum contributed during the scattering by the crystal. The period of the crystal along the x- and y-directions is $d_{x,y}$, (l, m) are integer numbers, $\boldsymbol{u}$ is the unit vector describing the SPP propagation direction before Bragg scattering, and $\boldsymbol{i}$ and $\boldsymbol{j}$ are the unit reciprocal lattice vectors of the periodic structure.

Figure 3 shows an application of **equation 3** for a one-dimensional (1D) SPC. For the calculations, a crystal made of Au with a period of $d_x = 550$ nm in the x-direction and $d_y \rightarrow \infty$ in the y-direction is considered. Experimental values from have been used for ε_{m} [42]. For illustrative purposes, an example of 1D crystal fabricated using focused ion beam (FIB) milling in a 220 nm thick Au film is shown in **Figure 3(a)** along with its measured dispersion in **Figure 3(b)**. To reproduce the experimental dispersion, the calculations accounted for nondegenerated SPP modes at the metal/glass ($n_{\mathrm{d}} = 1.5$) and the metal/air superstrate interfaces. This geometry is typical for SPCs. The different SPP-BMs are identified in **Figure 3** by the values of $(l, m)_i$ corresponding to the scattering direction in reciprocal space and subscript i denoting the supporting dielectric interface, g for glass and a for air. The dark gray region in **Figure 3(c)** represents SPP-BMs at the glass/metal interface not accessible to photons.

When using equation 3 one should be aware of the following. First, coherent interactions between counter-propagating SPP modes are not taken into account. Those lead to SPP-BM mixing at the Brillouin zone boundaries resulting in the formation of local zeros in the density of SPP states, usually referred to as band gaps. Two such band gaps are identified in **Figure 3(b)** and correspond to the mixing of (±1,0) SPP-BM. Secondly, equation 3 considers only Bragg-scattered smooth surface SPPs, and do not take into account properties of the lattice basis. SPCs are usually created by a regular arrangement of holes of different size and shapes, protrusions, slits or ridges milled in the metal film. This affects the scattering efficiency of SPP modes, resulting in more or less pronounced band-gap regions in the SPP-BM dispersion [43]. This also affects the effective index close to the interface.

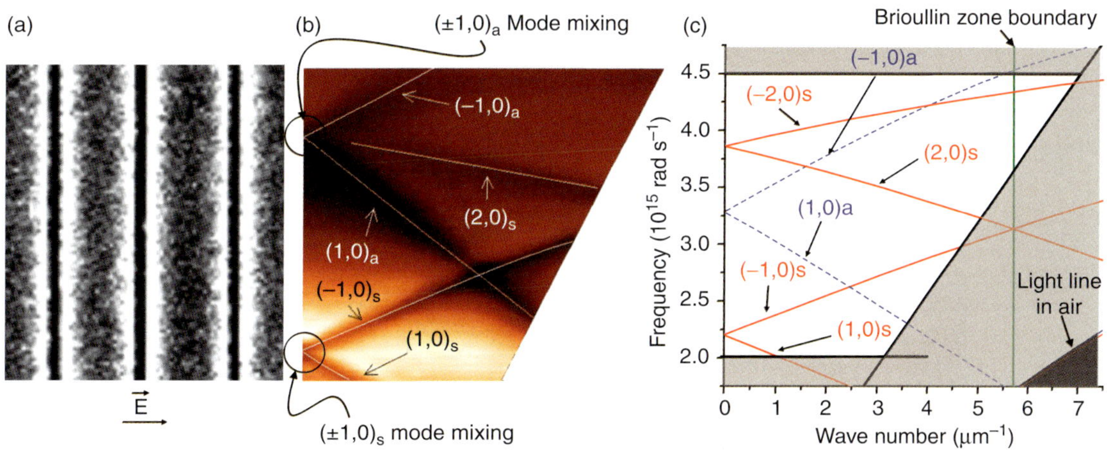

Figure 3 (a) SEM image of a 220 nm-thick Au surface plasmonic crystal made of an array of slits of width 150 nm. The period of the crystal is 500 nm; (b) zero-order transmission dispersion of the crystal shown in (a) measured for a TM-polarized incident field (illumination conditions given in (a)); (c) SPP dispersion of the crystal calculated using equation 3. The different bands are labeled according to the Bragg scattering process they encounter. In this 1D case, Bragg scattering is only present in the direction perpendicular to the slits, that is, the second index is always 0. The solid and broken lines correspond to scattering processes occurring at the glass–metal and air–metal interfaces, respectively. Three zones are defined in the dispersion: the gray and white regions correspond to modes coupled to photons in free space, that is, the radiative part of the dispersion. The white region corresponds to the part of the dispersion probed experimentally in (b), whereas the dark gray region defines surface modes, whose energy is confined to the glass–metal interface.

Because of the strong frequency dependence of the SPP field penetration in the dielectric, the presence of the holes in the film causes the SPP-BM dispersion to experience a frequency-dependent index of refraction resulting in a renormalization of the SPP dispersion [44]. Higher order (l, m) bands originating from higher momentum modes in the SPP dispersion have a stronger spatial confinement in a direction perpendicular to the metal/dielectric interface. These bands are, therefore, more sensitive to the presence of the holes in the film than lower-order bands. This can be observed in **Figure 3(b)** in comparison to **Figure 3(c)** for the higher-lying bands. A similar effect is discussed in Section 1.2.2.1. Finally, equation 3 is based on the assumption that the optical properties of SPCs are solely governed by SPP excitations. In fact, additional photonic contributions from the holes as well as from the film itself are also to be accounted for. These will be discussed in more detail below.

The formation of band gaps opens up opportunities to use SPC- it draws similarities to 2D photonic crystals-in applications dedicated to light guiding in the line defects of periodic structures, slow light tailoring, EM cavities, but with the advantage of intrinsic in-plane confinement of Bloch modes [45–47]. The most profound effect of Bloch mode formation is in the modification of conventional optical properties of SPC, such as light transmission, reflection, or absorption. This arises from the fact that some SPP-BMs are accessible to photons (**Figure 3**), in contrast to surface modes supported by smooth metal/dielectric interfaces.

1.2.1.3 Plasmonic cavities as an SPP crystal basis

The objective of this section is to familiarize the reader with some fundamental aspects of light interaction with plasmonic crystals and should not be regarded as an exhaustive study of plasmonic cavities. SPCs are created when the interface separating a metal from a dielectric material is periodically modulated at the scale of the free-space wavelength of interest [48]. They are generally made of a regular arrangement of holes in a metal film, constituting the basis of the crystal. The holes' first purpose is to generate SPP-BM, which to a large extent determines the dispersion of the crystal. In addition to this primary role, holes also bear optical properties on their own, which may impact the optical properties of the crystal. These properties govern the crystal's coupling to free-space radiation, the crystal's transmission and polarization properties, as well as the crystal's nonlinear optical response [49, 50]. SPCs formed by cavities of various cross-sections are found in the literature. These include circular, rectangular, and elliptical shapes, to name a few of the most common geometries studied experimentally.

1. *Longitudinal cavity modes.* To analyze the scattering cross-section of cavities, let us consider a single rectangular aperture of width w and length $l > w$ made in a metal film of thickness t and surrounding a dielectric of index n_d. The geometry is shown in **Figure 4(a)**.

In the situation where the metal is a perfect electric conductor (PEC) and $t \to \infty$, the geometry corresponds to a rectangular cross-section waveguide for which the lowest order supported mode is

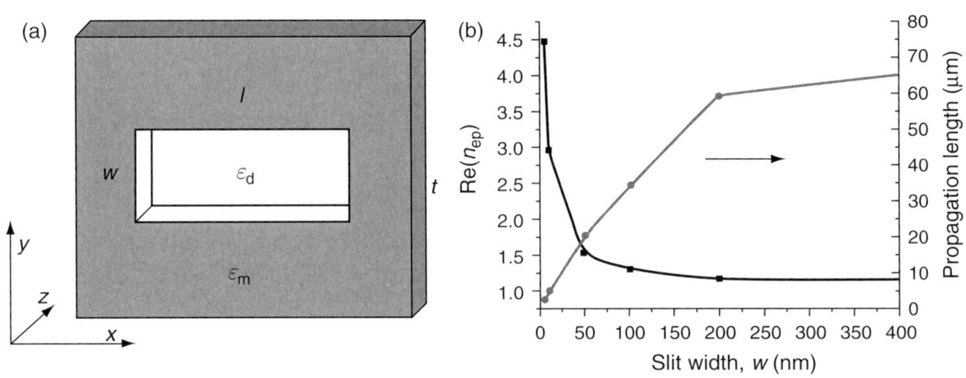

Figure 4 (a) General geometry of the plasmonic cavity considered in the study. The cavity has length l, width w, and depth t; t also represents the thickness of the metallic film in an SPC. (b) Re(n_{eff}) and propagation length for the lowest SPP mode in the z-direction supported by the structure in (a) as a function of slit width w, for $\lambda_0 = 650$ nm, $l \to \infty$, $t \to \infty$, and $\varepsilon_d = 1$.

transverse electric (TE) with an electric field polarized along the short axis of the aperture. This mode has a propagation constant along the guide $k_z = k_0\sqrt{1-\left(\frac{c\pi}{\omega l}\right)^2}$ corresponding to a mode wavelength $\lambda_g = \frac{2\pi}{k_z} > \frac{\lambda_0}{n_d}$, the wavelength in the unbounded dielectric. The cutoff wavelength λ_c for this guide is determined by the frequency ω at which k_z becomes imaginary and corresponds to $\lambda_c = 2l$ for a PEC. No EM energy will propagate along the guide for wavelengths exceeding λ_c.

The finite conductivity of real metals allows EM radiation to penetrate the metal over a distance on the order of the skin depth, which main effect is to red-shift the cutoff wavelength of photonic modes and allow SPP modes to exist [51,52]. For SPCs, the finite thickness t of the film, typically comprised between 20 and 400 nm, will allow exponentially decaying fields in the z-direction with $\lambda > \lambda_c$, or modes which are not eigenmodes of the structure, to be transmitted though the aperture and coupled to EM modes – bulk and surface modes – in the superstrate semi-space.

Another important consequence of the finite thickness of the metal film is the possibility for the hole to support cavity-like Fabry–Perot resonances. These standing-wave resonances result from waveguide modes reflected at the films interfaces and can be of photonic or plasmonic nature [53]. The resonance condition can be simply expressed as $\lambda_{\text{cavity}} = 2t_{\text{eff}}/m$, where m is a positive integer different from zero and $t_{\text{eff}} = n_{\text{eff}}t$ represents the frequency-dependent effective thickness of the metal film [53]. To discuss their impact on the cavity's optical properties, the transmission in particular, let us first consider the case when $l \rightarrow \infty$. Then the aperture becomes a slit, which constitutes the motif for the creation of a 1D SPC. Slits do not show a cutoff for radiation with an electric field polarized perpendicular to the slit's walls. For narrow slits, typically a few tens of nanometers, interface SPPs supported by the slit walls are spatially overlapping and can couple coherently to form mixed states. The dispersion of the new bound eigenmodes can be derived analytically and presents two branches corresponding to either symmetric or antisymmetric distributions of the field with respect to the center of the slit [54]. The spectral position of the branches depends on the dielectric's thickness w determining the wavelength, the damping, as well as the confinement δ_d of SPP modes in the dielectric within the slit. For a given frequency, the mode confinement increases with decreasing slit width. The slit width w can, therefore, be adjusted to increase the spatial density of EM energy in dielectric layer but also to tailor the spectral response of the Fabry–Perot cavity for a fixed film thickness. This is shown in **Figure 4(b)**, where the real part of the calculated effective index of refraction n_{eff} for the lower-frequency coupled SPP mode in the slit is plotted as a function of slit width w for $l,t \rightarrow \infty$. The calculations were performed for a free-space radiation wavelength of $\lambda_0 = 650$ nm. The quantity $\text{Re}(n_{\text{eff}})$, which determines the magnitude of the coupled SPP momentum, increases as the slit width decreases, an effect that is qualitatively similar to a increase in the permittivity of the dielectric core observed in **Figure 2**. The increase in the effective index translates into an increased confinement of the SPP field in the cavity but also affects the propagation length L of the mode, which decreases with slit width due to the relatively increased penetration of the SPP field in the metal resulting in increasing ohmic losses (**Figure 4(b)**). Typical SPP propagation lengths range from a few hundred nanometers to a few micrometers for the geometries of plasmonic cavities considered.

As observed in **Figure 4(b)**, $\text{Re}(n_{\text{eff}}) > n_d = 1$ reflects the bound nature of the mode. The finite depth of the slit, determined by the thickness t of the metal film, provides a scattering mechanism for mode coupling between coupled SPPs and free-space radiation. For slit widths on the order of 20 nm and below, their transmission spectrum is determined by cavity-mode coupled SPPs. This effect is shown in **Figure 5**, in which the normal incidence zero-order transmission of an isolated $w = 10$ nm wide slit in a Au film was calculated for various film thicknesses t. For the calculations, the substrate is glass ($n_g = 1.5$) and the superstrate is air ($n_a = 1$). The dielectric in the slit is air ($n_d = 1$) as well, while the permittivity for Au was taken from [42]. The schematic geometry for the calculations is shown in **Figure 5(a)**. The excitation field is a TM polarized plane wave, with a magnetic field oriented along the z-axis. The TE polarized incident field shows zero transmission related to the slit. For a film thickness of 100 nm, the transmission of the single slit is shown in **Figure 5(b)** and presents two main peaks. The higher-frequency transmission peak at $\lambda_0 = 500$ nm is a transmission of Au corresponding to interband electronic transitions from the d band to the sp conduction band. The second resonance around 730 nm is associated with the fundamental cavity resonance, $m = \frac{2t_{\text{eff}}n_{\text{eff}}}{\lambda_0} = 1$, supported by the slit between the

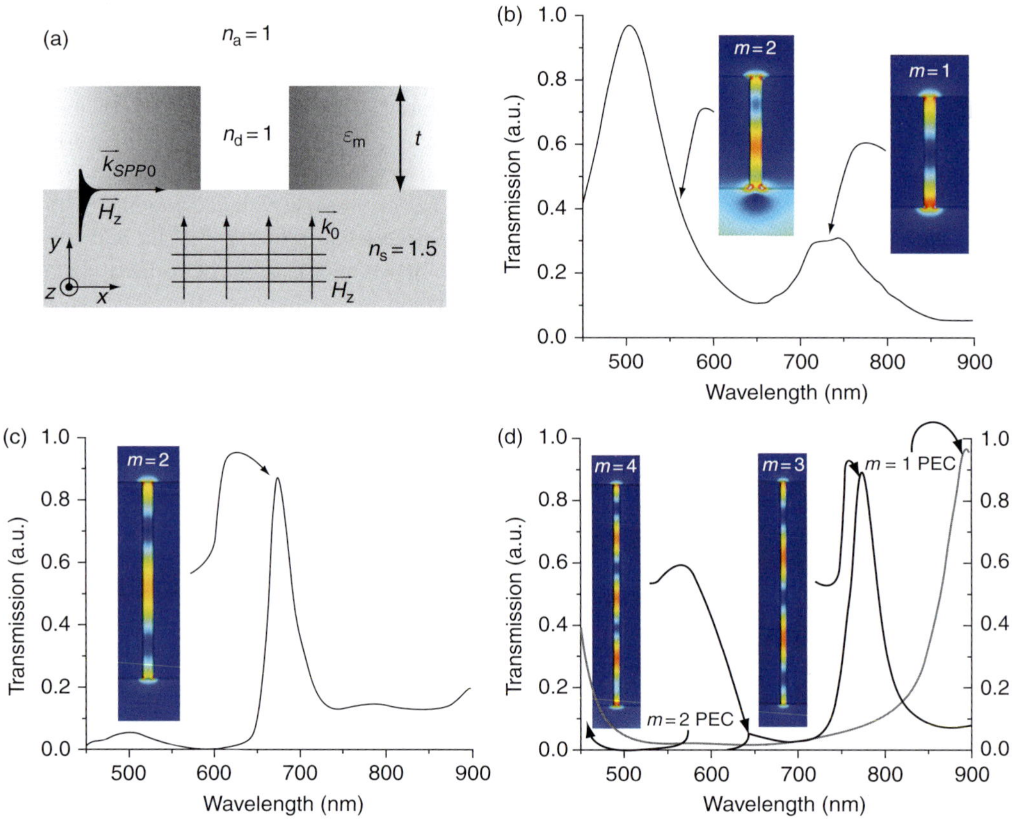

Figure 5 Optical properties of 2D plasmonic cavities. (a) Geometry considered. The cavity is a slit of width $w = 10$ nm made in a Au metal film of variable thickness t. The refractive index of the substrate and superstrate are $n_d = 1.5$ and $n_d = 1$, respectively. Two excitation conditions are shown, corresponding to either normal incidence plane wave excitation or SPP wave excitation propagating at the metal/substrate interface. (b–d) The zero-order transmission of the slit for the plane wave excitation conditions. (b) $t = 100$ nm Au film, Fabry–Perot resonances $m = 1$, and $m = 2$ contribute to the slit's transmission. (c) $t = 200$ nm Au film, Fabry–Perot resonance $m = 2$ contributes to the slit's transmission. (d) $t = 400$ nm Au film (black curve), Fabry–Perot resonances $m = 3$, and $m = 4$ contribute to the slit's transmission. The transmission for a PEC is shown for comparison (gray curve). The Fabry–Perot resonances $m = 1$, and $m = 2$ contribute to the slit's transmission at the free-space wavelengths of 895 and 442 nm, respectively.

glass substrate and the air superstrate. To account for the open cavity, the thickness t of the film is replaced by an effective thickness $t_{eff} > t$. The second-order resonance for which $m = 2$ is not apparent from the transmission spectrum of **Figure 5(b)** but is found from the field distribution to reveal itself around the wavelength of $\lambda_0 = 560$ nm. SPP damping at shorter wavelengths prevents this resonance to fully develop and to be observed in the transmission of the cavity. This higher-order mode can clearly be identified in **Figure 5(c)** for the 200 nm-thick Au film, where it appears around $\lambda_0 = 675$ nm. Further increase in the film thickness to $t = 400$ nm in **Figure 5(d)** shifts the subsequent $m = 3$ and $m = 4$ modes in the visible spectral range, each of those modes leading to a resonance in the transmission of the structure. These occur at free-space wavelengths of $\lambda_0 = 650$ and $\lambda_0 = 780$ nm for the modes $m = 4$ and $m = 3$, respectively. The increase in the resonance order with film thickness is merely the consequence of the condition for constructive interference in the cavity. For the geometry of **Figure 5** however, the spectral position of the peaks is determined by the coupled-mode SPPs dispersion in the cavity and not by photons. Indeed, replacing the Au film by a PEC in the calculation recovers the expected Fabry–Perot resonance position for photons. This is illustrated in **Figure 5(d)** for the 400 nm-thick film, for which the resonances appear at $\lambda_0 = 897$ nm ($m = 1$) and $\lambda_0 = 442$ nm ($m = 2$). Again, the boundary conditions at the dielectric interfaces shift the resonance positions to longer wavelengths because of the field penetration in the dielectrics. For the PEC, the effective thickness of the film is then estimated at about 442 nm and, to a good approximation, it can be considered to be independent of m. Using this thickness value for the cavity length allows one to calculate the refractive indices of the SPP modes

$$\mathrm{Re}(n_{\mathrm{eff}}) = \frac{m\lambda_0}{2t_{\mathrm{eff}}} = \frac{\lambda_0}{\lambda_{\mathrm{m}}}$$

For the 400 nm film, resonances correspond to $\mathrm{Re}(n_{\mathrm{eff}}) = 2.95$ and $\mathrm{Re}(n_{\mathrm{eff}}) = 2.63$ for the $m = 4$ and $m = 3$ modes, respectively. These results are in excellent agreement with the refractive index values estimated for the unrestricted ($t \to \infty$) coupled SPP mode at $\mathrm{Re}(n_{\mathrm{eff}}) = 2.94$ ($\lambda_0 = 650$ nm) and $\mathrm{Re}(n_{\mathrm{eff}}) = 2.67$ ($\lambda_0 = 780$ nm). These values show that material dispersion for Au in the visible spectral range is not predominant in modifying the effective thickness of the film, The boundary conditions at the substrate and superstrate are prevalent in determining the cavity length. The design of the cavity transmission can then be easily tailored by adjusting the film thickness. A continuous tuning of the film thickness will sweep the cavity modes through the spectral range of interest. Compared to the results presented in **Figure 5**, the 300 nm thick film for example shows the second $m = 2$ and third $m = 3$ order modes at $\lambda_0 = 873$ nm and $\lambda_0 = 660$ nm, respectively.

(a) *Wide slits.* The spectra of **Figure 5** were obtained for a 10 nm wide slit in a Au film illuminated by a plane wave of suitable polarization. The fact that the transmission is solely governed by SPP excitations emphasizes the strong free-space photon-to-SPP coupling at the slit. A similar behavior was observed for slit width up to about 20 nm. Larger slit widths show more complex transmission with both photons and SPPs contributing to the transmission of the slit. This behavior is illustrated in **Figure 6**, where the transmission for slits of different width in a $t = 400$ nm thick Au film is plotted. The larger 400 nm film was chosen to isolate the transmission from the slit from that of the unstructured film, that is, no significant transmission is observed in the absence of the slit. In **Figure 6**, the slits are under plane wave excitation at normal incidence and their zero-order transmission is shown. The geometry is that of **Figure 5(a)**.

Only slits with widths on the order of 50 nm or larger show a transmission that is a complex function of wavelength, with both photons and SPP modes responsible for the slit's transmission. For these slits, most if the incident photons are not coupled to SPP modes. The scattering cross-section for photonic mode becomes predominant and the transmission spectrum broader. As shown in **Figure 6**, an increase in slit width shifts the cavity orders to lower free-space wavelengths. In fact, we can observe in **Figure 6** that slit widths ranging from 5 to 50 nm support a third-order cavity mode $m = 3$ shifting across the visible spectrum from $\lambda_0 = 780$ nm ($w = 5$ nm) to $\lambda_0 = 550$ nm ($w = 50$ nm). To understand this behavior, let us

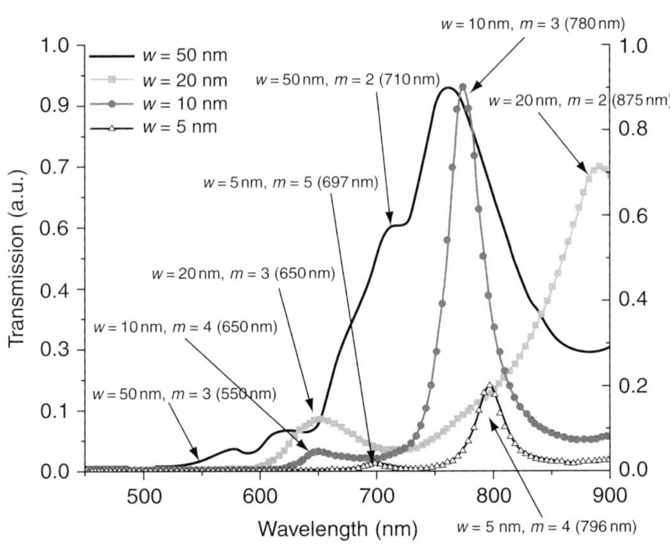

Figure 6 Transmission of a 2D cavity ($l, t \to \infty$) made in a $t = 400$ nm-thick Au film as a function of slit width w. The order for the Fabry–Perot resonance is indicated for the corresponding resonance.

recall that the cavity resonance condition relates the order of the mode to its index as $m \sim \mathrm{Re}(n_{\mathrm{eff}})$. A blue shift of the cavity orders is, therefore, a direct consequence of the decreased effective index of refraction of the coupled-SPP modes in the slits as the slit width increases. This behavior was discussed previously and is illustrated in **Figure 4(b)** for $\lambda_0 = 650\,\mathrm{nm}$. As for the film thickness, the slit width is, therefore, an effective way to tailor the resonant conditions for SPP modes supported by the cavity.

(b) *SPP excitation of cavity modes.* When considering the scattering properties of holes in a plasmonic crystal it is essential to account for two types of excitation processes. The first involves the excitation by a plane wave incident from the far-field. This scenario was just discussed for a single slit in a metallic film under normal incidence. An alternative excitation pathway by SPP waves shall be considered when dealing with SPCs. This is a secondary process originating from the scattering of light by neighboring cavities in the SPC leading to SPP waves on the interface acting as a secondary source for the neighboring cavities. For the moment we may restrict ourselves to analyzing the transmission of a single slit illuminated by an SPP mode incident from the glass/metal interface. The geometry is shown in **Figure 5(a)**, where the incident SPP field is characterized by ($\boldsymbol{H}_z$, $\boldsymbol{k}_{\mathrm{spp0}}$). Compared to the plane wave excitation, these excitation conditions are remarkable in two ways. First, the SPP-to-SPP coupling efficiency can be orders of magnitude larger than the direct photon-to-SPP coupling efficiency. The consequence of that observation is that the transmission process for narrow slits in an SPC involves a multi-order process with the following scattering sequence to couple photons to the slit through this mechanism: incident plane wave → SPP → slit mode. Second, the transmission spectrum obtained for SPP excitation is conditioned by the possibility of exciting SPP modes at the substrate/metal interface. Free-space wavelengths shorter than a certain cutoff wavelength can, therefore, not be coupled to the slit through this mechanism. Otherwise, the transmission spectrum for a narrow slit in an SPC is qualitatively similar irrespectively of whether the illumination field is a direct plane wave excitation or an excitation mediated by SPP originating from the scattering of neighboring slits. Slits wider than about 50 nm show differences in their transmission in the two excitation scenarios, but both cases result in the excitation of SPP and photonic modes in the cavity. Note that for the spectra of **Figures 5** and **6** only direct incident plane wave → slit mode or direct SPP → slit mode excitation processes are possible since the slit is isolated. Note also that the transmission spectrum reflects the decoupling of the cavity mode to free-space radiation in the air superstrate [55]. It is, therefore, conditioned by efficiency of alternative scattering processes accruing at the metal/air interdirectface and in particular to the excitation of SPPs at the output interface. Finally, for both plane wave and SPP wave excitations, it is found that the transmission spectrum is fairly isotropic with the main resonances present in both the bulk radiation in the far-field and the SPP radiation at the superstrate interface in the near-field.

(c) *Cavities with a two-dimensional (2D) cross-section.* For 2D crystals, cavities forming the lattice have a finite length l (**Figure 4(a)**). Considering the fundamental mode discussed thus far, the boundary conditions at the metal/dielectric interface along the x-direction for the EM field in the cavity lead to a cutoff wavelength of $\lambda/2$, where λ is the wavelength of the cavity mode. In practice, decreasing the length l of the cavity in an attempt to confine the cavity mode laterally impacts the effective index of the cavity modes resulting in a decrease of $\mathrm{Re}(n_{\mathrm{eff}})$. This effect is illustrated in **Figure 7**, where the dispersion for the lowest-order mode for rectangular waveguides ($t \to \infty$) of width $w = 50\,\mathrm{nm}$ is plotted. The figure shows that, as the cavity length l gradually decreases, the dispersion of the lowest mode comes closer to the smooth interface SPP dispersion, that is, the effective index of the mode decreases with l. Eventually, the cavity does not support this surface mode any longer as a cutoff wavelength λ_c appears. This cutoff was estimated to $\lambda_c = 650\,\mathrm{nm}$ for the waveguide of length $l = 250\,\mathrm{nm}$. For comparison, the dispersion of a cavity of same length ($l = 250\,\mathrm{nm}$) and width ($w = 50\,\mathrm{nm}$) in a PEC rather than in Au is also plotted in **Figure 8**. As can be observed, its dispersion is in the radiation zone because of the photonic nature of the mode and the cutoff wavelength is $\lambda_c = 500\,\mathrm{nm}$ as expected.

As a result, holes with low aspect ratio l/w do not sustain longitudinal surface plasmon resonances. In the simplest case, their optical response is then mainly governed by transverse cavity modes.

2. *Transverse cavity modes.* In three-dimensional (3D) dielectric cavities made in a metal film, transverse resonances due to the finite size of the hole may also contribute to the optical properties of the hole. A general understanding of such resonances can be

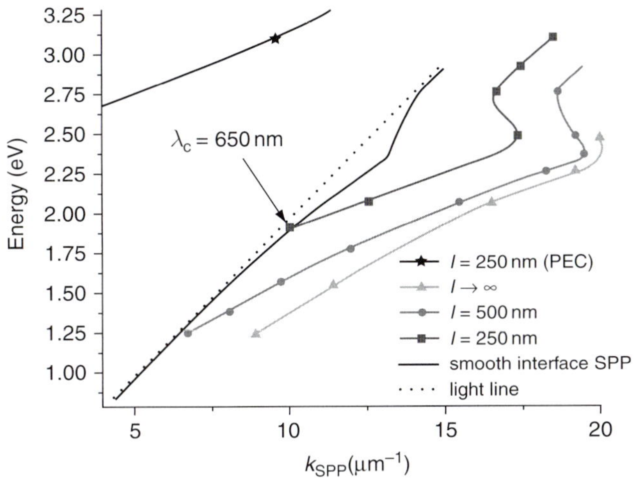

Figure 7 Dispersions of the fundamental SPP mode propagating in the z-direction supported by the rectangular cavity in **Figure 4(a)** with width $w = 10$ nm and infinite thickness t. The calculations were performed using finite-element method simulations for Au as a metal and air as a dielectric. The back-bending observed in the dispersions is due to the contribution of interband transitions in the dielectric constant of Au. The dispersions of a smooth interface SPP, the light line, and a rectangular cavity with infinite thickness made in a PEC are also shown for comparison. The variable λ_c indicates the cutoff wavelength for the SPP mode supported by the cavity of cross-section $lw = 250 \times 50$ nm^2.

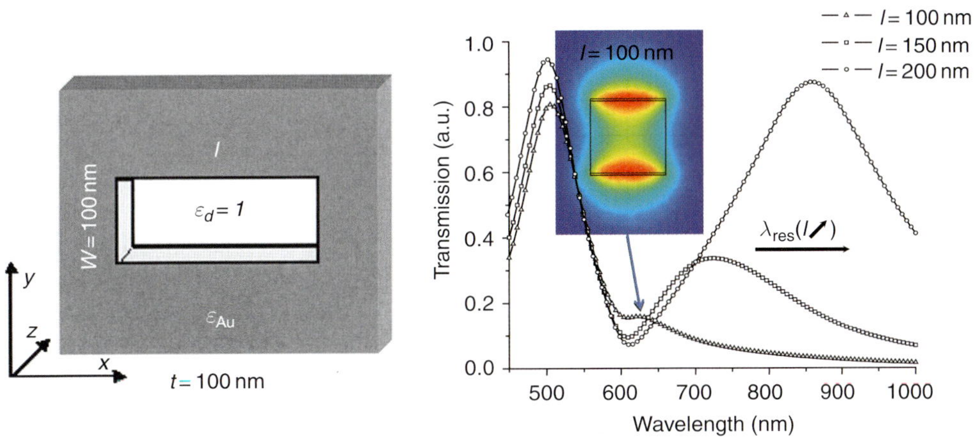

Figure 8 Optical properties of a rectangular hole with width $w = 100$ nm for different lengths l made in a Au film of thickness $t = 100$ nm. The dielectric in the hole and the superstrate is air and the substrate is glass. The optical spectrum shows the zero-order transmission of the cavity for $w = 100$ nm and $l = 100$ nm (triangles), $l = 150$ nm (squares), and $l = 200$ nm (circles). The spectra were obtained for an incident field at normal incidence from the substrate with an electric field linearly polarized along the y-direction. As the length, l, of the cavity increases, the dipolar transversal resonance wavelength, λ_{res}, red-shifts. The inset shows the norm of the electric-field distribution at a distance 10 nm above the hole in the air superstrate for the $l = w = 100$ nm square cross-section hole. The dipolar nature of the resonance is evident.

obtained by invoking the principle of complementarity between the optical response of ellipsoidal particles and voids supporting localized surface plasmon resonances. The complementarity principle can be expressed as follows [7]:

$$\omega^2_{p} = \omega^2_{particle} + \omega^2_{void} \quad (4)$$

where ω_p is the bulk plasmon frequency, and $\omega_{particle}$ and ω_{void} are the transverse localized resonance of the particle and the void, respectively. In the electrostatic approximation, and for a lossless Drude metal, the resonant wavelength of the lowest-order surface plasmon modes for an ellipsoidal particle in a field polarized along one of its principal axis j takes the form [56]:

$$\lambda_{\text{particle}} = \frac{2\pi c}{\omega_p}\sqrt{1-\varepsilon_d\left(1-\frac{1}{L_j}\right)}, \quad j=1,2,3 \qquad (5)$$

where L is the geometrical or depolarization factor that reflects the shape of the ellipsoid, c is the speed of light in vacuum, and ε_d denotes the dielectric constant of the embedding medium. In the general case, **equation 5** identifies three distinct localized surface plasmon resonances. The spectral position of the resonance along the axis j blue-shifts toward the plasma frequency when L_j increases between the limiting values of 0 and 1. This is the case, for example, for the long-axis resonance when the short-axis length increases. From **equations 4** and **5** the resonance condition as a function of aspect ratio of the void can be expressed as follows:

$$\lambda_{\text{void}} = \frac{2\pi c}{\omega_p}\frac{1}{\sqrt{1-\frac{1}{1-\varepsilon_d(1-1/L_j)}}} \qquad (6)$$

Equation 6 describes a complementary behavior to **equation 5** for the position of the surface plasmon modes of an elliptical void. This behavior is illustrated in **Figure 8**, which shows the transmission of an isolated hole (of rectangular cross-section) in a 100 nm-thick Au film. The incident field is a plane wave propagating along the z-axis and with an electric field linearly polarized along the y-axis. The wave is normally incident from the glass substrate and the zero-order transmission is computed in the air superstrate for different aperture lengths: $l = 100$, 150, 200 nm. This particular hole geometry was chosen to isolate the contribution of transverse resonances in the spectra.

The long-wavelength resonance peak in **Figure 8** corresponds to the transverse dipolar resonance of the hole. The field distribution in air 10 nm above the metal/air interface is shown in the inset for the square hole. The two-lobe characteristic of the dipolar response is evident. For the square hole the resonance is located at a wavelength of 625 nm. As the length of the cavity increases to 150 nm, and then to 200 nm, this resonance shifts to longer wavelengths, of about 725 and 860 nm, in agreement with the qualitative expectation of **equation 6**. Note that the actual shape of the hole does not affect the general behavior of this type of resonance.

1.2.1.4 Mechanisms of enhanced optical transmission (EOT) through SPCs

Having discussed the transmission properties of individual holes of the basis and Bloch modes of the periodic lattices, let us summarize the different mechanisms that may contribute to the transmission of photons through an SPC. **Figure 9** sketches three geometries for which specific transmission channels from the incident medium 1 to the transmission medium 2 are opened. These mechanisms may all contribute to the EOT of the SPC depending on the actual crystal geometry.

For an unstructured smooth metal film, the sole transmission mechanism for photons with frequency below the plasma frequency of the metal is via tunneling of an exponentially decaying electromagnetic wave through the metal as described by Beer's law (**Figure 9(a)**). The film thickness for which light can pass through the film is then limited approximately by the skin depth of the metal at the wavelength of interest. Typical values for Au are on the order of 30 nm [57]. An additional bulk transmission for Au is observed around 500 nm and corresponds to interband electronic transitions from the d band to the sp conduction band [58,59]. An example of such transmission is shown in **Figure 5(c)**. The second scenario (**Figure 9(b)**) involves tunneling via radiative SPP modes supported by the interfaces 1 and 2. This channel is opened when surface waves are coupled to bulk modes on one or both film interfaces. For

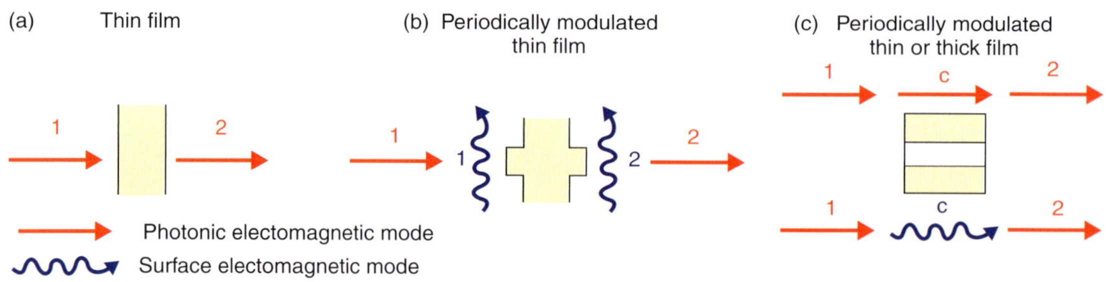

Figure 9 Different light transmission mechanisms between half-spaces 1 and 2 through (a) smooth metal film and (b, c) apertureless and apertured SPCs. The label C designates the cavity modes.

SPCs the coupling mechanism is Bragg scattering via the momentum generated by the periodic modulation of the interfaces. This transmission mechanism requires a spatial overlap of the field excited at interface 1 with the field at the interface 2 [60,61]. If the geometry of **Figure 9(b)** is symmetric with regard to the center of the metal film, SPP modes at both interfaces can interact coherently to form SPP film modes [54,62,63]. Then the excitation of the structure's eigenmode cannot be isolated to one particular interface, and the film as a whole contributes to the structure's transmission. Intermediate geometries, not symmetric in a geometrical sense, but supporting SPP at both interfaces of the film with incidental degeneracy may still strongly couple through the film at specific frequencies and contribute to the EOT [63]. In particular, this coupling can be modified by tuning the geometry of the interfaces and is maximized when the modes supported on both interfaces are degenerated, that is, when film SPP modes are formed [63–65]. In the thick-films scenario in **Figure 9(c)**, for which both half-spaces cannot be coupled electromagnetically directly through the film, evanescent or propagative (eigenmodes) modes in cavities created in the film are to be considered in the transmission of photons through the metal film [66]. These localized modes, in addition to providing coupling between the half-spaces, may be resonant with the cavity formed by the aperture (Fabry–Perot-type resonances) and may be of either plasmonic or photonic nature [67,68]. Finally, if the resonant frequencies of localized cavity modes overlap the frequency spectrum of the Bloch modes, the EM energy can be efficiently transferred between them, leading to strongly coupled modes so that their individual contribution to the transmission cannot be separated.

1.2.2 Sample Fabrication and Experimental Configuration

The SPCs were fabricated in Au films (40, 100, and 220 nm thick) sputtered onto glass substrates coated with a 10 nm-thick tantalum pentoxide (Ta_2O_5) adhesion layer using a magnetron sputtering system. The films were then nanostructured periodically with holes or slits, using focused ion beam milling to form 2D (hole arrays) and 1D (slit arrays) SPCs. The structural parameters of the crystals were characterized using atomic force microscopy (AFM) and scanning electron microscopy (SEM). The polarization properties of the SPCs were studied by analyzing the state of polarization of the zero-order transmission for different orientations of the linearly polarized normally incident light. The experimental setup used in the experiments is schematically described in **Figure 10**. The white-light source from a W-halogen lamp is linearly polarized and collimated before impinging the sample S at normal incidence. The transmitted light in analyzed and focused into a 100 μm core diameter optical fiber (OF) connected to a spectrometer equipped with a liquid-nitrogen-cooled charge-coupled device.

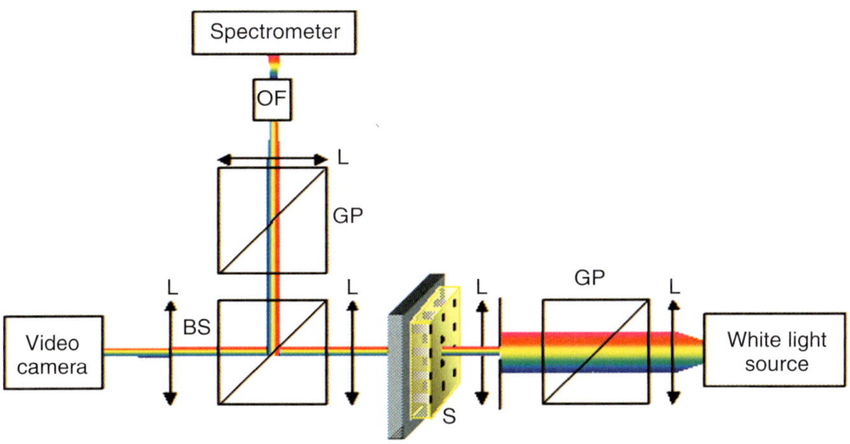

Figure 10 Typical Experimental setup. The white-light source from a W-halogen lamp is linearly polarized and collimated before impinging the sample S at normal incidence. The transmitted light in analyzed and focused into a 100 μm core optical fiber (OF) to be delivered to a liquid-nitrogen-cooled CCD. A beam splitter (BS) allows a routine view of the area considered for spectroscopy via a video camera. (L) lens and (GP) Glan–Thompson polarizer.

A beam splitter (BS) allows a routine view of the area considered for spectroscopy via a video camera.

1.2.3 Polarization Properties of SPCs

The polarization properties of SPCs are central to the controlled coupling between SPPs and photons, opening unique opportunities for the development of nanophotonic devices based on the SPP–light coupling [69–76]. In this section we present and discuss some aspects of the polarization properties of surface plasmon polaritonic crystals. The experiments were performed on SPCs made in gold films perforated periodically with holes or slits. The polarization properties of these crystals was studied in the zeroth-order transmission at normal incidence. The polarization properties of SPCs are determined by both the symmetry of the crystal and the plasmonic scattering mechanisms contributing to its optical transmission. In particular, the coupling of localized cavity modes and SPP-BM present at the metal interfaces is a mechanism that can lead to the mixing of the SPC eigenstates in different directions and strongly impacts its polarization properties. For an isotropic SPC with a fourfold symmetry, the state of polarization is not affected by the transmission of light through the structure at normal incidence. When the symmetry of the SPC is reduced by modifying the basis shape and/or orientation, the modified cavity modes as well as the lifted degeneracy of the surface polariton eigenmodes along different principal directions leads the SPC to exhibit wavelength-selective birefringence.

To assess the polarization properties of SPPCs, let us first consider a general case of an SPC with a film thickness of 100 nm for which all coupling mechanisms of **Figure 9** and consequent polarization transformations are possible. Because one or the other transmission mechanism is dominant in each case, for thinner as well as thicker films, the situation is somewhat simplified.

The SEM image of the SPC and its associated transmission spectrum for light polarized along the lattice axes are shown in **Figure 11**. The holes are 150 nm wide and the period of the lattice is 500 nm. Since the SPC is in a nonsymmetric environment, the

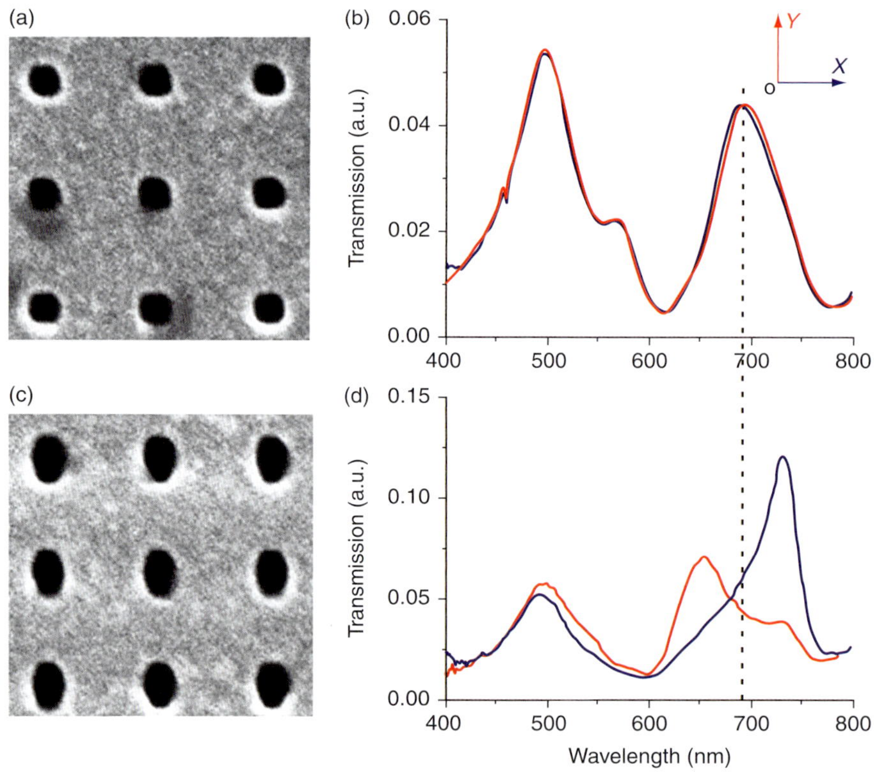

Figure 11 SEM images and zero-order transmission of a 100 nm-thick SPC in a glass-supported Au film. The SPCs have a square lattice with a 500 nm period. (a) An SPC with fourfold symmetry, 150 nm × 150 nm square holes, and (b) its optical transmission for light polarized along the principal axes. (c) An SPC with twofold symmetry, 150 nm × 210 nm elliptical holes (the symmetry axes coincide with the lattice axis) and (d) its optical transmission for light polarized along the principal axes.

resonant excitation of SPP-BM at each interface contributes to nondegenerate transmission peaks shifted in frequency. The SPC of **Figure 11(a)** exhibits transmission peaks around 690, 570, and 500 nm. For light polarized along the $\overrightarrow{ox}$ axis, the minima determining these resonances correspond to SPP-BM band edges along the (±1,0), (±1,1), and the (±2,0) directions at the substrate–metal interface, with the possible degeneracy of the (±2,0) excitation with the band edge associated to the SPP-BM at the air–metal interface along the (±1,0) directions. The crystal has a fourfold symmetry, which implies in particular that the SPP-BM along both the lattice axes directions are degenerate. This can be deduced from the spectra in **Figure 11(b)**, which were taken for light linearly polarized along the principal lattice directions of the crystal. Therefore, a linearly polarized incident field oriented, for example, at 45° from the lattice directions and coupled to SPP-BM along the (1,0) and (0,1) directions would excite two degenerate modes with the same amplitude along these axes. Photons normally incident on the crystal would, therefore, not undergo any change in their state of polarization after transmission through the crystal when these modes re-couple to photons in the superstrate. A similar reasoning holds for any other incident polarization at normal incidence and the field transmitted by the crystal.

The C_{4v} symmetry of the SPC leads to the isotropic polarization response of the crystal. When anisotropy is introduced in the SPC by increasing the size of the hole along the $\overrightarrow{oy}$ direction of the lattice as shown in **Figure 11(c)**, the resulting transmission for light polarized along the lattice directions is strongly modified, as can be observed in **Figure 11**. The modification in the holes' geometry only marginally affects the position of SPP-BM modes in frequency, and the lifted degeneracy in the transmission between the principal axes of the crystal is mainly determined by the different eigenmodes of the holes along these directions. The efficiency for one particular EM frequency to be coupled to SPP-BM will depend on the structure of the field inside the hole at this frequency. In fact, the transmission for the holes of the SPCs of **Figure 11** is mainly due to transverse dipolar cavity modes (see Section 1.2.1.3.2). Compared to **Figure 11(a)**, in the SPC of **Figure 11(c)**, the size of the hole was stretched along the $\overrightarrow{oy}$ direction from 150 to 212 nm while keeping its original width along the $\overrightarrow{ox}$ direction. For the modes supported by the hole in **Figure 11(c)**, this results in a splitting of the resonance frequency positions along the principal axes of the ellipse with regard to the position of the single resonance from the holes in **Figure 11(a)**. The resonances of the cavity mode polarized along both principal axes of the elliptical cross-section are, therefore, modified with a red-shifted cavity mode polarized along the short axis and a blue-shifted cavity mode polarized along the long axis of the ellipse. On the basis of this reasoning, and assuming the SPP-BM dispersion at the substrate/metal interface is still mainly degenerate along the principal axes of the ellipse, the modification of the transmission of the SPC in **Figure 12(d)** can be related to the coupling of the eigenmodes of the elliptical basis of the SPC to the Bloch modes. This is clearly demonstrated for the peak originally located at $\lambda = 690$ nm in **Figure 12(b)** that experiences a splitting into two new contributions at $\lambda = 660$ and 730 nm. These new contributions to the transmission spectrum correspond to the coupling of SPP-BM along the (1,1) directions with the mode supported by the elliptically shaped cavity along the long ($\lambda = 660$ nm) and to the coupling of SPP-Bloch modes along the (1,0) directions for the cavity resonance along the short axis ($\lambda = 730$ nm) of the ellipse. Comparing the relative amplitude of the resonances in **Figures 11(b)** and **11(d)**, it can be seen that the spectral overlap between the cavity modes of the hole and SPP-BMs at the substrate/metal interface is increased for the elliptical basis, leading to an increasing transmitted intensity near the plasmonic band-edges in **Figure 11(d)**.

Furthermore, SPP-BMs along the (1,1) directions at the substrate/metal interface represent a scattering pathways allowing for the holes eigenmodes to be coupled. Evidence of this coupling is given in **Figure 12** for a crystal of similar geometry. It can be observed that the transmission of the crystal shown in **Figure 12(b)** bears similar features to the one shown in **Figure 11(d)**. The small increased aspect ratio from 1.4 in **Figure 11(c)** to 2 in **Figure 12(a)** in the hole's cross-section leads to a larger splitting in the transmission peaks from 660 and 730 nm to, respectively, 630 and 760 nm. The more instructive behavior in **Figure 12(b)** is revealed as the polarization of the incident field is rotated in the plane of the crystal. Indeed, an isobestic I is observed at the wavelength of 660 nm, which reflects the coupling of the holes eigenmodes via SPP-BMs along the (1,1) directions. These modes, having orthogonal polarizations, would otherwise not be coupled. The amplitude of both peaks is shown to follow a $\cos^2\phi$ dependence on the in-plane polarization angle ϕ in **Figure 12(c)** emphasizing the cavity-related nature of these hybrid modes.

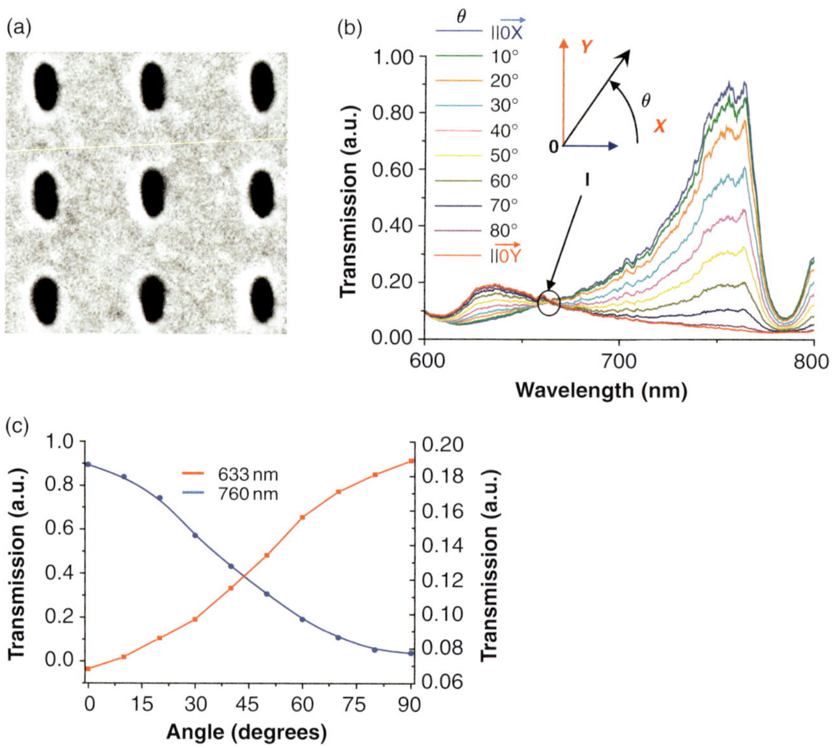

Figure 12 SPC with 110 nm × 230 nm holes basis made in 100 nm-thick SPC in a glass-supported Au film. The SPC has a square lattice with a 500 nm period with a twofold symmetry. (a) SEM image, (b) zero-order optical transmission as a function of in-plane polarization, and (c) transmitted intensity for the wavelengths of 633 and 760 nm as a function of in-plane polarization angle from 0° (*ox*) to 90° (*oy*), respectively.

Similarly to the SPC of **Figure 11(a)**, the transmission of light through the SPCs in **Figures 11(c)** and **12(a)** will not alter the state of polarization for light linearly polarized along one of the lattice axes. This is due to the fact that the crystal lattice symmetry axis coincides with the basis symmetry axis, that is, these are the principal axes of the crystal. However, the transmission of any other incident polarization will be modified according to the projected amplitude of the incident field intensity on the SPC eigenmodes excited along the principal axis of the crystal at the wavelength considered. In the particular case where this projection is the invariant point *I* in **Figure 12(b)** or zero, that is, the transmission for a particular wavelength is zero along one of the principal axes of the crystal, the polarization of the transmitted light at this wavelength will not be affected by the transmission through the crystal. This latter case is illustrated by the SPC of **Figure 13**. It is made of a collection of rectangular holes milled in a 220 nm thick Au film and placed in a square array with a 600 nm period. This crystal has the same symmetry as the crystals of **Figures 11(c)** and **12(a)** and the lattice axis in this case also coincide with the symmetry axis of the crystal. The transmission along these axes is shown in **Figure 13(b)**. Conversely to the SPCs considered so far, no spectral overlap exists between the modes with two principal transmitted polarizations (**Figure 13(b)**) and the light transmitted by the crystal is therefore always linearly polarized regardless of the wavelength or state of polarization of the incident light. Disregarding the bulk transmission of Au at a wavelength of 500 nm, it can be seen from **Figure 13(b)** that light incident upon the SPC in the spectral range comprised between 400 and 700 nm will be transmitted with a polarization along the $\overrightarrow{ox}$ axis and with an intensity proportional to the density of states of the crystal along this axis as revealed in **Figure 13(b)**. Similarly, at wavelengths above 600 nm, light will be linearly polarized along the $\overrightarrow{oy}$ axis. The SPC of **Figure 13(a)** is therefore acting as a spectrally sensitive linear polarizing filter.

A well-known limiting case for such a polarizing filter made from an array of slits with subwavelength width is shown in **Figure 14**. The boundary condition

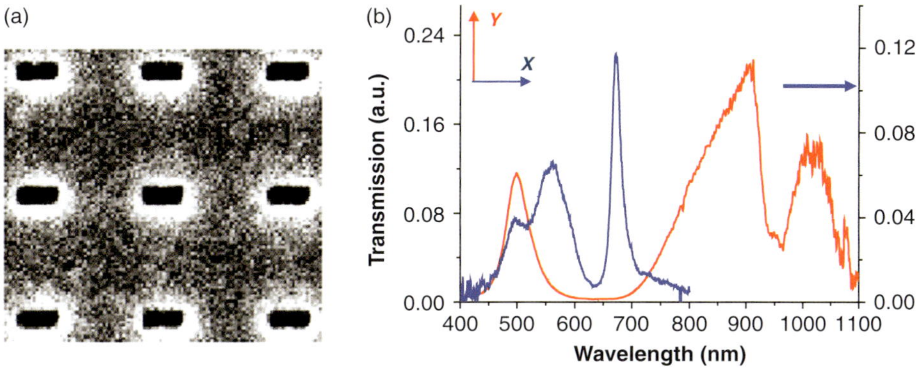

Figure 13 Glass-supported 220 nm-thick Au SPC made of 180 nm × 80 nm rectangular holes placed in a 600 nm period square array. (a) SEM image and (b) zero-order transmission for light polarized along the principal axis of the SPC. The SPC has a twofold symmetry with the symmetry axis coinciding with the lattice axis.

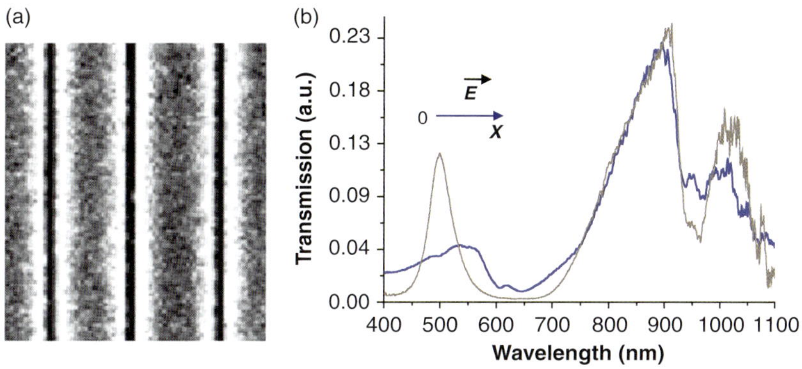

Figure 14 A 1D SPC made of 150 nm wide slits in a 1D array with 600 nm period. The crystal is made in Au and supported by glass. It has a twofold symmetry, with the symmetry axis coinciding with the lattice axis. (a) SEM image and (b) zero-order optical transmission (blue line) for light polarized along the *ox* axis. The transmission of the 2D SPC from **Figure 13(a)** is shown in gray for comparison.

at the metal interfaces – the continuity of the tangential component of the electric and magnetics fields – does not allow any propagating electromagnetic fields polarized along the metal interfaces in the slits and considering the thickness of the crystal only light with an electric field component polarized perpendicular to the slits will be detected in the far field transmitted by such a 1D SPC.

Let us briefly return to the discussion on the transmission of the 220 nm-thick Au crystal consisting of a square array of holes with a rectangular cross-section (**Figure 13(a)**). This crystal presents two peculiar features that deserve further comments. First, as a result of the absence of spectral overlap between eigenmodes polarized along either the short or the long axis of the cavity, the scattering through SPP-BMs along the (1,1) directions is not allowed. Indeed, a typical dispersion for a 2D SPC is shown in **Figure 16(j)**, where the (1,1) bands are identified according to **equation 3**. For the crystal shown in **Figure 13(a)**, this scattering event is prevented due to the scattering cross-section of the rectangular basis that shows a cutoff at a wavelength between 700 and 800 nm. The dispersion of this crystal then replicates that of the 1D crystal shown in **Figure 14**. Second, for light polarized along $\overrightarrow{oy}$, the spectrum in **Figure 13(b)** has the general characteristics of the transmission from a 1D grating and resembles strongly the transmission from the array of slits shown in **Figure 14(b)**. This emphasizes the primary role of SPP-BMs in the transmission mechanism of both SPPCs and suggests that the rectangular cavity's eigenmode along $\overrightarrow{oy}$ is coupled to an SPP-BM in the (0, ±1) directions at the substrate/metal interface and that this Bloch mode is not sensitive to the finite size of the cavity in these directions.

1.2.4 Dynamic Control of the Optical Properties of Plasmonic Crystals

The optical properties of plasmonic crystals can be tailored by varying their geometrical parameters. The ability to control their optical properties dynamically with various external stimuli, such as electric and magnetic field or all-optically, provides great flexibility in photonics, optoelectronics, and sensing applications enabling active functionalities and tunable optical response. In this section we review approaches to dynamically influence the optical properties of SPCs using static electric and magnetic as well as optical fields, via modification of SPP-BM.

1.2.4.1 Electronically controlled surface plasmon dispersion and optical transmission through metallic hole arrays using liquid crystal

The optical transmission through plasmonic crystals fabricated in a gold film can be controlled by electrical manipulation of the refractive index of a liquid crystal (LC) layer embedding the SPC. For practical purposes, an LC film is in direct contact with the SPC from the air superstrate, as shown in **Figure 15**. This allows one to use the SPC film as an electrode for electric field application. The electrically controlled variations of the director orientation of the LC in the layer adjacent to the nanostructured metallic film affects the plasmonic Bloch modes according to **equation 3**. This is a direct consequence of the anisotropy of the LC's refractive index whose ordinary and extraordinary refractive indices, perpendicular and along the LC orientation, are $n_0 = 1.525$ and $n_e = 1.734$, respectively, and assumed dispersionless in the spectral window considered. The refractive index variation influences both the state of polarization and the penetration depth of the Bloch mode's field in the LC layer [17].

Figure 16 shows examples of various SPCs and electric field-induced modifications of their transmission via their SPP-BM dispersion. These structures comprise both 1D and 2D arrays of subwavelength apertures having the same period (550 nm). In the 1D case, the array consists of slits having a width of approximately 100 nm. For the 2D structures, square arrays have been created using either a circular hole (diameter 180 nm) or a rectangle (200 nm length, 100 nm width) as the lattice basis. The dispersions plots $\omega(k)$ of the zero-order optical transmission of these nanostructures are shown in **Figure 16** for different values of the static electric field E_0 applied between the Au film and the ITO electrode to control the LC's orientation in the cell. The value of E_0 is $0\,\mathrm{kV\,cm^{-1}}$ (the OFF state) or $12.5\,\mathrm{kV\,cm^{-1}}$ (the ON state) for LC orientation along the (x,y) plane or the z-axis, respectively. All measurements were performed with the polarization of the incident light in the plane of incidence and perpendicular to the slits for the 1D arrays, and the long axis of the rectangles for the 2D arrays. The electric-field-induced changes of the transmission are highlighted in the differential dispersion plots in **Figures 16(d)**, **16(h)**, and **16(l)** that represent the differential ratio $(T_{\mathrm{ON}}\text{-}T_{\mathrm{OFF}})/T_{\mathrm{OFF}}$ between the transmission values when the static electric field is ON and OFF.

When the electric field is applied to the structure, both a modification in the magnitude of the transmission at the resonant wavelengths and a spectral shift of the resonances are observed. The changes in the transmission are most important close to the band-edges of the SPC. Illumination of the sample from the opposite side gives similar results, indicating that

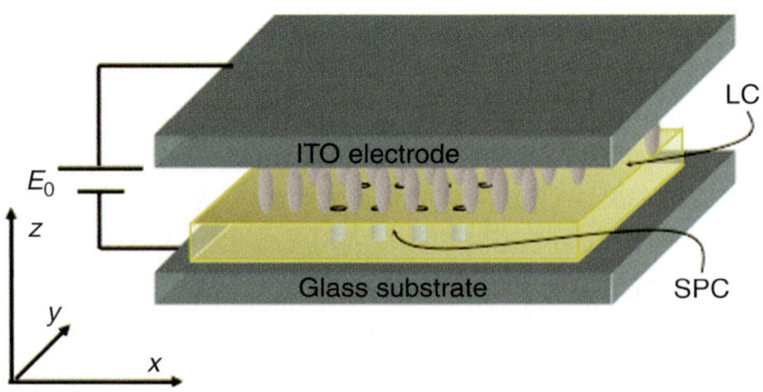

Figure 15 Geometry of an LC cell showing the SPC on a glass substrate in contact with the LC solution. The Au film and the ITO cover are used as the bias electrode to control the orientation of the LC in the cell. The optical measurements were performed in the configuration of **Figure 10**.

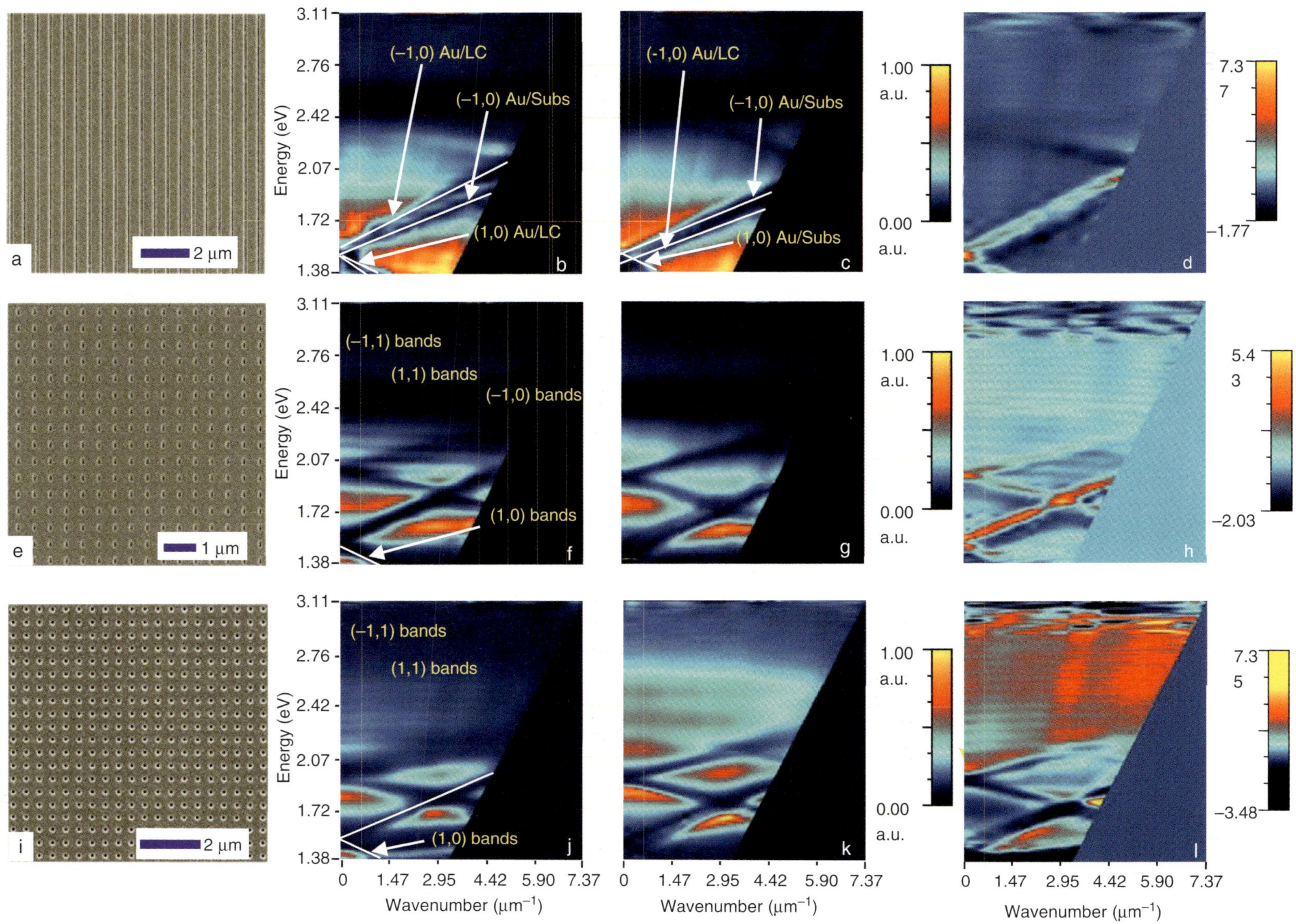

Figure 16 SEM images and optical transmission dispersion plots of SPC made in a 200 nm-thick Au film. SEM images for a periodic array of (a) slits (60 nm width, 550 nm period), (e) rectangles (200 nm × 50 nm, 550 nm period) and (i) holes (150 nm diameter, 550 nm period). Optical transmission dispersion plots for the structures in LC without and with an applied electric field (12.55 kV cm^{-1}) are shown in (b) and (c) for the slits; (f) and (g) for the rectangles and (j) and (k) for the holes, respectively. (d), (h), and (l) show the change in frequency dispersion associated with the application of the applied field for the slits, rectangles, and holes, respectively. The yellow arrow in (l) indicates where the dynamic response of the transmission was studied. The white lines are the dispersion calculated with **equation 3**. It should be noted that, in contrast to the slit array, the SPP bands on the opposite interface in the case of circular and rectangular basis overlap and cannot be distinguished in the plots.

pure polarization effects, which are responsible for the electric switching in the transmission of plasmonic nanorod arrays [77], are in this case negligible compared to the induced refractive index change, which results in the modification of the SPP-BM dispersion in the plasmonic crystal.

As discussed in Section 1.1, SPP Bloch modes as well as cavity modes of the basis are both sensitive to the spatial variations of the refractive index of the adjacent medium, which are changed in this experiment. The role of each of these effects will be exemplified when analyzing the response of the crystal to the applied static electric field. However, the general behavior of the optical properties of all SPCs studied to a reorganization of the adjacent LC layer is very similar. Therefore, we will focus the discussion on the case of the 1D crystal (**Figure 16(a)**) and then generalize the observed behavior to the 2D case.

As seen from **Figure 16(b)**, two SPP-BM intersect the $k=0$ ordinate at around 1.53 and 1.56 eV. These modes correspond to the $(\pm 1,0)$ SPP bands for the substrate and LC interfaces, respectively. The spectral position of these bands is related to the effective refractive index of these modes at each interface, where the substrate interface presents a higher effective index than the LC interface. Considering the refractive index of glass ($n=1.5$) and the expected orientation of the LC ($n_e=1.735$ along the metal/LC interface and $n_o=1.525$ perpendicular to it), the Au/LC band should lie at a lower frequency than the Au/glass band. However, the Au film is sputtered on a 10 nm-thick Ta_2O_5 base layer with a refractive index of 2, sufficient to displace the Au/substrate band below the Au/LC band. The effective indices of these low-frequency modes can be estimated using **equation 3** substituting $n_{\mathrm{eff}}^2 = \varepsilon_{\mathrm{d}}$ as a parameter. The values for the Au/substrate and Au/LC interfaces are then found to be $n_{\mathrm{eff\text{-}glass}}=1.42$ and $n_{\mathrm{eff\text{-}LC\text{-}no}}=1.39$, respectively.

On application of the static electric field, the upper band corresponding to the Au/LC SPP-BM, shifts to lower energy, sweeping across the substrate mode and crossing the center of the first Brillouin zone at around 1.4 eV. In this case, the corresponding effective index for this mode is $n_{\mathrm{eff\text{-}LC\text{-}ne}} \cong 1.55$ and corresponds to the LC being preferentially aligned with the static field, that is, perpendicular to the interface. This corresponds to a variation of 0.16 (1.55–1.39) in the effective refractive index of the LC layer, while the maximum refractive index variation corresponding to a 90° reorientation of the LC in the cell is 0.21 (n_e-$n_o=1.734$–1.525). Considering that the LC molecules are predominantly aligned with the applied field when the voltage is ON and principally aligned with the electrode interfaces when the voltage is OFF, and taking into account that the lower-frequency SPP-BM on the Au/LC interface has its electric field predominantly normal to the interface, this SPP-BM appears to be mostly sensitive to the variation of the refractive index perpendicular to the metal interface. Additionally, the measurements in **Figures 16(b)** and **16(c)** reflect the spatial sensitivity of the SPP-BM. This is evidenced in the differential dispersion plotted in **Figure 16(d)**. Indeed, this plot shows that the lower SPP-BM band (on the Au/LC interface) is most sensitive to changes of the refractive index in the bulk of the cell rather than those in the close proximity of the interface. The lower-lying bands at the Au/LC interface, that is the $(\pm 1,0)$ bands, have a stronger penetration depth in the LC region than the higher-lying bands, which are more confined to the Au/LC interface. The spectral position of the latter would then be more sensitive to changes in refractive index occurring in close proximity to the Au/LC interface, whereas the former are more sensitive to changes in the bulk, a few hundred nm into the LC medium.

Replacing the slits with holes of rectangular cross-section opens additional scattering channels, in particular along the $(\pm 1, \pm 1)$ directions (**Figures 16(f)–16(g)**). Compared to the dispersion of the 1D crystal, the dispersion of the array of rectangles also reflects the scattering cross-section of the holes. This is evidenced in the bare crystal dispersion (data not shown) and in the differential dispersion of **Figure 16(h)** showing that the LC reorganization affects not only the lower-frequency bands (as for the slits) but also leads to additional dispersion modification at normal incidence at around 1.75 eV. The latter feature is related to the changing SPP scattering properties of the basis subsequent to the LC reorganization in the hole when the static electric field E_0 is varied. However, the fact that both the dispersion, as evaluated from **equation 3**, as well as the SPP scattering properties of the holes are sensitive to the LC reorganization in a correlated manner makes it difficult to determine how the SPP scattering properties of the holes are changing as a function of the applied field. Additionally, LC geometries in confined structures can be extremely complicated requiring significant computational power in order to address them satisfactorily [78].

The contribution of the scattering cross-section of the basis to the differential optical response of the plasmonic crystal is further accentuated for the array of holes with circular cross-section. In this case, the change in the index of refraction due to LC reorganization leads primarily to the modification of a localized mode in the hole, which dominates the optical properties of the array at the frequency of about 1.84 eV at normal incidence. These results are shown in **Figures 16(j)–16(l)**. Returning briefly to the spatial sensitivity of the low-frequency SPP-BM to changes of the refractive index occurring in the bulk of the LC region, it is worth noting that the (1,1) band in **Figures 16(f)–16(h)** is not as sensitive to the LC reorganization as the (±1,0) bands. Although, these bands span the same spectral range in the dispersion, the field associated with the (1, 1) band is more confined to the Au/LC interface than the (±1,0) band, in agreement with our discussion in Section 1.2.1.

In order to investigate the dynamic response of the LC cell with a view on potential applications, the transmission of the array of circular holes (200 nm-thick film) was monitored as the cell was subjected to a time-modulated electric potential. The energy and momentum along the SPC surface of the incident light were chosen such as to probe the mode at $k = 0\ \mu m^{-1}$, $\hbar\omega = 1.85$ eV corresponding to a vacuum wavelength of 670 nm, where a large variation of transmitted light intensity was observed on application of the electric field (indicated by the yellow arrow in **Figure 16(d)**). The change in optical transmission with increasing applied electric field is illustrated by the plot shown in **Figure 17(a)**: the conformational change of the LC layer can be seen to take place between 0.4 and 5 kV cm^{-1} and, within this region, the transmission is unstable. Both the spectral shift and change in amplitude of the transmission peaks are also clearly visible with applied field (**Figure 17(b)**). The observed dynamic response of the transmission under the temporal modulation of the applied electric field at frequencies of 2.5 and 100 Hz demonstrate the ability of the structure to switch and modulate light at a frequency corresponding to that of today's modern LC displays with less than a 10 % reduction in amplitude. As the transmitted intensity is convoluted with the shape of the transmission spectrum at any particular wavelength, the possibility, therefore, exists to design plasmonic structures in order to perform complex optical signal processing and pulse shaping.

1.2.4.2 Magneto-optical control of surface plasmon polariton Bloch modes

The ability to actively influence plasmonic modes with external controlled magnetic fields has always been of paramount importance due to possible applications in magnetic and magneto-optical data storage as well as for the development of microscale optical isolators based on nonreciprocal photonic effects. The theory of the in-plane modification of the SPP behavior on smooth metal films by an applied magnetic field has recently been developed and some experiments have been carried out [79,80]. Magnetic fields have been proposed as a means to control the enhanced transmission of SPP crystals, but the required magnetic field strength is too high for the effect to be observed in good metals; this may, however, be achieved with doped semiconductors supporting plasmonic excitations in the THz spectral range [81–83]. It was predicted that the effect of magnetic fields on SPP modes can be enhanced via coupling to a neighboring ferromagnetic medium [80]. It was also demonstrated that the magneto-optical effects in magnetic medium can be enhanced via the resonant coupling to plasmonic modes, similarly to the resonant coupling in magneto-photonic crystals [84,85]. Additionally, magneto-optical effects due to cylindrical surface plasmons supported by arrays of plasmonic nanorods have also been investigated [86].

In this section we show how the optical transmission of magneto-plasmonic crystals, based on metal/ferroelectric layered structures, can be controlled and modulated under the application of an external static magnetic field. We show that the magneto-optical effects on the SPP field components induced by the magnetization of a dielectric layer with magneto-optical properties affects the optical properties of the SPC through both a net modulation of the transmitted intensity and a shift of the SPP-BM band edges add reference to New Journal of Physics.

In the absence of the external magnetic field, the iron garnet (IG) film exhibits an in-plane magnetic anisotropy (~2 kG). The optical properties of the IG in the visible spectral range are dispersive, but assuming an average refractive index $n \approx 2.34$ is a reasonable approximation. The magneto-optical response of an IG film in the polar (p) configuration, when the external magnetic field is normal to the interface, longitudinal (l) or transversal (t) configurations when the magnetic field is in the plane of the interface and in the direction of SPP propagation or

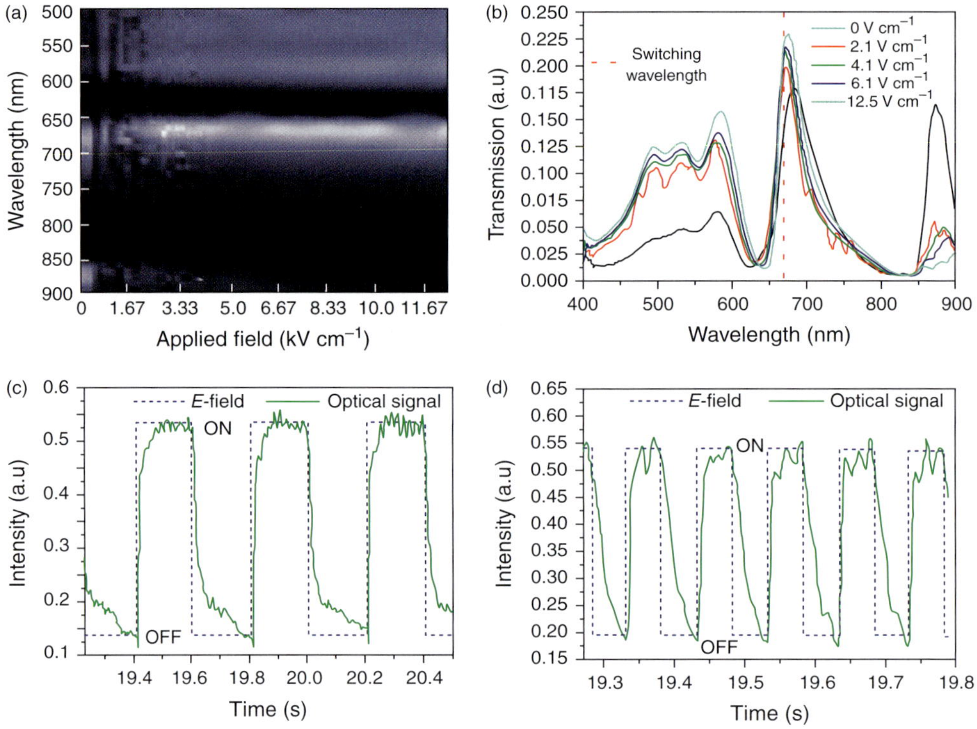

Figure 17 Electric-field dependence of the optical transmission for the 2D SPC of nanoholes shown in **Figure 16(i)**. (a) The change in transmission for increasing applied electric field. (b) Transmission spectra for several values of the applied electric field. The wavelength used to probe the switching dynamics is shown by a dashed orange line. (c,d) The transmitted light intensity at a wavelength of 670 nm through the periodic array of holes as the potential across the LC cell is modulated at 2.5 and 100 Hz, respectively. All measurements were performed at normal incidence.

perpendicular to it, respectively, is described by the permittivity tensor [87]:

$$\varepsilon = \begin{pmatrix} \varepsilon_0 & i\varepsilon_p & i\varepsilon_t \\ -i\varepsilon_p & \varepsilon_0 & i\varepsilon_l \\ -i\varepsilon_t & i\varepsilon_l & \varepsilon_0 \end{pmatrix} \qquad (7)$$

where only one pair of off-diagonal elements ε_i is nonzero for any given *p*-, *l*- or *t*-configuration. For an absorbing medium, the tensor components are complex. To the first order in magnetization, the diagonal components ε_0 do not depend on magnetization *M*. The off-diagonal components $\varepsilon_{p,l,t}$ vary linearly with *M* and give rise to magneto-optical effects, such as Faraday or Kerr rotation resulting in the ellipticity of the transmitted or reflected light, which initially is linearly polarized. The variation of the off-diagonal components of the permittivity tensor of the IG are typically on the order of $\Delta\varepsilon \sim 10^{-3}$.

The transmission spectra of the magneto-plasmonic crystals under investigation (**Figure 18**) are determined by both the transmission of the IG film and the transmission of the nanostructured Au film related to plasmonic resonances. Due to relatively high refractive index of the substrate compared to the air superstrate, SPP-BM bands on both the Au/air and the Au/IG interfaces are significantly spectrally separated at normal incidence. Additionally, as a result of the short period of the SPC crystal lattice, the low-order SPP-BM resonances associated with the Au/air interface are situated in the spectral range where the substrate absorption is significant and are neglected in the following considerations. Based on **equation 3**, the transmission resonances observed in the visible spectral range (**Figure 19(a)**) can be assigned to the scattering of photons due to the (1,1) and (2,0) SPP-BM on the Au/IG interface [17].

The application of the magnetic field leading to small variations of the off-diagonal elements of the permittivity tensor may affect the transmission spectra via the modifications of the effective refractive index of SPP-BM as well as via the polarization effects due to the enhanced Faraday rotation in the

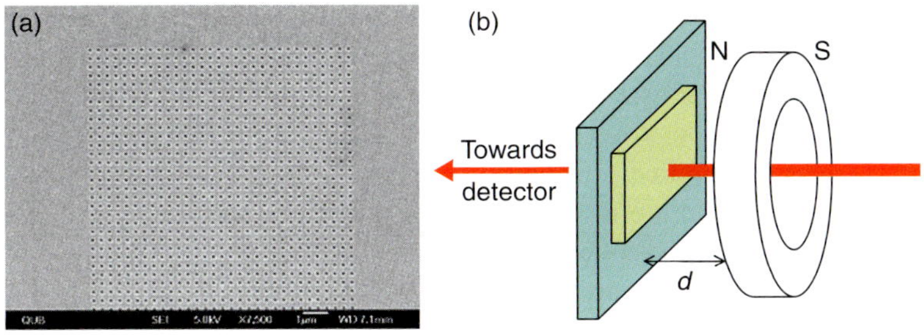

Figure 18 (a) SEM image of a 400 nm period 2D SPC made of 125 nm diameter circular holes in a 50 nm-thick Au film deposited onto a 3.5 μm-thick single crystalline IG of composition $Lu_{2.4}Bi_{0.6}Fe_{4.8}Ga_{0.2}O_{12}$. (b) Schematic of the magnetic-field-controlled transmission measurements. The spectroscopic measurements were performed in the configuration of **Figure 10**.

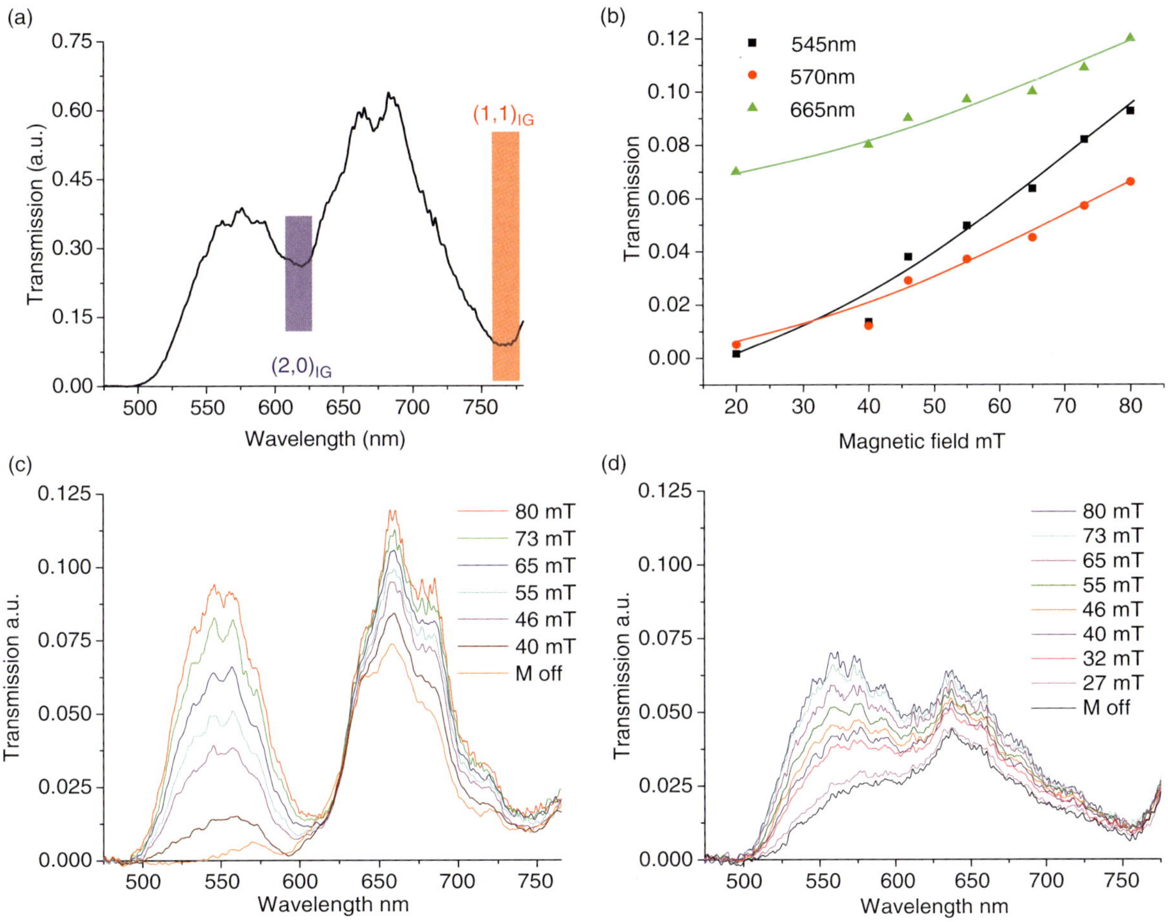

Figure 19 (a) Normal incidence zero-order transmission spectrum of the magneto-plasmonic crystal. The position of the SPP-BM estimated from **equation 3** is shown. (b) The dependence of the cross-polarized transmission at the wavelengths of 545, 570, and 645 nm on the applied magnetic field. (c,d) Cross-polarized transmission spectra of the magneto-plasmonic crystal in the varying external magnetic field (0–80 mT) for the polarization of the incident light along the two axes of the crystal lattice: (c) for (1,0) and (d) for (0,1) directions.

plasmonic resonances [85]. Both effects can be easily observed in a cross-polarized configuration with the analyzer probing the optical properties of the crystal along the (1,0) lattice vector, the incident field being polarized along the (0,1) lattice vector, and vice versa. The effect of the applied magnetic field on the cross-polarized transmission spectrum is shown in **Figures 19(b)–19(d)** for different polarizations of the incident light as a function of the distance between the sample and the magnet: a decreasing distance leads to an increasing applied magnetic field. The changes in the transmission are most pronounced in the 500–700 nm wavelength range where the magneto-optical effects in the IG film are strongest. The applied magnetic field impacts the optical properties of the SPC in two ways. First, the magnitude of the cross-polarized resonant transmission is strongly modified with varying magnetic field. This modification is dispersive, with a stronger effect at higher optical frequencies. Second, the transmission spectra also reveal the changes in the SPP-BM bandgap structure. In particular, the field-induced reorientation of magnetization shifts the band-edges of both the (1,1) and the (2,0) bands to longer wavelengths as the magnitude of the magnetic field is increased. Here again, the effect is dispersive, with a stronger shift observed for the (2,0) bands. For the SPP-BM existing on the interface of a metal and a dielectric with permittivity given by the tensor in **equation 7**, the electric field components E_x (along the SPP propagation direction) and E_z (normal to the interface) are coupled via off-diagonal tensor components. Thus, pure TE and TM surface electromagnetic modes cannot be separated and plasmonic excitations are not purely TM waves in longitudinal and polar configurations [80]. Significant changes of the effective refractive index of SPPs have been observed even for smooth metal films on a magneto-optically active substrate with nonreciprocal SPP propagation [80]. Even stronger effects should be expected near the SPP band gap, since the band-gap formation will be affected by the nonreciprocal effects that remove the directional degeneracy of the $(i,0)/(-i,0)$, $(i, i)/(-i,-i)$, etc. modes. At the same time, for the polarization of the incident field normal to the previous one, the observed magnetic-field dependence is slightly different (cf. **Figures 19(c)** and **19(d)**). This originates from the preferential orientation of the magnetic domains in the IG film leading to the situation where predominantly longitudinal or transverse configuration is realized for different polarizations of the incident light. Application of the external field results in the changes in all the off-diagonal elements of the permittivity tensor; thus, its influence on different SPP Bloch modes along different directions will be different.

To gain a better understanding of the experimental observations, full-vectorial finite-element modeling (FEM) of the effects of the magnetic field on the optical properties of magneto-plasmonic crystals have been performed. The model crystal consists of circular holes of diameter 150 nm placed in a square array with a period of 400 nm in a 50 nm-thick Au film. The crystal is supported by a 50 nm-thick IG substrate whose optical response is described by the dielectric tensor as in **equation 7**. The superstrate is air. The chosen thickness of the IG film enables us to analyze the behavior of the plasmonic crystal and minimize the magneto-optical effect that the substrate has on the polarization of the transmitted light. For technical reasons related to numerical simulations, a dispersionless and lossless permittivity of the IG film was assumed with the average values in the visible spectral range of $\varepsilon_0 = 5.5$ and $\varepsilon_{t,p,l} = 0.02$ for the diagonal and off-diagonal elements, respectively. In the simulations, a linearly polarized plane wave is normally incident on the crystal from the air superstrate. The incident field is polarized along one of the crystal's principal axes and the polarization of the light transmitted through the structure is analyzed in the zero-order direction as a function of the magnetization direction of the garnet substrate. To summarize these results, the introduction of nonzero off-diagonal elements in the dielectric tensor leads to the rotation of the plane of polarization of the transmitted light and is observed in the cross-polarized spectra where maxima in the transmission are observed near the band edges of the crystal. This demonstrates that the polarization rotation due to magneto-optical effects in the substrate is minimal in the calculated geometry, where the sensitivity of the SPP-BM dispersion to the dielectric tensor governs the transmission of the structure. A similar effect is observed for both in-plane and out-of-plane magnetization, with a difference originating from the sensitivity of the hybrid modes, responsible for the appearance of the band gaps.

1.2.4.3 Light-controlled optical transmission through nonlinear surface plasmonic crystals

All-optical signal processing in integrated photonic circuits and its applications in optical communications and computing necessitates the ability to control light with light [88]. Nonlinear optical effects required for this purpose can be achieved by applying the mechanisms of slow molecular motion (ms timescales), such as LCs or ultra-fast electronic nonlinearities (up to fs timescales) [17,89]. All-optical devices based on various types of optical nonlinearities have been considered in the past. The main drawbacks for the majority of such devices are twofold: the limitation on their minimum size required to provide a sufficient light pass-length to achieve a sizable nonlinear response; and the relatively high operational light intensities. To remedy these limitations, nonlinear optical devices based on photonic crystal defects have recently been proposed that utilize the electromagnetic field confinement and enhancement at the photonic defect location to enhance nonlinear effects [90–93]. Even stronger field enhancements may be achieved using plasmonic nanostructures to tailor electromagnetic modes with effective size well below the diffraction limit of the light. Significant field enhancements are routinely obtained in plasmonic structures of various geometries for surface-enhanced Raman scattering, stimulated emission, second-harmonic generation, and photon tunneling [2,9]. This enhancement is the result of an increase of the density of electromagnetic states near a metal surface corresponding to the spectrum of plasmonic excitations. Addressing such surface electromagnetic modes is, therefore, a possible way to enhance the effective nonlinear response of hybrid nanostructures formed when a nonlinear material is placed near a metal surface [94].

One example of such nonlinear optical metamaterials are bulk nonlinear materials doped with metallic nanoparticles [95,96]. In this metamaterials, the excitation of the hybrid structure at the wavelength of the localized surface plasmon resonances of the nanoparticles leads to an increase in the effective nonlinear susceptibility of the composite material compared to either materials taken separately. Similarly, the coverage of metallic nanoparticle arrays (e.g., nanospheres or nanorod assemblies) with nonlinear material results in the enhancement of the nonlinear response of the dielectric while also lowering the light intensities required for nonlinear action [97,98]. The utilization of field enhancement effects associated with localized surface plasmon modes has also been shown to enable controlled photon tunneling through nanoscale pinholes in a metal film covered with nonlinear polymer [99, 100]. In particular, measurements of photon tunneling through individual, naturally occurring, nanometer-scale pinholes have provided indication of 'photon blockade' effects, similar to Coulomb blockade phenomena observed in single-electron tunneling experiments [99].

The high sensitivity of SPP resonances to the structured surface on which they propagate could bring a viable solution to the problem of their all-optical control [1,9]. In this scheme, minute changes induced in the refractive index of a nonlinear material placed on the metal surface would significantly influence the surface plasmon resonant conditions and mode propagation length along the interface. This was first shown with smooth metal films in the Kretschmann geometry using LC phase transformations under the influence of light-induced thermal effects. In this configuration, both intensity switching of the reflected light and bistability of the incident light intensity were observed [89]. The latter was explained by a positive feedback due to the intensity-dependent refractive index of the LC placed on the metal film and illuminated by the evanescent field of the SPP wave. Nanostructured metal surfaces such as SPCs provide additional levels of flexibility [9,48,101,102]. Their geometry tailors the wavelength-dependent resonant conditions and the electromagnetic-field enhancement conditions in spatial regions a few wavelengths in size and are, therefore, suitable for the development of integrated all-optical devices.

The nonlinear optical response of hybridized SPCs can generally be observed throughout the transmission spectrum of the crystal. They originate mainly from SPP-BM excitations, cylindrical surface plasmons, as well as localized modes in the metal films [10,99,100,103]. An example of the zero-order normal incidence transmission spectrum from two SPP crystals is shown in **Figure 20**. The transmission spectrum of the bare (not coated with nonlinear polymer) crystal consists of a set of peaks corresponding to resonances in its spectrum of the electromagnetic modes that are coupled to photons. Coating the SPC originally in air ($n = 1$) with the nonlinear polymer ($n = 1.7$) modifies the optical response of both SPP-BM and localized modes of the holes. The analysis of the band-gap structure and SPP-BM position of the coated SPP crystal can be made in a good approximation using **equation 3**. This shows that the band-edges at around 720 and 660 nm correspond to the excitation of SPP-BM (ΓM and ΓX directions, respectively) on the polymer/metal interface.

The origin of the peak at around 700 nm is related to the coupling of localized surface plasmon

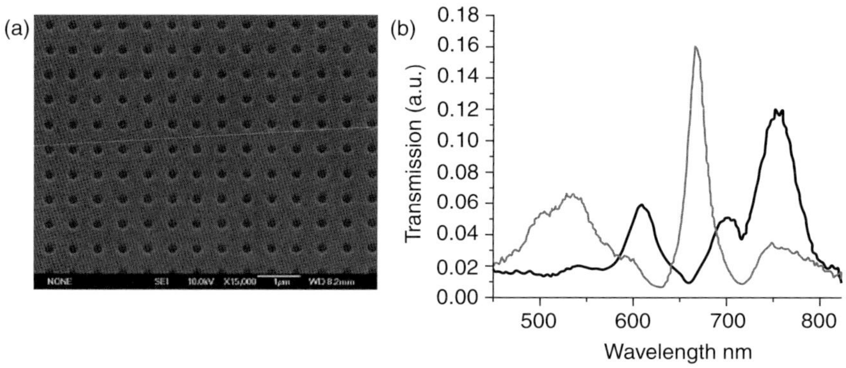

Figure 20 (a) SEM image of a 600 nm period 2D SPC made of 200 nm diameter circular holes in a 200 nm-thick Au film. (b) Normal incidence zero-order transmission of the SPC before (gray line) and after (black line) the deposition of the nonlinear polymer.

resonances of the holes with SPP-BM at the metal/polymer interface. Taking advantage of the nonlinear response of the polymer, a control illumination (at $\lambda_c = 488$ or 514 nm) is used to modify its refractive index of the polymer and, thus, the optical density of states of the nonlinear SPC. These modifications induce intensity-dependent changes in the optical transmission of the crystal. They are monitored in the zero-order transmitted light spectrum (**Figure 10**) using a probe beam whose low intensity does not affect the polymer's permittivity. **Figure 21(a)** shows the transmission of the SPC for two extreme values of control beam intensities. The control wavelength is $\lambda_0 = 514$ nm. A similar nonlinear transmission is observed for both control wavelengths, which affects the SPC transmission throughout the visible spectrum with different spectral regions exhibiting suppression (negative differential transmission) or enhancement (positive differential transmission) of the transmission (**Figure 21(b)**).

The observed transmission changes can reach up to 60% of the initial transmission of the SPC at some wavelengths (**Figure 21(b)**). In addition to the changes in the transmission magnitude, the position of both the maxima and the band-edges in the transmission spectra are also affected by the variation of the control intensity. This is an indication of the modification of the whole surface plasmon mode spectrum by the control light via induced changes in the polymer's dielectric constant. The nontrivial shifts in the SPP-BM band-edges observed in **Figure 21(a)** are an indication of the nonuniform changes in the dielectric constant of the polymer. These will not be discussed here because of space constrains. However, using the observed SPP-BM band-gap shift along with **equation 3**, the changes in the effective refractive index of the polymer induced by the control light can be estimated to be $\Delta n \sim 10^{-3}$–10^{-4}. This estimation assumes a uniform change of the polymer refractive index at the metal/polymer interface. In reality, however, the spatial variations of the induced refractive index changes follow the spatial variations of the field of the SPP-BM excited by the control light and are significantly nonuniform. Therefore, larger local Δn values (greater than 10^{-3}) may be expected in the maxima of the field distributions at the control light wavelength. This is several orders of magnitude larger than the Δn changes detectable in sensing experiments with surface plasmons, thus, further reduction of the control light intensity may be possible.

For low-intensity signal light, and neglecting the nonlinear response of the metal, the variations of the dielectric constant of the nonlinear material can be considered as solely induced by the control light of frequency ω_c:

$$\varepsilon(\boldsymbol{r}, \omega) = \varepsilon^{(0)}(\omega) + 4\pi\chi^{(3)}|E_L((\boldsymbol{r}, \omega), \omega_c, \boldsymbol{r})|^2 \quad (8)$$

where $\varepsilon^{(0)}(\omega)$ and $\chi^{(3)}$ are the linear dielectric constant and the third-order nonlinear susceptibility of the polymer, respectively, and $E_L((\boldsymbol{r}, \omega), \omega_c, \boldsymbol{r})$ is the local, position-dependent electric field of the control light, which is determined by the SPP crystal parameters. Thus, the field distribution described by $E_L((\boldsymbol{r}, \omega), \omega_c, \boldsymbol{r})$ depends on the induced permittivity changes in the nonlinear SPC and, in turn, the permittivity variations depend on the local SPP field. As a consequence, the crystal eigenmodes and associated

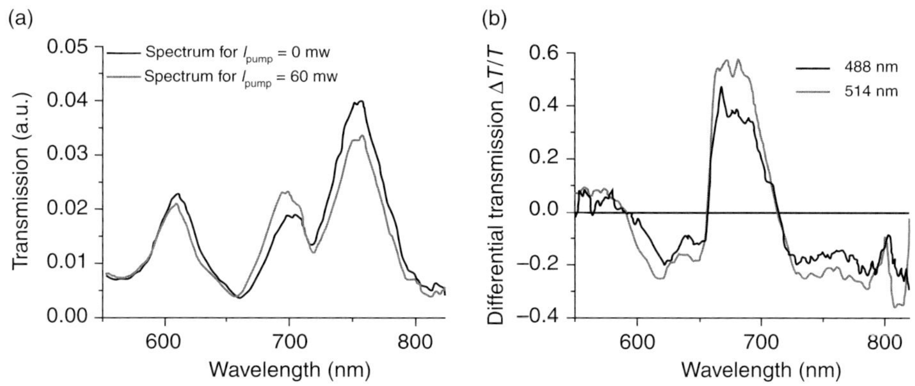

Figure 21 (a) Normal incidence zero-order transmission spectra of the nonlinear SPC measured at a control wavelength $\lambda_c = 488$ nm with incident power OFF (black line), and ON (gray line, $P = 0.75$ kW cm^{-2}). (b) Differential transmission for the same control power at control wavelengths of $\lambda_c = 488$ nm (black line), and $\lambda_c = 514$ nm (gray line).

field distribution for a given control intensity and wavelength depend self-consistently on the spatial variations of $\varepsilon(\boldsymbol{r}, \omega)$. The permittivity variations in turn modify the interaction of the signal light with the SPC, thus, changing the transmission spectrum, which is determined by the distribution of the dielectric constant around the nanostructured metal film [60,61]. This leads to a multivalued dependence of the transmission of SPC on the control light intensity in the case of the dielectric constant dependence described by **equation 8**. The changes in the spatial distribution of the polymer's permittivity $\varepsilon(\boldsymbol{r}, \omega)$and the magnitude of these changes, both being responsible for the SPC's eigenmode spectrum, provide a mechanism for bistable behavior of the crystal's optical transmission with the intensity of the control light. As in a typical configuration for optical bistability, the above-described process requires a nonlinear transmission dependence on the control light to achieve a transistor-type effect as well as a 'built-in' feedback mechanism (described by **equation 8**) to enable bistability [88].

In the case of the SPC considered in **Figure 20(a)**, for the control wavelength of 488 nm, and in the range of control intensities where only reversible changes of the transmission occur (below 5 kW cm^{-2}), bistability in the SPC's transmission has been observed for all main resonances of the transmission spectrum (**Figures 22(a)–22(c)**). However, the behavior of the hysteresis loops with control intensity is different at different probe wavelengths: the transmission increases at some resonances and decreases at the others for the same variations of the control light intensity. The complete switching occurs at similar control intensities for the transmission resonances considered, and the transmission returns to its original values when the control light is turned off.

The analysis shows that the different behavior of the bistability loops observed at different probe wavelengths is related to the origin of the transmission resonances at these wavelengths [10]. For these wavelengths the influence of the nonlinearity is chiefly a modification of the absolute value of the transmission at the resonance. The behavior of the peak at around 700 nm is particularly interesting: with an increased control intensity, not only does the transmission change, but the shape of the resonance also changes. This highlights the complex origin of this peak corresponding to the coupling of localized surface plasmon resonances from the holes with SPP-BM at the metal–polymer interface [10].

At the control wavelength of 514 nm (**Figures 22(d)–22(f)**), the nonlinear transmission is observed for all resonances as well. Compared to the experiments with the 488 nm control wavelength, the observed relative transmission changes in this case are up to 50% stronger at some probe wavelengths. The most prominent difference, however, comes from the very weak bistable behavior obtained at this wavelength, the exception being at around 690 nm [10]. The observed control wavelength dependence underlines the importance of both nonlinearity and feedback mechanism for the bistability to occur. At the 488 nm control wavelength, the nonlinear changes are smaller than at 514 nm, but the feedback is stronger, leading to bistability in the nonlinear transmission at all resonances of the SPC. At the same time, in spite of the stronger nonlinear

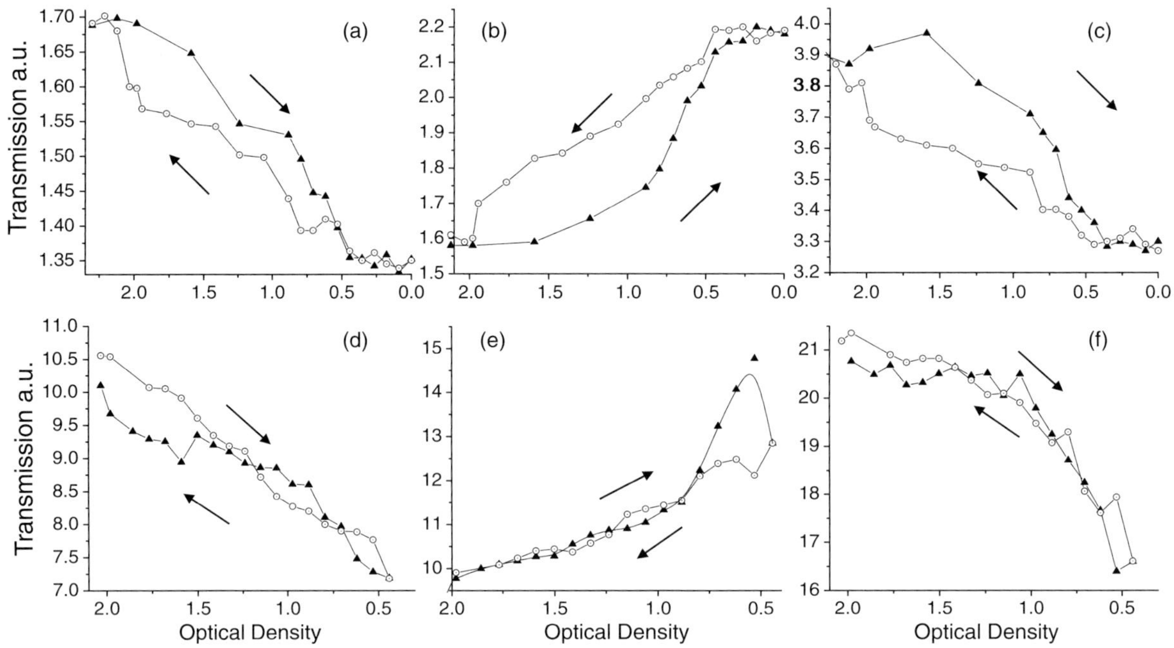

Figure 22 Normal incidence zero-order transmission of the nonlinear SPC of **Figure 20(a)** as a function of incident power for selected probe wavelengths: (a–d) 622 nm, (b–e) 690 nm, and (c–f) 750 nm. Increasing and decreasing incident power transmissions are labeled with triangles and circles, respectively. The plots are given as a function of optical density in the incident beam, that is, a smaller optical density corresponds to a higher incident power. Control wavelengths of $\lambda_c = 488$nm, and 514 nm have been used for (a–c) and (d–f), respectively.

dependence of the transmission, bistability is not always present under the 514 nm control illumination due to a weaker feedback. The dependence of the bistability type on the control light wavelength can be understood by taking into account that the SPP field distribution at this wavelength plays a crucial role in triggering bistability. In particular, under the 488 and 514 nm illumination, different resonances of the SPC are excited [10].

Peculiar all-optical control of the linear optical properties of SPP Bloch modes has also been demonstrated using acousto-optical effects [104]. In this case, a control optical pulse excites surface acoustic waves in the SPC, resulting in a mechanical modulation of the crystal's period and associated dispersion and optical properties. A coherent control of such acoustic vibrations can be achieved with an appropriate sequence of pulses.

1.3 Metallic Nanorod Arrays

Dielectric, semiconductor, and metallic nanowires have recently attracted significant attention due to their remarkable optical and electronic properties [105–110]. Among others, these include light guiding in sub-wavelength diameter metallic wires, negative permeability of pairs of closely spaced metallic rods, and optical plasmonic antenna [109,110]. All these effects rely on surface plasmon modes supported by a metallic nanowire or nanorod. Progress in nanofabrication techniques has opened up the possibility to fabricate metallic nanorods in macroscopic size arrays with their long axes aligned perpendicular to the substrate [111]. This geometry of oriented nanorods, with a much reduced influence of the underlying substrate, impacts the spectral optical properties of the array as well as the spatial field distribution in it, which both differ strongly from those of nanorods studied in colloidal solutions [112,113] or placed on substrates [109,114]. The optical properties of such arrays with their strongly anisotropic behavior can be beneficial for spectroscopic, sensing, and imaging applications [115–119]. One original and fundamental property of nanorod arrays is in the possibility of wide-range tuning of the geometrical parameters of the structure so as to control the electromagnetic interaction between the rods. This can be done, for example, by changing the inter-rod distance and is crucial for determining the reflection, transmission, and extinction of the array. It is also important for the development of new applications such as guiding electromagnetic energy on the nanoscale along the chains of nanorods.

In this section, a novel nanostructured material made of strongly interacting Au nanorods is presented. In contrast to widely studied plasmonic geometries involving nanorods on substrates or suspended in solutions, the nanorods studied here: have a much higher aspect ratio; are standing on a substrate with their long axes oriented perpendicularly to the substrate; and are placed in a quasi-regular lattice. This impacts their optical properties dramatically. In particular, based on experimental and numerical investigations, it is found that assemblies of closely spaced metallic nanorods support the formation of an extended plasmonic mode delocalized through the array. Both the spatial and spectral properties of this mode are discussed in detail. Furthermore, its controlled interaction to a neighboring molecular excitons is investigated in both weak- and strong-coupling regimes.

1.3.1 Plasmonic Nanorod Assembly

One can briefly describe a nanorod assembly as a set of aligned nanorods oriented perpendicularly to a supporting substrate and either embedded in a dielectric matrix or free-standing in air. Such a nanorod assembly can be grown electrochemically in a substrate-supported, porous anodized aluminum oxide template (AAO) [111]. The substrate is a multilayered structure comprising a 1 mm-thick glass slide, a 10 nm-thick tantalum pentoxide (Ta2O5) base layer and a 5 nm-thick Au film acting as the working electrode for the electrochemical reaction. Both layers are grown using magnetron sputter deposition. Tantalum pentoxide is deposited by sputtering tantalum using a 20% oxygen/80% argon mixture. An aluminum (Al) film of thickness up to 500 nm is then deposited onto the electrode using planar magnetron sputterring. This Al film is subsequently anodized at constant voltage with a platinum counter-electrode in a 0.3 M sulfuric acid electrolyte at $\sim$1 °C to produce the porous alumina template [111]. The diameter, separation, and the ordering of the Au rods in the assembly are determined by the geometry of the AAO template, that is, the anodization conditions and post-chemical etching processes. The rod diameter and spacing can vary from about 10 nm to a few hundred nanometers by altering the AAO growth conditions and post-growth chemical etching processes. The length of the rods is chosen by controlling the electrodeposition time, and can be varied from about 20 nm to the limiting value dictated by the thickness of the AAO template, typically a few hundred nanometers. All of these parameters – rod aspect ratio, rod length, rod separation – impact the optical response of the array. In contrast to the widely studied plasmonic geometries involving nanorods on substrates or suspended in solutions, the nanorods studied here have a much higher aspect ratio, are standing on a substrate with their long axes oriented perpendicularly to the substrate, and are placed in a quasi-regular lattice. An example of nanorod assembly is shown in **Figure 23** presenting an SEM image of the assembly of Au rods after removal of the AAO matrix to enable their observation. A cross-sectional TEM view, performed before AAO removal, is also shown in **Figure 23**.

1.3.2 Optical Properties

1.3.2.1 Eigenmodes of nanorod arrays: spectral properties

The optical properties of the arrays were investigated by measuring the transmittance T of the sample and plotting the extinction $-\log_{10}(T)$ as a function of wavelength for different angles of incidence and polarizations of the incident light. Considering the typical length of the rods in the (300–400 nm) range, both scattering and absorption contribute to T and, therefore, $-\log_{10}(T)$ represents the extinction of the sample. The optical response of the nanorods when forming such arrays is governed by a collective plasmonic mode resulting from the strong electromagnetic coupling between the dipolar longitudinal plasmon resonance supported by individual nanorods. Both the spectral position of this collective plasmonic resonance and the associated electromagnetic field distribution in the nanorod array strongly depend on the inter-rod coupling strength; this behavior differs from the dipolar response of isolated nanorods from which they originate. These unique properties therefore demonstrate strong potential for the use of oriented metallic nanorod assemblies in the manipulation of light in nanoscale waveguide applications, sensing and non-linearity enhancement applications, as well as in sub-wavelength imaging [38,120].

The experimental extinction spectra, measured for an array of Au nanorods embedded in an AAO matrix ($n \approx 1.6$) are shown in **Figure 24(a)** for different angles of incidence. At normal incidence or also with s-polarized light (the incident electric field

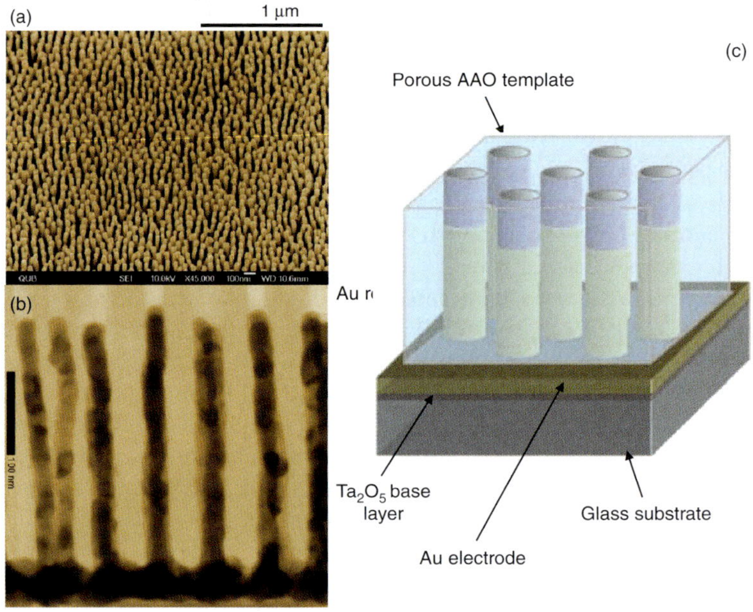

Figure 23 (a) SEM image of an array of Au nanorods in air. The sample is tilted to give the perspective view shown. (b) TEM cross-section of Au nanorods embedded in AAO (the scale bars are 100 nm). The bottom Au electrode is also visible in this image. (c) Schematics of the 3D geometry showing the assembly of Au nanorods standing on the transparent substrate and embedded in AAO. The AAO thickness exceeds the nanorod length.

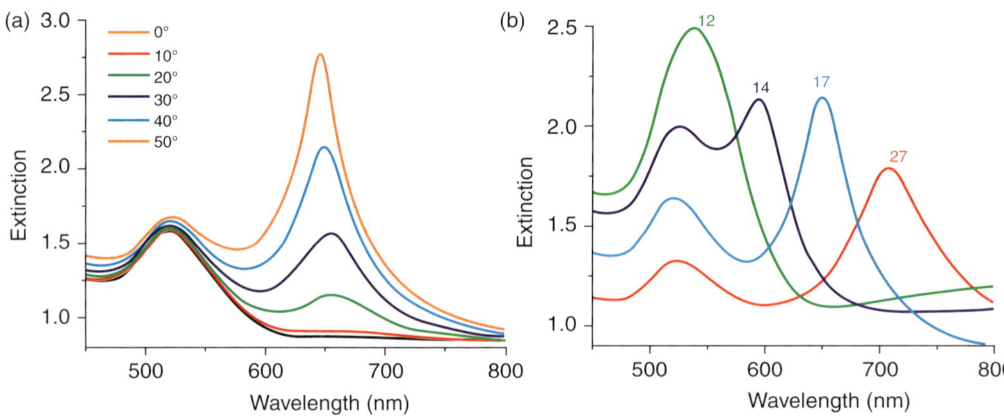

Figure 24 (a) Zero-order optical extinction spectra for different of angles of incidence. The spectra have been taken for nanorods embedded in an AAO matrix. The nanorods length is 300 nm, diameter is 30 nm, and the inter-rod distance (center-to-center) is about 100 nm. (b) Zero-order optical extinction spectra of arrays of Au nanorods in AAO as a function of rod aspect ratio. The nanorod length is 400 nm and the inter-rod distance (center-to-center) is about 100 nm. The curves are labeled according to the rod's aspect ratio corresponding to rods with diameter ranging from about 15 to 30 nm.

vibrates perpendicular to the nanorod long axis), the spectra reveal one single peak at around 520 nm wavelength. This resonance is associated with the transverse (T) plasmonic excitation in the direction normal to the nanorod long axes [115]. At oblique incidence and with p-polarized light (incident electric field has a component both along and perpendicular to the nanorod long axes), the spectra contain two peaks: the above-mentioned T mode as well as a longer-wavelength resonance, the L mode associated with a plasmonic excitation polarized along the nanorod long axes. This long-wavelength peak becomes more pronounced at larger angles of incidence for which this longitudinal plasmon resonance is excited more effectively. The angular sensitivity of the spectra of **Figure 24(a)** reflects the strong anisotropy of the structure defined by the orientation of the nanorods in the assembly. The general behavior of these two eigenmodes is consistent with the response of the dipolar modes supported by isolated or weakly interacting rods [121].

The L-mode resonance wavelength is strongly dependent on both the rod aspect ratio and the distance between the rods in the array. In accordance with the dipolar plasmonic response of nanorods, an increase in the nanorod aspect ratio causes the two resonances to split further apart spectrally, with the T mode undergoing a blue-shift while the L mode moves towards longer wavelengths. This behavior is illustrated in **Figure 24(b)** in the particular case of a varying rod diameter; however, it is also observed when the rod length is varied. The inter-rod distance can also be varied in order to tune the spectral position of the resonances in the array throughout the visible spectrum [38,120]. The spectral position of this resonance for a given nanorod aspect ratio is not, however, consistent with that of the localized surface plasmon resonances of an isolated rod or even for geometries involving dimers, for which the longitudinal resonance is expected to appear at a much longer wavelength [23,24,122,123]. For example, an isolated nanorod with a diameter of 30 nm and an aspect ratio of 10 embedded in AAO exhibits a longitudinal dipolar resonance at around 2 μm. The resonance demonstrating similar polarization properties in the array of nanorods of **Figure 24** is observed in the visible region of the spectrum at around 650 nm, therefore strongly blue-shifted compared to the expected position of the dipolar resonance. The direction of the L-mode shift as a function of aspect ratio is qualitatively identical to the behavior of the longitudinal dipolar resonance of an isolated nanorod, the value of the shift, being, however, much smaller than one would expect for an isolated dipolar resonance [122].

Figure 25(a) illustrates the spectral dependence of both the dipolar longitudinal plasmonic resonance supported by an isolated ellipsoid and the L mode of a nanorod assembly as a function of particle aspect ratio. It is clear from **Figure 25(a)** that both the spectral position of the resonances for any given aspect ratio and their qualitative variation as a function of aspect ratio are specific to the resonance considered. First, as mentioned earlier, the L mode is strongly blue-shifted compared to the dipolar resonance of the isolated particle, with a saturation in the red-shift occurring for rods with larger aspect ratios. At smaller aspect ratios the L mode is located around 550 nm, overlapping the T mode in the assembly, while an isolated particle of similar aspect ratio would show two very distinct dipolar resonances spectrally separated by a few hundred nanometers [122]. This spectral overlap is clearly observed in the spectrum obtained for nanorods with an aspect ratio of 12 (**Figure 24(b)**) and agrees with the observation of strong electromagnetic coupling [22, 24, 123, 124].

As can be seen form **Figure 24(a)**, in addition to its amplitude, the spectral position of the L mode in nanorod arrays depends on the angle of incidence of the probe light and, with an increase in the angle of incidence, shifts toward shorter wavelengths. This angular dispersion of the mode is shown in **Figure 26(a)** for the spectra of **Figure 24(a)** and contrasts with the dispersionless behavior of the localized plasmonic modes of isolated plasmonic modes. The angular dispersion of the L mode was shown to depend on the inter-rod coupling strength, as it increases for a decreasing effective inter-rod distance $\bar{n}d$, where $\bar{n}$ is the average refractive index of the medium embedding the rods and d is the distance between neighboring nanorods. This property is plotted in **Figure 26(b)** and was observed in the spectral dependence of nanorod arrays embedded in alumina with an additional air shell around the nanorods [120].

The experimental observations of the spectral position of the L mode as a function of rod aspect ratio, nanorod concentration, and inter-rod distance, as well as the dispersive behavior of the resonance, all indicate that although the L mode in the extinction spectra is related to the plasma oscillations along the long axis of the nanorods, the nature of this mode is in fact more complex. The observed behavior of the L mode can be explained by the formation of a spatially extended (collective) plasmonic resonance in the array caused by the interaction between the plasmonic modes supported by the nanorods. In particular, this extended mode can be considered as a linear combination of the plasmonic modes of each individual nanorod in the assembly experiencing electromagnetic interaction from neighboring nanorods if $k_p l < 2\pi$, where k_p is the field extension of the plasmonic mode outside the metal and l the distance between the nanorods in the array. Considering the interaction between two nanorods only, the coherent interaction between the

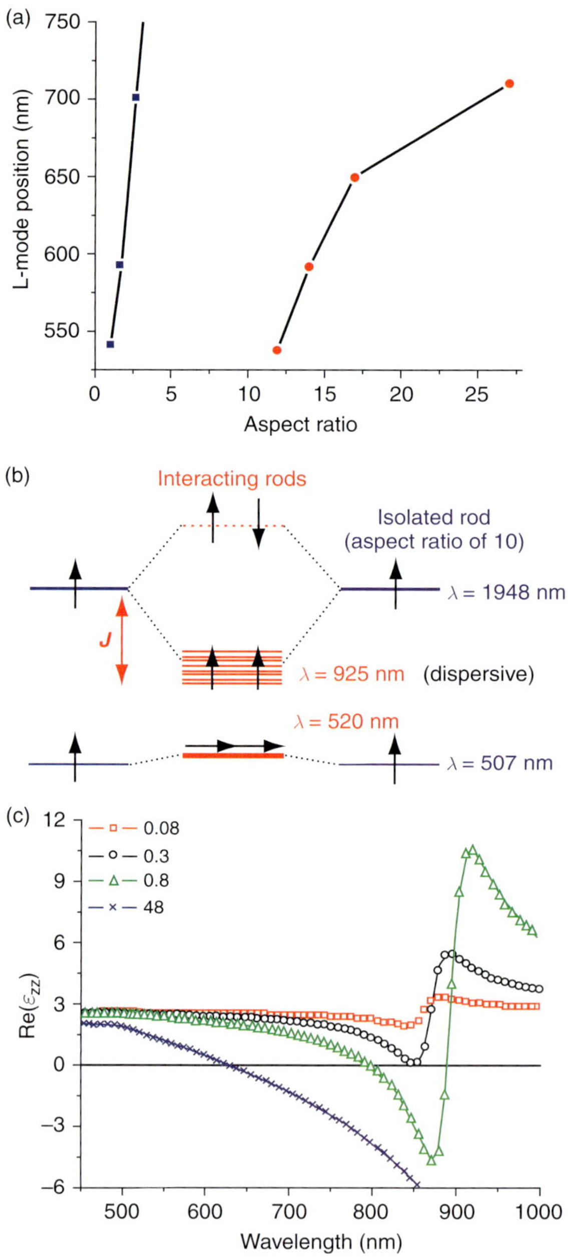

Figure 25 (a) The L-mode wavelength dependence on the aspect ratio for nanorods in the array (circles) and isolated ellipsoids (squares). The L-mode data are from **Figure 23(c)**. (b) Schematic energy diagram of the plasmonic resonances in the array of nanorods with regard to the isolated modes. Strong electromagnetic coupling of the plasmonic resonances supported by individual nanorods leads to the formation of a collective plasmonic mode. The *J* values measure the inter-rod coupling strength. See text for details. (c) Spectra of Re(ε_{zz}) calculated in the modified Maxwell–Garnet model for assemblies of Au ellipsoids with different concentrations: (squares) 0.08 rods per μm^2, (circles) 0.3 rods per μm^2, (triangles) 0.8 rods per μm^2, and (crosses) 48 rods per μm^2. The ellipsoids are embedded in AAO and have a major and minor axis length of 300 and 25 nm, respectively.

longitudinal dipolar modes supported by these rods would result in the formation of two eigenstates representing the symmetric and antisymmetric configurations of the plasmonic dipole moments of the coupled rod system [24]. Including more interacting nanorods will result in the formation of the band of plasmonic states formed in the nanorod assembly.

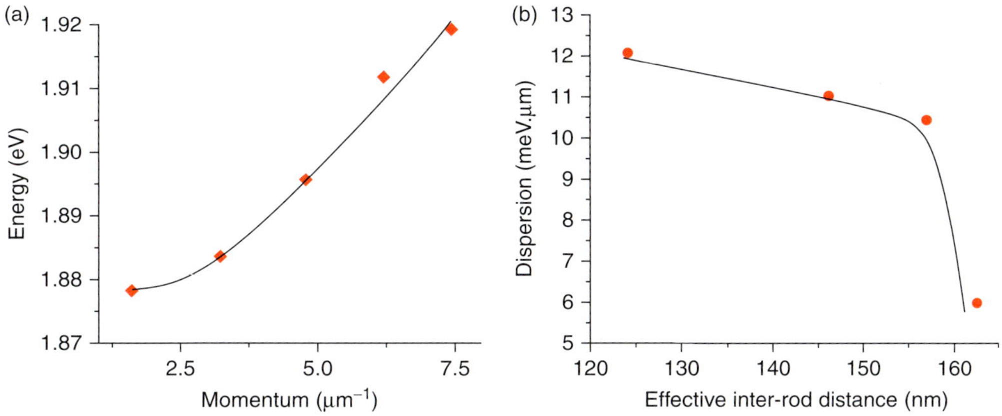

Figure 26 (a) Angular dispersion of the L mode as measured from **Figure 23(b)** for incidence angles varying from 10° to 50°. The momentum is the projection of the incident photon momentum perpendicular to the rod's long axis. (b) Dispersion $\hbar\Delta\omega/\Delta k_{//}$ for the L mode measured as a function of effective inter-rod distance between the incidence angles of 10° and 50°. The lines in (a) and (b) are a guide to the eye.

This interaction is schematically shown in the energy diagram of **Figure 25(b)**, where strong coupling between the nanorods shifts the symmetric state (parallel orientation of the dipole moments of the nanorods) in the short-wavelength spectral range compared to the longitudinal dipolar mode of the isolated nanorods. The mode associated with the latter state is the L mode studied above. The antisymmetric state corresponds to the antiparallel arrangement of the dipole moments and is not optically active under plane wave excitation.

The dependence of the spectral position of the L mode on the angle of incidence further emphasizes the collective behavior of this state and of its delocalized character in particular. The spectral dependence of the effective anisotropic permittivity of the nanorod array calculated in the Maxwell–Garnet model are presented in **Figure 25(c)** for different separations between nanorods in the array [115]. For clarity, only the real part of the dielectric constant ε_L seen by light polarized along the long axis of the nanorods, Re(ε_{zz}), is presented. The behavior of the ε_{xx} component is described by a single Lorentzian resonance related to the plasmonic mode associated with the short axis of nanorods. This resonance red-shifts with decreased inter-rod distance but, more importantly, its localized nature is independent of the nanorod separation. **Figure 25(c)** shows that, for larger inter-rod distances, when near-field interactions between the rods are negligible ($k_p l > 2\pi$), a Lorentzian resonance dominates the spectral behavior of the permittivity, which is associated with the longitudinal plasma resonance of the nanorods. In this regime the rods are weakly interacting and the nature of the resonance is that of an isolated nanorod embedded in an effective medium made of AAO and neighboring nanorods. With an increase in the nanorod concentration, the real part of the effective permittivity becomes negative indicating a metallic behavior of the nanorod array as a whole for light polarized along the nanorods' long axes. For the orthogonal polarization (along the short axis of the rods), however, the real part of the permittivity (not shown) remains positive as for a transparent dielectric. The frequency at which Re(ε_{zz}) changes sign can be considered as the effective plasma frequency $\tilde{\omega}_p$ of the metallic nanorod assembly. It corresponds to the excitation of the symmetric modes supported by the nanorod assembly with dipoles in adjacent nanorods excited in phase. Because of the different signs of epsilon zz and espilon xx observed at longer wavelentgh, these nanorod assemblies are showning negative refraction. (reference Podolski Nature Material,...).

1.3.2.2 Eigenmodes of nanorod arrays: spatial field distribution

To assess the spatial distribution of the electromagnetic field associated with the L-mode resonance of the array, we performed full 3D numerical calculations using an FEM. The geometry used is shown in **Figure 27**. The model consists of a periodic arrangement of Au rods placed in a square array and supported by a glass substrate. Considering that the spacing between the rods in the sample is smaller than k_0^{-1} (the wave vector of the incident light), the periodic geometry greatly diminishes the amount of memory required to perform the calculations, without affecting the generality of the results. The Au rods were modeled using a cylindrical shape with rounded extremities. Their diameter and length were set to 30 and 300 nm, respectively, with a dielectric response described by fitting the ellipsometric measurements on a sputtered Au film according to the Drude–Lorentz law [125]. The eigenmodes of the nanorods arrays were identified by computing the zero-order transmission spectra of the structures for a TM-polarized illumination field incident on the sample through the substrate with an angle in air varying between 0° (normal incidence) and 80°. The influence of the inter-rod coupling strength was investigated by changing the period of the array from 500 to 100 nm. The results are shown in **Figure 27**, in which both the magnitude and direction of the power flow, represented by the Pointing vector, are plotted alongside the spatial distribution of the norm of the electric field $\sqrt{\vec{E}^{*}.\vec{E}}$ for the L mode. For the larger inter-rod distance of 500 nm, the field distribution around the rods is dominated by the dipolar mode of the longitudinal plasmon resonance occurring at a wavelength of 1948 nm. In this case, the field is localized at the rods extremities with a small braking in symmetry due to the presence of the substrate and illumination conditions. The effect of a reduction in the inter-rod distance leads to a blue shift of the L-resonance, consistent with both the results from the effective-medium calculations and experimental observations. Note that in the conditions of the calculations, strong coupling is responsible for the optical response of the assemblies in **Figures 27(b)–27(d)**. The most surprising result comes from a dramatic change in both the spatial distribution of the electric field and the power flow in the array. In fact, the electric field concentrates within the assembly, in the middle part of the rods as the inter-rod separation decreases. This evolution corresponds to a simultaneous decrease of the field amplitude at the rods' extremities. This reconfiguration of the spatial distribution of the field in the array is supported by experimental observations that showed the lack of sensitivity in the spectral position of the array's extinction, and the L-mode position in particular, when the index of refraction of the superstrate is modified, that is, when only the rods extremities are subjected to a change in refractive index [120].

It is instructive to examine the behavior of the Pointing vector within the assembly as a function of inter-rod coupling strength (**Figures 27(a)–27(d)**). Going from the larger inter-rod distances to the strongly coupled regime, the electromagnetic energy flow around the rods evolves from being localized at the rod extremities to a net energy flux in the layer containing the nanorods in a direction perpendicular to their long axes. Thus, the electromagnetic energy coupled to the array, can effectively propagate from nanorod to nanorod within the assembly. This property is central to applications aimed at guiding light at the nanoscale [29,31,32,126,127]. In the geometry considered in the calculations, the maximum bandwidth of the L mode defined as $B_L = 2|\omega_0 - \omega_L|$, where ω_0 and ω_L are the resonant frequencies of the dipolar longitudinal mode of the isolated nanorods and the L-resonance, respectively, is obtained for the

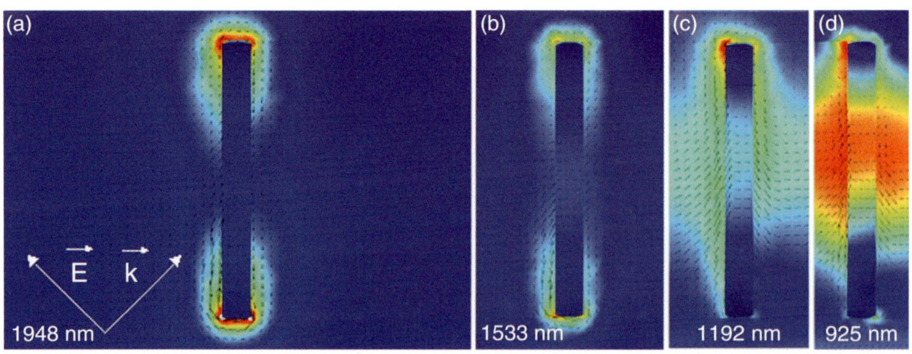

Figure 27 Norm of the Electric field distribution of L mode in the primitive cell of the nanorod array for different inter-rod distances: (a) 500 nm, (b) 200 nm, (c) 150 nm, and (d) 100 nm calculated using the FEM. The nanorod length and diameter are 300 and 30 nm, respectively. The nanorods are embedded in an AAO matrix ($n = 6$) and supported by a glass substrate ($n = 1.5$). The superstrate is air ($n = 1$). The angle of incidence of TM-polarized probe light is 45°. The arrows in (a)–(d) show the direction of the Poynting vector. The spectral position of the resonance is indicated in each situation.

smallest inter-rod distance and is estimated to be about 1.4 eV for an inter-rod distance of 100 nm. This corresponds to a maximum group velocity of about 1.7×10^7 m s^{-1} if a linear dependence of the group velocity of the L mode with bandwidth is assumed. The rate at which energy is lost in the assembly would then correspond to 1.2×10^{13} Hz, considering the 100 meV homogeneous broadening of the L mode [128]. The damping rate follows a monotonous and decreasing function of the inter-rod coupling strength, supporting the observation that the formation of the L mode enables the electromagnetic energy to propagate in the layer defined by the nanorod array.

The above properties were also investigated experimentally by gradually tuning the inter-rod coupling strength varying the inter-rod distance from about 170 to about 100 nm. This results in a shift of the L-mode resonance position from 1.74 to 2.25 eV [120]. Estimating the longitudinal resonance frequency of the isolated rod to be located around 0.54 eV, the effect of decreasing the inter-rod distance from 170 to 100 nm is to increase the bandwidth from 2.4 to 3.4 eV, respectively. Accounting for the bandwidth, the inter-rod distance, as well as the L-mode linewidth, we can estimate the group velocity and propagation length of electromagnetic energy in the nanorod assembly. These quantities are plotted in **Figure 28** as a function of inter-rod distance. The behavior of the group velocity is quite interesting as it undergoes a nonmonotonous variation as a function of inter-rod coupling strength with a maximum obtained for an inter-rod distance of about 155 nm. In this case, a group velocity of about 5×10^7ms^{-1} is reached, a value that is comparable to group velocities reported for other plasmonic guiding geometries [129,130]. The origin of this maximum was associated to an increasing confinement of electromagnetic energy in the rod assembly as near-field interactions become predominant in the system with decreasing inter-rod distances [125]. This observation suggests that the guiding properties of the structure are not obtained for maximum inter-rod coupling strength but rather for an optimal distribution of the electromagnetic field of the L mode. Turning back to **Figure 28**, the propagation length of electromagnetic energy in the rod assembly is shown to decrease from a value of about 2.7 μm estimated for coupling strengths varying between 1.2 and 1.4 eV, to a value of 1.66 μm for the maximum coupling strength of 1.7 eV. We attribute this increased damping to losses in the L mode due to an increased overlap between the L-mode resonance and the inter-band band edge of gold. This behavior was not observed in the numerical calculations for the array parameters, for which the L mode is always situated well in the infra-red spectral range.

Defining the inter-rod coupling strength for the L mode as $\mathcal{J} = \hbar|\omega_0 - \omega_L|$, coupling strengths on the order of 1 eV are typically observed in assemblies of nanorods but, more importantly, can be easily varied in order to control the degree of localization of electromagnetic energy within the assembly. For comparison, coupling strength on the order of 60 to 200 meV, depending on the polarization, have been measured for other plasmonic structures [22,24], while calculations show that coupling strengths comparable with our observations can be reached in more

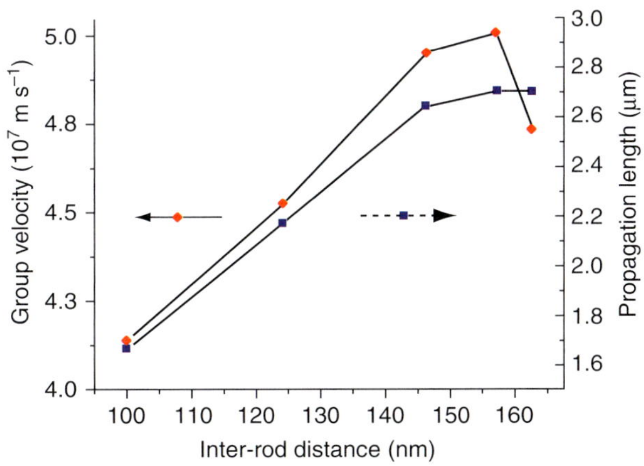

Figure 28 Group velocity (circles) and propagation length (squares) of electromagnetic energy of the L-mode as a function of inter-rod distance; a decreasing distance corresponds to an increasing coupling strength.

complex plasmonic systems made of chains of interacting plasmonic nanoparticles [131].

The flexible geometry of assemblies of nanorods open new opportunities to tailor the optical properties of plasmonic resonances both spectrally and spatially. The L mode obtained for strongly interacting nanorods shows particular potential in applications associated with guiding the electromagnetic energy and optical-manipulation applications on the nanometric scale [132,133]. These applications are particularly relevant since the recent demonstrations of the possibility to produce laterally confined assemblies of nanorods with single-nanorod resolution [119]. The optical properties of such assemblies should allow even further design enhancement. Assemblies of interacting nanorods also represent strongly anisotropic materials with either positive or negative values of the permittivity for different polarization directions. It may therefore find numerous applications in the design of meta-materials with specific sensing and imaging capabilities providing a route for the development of materials with indefinite dielectric tensor, highly phase-matched surfaces, as well as near-zero refractive index materials.

1.3.3 Assembly of Core–Shell Nanorods

This section describes the controlled fabrication of a dielectric shell of uniform thickness and tubular geometry around each nanorod composing the assembly. The optical properties of this unique core–shell system are investigated as a function of the shell thickness. The sensitivity of the structure as a chemical sensor based on the variation of the refractive index of the material in the nanoshell is discussed as one of the possible applications in a later section.

Following the fabrication procedure described in Section 1.3.2 for obtaining Au nanorods embedded in an AAO matrix, it is possible to create an air shell of tubular geometry and adjustable thickness along the sidewalls of the Au rods. This is simply done by subjecting the AAO-embedded assembly of Au nanorods to an additional aqueous sodium hydroxide (NaOH) etch. **Figure 29(a)** shows a typical TEM cross-section along with a schematic representation of the described geometry. The thickness of the shell can be controlled by varying the etch time and/or the concentration of NaOH from a fraction of a nanometer to a few tens of nanometers depending on the rod-to-rod distance in the assembly. Typical values are in the (10 s to 3 min) and (0.01–0.1 M) ranges for etching times and concentrations, respectively. The shell thickness variation as a function of etching time is shown in **Figure 29(b)** for a 0.75 mM aqueous solution of NaOH. As evidenced from **Figure 29(a)**, the uniformity obtained in the shell thickness is quite a remarkable characteristic of the process and reflects the dominance of the diffusion rate of NaOH along the rod walls, in the less-denser region of the alumina matrix, over the actual etching kinetics.

Figure 30(a) illustrates the optical response of the system for different shell thicknesses; **Figure 30(b)** shows the position of both the L and T resonances as a function of the ratio between the outer diameter of the shell and the diameter of the rod, this quantity being referred to as shell/core diametric ratio. The etching solution is a 0.75 mM aqueous solution of NaOH. The successive etching times are 15,75, and 90 s for shell/core diametric ratios of 1.25, 1.75, and 2.75, respectively. These diametric ratios correspond to an average shell thickness of 3, 8, and 19 nm, respectively.

The relationship between etching time and shell thickness for the investigated sample is shown in **Figure 29(b)**. In **Figure 30(a)**, the limiting values of 510 and 550 nm for the T and L resonances correspond to a geometry where the shell is etched away completely and only air separates the rods. The tendency for the T resonance to undergo a blue shift with increasing shell thickness can be explained by the resonance condition for the dipolar mode of the rods.

Qualitatively, if we consider a spherical lossless Mie scattering particle, this condition can be expressed as $\varepsilon'_{\mathrm{Au}} = -2\varepsilon_{\mathrm{d}}$, where $\varepsilon'_{\mathrm{Au}}$ and ε_{d} describe the real part of the dielectric constant of, respectively, Au and the surrounding dielectric. A Drude-like dielectric constant then leads to the following dependence of the resonant wavelength as a function of the dielectric constant of the surrounding dielectric:

$$\lambda_{\mathrm{resonance}} = \frac{2\pi c}{\omega_{\mathrm{p}}}\sqrt{1 + 2\varepsilon_{\mathrm{d}}}$$

where ω_{p} is the plasma frequency of the metal and c the speed of light in vacuum. Accounting for the net decrease in ε_{d} with increasing shell thickness leads to the observed blue shift of the resonance position. The blue shift associated with the L-mode results from the added effects of material dispersion and inter-rod electromagnetic coupling. The former has just been described for the T-mode and similarly displaces the isolated mode dipolar frequency to the blue with increasing shell thickness, while the latter

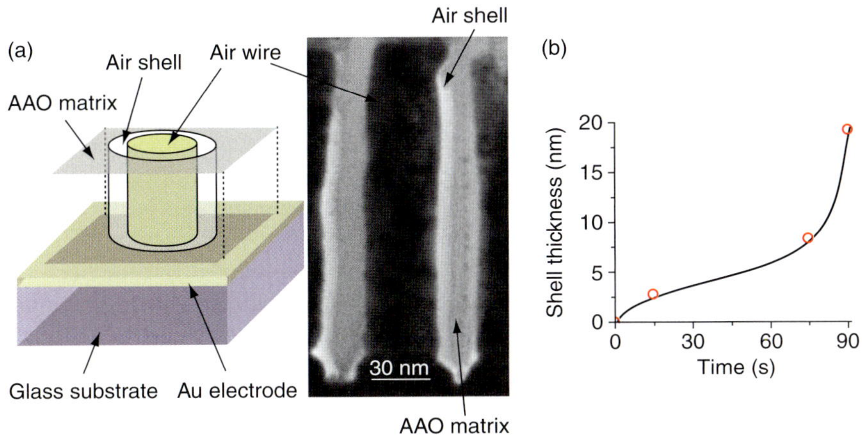

Figure 29 (a) Typical TEM cross-section of a Au nanorod surrounded by an air shell embedded in an alumina matrix. A schematic representation of the supported core–shell geometry is shown alongside. (b) Shell thickness variation as a function of etching time for a 0.75 mM aqueous solution of NaOH.

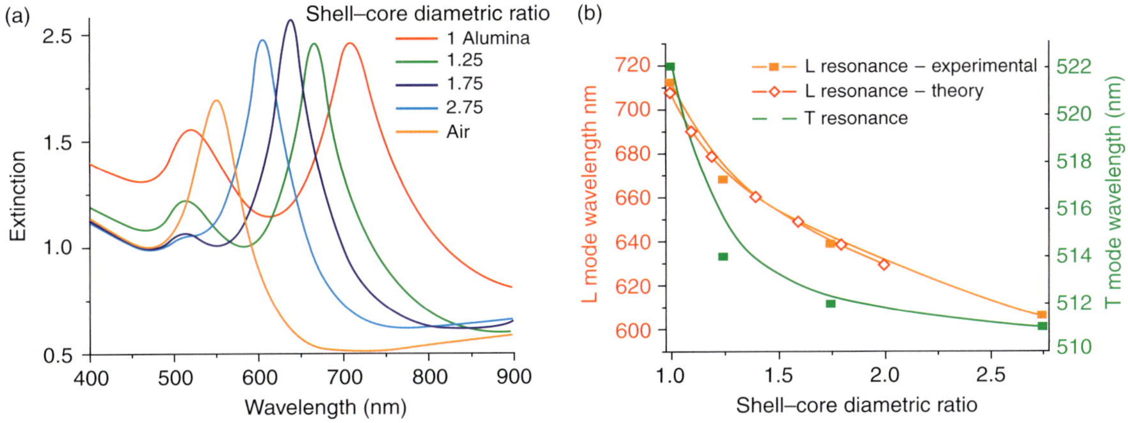

Figure 30 Optical properties of an array of Au core–shell nanorod array as a function of shell thickness. (a) Optical extinction of an array of free-standing core–shell nanorods embedded in an alumina matrix as a function of shell thickness. The zero-order extinction is shown for a TM-polarized beam incident on the array at an angle of 40°. The cores of the Au rods have an average diameter of 22 nm, length 300 nm, and separation of 120 nm. The diametric ratio is defined as the ratio between the shell's diameter and the rod's diameter of 22 nm and was characterized using SEM. The curves labeled alumina (diametric ratio of 1) and air, both, correspond to rods immersed in these respective environments, that is, no core–shell geometry. The insert shows the relationship between shell thickness and etching time for the sample considered. (b) Position of both the T and the L resonances as a function of shell/core diametric ratio according to (a). The lines are a guide for the eye. The results from the effective-medium calculations, accounting for the core–shell geometry, is also shown for the resonance of interest.

contribution originates from the increasing electromagnetic coupling strength between neighboring rods concomitant to the increasing shell thickness, that is, the decreasing effective optical distance $1/\sqrt{\varepsilon_d k_0}$ between rods, where k_0 is the free-space momentum of light, as ε_d decreases. An additional consequence of the increasing rod-to-rod coupling strength as the shell thickness is increased is deduced from the variation in the relative amplitude of both T and L modes: an increased shell thickness leads to a reduced polarizability for both T and L modes and should appear as a decreased amplitude for both resonances. This effect is observed for the T resonance only. For the L mode, the increased rod-to-rod coupling leads the resonance to carry an increased oscillator strength, which compensates for the decreased polarizability of the resonance. From a general point of view, the measurements in

Figure 30(a) provide a probe for the electric field of both the T and the L eigenmodes in the structure. In particular, they reveal the spatial delocalization of the L mode resulting from the strong interaction between the rods discussed previously. The results from calculations based on an anisotropic effective medium made of dipole–dipole interacting metallic-core/dielectric-shell ellipsoids, supporting only the fundamental dipolar resonance, are alongside the experimental data in **Figure 30(b)** and show reasonable quantitative agreement with the measurements [115]. The qualitative agreement between experiment and theory is broken if either the core–shell geometry or the inter-rod interaction is removed from the model. In the former case, omitting the core–shell structure in the calculation, and treating the increasing shell thickness as an effective medium decrease of the refractive index of the matrix, results in a simple linear dependence of the L mode spectral position versus shell thickness.

In the more general context of plasmonic devices, we will show in the next section that the core–shell plasmonic nanostructured system presented in this section demonstrates the spectral tunability and flexibility of geometry that are key to the efficient coupling to active materials with the aim of tailoring specific optical and optoelectronic functionalities. In the latter case, the inherent possibility to electrically address the nanostructured core–shell system through the supporting electrode along with the simplicity of the fabrication process make these structures possible candidates for the development of novel applications using active plasmonic devices.

1.3.4 Applications

1.3.4.1 Tunable exciton–plasmon coupling strength in J-aggregate hybridized Au nanorod assemblies

Controlling coherent electromagnetic interactions in molecular systems is a problem of both fundamental interest and important applicative potential in the development of photonic and optoelectronic devices. The strength of these interactions determines both the absorption and emission properties of molecules hybridized nanostructures, effectively governing the optical properties of such a composite meta-material. Here we discuss the observation of strong coupling between a plasmon–polariton supported by the assembly of gold nanorods and a molecular exciton. We show that the coupling is easily engineered and is deterministic as both spatial and spectral overlaps between the plasmonic structure and molecular aggregates are controlled [38]. These results in conjunction with the flexible geometry of the assembly of gold nanorods are of potential significance to the development of plasmonic molecular devices.

We discussed previously the promising prospects plasmonic modes offer in the development of nanodevices [1,9,134]. Particular efforts need to be taken toward designing active devices that demonstrate basic functionalities such as switching, lasing, energy storage, and conversion [10,134,135]. This can be achieved by coupling plasmonic modes with appropriately chosen active materials. A few examples were given for SPC earlier. Organic semiconductor materials, such as J-aggregates, are thought to offer great potential since they have the upper hand in low-cost manufacturing processes and possess attractive optical properties [136].

J-aggregates support excitonic states, which are electrically neutral electron–hole pairs, created by the absorption of photons. In photosynthetic processes, these states are used by plants to collect, store, and guide energy to the reaction center for energy conversion and could therefore drive artificial devices on identical principles. Excitonic states also show very strong nonlinear optical behavior that could be used to produce stimulated sources of photons and transistor-like action [137]. Experiments on colloidal solutions of Ag nanoparticles covered with J-aggregates demonstrated the possibility of using the strong scattering cross-section and the associated enhanced field concentrations of SPP to generate stimulated emission of excitons of J-aggregates with very low excitation powers [138]. Thus, their coupling with SPP appears as a particularly attractive approach to creating low-powered optical devices. In addition, the coupling of J-aggregates with SPP supporting geometries presents genuine fundamental interest in the creation of mixed plasmon/exciton states [35,36,139]. These quasiparticles may be created if the energy of the excitonic mode is resonant with a plasmonic transition. Either weak or strong coupling can then be achieved depending on whether the transition probabilities associated with the system's eigenmodes are still governed by the Fermi Golden Rule. Tuning the coupling strength is therefore of particular interest to tailor the absorption and emission characteristics of the molecular semiconductor. Strong coupling was observed in different systems including microcavities where a quasiparticles include cavity photons and molecular excitons [140,141], in plasmonic systems involving an

SPP and a waveguide photon [142], and between an SPP and an exciton [35,36,143]. The latter were observed in conventional plasmonic geometries enabling none or little tunability and therefore offering minimal prospect for practical interests. In this section, we focus our observations on the formation of hybrid plasmonic/excitonic states in assemblies of Au nanorods surrounded by a shell of J-aggregates supporting a molecular excitonic transition dipole moment. The flexible geometry of the assembly of nanorods was used to control the positioning of the J-aggregate in the plasmonic structure as well as the mixing of the hybrid system's eigenstates, offering a unique possibility in the design of molecular plasmonic nanodevices with tailored optoelectronic functionalities.

Figure 31 shows the extinction spectra for both the assembly of nanorods and the J-aggregate when formed on a smooth Au film. The J-aggregate is obtained from Benzothiazolium 5-chloro-2-(2-((5-chloro-3-(3-sulfopropyl)-2(3H)-benzothiazolylidene)-methyl)-1-butenyl)-3-(sulfopropyl), inner salt, compd. N,N-diethylethanamine (1:1) dye (CAS RN 27268-50-4); its molecular structure is also shown in **Figure 31**.

The assembly of nanorod spectra is dominated by two resonances around 500 and 750 nm, as discussed earlier. Because of the peculiar electric field

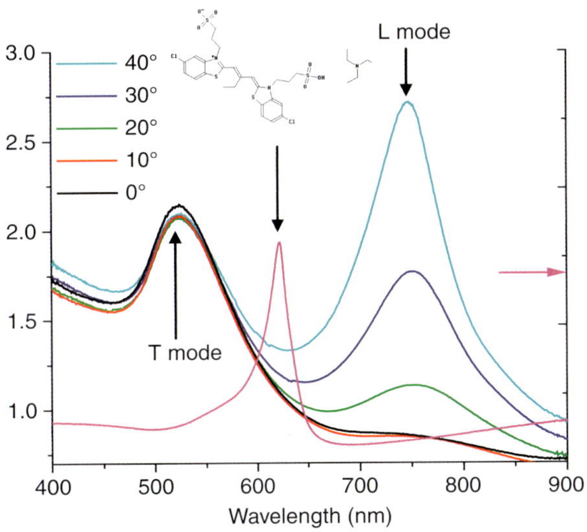

Figure 31 Optical extinction spectrum of the assembly of nanorods embedded in the porous alumina template as a function of angle of incidence. The inset shows the molecular structure of the dye used to hybridize the assembly of nanorods along with its extinction spectrum shown in a magenta line when J-aggregated on a 50 nm-thick smooth Au film magnetron sputtered on a glass substrate.

distribution of the L mode showing a maximum in the inter-rod spacing rather than around the rod extremity, coupling of the excitonic state from the molecular aggregate to either or both the plasmonic modes of the assembly then requires a shell around the rods in which to introduce the dye. This opens up unique opportunities to couple plasmonic structures with their environment in a controlled manner. Indeed, the thickness of the shell represents an additional geometrical parameter allowing one to tune the optical properties of the plasmonic structure over the 500–900 nm range, while it simultaneously offers spatial selectivity to couple the plasmonic resonances with active molecules. For this study, shells of various thicknesses were created successively in the same assembly of nanorods to create shell thicknesses of 2.5, 3.8, 5.6, 7.7, 8.9, and 20 nm. The hybrid plasmonic/excitonic system is obtained by depositing a 25 μl droplet from a 5% (by weight) dye methanol solution onto the core–shell structure. Upon methanol evaporation, the dye adsorbs onto the Au nanorods within the shell where it spontaneously J-aggregates. **Figures 32(a)** and **32(b)** show the extinction of the coupled system for two different shell thicknesses of 2.5 and 20 nm, respectively. Varying the shell thickness effectively tunes the plasmonic resonances of the assembly of nanorods with no measurable effect on the excitonic transition energy located around the wavelength of 622 nm. In the following, the shell thickness is therefore used as a parameter to control the spectral overlap between the plasmonic modes of the assembly of nanorods and the excitonic state of the J-aggregate.

From **Figure 32(a)** it can be deduced that when the overlap between the L mode ($\lambda_L = 737$ nm) and the excitonic transition ($\lambda_{J\text{-agg}} = 622$ nm) is small, then the spectral response of the extinction associated with the hybrid system ($\lambda_{hybrid+} = 745$ nm, $\lambda_{hybrid-} = 622$ nm) is essentially determined by the resonances of the isolated systems, that is, weak coupling is observed. Conversely, **Figure 32(b)** illustrates the regime where this overlap is strong ($\lambda_L = 610$ nm, $\lambda_{J\text{-agg}} = 622$ nm). The resonances of the hybrid system ($\lambda_{hybrid+} = 654$ nm and $\lambda_{hybrid-} = 593$ nm) then reflect the hybridization of the original resonances into mixed states with shared plasmon/exciton character, that is, strong coupling is observed.

The general behavior observed in **Figures 32(a)** and **32(b)** can be illustrated following the procedure described in Ref. [35] and calculating the extinction cross-section for an isolated ellipsoidal nanoparticle with a Au core surrounded by an excitonic shell and embedded in a homogeneous medium of dielectric constant ε_d [38]. This quantity can be expressed as

$$\sigma_{\mathrm{Extinction}_i} = \frac{1}{3}\sum_i 2\pi/\lambda\,\mathrm{Im}\{\alpha_i\} + \frac{1}{3}\sum_i 1/6\pi\,(2\pi/\lambda)^4|\alpha_i|^2$$

where λ is the wavelength in vacuum and α_i the dipolar polarizability of the core–shell ellipsoid along axis *i* expressed in the electrostatic approximation [56]. The dielectric constant ε_d of the embedding medium alumina is taken to be 2.56, whereas a polynomial fit of the data published in [42] is used to describe the dielectric constant of Au, $\varepsilon_{\mathrm{Au}}$. The response of the shell was modeled by an excitonic resonance with a Lorentzian response in a dielectric background ε_∞. The anisotropy of the exciton's transition dipole moment was approximated by

$$\varepsilon_{\text{J-agg}//} = \varepsilon_\infty + \frac{\omega_0^2 \cdot f}{\omega_0^2 - \omega^2 - i\omega\gamma} \quad \text{and} \quad \varepsilon_{\text{J-agg}\perp} = \varepsilon_\infty$$

for the contributions along and perpendicular to the aggregate's transition dipole moment, respectively. The high-frequency dielectric constant ε_∞ was extracted from ellipsometric measurements made on a thin J-aggregate film formed on a smooth Au surface, whereas the transition energy, the reduced oscillator strength, and the damping of the excitonic state were determined by fitting the J-aggregate's

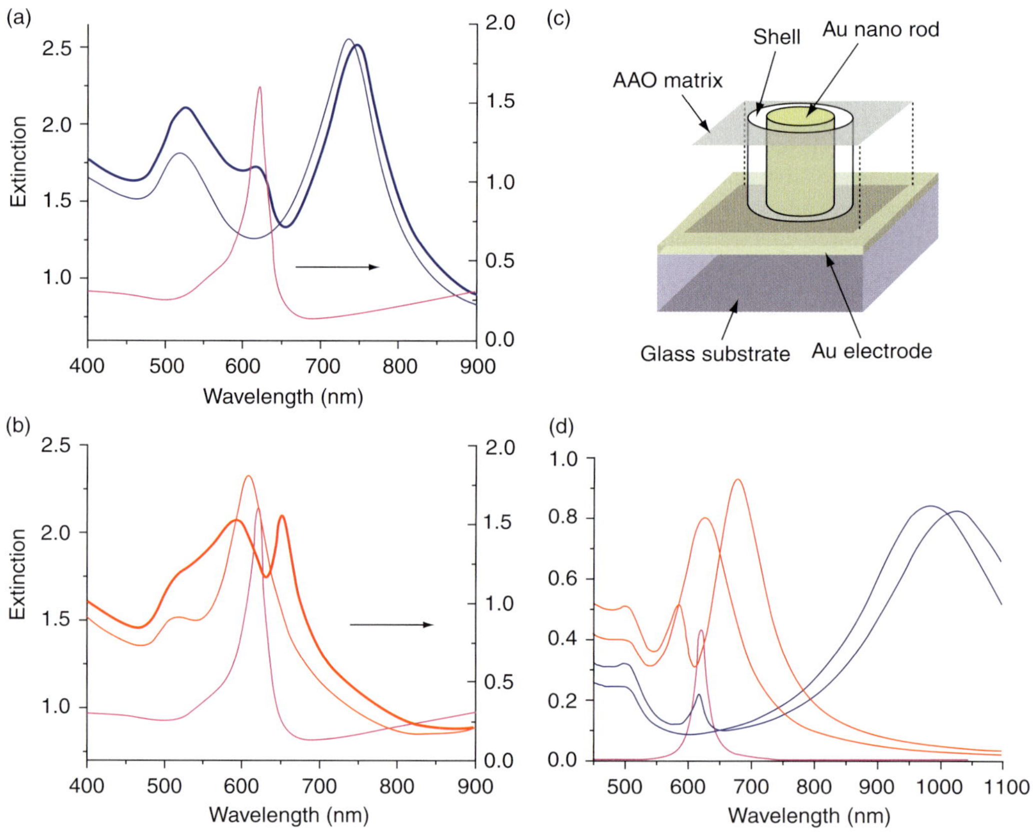

Figure 32 (a) Extinction spectrum of the hybrid assembly of nanorods (ANRs) in the weak plasmon–exciton coupling regime (thick blue line) along with the extinction of the isolated ANR (thin blue line) and the extinction of the J-aggregate on a Au film (magenta line). The shell thickness is 2.5 nm. (b) Extinction spectrum of the hybrid ANR in the strong plasmon–exciton coupling regime (thick red line) along with the extinction of the isolated ANR (thin red line) and the extinction of the J-aggregate on a Au film (magenta line). The shell thickness is 20 nm. The spectra are taken for an angle of incidence of 40°. (c) Schematic illustration of the structure. The red arrow shows the orientation of the molecular transition dipole moment with respect to the long axis of the Au nanorod when introduced in the shell. (d) Extinction from a coated Au ellipsoid calculated in the dipolar limit and illustration of the general behavior observed experimentally in (a) and (b). Strong and weak coupling regimes are plotted in red and blue colors, respectively. The thin lines correspond to the isolated case, whereas thick lines describe the response of the hybrid system. An homogeneously broadened J-aggregate extinction is shown in magenta.

absorption spectrum as measured on the smooth Au film in **Figure 31**. These parameters then take the respective values of $\varepsilon_\infty = 1.21$, $\hbar\omega_0 = 1.99$ eV ($\lambda = 622$ nm), $f = 0.054$, and $\hbar\gamma = 66$ meV. **Figure 32(d)** shows the calculated extinction cross-section for core–shell ellipsoids with aspect ratios of 2.4 and 6 with a constant shell thickness of 2 nm. The thick lines correspond to the coupled plasmon/exciton system, whereas the thin lines describe the extinction of the Au ellipsoids but surrounded by an air shell. The extinction of the J-aggregate is also shown alongside by the magenta line spectrum. The aspect ratio of the ellipsoid was parametrized to modify the spectral overlap between the long-axis plasmonic resonance of the ellipsoid and the transition energy of the exciton and, therefore, the coupling strength between the two transition dipole moments. In **Figure 32(d)**, only the long-axis plasmonic mode was coupled to the excitonic transition, reflecting the experimental observations. In a situation where both transverse and longitudinal modes from the ellipsoid were to be coupled with the excitonic transition, four modes would be observed in the hybrid's extinction spectrum: two transverse modes as well as two longitudinal modes.

In order to gain a better insight into the coupling mechanism, let us consider a 3D physical system whose Hamiltonian is H_0. The eigenstates of H_0 are $|\phi_T>$, $|\phi_L>$ and $|\phi_{J\text{-agg}}>$ associated with the rods dipolar transverse plasmonic resonance, the ANR's longitudinal resonance, and the J-aggregtate excitonic state. The corresponding eigenvalues are E_T, E_L and $E_{J\text{-agg}}$ satisfying $H_0|\phi_T> = E_T|\phi_T>$, $H_0|\phi_L> = E_L|\phi_L>$, and $H_0|\phi_{J\text{-agg}}> = E_{J\text{-agg}}|\phi_{J\text{-agg}}>$, with $<\phi_i|\phi_j> = \delta_{ij}$. Introducing coupling between these levels and searching for the eigenvalues of the coupled system in the $\{|\phi_T>, |\phi_L>, |\phi_{J\text{-agg}}>\}$ basis, we rewrite the Hamiltonian as $H = H_0 + V$, where V is a time-independent perturbation accounting for the coupling between the different eigenstates of the isolated system. Assuming that the J-aggregate's transition dipole moment is oriented along the long axis of the nanorods only, the Hamiltonian takes the following simplified form, where V_{hybrid} couples $|\phi_L>$ and $|\phi_{J\text{-agg}}>$:

$$H = \begin{bmatrix} E_T + V_{11} & 0 & 0 \\ 0 & E_L + V_{22} & V_{\text{hybrid}} \\ 0 & V_{\text{hybrid}} & E_{\text{Jagg}} + V_{33} \end{bmatrix} \tag{9a}$$

from which the eigenvalues are readily obtained:

$$E'_T = E_T + V_{11} \tag{9b}$$

$$E_{\text{hybrid}+} = \frac{1}{2}(E'_L + E'_{\text{Jagg}}) + \frac{1}{2}\sqrt{(E'_L - E'_{\text{Jagg}})^2 + 4V^2_{\text{hybrid}}} \tag{9c}$$

$$E_{\text{hybrid}-} = \frac{1}{2}(E'_L + E'_{\text{Jagg}}) - \frac{1}{2}\sqrt{(E'_L - E'_{\text{Jagg}})^2 + 4V^2_{\text{hybrid}}} \tag{9d}$$

where $E'_L = E_L + V_{22}$ and $E'_{\text{Jagg}} = E'_{\text{Jagg}} + V_{33}$. The eigenvector associated with E_T' is $|\psi_T> \sim |\phi_T>$, reflecting the fact that the J-aggregate's transition dipole moment and the T mode were taken to be perpendicular to each other in our expression of the coupling Hamiltonian V. Consequently, for the transverse mode in the hybrid structure, the presence of the J-aggregate along the Au nanorod results in a net change in the energy E_T' through a change in the dielectric constant in the shell, with an eigenvector that is essentially unchanged from its original dipolar form. Neglecting the core–shell geometry as well as the presence of the substrate, an approximate expression for V_{11} can be found by assuming a lossless Drude-like dielectric response of the form $\varepsilon(\omega) = 1 - \omega_p^2/\omega^2$ for an Au ellipsoid embedded in an homogeneous medium of effective dielectric constant ε_{eff}:

$$V_{11} = \frac{\hbar\sqrt{L}\omega_p(L-1)}{2[\varepsilon_{\text{eff}}(1-L) + L]^{3/2}}\Delta\varepsilon_{\text{eff}}$$

where ω_p is the bulk plasma frequency of Au, L is the geometric factor defined earlier, and $\Delta\varepsilon_{\text{eff}}$ reflects the change in dielectric constant induced by the introduction of the J-aggregate in the shell. The function V_{11} is an increasing function of the shell thickness through $\Delta\varepsilon_{\text{eff}}$ with an upper value of $V_{11} \sim -20$ meV measured for a complete removal of the AAO matrix ($\Delta\varepsilon_{\text{eff}} \sim \varepsilon_{air} - \varepsilon_{\text{Jagg}\perp} = -0.21$). The negative value of V_{11} reflects the increase in the index of refraction of the shell when introducing the dye. Similarly to V_{11}, V_{22} reflects the change in the L-mode position upon J-aggregation in the shell via the off-resonance high-frequency dielectric constant of the J-aggregate. However, V_{22} is more difficult to express analytically, since the field associated with the L mode within the assembly of nanorods is no longer following the dipolar response of the isolated rod's long-axis resonance. An estimate value of V_{22} can be obtained by measuring the sensitivity of the L mode as a function of shell thickness for an index of refraction of 1.1 in the shell. This allows us to evaluate the variation ΔV_{22} of V_{22} over the shell thickness range considered in this study of about 65 meV, whereas V_{22} at 736 nm is about −20 meV and ΔE_L is about 570 meV. The function V_{33} is a measure of the variation of the intermolecular coupling $\mathcal{J}$ within the aggregate when it adsorbs on the Au nanorods with respect to a reference state measured in water: $E_{JAgg}{'} = E_{J\text{-}agg} + V_{33}$. The loss in potential energy of the highest occupied molecular orbital (HOMO) upon J-aggregation determines the coupling energy $\mathcal{J}$, which can then be defined from the transition energy $\Delta E_{J\text{-}agg}$ of the aggregate as $2\mathcal{J} = \Delta E - \Delta E_{J\text{-}agg}$, where ΔE is the transition energy of the monomer and $\mathcal{J}$ is the positive coupling energy. To evaluate V_{33}, J-aggregation has been studied on a smooth Au film, a borosilicate glass substrate and on porous AAO from the dye solution in methanol. For all these situations the excitonic transition was observed at an identical wavelength of 622 nm, slightly blue-shifted from its value in water. This underlines that, in the experimental configurations considered, it is the interfacial geometry that governs the aggregation energy rather than the material properties of the substrate itself. Comparing **Figure 31** with **Figure 32(a)**, it is deduced that J-aggregation in the shell around the rods leads to an intermolecular coupling energy similar to the one measured on planar interfaces, suggesting that the core–shell geometry does not strongly affect the value of $\mathcal{J}$ in the aggregate even for the smallest shell thicknesses studied (~2 nm). Consequently, in what follows, the value of V_{33} is assumed to be shell-thickness independent. Considering the absorption frequency of the J-aggregate in both water ($\lambda = 650$ nm, i.e., $2\mathcal{J} = 230$ meV) and when adsorbed on Au nanorods ($\lambda = 622$ nm, i.e., $2\mathcal{J} = 150$ meV) we can deduce that $V_{33} = E_{Jagg}{'} - E_{Jagg} = \Delta E - 2\mathcal{J}' - (\Delta E - 2\mathcal{J}) = 80$ meV.

The eigenvectors $|\psi_+>$ and $|\psi_->$, associated, respectively, with $E_{hybrid+}$ and $E_{hybrid-}$, can be expressed as linear combinations of the longitudinal plasmonic orbital $|\phi_L>$ and the J-aggregate excitonic orbital $|\phi_{JAgg}>$as

$$|\psi_+\rangle = \cos\frac{\alpha}{2}|\phi_L\rangle + \sin\frac{\alpha}{2}|\phi_{Jagg}\rangle \quad \text{and}$$
$$|\psi_-\rangle = -\sin\frac{\alpha}{2}|\phi_L\rangle + \cos\frac{\alpha}{2}|\phi_{Jagg}\rangle$$

where

$$\tan\alpha = \frac{V_{hybrid}}{\frac{1}{2}(E'_L - E'_{Jagg})}$$

The effect of the coupling matrix V on H_0 is to hybridize the longitudinal and excitonic states into two new eigenstates with a mixed plasmon–exciton character. It follows that, if $|E_L{'} - E_{J\text{-}agg}{'}| >> 2V_{hybrid}$, then $\alpha \sim 0$, cos $\alpha/2 \sim 1$ and sin $\alpha/2 \sim V_{hybrid}/|E_T{'} - E_{J\text{-}agg}{'}|$, from which, if we assume in this instance that $E_L{'} > E_{J\text{-}agg}{'}$, we deduce the wave functions of the system as

$$|\psi_+\rangle = |\phi_L\rangle + \frac{V_{hybrid}}{E'_L - E'_{Jagg}}|\phi_{Jagg}\rangle \sim |\phi_L\rangle \quad \text{and}$$
$$|\psi_-\rangle = -\frac{V_{hybrid}}{E'_L - E'_{Jagg}}|\phi_L\rangle + |\phi_{Jagg}\rangle \sim |\phi_{Jagg}\rangle$$

These hybrid orbitals then resemble the plasmonic and excitonic orbitals from the uncoupled system; in this case these orbitals represent a weakly coupled system as they retain their original character. This case is illustrated in **Figure 32(a)** for $E_{J\text{-}agg}{'} > E_L{'}$. Similarly, if $|E_L{'} - E_{J\text{-}agg}{'}| << 2V_{hybrid}$, then $\alpha \sim \pi/2$, cos α = sin $\alpha \sim 1/\sqrt{2}$ and

$$|\psi_\pm\rangle = \frac{1}{\sqrt{2}}\left[|\phi_L\rangle \pm |\phi_{Jagg}\rangle\right]$$

These functions are hybrid orbitals of the system of **Figure 32(b)** where the two original uncoupled orbitals overlap spectrally; in this case strong coupling is achieved.

The unique characteristic of the core–shell assembly of nanorods geometry is to enable a continuous tuning of the plasmonic resonances by controlling the thickness of the shell created around the rods. The spectral position of the L mode can therefore be scanned through most of the visible spectrum and across the excitonic resonance to modulate the mixing of these states. The result is shown in **Figure 33**, where the spectrum of the hybrid assembly of nanorods is plotted for different positions of the L mode in air.

The occurrence of the strong coupling regime in the structure is evident, since no resonances appear in the spectral range from 622 to 658 nm, whereas the L-mode resonance was continuously tune from 736 to 550 nm. The anticrossing between the L mode and the excitonic transition is illustrated in **Figure 34(a)**. Deriving V_{hybrid} from **equations 9c** and **9d**, we use **Figure 33** to estimate the coupling strength within the hybrid as a function of the overlap between the L mode and the excitonic states as

$$V_{hybrid} = \sqrt{\frac{(E_{hybrid+} - E_{hybrid-})^2 - (E_L - E'_{Jagg})^2}{4}} \qquad (10)$$

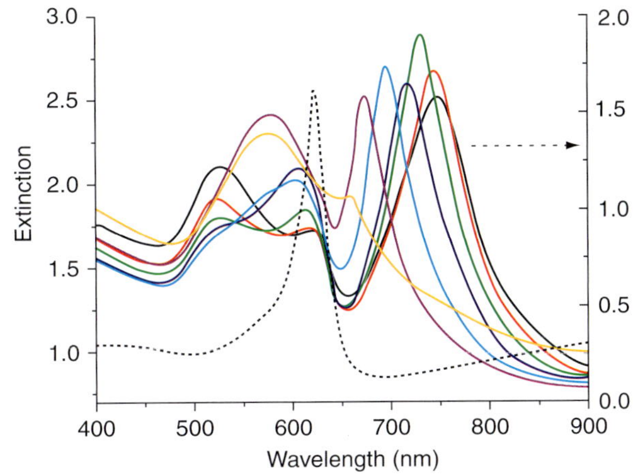

Figure 33 Extinction from the hybrid assembly of nanorods varying the coupling strength between the plasmonic L mode and the molecular excitonic transition. The spectra are taken for an angle of incidence of 40°. The plasmon–exciton coupling strength has been adjusted by tuning the spectral position of the L mode from 730 to 550 nm by increasing the thickness of the shell around the Au rods. The L-mode resonance position in air for the hybrids represented by the black, red, green, blue, cyan, magenta, and yellow spectra was located at a wavelength (and corresponding shell thickness) of 740 nm (2.5 nm), 730 nm (3.8 nm), 720 nm (5.6 nm), 700 nm (7.7 nm), 650 nm (8.9 nm), 600 nm (20 nm), 550 nm (no shell-matrix completely removed), respectively. The black broken line shows the excitonic extinction on a 50 nm-thick smooth Au film.

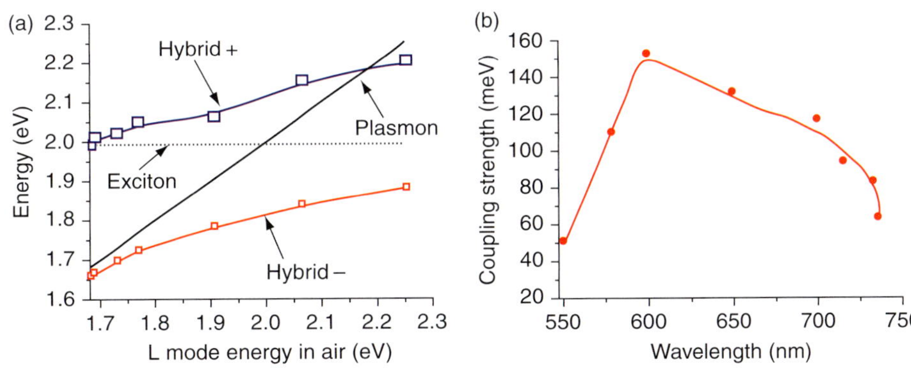

Figure 34 (a) Energy diagram showing the anti-crossing of the plasmonic (diagonal solid black line) and the excitonic (horizontal dotted black line). The crossing of the high-energy branch and the plasmon line is due to V_{22} upon hybridization (see text for details). The size of the square markers represents the error bars. (b) Coupling strength calculated from **equation 10**. The data are represented by circular dots and the red curve is a guide for the eye.

where we assumed that $E_L' \sim E_L$. The result is plotted in **Figure 34(b)** as a function of the original position of the L mode. A similar relationship derived from an analytical theory for the optical properties of ellipsoidal plasmonic particles covered by anisotropic molecular layers leads to very similar values for V_{hybrid} as a function of the L-mode wavelength with a small overestimation of the coupling strength when using **equation 10** [144]. This overestimation reflects the effect V_{22} has on E_L': the absolute value of V_{22} increases with shell thickness and is negligible at maximum coupling strength. This maximum is reached for an L-mode resonance in air located at 600 nm and corresponds to a Rabi splitting of 155 meV. This value is commensurate with the largest Rabi splitting observed in planar organic microcavities [145] and similar to recent observations for an excitonic state mixed with a planar SPP [36].

The ability to control the coupling strength between a plasmon–polariton and a molecular exciton both spatially and spectrally is a major step toward implementing active and passive plasmonic devices. Based on the hybrid nature of the coupled states, the nonlinear optical properties of such a nanostructured material should demonstrate extreme sensitivity over a large spectral range, that is, by addressing one of the hybrid resonances one would effectively affect both coupled resonances and therefore in turn the extinction of the device from 550 to 700 nm. This could be achieved optically in a pump–probe configuration for example, and possibly also electrically by dynamically controlling the orientation/ordering of the molecular aggregate. Because of the coupled nature between the plasmonic and excitonic states, future investigations on the emission properties of this hybrid structure should stimulate considerable interest as well [36]. Finally, we emphasize the practical interest of such device, as the ANR's fabrication process is simple and scalable [111] while also demonstrating strong potential in the fabrication of nanodevices [132].

1.3.4.2 Refractive index sensing

The nanorod core–shell assemblies represent a suitable nanostructured system for applications based on monitoring refractive index (RI) changes of a substance incorporated in the sub-attoliter volume created by the shell [120]. Importantly, it is to be pointed out that no spectral shifts in the position of the resonances are observed when a change in the RI of the superstrate takes place, that is, when only the apex of the rods is exposed to the changing RI. This reflects the unique properties of the new electromagnetic field distribution, which has a maximum in the middle of the rods (**Figure 27(d)**) and develops when the rods are brought closer together to form an array. In contrast, for an isolated rod the field distribution associated with the longitudinal dipolar resonance is confined to the apexes of the rod, which consequently are the locations the most sensitive to changes in the surrounding RI [146]. For the rod assembly, only a varying RI in the shell surrounding the rod walls affects the plasmonic resonance position.

The presence of the shell is therefore necessary to achieve RI sensing with arrays of nanorods. The shell, in combination with other geometric parameters, such as rod aspect ratio, separation, and length, determines the sensitivity of the structure to RI changes. Examples illustrating this RI sensitivity are presented in **Figure 35** for two sets of rod assemblies with different rod aspect ratios and shell thicknesses as a function of shell RI. The shell RI is assumed to vary between $n=1.6$ and $n=1$, which values correspond, respectively, to shells containing bulk alumina (i.e., shell not yet created) and air (once the shell is formed and empty).

The sensitivity calibration curves for different array geometries is presented in **Figure 36**. The shift of the L-mode resonance for different structures follows a fairly linear dependence on the shell's RI. This behavior is reproduced using the

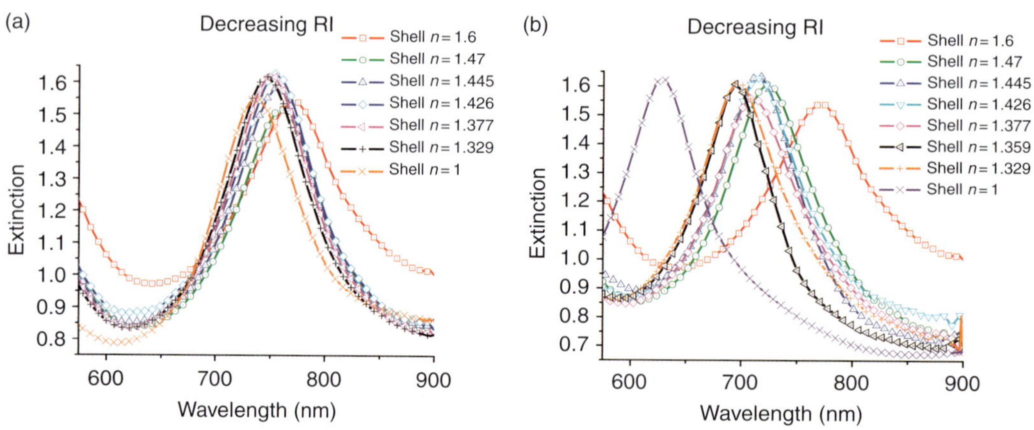

Figure 35 Optical extinction of an array of free-standing core–shell nanorods embedded in an alumina matrix plotted for a varying index of refraction of the shell. The zero-order extinction is shown for a TM-polarized beam incident on the array at an angle of 40°. The rod aspect ratio is 27, the rod length is 400 nm, and the shell thickness is (a) 8 nm, and (b) 20 nm. The values of $n=1.6$ and $n=1$ correspond to rods surrounded by a homogeneous environment made of alumina and air, respectively, that is, no core–shell geometry. The shell's RI has been controlled using the following solvents: methanol ($n=1.329$), acetone ($n=1.359$), isopropanol ($n=1.377$), cyclohexane ($n=1.426$), chloroform ($n=1.445$), 1,1,2-trichloroethane ($n=1.47$), and aqueous solutions of sucrose with concentrations in weight ranging from 20% ($n=1.364$) to 80% ($n=1.459$). The index of refraction of the different solutions was measured using Abbe's refractometer.

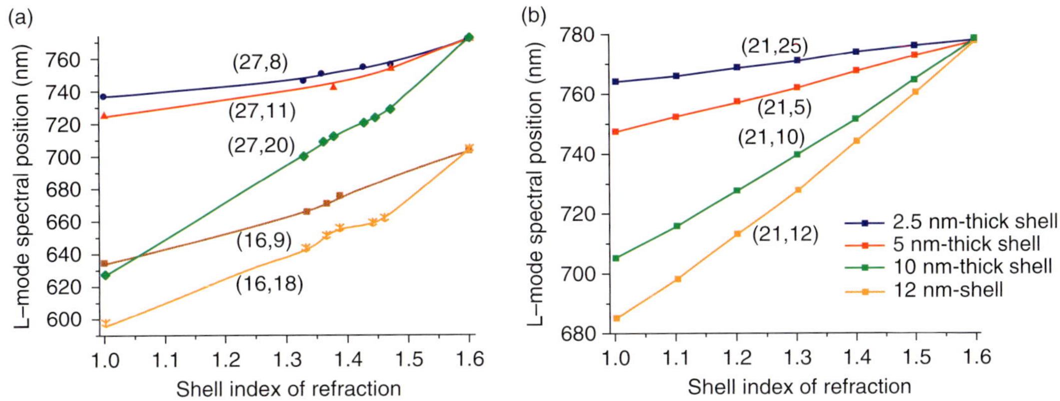

Figure 36 Spectral position of the L resonance for an array of free-standing Au core–shell nanorods embedded in an alumina matrix plotted for different couples (α, β), where α is the rod aspect ratio and β is the shell thickness, as a function of the shell's refractive index. (a) Experimental results. The values of $n = 1.6$ and $n = 1$ correspond to rods surrounded by a homogeneous environment made of alumina and air, respectively, that is, no core–shell geometry. (b) Theoretical results from the effective-medium model calculations exemplifying the experimental behavior.

effective-medium model calculation and the results are plotted in **Figure 36(b)** for four different shell thicknesses. The ellipsoidal geometry assumed for the nanorods in the calculations leads to the observed discrepancy in the array parameters between experimental and theoretical results [122]. The slope of these curves determines the spectral sensitivity of the structure per refractive index unit (RIU) and is a function of both the shell thickness and the rod aspect ratio. For example, in **Figure 36(a)**, for an aspect ratio of 27 and a shell thickness of 20 nm, a typical sensitivity of around 100 nm/RIU is obtained. This value is to be compared to previously reported sensitivities of 100–400 nm/RIU for single gold nanoholes and nanohole arrays [147,148], 191–230 nm/RIU for nanosphere lithographically produced Ag nanoparticles [149, 150], 328.5 nm/RIU for gold nanoshells [151], and 538 nm/RIU for gold nanowells [152].

As can be deduced from **Figure 36(a)**, for a fixed value of the rod aspect ratio, increasing the shell thickness will increase the spectral window covered by the shift of the L peak for a given change of RI in the shell. A linear fit reveals a slope scaling as $\sim 12e$, where e is the thickness of the shell, a value that was found to be rather insensitive to the rod aspect ratio. The 80 nm bandwidth of the L resonance would, however, allow for an enhanced sensitivity to changes in RI by measuring the extinction of the L resonance at a wavelength demonstrating high $\Delta OD/\Delta\lambda$. For example, the structure of **Figure 35(b)** could be used at a probing wavelength of 670 nm to sense RI changes in the range 1.2–1.5, with a sensitivity on the order of 2×10^{-2} [120]. A fourfold improvement in the sensitivity is expected from the use of thermally annealed Au rods [120], which can show a 50% bandwidth narrowing for the L resonance compared to their un-annealed counterparts due to a decrease in electron scattering in the rods [115]. The use of other plasmonic supporting metals (e.g., Ag) or the use of ordered templates shows a reduced inhomogeneous broadening in the plasmonic resonances, and can also be explored to achieve improved sensitivity [153].

The originality and advantage of the considered structure is in its flexible geometry and associated optical properties. In particular, it allows controlling the volume of the sensed substance with a resolution on the order of a fraction of an attoliter (10^{-18} l) by controlling the thickness of the nanoshell, and independently enables tuning the spectral range of operation over the visible spectrum by varying the aspect ratio of the Au rod core.

1.3.4.3 Electrically switchable nonreciprocal transmission

In contrast to the electro-optical tunability of plasmonic crystals incorporating an LC layer, the electrical reorientation of the anisotropic LC molecules do not provide the same spectral tunability for the nanorod array. This is due to the field distribution of the eigenmodes, concentrated inside the AAO matrix and therefore insensitive to the RI changes of the superstrate. Nevertheless, polarization effects, associated with LC reorientation in an applied electric field, combined with the polarization properties of the nanorod arrays, can be used for light switching and modulation applications [77].

Figure 37(a) shows the effect of the LC and electric applied field on the extinction spectra of 20 nm-diameter and 400 nm-long gold nanorods embedded in alumina. When no LC is in the cell, then both T and L modes are observed in the extinction spectra. From **Figure 37(a)** it can be seen that when the LC is inserted into the cell it completely quenches the L mode at 712 nm if no electric field is applied. A small increase in the extinction at the T-mode resonance wavelength of 520 nm is also observed. Applying a small field of 1 V μm^{-1} leads to the recovery of the L-mode extinction and a small decrease in the extinction at the T mode to the same absorbance that was observed with no LC. With the application of a small field (0.05 V μm^{-1}) there is an initial increase in the extinction for the T mode and a decrease in the extinction at the L-mode resonance position. As the voltage is increased (to 0.15 V μm^{-1}) the extinction for the T mode decreases and the L mode starts to appear, which then rapidly increases as the voltage is increased to +0.5 V μm^{-1}. As the field is increased, further there is a steady increase in the L-mode amplitude, which eventually saturates. At a field of 1.5 V μm^{-1}, a well-defined L mode is observed. **Figure 37(b)** shows the change in extinction for the L mode with applied field. This process is completely reversible and shows good reproducibility as the field is reduced back to zero. The above data are obtained for illumination through the LC first. When light is incident through the nanorod side of the cell first, the L-mode extinction is unchanged by the presence of the LC and applying a field was found to have little effect. The observations of electric-field-dependent extinction and nonreciprocity with respect to the illumination side strongly suggest that the important effect in our system is that of the LC on the polarization of the incident light. The LC layer conformation has molecules oriented in the plane of the substrate (perpendicularly to the nanorod axes) when no electric field is applied to the cell. This rest orientation results from the anisotropic LC/surface interaction and inter-molecular interaction within the cell. The TM-polarized incident light is then converted into TE-polarized light when transmitted through the LC molecular layer. In the configuration when light is first incident through the LC, before interacting with the nanorod layer, the L mode cannot be addressed and is not present in the extinction of the sample. Applying the voltage causes the LCs to orient along the field, that is, perpendicular to the substrate and along the direction of the nanorod long axes. In this orientation, they no longer change the polarization of the incident TM-polarized light and hence the longitudinal extinction peak is observed again.

A comparison with previous work on gold nanorods [154], nanoparticles [155] and nano dots [156] shows that we observe much larger changes in the extinction spectra for smaller applied electric fields and without the need for a rubbing layer. However, these works relied on a shift in the plasmon resonance wavelength due to the changes of the dielectric constant of differently oriented LCs.

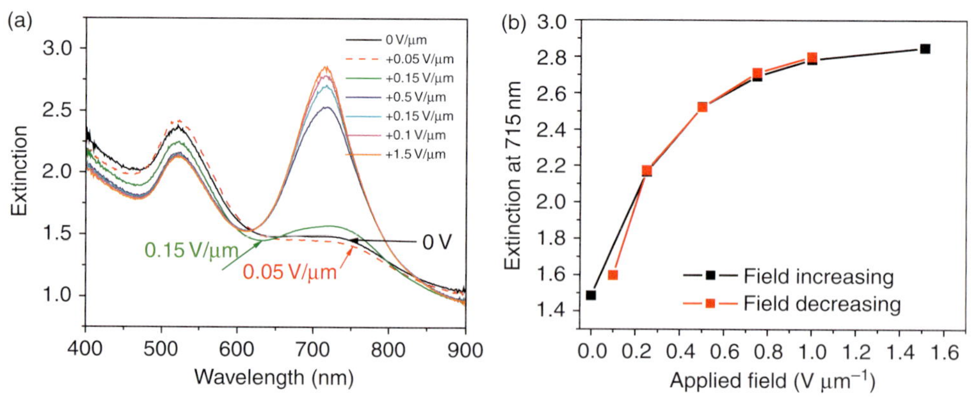

Figure 37 (a) Extinction spectra of gold nanorods, 20 nm in diameter and 400 nm long, embedded in alumina with a 100 μm top liquid crystal layer as a function of applied electric field. The incident light is TM-polarized and the angle of incidence is 40°. (b) Dependence of the extinction at 715 nm for gold nanorods embedded in alumina with an LC top layer as a function of applied electric field.

1.3.4.4 Light-controlled transmission through nanorod arrays embedded in a nonlinear dielectric

Similar to the nonlinear optical effects in plasmonic crystals described in Section 1.2, the optical properties of arrays of nanorods can also be controlled using an optical field. This control can be achieved using the third-order nonlinear optical properties of a Kerr-type dielectric matrix to modify the array's eigenmodes when the RI of the matrix is changed upon optical excitation. This has been demonstrated by replacing the AAO matrix with a nonlinear polymer (poly-3BCMU) as a dielectric embedding the Au nanorods [97,157]. The polymer matrix was made by spin-coating mono-3BCMU on the bare array of rods and subsequently polymerizing the monomer under UV radiation [158]. To characterize the nonlinear optical response of the hybrid structures, a two-color continuous-wave pump–probe experimental configuration was used [97]. The control light was coupled to the T mode of the nanorod array. The changes in RI of the polymer, induced by the electric field associated with the T mode, effectively alter the coupling between the nanorods in the array, which affects the spectral position of the array eigenmodes. Considering the sensitivity of both the T and L modes to changes in the adjacent RI [120], only the L mode will show substantial spectral shifts as a function of pump power. This guarantees an efficient and mostly power-insensitive coupling of the control beam with the T mode. The pump-induced spectral modifications of the L mode are then monitored by a broadband probe light.

Figure 38 shows the differential changes in the extinction of the L mode for a pump power of 23 W cm^{-2}. The changes are normalized with respect to the extinction of the array when the control beam is absent. A blue shift of the L mode is observed as the control light is turned on. Considering that poly-3BCMU is a nonlinear optical material with a positive Kerr coefficient ($d\varepsilon_d > 0$), one would expect the L-mode resonance peak to shift to longer wavelengths as $d\lambda_{resonance} \propto \varepsilon_d^{-1/2} d\varepsilon_d$ with increasing pump light intensity [56]. Instead, the resonant excitonic behavior of poly-3BCMU in the spectral range overlapping the L-mode resonance affects both the real and imaginary parts of the polymer's refractive index, leading to a blue shift of the L mode with increasing control beam power.

Due to these competing effects, the effective response of the nonlinear nanorod arrays is weaker than that of the plasmonic crystals, leading to less pronounced variations in its optical properties and requiring much higher control light intensities for equivalent extinction changes. The upper-limit estimate of the sensitivity of the L mode to the changes of RI of the matrix can be made using the data presented in Section 1.03.5. This 'pure' Kerr-type sensitivity, neglecting absorption changes, is on the order of

$$\frac{d\lambda_{resonance}}{d\varepsilon_d} \propto \frac{\Delta\lambda_{resonance}}{\Delta\varepsilon_d} = 400\ \mathrm{nm}/\varepsilon_d$$

Again, compared to nonlinear plasmonic crystals, where nonlinear effects depend more strongly on both the absolute value of RI modification and its spatial variations, arrays of closely spaced nanorods, which are well described by the effective-medium

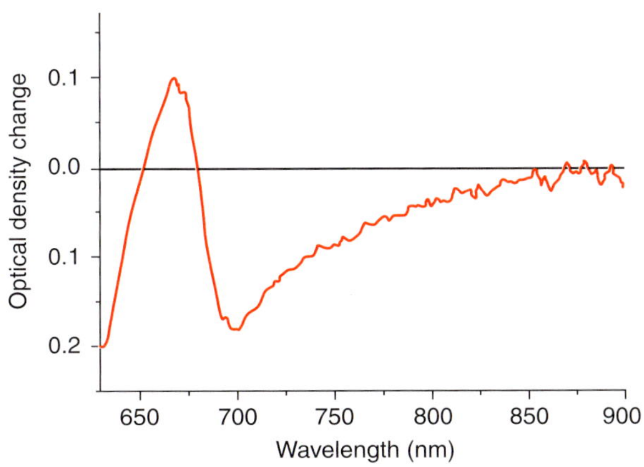

Figure 38 Nonlinear nanorod array. Differential extinction spectrum for a control wavelength of 514 nm coupled to the T mode of the array and a control power of 23 W cm^{-2}. The white-light probe monitors the control-intensity-dependent position of the L mode. The nanorod length, diameter, and separation are 300, 20, and 40 nm, respectively.

approximation, the spatial variations of RI changes are less important than their absolute values.

1.4 Conclusion

The coupling of light to plasmonic excitations, that are collective electronic excitations in metallic nanostructures, allows one to confine electromagnetic fields on the sub-wavelength scale and manipulate these fields extensively. This article provides an overview of the optical properties of two important classes of nanostructured metallic systems: surface plasmon polaritonic crystals and assemblies of strongly coupled metallic nanorods. The consideration of these geometries enables an overview of different types of plasmonic modes, such as surface plasmon polaritons, cavity modes, localized surface plasmons, and coherent effects associated with their mutual coupling. Both nanostructured systems allow for an extensive tailoring of their eigenmodes through geometric design, including mode dispersion, spectral properties, and mode lifetime, as well as spatial field distribution. All these properties influence the optical properties of nanostructured materials, including transmission, reflection, and absorption. Moreover, these optical properties can be controlled dynamically by influencing the adjacent dielectric RIs with external static electric and magnetic fields, or even optical fields. This is the fundamental strength of these materials, as one seeks to manipulate electromagnetic signals at the nanoscale in a variety of applications, such as waveguiding, ultralow-light-intensity nonlinear effects, as well as all-optical modulators and transistors.

Acknowledgments

The authors thank all their collaborators, whose contribution to this work was extensive. In particular, they acknowledge R. Atkinson, W. Dickson, P. Evans, W. Hendren, and D. O'Connor from the Queen's University of Belfast, C. Bower from Kodak Ltd, W. Harrison from Kodak Inc., and L. Le Guyader, A. Kirilyuk, and Th. Rasing from University of Nijmegen. The authors acknowledge support by the EPSRC (UK), the EU FP6 Network of Excellence Plasmo-Nano-Devices, and by Invest Northern Ireland for the provision of funding under Nanotec NI, facilitating the formation of the Centre for Nanostructured Media at the Queen's University of Belfast.

References

1. Barnes WL, Dereux A, and Ebbesen TW (2003) Surface plasmon subwavelength optics. *Nature* 424: 824.
2. Maier SA (2007) *Plasmonics: Fundamentals and Applications*. New York: Springer.
3. Atwater HA (2007) The promise of plasmonics. *Scientific American* 296: 56.
4. Ebbesen TW, *et al.* (2008) Surface-plasmon circuitry. *Physics Today* 61: 44.
5. Bozhevolnyi SI (2009) *Plasmonic nanoguides and circuits*. Singapore: World Scientific.
6. Raether H (1987) *Surface Plasmons on Smooth and Rough Surfaces and on Gratings*. Berlin: Springer.
7. Boardman AD (1982) *Electromagnetic Surface Modes* Chichester: Wiley.
8. Zayats AV and Smolyaninov II (2003) Near-field photonics: Surface plasmon polaritons and localized surface plasmons. *Journal of Optics A: Pure and Applied Optics* 5: S16.
9. Zayats AV, *et al.* (2005) Nano-optics of surface plasmon polaritons. *Physical Reports* 408: 131.
10. Wurtz GA, *et al.* (2006) Optical bistability in nonlinear surface-plasmon polaritonic crystals. *Physical Review Letters* 97: 057402.
11. Nie, S and Emory SR (1997) Probing single molecules and single nanoparticles by surface-enhanced raman scattering. *Science* 275: 1102.
12. Anker JN, *et al.* (2008) Biosensing with plasmonic nanosensors. *Nature Materials* 7: 442.
13. Cai W, *et al.* (2007) Optical cloaking with metamaterials. *Nature Photon* 1: 224.
14. Shalaev VM (2007) Optical negative-index metamaterials. *Nature Photon* 1: 41.
15. Litchinitser NM and Shalaev VM (2009) Metamaterials: Loss as a route to transparency. *Nature Photon* 3: 75.
16. Sandtke M and Kuipers L (2007) Slow guided surface plasmons at telecom frequencies. *Nature Photon* 1: 573.
17. Dickson W, *et al.* (2008) Electronically controlled surface plasmon dispersion and optical transmission through metallic hole arrays using liquid crystal. *Nano Letters* 8: 281.
18. Wurtz GW and Zayats AV (2008) Nonlinear surface plasmon polaritonic crystals. *Laser and Photon Review* 2: 125.
19. Kottmann JP, *et al.* (2001) Plasmon resonances of silver nanowires with a nonregular cross section. *Physical Review B* 64: 2354025.
20. Noguez C (2007) Surface plasmons on metal nanoparticles: The influence of shape and physical environment. *Journal of Physical Chemistry C* 111: 3806.
21. Prodan E, *et al.* (2003) A hybridization model for the plasmon response of complex nanostructures. *Science* 302: 419.
22. Rechberger W, *et al.* (2003) Optical properties of two interacting gold nanoparticles. *Optics Communications* 220: 137.
23. Xiao JJ, *et al.* (2005) Optical response of strongly coupled metal nanoparticles in dimer arrays. *Physical Review B* 71: 045404.
24. Jain PK, *et al.* (2006) Plasmon coupling in nanorod assemblies: Optical absorption, discrete dipole approximation simulation, and exciton-coupling model. *Journal of Physical Chemistry B* 110: 18243.
25. Hao F, *et al.* (2007) Plasmon resonances of a gold nanostar. *Nano Letters* 7: 729.
26. Jain PK and El-Sayed MA (2007) Universal scaling of plasmon coupling in metal nanostructures: Extension from particle pairs to nanoshells. *Nano Letters* 7: 2854.

27. Dionne JA, *et al.* (2005) Plasmon slot waveguides: Towards chip-scale propagation with subwavelength-scale localization. *Physical Review B* 73: 035407.
28. Krenn JR, *et al.* (1999) Squeezing the optical near-field zone by plasmon coupling of metallic nanoparticles. *Physical Review Letters* 82: 2590.
29. Brongersma ML, *et al.* (2000) Electromagnetic energy transfer and switching in nanoparticle chain arrays below the diffraction limit. *Physical Review B* 62: R16356.
30. Maier SA, *et al.* (2003) Local detection of electromagnetic energy transport below the diffraction limit in metal nanoparticle plasmon waveguides. *Nature Materials* 2: 229.
31. Citrin DS (2004) Coherent excitation transport in metal-nanoparticle chains. *Nano Letters* 4: 1562.
32. Weber WH and Ford GW (2004) Propagation of optical excitations by dipolar interactions in metal nanoparticle chains. *Physical Review B* 70: 125429.
33. Kelf TA, *et al.* (2005) Plasmonic band gaps and trapped plasmons on nanostructured metal surfaces. *Physical Review Letters* 95: 116802.
34. Kelf TA, *et al.* (2006) Localized and delocalized plasmons in metallic nanovoids. *Physical Review B* 74: 245415.
35. Wiederrecht GP, *et al.* (2004) Coherent coupling of molecular excitons to electronic polarizations of noble metal nanoparticles. *Nano Letters* 4: 2121.
36. Bellessa J, *et al.* (2004) Strong coupling between surface plasmons and excitons in an organic semiconductor. *Physical Review Letters* 93: 036404.
37. Zhang W, *et al.* (2006) Semiconductor-metal nanoparticle molecules: Hybrid excitons and the nonlinear Fano effect. *Physical Review Letters* 97: 146804.
38. Wurtz GA, *et al.* (2007) Molecular plasmonics with tunable exciton–plasmon coupling strength in J-aggregate hybridized Au nanorod assemblies. *Nano Letters* 7: 1297.
39. Vasa P, *et al.* (2008) Coherent exciton – surface-plasmon-polariton interaction in hybrid metal-semiconductor nanostructures. *Physical Review Letters* 101: 116801.
40. Fofang NT, *et al.* (2008) Plexcitonic nanoparticles: Plasmon–exciton coupling in nanoshell J-aggregate complexes. *Nano Letters* 8: 3481.
41. Barnes WL (2006) Surface plasmon-polariton length scales: A route to sub-wavelength optics. *Journal of Optics A: Pure and Applied Optics* 8: S87.
42. Johnson PB and Christy RW (1972) Optical constants of the noble metals. *Physical Review B* 6: 4370.
43. Abajo FJG (2007) Light scattering by particle and hole arrays. *Reviews of Modern Physics* 79: 1267.
44. Karalis A, *et al.* (2005) Surface-plasmon-assisted guiding of broadband slow and subwavelength light in air. *Physical Review Letters* 95: 063901.
45. Bozhevolnyi SI, *et al.* (2001) Waveguiding in surface plasmon polariton band gap structures. *Physical Review Letters* 86: 3008.
46. Zayats AV, *et al.* (2003) Polarization superprism effect in surface polaritonic crystals. *Applied Physics Letters* 82: 4438.
47. Weeber J-C, *et al.* (2007) Surface-plasmon hopping along coupled coplanar cavities. *Physical Review B* 76: 113405.
48. Barnes WL, *et al.* (1996) Physical origin of photonic energy gaps in the propagation of surface plasmons on gratings. *Physical Review B* 54: 6267.
49. Schouten HF, *et al.* (2005) Plasmon-assisted two-slit transmission: Young's experiment revisited. *Physical Review Letters* 94: 053901.
50. Nieuwstadt JAHV, *et al.* (2006) Strong modification of the nonlinear optical response of metallic subwavelength hole arrays. *Physical Review Letters* 97: 146102.
51. Garcia-Vidal FJ, *et al.* (2006) Transmission of light through a single rectangular hole in a real metal. *Physical Review B* 74: 153411.
52. Gordon R and Brolo AG (2005) Increased cut-off wavelength for a subwavelength hole in a real metal. *Optics Express* 13: 1933.
53. Ung B and Sheng Y (2007) Interference of surface waves in a metallic nanoslit. *Optics Express* 15: 1182.
54. Economou EN (1969) Surface plasmons in thin films. *Physical Review* 182: 539.
55. Liu H and Lalanne P (2008) Microscopic theory of the extraordinary optical transmission. *Nature* 452: 728.
56. Bohren CF and Huffmann DR (1983) *Absorption and Scattering of Light by Small Particles*. New York: Wiley Science.
57. Palik ED (1985) *Handbook of Optical Constants of Solids*. New York: Academic Press.
58. Mooradian A (1969) Photoluminescence of metals. *Physical Review Letters* 22: 185.
59. Khriachtchev L, *et al.* (2001) Red photoluminescence of gold island films. *Applied Physics Letters* 78: 1994.
60. Darmanyan SA and Zayats AV (2003) Light tunneling via resonant surface plasmon polariton states and the enhanced transmission of periodically nanostructured metal films: An analytical study. *Physical Review B* 67: 035424.
61. Darmanyan SA, *et al.* (2004) Analytical theory of optical transmission through periodically structured metal films via tunnel-coupled surface polariton modes. *Physical Review B* 70: 075103.
62. Dragila R, *et al.* (1985) High transparency of classically opaque metallic films. *Physical Review Letters* 55: 1117.
63. Gerard D, *et al.* (2004) Analysis of the Bloch mode spectra of surface polaritonic crystals in the weak and strong coupling regimes: Grating-enhanced transmission at oblique incidence and suppression of SPP radiative losses. *Optics Express* 12: 3652.
64. Lalanne P, *et al.* (2005) Theory of surface plasmon generation at nanoslit apertures. *Physical Review Letters* 95: 263902.
65. Barbara A, *et al.* (2002) Optical transmission through subwavelength metallic gratings. *Physical Review B* 66: 161403.
66. Porto JA, *et al.* (1999) Transmission resonances on metallic gratings with very narrow slits. *Physical Review Letters* 83: 2845.
67. Marquier F, *et al.* (2005) Resonant transmission through a metallic film due to coupled modes. *Optica Express* 13: 70.
68. Shin H, *et al.* (2005) Effect of the plasmonic dispersion relation on the transmission properties of subwavelength cylindrical holes. *Physical Review B* 72: 085436.
69. Altewischer E, *et al.* (2003) Depolarization induced by subwavelength metal hole arrays. *Journal of Optical Society of America B* 20: 1927.
70. Papakostas A, *et al.* (2003) Optical manifestations of planar chirality. *Physical Review Letters* 90: 107404.
71. Moreno E, *et al.* (2004) Theory of plasmon-assisted transmission of entangled photons. *Physical Review Letters* 92: 236801.
72. Elliott J, *et al.* (2004) Wavelength dependent birefringence of surface plasmon polaritonic crystals. *Physical Review B* 90: 233403.
73. Kats AV, *et al.* (2005) Polarization properties of a periodically-modulated metal film in regions of anomalous optical transparency. *Physical Review B* 72: 193405.
74. Genet C, *et al.* (2005) Optical depolarization induced by arrays of subwavelength metal holes. *Physical Review B* 71: 033409.

75. Gordon R, *et al.* (2004) Strong polarization in the optical transmission through elliptical nanohole arrays. *Physical Review Letters* 92: 037401.
76. Altewischer E, *et al.* (2005) Fano-type interference in the point-spread function of nanohole arrays. *Optics Letters* 30: 2436.
77. Evans PR, *et al.* (2007) Electrically switchable nonreciprocal transmission of plasmonic nanorods with liquid crystal. *Applied Physics Letters* 91: 043101.
78. Priezjev N and Pelcovits RA (2000) Surface extrapolation length and director structures in confined nematics. *Physical Review E* 62: 6734.
79. Khurgin JB (2006) Optical isolating action in surface plasmon polaritons. *Applied Physics Letters* 89: 251115.
80. Sepulveda B, *et al.* (2006) Magnetooptic effects in surface-plasmon-polaritons slab waveguides. *Journal of Lightwave Technology* 24: 945.
81. Strelniker YM and Bergman DJ (1999) Optical transmission through metal films with a subwavelength hole array in the presence of a magnetic field. *Physical Review B* 59: R12763 LP.
82. Battula A, *et al.* (2007) Tuning the extraordinary optical transmission through subwavelength hole array by applying a magnetic field. *Optics Letters* 32: 2692.
83. Lan Y-C (2008) Magnetic-field-induced frequency shifts for resonant tunneling effects caused by surface plasmons excitation. *IEEE Quantum Electron* 14: 1559.
84. Inoue M, *et al.* (2006) Magneto-optical crystals. *Journal of Physics D: Applied Physics* 39: R151.
85. Belotelov VI, *et al.* (2007) Extraordinary magneto-optical effects and transmission through metal-dielectric plasmonic systems. *Physical Review Letters* 98: 077401.
86. García-Martín A, *et al.* (2005) Light transport in photonic crystals composed of magneto-optically active materials. *Physical Review B* 71: 205116.
87. Landau LD and Lifshits EM (1984) *Electrodynamics of Continuous Media*. Oxford: Pergamon.
88. Gibbs HM (1985) *Optical Bistability: Controlling Light with Light*. New York: Academic.
89. Innes RA and Sambles JR (1989) Optical nonlinearity in liquid crystals using surface plasmon-polaritons. *Journal of Physics: Condensed Matter* 1: 6231.
90. John S and Florescu M (2001) Photonic band gap materials: Toward an all-optical micro-transistor. *Journal of Optics A: Pure Applied Optics* 3: S103.
91. Centeno E and Felbacq D (2000) Optical bistability in finite-size nonlinear bidimensional photonic crystals doped by a microcavity. *Physical Review B* 62: R7683 LP.
92. Mingaleev SF, *et al.* (2006) All-optical switching, bistability, and slow-light transmission in photonic crystal waveguide-resonator structures. *Physical Review E* 74: 046603.
93. Yanik MF, *et al.* (2003) All-optical transistor action with bistable switching in a photonic crystal cross-waveguide geometry. *Optics Letters* 28: 2506.
94. Smolyaninov II (2005) Quantum fluctuations of the refractive index near the interface between a metal and a nonlinear dielectric. *Physical Review Letters* 94: 057403.
95. Liu T-M, *et al.* (2006) Measuring plasmon-resonance enhanced third-harmonic chi[sup (3)] of Ag nanoparticles. *Applied Physics Letters* 89: 043122.
96. Shen Y and Prasad PN (2002) Nanophotonics: A new multidisciplinary frontier. *Applied Physics B* 74: 641.
97. Dickson W, *et al.* (2008) Towards nonlinear plasmonic devices based on metallic nanorods. *Journal of Microscopy* 420: 415.
98. Marinica DC, *et al.* (2007) Second harmonic generation from arrays of subwavelength cylinders. *Physical Review B* 76: 085311.
99. Smolyaninov II, *et al.* (2002) Single-photon tunneling via localized surface plasmons. *Physical Review Letters* 88: 187402.
100. Smolyaninov II, *et al.* (2002) Light-controlled photon tunneling. *Applied Physics Letters* 81: 3314.
101. Salomon L, *et al.* (2001) Near-field distribution of optical transmission of periodic subwavelength holes in a metal film. *Physical Review Letters* 86: 1110.
102. Kretschmann M and Maradudin AA (2002) Band structures of two-dimensional surface-plasmon polaritonic crystals. *Physical Review B* 66: 245408.
103. Smolyaninov II, *et al.* (2002) Optical control of photon tunneling through an array of nanometer-scale cylindrical channels. *Physical Review Letters* 66: 205414.
104. Le Guyader L, *et al.* (2008) Coherent control of surface plasmon polariton mediated optical transmission. *Jounal of Physics D: Applied Physics* 41: 195102.
105. Friedman RS, *et al.* (2005) High-speed integrated nanowire circuits. *Nature* 434: 1085.
106. Tong LM, *et al.* (2003) Subwavelength-diameter silica wires for low-loss optical wave guiding. *Nature* 426: 816.
107. Xiang J, *et al.* (2006) Ge/Si nanowire heterostructures as high-performance field-effect transistors. *Nature* 441: 489.
108. Gudiksen MS, *et al.* (2002) Growth of nanowire superlattice structures for nanoscale photonics and electronics. *Nature* 415: 617.
109. Mühlschlegel P, *et al.* (2005) Resonant optical antennas. *Science* 308: 1607.
110. Shalaev VM, *et al.* (2005) Negative index of refraction in optical metamaterials. *Optics Letters* 30: 3356.
111. Evans P, *et al.* (2006) Growth and properties of gold and nickel nanorods in thin film alumina. *Nanotechnology* 17: 5746.
112. Lee K-S and El-Sayed MA (2006) Gold and silver nanoparticles in sensing and imaging: Sensitivity of plasmon response to size, shape, and metal composition. *Journal of Physical Chemistry B* 110: 19220.
113. Pelton M, *et al.* (2006) Ultrafast resonant optical scattering from single gold nanorods: Large nonlinearities and plasmon saturation. *Physical Review B* 73: 155419.
114. Schider G, *et al.* (2003) Plasmon dispersion relation of Au and Ag nanowires. *Physical Review B* 68: 155427.
115. Atkinson R, *et al.* (2006) Anisotropic optical properties of arrays of gold nanorods embedded in alumina. *Physical Review B* 73: 235402.
116. Ono A, *et al.* (2005) Subwavelength optical imaging through a metallic nanorod array. *Physical Review Letters* 95: 265407.
117. Podolskiy VA, *et al.* (2005) Resonant light interaction with plasmonic nanowire systems. *Journal of Optics A: Pure Applied Optics* 7: S32.
118. Rahachou AI and Zozulenko IV (2007) Light propagation in nanorod arrays. *Journal of Optics A: Pure Applied Optics* 9: 265.
119. Reinhardt C, *et al.* (2006) Rapid laser prototyping of plasmonic components. *Applied Physics Letters* 89: 231117.
120. Evans PR, *et al.* (2007) Plasmonic core/shell nanorod arrays: Subattoliter controlled geometry and tunable optical properties. *Journal of Physical Chemistry C* 111: 12522.
121. Link S and El-Sayed MA (1999) Spectral properties and relaxation dynamics of surface plasmon electronic oscillations in gold and silver nanodots and nanorods. *Journal of Physical Chemistry B* 103: 8410.
122. Prescott SW and Mulvaney P (2006) Gold nanorod extinction spectra. *Journal of Applied Physics* 99: 123504.

123. Aizpurua J, *et al.* (2005) Optical properties of coupled metallic nanorods for field-enhanced spectroscopy. *Physical Review B* 71: 235420.
124. Smythe EJ, *et al.* (2007) Optical properties of surface plasmon resonances of coupled metallic nanorods. *Optics Express* 15: 7439.
125. Wurtz GA, *et al.* (2008) Guided plasmonic modes in nanorod assemblies: Strong electromagnetic coupling regime. *Optics Express* 16: 7460.
126. Maier SA, *et al.* (2001) Plasmonics – a route to nanoscale optical devices. *Advanced Materials* 13: 1501.
127. Waele RD, *et al.* (2007) Tunable nanoscale localization of energy on plasmon particle arrays. *Nano Letters* 7: 2004.
128. Maier SA, *et al.* (2002) Observation of coupled plasmon-polariton modes of plasmon waveguides for electromagnetic energy transport below the diffraction limit. *SPIE* 4810: 71.
129. Maier SA, *et al.* (2003) Optical pulse propagation in metal nanoparticle chain waveguides. *Physical Review B* 67: 205402.
130. Park SY and Stroud D (2004) Surface-plasmon dispersion relations in chains of metallic nanoparticles: An exact quasistatic calculation. *Physical Review B* 69: 125418.
131. Koenderink AF and Polman A (2006) Complex response and polariton-like dispersion splitting in periodic metal nanoparticle chains. *Physical Review B* 74: 033402.
132. Vlad A, *et al.* (2006) Controlled growth of single nanowires within a supported alumina template. *Nanotechonolgy* 17: 4873.
133. Wolfrum B, *et al.* (2006) Fabrication of large-scale patterned gold-nanopillar arrays on a silicon substrate using imprinted porous alumina templates. *Small* 2: 1256.
134. Andrew P and Barnes WL (2004) Energy transfer across a metal film mediated by surface plasmon polaritons. *Science* 306: 1002.
135. Seidel J, *et al.* (2005) Stimulated emission of surface plasmons at the interface between a silver film and an optically pumped dye solution. *Physical Review Letters* 94: 177401.
136. Kobayashi J (1996) *J-Aggregates*. Singapore: World Scientific.
137. Wiederrecht GP, *et al.* (2007) Control of molecular energy redistribution pathways via surface plasmon gating. *Physical Review Letters* 98: 083001.
138. Ozcelik S, *et al.* (1998) Superradiant lasing from J-aggregated molecules adsorbed onto colloidal silver. *Applied Physics Letters* 73: 1949.
139. Dintinger J, *et al.* (2005) Strong coupling between surface plasmon-polaritons and organic molecules in subwavelength hole arrays. *Physical Review B* 71: 035424.
140. Lidzey DG, *et al.* (2000) Photon-mediated hybridization of frenkel excitons in organic semiconductor microcavities. *Science* 288: 1620.
141. Hobson PA, *et al.* (2002) Strong exciton–photon coupling in a low-Q all-metal mirror microcavity. *Applied Physics Letters* 81: 3519.
142. Christ A, *et al.* (2003) Waveguide-plasmon polaritons: Strong coupling of photonic and electronic resonances in a metallic photonic crystal slab. *Physical Review Letters* 91: 183901.
143. Pockrand I, *et al.* (1980) Nonradiative decay of excited molecules near a metal surface. *Chemical Physics Letters* 69: 499.
144. Ambjörnsson T, *et al.* (2006) Resonant coupling between localized plasmons and anisotropic molecular coatings in ellipsoidal metal nanoparticles. *Physical Review B* 73: 085412.
145. Agranovich VM, *et al.* (2003) Cavity polaritons in microcavities containing disordered organic semiconductors. *Physical Review B* 67: 085311.
146. Imura K, *et al.* (2005) Near-field optical imaging of plasmon modes in gold nanorods. *Journal of Chemical Physics* 122: 154701.
147. Prikulis J, *et al.* (2004) Optical spectroscopy of single trapped metal nanoparticles in solution. *Nano Letters* 4: 115.
148. Brolo A, *et al.* (2004) Surface plasmon sensor based on the enhanced light transmission through arrays of nanoholes in gold films. *Langmuir* 20: 4813.
149. McFarland AD and Duyne RPV (2003) Single silver nanoparticles as real-time optical sensors with zeptomole sensitivity. *Nano Letters* 3: 1057.
150. Malinsky MD, *et al.* (2001) Chain length dependence and sensing capabilities of the localized surface plasmon resonance of silver nanoparticles chemically modified with alkanethiol self-assembled monolayers. *Journal of American Chemical Society* 123: 1471.
151. Sun YG and Xia YN (2003) Gold and silver nanoparticles: A class of tunable in the range from 400 to 750 nm. *Analyst* 128: 686.
152. Hicks EM, *et al.* (2005) Plasmonic properties of film over nanowell surfaces fabricated by nanosphere lithography. *Journal of Physical Chemistry B* 109: 22351.
153. Haes AJ, *et al.* (2004) A nanoscale optical biosensor: The long range distance dependence of the localized surface plasmon resonance of noble metal nanoparticles. *Journal of Physical Chemistry B* 108: 109.
154. Chu KC, *et al.* (2006) Electrically controlled surface plasmon resonance frequency of gold nanorods. *Applied Physics Letters* 89: 103107.
155. Müller J, *et al.* (2002) Electrically controlled light scattering with single metal nanoparticles. *Applied Physics Letters* 81: 171.
156. Kossyrev PA, *et al.* (2005) Electric field tuning of plasmonic response of nanodot array in liquid crystal matrix. *Nano Letters* 5: 1978.
157. Dickson W, *et al.* (2007) Dielectric-loaded plasmonic nanoantenna arrays: A metamaterial with tunable optical properties. *Physical Review B* 76: 115411.
158. Sottini S, *et al.* (1996) Efficient UV polymerisation of 3BCMU: Optical and waveguiding properties of the material. *Optical Materials* 5: 285.

2 Surface Nanophotonics Theory

J M McMahon, Northwestern University, Evanston, IL, USA

S K Gray, Argonne National Laboratory, Argonne, IL, USA

G C Schatz, Northwestern University, Evanston, IL, USA

2.1 Introduction

Nanostructured metal films have recently attracted a lot of attention with the discovery that periodic arrays of sub-wavelength diameter holes can transmit a relatively large amount of light, a phenomenon known as extraordinary optical transmission (EOT) [1]. Another source of interest is that a number of distinct mechanisms for the interaction of light with them are possible, including surface plasmon polaritons (SPPs), localized surface plasmon resonances (LSPRs), Rayleigh anomalies (RAs), and waveguide modes. Often it is possible to vary these interactions independently, leading to a variety of complex optical properties. The popularity of these systems has been fueled by the fact that they can be fabricated by a variety of techniques, such as electron beam lithography [2], nanosphere lithography [3], and soft lithography [4, 5], allowing precise control over the important dimensions of the problem. These systems have found many applications, as has recently been reviewed [6], including sensing applications based on spectral features that are sensitive to the refractive index (RI) of materials interfacing with the metal surface [7–9] and substrates for surface enhanced Raman scattering [10, 11] and enhanced fluorescence [12].

Nanostructured metal films have attracted extensive interest among theorists, as computational methods for solving Maxwell's equations have progressed to the point where it is possible to provide accurate solutions for many of the important experiments that have been done. Thus, there have been extensive simulations of the transmission of light through hole arrays and related structures, including isolated apertures and particle arrays in metal films [7, 8, 13–16]. There have even been simulations of near-field scanning optical microscope (NSOM) measurements of hole arrays, characterizing the details of the electromagnetic fields associated with propagating plasmon polaritons [16, 17].

In this chapter, we review a number of computational electromagnetics methods that have been applied to nanostructured metal films in the last few years, and present a select set of applications of them. We begin by reviewing the pertinent equations of electromagnetics. The qualitative aspects of metal films, SPPs, and periodic systems are then discussed. This is followed by a discussion of the primary computational tool that has been used for numerical studies, the finite-difference time-domain (FDTD) method, as well as alternative methods, such as the rigorous coupled-wave analysis (RCWA) and the modal expansion method. After this, we present results from applications of these methods to isolated slits and holes, hole and pillar arrays, and other periodic systems. We conclude by giving an outlook on the future of nanostructured metal films and the computational tools used to study them.

2.2 Background

2.2.1 Basic Electromagnetic Theory

Electromagnetics is governed by Maxwell–Ampere's law, Faraday's law, and Gauss' laws, collectively known as Maxwell's equations. In differential form, these equations are

$$\varepsilon \frac{\partial}{\partial t} \boldsymbol{E} = \nabla \times \boldsymbol{H} - \boldsymbol{\mathcal{J}} \tag{1}$$

$$\mu \frac{\partial}{\partial t} \boldsymbol{H} = -\nabla \times \boldsymbol{E} \tag{2}$$

$$\nabla \cdot (\varepsilon \boldsymbol{E}) = \rho \tag{3}$$

$$\nabla \cdot (\mu \boldsymbol{H}) = 0 \tag{4}$$

where, respectively, $\boldsymbol{E}$ and $\boldsymbol{H}$ are the electric and magnetic fields, ε and μ are the permittivity and permeability of the medium, and $\boldsymbol{\mathcal{J}}$ and ρ are current and charge densities. For a time-harmonic field (i.e., the field harmonically oscillates at a single frequency, ω), **equations 1** and **2** become (where the conventional time dependence, $\exp(i\omega t)$, is assumed here, although later we also implicitly use $\exp(-i\omega t)$)

$$i\omega\varepsilon \boldsymbol{E} = \nabla \times \boldsymbol{H} - \boldsymbol{\mathcal{J}} \tag{5}$$

$$i\omega\mu \boldsymbol{H} = -\nabla \times \boldsymbol{E} \tag{6}$$

Equations 5 and **6**, as well as 1 and 2, can be rearranged into equations in terms of only $\boldsymbol{E}$ or $\boldsymbol{H}$,

$$\nabla \times \left(\frac{1}{\mu}\nabla \times \boldsymbol{E}\right) - \omega^2 \varepsilon \boldsymbol{E} = -\mathrm{i}\omega \boldsymbol{\mathcal{J}} \quad (7)$$

$$\nabla \times \left(\frac{1}{\varepsilon}\nabla \times \boldsymbol{H}\right) - \omega^2 \mu \boldsymbol{H} = \nabla \times \left(\frac{1}{\varepsilon}\boldsymbol{\mathcal{J}}\right) \quad (8)$$

called the vector wave equations. If the system is invariant along an axis, which we will take to be the z-axis, **equations 7** and **8** simplify to scalar wave equations,

$$\left[\frac{\partial}{\partial x}\left(\frac{1}{\varepsilon_r}\frac{\partial}{\partial x}\right) + \frac{\partial}{\partial y}\left(\frac{1}{\varepsilon_r}\frac{\partial}{\partial y}\right) + k_0{}^2 \mu_r\right] H_z = -\frac{\partial}{\partial x}\left(\frac{1}{\varepsilon_r}\mathcal{J}_y\right) + \frac{\partial}{\partial y}\left(\frac{1}{\varepsilon_r}\mathcal{J}_x\right) \quad (9)$$

$$\left[\frac{\partial}{\partial x}\left(\frac{1}{\mu_r}\frac{\partial}{\partial x}\right) + \frac{\partial}{\partial y}\left(\frac{1}{\mu_r}\frac{\partial}{\partial y}\right) + k_0{}^2 \varepsilon_r\right] E_z = \mathrm{i}k_0 Z_0 \mathcal{J}_y \quad (10)$$

where $\varepsilon_r = \varepsilon/\varepsilon_0$ and $\mu_r = \mu/\mu_0$ are relative permittivity and permeability values, respectively, $Z_0 = \sqrt{\mu_0/\varepsilon_0}$ is the impedance of free space, and

$$k_0 = \frac{\omega}{c} = \frac{2\pi}{\lambda} \quad (11)$$

is the wavevector magnitude in free space, with c being is the speed of light and λ the wavelength. Note that in a dielectric medium, the wavevector magnitude becomes

$$k = \frac{\omega}{c}\sqrt{\varepsilon_r} \quad (12)$$

Equations 9 and **10** correspond to two distinct sets of polarization, one where $\boldsymbol{H}$ is transverse to the z-axis, called TE_z polarization (**equation 9**), and one where $\boldsymbol{E}$ is transverse to the z-axis, called TM_z polarization (**equation 10**).

At the interface between two media, which we will denote by subscripts 1 and 2, the electromagnetic fields must satisfy four conditions, known as boundary conditions,

$$\hat{n} \times (\boldsymbol{E}_1 - \boldsymbol{E}_2) = 0 \quad (13)$$

$$\hat{n} \cdot (\varepsilon_1 \boldsymbol{E}_1 - \varepsilon_2 \boldsymbol{E}_2) = \rho_s \quad (14)$$

$$\hat{n} \times (\boldsymbol{H}_1 - \boldsymbol{H}_2) = \mathcal{J}_s \quad (15)$$

$$\hat{n} \cdot (\mu_1 \boldsymbol{H}_1 - \mu_2 \boldsymbol{H}_2) = 0 \quad (16)$$

where $\hat{n}$ is a unit vector normal to the interface pointing from medium 2 to 1, and ρ_s and $\mathcal{J}_s$ are the surface charge and current density, respectively. The boundary conditions are greatly simplified when one of the media cannot sustain internal fields, known as a perfect electric conductor (PEC),

$$\hat{n} \times \boldsymbol{E} = 0 \quad (17)$$

$$\hat{n} \cdot (\mu \boldsymbol{H}) = 0 \quad (18)$$

Note that in this case there is always a surface charge and current density,

$$\hat{n} \cdot (\varepsilon \boldsymbol{E}) = \rho_s \quad (19)$$

$$\hat{n} \times \boldsymbol{H} = \mathcal{J}_s \quad (20)$$

2.2.2 Metal Films, Apertures, and Periodic Structures

Many mechanisms for the interaction of light with nanostructured metal films are possible. One of these is through surface plasmons (SPs), which are oscillations of electron density on the metal–dielectric interface. Under certain circumstances, light can be efficiently coupled into SPs, resulting in a surface plasmon polariton (SPP) [18], which can lead to EOT relative to transmission through a metal film without SPPs [1]. For future reference, the transmission through 30 and 50 nm thick gold (Au) films (obtainable from a simple analytical formula) on a glass substrate (which has an RI of 1.5) in air are shown in **Figure 1**.

The peak structure in **Figure 1** arises from direct transmission of light through the Au film, which occurs when the real part of the relative permittivity is +1 (see below), leading to transparency in the absence of damping. On a flat metal film–dielectric interface, SPs propagate in the plane with a wavevector magnitude [18]

$$k_{SP} \approx \frac{\omega}{c}\sqrt{\frac{\varepsilon_M \varepsilon_r}{\varepsilon_M + \varepsilon_r}} \quad (21)$$

where ε_M is the relative permittivity of the metal.

For most metals, ε_M is dispersive (i.e., its value depends on the frequency), and is often well approximated using a Drude plus 2 Lorentz pole function,

$$\varepsilon(\omega) = \varepsilon_\infty - \frac{\omega_D{}^2}{\omega^2 - \mathrm{i}\omega\gamma_D} + \sum_{L=1}^{2} \frac{\Delta\varepsilon_L \omega_L{}^2}{\omega_L{}^2 + 2\mathrm{i}\omega\delta_L - \omega^2} \quad (22)$$

where ε_∞ is the relative permittivity at infinite frequency, ω_D is the Drude pole frequency, γ_D the inverse of the Drude pole relaxation time, ω_L the Lth Lorentz pole frequency, δ_L a damping factor of the Lorentz pole, and $\Delta\varepsilon_L$ the change in relative

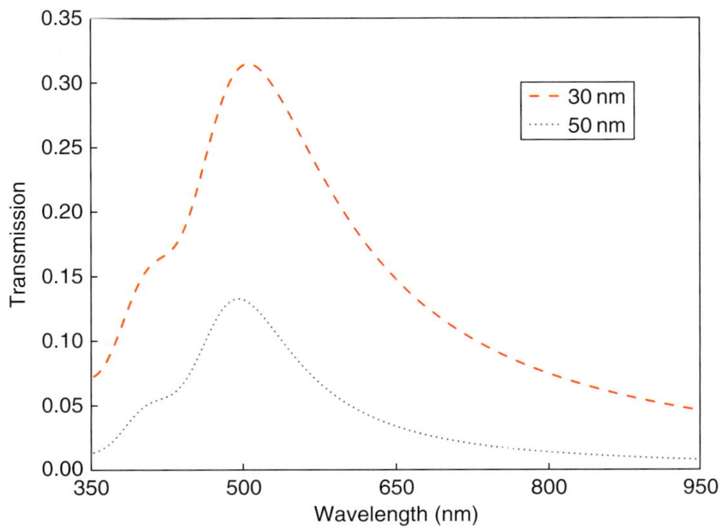

Figure 1 Transmission through Au films on a glass substrate in air.

permittivity at ω_L. **Figure 2** shows $\varepsilon(\omega)$ for Au and silver (Ag) over the range 300–800 nm calculated using **equation 22** with the parameters in **Table 1**, consistent with the Lynch and Hunter (Au) [19] and Johnson and Christy (Ag) [20] empirically determined dielectric data, respectively.

In **Figure 2**, Real(ε) is seen to decrease over the range 300–800 nm, becoming large and negative. This corresponds to the ability of the metal to easily polarize against the incident field. In addition, for a planar metal film this leads to an increase in reflection and decrease in transmission, as can be seen by comparison with **Figure 1**. **Figure 2** also shows that both Au and Ag have a nonzero Imag(ε), which corresponds to absorption in the metal. As a side note, the structure near 400 nm for Au arises from an interband transition, and is not directly responsible for any plasmonic effects. The use of models for $\varepsilon(\omega)$ is common with time-domain computational methods, such as the FDTD method (see Section 2.3.1). For frequency-domain methods, however, the empirically determined dielectric data can be used directly.

Figure 3 shows the SP wavevector magnitude, **equation 21**, for Au and Ag metal film–air interfaces, as well as the free space wavevector magnitude, **equation 11**, known as a dispersion diagram. For most ω, it is seen that $k_{SP} > k_0$, which is a result of $\varepsilon_M < 0$. In addition, near $\omega = 3.8 \times 10^{15}\ s^{-1}$ ($\lambda = 496$ nm) for Au and $\omega = 5.5 \times 10^{15}\ s^{-1}$ ($\lambda = 342$ nm) for Ag a backbending of ck is seen. This is due to the complex nature of ε_M, which causes $dk_{SP}/d\omega$ to change sign with Real(ε) (see **Figure 2**).

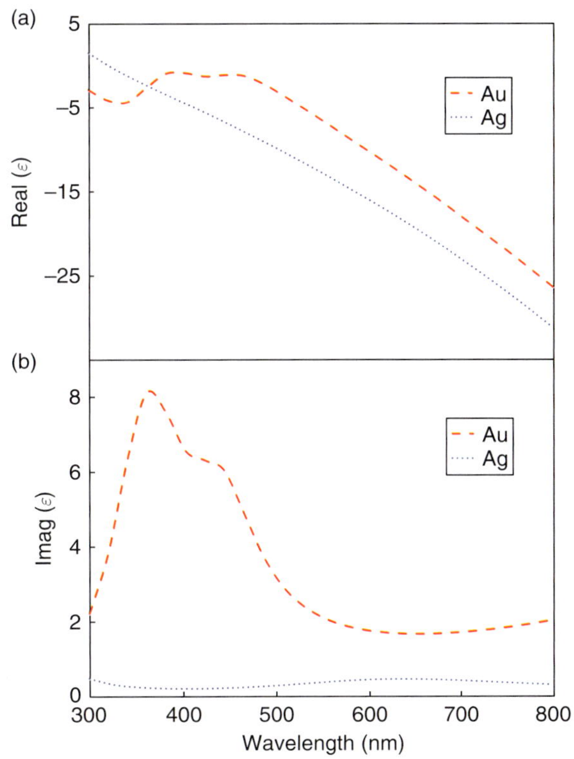

Figure 2 Relative permittivities of Au and Ag modeled using **equation 22** with parameters from **Table 1**.

Because $k_{SP} > k_0$, it is not possible to excite SPPs on a metal film–dielectric interface without an additional source of momentum. Some common methods to provide additional momentum are prisms [21, 22],

Table 1 Drude plus 2 Lorentz pole parameters used in **equation 22** for Au and Ag modeling

Material	ε_∞	ω_D *(rad s^{-1})*	γ_D *(s^{-1})*
Au	5.39833	1.39783×10^{16}	1.03337×10^{14}
Ag	1.17152	1.39604×10^{16}	12.6126
Material	$\Delta\varepsilon_L$	ω_L (rad s^{-1})	δ_L (s^{-1})
Au	0.68086	$4.27392\cdot10^{15}$	4.35335×10^{14}
	1.86089	$5.22540\cdot10^{15}$	6.60773×10^{14}
Ag	2.23994	$8.25718\cdot10^{15}$	1.95614×10^{14}
	0.22265	$3.05707\cdot10^{15}$	8.52675×10^{14}

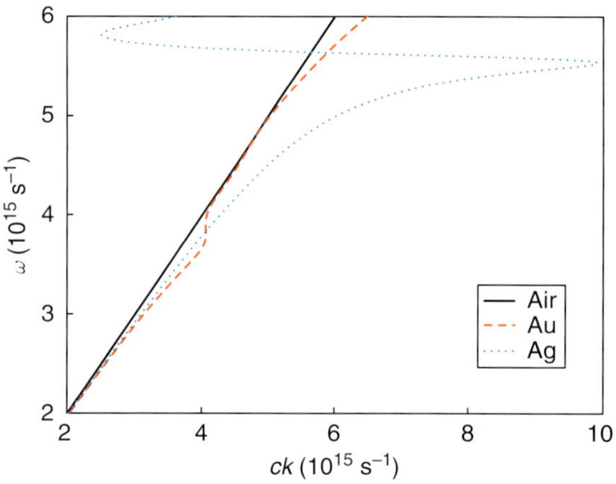

Figure 3 Dispersion diagrams of light in air compared to those of SPs on Au and Ag metal film–air interfaces.

isolated holes [15, 23], and diffraction gratings, such as hole arrays [1]. For example, a two dimensional (2D) diffraction grating generates x- and y-wavevector components with magnitudes

$$k_x = \frac{2\pi n_x}{P_x} \tag{23}$$

$$k_y = \frac{2\pi n_y}{P_y} \tag{24}$$

where n_x and n_y are diffraction orders and P_x and P_y are the periodicities in the x- and y-directions, respectively. For a square periodicity, $P = P_x = P_y$, **equations 23** and **24** combine to give a total wavevector magnitude of

$$k = \frac{2\pi}{P}\sqrt{n_x^2 + n_y^2} \tag{25}$$

Equating equations 25 and 21 shows that the wavelength of an SPP excited by a square grating, λ_{SPP}, is given by

$$\lambda_{\mathrm{SPP}} = \frac{P}{\sqrt{s_x^2 + s_y^2}}\sqrt{\frac{\varepsilon_{\mathrm{M}}\varepsilon_{\mathrm{r}}}{\varepsilon_{\mathrm{M}} + \varepsilon_{\mathrm{r}}}} \tag{26}$$

where (s_x, s_y) is an integer pair that defines the diffraction orders that excited the SPP (where the symbol s has replaced n in **equation 25** in order to distinguish it from diffraction orders that excite other effects; see below). Because of the system periodicity, the excited SPP is sometimes referred to as a SPP–Bloch wave (SPP–BW).

Besides exciting SPPs, other mechanisms of light interaction with nanostructured metal films are possible. For example, by equating **equations 25** and **12** one finds that

$$\lambda_{\mathrm{RA}} = \frac{P}{\sqrt{w_x^2 + w_y^2}}\sqrt{\varepsilon_{\mathrm{r}}} \tag{27}$$

where (w_x, w_y) is an integer pair indicating the diffraction order. **Equation 27** corresponds to the light diffracted at an angle parallel to the grating, a RA. A RA occurs at the threshold of a diffraction order, and is sometimes also referred to as a Wood anomaly [2, 15]. However, it is important to keep in mind that, following the detailed analysis and classification of Wood anomalies by Hessel and Oliner [24], a RA is one of the two types of Wood anomalies. The other type of Wood anomaly identified by them is a 'resonant' anomaly, which could be an SPP, waveguide mode, or possibly a hybrid mode.

It is important to note that the actual wavelengths associated with **equation 26** are approximate because coupling between resonances on the top and bottom film interfaces is ignored. Therefore, one expects this equation to be most accurate when the metal film is thick or the resonances occur significantly far from one another (e.g., when ε_{I} is significantly different from $\varepsilon_{\mathrm{III}}$, where I and III refer to the two sides of the metal film). In thin metal films, when ε_{I} and $\varepsilon_{\mathrm{III}}$ are similar (ca. $\Delta\varepsilon < 0.2$), which typically occurs when a metal film is sandwiched between similar dielectric materials, the actual positions of the SPP resonance features are red- or blue-shifted slightly relative to **equation 26**. This is due to SPP–SPP coupling, which leads to a set of hybridized SPPs: (1) a high-energy antisymmetric combination and (2) a low-energy symmetric combination [25]. The long-wavelength mode (symmetric) is red-shifted, while the short-wavelength mode (antisymmetric) is blue-shifted, relative to their predicted positions. In addition, the amplitude of the symmetric mode increases, and that of the antisymmetric mode decreases. Interaction between SPPs and RAs can also occur, leading to narrow spectral features that are particularly sensitive to the RI, a phenomenon known as the

RA–SPP effect (see Section 2.5.1) [7]. RAs, however, are nonresonant spectral features associated with a grating, so the RI sensitivity in this case is associated with the SPP, while the narrow linewidth is associated with the RA.

Some of the most popular types of diffraction gratings in metal films are hole and slit arrays [1, 7]. Besides generating additional momentum to couple into SPPs, the holes and slits themselves provide a waveguide mechanism for the transmission of light from the lower surface to the upper one [13, 14]. However, Bethe showed [26] that the amount of transmission through a single circular hole in an infinitely thin PEC film (which does not support SPPs) scales as $(2r/\lambda)$, where r is the hole radius. Therefore, the transmission through sub-wavelength holes alone is very low, and other effects, such as SPP excitations, are primarily responsible for EOT. However, this waveguide mechanism itself can result in complex optical properties, such as the dephasing of electromagnetic fields (see Section 2.4.1) [27].

2.3 Theoretical and Computational Methods

Many computational electromagnetics methods are well-suited to model nanostructured metal films. One of the most general, and popular techniques to model both nonperiodic and periodic systems is the FDTD method. However, FDTD calculations can be very computationally demanding, and other, more efficient methods are also sometimes used. For example, for periodic systems RCWA can be used, and for nonperiodic systems an electrodynamic mode (modal) expansion method can be used. In passing, we mention that another powerful, yet less-often used method in the present context, is the finite-element method [28]. In this section, we discuss FDTD, RCWA, and the modal expansion method.

2.3.1 The Finite-Difference Time-Domain (FDTD) Method

To computationally model nanostructured metal films, the FDTD method [29, 30] is often used [7, 8, 15, 31]. The FDTD method explicitly solves Maxwell–Ampere's law and Faraday's law in differential form, **equations 1** and **2**, by approximating the partial derivatives using central finite-difference expressions (derived using Taylor expansions) over

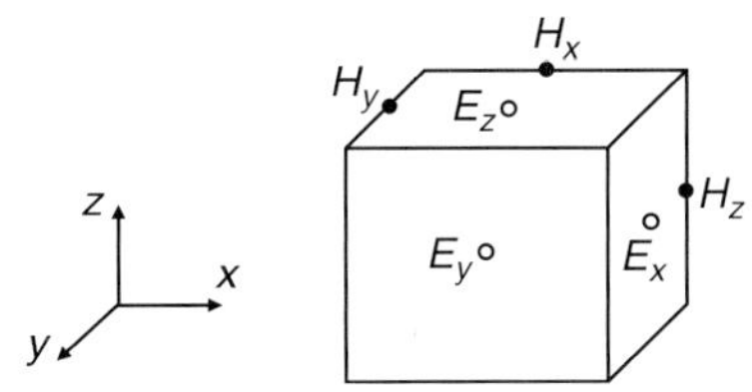

Figure 4 Yee spatial lattice used in the FDTD method.

a grid-based domain, which has second-order accuracy (i.e., the truncation errors are of higher order than 2). Even though only **equations 1** and **2** are explicitly solved, special care must be taken in order to satisfy Gauss' laws, **equations 3** and **4**. One way to do this, suggested by Yee in his seminal paper [30], is to use a Cartesian spatial grid with all of the field components shifted by half of a grid spacing relative to each other, known as the Yee spatial lattice (**Figure 4**). For a proof of this result see Ref. [29].

To obtain second-order accuracy in the temporal derivatives of **equations 1** and **2**, $\boldsymbol{E}$ and $\boldsymbol{H}$ are defined on time grids shifted by Δt. For example, in **equation 1** the time derivative of $\boldsymbol{E}$ 'leaps over' the spatial derivatives of $\boldsymbol{H}$, known as a 'leap-frog algorithm' (see below).

Using the Yee spatial lattice (**Figure 4**) and second-order accurate central finite-differences leads to approximations of **equations 1** and **2** as

$$\varepsilon_{i+1/2,j,k}\frac{E_x^{n+1/2}{}_{i+1/2,j,k}-E_x^{n-1/2}{}_{i+1/2,j,k}}{\Delta t} = \frac{H_z^n{}_{i+1/2,j-1/2,k}-H_z^n{}_{i-1/2,j+1/2,k}}{\Delta y} - \frac{H_y^n{}_{i+1/2,j,k+1/2}-H_y^n{}_{i+1/2,j,k-1/2}}{\Delta z} - \mathcal{J}_x^n \tag{28}$$

$$\varepsilon_{i,j+1/2,k}\frac{E_y^{n+1/2}{}_{i,j+1/2,k}-E_y^{n-1/2}{}_{i,j+1/2,k}}{\Delta t} = \frac{H_x^n{}_{i,j+1/2,k+1/2}-H_x^n{}_{i,j+1/2,k-1/2}}{\Delta z} - \frac{H_z^n{}_{i+1/2,j+1/2,k}-H_z^n{}_{i-1/2,j+1/2,k}}{\Delta x} - \mathcal{J}_y^n \tag{29}$$

$$\varepsilon_{i,j,k+1/2}\frac{E_z^{n+1/2}{}_{i,j,k+1/2}-E_z^{n-1/2}{}_{i,j,k+1/2}}{\Delta t} = \frac{H_y^n{}_{i+1/2,j,k+1/2}-H_y^n{}_{i-1/2,j,k+1/2}}{\Delta x} - \frac{H_x^n{}_{i,j+1/2,k+1/2}-H_x^n{}_{i,j-1/2,k+1/2}}{\Delta y} - \mathcal{J}_z^n \tag{30}$$

$$\mu_{i,j+1/2,k+1/2}\frac{H_x^{n+1}{}_{i,j+1/2,k+1/2}-H_x^{n}{}_{i,j+1/2,k+1/2}}{\Delta t}=\frac{E_y^{n+1/2}{}_{i,j+1/2,k+1}-E_y^{n+1/2}{}_{i,j+1/2,k}}{\Delta z}-\frac{E_z^{n+1/2}{}_{i,j+1,k+1/2}-E_z^{n+1/2}{}_{i,j,k+1/2}}{\Delta y} \tag{31}$$

$$\mu_{i+1/2,j,k+1/2}\frac{H_y^{n+1}{}_{i+1/2,j,k+1/2}-H_y^{n}{}_{i+1/2,j,k+1/2}}{\Delta t}=\frac{E_z^{n+1/2}{}_{i+1,j,k+1/2}-E_z^{n+1/2}{}_{i,j,k+1/2}}{\Delta x}-\frac{E_x^{n+1/2}{}_{i+1/2,j,k+1}-E_x^{n+1/2}{}_{i+1/2,j,k}}{\Delta z} \tag{32}$$

$$\mu_{i+1/2,j+1/2,k}\frac{H_z^{n+1}{}_{i+1/2,j+1/2,k}-H_z^{n}{}_{i+1/2,j+1/2,k}}{\Delta t}=\frac{E_x^{n+1/2}{}_{i+1/2,j+1,k}-E_x^{n+1/2}{}_{i+1/2,j,k}}{\Delta y}-\frac{E_y^{n+1/2}{}_{i+1,j+1/2,k}-E_y^{n+1/2}{}_{i,j+1/2,k}}{\Delta x} \tag{33}$$

Equations 28–33 can be solved by discretizing a domain, specifying ε and μ at the discrete grid points, and sequentially solving **equations 28–30** and 31–33.

The FDTD equations given above only apply to regions of nondispersive materials. Dispersive materials, such as metals, can be modeled using a modified set of FDTD equations, where the frequency-dependent response is modeled in the time domain using equivalent current densities which are updated self-consistently with the normal FDTD equations. This is known as the auxiliary differential equation (ADE) method [29, 32, 33]. In this approach, a model for the dispersive material is chosen, inserted into **equation 5**, and the resulting equation is then Fourier-transformed to the time domain to obtain a modified Maxwell–Ampere's law. For the dispersive dielectric model in **equation 22**, the modified Maxwell–Ampere's law in time-discretized form is

$$C_1\boldsymbol{E}^{n+1/2}=C_2\boldsymbol{E}^{n-1/2}+C_3\boldsymbol{E}^{n-3/2}+\nabla\times\boldsymbol{H}^n-\frac{1}{2}\left\{(1+\kappa_D)\mathcal{J}_D^{\,n-1/2}+\sum_{L=1}^{2}\left[(1+\alpha_L)\mathcal{J}_L^{\,n-1/2}+\zeta_L\mathcal{J}_L^{\,n-3/2}\right]\right\} \tag{34}$$

where

$$C_1=\frac{\varepsilon_0\varepsilon_\infty}{\Delta t}+\frac{1}{2}\beta_D+\frac{1}{4\Delta t}\sum_{L=1}^{2}\gamma_L \tag{35}$$

$$C_2=\frac{\varepsilon_0\varepsilon_\infty}{\Delta t}-\frac{1}{2}\beta_D \tag{36}$$

$$C_3=\frac{1}{4\Delta t}\sum_{L=1}^{2}\gamma_L \tag{37}$$

$$\kappa_D=\frac{1-\gamma_D\,\Delta t/2}{1+\gamma_D\,\Delta t/2} \tag{38}$$

$$\beta_D=\frac{\omega_D^2\,\varepsilon_0\,\Delta t/2}{1+\gamma_D\,\Delta t/2} \tag{39}$$

$$\alpha_L=\frac{2-\omega_L^2\,\Delta t^2}{1+\delta_L\,\Delta t} \tag{40}$$

$$\zeta_L=\frac{\delta_L\,\Delta t-1}{\delta_L\,\Delta t+1} \tag{41}$$

$$\gamma_L=\frac{\varepsilon_0\Delta\varepsilon_L\omega_L^2\,\Delta t^2}{1+\delta_L\,\Delta t} \tag{42}$$

The equivalent current densities, $\mathcal{J}_D$ and $\mathcal{J}_L$, are then updated using discretized ADEs,

$$\mathcal{J}_D^{\,n+1/2}=\kappa_D\mathcal{J}_D^{\,n-1/2}+\beta_D\left(\boldsymbol{E}^{n+1/2}+\boldsymbol{E}^{n-1/2}\right) \tag{43}$$

$$\mathcal{J}_L^{\,n+1/2}=\alpha_L\mathcal{J}_L^{\,n-1/2}+\zeta_L\mathcal{J}_L^{\,n-3/2}+\gamma_L\left(\frac{\boldsymbol{E}^{n+1/2}-\boldsymbol{E}^{n-3/2}}{2\Delta t}\right) \tag{44}$$

The FDTD method can be used to study periodic systems by modeling only a single unit cell and truncating the domain using periodic boundary conditions. Even when simulating infinite domains, the computational domain must be truncated. However, many techniques have been developed to mimic open regions of space. One of the most successful techniques is to truncate the domain with artificial materials that absorb nearly all incident waves, called perfectly matched layers (PMLs) [34]. An efficient and accurate way of implementing PML with the Yee spatial lattice is to use convolutional PML (CPML) [35]. The implementation of CPML involves stretching the spatial derivatives and superimposing time-dependent, two-component functions onto them. For a complete discussion of PML see Ref. [29].

Given suitable initial conditions defined everywhere in the computational domain, **equations 28–33** will properly evolve the fields according to **equations 1** and **2**. However, defining computational domains for initial conditions with large spatial extent is often inefficient and unnecessary. A more efficient technique to introduce fields into the computational domain, particularly with the Yee spatial lattice, is to use the total field–scattered field (TF–SF) technique [36–38]. The implementation of this involves splitting the domain into two regions, an interior total field region and an exterior scattered field region. The FDTD method is applied directly in each region without modification. However, near the boundaries, where the spatial derivatives extend into both regions, the fields are modified using the (known) incident field so that all equations are consistent.

Some of the main interests in nanostructured metal films have to do with transmission, reflection, and absorption spectra. Often, but not always, the measured spectra are for zero-order transmission, that is the measurements are such that they do not include contributions from light scattered away from the incident direction. To obtain this result computationally, it is necessary to Fourier transform the time-domain E and H fields on xy-planes above or below the surfaces of the film. Since the wavevector for zero-order scattering is $\mathbf{k} = \frac{\omega}{c}\hat{z}$, and z is fixed for a given xy-plane of interest, the relevant projections of the fields are simply proportional to integrating the fields over x and y [7]. If $\boldsymbol{E}_0(\omega)$ and $\boldsymbol{H}_0(\omega)$ denote these projections, then the zero-order transmission is given by $T(\omega) = P_T(\omega)/P_{\text{inc}}(\omega)$, with

$$P_T(\omega) = (1/2)\text{Re}[\boldsymbol{E}_0(\omega) \times \boldsymbol{H}_0(\omega)^*]$$

and the quantity $P_{\text{inc}}(\omega)$ is the analogous incident flux.

For all of the FDTD calculations discussed herein, grid spacings of $\Delta x = \Delta y = \Delta z = 4$ or 5 nm are used, and the computational domains are truncated with CPML. Total simulation times of at least 150 fs are used to obtain accurate Fourier-transformed fields. The TF–SF technique is used to introduce incident light into the computational domains, and typically the incident field is a sinusoidal pulse that is Gaussian damped, with parameters chosen such that the field contains wavelength content over the range of interest ($\lambda = 300$–1000 nm).

2.3.2 The Rigorous Coupled-Wave Analysis (RCWA)

A computationally efficient approach to study periodic nanostructured metal films is RCWA [39, 40]. An additional benefit of RCWA is that specific diffraction orders can be independently studied. A schematic diagram of the 2D system under consideration with RCWA is shown in **Figure 5** for TE_z polarization. The system consists of a 1D grating with slits of diameter d, permittivity ε_s, and periodicity P in a metal film with thickness h and permittivity ε_{II}, which is surrounded by dielectric media with permittivities ε_{I} and ε_{III}. The grating is illuminated from above with TE_z polarized light at an angle θ with respect to the film normal.

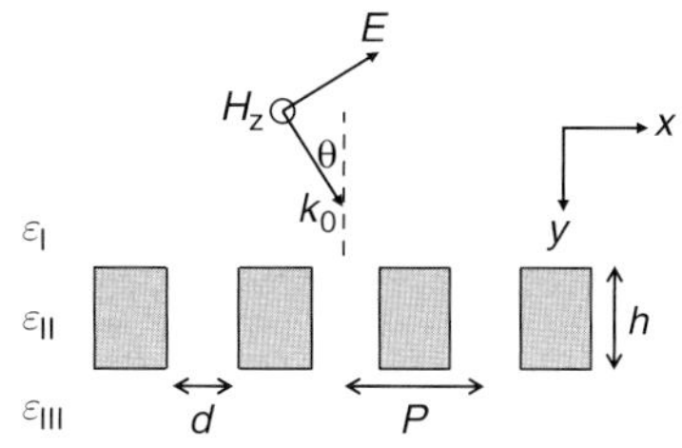

Figure 5 Schematic diagram of a 1D metallic grating illuminated with TE_z polarized light. The parameters in the figure are defined in the text.

Inside the grating region, the permittivity is expanded using a Fourier series,

$$\varepsilon_{\text{II}}(x) = \sum_n \varepsilon_n \exp(\text{i}ngx) \tag{45}$$

where n is an integer and $g = 2\pi/P$ is the period of the reciprocal lattice. The Fourier components of **equation 45** are

$$\varepsilon_n = (\varepsilon_s - \varepsilon_m)\frac{\sin(n\pi f)}{n\pi} \tag{46}$$

for $n \neq 0$ and

$$\varepsilon_0 = f\varepsilon_s + (1-f)\varepsilon_m \tag{47}$$

for $n = 0$, where $f = d/P$ is a filling factor. In **equation 47,** ε_0 corresponds to the average value of the dielectric function of the grating region, and should not be confused with the vacuum permittivity. For TE_z polarization, the incident magnetic field, H_z^{Inc}, is

$$H_z^{\text{Inc}} = \exp\left(-\text{i}\left(k_{Ix}x + k_{Iy}y\right)\right) \tag{48}$$

where $k_{\text{I}x} = k_{\text{I}} \sin\theta$ and $k_{\text{I}y} = k_{\text{I}} \cos\theta$, with $k_{(\text{I,III})} = k_0\sqrt{\varepsilon_{(\text{I,III})}}$. Diffraction by the grating gives rise to a set of diffraction orders, where the in-plane wavevector magnitude of the nth diffracted wave is given by $k_{y_n} = k_{\text{I}x} + ng$, where n is an integer. Rayleigh expansions are used to express the reflected and transmitted magnetic fields in regions I and III as

$$H_z^{\text{I}} = H_z^{\text{Inc}} + \sum_n r_n \exp\left(-\text{i}\left(k_{x_n}x - k_{\text{I}y_n}y\right)\right) \tag{49}$$

$$H_z^{\mathrm{III}} = \sum_n t_n \exp\left(-\mathrm{i}\left(k_{x_n}x + k_{\mathrm{III}y_n}(y-h)\right)\right) \quad (50)$$

where $k^2_{(I,III)z_n} = k^2_{(I,III)} - k^2_{x_n}$. The coefficients r_n and t_n correspond to the amplitudes of the nth reflected and transmitted waves, respectively. Inside region II, Bloch's theorem is used to express H_z and E_x using Fourier expansions,

$$H_z^{\mathrm{II}} = \sum_n H_{\mathrm{II}_n}(y)\exp\left(-\mathrm{i}k_{x_n}x\right) \quad (51)$$

$$E_x^{\mathrm{II}} = \mathrm{i}\sqrt{\frac{\mu}{\varepsilon}}\sum_n E_{\mathrm{II}_n}(y)\exp\left(-\mathrm{i}k_{x_n}x\right) \quad (52)$$

where $H_{\mathrm{II}_n}(y)$ and $E_{\mathrm{II}_n}(y)$ are the normalized amplitudes of the spatial harmonics and μ and ε are free space values. It can be shown that $H_{\mathrm{II}_n}(y)$ and $E_{\mathrm{II}_n}(y)$ are related by a set of coupled equations, and by projecting them onto a set of plane waves and matching H_z and E_x at the metal interfaces (**equations 13** and **16**), they can be used to solve for the coefficients r_n and t_n [41]. Once r_n and t_n are determined, the diffraction efficiencies can be calculated from

$$R_n = r_n r_n^* \operatorname{Re}\left(\frac{k_{\mathrm{I}y_n}}{k_0\sqrt{\varepsilon_{\mathrm{I}}}\cos\theta}\right) \quad (53)$$

$$T_n = t_n t_n^* \operatorname{Re}\left(\frac{\sqrt{\varepsilon_{\mathrm{I}}}k_{\mathrm{III}y_n}}{k_0\varepsilon_{\mathrm{III}}\cos\theta}\right) \quad (54)$$

In practice, RCWA is often less rigorous than FDTD because of the approximation of the dielectric function of the grating region, **equation 45**, with a finite number of terms. In addition, convergence issues exist for RCWA for TE_z polarization due to material discontinuities inside the grating, however, with care they are possible to avoid [42].

For the RCWA results presented herein, we have used the model of Ref. [41], which presented analytical expressions for the diffraction efficiencies for normal incident light and only retaining the first two diffraction orders and Fourier components in the grating region. In addition, we apply this 2D method to 3D problems, which we have previously found captures the important features [7, 31].

2.3.3 The Modal Expansion Method

To describe the transmission through an isolated slit, a modal expansion of the electromagnetic fields in the presence of the slit can be used. A schematic diagram of the system under consideration is shown in **Figure 6**. The system consists of an isolated slit

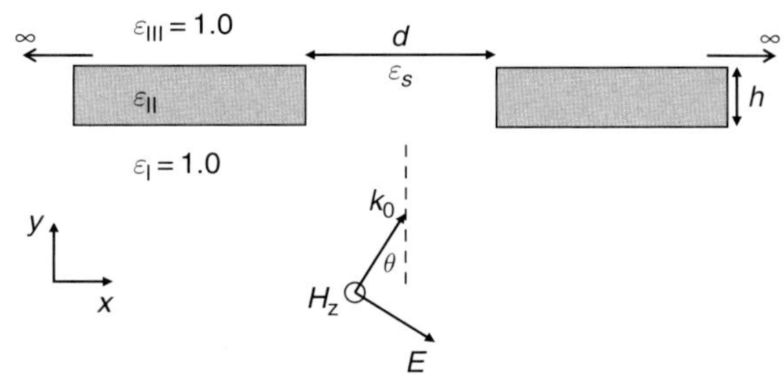

Figure 6 Schematic diagram of an isolated slit in a metal film illuminated using $\mathrm{TE_z}$ polarized light. The parameters in the figure are defined in the text.

with diameter d in a metal film (taken here to be a PEC) with thickness h in air. Even though the method is applied to a PEC film, we have previously found that the results accurately describe many aspects of a real metal [27]. The slit is illuminated from below with TE_z polarized light at an angle θ with respect to the film normal.

For TE_z polarization, H_z is expressed in regions I and III as a superposition of propagating and evanescent plane waves,

$$H_z^{\mathrm{I}} = H_z^{\mathrm{inc}} + \int_{-\infty}^{\infty} \mathrm{d}k\,\rho(k)\exp\left[\mathrm{i}\left(kx - k_y y\right)\right] \quad (55)$$

$$H_z^{\mathrm{III}} = \int_{-\infty}^{\infty} \mathrm{d}k\,\tau(k)\exp\left[\mathrm{i}\left(kx + k_y y\right)\right] \quad (56)$$

where $\rho(k)$ and $\tau(k)$ are the reflection and transmission amplitudes, respectively, $k_y = \sqrt{k_0{}^2 - k^2}$, and the incident magnetic field, $H_z{}^{\mathrm{inc}}$, is given by

$$H_z^{\mathrm{inc}} = \exp\left[\mathrm{i}\left(k_{0x}x + k_{0y}y\right)\right] \quad (57)$$

where k_{0x} and k_{0y} are the projections of the incident wavevector magnitude along the x- and y-axes. In region II, H_z is expanded in terms of the eigenmodes of a waveguide,

$$H_z^{\mathrm{II}} = \sum_{\mathrm{m}=0}^{\infty}\left[A_{\mathrm{m}}\exp\left(\mathrm{i}\beta_{\mathrm{m}}y\right) + B_{\mathrm{m}}\exp\left(-\mathrm{i}\beta_{\mathrm{m}}y\right)\right]\phi_{\mathrm{m}}(x) \quad (58)$$

where m is the order of the waveguide mode, $\beta_{\mathrm{m}} = \left[k_0{}^2\varepsilon_{\mathrm{s}} - (m\pi/d)^2\right]^{1/2}$ and $\phi_{\mathrm{m}} = \left(2/\sqrt{d}\right)\cos\left[m\pi(x + d/2)/d\right]$ is the solution to **equation 9** within the slit subject to PEC boundary conditions at $x = \pm d/2$. By matching H_z and E_x at the film interfaces (**equations 13** and **16**) and projecting the resulting equations onto plane waves, the modal amplitudes A_{m} and B_{m} can be determined via numerical solution of a set of linear equations.

2.4 Isolated Apertures in Metal Films

2.4.1 Isolated Slits

In this section, we perform FDTD calculations on an isolated slit with diameter $d = 2.23\ \mu m$ in a Au film with thickness $h = 60$ nm in air to study the behavior of confined electromagnetic fields for 543-nm TE_z, TM_z, and circular polarizations of incident light. Note that for circular polarization, both TE_z and TM_z polarizations exist, but are out of phase. The system parameters were chosen to match those of a previous study [19]. **Figure 6** shows a schematic diagram of the system under consideration for TE_z polarization.

The electric and magnetic field intensities, $E^2 = E_x{}^2 + E_y{}^2 + E_z{}^2$ and $H^2 = H_x{}^2 + H_y{}^2 + H_z{}^2$, respectively, inside the slit calculated with FDTD for TE_z polarization are shown in **Figure 7(a)**. It is seen that the fields have oscillatory behavior within the slit, with a peak-to-peak separation of 515 and 545 nm for E^2 and H^2, respectively, which is very close to the free-space wavelength of 543 nm. In addition, it is seen that E^2 and H^2 are out of phase.

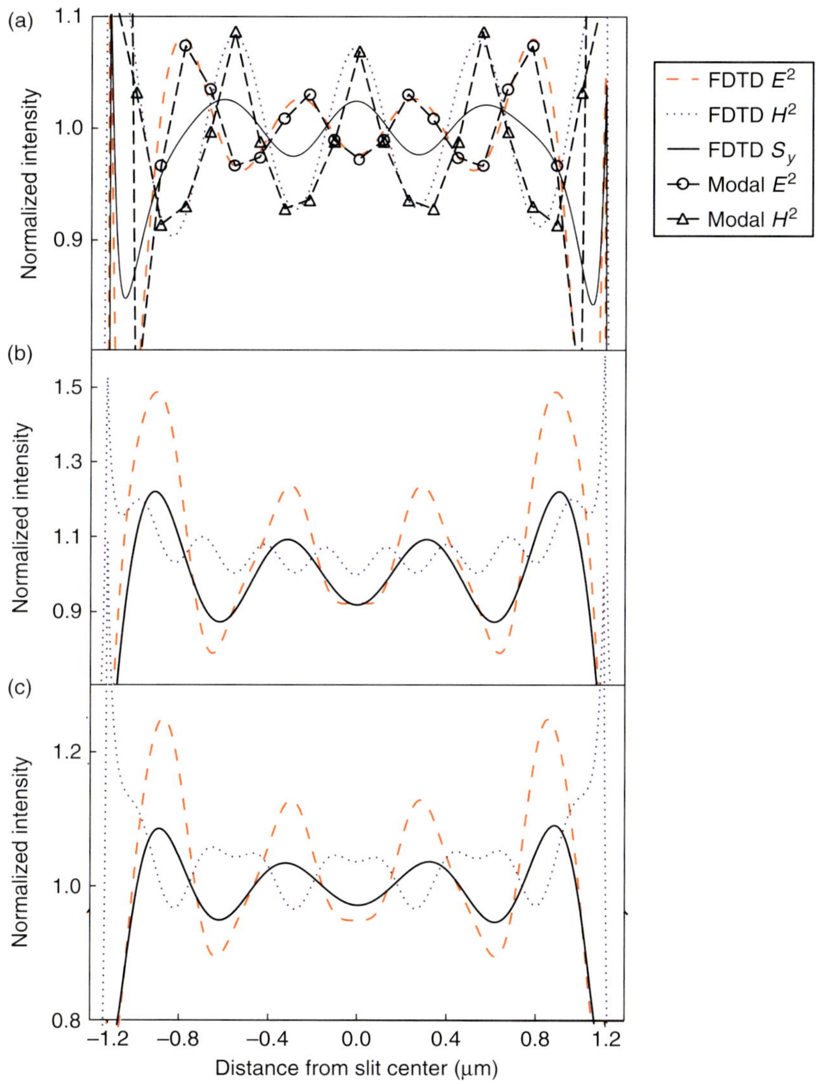

Figure 7 FDTD calculated field intensities and power flow inside a $d = 2.23\ \mu m$ diameter slit in a $h = 60$ nm thick Au film for 543 nm incident light: (a) TE_z, (b) TM_z, and (c) circularly polarized. Modal expansion method results are also shown for TE_z polarization. Reprinted from McMahon JM, Gray SK, and Schatz GC (2008) Dephasing of electromagnetic fields in scattering from an isolated slit in a gold film. In: *Proceedings of SPIE: Plasmonics: Nanoimaging, Nanofabrication, and Their Applications IV*, p. 703311. San Diego, CA, USA. Bellingham, WA: SPIE Press.

To gain physical insight into this dephasing, the modal expansion method was also used to calculate the field intensities, **Figure 7(a)**. The results are seen to agree very well with the FDTD results, confirming that the behavior of the fields inside a slit in a Au film and a PEC film are similar. The amplitudes of the waveguide modes, A_m and B_m in **equation 58,** obtained from the modal analysis are shown in **Figure 8**. It is seen that coefficients are mainly composed of the $m=0$ and $m=8$ modes, which means that scattering by the slit induces a modal transition from propagating mode to evanescent mode. The latter corresponds to light propagating parallel to the surface with a wavelength that is just slightly longer than the free-space wavelength, which explains the similarity between this and the peak-to-peak separation. In addition, the superposition of the higher-order evanescent modes with the propagating modes explains the dephasing of the fields. FDTD was also used to calculate the normal component of the Poynting vector, S_y, inside the slit, **Figure 7(a)**. It is seen that power flow through the slit is governed by $\boldsymbol{H}$, except near the slit edges.

E^2, $\boldsymbol{H}^2$, and S_y were also calculated using FDTD for TM_z polarization, and are shown in **Figure 7(b)**. It is seen that the peak-to-peak separation of E^2, 590 nm, is again very close to the free-space wavelength of 543 nm. However, the peak-to-peak separation of H^2 is seen to be 270 nm, close to half of the free-space wavelength. This can be attributed to $\boldsymbol{H}$ components composed almost entirely of the $m=8$ waveguide mode. This result is also seen for the y-component of E for TE_z polarization (not shown), except that it is overshadowed by the x-component, which contains both $m=0$ and $m=8$ modes. For this polarization, it is also seen that the power flow throughout the entire slit is governed by $\boldsymbol{E}$.

E^2, $\boldsymbol{H}^2$, and S_y were also calculated using FDTD for circular polarization, and are shown in **Figure 7(c)**. It is seen that the results are essentially a superposition of the TE_z and TM_z polarized results. However, the power flow throughout the entire slit is seen to be governed by E, similar to TM_z polarization.

The results in **Figures 7(a)–7(c)** are important to the interpretation of NSOM imaging measurements [17], as they show that the power flow (which is the property that is directly probed) provides a measurement of the local $\boldsymbol{E}$ field in some cases and the local $\boldsymbol{H}$ field in others. While this can complicate the use of NSOM to experimentally study interference effects, FDTD and the modal expansion method provide simple ways to model the results.

2.4.2 Isolated Holes

Besides waveguide effects associated with isolated apertures, SPPs and LSPRs can be excited by, and in, them, respectively [15, 23]. Consider a thin metal film with an isolated circular hole supported by a glass substrate in air, similar to **Figure 6**. If the film is illuminated from the glass side by incident light with a wavelength greater than the hole diameter, several

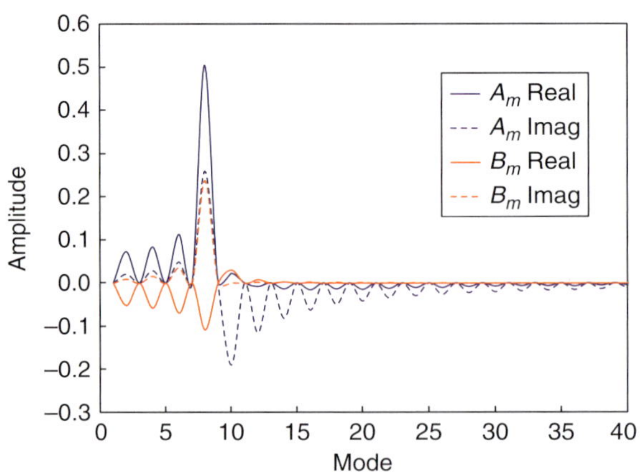

Figure 8 Amplitudes of the waveguide modes inside a $d=2.23$ μm diameter slit in an $h=60$ nm thick PEC film for 543 nm TE_z polarized incident light. The $m=0$ mode, which has amplitudes of $A_m=0.658-i0.846$ and $B_m=0.601+i0.858$, is not displayed. Adapted from McMahon JM, Gray SK, and Schatz GC (2008) Dephasing of electromagnetic fields in scattering from an isolated slit in a gold film. In: *Proceedings of SPIE: Plasmonics: Nanoimaging, Nanofabrication, and Their Applications IV*, p. 703311. San Diego, CA, USA. Bellingham, WA: SPIE Press.

interesting phenomena can occur. (A system of randomly distributed, but not too dense, holes also will have similar behavior.)

First, even if the light is propagating normal to the film, it is possible to change SPP excitations on the metal–air interface [16, 43]. This is because the hole itself acts as a scattering defect that can generate the additional momentum required for SPP excitation (see Section 2.2.2). A combined FDTD and experimental study by Yin *et al.* [16] of 100 nm-thick Au films with an isolated hole showed that SPPs tend to be launched from the holes in the direction of the incident polarization. The SPP near fields were experimentally measured with an NSOM probe, and they showed periodic intensity variations, with peak-to-peak separations approximately equal to $\lambda_{SPP} = 2\pi/k_{SPP}$. It was at first surprising that there would be periodic oscillations in the intensity, since that of a pure SPP excitation should decay exponentially along the propagation direction (if the metal is lossy). However, the theoretical analysis showed that the near fields are composed of two interfering terms, the excited SPP and directly penetrating (or tunneling) light through the film. The resulting interference pattern is easily seen to have the observed oscillations. Subsequently carried out more detailed calculations and analyses, showing more explicitly that an isolated hole can act as a point source of SPPs.

When SPPs can be generated by an isolated hole, there can also be EOT [15, 43]. This was shown in a theoretical study by Wannemacher [43], who used a numerically rigorous multiple multipole [44] approach. In addition, while not emphasized in this work, contributions to EOT can also arise from LSPR excitations, which are dipolar (and possibly higher—order) charge oscillation excitations associated with the holes themselves [15, 45, 46]. LSPR excitations associated with holes in metal films are analogous to those in metal nanoparticles, as shown by Kall and coworkers [45, 46]. For example, light transmission through holes can mimic the corresponding optical scattering spectra of nanoparticles, displaying resonance features that red-shift with hole size. (However, some interesting differences regarding elongated holes and particles have recently been pointed out [47].) It should again be mentioned (see Section 2.2.2) that these effects are not completely isolated from each other, and in reality, LSPRs and SPPs are inherently coupled.

2.5 Periodic Nanostructured Metal Films

2.5.1 Hole Arrays

Light transmission through arrays of sub-wavelength diameter holes in metal films has been the focus of much research since the discovery by Ebbesen *et al.* [1] of EOT. A schematic diagram of a typical hole array system in a Au film is shown in **Figure 9**. The system consists of a square array of holes with diameter d and periodicity P in a Au film of thickness h, sandwiched between dielectric materials I and III with RIs $n_{\mathrm{I}} = \sqrt{\varepsilon_{\mathrm{I}}}$ and $n_{\mathrm{III}} = \sqrt{\varepsilon_{\mathrm{III}}}$. Material I (bottom) is taken to be the substrate or supporting material for the film (e.g., glass), with a fixed RI, and material III (top) is a variable RI superstrate. Light is incident from region I, and the transmission spectrum is obtained in region III.

The mechanisms of EOT have been the subject of much discussion, and important contributions arise

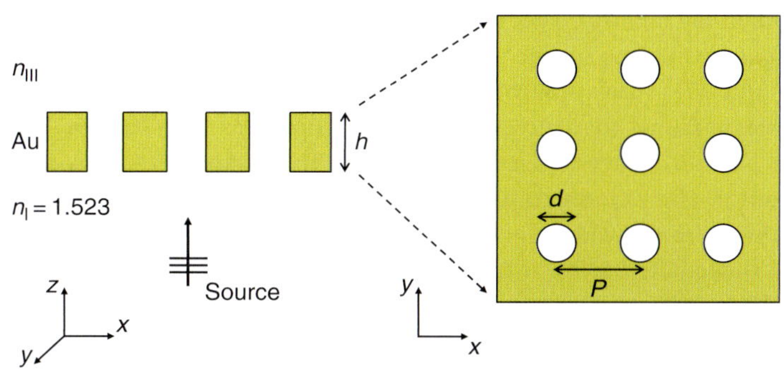

Figure 9 Schematic diagram of a typical hole array in a Au film. The parameters in the figure are defined in the text. Reprinted from Schatz GC, McMahon JM, and Gray SK (2007) Tailoring the parameters of nanohole arrays in gold films for sensing applications. In: *Proceedings of SPIE: Plasmonics: Metallic Nanostructures and Their Optical Properties V*, p. 664103. San Diego, CA, USA. Bellingham, WA: SPIE Press.

from SPPs, diffractive effects including RAs, and localized waveguide modes [1, 2, 15, 48–51]. For example, all of these features play a role in the large RI sensitivity observed in Au films deposited on arrays of 500-nm wells molded in a polymer substrate (see Section 2.5.3) [8]. We mention once more (see Section 2.2.2) that these effects are not isolated from one another, and can interact leading to additional effects.

The details of EOT vary significantly with parameters of the hole array structure, including size [52] and shape [53, 54] of the holes, metal thickness [55, 56], and periodicity. As an example of these concepts, **Figure 10** shows FDTD calculated zero-order transmission spectra for a hole array consisting of $d = 100$ nm holes separated by $P = 400$ nm in a Au film on a glass ($n_I = 1.523$) substrate and $n_{III} = 1.35$ superstrate as h is increased from 25 to 100 nm.

All of the spectra in **Figure 10** show a considerable amount of transmission (EOT) relative to the thin film transmission (**Figure 1**). Initially, one may suspect that this is due to direct transmission through the holes. However, the geometric optics prediction for these parameters is <5%, and that predicted by Bethe's analysis [26] (see Section 2.2.2) is even less. The EOT can be primarily attributed to coupling of the incident light into SPPs. A broad maximum near $\lambda = 490$ nm is observed in all spectra, with a peak transmission of about 15%. This peak is from direct transmission of light through the Au film, and is not related to any hole or array properties, as is confirmed by comparison with **Figure 1**. The peak structure between $\lambda = 560$ and 580 nm can be attributed to a combination of a LSPR (see Section 2.4.2) in each hole and a (1, 1) n_I SPP-BW (i.e., the solution to **equation 26** with ($s_x = 1$, $s_y = 1$) and $\varepsilon_r = n_I^2$, associated with the metal–material I interface; see Section 2.2.2). The minima in the spectra for $\lambda > 580$ nm can be assigned using the simple predictions of **equations 26** and **27**. (Note that the Fano resonance feature corresponding to a given SPP occurs as a minimum and an adjacent maximum, with the minimum position often closer to the prediction [15, 49, 51].) For example, when $h = 50$ nm, the (1, 0) n_{III} SPP-BW is predicted to be at 601 nm, and the (1, 0) n_I SPP-BW at 662 nm. These positions are in relatively good agreement with the calculated FDTD minima of 610 and 683 nm, respectively, **Figure 10(a)**. It is important to note that, as the film thickness is decreased for $h < 65$ nm, the peak positions shift considerably. This is attributed to an increase in coupling between SPP resonances on the top and bottom film surfaces (see Section 2.2.2).

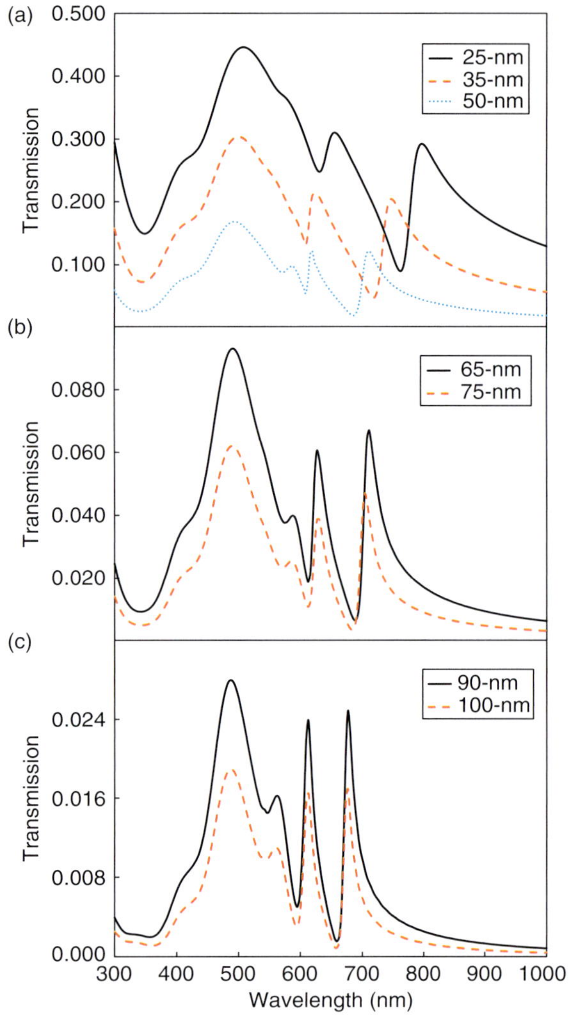

Figure 10 FDTD calculated zero-order transmission through a hole array with $d = 100$ nm holes separated by $P = 400$ nm in a Au film on a glass substrate and $n_{III} = 1.35$ superstrate as h is increased from 25 to 100 nm.

Under certain circumstances, SPP-BWs and RAs on opposite sides of the film can strongly interact [7, 31]. **Figure 11** shows how the positions of SPP-BWs and RAs, predicted by **equations 26** and **27**, evolve as n_{III} is varied while $n_I = 1.523$ and $P = 400$ nm remain fixed. Only the first-order cases are shown in **Figure 11** since they are the strongest features in this wavelength range and for this P. The crossing point labeled as RA-SPP corresponds to where the (1, 0) n_{III} RA (i.e., the solution to **equation 27** with ($w_x = 1$, $w_y = 0$) and $\varepsilon_r = n_{III}^2$, associated with the material III region; see Section 2.2.2) and

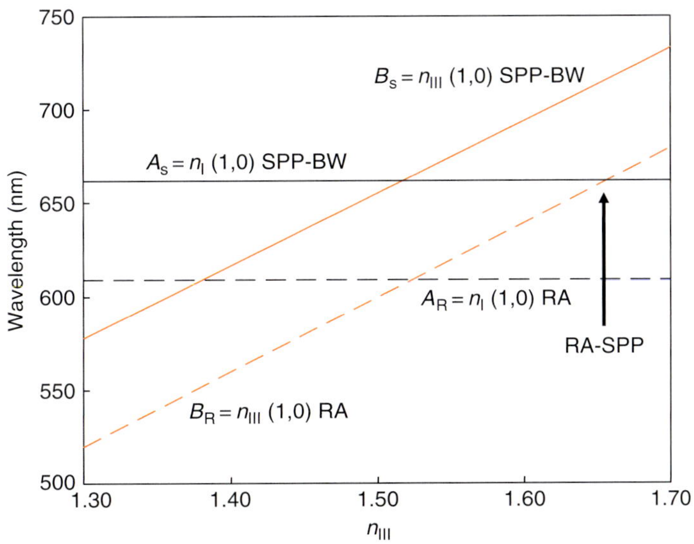

Figure 11 Predicted positions of (1, 0) SPP-BWs and (1, 0) RAs (**equations 26** and **27**) as a function of n_{III} for $n_I = 1.523$ and $P = 400$ nm. The n_I SPP-BW and RA are denoted as A_S and A_R, and the n_{III} SPP-BW and RA as B_S and B_R, respectively. Reprinted from McMahon JM, Henzie J, Odom TW, Schatz GC, and Gray SK (2007) Tailoring the sensing capabilities of nanohole arrays in gold films with Rayleigh anomaly – surface plasmon polaritons. *Optics Express* 15: 18119.

the (1, 0) n_I SPP-BW coexist and can interact if the Au film is thin enough, leading to a strong variation in transmission amplitude for $h < 65$ nm [31]. The value of n_{III} at the RA-SPP crossing point for a general hole array can be found by equating **equation 26** for n_I with **equation 27** for n_{III} for the first diffraction orders, $(s_x = 1, s_y = 0)$ and $(w_x = 1, w_y = 0)$,

$$n_{III} = \sqrt{\frac{\varepsilon_M(\lambda_{RA-SPP})\varepsilon_I}{\varepsilon_M(\lambda_{RA-SPP}) + \varepsilon_I}} \tag{59}$$

Equation 59, like **equation 18,** is an implicit relation for a wavelength. In this case, the wavelength $\lambda_{RA\text{-}SPP}$ is the incident wavelength where the (1, 0) n_I SPP-BW and (1, 0) n_{III} RA conditions are the same. The required P for the hole array at this crossing point is subsequently found from

$$P = \frac{\lambda_{RA-SPP}}{n_{III}} \tag{60}$$

To achieve the RA-SPP effect near some specific n_{III} value, **equation 59** is first rewritten as a root equation, $f(\lambda_{RA-SPP}) = 0$, and solved using bisection for λ_{RA-SPP}. Then, **equation 60** is used to determine the necessary P. For example, $n_I = 1.523$ corresponds to $n_{III} = 1.65$ and $\lambda_{RA-SPP} = 662$ nm with $P \approx 400$ nm. It turns out that RA-SPP solutions exist only for n_{III} values in a small range, $n_I < n_{III} < n_I + \Delta$, where Δ depends on the nature of the metal, n_I, and h (e.g., $\Delta \approx 0.35$ for an $h = 50$ nm Au film with $n_I = 1.523$). In what follows, we show both FDTD calculations and experimental results that confirm this expectation for a hole array in an $h = 50$ nm thick Au film with $d = 100$ nm and $P = 400$ nm on a glass ($n_I = 1.523$) substrate [7]. **Figures 12** and **13** show the FDTD calculated and experimental zero-order transmission spectra for n_{III} ranging from 1.00 to 1.70.

Similar to **Figure 10**, all spectra in **Figures 12** and **13** display a broad maximum near $\lambda = 490$ nm with peak transmission of about 15% due to direct transmission of light through the Au film, and the peak structure between $\lambda = 560$ and 580 nm can be attributed to a combination of an LSPR in each hole and a (1, 1) n_I SPP-BW. The minima in the spectra for $\lambda > 580$ nm can be assigned using **equations 26** and **27** as well as **Figure 11**. For example, when $n_{III} = 1.70$ (**Figure 13**), the $B_S = (1, 0)$ n_{III} SPP-BW is predicted to be at 733 nm, while the calculated and experimental wavelengths are 757 and 750 nm, respectively. In addition, the $A_S = (1, 0)$ n_I SPP-BW is predicted to be at 662 nm, and is calculated and observed at 669 and 660 nm, respectively, and the (1, 1) n_{III} SPP-BW is predicted to be at 580 nm, and is calculated and observed at 614 and 609 nm, respectively. The result of coupling between SPPs on the top and bottom film interfaces can be seen as n_{III} is increased for $n_{III} < n_I$, where the hybridized $B_S = (1, 0)$ n_{III} SPP-BW increases in wavelength from 574 to 620 nm and decreases in

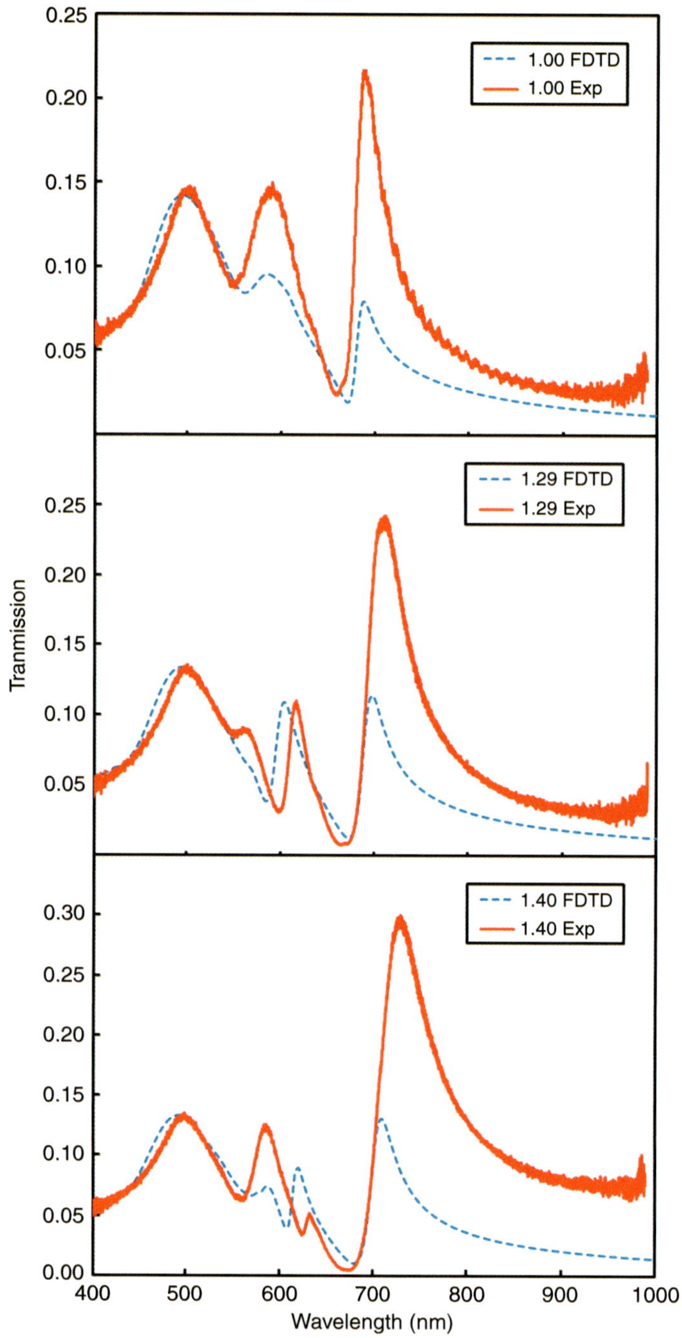

Figure 12 FDTD calculated and experimental zero-order transmission spectra for an $h = 50$ nm thick Au film hole array with $d = 100$ nm and $P = 400$ nm for $n_{III} < n_I$, where $n_I = 1.523$. The letters are assignments based on **Figure 11**. Adapted from McMahon JM, Henzie J, Odom TW, Schatz GC, and Gray SK (2007) Tailoring the sensing capabilities of nanohole arrays in gold films with Rayleigh anomaly – surface plasmon polaritons. *Optics Express* 15: 18119.

amplitude, while the hybridized $A_S = (1, 0)$ n_I SPP-BW peak near $\lambda = 708$ nm increases in amplitude (**Figure 12**). In this discussion so far, RAs have not appeared prominently in the calculated or measured transmission spectra because they are generally narrow spectral features (or possibly just kinks) that can be overwhelmed by other processes. For $n_{III} > n_I$, the maximum to the immediate red of the minimum

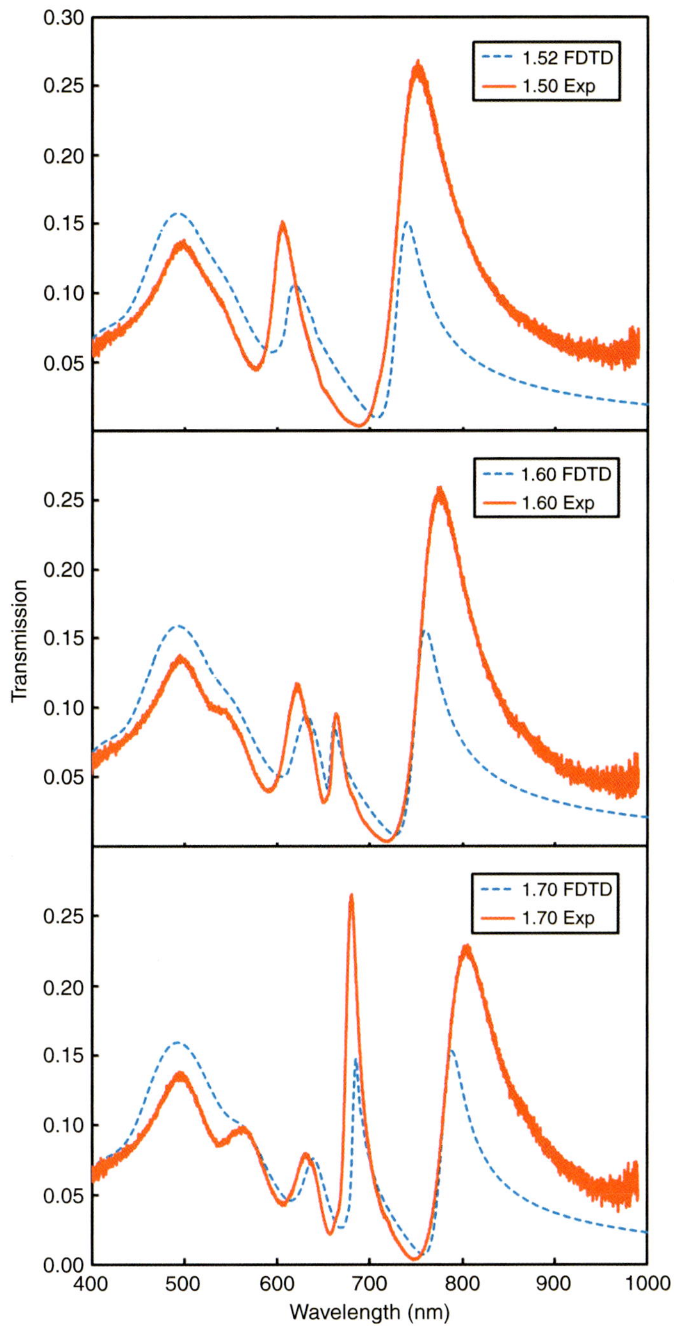

Figure 13 FDTD calculated and experimental zero-order transmission spectra for an $h = 50$ nm thick Au film hole array with $d = 100$ nm and $P = 400$ nm for $n_{III} \geq n_I$, where $n_I = 1.523$. The letters are assignments based on **Figure 11**. Adapted from McMahon JM, Henzie J, Odom TW, Schatz GC, and Gray SK (2007) Tailoring the sensing capabilities of nanohole arrays in gold films with Rayleigh anomaly – surface plasmon polaritons. *Optics Express* 15: 18119.

associated with $A_s = (1, 0)$ n_I SPP-BW shows a rapid rise and narrowing (**Figure 13**). This rapid amplitude increase is a direct result of the RA-SPP effect, and the evolution of the peak associated with this is shown in **Figure 14** on a finer RI scale, calculated with both FDTD and RCWA.

Consistent with **Figure 11**, the $B_R = (1, 0)$ n_{III} RA passes through the wavelength region of the $A_S = (1,$

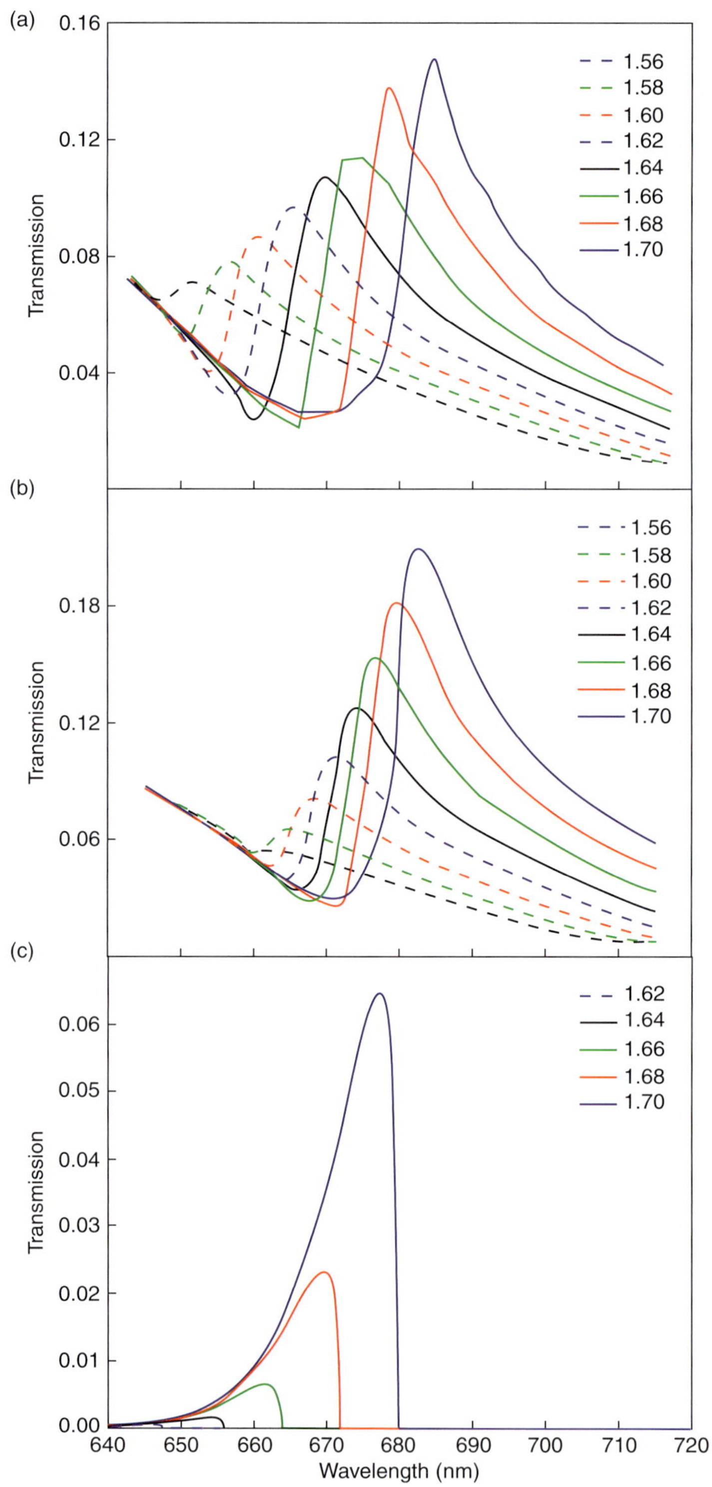

Figure 14 (a) FDTD calculated zero-order transmission spectra for hole arrays in $h = 50$ nm thick Au films with $d = 100$ nm, $P = 400$ nm, and $n_{\text{I}} = 1.523$, showing a rapid increase in amplitude of the peak associated with the RA-SPP effect as n_{III} is varied. (b) RCWA calculated results consistent with (a). (c) RCWA calculated first-order transmission as n_{III} is varied through the region of the RA-SPP effect. Adapted from McMahon JM, Henzie J, Odom TW, Schatz GC, and Gray SK (2007) Tailoring the sensing capabilities of nanohole arrays in gold films with Rayleigh anomaly – surface plasmon polaritons. *Optics Express* 15: 18119.

0) n_I SPP-BW, coupling with it to produce a narrow, n_III-sensitive spectral feature. Note that owing to partial RA character, the RA-SPP feature occurs close (but still red-shifted due to SPP-SPP coupling; see Section 2.2.2) to the prediction of **equation 59**. The FDTD calculated amplitude variation of the RA-SPP peak in the n_III range 1.55–1.70 is ≈65%/RIU (RI unit); **Figure 14**. The RCWA method reveals that first-order diffraction is also significantly enhanced by the RA-SPP effect, **Figure 14(c)**. As expected, first-order diffraction, dominated by the (1, 0) n_III RA, exhibits a wavelength sensitivity equal to the theoretical upper limit, P/RIU.

FDTD calculated electric field plots also verify coupling between the (1, 0) n_I SPP-BW and (1, 0) n_III RA. **Figure 15** shows the frequency-resolved $|E_z|^2$ at $\lambda = 679$ nm for $n_\text{III} = 1.70$ (near the RA-SPP crossing point). **Figure 15(a)** shows that the (1, 0) n_I SPP-BW has greatest intensity on the n_I side, but also some noticeable intensity on the n_III side. The (1, 0) n_III RA is present, but is not visible in **Figure 15(a)** since it's intensity relative to the (1, 0) n_I SPP-BW is small. However, **Figure 15(b)** shows the field for $z > 200$ nm above the film surface, where characteristic RA features are visible, that is, an extended propagating plane wave [57]. The high intensity of the (1, 0) n_III RA near the film interface, which decreases in strength as the distance from the film increases, suggests that the (1, 0) n_III RA is coupled to the (1, 0) n_I SPP-BW. Further RCWA calculations revealed that the RA-SPP effect persists strongly for film thicknesses up to $h = 65$ nm and hole (or slit) diameters $d > 80$ nm, with optimum values of $h = 45$ nm and $d = 175$ nm, respectively [31]. For example, using the optimum hole diameter, the amplitude variation with RIU almost doubles. RCWA calculations were carried out over a wide range of P and n_I values to confirm that the rapid variations in transmission amplitude are always consistent with the predictions of **equations 59** and **60** (not shown).

The other crossing points in **Figure 11** lead to couplings that are less relevant for obtaining narrow spectral features that are especially sensitive to variations of n_III. The coupling of the two SPP-BW features near $n_\text{III} = 1.523$ will lead to a broader peak than the RA-SPP peak since SPP-BWs are intrinsically broader than RAs due to their finite lifetimes. (The hybridization that results from this interaction, however, plays a role in the RA-SPP effect [7].) The crossing of the two RA features could produce narrow spectral features, but this effect is overwhelmed by SPP-BW coupling near the same wavelength. Finally, the n_III SPP-BW and n_I RA crossing near $n_\text{III} = 1.35$ occurs in a wavelength region where higher order SPP-BWs, LSPRs, and direct transmission are all important.

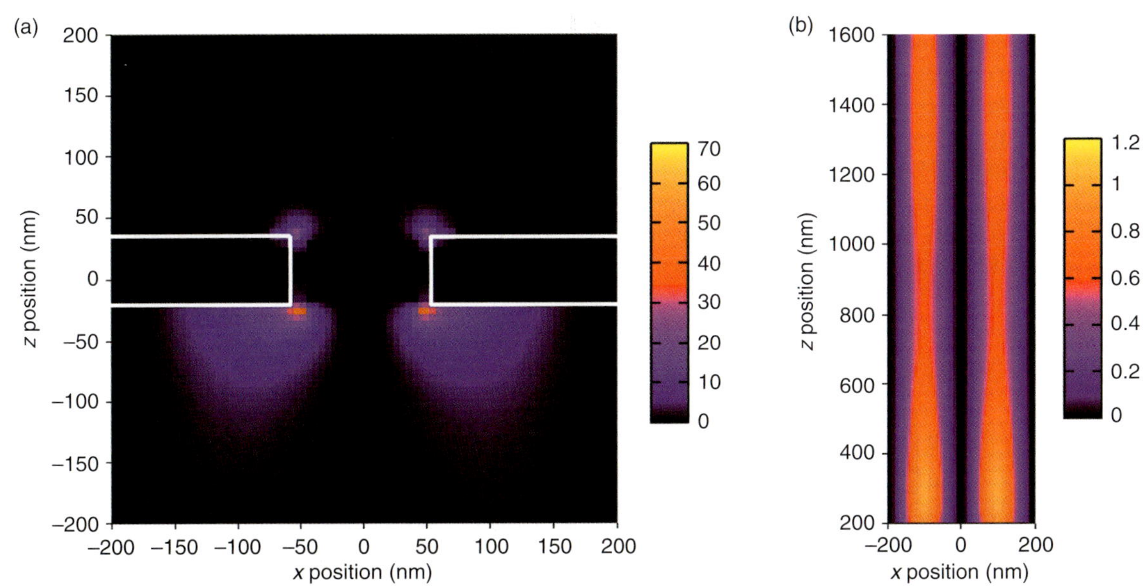

Figure 15 FDTD calculated frequency-resolved $|E_z|^2$ at $\lambda = 679$ nm for $n_\text{III} = 1.70$, showing (a) the region near the hole, and (b) 200 nm above the Au film. The hole is centered at the origin, and the film boundaries are outlined in white. Reprinted from McMahon JM, Henzie J, Odom TW, Schatz GC, and Gray SK (2007) Tailoring the sensing capabilities of nanohole arrays in gold films with Rayleigh anomaly – surface plasmon polaritons. *Optics Express* 15: 18119.

These results show the richness of phenomena associated with light transmission through arrays of sub-wavelength holes in metal films. Of particular interest for sensing applications is the discussed RA-SPP effect, which leads to narrow transmission peaks that have strong RI sensitivity. Simple models were presented that accurately predict the system parameters, RI, and wavelengths where RA-SPP effects occur. However, more quantitative full electrodynamics calculations are needed to determine the precise positions, intensities, and widths of the transmission peaks.

2.5.2 Pillar Arrays

While somewhat counter-intuitive, the existence of holes in a metal film is not required to observe EOT [58]. All that is necessary is to couple into SPPs (or other effects) associated with the metal–dielectric interface, and for the metal film to be optically thick [2, 59]. As an example of this, consider a 400 nm periodic array of 60 nm high × 120 nm diameter Ag disks on top of a flat 30 nm thick Au metal film on a glass ($n_{\text{I}} = 1.523$) substrate with a variable index superstrate (n_{III}); **Figure 16**.

FDTD calculated the zero-order transmission spectra for superstrate indices of $n_{\text{III}} = 1.0$ to 1.65 are shown in **Figure 17**. As can be seen, even without holes, a large amount of transmission through these Au film/pillar arrays occurs compared to a flat Au film (**Figure 1**). This EOT arises because the Ag pillars act as a diffraction grating that couples incident light into SPPs. As was the case for hole arrays (see Section 2.5.1), most peaks not due to film transmission can be explained using **equations 26** and **27**. For example, for $n_{\text{III}} = 1.65$, the small minimum at 742 nm is close to the predicted (1, 0) n_{III} SPP-BW position of 712 nm. In addition, in all spectra a slight kink is seen near 603 nm, which can be attributed to a (1, 0) n_{I} RA, expected to occur at 609 nm. A large minimum is also observed in all of the spectra from the LSPR of the Ag disks (e.g., at 550 nm for $n_{\text{III}} = 1.35$, and 670 nm for $n_{\text{III}} = 1.65$), which is confirmed by plots of the frequency-resolved $|E_z|^2$ at 666 nm for $n_{\text{III}} = 1.65$, **Figure 18(c)**, and calculations not involving the Au film (not shown).

Coupling between SPP-BWs on both sides of the film is strong when $n_{\text{I}} = 1.523$ and $n_{\text{III}} = 1.5$ are similar, as can be seen in the plot of $|E_z|^2$ at 619 nm, **Figure 18(a)**. RAs and SPP-BWs can also interact in these systems, giving the RA-SPP effect (see Section 2.5.1). For the system under consideration, this effect is seen in the spectra near 638 nm for $n_{\text{III}} = 1.65$, **Figure 17**, and is confirmed by a plots of $|E_z|^2$ at 633 nm, **Figure 18(b)**, where a (1, 0) n_{I} SPP-BW and (1, 0) n_{III} RA are seen.

The LSPR of the Ag disks provides an additional parameter that can be used to control the EOT in periodic systems. For example, by using elliptical Ag disks, the large minimum in the transmission spectra due to the LSPR can be red-shifted by switching the polarization direction from the short to the long axis. For example, **Figure 19** shows the transmission through a Ag pillar array with $n_{\text{III}} = 1.65$ for 120 nm × 240 nm pillars (2:1 aspect ratio), for both short- (x) and long-axis (y) polarizations. As can be seen, the Ag LSPR is red-shifted from 662 to 798 nm by switching the polarization direction. For circular polarization, both short- and long-axis LSPR resonances are seen to exist.

2.5.3 Other Periodic Systems

In addition to periodic arrays of holes and pillars in metal films, other periodic structures have been investigated. For example, Rogers and co-workers [8, 60, 61] have fabricated quasi-3D plasmonic

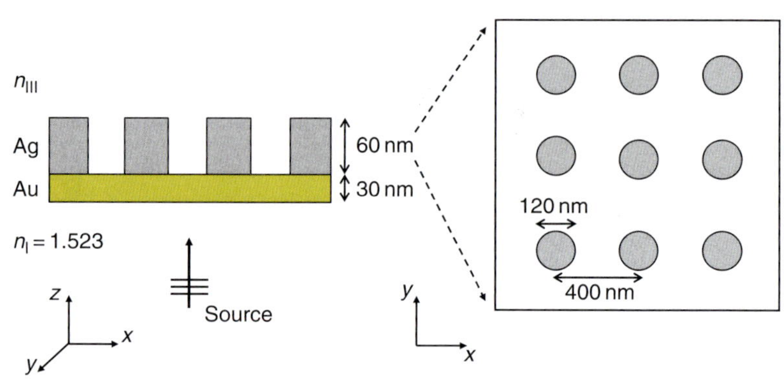

Figure 16 Schematic diagram of a periodic Ag pillar array on a Au film.

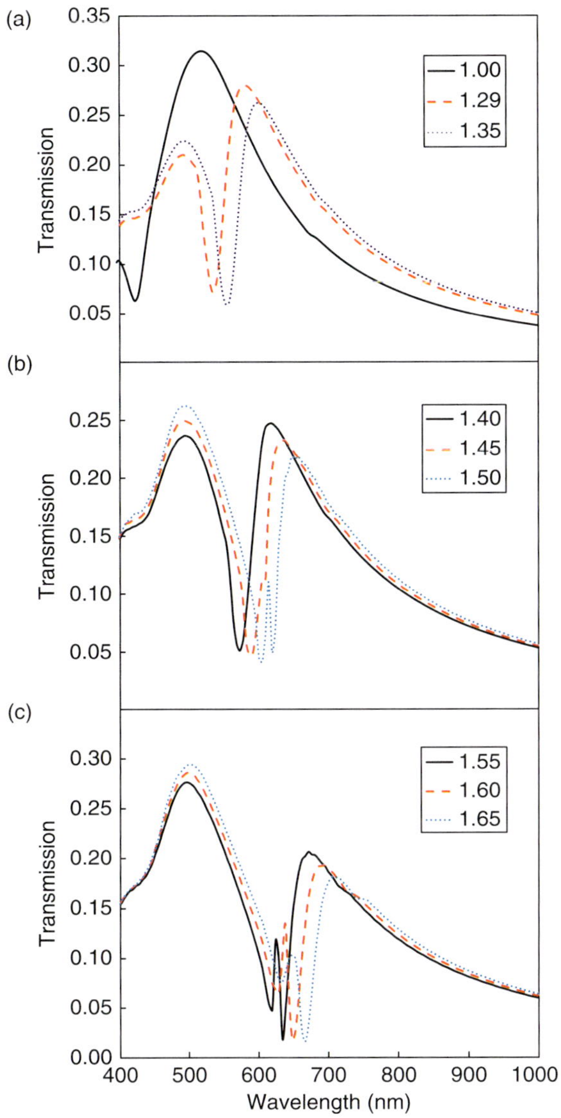

Figure 17 FDTD calculated zero-order transmission through periodic Ag pillars on a Au film with parameters shown in **Figure 16**.

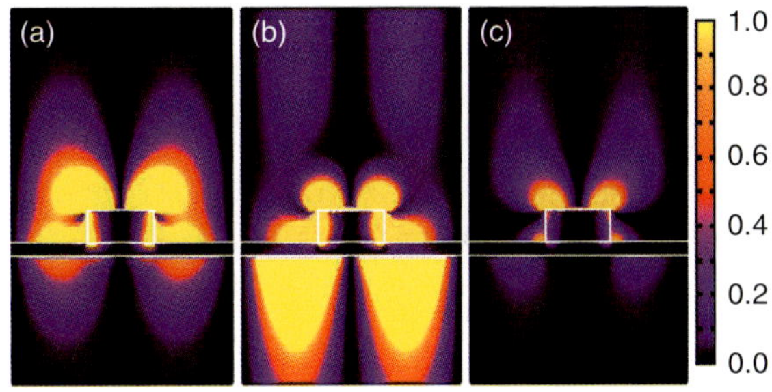

Figure 18 FDTD calculated frequency-resolved $|E_z|^2$ (normalized) at (a) $\lambda = 619$ nm for $n_{III} = 1.5$, showing coupling of SPP-BWs on both sides of the film, (b) $\lambda = 633$ nm for $n_{III} = 1.65$, showing the RA-SPP effect, and (c) $\lambda = 666$ nm for $n_{III} = 1.65$, showing the LSPR of the Ag pillars.

crystals by depositing a Au film onto periodic arrays of finite-depth wells stamped in a polymer. Depending on the metal deposition technique, the resulting systems have either holes in the top film and Au disks in the bottom of the wells [8], or well structures conformally covered with a Au film [61] (i.e., the well side-walls are also coated). One advantage of these structures is that their greater complexity relative to simple hole arrays can be used to obtain extra sensitivity to RI. FDTD simulations have been used to identify the physical origins of the RI-sensitive transmission peaks, and in general they involve SPP-BWs, RAs, and LSPRs [8]. A related structure based on nanosphere lithography was studied by Hicks *et al.* [3], who looked at reflected light rather than the transmitted light.

Another interesting periodic system is an array of coaxial apertures in a metal film; these arrays are essentially hole arrays with metal pillars inside the holes [62–65]. These systems may alternatively be viewed as dielectric (or possibly air) ring structures in a metal film, where the interior of the dielectric ring is the metal pillar and the exterior is the metal film. These systems are able to strongly support waveguide modes (see Sections 2.3.3 and 2.4.1) that can travel up through the coaxial structures, which can also lead to EOT. In addition, when the wavelengths are such that SPPs can be excited in the metal, FDTD calculations show that the waveguide modes can significantly red-shift and exhibit unusual cutoff properties [62, 64]. Experimental results have been found to be in remarkably good accord with these theoretical predictions [63, 65], and Haftel and co-workers [64, 65] refer to these modes as cylindrical SPPs.

2.6 Summary and Outlook

In this chapter, we reviewed some common theoretical and computational methods for modeling nanostructured metal films, and discussed a select set of applications of them. We focused mainly on the FDTD method, but also discussed the RCWA and modal expansion methods. Many of the phenomena associated with nanostructured metal films were discussed, such as SPPs, RAs, and LSPRs, all which lead to complex optical properties, including EOT. It

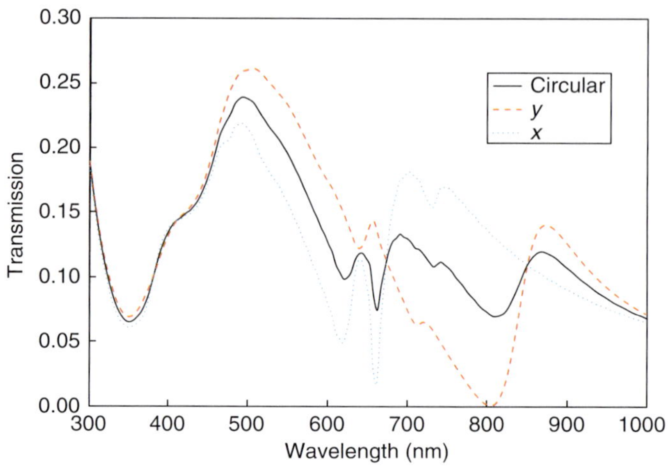

Figure 19 FDTD calculated zero-order transmission spectra for 120 nm × 240 nm Ag pillars, oriented along the *y*-axis, on a 30 nm thick Au film on a glass ($n_{\text{I}} = 1.5$) substrate with $n_{\text{III}} = 1.65$ for linear (*x*- or *y*-axis) and circular incident polarizations of light.

was emphasized that these effects are not isolated from one-another, and FDTD and RCWA calculations were presented that show that SPPs and RAs can strongly interact (the RA-SPP effect), producing especially narrow EOT features. We also demonstrated that waveguide mechanisms lead to interesting effects in apertures in metal films, such as the dephasing of electromagnetic fields inside an isolated gold slit. For these systems, FDTD calculations and the modal expansion method showed that scattering by the aperture induces a modal transition from propagating to evanescent of the incident light, leading to dephasing.

The methods and applications that were presented are only a small selection of the work that has been done. In addition, as the experimental techniques for making nanostructured metal films becomes more sophisticated, leading to more complex structures, these theoretical and computational tools will continue to play a large part in modeling and describing the effects associated with them.

Acknowledgments

GCS and JMM were supported by the US Department of Energy under grant No. DEFG02-03-ER15487 and the Northwestern Materials Research Center, sponsored by the National Science Foundation (DMR-0520513). SKG was supported by the US Department of Energy, Office of Basic Energy Sciences under contract DE-AC02-06CH11357. This research used resources of the National Energy Research Scientific Computing Center, which is supported by the Office of Science of the US Department of Energy Contract DE-AC02-05CH11231.

References

1. Ebbesen TW, Lezec HJ, Ghaemi HF, Thio T, and Wolff PA (1998) Extraordinary optical transmission through sub-wavelength hole arrays. *Nature (London)* 391: 667–669.
2. Ghaemi HF, Thio T, Grupp DE, Ebbesen TW, and Lezec HJ (1998) Surface plasmons enhance optical transmission through subwavelength holes. *Physical Review B* 58: 6779–6782.
3. Hicks EM, Zhang X, Zou S, *et al.* (2005) Plasmonic properties of film over nanowell surface fabricated by nanosphere lithography. *Journal of Physical Chemistry B* 109: 22351–22358.
4. Lee MH, Gao H, Henzie J, and Odom TW (2007) Microscale arrays of nanoscale holes. *Small* 3: 2029–2033.
5. Henzie J, Lee MH, and Odom TW (2007) Multiscale patterning of plasmonic metamaterials. *Nature Nanotechnology* 2: 549–554.
6. Gordon R, Sinton D, Kavanagh KL, and Brolo AG (2008) A new generation of sensors based on extraordinary optical transmission. *Accounts of Chemical Research* 41: 1049–1057.
7. McMahon JM, Henzie J, Odom TW, Schatz GC, and Gray SK (2007) Tailoring the sensing capabilities of nanohole arrays in gold films with Rayleigh anomaly – surface plasmon polaritons. *Optics Express* 15: 18119.
8. Stewart ME, Mack NH, Malyarchuck V, *et al.* (2006) Quantitative multispectral biosensing and 1D imaging using quasi-3D plasmonic crystals. *Proceedings of the National Academy of Sciences of the United States of America* 103: 17143–17148.
9. Brolo AG, Gordon R, Leathem B, and Kavanagh KL (2004) Surface plasmon sensor based on the enhanced light transmission through arrays of nanoholes in gold films. *Langmuir* 20: 4813–4815.

10. Reilly TH, Chang S-H, Corbman JD, Schatz GC, and Rowlen KL (2007) Quantitative evaluation of plasmon enhanced Raman scattering from nanoaperture arrays. *Journal of Physical Chemistry C* 111: 1689–1694.
11. Brolo AG, Arctander E, Gordon R, Leathem B, and Kavanagh KL (2004) Nanohole-enhanced Raman scattering. *Nano Letters* 4: 2015–2018.
12. Brolo AG, Kwok SC, Moffitt MG, Gordon R, Riordon J, and Kavanagh KL (2005) Enhanced fluorescence from arrays of nanoholes in a gold film. *Journal of the American Chemical Society* 127: 14936–14941.
13. Porto JA, Garcia-Vidal FJ, and Pendry JB (1999) Transmission resonances on metallic gratings with very narrow slits. *Physical Review Letters* 83: 2845–2848.
14. Bravo-Abad J, Martín-Moreno L, and García-Vidal FJ (2004) Transmission properties of a single metallic slit: From the subwavelength regime to the geometrical-optics limit. *Physical Review E* 69: 026601.
15. Chang S-H, Gray SK, and Schatz GC (2005) Surface plasmon generation and light transmission by isolated nanoholes and arrays of nanoholes in thin metal films. *Optics Express* 13: 3150–3165.
16. Yin L, Vlasko-Vlasov VK, Rydh A, *et al.* (2004) Surface plasmons at single nanoholes in Au films. *Applied Physics Letters* 85: 467–469.
17. Babayan Y, McMahon JM, Li S, Gray SK, Schatz GC, and Odom TW (2009) Confining standing waves in optical corrals. *ACS Nano* 3: 615–620.
18. Raether H (1988) *Surface Plasmons on Smooth and Rough Surfaces and on Gratings*. New York: Springer.
19. Lynch DW and Hunter WR (1985) Comments on the optical constants of metals and an introduction to the data for several metals. In: Palik ED (ed.) *Handbook of Optical Constants of Solids*, pp. 275–368. Orlando, FL: Academic Press.
20. Johnson PB and Christy RW (1972) Optical constants of the noble metals. *Physical Review B* 6: 4370–4379.
21. Kretschmann E (1971) Die Bestiemmung optisher Konstanten von Metallen durch Anregung von Obelflachenplasmaschwingungen. *Zeitschrift fur Physik* 241: 313–324.
22. Otto A (1968) Excitation of nonradiative surface plasma waves on silver by the method of frustrated total reflection. *Zeitschrift fur Physik* 216: 398–410.
23. Degiron A, Lezec HJ, Yamamoto N, and Ebbesen TW (2004) Optical transmission properties of a single subwavelength aperture in a real metal. *Optics Communications* 239: 61–66.
24. Hessel A and Oliner AA (1965) A new theory of Wood's anomalies on optical gratings. *Applied Optics* 4: 1275–1297.
25. Darmanyan S, Neviere M, and Zayats A (2004) Analytical theory of optical transmission through periodically structured metallic films via tunnel-coupled surface polariton modes. *Physical Review B* 70: 075103.
26. Bethe HA (1944) Theory of diffraction by small holes. *Physical Review* 66: 163–182.
27. McMahon JM, Gray SK, and Schatz GC (2008) Dephasing of electromagnetic fields in scattering from an isolated slit in a gold film. In: *Proceedings of SPIE: Plasmonics: Nanoimaging, Nanofabrication, and Their Applications IV*, p. 703311. San Diego, CA, USA. Bellingham, WA: SPIE Press.
28. Hendry E, Garcia-Vidal FJ, Martin-Moreno L, Gómez Rivas J, Bonn M, Hibbins AP, and Lockyear MJ (2008) Optical control over surface-plasmon-polariton-assisted THz transmission through a slit aperture. *Physical Review Letters* 100: 123901.
29. Taflove A and Hagness SC (2005) *Computational Electrodynamics: The Finite-Difference Time-Domain Method*, 3rd edn. Norwood, MA: Artech House.
30. Yee SK (1966) Numerical solution of initial boundary value problems involving Maxwell's equations in isotropic media. *IEEE Transactions on Antennas and Propagation* 14: 302–307.
31. Schatz GC, McMahon JM, and Gray SK (2007) Tailoring the parameters of nanohole arrays in gold films for sensing applications. In: *Proceedings of SPIE: Plasmonics: Metallic Nanostructures and Their Optical Properties V*, p. 664103. San Diego, CA, USA. Bellingham, WA: SPIE Press.
32. Kashiwa T and Fukai I (1990) A treatment by FDTD method of dispersive characteristics associated with electronic polarization. *Microwave and Optical Technology Letters* 3: 203–205.
33. Joseph RM, Hagness SC, and Taflove A (1991) Direct time integration of Maxwell's equations in linear dispersive media with absorption for scattering and propagation of femtosecond electromagnetic pulses. *Optics Letters* 16: 1412–1414.
34. Berenger JP (1994) A perfectly matched layer for the absorption of electromagnetic waves. *Journal of Computational Physics* 114: 185–200.
35. Roden JA and Gedney SD (2000) Convolutional PML (CPML): An efficient FDTD implementation of the CFS-PML for arbitrary media. *Microwave and Optical Technology Letters* 27: 334–339.
36. Merewether DE, Fisher R, and Smith FW (1980) On implementing a numeric Huygen's source scheme in a finite difference program to illuminate scattering bodies. *IEEE Transactions on Nuclear Science* 27: 1829–1833.
37. Mur G (1981) Absorbing boundary conditions for the finite-difference approximation of the time-domain electromagnetic field equations. *IEEE Transactions on Electromagnetic Compatibility* 23: 377–382.
38. Umashankar KR and Taflove A (1982) A novel method to analyze electromagnetic scattering of complex objects. *IEEE Transactions on Electromagnetic Compatibility* 24: 397–405.
39. Moharam MG, Grann EB, Pommet DA, and Gaylord TK (1995) Formulation for stable and efficient implementation of the rigorous coupled-wave analysis of binary gratings. *Journal of the Optical Society of America A* 12: 1068–1076.
40. Lalanne P and Morris GM (1996) Highly improved convergence of the coupled-wave method for TM polarization. *Journal of the Optical Society of America A* 13: 779–784.
41. Benabbas A, Halté V, and Bigot J-Y (2005) Analytical model of the optical response of periodically structured metallic films. *Optics Express* 13: 8730–8745.
42. Li L (1996) Use of Fourier series in the analysis of discontinuous periodic structures. *Journal of the Optical Society of America A* 13: 1870–1876.
43. Wannemacher R (2001) Plasmon – supported transmission of light through nanometric holes in metallic thin films. *Optics Communications* 195: 107–118.
44. Hafner C (1990) The generalized multipole technique for computational electromagnetics, ch. 7–8, pp. 157–266. Boston, MA: Artech House.
45. Prikulis J, Hanarp P, Olofsson L, Sutherland D, and Kall M (2004) Optical spectroscopy of nanometric holes in thin gold films. *Nano Letters* 4: 1003–1007.
46. Rindzevicius T, Alaverdyan Y, Sepulveda B, *et al.* (2007) Nanohole plasmons on optically thin gold films. *Journal of Physical Chemistry C* 111: 1207–1212.
47. Sepulveda B, Alaverdyan V, Alegret J, Kall M, and Johansson P (2008) Shape effects in the localized surface

plasmon resonance of single nanoholes in thin metal films. *Optics Express* 16: 5609–5616.

48. Treacy MMJ (2002) Dynamical diffraction explanation of the anomalous transmission of light through metallic gratings. *Physical Review B* 66: 195105.
49. Sarrazin M, Vigneron J-P, and Vigoureux J-M (2003) Role of Wood anomalies in optical properties of thin metallic films with a bidimensional array of subwavelength holes. *Physical Review B* 67: 085415.
50. Ruan Z and Qiu M (2006) Enhanced transmission through periodic arrays of subwavelength holes: The role of localized waveguide resonances. *Physical Review Letters* 96: 233901.
51. Genet C, van Exter MP, and Woerdman JP (2003) Fano-type interpretation of red shifts and red tails in hole array transmission spectra. *Optics Communications* 225: 331–336.
52. van der Molen KL, Segerink FB, van Hulst NF, and Kuipers L (2004) Influence of hole size on the extraordinary transmission through subwavelength hole arrays. *Applied Physics Letters* 85: 4316–4318.
53. Klein Koerkamp KJ, Enoch S, Segerink FB, van Hulst NF, and Kuipers L (2004) Strong influence of hole shape on extraordinary transmission through periodic arrays of subwavelength holes. *Physical Review Letters* 92: 183901.
54. van der Molen KL, Koerkamp KJK, Enoch S, Segerink FB, van Hulst NF, and Kuipers L (2005) Role of shape and localized resonances in extraordinary transmission through periodic arrays of subwavelength holes: Experiment and theory. *Physical Review B* 72: 045421.
55. Shou X, Agrawal A, and Nahata A (2005) Role of metal film thickness on the enhanced transmission properties of a periodic array of subwavelength apertures. *Optics Express* 13: 9834–9840.
56. Kim JH and Moyer PJ (2006) Thickness effects on the optical transmission characteristics of small hole arrays on thin gold films. *Optics Express* 14: 6595–6603.
57. Steele J, Moran C, Aguirre A, Lee A, and Halas N (2003) Metallodielectric gratings with subwavelength slots: Optical properties. *Physical Review B* 68: 205103.
58. Bonod N, Enoch S, Li L, Evgeny P, and Neviere M (2003) Resonant optical transmission through thin metallic films with and without holes. *Optics Express* 11: 482–490.
59. Degiron A and Ebbesen TW (2004) Analysis of the transmission process through single apertures surrounded by periodic corrugations. *Optics Express* 12: 3694–3700.
60. Stewart ME, Anderton CR, Thompson LB, Maria J, Gray SK, Rogers JA, and Nuzzo RG (2008) Nanostructured plasmonic sensors. *Chemical Review* 108: 494–521.
61. Yao JM, Stuart ME, Maria J, *et al.* (2008) Seeing molecules by eye: Surface plasmon resonance imaging at visible wavelengths with high spatial resolution and submonolayer sensitivity. *Angewandte Chemie International Edition* 47: 5013–5017.
62. Baida FI, Belkhir A, and Van Labeke D (2006) Subwavelength metallic coaxial waveguides in the optical range: Role of the plasmonic modes. *Physical Review B* 74: 205419.
63. Poujet Y, Salvi J, and Baida FI (2007) 90% Extraordinary optical transmission in the visible range through annular aperture metallic arrays. *Optics Letters* 32: 2942–2944.
64. Haftel MI, Schlockermann C, and Blumberg G (2006) Enhanced transmission with coaxial nanoapertures: Role of cylindrical surface plasmons. *Physical Review B* 74: 235405.
65. Orbons SM, Roberts A, Jamieson DN, *et al.* (2007) Extraordinary optical transmission with coaxial apertures. *Applied Physics Letters* 90: 251107.

3 Second Harmonic Generation in Nanostructures

P-F Brevet, Université Claude Bernard Lyon 1, Villeurbanne, France

3.1 Introduction

The study of the optical properties of nanostructures is a field in rapid expansion, driven by fundamental questions such as the influence of the characteristic size of the systems on the optical response and their potential applications in many diverse areas, from optical telecommunications to the life sciences [1]. It has also given birth to a whole new field known as 'plasmonics' [2]. This interest originally stems from the unique optical features of the visible optical spectra of nanostructured metallic nano-objects, more particularly those made from gold and silver. Their optical spectra are dominated in the visible frequency range by the surface plasmon resonance (SPR), that is, the collective excitation of the conduction band electrons. Many studies have since demonstrated that the frequency of the SPR is dramatically dependent upon many parameters like the size, the shape, and the morphology or the environment of the particles [3,4]. Hence, in recent years, nanostructure fabrication methods have been developed and a great versatility in the morphology of the systems has been achieved. Two routes are currently proposed. The first one is based on a bottom-up approach, where the nanostructures are built starting from the individual nanoparticles used as building blocks of more elaborated architectures. This approach requires a good knowledge of the optical properties of the particles and their modifications when the particles are integrated in the nanostructures. The second route is a top-down approach, which finds its origin in the development of the standard microfabrication techniques at the nanometre scale. One of the techniques now routinely used in laboratories is electron-beam lithography (EBL). Characterization has also become an important issue, but in both the bottom-up and the top-down routes the method of choice is electron microscopy – either scanning (SEM) or transmission (TEM) electron microscopy. However, optical methods, for example, absorption and reflectance or photoluminescence spectroscopy, have been able to reveal many important features. Besides characterization, applications based on the use of the linear optical techniques of nanostructures have regularly appeared owing to the sensitivity of the optical response to the nearby environment for example. In this context, it is interesting to ask whether nonlinear optical techniques could offer similar perspectives. In nonlinear optics, the interaction processes between the exciting electromagnetic waves and the material are no longer linear in the field amplitude but involve higher orders [5,6]. New features can be observed and will eventually lead to new applications. It is the purpose of this work to focus on one of the simplest nonlinear optical processes, namely second-harmonic generation (SHG). Several experimental approaches have been reported, from incoherent to coherent SHG using different types of nanostructures: dispersed particles in a solid or a liquid matrix, arrays of particles deposited on a substrate, nanostructured semicontinuous films, etc.

Second-harmonic generation, the phenomenon whereby two photons at the fundamental frequency ω are converted into one photon at the harmonic frequency $\Omega = 2\omega$, is a process forbidden in media possessing a center of inversion. As a result, only a reduced subset of the crystal group symmetries yield a nonvanishing quadratic susceptibility $\chi^{(2)}$ [5,6]. Many studies have made use of this general selection rule to investigate surfaces of, or interfaces between, media possessing a center of inversion. Indeed, no coherent SHG signal is generated in the volume of both centrosymmetrical media, whereas one is produced at the interface where the inversion symmetry is broken. Surface SHG has been used to investigate planar surfaces, such as air/solid or air/liquid surfaces as well as liquid/solid or liquid/liquid interfaces [7–12]. The role of nanometer scale has not been the primary focus in these studies, although some authors have pointed out features arising from the nanoscale dimensions of the samples. This was the case, for example, for the investigation of the enhancement of the SHG intensity from rough metal surfaces [13]. The problem must be reconsidered in a large part when dealing with structures where the characteristic dimensions of the morphological features are no longer of the order of or larger than the wavelength of light. For systems with smaller characteristic sizes made from noncentrosymmetrical material, the whole volume of the structure participates in the frequency doubling process. As the dimensions of the structure decrease further, the surface contributions start to dominate as the number of surface atoms compared to the number of volume atoms increases. For structures made from a few atoms only, quantum effects such as quantum confinement would also need to be considered. For systems made of centrosymmetrical material, the SHG phenomenon is forbidden in the volume of the structure within the electric dipole approximation. The problem must then be fully reconsidered taking a careful account of the surface of the structure and the spatial inhomogeneity of the electromagnetic fields. Retardation of the electromagnetic fields in particular, namely the dependence of the amplitude of the electromagnetic fields on the spatial coordinates, must be taken into account correctly. This often introduces a higher degree of complexity in the experimental results.

In the first part of this chapter, we briefly review the basics of SHG in continuous media. In particular, we will underline the role of phase matching and introduce the coherence length in discussing the SHG intensity. Although this part does not bring any new material, it provides a smooth introduction to the main features that have to be reconsidered at the nanometer scale. Then in a second part, we discuss the case of particles dispersed in a homogenous matrix. A large part of the literature is devoted to this type of nanostructures. We will, therefore, discuss separately the case of particles made from a noncentrosymmetrical material and those made from a centrosymmetrical one. The section devoted to the particles made from a noncentrosymmetrical material is essentially concerned with semiconducting particles. The case of centrosymmetrical particles reduces to that of metallic particles. Results on amorphous materials and centrosymmetrical dielectrics have been scarcely reported. At the microscopic level, the efficiency of nanoobjects to convert fundamental-frequency photons into harmonic photons is determined by the quadratic hyperpolarizability. The method of choice to determine this quantity is the technique of hyper Rayleigh scattering (HRS), where incoherent SHG is detected. This method is appropriate when the particles are randomly distributed in a solid or a liquid matrix. Despite the weak signals measured, often obtained in the photon-counting regime, the technique allows to avoid the use of a substrate to support the particles. Comparison with theoretical models is also easier because the environment of the particle is homogeneous. In a third part, we address the case of nanoparticles regularly arranged on a substrate. These structures are produced by EBL and a wide range of shapes is available, whether centrosmmetrical, noncentrosymmetrical, or even chiral. Finally, we conclude this work with several considerations on random nanostructures, opening the scope of the previous sections to less-controlled systems.

3.2 Fundamentals of Second Harmonic Generation

We first start with a brief introduction to the basic theory of SHG. Fundamental issues are presented, in particular those of wave mixing, phase matching, and coherence length. These issues will have to be reconsidered in light of the nanometer dimensions of the structures where the frequency-doubling process takes place.

Upon excitation by the electric and magnetic fields of an intense electromagnetic wave, a medium is polarized and magnetized, and this response may not be linear in the excitation fields. In particular, it may be nonlinear and nonlocal [5,6,14]. Hence, considering the perturbation of the medium due to the electric field, the polarization vector may be written as a power series in the electric field amplitude:

$$\boldsymbol{P} = \boldsymbol{P}^{(1)} + \boldsymbol{P}^{(2)} + \boldsymbol{P}^{(3)} + \cdots \tag{1}$$

where the first term $\boldsymbol{P}^{(1)}$ is the linear polarization proportional to the field amplitude, the second term $\boldsymbol{P}^{(2)}$ is the first nonlinear polarization proportional to the second power of the field amplitude, and so on. In particular, the linear polarization writes as follows:

$$\boldsymbol{P}^{(1)}(\omega) = \varepsilon_0 \boldsymbol{\chi}^{(1)}(\omega)\boldsymbol{E}(\omega) \tag{2}$$

for a monochromatic electric field. The linear susceptibility $\boldsymbol{\chi}^{(1)}(\omega)$ is defined as the Fourier transform component (at the frequency ω) of the tensor representing the response of the medium to the electric field, and is directly related to the dielectric function of the material through

$$\varepsilon(\omega) = 1 + \boldsymbol{\chi}^{(1)}(\omega) \tag{3}$$

In **equation 2**, we have omitted the spatial dependence of the field, assuming a plane wave propagating in a defined arbitrary direction. We now focus on the first nonlinear polarization $\boldsymbol{P}^{(2)}$, the general expression of which is

$$\boldsymbol{P}^{(2)}(\omega_3) = \varepsilon_0 \boldsymbol{\chi}^{(2)}(\omega_3;\omega_1,\omega_2)\boldsymbol{E}_1(\omega_1)\boldsymbol{E}_2(\omega_2) \tag{4}$$

assuming that the excitation is performed with two incoming waves $E_1(\omega_1)$ and $E_2(\omega_2)$ with respective frequencies ω_1 and ω_2. In the particular case of SHG generated with a single beam, we simply have $\omega_3 = 2\omega$ and $\omega_1 = \omega_2 = \omega$ along with $E_1(\omega_1) = E_2(\omega_2)$. However, as we shall see later, **equation 4** allows the possibility to produce a nonlinear polarization oscillating at the second-harmonic (SH) frequency $\omega_3 = 2\omega$ through excitation with two different beams with $\omega_1 = \omega_2 = \omega$. This two-beam SHG configuration yields more flexibility in terms of the selection of the nonlinear sources at the microscopic level participating to the conversion process. In **equation 4**, the nonlinear susceptibility tensor $\boldsymbol{\chi}^{(2)}(\omega_3; \omega_1, \omega_2)$ is a 27-component complex-valued tensor. The form of **equation (4)** also assumes the electric dipole approximation, where the nonlinear polarization at the location $\boldsymbol{r}$ depends only on the field amplitude at the same location. We shall see below how this approximation must be revised in order to discuss field gradient sources. The quadratic susceptibility tensor $\boldsymbol{\chi}^{(2)}(\omega_3; \omega_1, \omega_2)$ embeds all the physical properties of the medium and, in particular, is left invariant through the same-space operations leaving the nonlinear medium invariant. These symmetry operations lead to a dramatic reduction of the number of nonvanishing and independent tensor components. In media possessing a center of inversion, for instance, the centrosymmetry leads to the vanishing of all the tensor elements. As mentioned in the introduction, this is an interesting property to investigate interfaces between two centrosymmetrical media.

In infinite media, for a single plane wave of amplitude $E^{(\omega)}$ with frequency ω, linearly polarized and propagating along the Oz direction, the wave equation writes [15]:

$$\frac{\partial E^{(2\omega)}}{\partial z} = -\frac{\mathrm{i}\omega}{2n^{(2\omega)}c}\chi^{(2)}\left(E^{(\omega)}\right)^2 \qquad (5)$$

assuming the slowly varying envelope approximation. Integration of this equation for a nonlinear medium with length L yields the following intensity of the second-harmonic wave in the case of the nondepletion of the fundamental wave:

$$I^{(2\omega)} = \frac{\omega^2 L^2}{8(\varepsilon_0 c)^3 n^{(2\omega)}\left(n^{(\omega)}\right)^2}\left|\chi^{(2)}\right|^2\left(I^{(\omega)}\right)^2\left(\frac{\sin \Delta kL/2}{\Delta kL/2}\right)^2 \qquad (6)$$

The exact expression of the modulus of the quadratic susceptibility tensor $\left|\chi^{(2)}\right|$ depends on the exact choice of the linear polarization with respect to the crystal orientation. We observe in **equation 6** that the second-harmonic intensity is an oscillating function of the quantity $\Delta kL/2$. This oscillatory behavior has been clearly demonstrated in the Maker fringe experiment, one of the fundamental experiments in SHG (**Figure 1**) [16]. From **equation 6**, it is obvious that the maximum intensity will be obtained for a vanishing

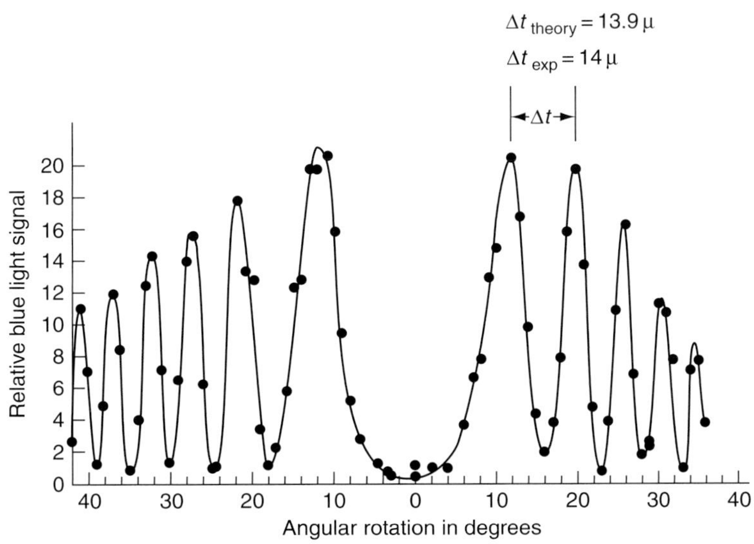

Figure 1 SH light generated from a 0.0308 inch-thick quartz platelet as a function of the inclination angle. The fundamental beam is unfocused and linearly polarized parallel to the rotation axis, which is normal to the beam propagation direction. Reproduced with permission from Maker PD, *et al.* (1962) *Physical Review Letters* 8: 21.

value of $\Delta kL/2$, whereas the first minimum will be obtained when this value equals π. It is common to define an optimum length L_c for the nonlinear efficiency of the material corresponding to the distance over which the polarization wave and the harmonic wave are out of phase with each other by an amount of π. Such a condition lead to the maximum transfer of energy from the fundamental wave to the second-harmonic wave. The expression of this optimum length L_c, also known as the coherence length, is

$$L_c = \frac{\lambda}{4\Delta n} \tag{7}$$

and is usually of the order of a few tens of microns. For example, taking a fundamental wavelength of 1 μm and an index mismatch of $\Delta n = n^{(2\omega)} - n^{(\omega)} \approx 10^{-2}$, the coherence length is 25 μm. Since $L_c \propto 1/\Delta k$, the best conversion efficiency will be obtained for a vanishing Δk, a condition called the 'phase-matching condition'. Conversion efficiencies approaching 100% have been reached when the phase-matching condition is fulfilled but in this case the energy depletion of the fundamental wave must be accounted for. Therefore, **equations 5** and **6** are no longer sufficient to describe the problem, a development out of the scope of the present work. In nanostructures, the characteristic size of the nonlinear medium is much smaller than the coherence length and, therefore, the polarization wave and the harmonic wave do not go out of phase by much. It is therefore said that, in nanostructures, SHG cannot be phase matched. As a result, the orientation of nonlinear nanocrystals with respect to the propagation of the fundamental wave is no longer critical. In regularly organized nanostructures though, like arrays of nanoparticles or nanocrystals deposited on a substrate, grating effects can still be observed owing to the coherent superposition of the SH wavelets radiated by each particle in a particular direction. This is the reproduction of the arrangement of the nanoparticles in the array as that of the atoms in a crystal cell. Hence, the coherent SHG signal radiated in specific directions closely resembles phase-matched SHG [17,18].

3.3 Particles from Noncentrosymmetrical Material

3.3.1 General Theory

In this section we describe the case of particles made from noncentrosymmetrical material. The nonlinear optical response from these systems is allowed within the electric dipole approximation and arises from the whole volume of the nanoobject or nanostructure. Approaches based on an effective medium have been proposed in the past to address this problem where the nanostructures are embedded in a matrix. This matrix may or may not have a quadratic susceptibility. The theory reproduces the well-known Maxwell–Garnett effective-medium approach for the linear properties of composite media [19–22]. Since the nanometre-scale structure of the nonlinear medium is not properly described, here we consider the SH generation at the level of a single particle and sum the response over all particles. We describe the particle made from the nonlinear medium irradiated by the fundamental plane wave linearly polarized with an angle of polarization γ in the laboratory frame as seen in **Figure 2**. Because the ratio between the characteristic size a of the particle and the coherence length L_c is much smaller than 1, we can

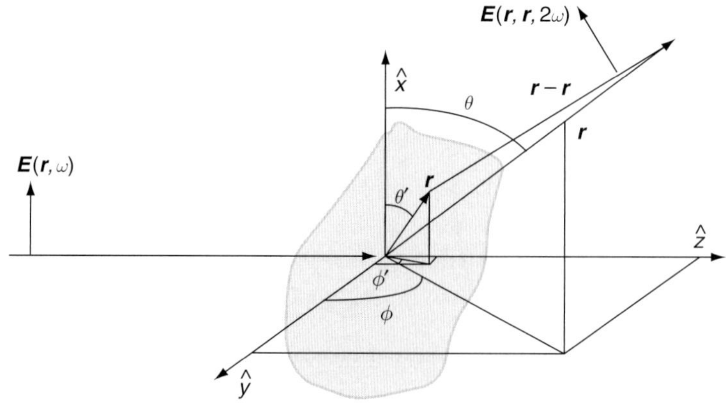

Figure 2 Schematics of the geometrical configuration used for the description of SHG from small particles. The direction to the detector is $\boldsymbol{r}$ and the position of any point located within the particle volume is defined by $\boldsymbol{r}'$.

disregard the orientation of the particle at the atomic level with respect to the propagation direction and the polarization state of the incoming fundamental wave in the linear optical regime. This means that the anisotropic nature of the nanoparticles is disregarded for the linear propagation of the fundamental and harmonic wave. For uniaxial crystalline particles, we therefore consider an optical index at the fundamental and the harmonic wavelength but do not discuss the ordinary or extraordinary character of these indices. Hence $\Delta kL/2 \propto a/L_c << 1$ and $\sin x/x \approx 1$ in **equation 6**. However, the orientation of the particle will be considered in the nonlinear frequency conversion process. We assume that the harmonic light is collected along a direction $\hat{n} = \boldsymbol{r}/r$ defined with the usual scattering angles (θ,ϕ). The incoming plane wave electric field is thus given by

$$\boldsymbol{E}(\boldsymbol{r}',\omega) = E_0\hat{e}\exp\left[-\mathrm{i}\left(\omega t + \boldsymbol{k}^{(\omega)}\cdot\boldsymbol{r}'\right)\right] \tag{8}$$

at any location $\boldsymbol{r}'$. The unit vector $\hat{e}$ is defined simply as $\hat{e} = \cos\gamma\,\hat{x} + \sin\gamma\,\hat{y}$ since we assume that the incident wave propagates along the Oz axis and the wave is linearly polarized with the angle γ. The Ox axis is arbitrarily defined as the vertical direction in the laboratory frame.

In an approach that reproduces the discrete dipole approximation (DDA) [23], a nanoparticle is described as an ensemble of induced nonlinear dipoles excited at the fundamental frequency and radiating at the harmonic frequency. Because the medium is noncentrosymmetrical, the electric dipole approximation applies. It yields

$$\boldsymbol{p}(\boldsymbol{r}',\Omega) = (E_0)^2\mathsf{T}(\hat{r}')\boldsymbol{\beta}':\hat{e}\hat{e}\exp\left[-2\mathrm{i}\left(\omega t - \boldsymbol{k}^{(\omega)}\cdot\boldsymbol{r}'\right)\right] \tag{9}$$

for a small nonlinear dipole located at position $\boldsymbol{r}'$. In **equation 9**, the small nonlinear dipole $\boldsymbol{p}(\boldsymbol{r}',\Omega)$ is given in the laboratory frame and makes use of the microscopic counterpart of the quadratic susceptibility, namely the quadratic hyperpolarizability $\boldsymbol{\beta}'$. Therefore, the frame transformation tensor $\mathsf{T}(\hat{r}')$ accounts for the transformation from the microscopic frame attached to the particle to the laboratory frame. It is indeed necessary to account for the orientation of the nanoparticle in the nonlinear regime to introduce the crystal symmetry properties of the particles. This is in contrast to the assumption made in the linear optical regime (see above), although nothing prevents the incorporation of the anisotropic linear optical properties of the wave propagation in the particle. The nonvanishing and independent elements of the quadratic hyperpolarizability tensor used in **equation 9** and reported in several textbooks are in effect given according to the crystal axes of the medium [5, 6]. The elements of the transformation tensor $\mathsf{T}(\hat{r}')$ are sine and cosine functions of the three Euler angles required to make the frame change. The total nonlinear dipole radiated by the particle is simply the superposition of all the small dipoles present within the volume of the particle. Hence,

$$\boldsymbol{p}(\Omega) = \int_{V'}\boldsymbol{p}(\boldsymbol{r}',\Omega)\mathrm{d}V' = (E_0)^2\exp[-2\mathrm{i}\omega t] \times \int_{V'}\mathsf{T}(\hat{r}')\boldsymbol{\beta}'\mathrm{d}V':\hat{e}\hat{e} = (E_0)^2\exp[-2\mathrm{i}\omega t]\boldsymbol{\beta}_{L,m}:\hat{e}\hat{e} \tag{10}$$

with the introduction of the hyperpolarizability of the particle integrated over the volume V' of the particle. It is easy to show that with a fully coherent superposition of all the individual nonlinear dipoles, this microscopic quadratic hyperpolarizability $\boldsymbol{\beta}_{L,m}$ will scale with the volume of the particle. A fully coherent superposition is defined as $\boldsymbol{\beta}_{L,m} = N\boldsymbol{\beta}'$, where N is the number of individual nonlinear dipoles contained in the particle. It is difficult to define such individual dipoles $\boldsymbol{p}(\boldsymbol{r}',\Omega)$ in a particle. It is more convenient to consider the quadratic hyperpolarizability $\boldsymbol{\beta}_{L,m}$ although **equations 9** and **10** demonstrate its scaling with the volume of the particle. The full coherence is obtained for a single nanocrystal, whereas defects in the crystalline structure would lead to an expression of the form of $\boldsymbol{\beta}_{L,m} = \alpha\boldsymbol{\beta}'$, with $\alpha < N$. In **equation 10**, we have omitted the spatial dependence of the electric field because we assume a nanoparticle, the characteristic size a (defined as $a = V'^{1/3}$) of which is much smaller than the fundamental wavelength λ, namely $a/\lambda << 1$. Also, we have directly used the fundamental electric field instead of the local fields, rejecting the local field factors in the hyperpolarizability itself. More refined models are accessible if the strict coherent superposition is not assumed. In particular, an account of a surface contribution is easily incorporated if the volume integration performed in **equation 10** is decomposed into a volume and a surface contribution:

$$\int_{V'}\boldsymbol{p}(\boldsymbol{r}',\Omega)\,\mathrm{d}V' = \int_{V_0'}\boldsymbol{p}(\boldsymbol{r}',\Omega)\,\mathrm{d}V' + \int_{S'}\boldsymbol{p}(\boldsymbol{r}',\Omega)\,\mathrm{d}V' \tag{11}$$

The surface integral can be formally taken as the integration on a small surface layer the thickness of which will likely depend on the microscopic structure of the interface, in particular on the compounds adsorbed at the surface. It is indeed known that in the chemical synthesis of particles, addition of stabilizing agents is required to prevent aggregation.

The cross-section of a particle for SHG is defined as $\sigma_{SHG} = I^{(2\omega)}/(I^{(\omega)})^2$ and is easily related to the quadratic hyperpolarizability. The quadratic hyperpolarizability is regularly reported in the literature in the old cgs system of units for comparison with previous data and is of the order of 10^{-26} to 10^{-30} esu (1 esu $= 3.7133 \times 10^{-21}\,C^3\,m^3\,J^{-3}$). This yields very weak signals for single particle experiments although there has been one report in the literature of a single particle SHG measurement [24, 25]. This experimental achievement was obtained with silver particles deposited on a substrate. The role of the substrate cannot be easily incorporated in the formalism given above and may break the symmetry of the geometrical configuration through the static polarization of the particle. One theoretical work addressed this configuration, though with the introduction of inhomogeneous fields [26]. Most of the experiments have been therefore devoted to the study of large ensembles of particles randomly distributed into a matrix, a solid matrix like alumina or silica, or a liquid phase. In principle, there is no real practical difference in the two samples apart from the shorter translation and rotational diffusion times of the particles in the liquid solutions as compared to the solid matrices. This is not a problem since usually the pulse duration provided by the laser sources is much shorter than this characteristic time. For large ensembles, the experiment is therefore performed on a random distribution in position and orientation of the particles within the volume sampled. This configuration leads to the detection of incoherent scattering also known as nonlinear scattering or hyper Rayleigh scattering, and such experiments have been performed on many different types of particles, from semiconducting to metallic and dielectric particles. The description of a large ensemble of particles randomly distributed requires the averaging over all their orientation to compute the intensity collected in a direction $\hat{n}$. The details of this procedure have been given in the literature for several symmetries [27, 28].

The harmonic wave amplitude of the SH field is that of a radiating dipole. In the far field, it is given by [29]

$$E(r,\Omega) = \frac{\exp\left[ik^{(\Omega)}r\right]}{r}[\hat{n} \times p(\Omega)] \times \hat{n} \qquad (12)$$

for a detector located at position r far from all particles, that is, $|r| >> |r'|$. **Equation 12** yields the field amplitude of the SH wave scattered by a particle in the direction of observation $\hat{n} = r/r$. For a collection of particles, we must sum up all the field amplitudes of the form of **equation 12**, taking into account the incoherent nature of the HRS signal arising from a collection of particles freely moving in position and orientation in the liquid solution or randomly distributed in the solid matrix. As a result, the HRS intensity stems from the orientation correlation between the particles and is a mere superposition of the intensities radiated by each single particle, namely [30]:

$$I_{HRS} = GN\langle E(r,\Omega))E^*(r,\Omega)\rangle \qquad (13)$$

for a collection of N particles. Here G is a constant embedding all constants and factors depending on the geometry of the experimental setup and the brackets $\langle \ldots \rangle$ stand for the averaging procedure over all possible orientations of the particle since the matrix is isotropic in all cases. Setting $\theta = \pi/2$ and $\phi = \pi/2$, corresponding to a conventional geometrical configuration where the detection direction is at right-angles to the excitation direction (Figure 2), we finally obtain the following expressions for the HRS intensity collected with output polarization state $\Gamma = V\,(\Gamma = 0)$:

$$I^{V}_{HRS} = G|p_x|^2 \qquad (14)$$

irrespective of the input polarization angle γ. For the other polarization direction $\Gamma = H\,(\Gamma = \pi/2)$, we get:

$$I^{H}_{HRS} = G|\sin\theta\, p_z - \cos\theta\, p_y|^2 \qquad (15)$$

The developments given above are identical to those describing the HRS measurements of the quadratic hyperpolarizability of molecules in solution [31]. Indeed, the particles behave similarly to molecules in the approximation of small particles compared to the wavelength of light. As an example, the angular dependence of the HRS intensity obtained from a chloroform solution of maghemite particles (a non-centrosymmetrical cubic crystalline structure of Fe_2O_3) is reported in **Figure 3** [32].

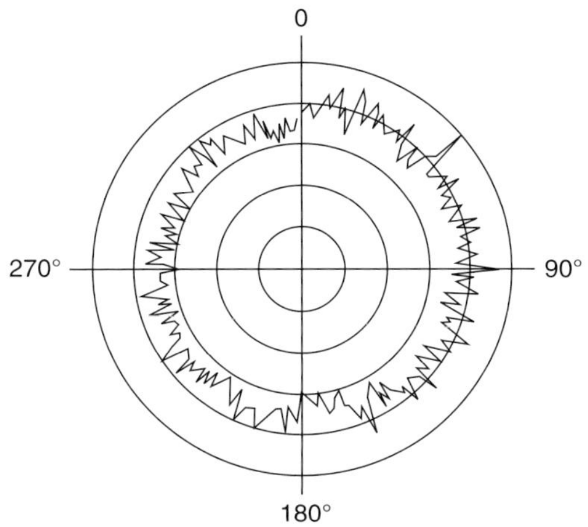

Figure 3 Angular dependence of the HRS signal from maghemite particles with the fundamental input vertically polarized ($\gamma = 0$) and $\phi = \pi/2$. Reproduced with permission from Gallet S, *et al.* (2003) *Chemical Physics Letters* 378: 101.

In most cases, one is interested in the absolute value of the quadratic hyperpolarizability tensor elements. To extract these values from the experimental data, **equation 13** must be rewritten with explicit dependence on the hyperpolarizability tensor elements, including the possible contribution of the surrounding matrix, so far neglected. Thus, **equation 15** rewrites as

$$I_{\mathrm{HRS}} = G\left\langle N_{\mathrm{S}}\beta_{Lm,S}^2 + N\beta_{Lm}^2\right\rangle I^2 \exp[-(2A_\omega + A_\Omega)] \quad (16)$$

where we specifically introduce the number densities N_{S} and N of, respectively, solvent molecules and particles and their respective quadratic hyperpolarizabilities $\beta_{Lm,S}$ and β_{Lm}. The exponential factor accounts for the absorption of either the incoming fundamental or the outgoing harmonic waves or both through the corresponding absorbance A_ω and A_Ω. Usually, in these experiments, one disregards the polarization state of the outgoing HRS intensity. Hence, assuming that the fundamental wave is vertically polarized (i.e., $\gamma = 0°$), the quadratic hyperpolarizabilities as given in **equation 16** is simply related to its elements in the laboratory frame: $\beta_{Lm} = \sqrt{\beta_{Lm,xxx}^2 + \beta_{Lm,zxx}^2}$. Therefore, this geometrical configuration only yields two tensor elements. A careful analysis of all the details of the experiment would reveal that at the most only six invariants are accessible [31].

Normalization of the hyperpolarizability tensor elements into absolute units is an important procedure in order to yield comparable quantities, from one experiment to the other. Considering **equation 16**, two different approaches are available, also known as the internal and external reference methods. The former consists in using the known value of the solvent hyperpolarizability to normalize the HRS intensity in absence of the compound studied. In a dilution experiment where the concentration of the compound with unknown hyperpolarizability is varied, a linear plot is obtained for the HRS intensity with the concentration of the compound, in agreement with **equation 16** (**Figure 4**).

The HRS intensity for the vanishing concentration of the compound is therefore perfectly determined and allows the normalization of the slope. The external reference method, on the other hand, consists in producing two different linear plots

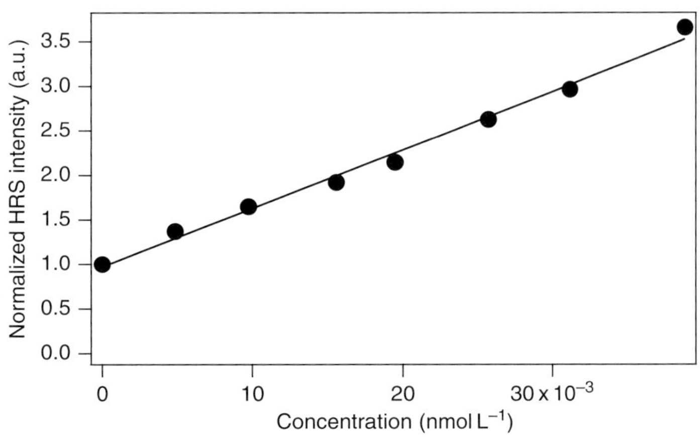

Figure 4 Dilution plot of the HRS intensity for 47 nm diameter mixed gold–silver metallic particles dispersed in an aqueous suspension as a function of the particle concentration. The stoechiometry of the particles is $Au_{0.5}Ag_{0.5}$.

in two dilution experiments and comparing the two slopes. One of these two slopes is obtained from a dilution experiment of a compound with a known hyperpolarizability. In these measurements, several caveats must be avoided. It is thus common practice to ensure that the HRS signal is monochromatic and scales with the square of the fundamental intensity and linearly with the number density of the particles. Any deviation from these observations would, in contrast, indicate that some other phenomena are occurring and are possibly affecting the measurements. For instance, one of the major problems encountered is the removal of a broadband background radiation that could arise from a luminescence or fluorescence signal. Such a problem is circumvented with the recording of the HRS band and subtracting the broadband background.

3.3.2 Volume Contribution

We turn our attention to a discussion on the quadratic hyperpolarizability for particles made from noncentrosymmetrical material. A comparison of this quantity as a function of the size of the particle is interesting in order to observe its scaling with the volume of the particle. For such media, the main contribution to the signal arises from the volume of the particle and, therefore, the absolute value of the hyperpolarizability per unit volume $\boldsymbol{\beta}_m/V$ should be comparable with the first-order nonlinear susceptibility of the corresponding bulk material $\boldsymbol{\chi}^{(2)}$. This assumption necessarily assumes that the crystalline structure is not modified in particles as compared to the bulk crystalline structure. This hypothesis is, however, not necessarily true for systems with nanometre-scale sizes. Inorganic crystals like $LiNbO_3$ have nonlinear susceptibility in the range of about 10^{-7} esu [6]. Therefore, the hyperpolarizability magnitude for particles with a diameter of about 10 nm should be about 10^{-25} to 10^{-26} esu. Several noncentrosymmetrical materials have been investigated, for example CdS, CdSe, ZnS and Fe_2O_3 [32]. All the measured hyperpolarizabilities indeed fall in the range of these expected magnitudes with absolute values of about 1000×10^{-30} esu for particles with a diameter of 2–3 nm. Nevertheless, a large dispersion of the values with respect to the size of the particles is observed. This clearly indicates that other phenomena contribute to the overall magnitude of hyperpolarizability. One contribution to consider is linked to the finite-size effects observed for small particles when their diameter is only a few nanometres large. This is especially true for semiconducting quantum dots for which the bandgap energy is directly related to the size of the particle [33]. In contrast, the quantum confinement in the particle has a direct effect on the oscillator strength of the transition between the valence and the conduction bands. This point can be emphasized by introducing a two-level model, originally proposed for molecular systems [34]. In this simple molecular picture, the hyperpolarizability is given as a function of the transition dipole moment $\boldsymbol{\mu}$, the change in the dipole moment between the ground and the excited states $\Delta\boldsymbol{\mu}$, the bandgap energy E_g and the energy of the incident photon $E_0 = \hbar\omega$. If $\boldsymbol{\mu}$ and $\Delta\boldsymbol{\mu}$ are collinear, the tensor element along this common direction is

$$\beta' = \frac{3\mu^2\,\Delta\mu E_g^2}{2\left(E_g^2 - E_0^2\right)\left(E_g^2 - 4E_0^2\right)} \tag{17}$$

If the fundamental and the harmonic frequencies are both far away from the valence band-to-conduction band resonance frequency, the denominator in **equation 17** is almost constant, independent of the photon energy. This is true if the fundamental and the harmonic energies are both below the bandgap energy. The conversion process is thus essentially nonresonant. In contrast, if the harmonic frequency is above the bandgap energy, the hyperpolarizability is resonantly enhanced, and a dramatic intensity change is expected. In particular, with the fundamental frequency scanned around half the bandgap frequency, the harmonic frequency will be tuned through the resonance, and dramatic changes in intensity will be observed. This latter case has been produced for TiO_2 particles, in agreement with **equation 17**, although the assignment of the corresponding resonances in the SH spectrum was unclear [35]. With the decrease in the size of the particle, the transition dipole moment $\boldsymbol{\mu}$ and the change in the dipole moment $\Delta\boldsymbol{\mu}$ increase owing to the enhanced overlap between the electron and hole wave functions and because of the coherent motion of the exciton. Experimentally, an increase of the hyperpolarizability per unit volume with the decreasing size of the particle has been observed for CdS and CdSe at small diameters [36,37].

3.3.3 Surface Contribution

For very small particles though, with diameters of only a couple of nanometres, the increase of the hyperpolarizablity per unit volume was observed to be larger than expected. This was attributed to a surface contribution arising from the combination of the breaking of the centrosymmetry at the surface of the particle, similarly to macroscopic planar interfaces, and a deviation from a centrosymmetrical shape. At the surface of the particle, the environment of the surface atoms is intrinsically noncentrosymmetrical and, therefore, yields a polarization of the bonds between the atoms. A surface hyperpolarizability results, the value of which is different from that for the bulk. The hyperpolarizability per atomic unit, that is, the hyperpolarizability of the particle divided by the number of atomic units in the particle, is therefore different at the surface of the particle, both in terms of symmetry and magnitude, as compared to that in the volume of the particle. However, if the particle has a centrosymmetrical shape, this surface contribution should vanish. This particular point will be discussed in greater detail below. This contribution can therefore only contribute to the total SH signal if the particle is noncentrosymmetrical. This effect is more prominent for smaller sizes, but it may be of considerable strength. For example, the hyperpolarizability of ZnS particles was determined for particles of diameter 2.5 nm. A value of 2.34×10^{-27} esu was determined corresponding to a value of 1.63×10^{-28} esu for the hyperpolarizability per ZnS unit. In contrast, the corresponding value per ZnS unit determined for the bulk ZnS material is only 0.36×10^{-30} esu, revealing a considerable surface contribution [38]. Another example given in **Figure 5** is the hyperpolarizability per CdSe units

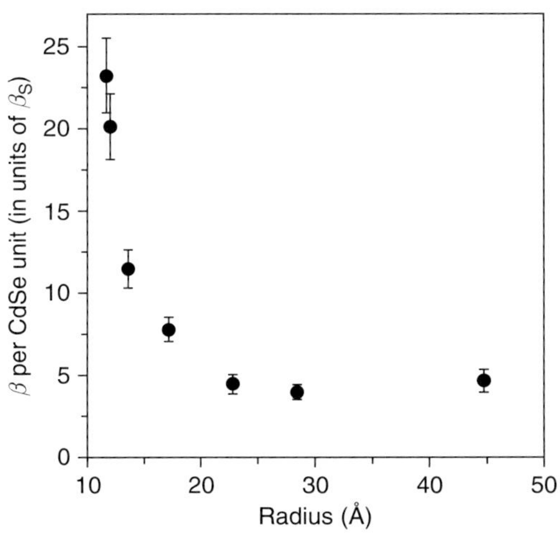

Figure 5 Size dependence of the hyperpolarizability per CdSe units. Reproduced with permission from Jacobsohn M and Banin U (2000) *Journal of Physical Chemistry B* 104:1.

observed for particles with diameters between 1.0 and 5.0 nm [39]. The enhancement of the hyperpolarizability per CdSe units was attributed to a surface contribution, the magnitude of which increases as the particle diameter decreases.

The surface origin to the hyperpolarizability for very small particles is thus often the main contribution and has been discussed in the past in terms of the noncentrosymmetry of the shape of the particle [40]. In the latter approach, the theoretical model was based on the chaotic motion of the electrons in the particle, similarly to a three-dimensional random billiard, and on the asymmetry of the surface of the particle.

Owing to the increase of the surface-to-volume ratio with the decrease of the particle size, a large sensitivity to the environment and the surface properties of the particles are observed through the surface contribution of the hyperpolarizability despite the noncentrosymmetry of these particles. This sensitivity has been the focus of interest of further studies, principally on CdS, ZnS, and CdSe particles. For CdSe, substitution of different adsorbates was performed, nitrothiophenol (NTP) substituting tris-*n*-octylphosphineoxide (TOPO) [37]. Larger HRS signals were observed for NTP-coated particles as compared to TOPO-coated particles for particles with similar diameters. Both compound, NTP and TOPO, possess a nonvanishing hyperpolarizability and, therefore, directly introduce a surface contribution to the overall SH response of the particle. A linear relationship was observed for the difference between the hyperpolarizabilities of the NTP- or the TOPO-coated particles as a function of the number of adsorbing sites. This latter number was taken as half the number of surface atoms of the particle. The linear relationship observed thus supported a surface contribution scaling directly with the number of ligand molecules adsorbed at the particle surface. Because of the steric hindrance, a larger number of NTP molecules were accommodated by the particle surface, possibly all the available sites, which does not occur for the bulkier TOPO ligand. Despite the quantitative determination of the change in the particle hyperpolarizability and the estimated number of surface sites, the exact number of adsorbed ligands could not be inferred. Complete substitution of TOPO by NTP cannot usually be obtained as extensive ligand exchange reactions have shown. Notably, the ligand hyperpolarizabilities could not be determined for the ligand adsorbed on the particle but the order of magnitude of the difference of hyperpolarizabilites between the two ligands could be estimated and compared to the hyperpolarizabilities of the free ligands. The determination of a larger hyperpolarizability for free NTP as compared to free TOPO further supported the analysis. Note also that symmetry considerations had to be introduced to prevent the cancellation of the surface contribution altogether for a centrosymmetrical surface.

Along the same line, a related work was performed with the adsorption of *n*-butylamine at the surface of CdSe particles, a study performed as a function of the CdSe particle size [41]. The interesting feature here is that, despite the vanishing hyperpolarizability of the compound, the adsorption at the surface of CdSe particles introduced a decrease of the particle hyperpolarizability. These experiments demonstrated that even if a direct surface molecular contribution may be vanishing, a contribution arising from an interaction between the ligand and the surface may still appear. Hence, one may write for the microscopic hyperpolarizability

$$\boldsymbol{\beta}_m = \boldsymbol{\beta}_{m,V} + \boldsymbol{\beta}_{m,S} + \boldsymbol{\beta}_{m,I} \tag{18}$$

where the total particle hyperpolarizability $\boldsymbol{\beta}_m$ is the sum of a volume, a surface, and an interaction contribution, respectively labeled $\boldsymbol{\beta}_{m,V}$, $\boldsymbol{\beta}_{m,S}$ and $\boldsymbol{\beta}_{m,I}$, the latter two being in effect two surface contributions. In the former case of the NTP and TOPO adsorption, $\boldsymbol{\beta}_{m,S} \neq 0$ and $\boldsymbol{\beta}_{m,I} = 0$, whereas in the latter case of the *n*-butylamine adsorption at the surface of the CdSe particles, $\boldsymbol{\beta}_{m,S} = 0$ and $\boldsymbol{\beta}_{m,I} \neq 0$. The decrease of the overall hyperpolarizability was attributed to the increase of the local crystalline order upon adsorption of the ligand molecule. A possible phase change can also occur. Hence, in that particular case, the subsequent decrease in hyperpolarizability was attributed to a change from the four-coordinated wurtzite structure for large particles to the six-coordinated zinc blende structure for smaller particles, which introduced a larger order and an increase in symmetry. A similar work has been performed on CdS particles. For this noncentrosymmetrical material, the hyperpolarizability arises from the bulk of the particle. The hyperpolarizability per CdS units is measured to be 16×10^{-30} esu for particles of diameter 2 nm, a value much larger than the estimated value of the bulk susceptibility, which is only 1.3×10^{-30} esu. This difference was attributed to the quantum confinement and the enhanced overlap of the electron and hole wave functions [36]. However, surface effects were also observed for CdS particles, owing to the noncentrosymmetrical nature of the particle surface at the atomic level and its lack of overall centrosymmetry at the level of the particle. This surface contribution was modified through aging of the particles, inducing crystalline re-arrangements, and upon thiourea adsorption, which suppresses any possible dangling bonds [42]. Further evidence was obtained by adsorption of pyridine and of the anion bis-(2-ethylhexyl) sulfosuccinate (AOT). In the case of pyridine, a σ-donation occurs from an adsorbed pyridine towards the Cd^{2+}-rich particle surface, whereas an electrostatic interaction exists for AOT. This difference in the nature of the adsorption process explains the observed smaller hyperpolarizability measured for pyridine-coated CdS particles as compared to AOT-coated CdS particles, although this point requires the assumption of an absence of surface reconstruction. Furthermore, an increase of the surface contribution can also be optimized during the synthesis, as has been shown for $CdS/PW_{12}O_{40}{}^{3-}$ composite particles.

The surface hyperpolarizability as introduced in **equation 18** should also obey the two level model as given in **equation 17**. The validity of this two level model arises from the molecular description of the adsorbates. This feature was indeed observed by modification of the protonation state of the surface layer with a pH change of the liquid solution. This was achieved by coating the CdS particles with thiourea. Cadmium-containing molecular compounds derived from thiourea are known to be rather efficient second-order nonlinear materials. One such example is bis(thiourea) cadmium chloride [43]. Thus, the CdS particles were covered with efficient nonlinear scatterers and these compounds could be described with a simple two level model. The transition energy E_g was displaced by the pH change of the environment leading to a maximum HRS intensity collected at a pH of 4.1. A similar two-level model was also very efficient for the description of other ligand molecules adsorbed at the surface of small particles [44].

3.3.4 Magnetic Particles

Interest in magneto-optical properties of materials has arisen due to the fact that nonlinear magneto-optical Kerr effect, also known as NOMOKE or magnetization-induced second-harmonic generation, exists in magnetic granular films along with giant magneto-resistance (GMR). In the past, nonlinear magneto-optics was mainly restricted to regular films but recent studies have turned their attention to random structures. Magnetization-induced HRS has thus been observed. Static (DC) magnetic fields do not break inversion symmetry in centrosmmetrical bulk materials because they are effectively axial vectors. Instead they break time reversal symmetry. Centrosymmetrical materials therefore do not participate in magnetization-induced SHG within the dipole approximation. Hence, a combination of magnetization and surface inversion symmetry breaking confers to the technique an extreme surface sensitivity. Magnetic granular films (e.g., of the composition Co_xAg_{1-x}, where x is the volume fraction with Co granules with average sizes between 2 and 6 nm) have received a lot of attention recently [45, 46]. This type of material exhibits GMR and SHG for DC magnetic fields below 2.5 kOe. In Co_xAg_{1-x} films, NOMOKE was observed as a function of the volume content x and the magnetic contrast $\rho_{2\omega}$, defined as

$$\rho_{2\omega} = \frac{I_+^{2\omega} - I_-^{2\omega}}{I_+^{2\omega} - I_-^{2\omega}} \tag{19}$$

where $I_+^{2\omega}$ and $I_-^{2\omega}$ are the SHG intensities collected for the two opposite directions of the magnetization, was found to be equal to 0.1 at the fundamental wavelength of 1064 nm. Furthermore, the maximum of the magnetic contrast was observed below the percolation threshold at a volume content of about 0.2–0.25, as for GMR. The contrast was found to have a sharp rise above the percolation threshold obtained at 0.45–0.5 volume fraction, whereas the GMR showed a regular decrease in this regime.

Another granular system investigated was that of yttrium iron garnet $Y_3F_5O_{12}$ (YIG), which presents one of the largest Verdet constant known [47]. The Verdet constant is responsible for the rotation of light in the Faraday effect as well as the magneto-optical Kerr effect. Hence, this material is extensively investigated for potential applications in optoelectronics. In order to define the coherent nature of the response, several experiments were performed (**Figure 6**). First, in-plane isotropy was established for granular YIG films, by recording the intensity as a function of the rotation of the sample around the direction normal to the surface. Then, the granular structure of the material was observed through the diffuse SHG response. Indeed, an SH intensity as large as that observed in specular reflectance was recorded in a direction normal to the surface. The incoherent nature of the SHG response, that is, its HRS character, was assessed by recording the SH intensity as a function of the number of layers of YIG deposited on the substrate. This was observed for both the p-in/p-out and p-in/s-out polarization configurations where -in indicates the polarization state of the incoming fundamental wave and -out the polarization state of the outgoing harmonic wave. As seen in **Figures 6(c)** and **6(d)**, a linear dependence is observed which indicates incoherence – and not coherence. Hence, these results suggest that the origin of the response stems from the breaking of the centrosymmetry at the surface of the particles. The shape of these particles being noncentrosymmetrical and these particles being randomly distributed, the SH intensity scales linearly with the number of particles, or layers in the experiments described, a signature of HRS. The diffuseness and depolarization character of the SH response is also in agreement with this general picture. Finally, **Figure 6(e)** underlines the mixed character of the total response from these YIG granular films. Inversion of the direction of the applied DC magnetic field reverts the sign of the magnetization-induced contribution to the field amplitude but does not affect the remaining contributions arising from the matrix or the YIG particles.

The total HRS intensity is therefore of the general form

$$I_\alpha^{2\omega} \propto \left| E_\alpha^{2\omega} + E^{2\omega} \right| \tag{20}$$

where $\alpha = 0, +$ or $-$ depending on the direction of the applied DC magnetic field. The first contribution is induced by the magnetization, whereas the second one is not. As observed in **Figure 6(e)**, the slope of the linear plots of the SH intensity as a function of the YIG layers depends on the value of the parameter α. This feature indicates that both contributions in **equation 20** are linear functions of the number of YIG particles. Hence, interferences must occur at the level of each single particle with a mean diameter of 32 nm. This is the reason why this behavior has been termed 'internal homodyne detection' of the magnetization-induced contribution.

These systems where a DC magnetic field may be applied are interesting to investigate the transition between incoherent and coherent SHG. In a fluid where the magnetic particles are initially randomly distributed, the isotropy may be broken upon application of a DC magnetic field if the

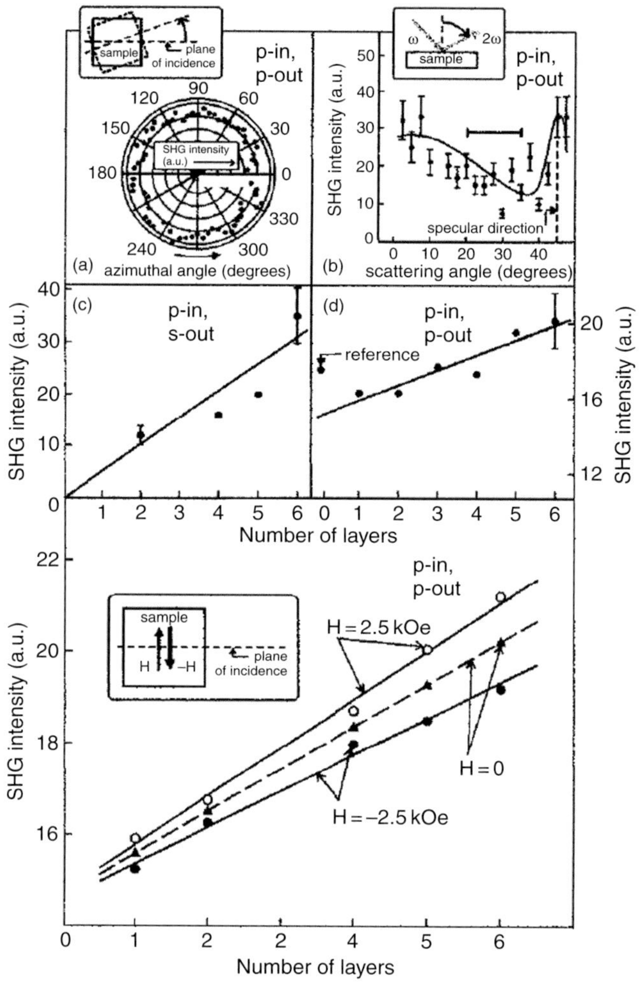

Figure 6 (a) Dependence of the SHG intensity from a six-layer film on the azimuthal angle measured as shown in the inset. (b) Dependence of the SHG intensity from a six-layer film on the scattering (polar) angle measured as shown in the inset. (c–d) Dependence of SHG intensity on the number of YIG-containing layers in the (c) p-in, s-out and (d) p-in, p-out geometries. (e) Dependencies of the SHG intensity on the number of YIG-containing layers at three values of the DC magnetic field applied as shown in the inset, circles and triangles: experimental data, lines approximations, see text. Reproduced with permission from Murzina TV, *et al.* (2001) *Applied Physics Letters* 79: 1309.

particles are sufficiently free to move, as opposed to the previous case where the particles embedded in the matrix possess fixed positions and orientations. These criteria were achieved in an aqueous suspension of cobalt ferrite particles $CoFe_2O_4$ with a mean diameter of 14.1 nm, the volume fraction being about 6% [48]. The breaking of the centrosymmetry was achieved with reasonable magnetic fields of less than 0.3 T. The observation of the transition was made possible by the large value of the product KV, where K is the magnetic anisotropy constant and V the particle volume. This product was as large as 100 times kT at room temperature, kT being the product of the Boltzmann constant and the temperature. This high-magnetic energy barrier prevented the particles to revert their magnetic moment without rotating their orientation, (**Figure 7**).

No incoherent signal was observed from the cell at vanishing DC magnetic field applied, as indicated by the vanishing SHG intensity when no magnetic fields were applied (see **Figure 7**).

Furthermore, the number of particles in the solution was kept constant, and thus saturation of the SH intensity was observed once the alignment of all the magnetic dipoles supported by the

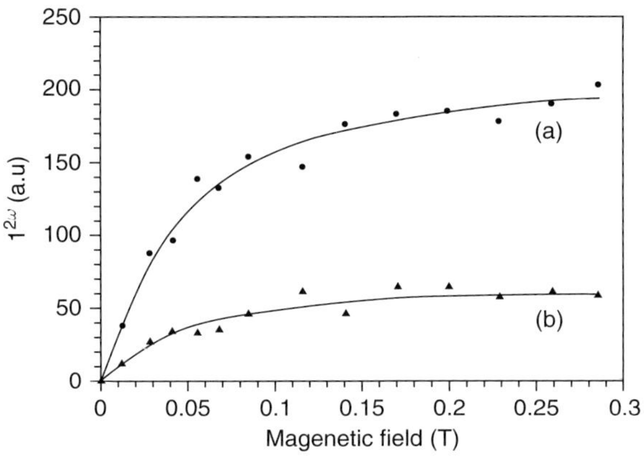

Figure 7 Dependence of the SHG intensity on the applied magnetic field for two different geometries: $\theta_\omega = \theta_{2\omega} = 0°$ and $\theta_\omega = 45°$, $\theta_{2\omega} = 90°$. Reproduced with permission from Lenglet J, *et al.*, (1996) *Physical Review* 53: 14941.

particles was achieved. Not surprisingly, these data were supported by a simple Langevin-like model.

3.4 Particles from a Centrosymmetrical Material

3.4.1 Particles with Noncentrosymmetrical Shape

We now focus on the description of the SH response from particles of centrosymmetrical materials. Metallic particles like silver and gold particles fall in this class of particles because gold and silver crystalline lattice structures are both cubic face-centered and, therefore, centrosymmetrical. Other metals have also been investigated like copper and platinum. In these considerations, we do not consider crystalline defects that would lead to the breaking of the symmetry of inversion in the bulk of the particle. We assume that these defects are either absent or present in a large number and randomly oriented in the volume. In these media possessing the property of inversion symmetry, no SHG takes place in the electric dipole approximation. We are led to consider that the nonlinear response arises from higher orders of the approximation: an electric quadrupole and magnetic dipole bulk response and an electric dipole surface response. In this section, we assume that the bulk volume response is negligible and that the electric dipole surface response dominates. In the next section, we will consider the opposite case before investigating real cases, where both contributions are present.

At the surface of the particle, the inversion symmetry of the bulk material is broken and, therefore, an electric dipole surface second-harmonic response is allowed. This case has already been mentioned above, and is very similar to previous studies performed at metal surfaces. These studies have essentially been performed so far at planar surfaces or surfaces with large macroscopic radii of curvature [7–9]. In the case of small particles, the radii of curvature and the exact shape of the particles must be considered, especially from the point of view of the ratio of the characteristic size of the structure with respect to the wavelength of light and the symmetry of the shape of the structure. For curvatures of the order of the wavelength of light, retardation effects may play a nonnegligible role. Here, retardation is defined as the spatial dependence of the electromagnetic fields on a distance of a wavelength. Furthermore, the symmetry of the shape of the particle is critical, because it may lead to an overall cancelation of the different surface contributions of the nonlinear polarization. For example, spherical surfaces with a curvature of the order of or close to the micron size have been investigated recently for molecular systems like liposomes and exhibit coherent SHG [49]. The general expression for the field amplitude, as given above in **equation 12**, is no longer adequate and must be rewritten as

$$E(\boldsymbol{r},\Omega) = (E_0)^2 \int_V \frac{\exp\left[ik^{(\Omega)}|\boldsymbol{r}-\boldsymbol{r}'|\right]}{|\boldsymbol{r}-\boldsymbol{r}'|}[\hat{n}\times(\mathsf{T}(\hat{r}')\boldsymbol{\beta}' :\hat{e}\hat{e})] \times \hat{n}\exp\left[-2i\left(\omega t-\boldsymbol{k}^{(\omega)}\cdot\boldsymbol{r}'\right)\right]\mathrm{d}V \quad (21)$$

Expansion of the exponential functions when assuming particle sizes much smaller than the wavelength of light yields a factor of the form $1 + i\Delta\boldsymbol{k}\cdot\boldsymbol{r}'$ in the integral over the particle volume, where the wave vector mismatch between the fundamental and the harmonic waves is given by $\Delta \boldsymbol{k} = 2\boldsymbol{k}^{(\omega)} - \boldsymbol{k}^{(\Omega)}\hat{\boldsymbol{n}}$. **Equation 12** is easily recovered when the second term of the exponential expansion is neglected. It is straightforward to see that, for small particles, $\Delta \boldsymbol{k} \cdot \boldsymbol{r}' << 1$. The microscopic nonlinear dipole $\boldsymbol{p}(\Omega)$ induced in the particle and already given in **equation 8** reduces to its surface integral as given in **equation 9**. Within the volume, the integral vanishes owing to the centrosymmetry property of the medium and we neglect higher-order contributions. Keeping the electric dipole approximation, retardation is simply introduced by reinstating the second term $\mathrm{i}\Delta \boldsymbol{k} \cdot \boldsymbol{r}'$ in **equation 21**. This term may only start to take nonnegligible amplitudes for particles with diameters no longer small compared to the fundamental wavelength. Without field retardation, the surface response can only subsist if the particle shape is noncentrosymmetrical. It is added here that a centrosymmetrical shape combined with an inhomogeneous surface distribution of the microscopic nonlinearity also yields a noncentrosymmetrical shape. This case may practically be achieved, for example, with inhomogeneous adsorption at the surface of particles. Assuming that the electric dipole surface response dominates, and neglecting the field retardation because the particles are small (i.e., the limit $\Delta \boldsymbol{k} \cdot \boldsymbol{r}' << 1$ holds), the problem is very similar to that of the particles made from noncentrosymmetrical material apart from the surface origin of the response. The SHG response of the particle will thus scale with the surface area of the particle rather than its volume. With the field retardation included, a dependence with the volume of the particle starts to appear. Before discussing real cases, we discuss the case of particles with a centrosymmetrical shape, where the electric dipole approximation for the response is no longer sufficient.

3.4.2 Particles with Centosymmetrical Shape

For small particles of a centrosymmetrical material and possessing a centrosymmetrical shape, the origin of the second-harmonic response can no longer be of pure electric dipole origin without incorporation of retardation. Indeed, within the electric dipole approximation, the volume nonlinear polarization vanishes because of the centrosymmetry of the crystal structure of the material, and the surface nonlinear polarization as discussed above also vanishes because of the centrosymmetry of the shape of the particle. The theoretical developments must then be taken to the next order of expansion of the intensity of the harmonic light in terms of the ratio of the particle radius and the wavelength of light, namely a/λ, where a is the characteristic size of the particle. For spherical particles, this quantity is simply the diameter of the particle.

In going beyond the electric dipole approximation, we have to disentangle carefully the multipole expansion of the fields for both the exciting incoming fundamental field and the radiated outgoing harmonic field and the multipole expansion of the nonlinear polarization. For the fields, the multipole expansion may be understood, within the spherical symmetry for instance, as the expansion of the fields as derived by G. Mie in the case of scattering by small spherical particles [3, 4]. For small particles, the zeroth-order term will be sufficient in most cases: this is the quasi-static approximation. For larger particles, an increasing number of terms will have to be added in the expansion as the ratio a/λ increases, and we will always consider only the first-order correction term. For the nonlinear polarization, a multipole expansion must be performed as well. At the excitation, the nonlinear polarization involves a local and a nonlocal contribution. The local excitation contribution depends on the field value at the location considered and corresponds to the electric dipole approximation expression. The nonlocal one arises from the excitation through field gradients occurring either at the surface of the particle, owing to the passage across the interface from the outer surrounding medium into the particle medium, or within the bulk of the particle owing to the field retardation. The surface field gradient contributions may be recast within the surface quadratic susceptibility and, therefore, vanish for a centrosymmetrical shape without retardation. We are left to consider the possibility of volume field gradients. A nonlinear polarization arising from these gradients has already been recognized in the past when considering metal spheres. Agarwal and Jha [50] were among the first to investigate the problem of the SHG response from metallic spheres in 1982, in relation with the contemporary observations of surface-enhanced Raman scattering (SERs) measurements. Their treatment was based on a nonlinear polarization earlier developed for planar metal surfaces within a free-electron model [26, 51, 52]. They introduced the following nonlinear polarization, consisting of two contributions:

$$p(r', 2\omega) = \gamma' \nabla \left[E^2(r', \omega)\right] + \delta'(E(r', \omega)\cdot\nabla)E(r', \omega) \tag{22}$$

A third one incorporating the divergence of the fundamental field vanishes owing to Maxwell equation in a homogeneous medium.

At the harmonic frequency, we must consider the radiation phenomenon. Besides the electric dipole contribution to the radiation, we add the electric quadrupole and the magnetic dipole contributions to incorporate the next terms of the expansion with the parameter a/λ. Agarwal and Jha found a power radiated at the harmonic frequency over all solid angles scaling with the square of the volume of the particle [50]. They also observed that the intensity could be resonantly enhanced through an electric dipole and an electric quadrupole resonance at the fundamental and the harmonic frequency. An l-type resonance frequency is determined by

$$l\varepsilon(\omega) + (l+1)\varepsilon_m = 0 \tag{23}$$

where $\varepsilon(\omega)$ is the complex dielectric function of the nonlinear material and ε_m that of the surrounding medium. The electric dipole resonance is obtained with $l = 1$ and the electric quadrupole resonance with $l = 2$. Hence, the resonances directly arise from the retarded and nonretarded contributions appearing at both frequencies. In metallic particles, such as gold and silver particles, this result is expected to yield large enhancements through the SP resonances, in a way similar to the linear optical case. Hua and Gersten [51] developed a similar model starting with the same expression of the nonlinear polarization, but they solved the problem within the hydrodynamic model of the electronic density in metals. Östling *et al.* [52] proposed an approach with the accent put on the oscillating surface charges. In the description of the response, a formalism introducing the effective dipole and quadrupole contributions has been proposed by Dadap *et al.* [53, 54]. The first contribution is an effective electric dipole, p_{eff}, and the second one an electric quadrupole $\mathsf{Q}_{\mathrm{eff}}$. The SH intensity then takes the form

$$E(\Omega, r) = \frac{k_1^2(\Omega)\exp[ik_1(\Omega)r]}{\varepsilon_1(\Omega)r} \times \left[\hat{n} \times \left(p_{\mathrm{eff}} - \frac{ik_1(\Omega)}{6}\mathsf{Q}_{\mathrm{eff}}(\hat{n})\right)\right] \times \hat{n} \tag{24}$$

where $\mathsf{Q}_{\mathrm{eff}}(\hat{n})$ stands for the quadrupole projected onto the direction of observation $\hat{n}$. For the effective dipole p_{eff}, an unusual orientation along the direction of propagation is observed, which yields a well-defined pattern as a function of the angle of polarization of the incoming fundamental wave. Furthermore, because of its nature, the contribution to the second-harmonic response originating from p_{eff} has a dipole-type resonance defined by the condition $\varepsilon(\Omega) + 2\varepsilon_m = 0$ at the harmonic frequency. The contribution arising from the electric quadrupole $\mathsf{Q}_{\mathrm{eff}}(\hat{n})$ has a resonance frequency determined by the condition $2\varepsilon(\Omega) + 3\varepsilon_m = 0$ at the harmonic frequency. SP resonances also appear at the fundamental frequency, but they are often avoided in order to preserve the samples from laser-induced degradation. However, they have to be incorporated into the discussion in some cases, for example, that of metallic particle aggregation [55]. Also, p_{eff} and $\mathsf{Q}_{\mathrm{eff}}(\hat{n})$ both exhibit an a^3 dependence.

The patterns of the harmonic signal intensities with the input polarization angle are unusual owing to the orientation of the effective dipole p_{eff} and quadrupole $\mathsf{Q}_{\mathrm{eff}}(\hat{n})$. Using an experimental geometry where the fundamental wave propagates along the Oz axis and the direction of collection of the scattered light is taken along the $\hat{n} = \hat{y}$ axis, the model yields [53, 54]

$$p_{\mathrm{eff}} = \alpha_1 \hat{z} \tag{25a}$$

$$\mathsf{Q}_{\mathrm{eff}}(\hat{y}) = \alpha_2 \sin\gamma(\cos\gamma\hat{x} + \sin\gamma\hat{y}) \tag{25b}$$

where α_1 and α_2 are two complex numbers. Using **equation 24**, the HRS intensity becomes [56]

$$I_{\mathrm{HRS}}^{\mathrm{V}} = K|\alpha_2|^2 \sin^2 2\gamma\, I^2 \tag{26a}$$

$$I_{\mathrm{HRS}}^{\mathrm{H}} = K|\alpha_1|^2\, I^2 \tag{26b}$$

where K is a constant and I is the intensity of the fundamental incoming wave. For harmonic light polarized vertically, the HRS intensity $I_{\mathrm{HRS}}^{\mathrm{V}}$ exhibits a bell shape that vanishes for $\gamma = 0$ and for $\gamma = \pi/2$. For harmonic light polarized horizontally, the HRS intensity $I_{\mathrm{HRS}}^{\mathrm{H}}$ is constant, irrespective of the input polarization angle. We note further that, because the two contributions arise from the same order of the multipole expansion in a/λ, the two parameters α_1 and α_2 should be of similar magnitude, except near resonances since they are not enhanced for the same resonance frequencies. A general picture of the radiation patterns is given in **Figure 8**.

With the last two sections devoted to particles with noncentrosymmetrical and centrosymmetrical shapes, two perfect cases have been described. It is

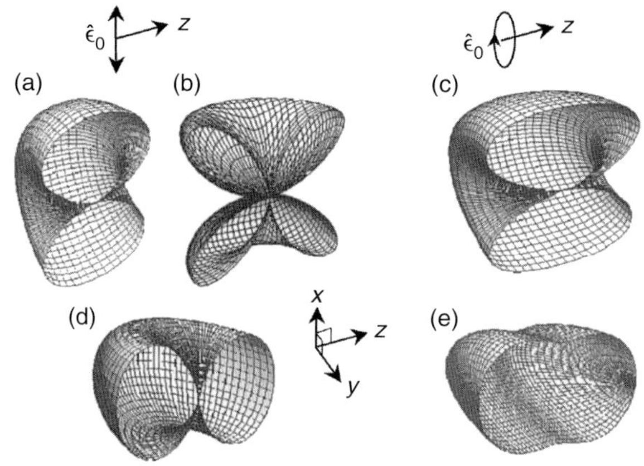

Figure 8 Graphical representation of the scattering pattern for (a–c) hyper Rayleigh scattering and (d, e) linear Rayleigh scattering: (a) pure electric dipole and (b) pure electric quadrupole radiation pattern upon linear polarization excitation; (c) pure electric quadrupole radiation pattern upon circular polarization excitation. For comparison the radiation patterns for (d) linear and (e) circular polarized excitations are also shown. Reproduced with permission from Dadap JI, *et al.* (2004) *Journal of the Optical Society of America* B21: 1328.

rather obvious that it is difficult to synthesize particles with perfectly centrosymmetrical shapes, without defects. Hence, if the shape is close to the centrosymmetrical one, it will not dominate and a competition between the two cases, pure electric dipole arising from the surface of the particle and higher orders arising from both the surface and the volume, will occur. Small gold and silver metallic particles exhibit such a competitive behavior in their SHG response.

3.5 Metallic Particles

3.5.1 Theoretical Approach

Metallic particles constitute the main class of centrosymmetrical particles studied. Their response has been investigated in great detail owing to the possibility of resonantly enhancing the SHG response in the visible range of the spectrum through the SP resonances. The origin of the response as well as the size and composition dependence are now discussed in light of the previous sections and the large availability of data in the literature. We will close the section with a discussion of the applications developed by using their nonlinear optical response, principally in analytical sciences.

We are interested in a first stage in the description of the SH response from metallic particles with minimum restrictions on their shape or their size. Therefore, in agreement with the previous sections, we assume that a surface and a bulk term contribute to the overall response. We first argue that the surface contribution is derived from a local surface hyperpolarizability tensor. Here, we disregard the possibility of surface anisotropy arising from crystalline faces like the common (001), (110), and (111) single-crystal faces of metal surfaces. The elements of the local tensor vanish according to the local surface symmetry of the particle surface, and can be found in the tables provided by classical textbooks [5,6]. If an isotropic surface is considered, the local direction normal to the surface must be distinguished from the two in-plane axes, the latter two axes being undistinguishable. Thus, the local hyperpolarizability tensor possesses only three independent elements, namely β'_{zzz}, β'_{zxx} and β'_{xzx}, and seven nonvanishing elements, namely β'_{zzz}, β'_{zxx}, $\beta'_{xxz} = \beta'_{xzx}$, β'_{zyy}, $\beta'_{yyz} = \beta'_{yzy}$, with $\beta'_{xxz} = \beta'_{yyz}$ and $\beta'_{zxx} = \beta'_{zyy}$. For metallic surfaces, those nonvanishing and independent elements have been recast into three microscopic parameters, also named the three Rudnick and Stern parameters *a*, *b*, and *d*, corresponding to the three nonlinear currents induced at the harmonic frequency, respectively, two surface currents perpendicular and parallel to the surface and one volume current perpendicular to the surface [57]. Theoretical expressions for these parameters are known within some approximations and for perfect surfaces [58]. It is also debatable whether the surface nonlinear polarization is located inside or outside the particle. This problem has been discussed for planar interfaces and a similar discussion could be developed here. Besides, for metallic particles, the polarization field inside the particle is nonnegligible and the local exciting field should be taken as the superposition of the incoming and the polarization fields.

The problem of the SH radiation from small metallic particles is, therefore, split into two fundamental steps: excitation at the fundamental frequency and radiation at the harmonic frequency, the two being linked through the nonlinear polarization. At the excitation stage, the fundamental field inside the particle is the superposition of the exciting field and the polarization field. For spherical shapes, the problem has been solved by G. Mie using harmonic spherical vectors, making use of the spherical symmetry of the problem[3,4]. The electromagnetic fields can be expanded in a series of spherical harmonic vectors $Y_{llm}(\hat{r}')$ of increasing orbital numbers l and m, where $\hat{r}'$ indicates the angular coordinates only. The problem does not present any difficulties in the spherical geometry, but for other shapes of particle, the problem gets more complicated and possibly no longer analytical. Formally, the incoming fundamental plane wave can be expanded according to its multipole expansion into spherical waves. For the first two terms, this yields

$$E(r', \omega) = D(\omega) + xQ(r', \omega) \tag{27}$$

to first order in $x = a/\lambda$. The first term of **equation 27** is the electric dipole field. This field is homogeneous over the whole particle volume, which is the reason why the spatial dependence has been dropped. The second term, linear in the parameter $x = a/\lambda$, corresponds to the electric quadrupole term. Once we have defined the field expansions, the nonlinear polarization may be built. Since we wish to incorporate both the surface and bulk contributions, according to the previous sections, we write

$$p(r', 2\omega) = \beta' E(r', \omega)E(r', \omega) + \gamma' \nabla\left[E^2(r', \omega)\right] + \delta'(E(r', \omega)\cdot\nabla)E(r', \omega) \tag{28}$$

The first term is the usual electric dipole excitation term. It vanishes within the bulk volume of the particle but may subsist at the surface if the spherical symmetry is broken. A careful description of the radiation fields in terms of the different terms in **equation 28** is cumbersome. Several authors have already introduced more formal descriptions in the past, and we follow this approach. The nonlinear polarization is developed according to the fundamental field expansion but we take into account the electric dipole, electric quadrupole, and magnetic dipole contributions. Hence, the nonlinear polarization writes as

$$p(r', 2\omega) = \beta^{dd\prime} D(\omega)D(\omega) + x\beta^{dq\prime} D(\omega)Q(r', \omega) + x\beta^{dm\prime} D(\omega)M(r', \omega) \tag{29}$$

to first order in x, where effective quadratic hyperpolarizabilities $\beta^{dd\prime}$, $\beta^{dq\prime}$, and $\beta^{dm\prime}$ have been introduced. At the radiation stage, each single point r' radiates a spherical SH wave and the total field amplitude collected at the harmonic frequency at a position r is given by the coherent superposition of these spherical waves. Hence, the radiated field can also be decomposed into a multipole series expansion to account for the nonvanishing size of the particle. At the first order in x, it will be decomposed into an electric dipole, an electric quadrupole, and a magnetic dipole term. Also, the polarization field at the SH frequency must not be forgotten. Hence, the scattered field at the harmonic frequency far from the particle is linked through the boundary conditions to the superposition of the radiated field and the polarization field inside the particle. Finally, the total field amplitude at the harmonic frequency formally writes

$$\begin{aligned} E(\Omega, r) = {} & E_{\mathrm{Lm}}^{\mathrm{Ddd}} D(\Omega)D(\omega)D(\omega) + x\Big[E_{\mathrm{Lm}}^{\mathrm{Ddq}} D(\Omega)D(\omega)Q(\omega) \\ & + E_{\mathrm{Lm}}^{\mathrm{Ddm}} D(\Omega)D(\omega)M(\omega) \\ & + E_{\mathrm{Lm}}^{\mathrm{Qdd}} Q(\Omega)D(\omega)D(\omega) \\ & + E_{\mathrm{Lm}}^{\mathrm{Mdd}} M(\Omega)D(\omega)D(\omega)\Big] \end{aligned} \tag{30}$$

where the E_{Lm} parameters stand for the corresponding field amplitudes. **Equation 30** is general, although formal, and does not provide clear insights into the detailed microscopic origin of the response. Nevertheless, it allows us to discuss the data available for metallic particles within a general framework. It is made of a first contribution, purely local, and a second contribution, itself made of several terms where retardation is only included once, either at the fundamental or the harmonic frequency. For perfectly centrosymmetrical particles, centrosymmetrical both in their crystal structure and their shape, the first term in **equation 30** vanishes. In all other case, a balance between these purely local and the nonlocal contributions is established. Finally, it is also emphasized that the case of particles of noncentrosymmetrical material, as discussed above, is described by keeping only the first term since only this term dominates. This term would, however, be volume dependent. The model developed above is similar to that proposed by Dadap *et al.* [54]. Contributions with retardation effects at both the fundamental and the harmonic frequencies would be of higher orders in the parameter x. As we shall see later, the different contributions present in **equation 30** can be distinguished, to some extent, through their angular and polarization patterns.

3.5.2 Origin of the SH Response

Experimental work on the SH response of metallic particles has been performed in the past, with the report of large quadratic hyperpolarizabilities, in the range of 10^{-23} esu for silver particles with diameters of about 50 nm [59–61]. Recently, experiments have focused on the origin of their SH response. One of the first challenging questions relates to the possibility of distinguishing the purely local contribution from the retarded contributions. This question was answered through a set of experiments where the dependence of the logarithm of the square root of the HRS intensity was reported with respect to the particle diameter. The reason for this analysis is that a purely local response, as given first term in **equation 30**, scales with the surface area of the particle as opposed to its volume, as initially proposed. On the opposite, because of retardation, any response stemming from the nonlocal contributions scales with the volume of the particle. These experiments, as reported in **Figure 9**, were performed through HRS because of the requirement of a homogeneous environment to prevent any influence of a substrate [62]. The presence of a substrate can indeed modify the response of the particles with the breaking of the centrosymmetry. The slope of the line fit of the square root of the HRS intensity, effectively a quantity scaling linearly with the quadratic hyperpolarizability, with the radius of gold particles suspended in an aqueous solution and the diameter of which was ranging between 20 and 150 nm, was found to be 1.9 ± 0.2. This result demonstrates that the origin of the SH process in these particles is of electric dipole origin, stemming from the breaking of the centrosymmetry of the particle shape. The analysis of the HRS intensity as a function of the angle of polarization of the fundamental wave was also performed for these particles. **Figure 10** reports the polar plot of the HRS intensity as a function of this angle of polarization for aqueous suspensions of 20 and 80 nm-diameter gold particles for the output SH light polarized vertically. The observed pattern on **Figure 10(a)** for the 20 nm-diameter particles is in agreement with the expected pattern of a suspension of point-like sources, like noncentrosymmetrical molecules. Hence, small particles, since they are not fully centrosymmetrical because of their shape, behave like noncentrosymmetrical objects.

For larger diameters, the retarded contributions are no longer negligible. As a result, these contributions start to dominate, as seen on **Figure 10(b)**. The dominating retarded contribution corresponds to an electric quadrupole radiation pattern. These results confirm that the origin of the SH process in small metallic particles arises from the surface of the particle owing to the deviation of the particle shape from that of a perfect sphere and that, for larger diameters, retardation must be considered. These results underline in particular the sensitivity of the SH process to

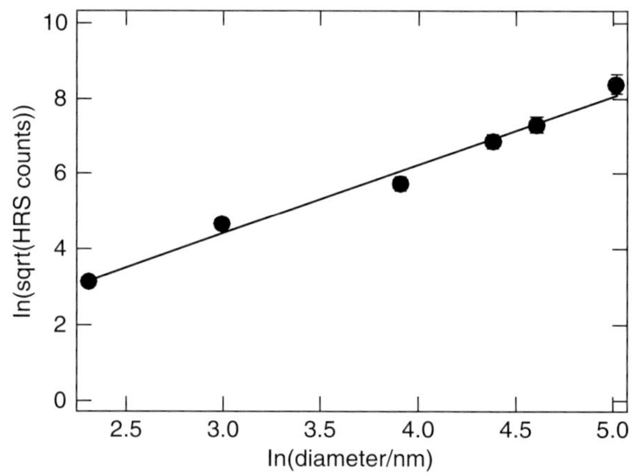

Figure 9 Logarithmic plot of the square root of the HRS intensity, recorded for both the input fundamental and the output harmonic intensities polarized along the *Ox* direction, versus the diameter of the particles. The slope is 1.9 ± 0.2, in agreement with a surface origin of the signal. Reproduced with permission from Nappa J, *et al.* (2005) *Physical Review B* 71: 165407.

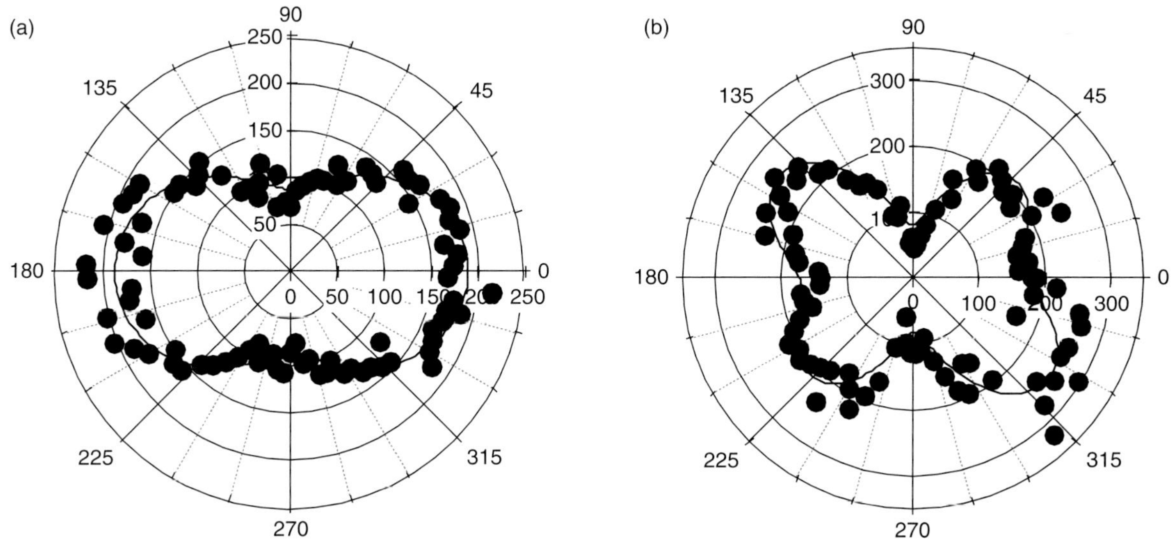

Figure 10 Polarization pattern, given as a polar plot of the HRS intensity as a function of the angle of polarization of the fundamental wave for an aqueous suspension of (a) 20 nm-diameter and (b) 80 nm-diameter gold particles for vertically polarized SH scattered light. Reproduced with permission from Nappa J, *et al.* (2005) *Physical Review B*, 71: 165407.

the breaking of the centrosymmetry. The polarization patterns reported for HRS intensities collected at right-angles from the excitation can be recast in general as

$$I_{\mathrm{HRS}}^{\Gamma} = a^{(\Gamma)}\cos^4\gamma + b^{(\Gamma)}\cos^2\gamma\sin^2\gamma + c^{(\Gamma)}\sin^4\gamma \qquad (31)$$

as a function of the polarization angle of the incoming fundamental wave. In **equation 31**, Γ stands for the polarization of the scattered harmonic intensity. The pure case of noncentrosymmetrical particles with vanishing size requires $b^{(\mathrm{V})} = a^{(\mathrm{V})} + c^{(\mathrm{V})}$, whereas the case of pure centrosymmetrical particles requires $a^{(\mathrm{V})} = c^{(\mathrm{V})} = 0$ for the SH intensity polarized vertically. Hence, defining the sparameter $\zeta^{(\mathrm{V})}$ as

$$\zeta^{\mathrm{V}} = \left|\frac{b^{(\mathrm{V})} - a^{(\mathrm{V})} + c^{(\mathrm{V})}}{b^{(\mathrm{V})}}\right| \qquad (32)$$

we build a parameter that quantitatively assesses the weight of the dipole or quadrupole character of the SH response, (see **Figure 11** for gold and silver particles).

Interestingly, it appears that, for silver particles, the weight of the quadrupole response is larger than in the case of gold particles. The reason for this is that

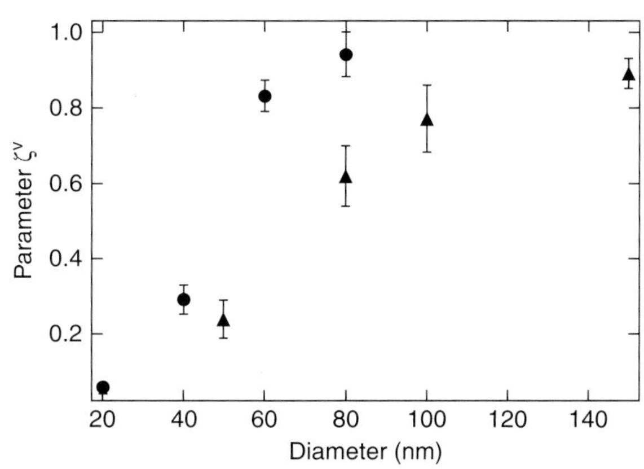

Figure 11 Plot of the parameter ζ^{V} as a function of the diameter of the (circles) silver metallic particles at 390 nm and (triangles) gold metallic particles at 400 nm. Reproduced with permission from Russier-Antoine I, *et al.* (2007) *Journal of Physical Chemistry C* 111: 9044.

retardation scales with the parameter $x=a/\lambda$. Strictly, the wavelength, and not only the diameter of the particles, must be considered within the medium. Hence, retardation appears at smaller particles sizes in the case of silver as compared to the case of gold.

3.5.3 Resonance Enhancement

Gold and silver metallic particles are expected to have rather large quadratic hyperpolarizability magnitudes because of the possibility of resonance enhancements through SP excitations. This possibility has been clearly identified and discussed above. Electric dipole and electric quadrupole fields are therefore, eventually enhanced through the corresponding resonances, the conditions for which are given by **equation 23**. These resonances can appear at the fundamental or the harmonic frequency. Such resonance enhancements have indeed been observed for the SH intensity collected from particles deposited on surfaces [63]. However, in that particular case, a breaking of the particle symmetry is induced by the substrate and the weak static polarization of the particles leads to a pure electric dipole origin for the signal. In solutions, the surrounding medium can be considered as homogeneous and modeled through a continuous dielectric function ε_m.

Primarily, gold and silver metal particles have been investigated, although some authors have reported hyperpolarizabilities for other metal particles [64,65]. The magnitude of their hyperpolarizability has been reported for several sizes [59]. Not surprisingly, platinum particles did not give any detectable signal, a feature attributed to the absence of an SP resonance in the vicinity of the excitation frequency. Because the HRS intensity from metallic particles may be enhanced through SP resonances, wavelength-dependence measurements were also performed. These experiments allowed for a direct confrontation of the theoretical models with the experimental data. Wavelength dependence of the HRS intensity for metallic particles dispersed in a homogeneous environment has been reported for 32 and 40 nm-diameter silver particles (see, e.g., **Figure 12**) [66,67]. The spectrum clearly displays the electric dipole and the electric quadrupole SP resonances. According to the above discussion, this indicates that, in these 32 nm-diameter particles, retardation effects are already present but nonnegligible. Furthermore, the spectral location of the two SP resonances is in agreement with the conditions $\varepsilon(\Omega)+2\varepsilon_m=0$ and $2\varepsilon(\Omega)+3\varepsilon_m=0$. In **Figure 12**, the continuous curves are results from theoretical models adjusted to the experimental data points for two different ratios of the weights A and B determined by [66,67]

$$I_{HRS}=\frac{AC_A(\gamma)}{|\varepsilon(\Omega)+\varepsilon_m|^2}+\frac{BC_B(\gamma)}{|2\varepsilon(\Omega)+3\varepsilon_m|^2} \tag{33}$$

where the functions $C_A(\gamma)$ and $C_B(\gamma)$ of the fundamental polarization angle must be introduced to account for the integration of the total intensity

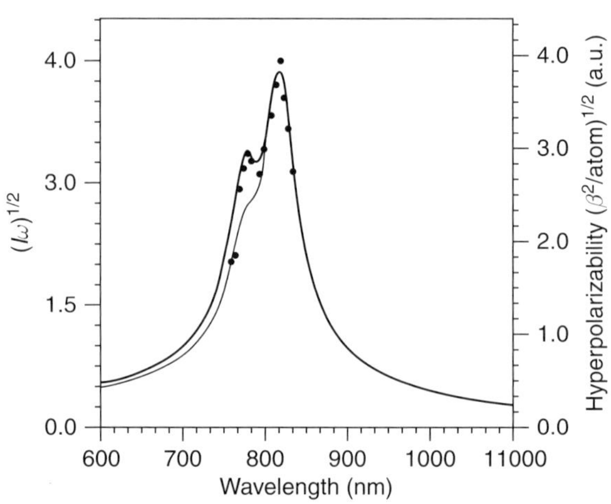

Figure 12 Wavelength dependence of the HRS intensity for an aqueous solution of silver particles the diameter of which is 32 nm for a disk solid angle. Solid lines are (1) fit to the model with $A=1$ and $B=6.2$ exhibiting two resonances, and (2) Agarwal and Jha's model exhibiting only a shoulder to the main resonance. Reproduced with permission from Hao EC, *et al.* (2002) *Journal of Chemical Physics*, 117: 5963.

detected over the solid angle. The adjustment of the model yields an idea of the different contributions to the total SH scattered intensity. However, it has to be noted that according to the discussion above, the first term contains the purely local contribution as well as the nonlocal contribution at the fundamental frequency, whereas the second term only contains the nonlocal contribution at the harmonic wavelength.

In **equation 33**, the expression is valid for a detection at right-angles from the direction of propagation of the incoming exciting wave. The ratio of the two parameters A and B can, however, be modified using a slit arrangement. This possibility arises from the different angular patterns exhibited by the electric dipole and electric quadrupole contributions to the total SH intensity. The slit indeed modifies the solid angle of the collection.

A similar wavelength dependence of the HRS intensity was reported for an aqueous suspension of the smaller 12 nm-diameter gold particles and has yielded only one single electric dipole SP resonance [67]. For these particles, only the electric dipole SP resonance is expected, because no retardation occurs for such small sizes. Hence, in **equation 33**, the coefficient B vanishes and the first term involving the purely local contribution remains. It is interesting to note also that the SP resonance frequency differs if it is of electric dipole or quadrupole origin. Hence, for particles large enough, it is possible to select the type of enhancement through the tuning of the fundamental frequency. This was achieved as reported in **Figure 13**, where for 40 nm-diameter silver particles, an electric dipole enhancement was obtained with the fundamental wavelength of 820 nm, whereas an electric quadrupole enhancement was obtained with a fundamental frequency of 780 nm. Interestingly, this yielded two different polarization patterns [68].

In experiments investigating resonance enhancement, an analysis of the intensity as a function of the fundamental frequency or wavelength is highly desirable. This is, however, not easily possible if the laser source does not provide tunability or provides too narrow a range of frequencies. This problem was circumvented in a set of experiments where instead the SP resonance of the particles was tuned by changing the metal composition. Gold particles possess their SP resonance at a wavelength of about 520 nm, whereas silver ones have their resonance at about 400 nm when dispersed in aqueous suspensions. Synthesis of bimetallic gold–silver particles allowed the linear tuning of the SP resonance from 520 nm down to 400 nm by adjusting the metal composition of the particles, see **Figure 14(a)** and inset therein [69]. Similar experiments were then performed in HRS to investigate the behavior of the hyperpolarizability as a function of the metal composition. This work was motivated by the surface origin of the response from small metallic particles. For small enough particles, whereas the absorption cross-section scales with the particle volume, the quadratic hyperpolarizability scales

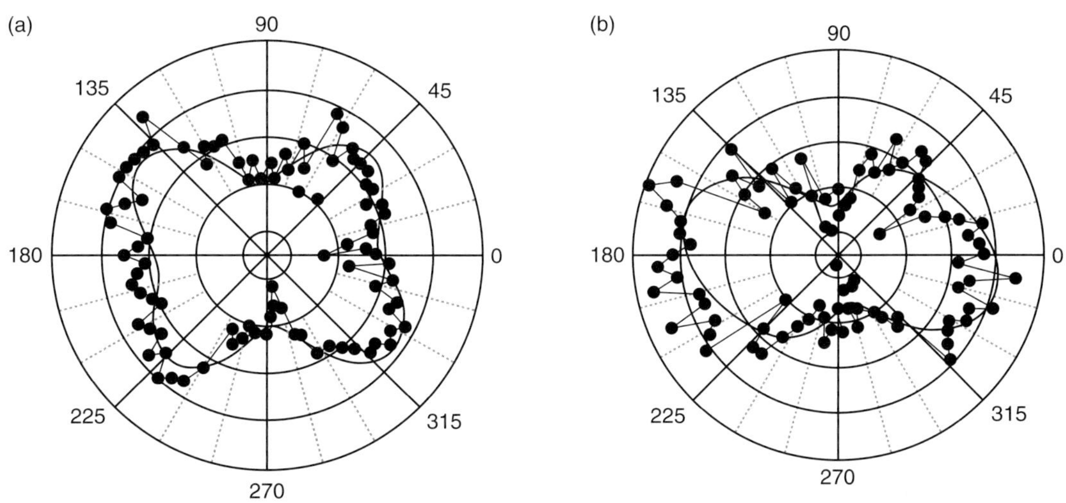

Figure 13 Wavelength dependence of the polarization pattern of the HRS intensity for an aqueous solution of silver particles with diameter 40 nm. The fundamental wavelength is 780 nm (a) and 820 nm. (b) Reproduced with permission from Nappa J, *et al.* (2005) *Chemical Physics Letters* 415: 246.

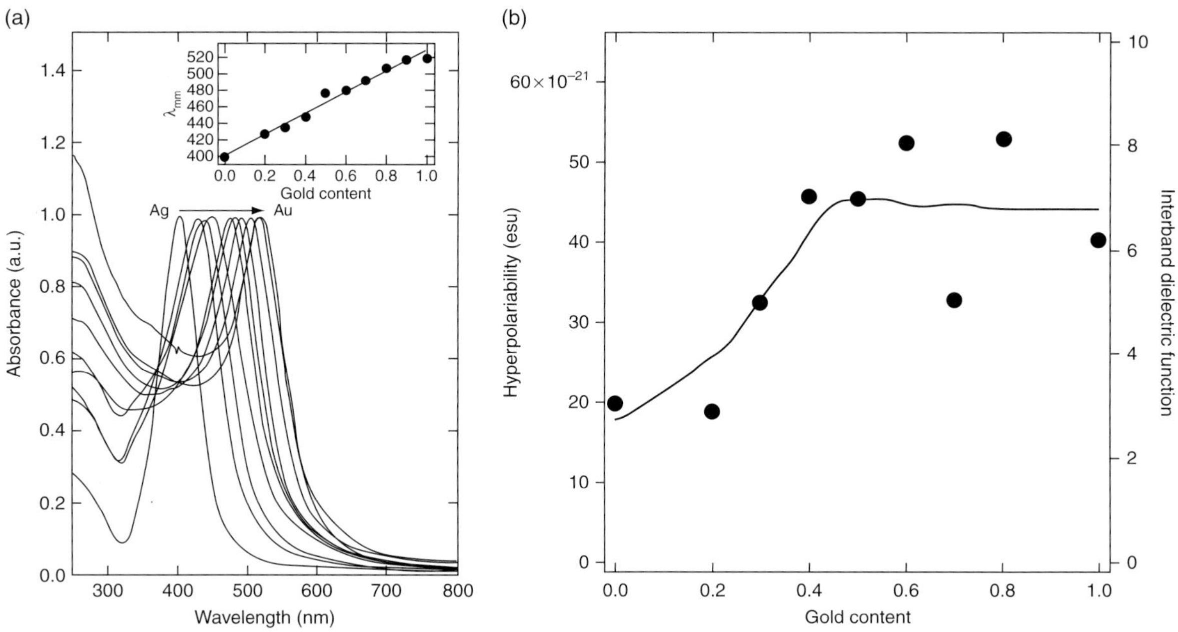

Figure 14 (a) UV–visible absorption spectra of gold–silver alloyed metallic nanoparticles, with the dependence of the wavelength of maximum absorption on the gold content (inset). (b) Dependence of the corrected hyperpolarizability of gold–silver alloyed metallic nanoparticles on the gold content. Reproduced with permission from Russier-Antoine I, *et al.* (2008) *Physical Review B* 78: 035436.

with the surface. The results are reported on **Figure 14(b)**, where the hyperpolarizability magnitude, once corrected for size and resonance field factors, is reported as a function of the silver content of the particles. A clear nonlinear behavior is observed. These results demonstrate the ability of the method to address the surface of the particles instead of their volume, an important difference in many studies like catalysis.

The dispersion of the corrected hyperpolarizabilities as a function of the silver content of the particles also underlines the difficulty to prepare a surface with a similar control as that obtained for the volume. Furthermore, considering the miscibility of gold and silver metals, it appears that the main difference between the two atoms stems from the interband transitions only, the latter transitions being present for gold, and not for silver, at the harmonic wavelength of 400 nm. The dispersion of the corrected hyperpolarizabilities as a function of the silver content in **Figure 14(b)** is observed to follow the interband dielectric function of the bimetallic system. These results point to an atomic heterogeneity of the bimetallic particle surface. This heterogeneity is another contribution to the breaking of the centrosymmetry of the particle at the surface.

3.5.4 Aggregation

Monodispersed solutions of metallic nanoparticles are not very stable if no care is devoted to the stabilization of the solution. This is achieved by incorporating stabilizing agents or by passivating the particle surface. A large number of chemical procedures have been devised and are available in the literature. In the simplest case, the synthesis procedure leads to charged particles, and thus repulsive coulombic forces prevent the aggregation of the particles. However, aggregation may be initiated in these particle solutions by simply modifying the ionic strength of the solution, thereby decreasing the screening Debye length. This has recently been performed by Vance *et al.* [60] by adding sodium chloride to a solution of gold particles. Upon addition of NaCl, aggregation was quickly initiated and followed by UV–visible photoabsorption spectroscopy and HRS. A colorful change of the solution from a red to a blue hue was observed. This color change is induced by the appearance of a longitudinal SP resonance mode in the aggregates at longer wavelengths between 550 and 800 nm as compared to the initial SP resonance mode of the sphere located at 525 nm. In the meantime, the transverse SP resonance mode remaining at 525 nm is damped. The HRS intensity during the NaCl-induced aggregation was reported for a harmonic wavelength of 410 nm and a fundamental wavelength of 820 nm. The onset of the enhancement of the HRS intensity, an enhancement reaching eventually an order of magnitude, was observed for a concentration of 20 mM NaCl, a concentration at which the one-photon Rayleigh scattering signal did not show any sign of enhancement at all. This feature was attributed to the greater sensitivity of the HRS signal to the small aggregates appearing at small NaCl concentrations. Many questions are raised by these experiments in relation to the exact origin of the HRS intensity. Indeed, one possibility is the occurrence of an SP resonance enhancement with the SP longitudinal mode at the fundamental frequency at 820 nm. Another one stems from a change of regime where the noncentrosymmetrical shape of the aggregates plays a major role as compared to the case of isolated particles and the centrosymmetrical shape may dominate.

In order to investigate these questions further, similar experiments were conducted on the aggregation process of gold particles by pyridine. In this case, pyridine displaces charged organic compounds at the surface of the particles, such as citrate, thereby reducing the repulsive coulombic barrier. Aggregation was, therefore, quickly initiated but minute amounts of pyridine were required in order to yield a very slow kinetics allowing the HRS measurements. The latter measurements indeed usually require long acquisition times. Large enhancements, by a factor of about 10, of the HRS intensity were observed for a harmonic wavelength in resonance with the transverse SP mode at 532 nm (**Figure 15**) [55].

A simple model based on the formation of ellipsoidal particles was used to support the data, noting that, at the early stages of the aggregation process, only small linear aggregates were formed. Obviously, as the aggregation proceeds, larger aggregates are formed that eventually reach radii of gyration in excess of the size of the wavelength of light. An enhancement ratio can be defined for small aggregates with respect to the intensity obtained for spherical particles:

$$\eta = \frac{\left|f_{\mathrm{agg}}^2(\omega) f_{\mathrm{agg}}(\Omega)\right|^2}{\left|f_{\mathrm{sph}}^2(\omega) f_{\mathrm{sph}}(\Omega)\right|^2} \tag{34}$$

where the field enhancement factors of the aggregates and the spherical particles at the fundamental and the harmonic frequencies are defined by using the longitudinal and the transverse polarizabilities, respectively, $\alpha_i^l(\omega)$ and $\alpha_i^t(\omega)$, of the ellipsoidal particles:

$$f(\omega) = \frac{4\pi}{V} \sum_i w_i \frac{\mathrm{Im}\left[\alpha_i^l(\omega) + 2\alpha_i^t(\omega)\right]}{\mathrm{Im}\,\varepsilon(\omega)} \tag{35}$$

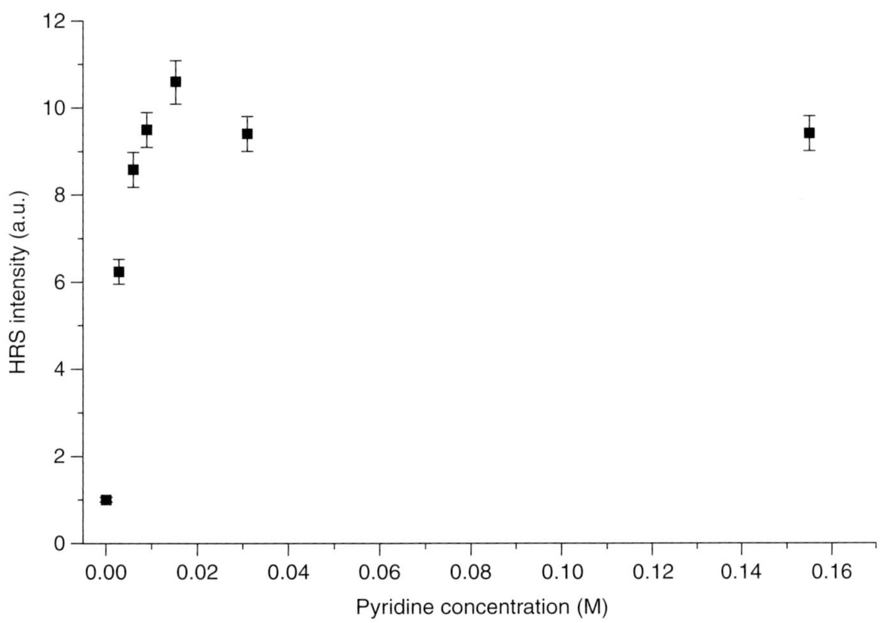

Figure 15 HRS intensity as a function of pyridine concentration for an aqueous solution of gold particles of diameter 22 nm. Reproduced with permission from Galletto P, *et al.* (1999) *Journal of Physical Chemistry B*, 103: 8706.

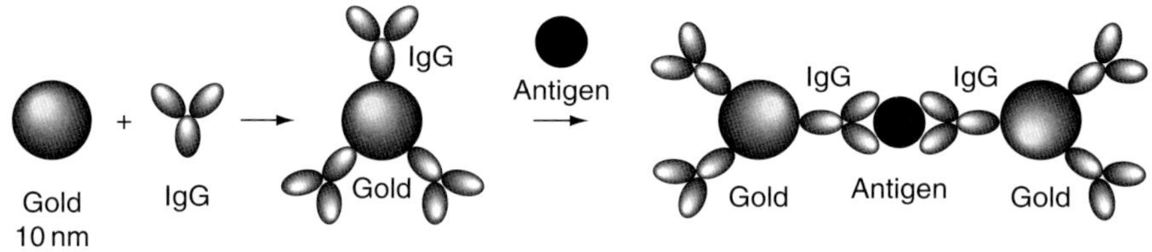

Figure 16 Graphical scheme of gold–IgG conjugation and antigen-induced aggregation. Reproduced with permission from Zhang CX, *et al.* (2003) *Analytical Biochemistry* 320: 136.

In this model, a weighted distribution with weights w_i of linear aggregates, with different lengths was used. The model is in good agreement with the experimental data for the initial times of the aggregation, and it was concluded that the enhancement of the HRS intensity could be attributed to the excitation of the longitudinal SP resonance mode at the fundamental frequency. These results sparked a renewed interest in the controlled design of dense solutions of particles and ultimately aggregated particles. Synthesis of molecularly bridged dimers and trimers of gold particles was achieved and the HRS intensity from aqueous solutions of these compounds compared for particle diameters ranging between 5 and 10 nm [70]. Dimers had larger hyperpolarizabilities than monomers but trimers had much larger hyperpolarizabilities. This was attributed to the symmetry change from the centrosymmetrical monomers and dimers to the noncentrosymmetrical trimers. However, the distance between the different particles of an aggregate was also an important parameter determining the hyperpolarizability of the ensemble.

Recently, HRS from aggregated particles has been used as a possible route to develop new techniques for bio-assay. Protein-modified gold particles were dispersed in solution, and their aggregation state was modified by seeding the solution with antigens (**Figure 16**) [71]. The resulting advancement of the aggregation was monitored by HRS (**Figure 17**).

The gold particles were first coated by goat-anti-human IgG proteins and, as expected, the HRS signal from the protein-aggregated particle solution was larger than that of the nonaggregated particle solution. Interestingly, the aging of the solution was described in terms of the time evolution of the HRS intensity after bio-conjugation of the particles by proteins, a phenomenon attributed to the rise of a contribution from the surface of the particles. Upon addition of human IgG, aggregation of the gold particles was further induced, in a manner very close to the aggregation reaction observed with addition of NaCl or pyridine.

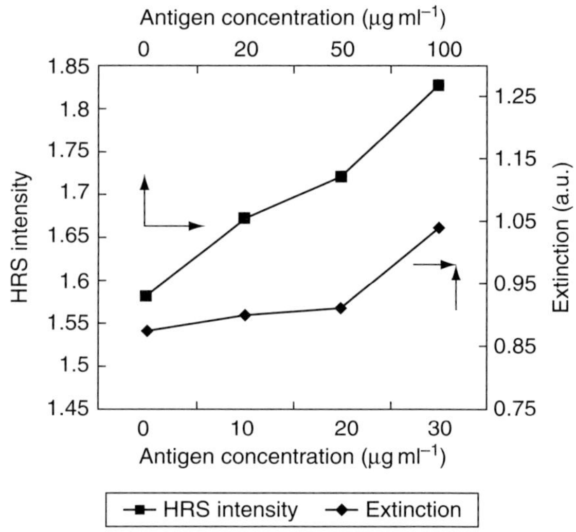

Figure 17 HRS intensity and extinction of gold–IgG conjugates as a function of antigen concentration for an aqueous solution of gold particles of diameter 22 nm. Reproduced with permission from Zhang CX, *et al.* (2003) *Analytical Biochemistry* 320: 136.

Comparison with UV–visible photoabsorption spectroscopy showed that HRS was a more sensitive technique revealing the aggregation process at smaller antigen concentration. For instance, at an antigen concentration of 20 μg ml^{-1}, the HRS signal had already increased by 6% from its initial value, whereas the extinction coefficient did not increase by more than 1%.

The reason for such a sensitivity of HRS as compared to a linear optical method was attributed to the weak protein–protein interactions. Aggregation was so weak between the particles that linear spectroscopy could not resolve any change in the SP resonances due to the weak electromagnetic coupling between the particles. The possibility of sensing very small aggregates with HRS was thus proved. HRS is then potentially a sensitive method to follow the antigen concentration by conjugation in aqueous samples. Another recent work pointed out the care required though in interpreting the results in HRS [72]. Indeed, since particles coated with organic compounds give strong luminescence, a first step where the photoluminescence is removed must be incorporated into the analysis process.

3.5.5 Other Metallic Nanostructures

Thus far, we have given a general approach of systems with either a centrosymmetrical or a noncentrosymmetrical shape. However, in the latter case, we have not discussed particular regular shapes. As opposed to spheres where the selection rules for SHG have been discussed in detail, more complicated shapes such as prisms, cubes, or decahedra have been synthesized in solution. These particles can be centrosymmetrical like prisms or noncentrosymmetrical like cubes. These particular shapes retain a high degree of symmetry, and selection rules may be derived for them. This has been at the focus of works, for instance, of Finazzi *et al.* [73] who have derived such selection rules for several shapes. The problem has been treated for a plane-wave illumination and also for a nonplane-wave illumination such as a converging wave. The latter case is relevant for SH microscopy, where the illumination is so stringent that the particle size cannot be neglected compared to the spatial extension of the beam waist. In this case, higher-order multipoles can be found more prominently in the SH response. Other more refined geometries have also been investigated. Self-similar chains of metal spheres for example have been given a particular attention because of the possibility of producing an efficient nanolens, where the SH local fields may be greatly enhanced at the focus [74]. The chains consisted of a series of three metal silver nanospheres with decreasing diameter. An enhancement by a factor of about 10^6 was predicted for the SH field amplitude as compared to a flat silver surface. Since the whole work was derived for perfect spheres, the first term of **equation 30** was omitted. Considering the above results, incorporation of this contribution too would be highly desirable.

Addressing the SH response from single particles in great detail is still a way forward, although several works have reported the detection of an SH response from single particles [24,25]. Since the response is dramatically modified by the surface of the particles, observing the problem at the single particle and correlating it to its true geometry could give some answers. As a result, near-field optical studies have also been undertaken, despite the technical challenge of such experiments. In near-field optical SH generation, the easiest route is to illuminate the sample, in the present case the particle, and collect the signal in the far-field and perturb it with a metal tip in the near-field. Hence, the response from a metal tip has also been investigated since this is an intrinsically noncentrosymmetrical nanostructure, (**Figure 18**) [75].

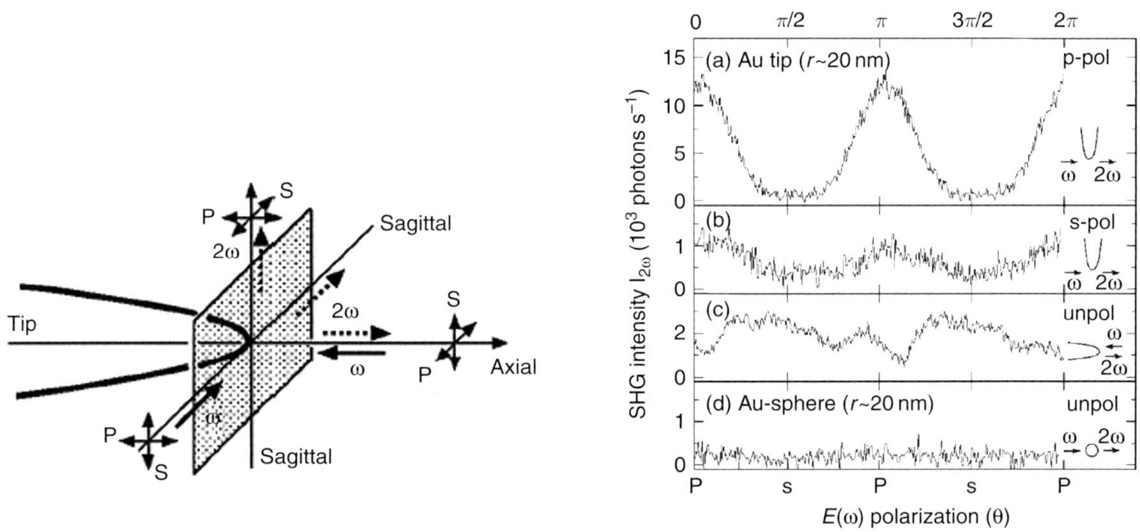

Figure 18 (Left) Experimental geometry for a metal tip. (Right) (b–c) Polarization dependence of the SH intensity for a gold tip with radius ~20 nm. (d) Polarization dependence of the SH intensity for a gold nanosphere with radius ~40 nm deposited on a glass substrate. Reproduced with permission from Neacsu CC, *et al.* (2005) *Physical Review B* 71: 201402.

The breaking of the selection rules observed for metallic nanospheres was clearly observed. In **Figure 18(d)**, the polarization pattern for a gold nanosphere is clearly unpolarized, whereas, for the gold tip, the SH response presents strongly polarization-dependent patterns. If the s-polarized SH radiation polarization pattern is attributed to the rotation invariance of the tip around its axis, the strongly modulated p-polarized SH polarization pattern is attributed to the observation of the local electric dipole contribution owing to the loss of the centrosymmetry of the structure. Other geometries, for instance, axial excitation and sagittal radiation, yield the possibility of observing both the local and the nonlocal contributions.

3.6 Arrays of Metallic Particles

3.6.1 Regular Arrays of Metallic Nanoparticles

With the advance of microstructuring techniques into the nanometer-scale domain, it has been possible to develop a whole new set of samples where the shape of the particles and their distribution on the substrate are better controlled. Hence, UHV shadow deposition or electron-beam lithography, for instance, have received great attention to investigate, and these techniques can now routinely produce arrays of metallic structures the characteristic size of which is in the range of several tens of nanometers. For example, arrays of gold nanowires distributed on a substrate have been investigated [76]. One of the main results of these works is the possibility of selecting the excitation of longitudinal and transverse local fields in the gold structure and observe the corresponding enhancement of the SH response (**Figure 19**).

For example, in **Figure 20**, the SH intensity for the gold nanowires array described in **Figure 19** is reported as a function of the angle of rotation of the sample around the surface normal. A model assuming C_s symmetry could account for the different plots. Ten independent nonlinear elements of susceptibility were incorporated into the theoretical model developed within the reference frame given in **Figure 19**.

Considering the above discussion on the origin of the SH radiation from metallic nanostructures, a new degree of complexity is introduced with a substrate in these systems. The substrate will indeed introduce a geometrical deformation of the nanostructure, for instance, a sphere will be flattened at the interface with the substrate or weakly polarized. The difficulty then stems from the extent of symmetry breaking introduced by the substrate. In their study of an array of gold particles with planar inversion symmetry, McMahon *et al.* [17] assumed an origin of the SH response principally stemming from the substrate–particle interaction (**Figure 21**). In the plane-wave excitation approximation, the nonlinear polarization at the harmonic frequency was, therefore, oriented toward the normal to the substrate surface, preventing any SH radiation in a transmission configuration. Thus, they made use of the diffractive properties of the array and investigating the SH intensity radiated at some angular direction. Their experimental observations were in agreement with the expected behavior of a grating with the corresponding spacing.

The observation of an SH intensity in a forward direction determined by the diffractive properties of the grating formed by the array of nanoparticles is rather general. However, considering the microscopic

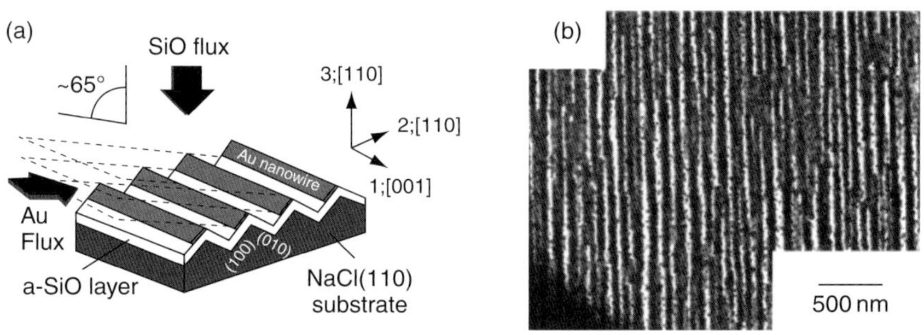

Figure 19 (a) Schematic topography of the surface of the NaCl (110) template with Au nanowires created by the shadow deposition method. (b) The Au nanowire array deposited on the stepped NaCl (110) template as observed through the bright field images. The Au nanowires seen in the dark contrast run along the [001] azimuth of the NaCl (110) substrate. Reproduced with permission from Kitahara T, *et al.* (2004) *Journal of Applied Physics* 95: 5002.

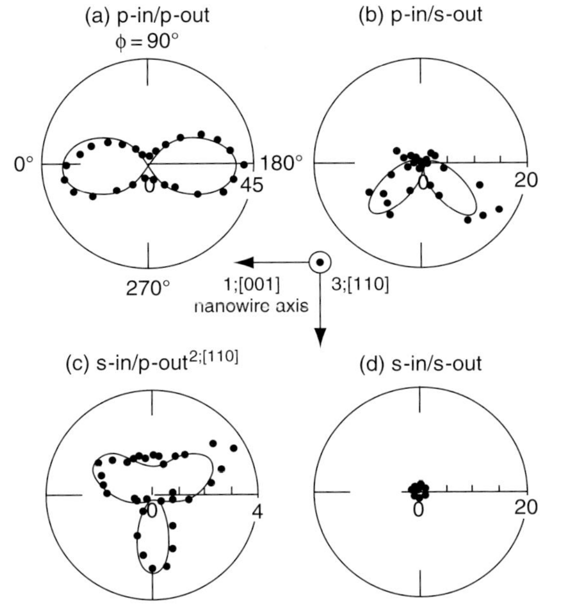

Figure 20 SH intensity from the quasi-C_s symmetric Au nanowire arrays/NaCl (110) system at a second-harmonic energy of 2.33 eV as a function of the sample rotation angle defined as the angle between the plane of incidence and the [001] direction (parallel to the nanowire axis) for four different polarization combinations. Reproduced with permission from Kitahara T, *et al.* (2004) *Journal of Applied Physics* 95: 5002.

origin of the SH response, it is surprising that no SH intensity was detected in the transmission geometry. Indeed, since perfectly centrosymmetrical particles are difficult to achieve, a weak breaking of the planar inversion would have yielded a nonvanishing SH signal in transmission. In fact, this requires a nonlinear radiated dipole oriented in a direction other than the transmission direction, otherwise, a node is observed in the radiation pattern. This condition may be obtained at the level of a single particle, but, if it arises from defects in the morphology of the particles, the random distribution in orientation of this nonlinear radiating dipole from one particle to the other should cancel its existence. The possibility of a bias in the sample preparation may, however, produce defects with rather similar orientation. This could, for example, be produced by the raster mode used in the EBL fabrication. Hence, experimental work has been undertaken to determine the exact origin of the SH response in arrays of particles.

In order to disentangle the different contributions already discussed in the previous sections, two schemes have been proposed. The first one involves the use of a two-beam SHG configuration. In this case, contributions to the nonlinear polarization of the form $(\boldsymbol{E}\cdot\nabla)\boldsymbol{E}$ may be excited as $(E_1\cdot\nabla)E_2$ or $(E_2\cdot\nabla)E_1$, yielding a larger freedom to select the weight of the different contributions [77,78]. This method has been used, in particular, for the investigation of the SH response from diamond nanocrystals and metallic surfaces. The authors were then able to partly separate surface and bulk contributions. The second method involves the comparison of reflection and transmission geometries. This method has been extensively investigated in the

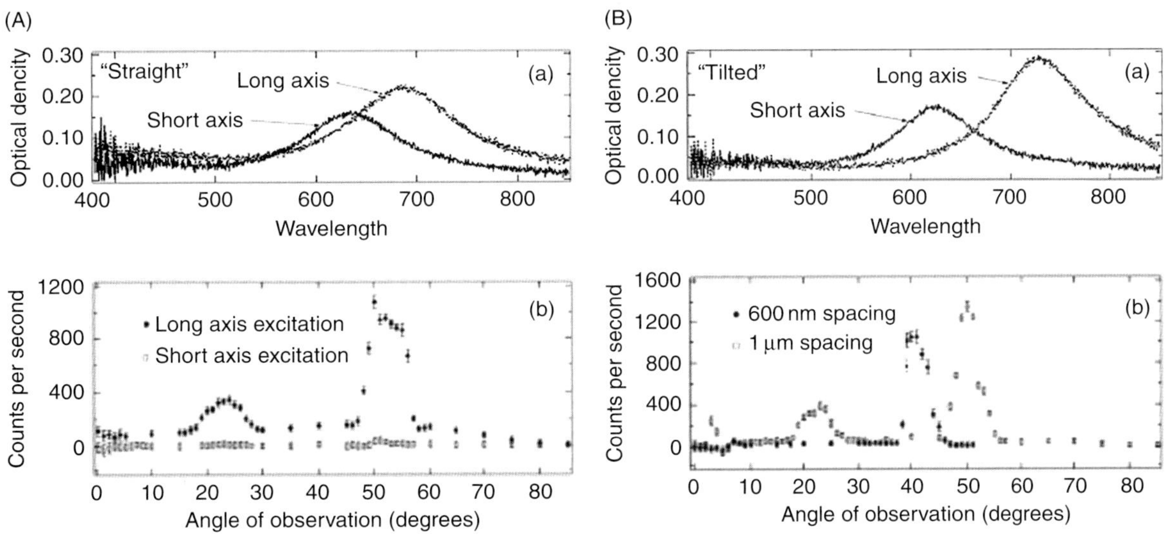

Figure 21 (A) (a-b) angular dependence of the SH emission from an array of centrosymmetrical rods. (B) Experimental SH emission pattern recorded for two array spacings. Reproduced with permission from McMahon MD, *et al.* (2006) *Physical Review B* 73: 041401.

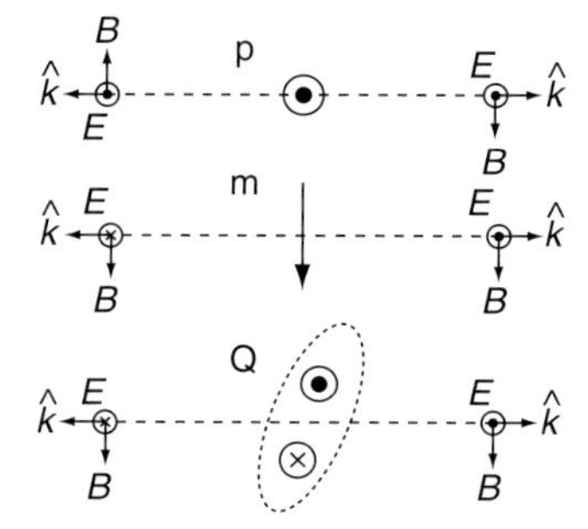

Figure 22 Simple schematics of the radiative properties of an electric dipole (p), magnetic dipole (m) and electric quadrupole (Q) in reflection and transmission geometries. Reproduced with permission from Kujala S, *et al.* (2007) *Physical Review Letters* 98: 167403.

group of Kujala *et al.*, [79] and is based on the different radiation properties of the electric dipole, electric quadrupole, and magnetic dipole (**Figure 22**). These properties are given for the reflection and transmission geometries and demonstrate that a different interference pattern between these contributions will be built in reflection and transmission.

Using a regular array of noncentrosymmetrical gold particles deposited on a glass substrate by EBL, Kujala *et al.* [79] manage to demonstrate this property (**Figure 23**). Since it is difficult to describe the SH response of the array in terms of the microscopic origin of the nonlinear polarization, they discussed their results in terms of a nonlinear response tensor linking the exciting and the radiating fields directly. They also observed that the dominating element of this tensor was the allowed electric dipole, in agreement with our discussion for particles made from a centrosymmetrical material but possessing a noncentrosymmetrical shape (an L-shape). The estimation weight of the higher-order multipoles compared to the electric dipole contribution was about 20%, the phase of which was returned upon going from the transmission to the reflection geometry.

In a detailed analysis of these arrays of L-shaped particles, Canfield *et al.* [80] also investigated the selection rules for the different elements of the nonlinear response tensor. They could not observe any dominating elements and, besides, they observed that elements of the form A_{zxy} were also allowed, with strong amplitude as compared to the allowed electric dipole one. This result was attributed to the chirality of the particles, indeed achieved in their sample through either the chirality of the particles or that of the organization of the particles on the substrate (**Figure 24**) .

Experiments using circularly polarized light clearly supported the analysis of the data in terms of chirality. Nevertheless, these systems call for more refined studies. It appears, indeed, that the SH response arises from defects in the EBL fabrication of the samples, with a similar orientation of the nonlinear radiating dipole of the defect for all the particles. In contrast, the chiral contribution of the SH response, built from a designed fabrication of an array of nanoparticles with chiral properties, did not yield tensor elements with dominating amplitudes. Work on defect-induced SH scattering is still being investigated theoretically and it is possible that, in the future, these approaches will converge with the experimental geometries (**Figure 25**) [81].

3.6.2 Random Metallic Structures

Before closing this section, we wish to mention a set of nanostructures where organization is lacking. This type of structures has already been investigated mainly because they provide geometries leading to

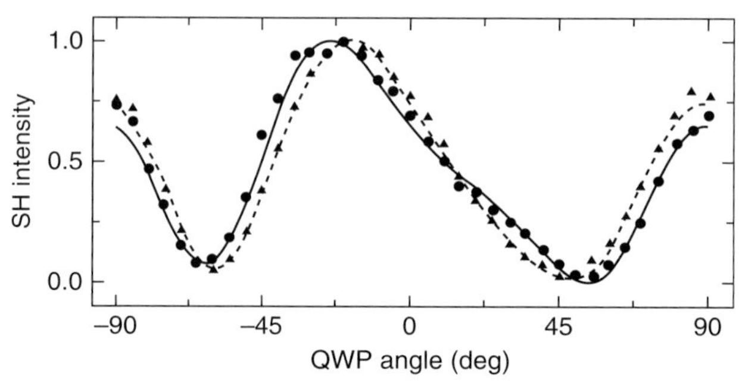

Figure 23 Measured SH intensity as a function of the angle of the quarter-wave plate for a transmission (triangles) and a reflection (circles) geometry. Reproduced with permission from Kujala S, *et al.* (2007) *Physical Review Letters* 98: 167403.

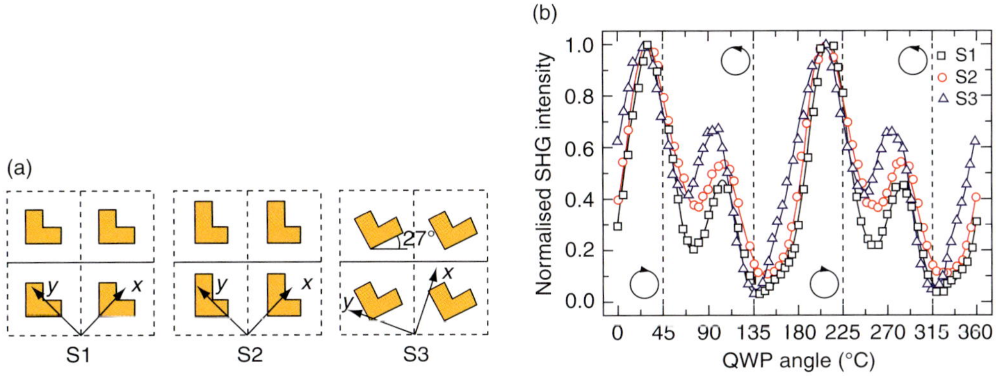

Figure 24 (a) Nanoparticle array designs. S1: achiral array of achiral particles; S2: achiral array of chiral particles; and S3 : chiral array of achiral particles. (b) SH intensity as a function of the angle of the quarter-wave plate (circular polarization is indicated by vertical bars). Reproduced with permission from Canfield BK, *et al.* (2006) *Optics Express* 14: 951.

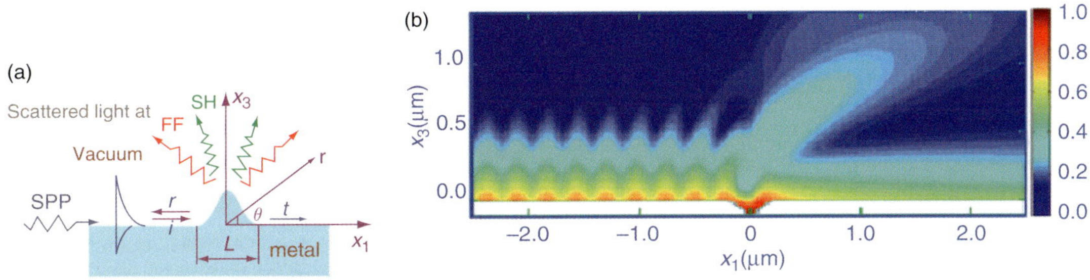

Figure 25 (a) Schematics of the scattering geometry and the radiated waves in the nonlinear scattering process. (b) Calculated near-field distribution in vacuum at far-field computed for a Gaussian-shaped nanogroove with a width of 100 nm, a depth of 100 nm, and for an incident surface plasmon polariton wavelength of 560 nm. Reproduced with permission from Cao L, *et al.* (2007) *Physical Review B* 75: 205401.

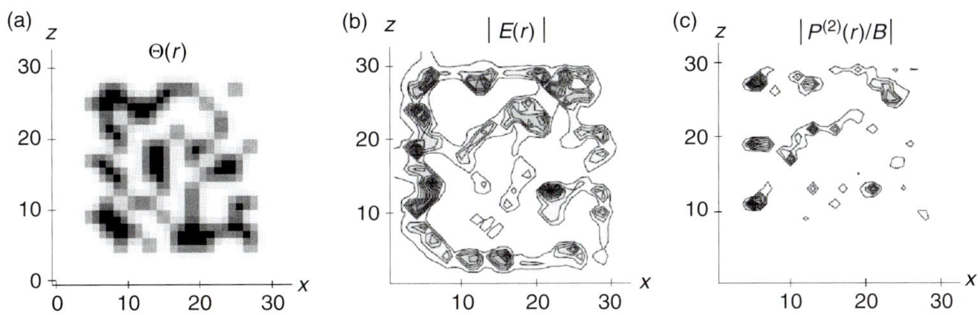

Figure 26 (a) Geometry of a random planar composite (cross-section through the plane of symmetry): characteristic function Θ(r) displayed as a density plot. The axis unit is the grid step; one unit may correspond to a length between 2 and 5 nm. Fill factor $p = 0.5$. (b) Spatial distribution of the magnitude of the linear field $|\boldsymbol{E}(\boldsymbol{r})|$ in the plane of the RPC shown as a contour map, max $|\mathbf{E}| = 35$. (c) Same as (b), but for the SH polarization enhancement, $|\boldsymbol{P}^{(2)}(\boldsymbol{r})/B|$, max $|\boldsymbol{P}^{(2)}(\boldsymbol{r})/B| = 250$. Reprinted from Stockman MI, *et al.* (2004) *Physical Review Letters* 92: 057402.

field localization and large SH enhancement. With random structures produced in semicontinuous metallic films for instance, roughness is obtained with characteristic length scale in the nanometre scale domain [82]. Computation in these systems has shown that electric field singularities are produced in the linear and the nonlinear regime, forming areas where the local field enhancement can be rather large (**Figure 26**). The characteristic lengths of these areas remain much smaller than the wavelength of light and the SH fields are depolarized and even dephased between them.

Hence, it has been proposed to observe these SH field singularities using near-field optical SH microscopy, and this was experimentally achieved. Nevertheless, confocal microscopy has also been demonstrated to be a viable experimental configuration to observe these singularities. One of the features of these singularities is their appearance near the percolation threshold in these films and their being closely linked to localized plasmon excitation [83]. Because of the large distribution of the metallic structures, the SP resonance excitation frequencies strongly vary from one singularity to the other, preventing a direct control on the SH field enhancement. Nevertheless, these investigations closely resemble those performed in SERS to develop very active substrates for high sensitivity. It is possible that, with the future development of the fabrication techniques, a better control on these enhancements through field localization will be achieved.

3.6.3 Conclusions

In this review, the SH response from nanostructures has been investigated. The SH response can be separated according to the symmetry properties of the structures. The centrosymmetry property in particular, which yields the strongest selection rule since SHG is forbidden in media with inversion symmetry in the electric dipole approximation, must be investigated from the point of view of both the crystal structure of the material and the shape of the nanostructure. This is the main difference with macroscopic media, where only the crystal structure of the medium matters. As a result of the shape dependence, the problem of the origin of the SH response is usually difficult to answer in a general manner. In most cases, the response is defined by the exact nature of the system and possibly its interaction with the surrounding medium. This is especially true in the case of a nanostructure deposited on a substrate.

Applications have already appeared, although only at the demonstration level. This is certainly the case in analytical chemistry or biochemistry. However, in the design of nanostructures for applications, it appears that a careful control of the fabrication stage will be required in order to design the SH response. This is perfectly demonstrated with arrays of metallic nanostructures the response of which is dominated by defects. Nevertheless, it appears that a large freedom is available owing to the number of contributions to the SH response. Other geometries, less conventional, have also been proposed, such as random structures, and it is difficult to predict where the next advances will result.

Finally, it must be underlined that this work has deliberately focused on semiconducting and metallic nanostructures. Other nanostructures do exist, such as those based on molecules or biomolecules, (e.g., liposomes and micelles) or large biomolecules (e.g., proteins). Recent works have appeared on this topic, but the approach would be rather different, based on the response from single molecular units arranged in larger coherent structures. This opens up a whole new world.

References

1. Dupas C, Houdy Ph, and Lahmani M (eds.) (2007) *Nanosciences: Nanotechnology and Nanophysics*. Berlin: Springer.
2. Maier SA (2006) *Plasmonics: Fundamentals and Applications*. Berlin: Springer.
3. Kreibig U and Vollmer M (1995) *Optical Properties of Metal Clusters*. Berlin: Springer.
4. Bohren CF and Huffman DR (1998) *Absorption and Scattering of Light by Small Particles*. New York: Wiley.
5. Boyd RW (1984) *Nonlinear Optics*. New York: Academic Press.
6. Shen YR (1984) *The Principles of Nonlinear Optics*. New York: Wiley.
7. Heinz TF (1991) In: Agranovich VM and Maradudin AA (eds.) *Nonlinear Surface Electromagnetic Phenomena*. Amsterdam: North Holland.
8. Brevet PF (1997) *Surface Second Harmonic Generation*. Lausanne: Presses Polytechniques Universitaires Romandes.
9. Shen YR (1989) *Annual Review of Physical Chemistry* 40: 329.
10. Eisenthal KB (1992) *Annual Review of Physical Chemistry* 43: 627.
11. Higgins DA and Corn RM (1994) *Chemical Reviews* 94: 107.
12. Brevet PF (2001) Nonlinear optics at liquid/liquid interfaces. In: Volkov AG (ed.). *Liquid Interfaces in Chemical, Biological and Pharmaceutical Applications, Surface Science Series*, vol95, pp. 123–151. New York: Marcel Dekker.
13. Boyd GT, *et al.* (1984) *Physical Review B* 30: 519.
14. Butcher PN and Cotter D (1991) *The Elements of Nonlinear Optics, Cambridge Studies in Modern Optics*, vol. 9 Cambridge: Cambridge University Press.
15. Svelto O (1998) *Principles of Lasers,* 4th ed.; New York: Plenum Press.
16. Maker PD, *et al.* (1962) *Physical Review Letters* 8: 21.
17. McMahon MD, *et al.* (2006) *Physical Review B* 73: 041401(R).
18. Zheludev NI and Emelyanov VI (2004) *Journal of Optics A: Pure and Applied Optics* 6: 26.
19. Maxwell-Garnett JC (1904) *Philosophical Transactions of the Royal Society of London* 203: 385.
20. Maxwell-Garnett JC (1906) *Philosophical Transactions of the Royal Society of London* 205: 237.

21. Kothari NC (1990) *Physical Review A* 41: 4486.
22. Sipe JE and Boyd RW (1992) *Physical Review B* 46: 1614.
23. Draine BT and Flatau PJ (1994) *Journal of the American Optical Society A* 11: 1491.
24. Jin R, *et al.* (2005) *Journal of the American Chemical Society* 127: 12482.
25. Lippitz M, *et al.* (2005) *NanoLetters* 5: 799.
26. Brudny VL, *et al.* (2000) *Physical Review B* 62: 11152.
27. Bersohn R, *et al.* (1966) *Journal of Chemical Physics* 45: 3184.
28. Cyvin SG, *et al.* (1965) *Journal of Chemical Physics* 43: 4083.
29. Jackson JD (1975) *Classical Electrodynamics*. New York: Wiley.
30. Clays K and Persoons A (1991) *Physical Review Letters* 91: 2980.
31. Brasselet S and Zyss J (1998) *Journal of the Optical Society of America B* 15: 257.
32. Gallet S, *et al.* (2003) *Chemical Physics Letters* 378: 101.
33. Alivisatos AP (1996) *Journal of Physical Chemistry* 100: 13226.
34. Chemla DS and Zyss J (1987) *Nonlinear Properties of Organic molecules and Crystals*, Vol. 2. Orlando: Academic Press.
35. Sahyun MRV (2002) *Spectrochimica Acta A* 58: 3149.
36. Santos BS, *et al.* (2000) *Optics Communications* 178: 187.
37. Jacobsohn M, *et al.* (2001) *Journal of Physical Chemistry B* 105: 12726.
38. Zhang Y, *et al.* (2002) *Journal of Physical Chemistry of Solids* 63: 2115.
39. Jacobsohn M and Banin U (2000) *Journal of Physical Chemistry B* 104: 1.
40. Aktsipetrov OA, *et al.* (1995) *Physical Review B* 51: 17591.
41. Landes C, *et al.* (2002) *Chemical Physics Letters* 363: 465.
42. Zhang Y, *et al.* (2001) *Journal of Physical Chemistry of Solids* 62: 903.
43. Fu D, *et al.* (2002) *Materials Chemistry and Physics* 77: 285.
44. Zhang Y, *et al.* (2003) *Journal of Physical Chemistry of Solids* 64: 927.
45. Aktsipetrov OA, *et al.* (2005) *Journal of the Optical Society of America B* 22: 138.
46. Aktsipetrov OA (2002) *Colloids and Surfaces A* 202: 165.
47. Murzina TV, *et al.* (2001) *Applied Physics Letters* 79: 1309.
48. Lenglet J, *et al.* (1996) *Physical Review B* 53: 14941.
49. Eisenthal KB (2006) *Chemical Reviews* 106: 1462.
50. Agarwal GS and Jha SS (1982) *Solid State Communication* 41: 499.
51. Hua XM and Gersten JI (1986) *Physical Review B* 33: 3756.
52. Ostling D, *et al.* (1993) *Zeitschrift für* Physik *D* 28: 169.
53. Dadap JI, *et al.* (1999) *Physical Review Letters* 83: 4045.
54. Dadap JI, *et al.* (2004) *Journal of the Optical Society of America B* 21: 1328.
55. Galletto P, *et al.* (1999) *Journal of Physical Chemistry B* 103: 8706.
56. Nappa J, *et al.* (2004) *Faraday Discussions* 125: 145.
57. Rudnick J and Stern EA (1971) *Physical Review B* 4: 4274.
58. Schmickler W and Urback M (1993) *Physical Review B* 47: 6644.
59. Russier-Antoine I, *et al.* (2007) *Journal of Physical Chemistry C* 111: 9044.
60. Vance FW, *et al.* (1998) *Journal of Physical Chemistry B* 102: 10091.
61. Galletto P, *et al.* (1999) *Chemical Communications.*581.
62. Nappa J, *et al.* (2005) *Physical Review B* 71: 165407.
63. Antoine R, *et al.* (1997) *Chemical Communications* 1901.
64. Chandra M, *et al.* (2007) *Journal of Physical Chemistry C* 111: 10652.
65. Johnson RC, *et al.* (2002) *Chemical Physics Letters* 356: 534.
66. Hao EC, *et al.* (2002) *Journal of Chemical Physics* 117: 5963.
67. Russier-Antoine I, *et al.* (2004) *Journal of Chemical Physics* 117: 5963.
68. Nappa J, *et al.* (2005) *Chemical Physics Letters* 415: 246.
69. Russier-Antoine I, *et al.* (2008) *Physical Review B* 78: 035436.
70. Novack JP, *et al.* (2000) *Journal of the American Chemical Society* 122: 12029.
71. Zhang CX, *et al.* (2003) *Analytical Biochemistry* 320: 136.
72. Russier-Antoine I, *et al.* (2008) *Chemical Physics Letters* 450: 345.
73. Finazzi M, *et al.* (2007) *Physical Review B* 76: 125414.
74. Li K, *et al.* (2005) *Physical Review B* 72: 153401.
75. Neacsu CC, *et al.* (2005) *Physical Review B* 71: 201402(R).
76. Kitahara T, *et al.* (2004) *Journal of Applied Physics* 95: 5002.
77. Sun L, *et al.* (2005) *Optics Letters* 30: 2287.
78. Cattaneo S, *et al.* (2002) *Physical Review Letters* 88: 243901.
79. Kujala S, *et al.* (2007) *Physical Review Letters* 98: 167403.
80. Canfield BK, *et al.* (2006) *Optics Express* 14: 950.
81. Cao L, *et al.* (2007) *Physical Review B* 75: 205401.
82. Stockman MI, *et al.* (2004) *Physical Review Letters* 92: 057402.
83. Breit M, *et al.* (2001) *Physical Review B* 64: 125106.

4 Organic Electronic Devices with Water-Dispersible Conducting Polymers

J Tarver, Princeton University, Princeton, NJ, USA

J E Yoo, University of Texas at Austin, Austin, TX, USA

Y-L Loo, Princeton University, Princeton, NJ, USA

4.1 Introduction

Organic electronics represent a radical shift in modern technology; once purely in the realm of academic interest, the field of organic electronics has matured in the past decade and is now primed for commercialization. Already, organic light-emitting diodes (OLEDs) are in production and are gaining market share as active displays for cellular telephones, cameras, and televisions. Organic thin-film transistors (OTFTs) and organic solar cells (OSCs) currently receive levels of attention rivaling that received by their silicon-based analogs, while organic-based electrochromic devices (ECDs) can be readily functionalized to fit almost any designer niche.

Although organic materials are unlikely to outperform inorganic conductors or semiconductors, a number of qualities make them desirable alternatives. Organic molecules are produced through versatile synthetic routes involving inexpensive reactants and reaction conditions that only deviate slightly from ambiance; by harnessing a vast array of well-established organic chemistries, the properties of these molecules can also be tuned systematically via postsynthesis derivatization and functionalization [1, 2]. Unlike materials traditionally used in electronics, organic molecules are often processed at or near room temperature. The promise of facile processing obviates the need for costly, high-temperature, high-vacuum deposition, and patterning procedures that inorganics require and could translate into immense capital and operational savings [3]. These processing advantages also permit large-area, roll-to-roll manufacturing and thus free devices from the limited size constraints of silicon wafers. In addition to economic benefits, organic electrically active materials can be lightweight and often possess the mechanical flexibility and moldability associated with plastics. The promise of this unique combination of mechanical and electrical properties permits novel implementation in a number of specialty applications.

Electrically active organic materials can be divided into two classes: small molecules and polymers. Each set has its own innate strengths and drawbacks. The chemistry of small molecules, for instance, can be tuned to suit either evaporation or solution deposition. In addition, small molecules often achieve high degrees of crystallinity and consequently exhibit remarkable electrical properties. The corollary of this strength, however, is that small molecules are vulnerable to defect formation during crystallization and can have poor film formability. Unlike electrically active small-molecule organics, polymers frequently form viscous solutions that can be uniformly coated across large dimensions [1]. By doping electrically active polymers, the electrical characteristics can span insulating, semiconducting, and conducting regimes. For the past few decades, the goal of conducting polymer research has thus been to identify candidates within the group of stable systems that exhibit electrical performance comparable to, or exceeding that of, small molecules. This chapter details the history, synthesis, characterization, and application of conducting polymers, in particular two of its most robust and successful varieties: poly(3,4-ethylenedioxythiophene) and polyaniline.

4.1.1 History of Conducting Polymers

Following a breakthrough work by Heeger [4], MacDiarmid [5], and Shirakawa [6] during the late 1970s, the study of conducting polymers has emerged as a thriving field within organic electronics. The trio was awarded the Nobel prize in Chemistry in 2000 for both the discovery of metal-like conductivities in polyacetylene films that were oxidatively doped with iodine vapors, as well as for the subsequent development and investigation of new conducting polymer systems [7]. They were the first to investigate the mechanisms and physical origins that underlie the charge transport in this new class of materials [7].

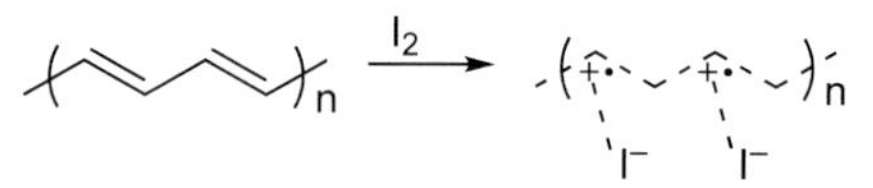

Scheme 1 Iodine-doping of *trans*-polyacetylene to its highly conductive form. (·) represents radicals along the conjugated backbone.

Much success was met initially with the discovery that polyacetylene can be oxidatively doped with iodine to induce metal-like conductivities. Free radicals produced during the oxidation of polyacetylene can delocalize along the backbone of the conjugated polymer, as shown in **Scheme 1**, to give rise to electrical conductivities in the order of $500\,S\,cm^{-1}$ [8]. This phenomenon, however, is short lived at ambient conditions. The free radicals are quickly quenched when exposed to oxygen and the material's conductivity reverts to its initial insulating level [8]. Although useful as an experimental model, it was quickly realized that polyacetylene's environmental instability severely limits its use in practical applications.

Equipped with a working example of electrical conductivity in polymers, researchers turned their attention toward other conjugated systems in the hopes of identifying more stable candidates for use in electronic devices. Conductivity was predicted, and soon after measured, in a number of polymers, most notably poly(3,4-ethylenedioxythiophene) and polyaniline – often abbreviated as PEDOT and PANI, respectively. The unique stability possessed by these systems has generated significant amounts of research. Between 1990 and 2008 alone, nearly 13 000 papers describing the synthesis, characterization, or applications of PEDOT or PANI have been published.

4.1.1.1 History of PEDOT

The discovery of electrical conductivity in oxidized polythiophenes was simultaneously reported in 1980 by Yamamoto *et al.* [9] and Lin and Dudek [10]. The mechanism underlying polythiophene's conductivity was found to be similar to that of oxidatively doped polyacetylene: generation of delocalized radicals by an oxidizing dopant, and subsequent stabilization via ionic interactions between the charged polymer and spent dopant. Although the observed conductivities were low (10^{-3}–10^{-1} S cm^{-1}) relative to doped polyacetylene, thiophene monomers offer a more diverse opportunity for functionalization than bare acetylene. Through proper modification of the base thiophene monomer, free radicals that are generated on oxidation can thus be stabilized. During the next decade, poly(cycloalkylenedioxythiophenes) offered the best combination of conductivity and environmental stability [11,12], culminating in the first-reported synthesis of PEDOT via a patent application by Bayer [13] in 1989. A particularly conductive form of PEDOT, synthesized along, and stabilized by, poly(styrene sulfonic acid), or PSS, has been commercially available as an aqueous dispersion under the brand-name Baytron (though recently changed to Clevios and colloquially known as PEDOT–PSS) since the early 1990s. A variety of forms are produced with conductivities ranging from 10^{-6} to $300\,S\,cm^{-1}$.

4.1.1.2 History of PANI

The first-reported synthesis of PANI dates back to an 1862 communication by H. Letheby detailing what is now clearly the electrochemical polymerization of aniline in sulfuric acid at oxidizing potentials [14]. Due to the scientific community's incomprehension of polymers at the time, no effort was made to characterize the structure of the product – much less probe its electrical properties. Nearly 40 years later, Green and Woodhead [15,16] postulated that oxidation of aniline yielded a molecule comprising alternating amine/imine + benzenoid/quinoid segments that is capable of oscillating between distinct oxidation states. Although their assessment of PANI's chemical structure was accurate, they failed to recognize the extension of this sequence beyond an octameric form. Conductivity in polyaniline was first detected by de Surville and Jozefowicz in 1968 [17]. They reported conductivities near $20\,S\,cm^{-1}$ for pellets comprising PANI's intermediate oxidation state (emeraldine) that is hydrated at pH 0. Hydration at elevated pH decreased conductivity, but no attempt was made to explain the conduction mechanism or its unusual dependence on proton concentration.

Following the discovery of the conduction mechanism of polyacetylene in the late 1970s by Heeger and coworkers, the work of de Surville and Jozefowicz was revisited and interest in PANI's unique pH dependence grew [18–20]. Unlike other conducting polymer systems, namely PEDOT and polyacetylene, delocalization of radicals along PANI occurs via protonation by an acid rather than by oxidation. This unique doping mechanism presented unprecedented environmental stability in PANI, which subsequently spurred a rush of research. Studies on PANI gained additional momentum in the 1990s when Heeger and coworkers [21] reported conductivities as high as 400 S cm^{-1} in PANI films treated with 'secondary dopants' [22].

Several major problems have historically hindered the use of PANI in practical applications. The volatile small-molecule acids initially used to dope PANI would evaporate, rendering PANI electrically insulating upon long-term storage. In addition, the doped form of PANI is insoluble in virtually all solvents and is not melt processable. The undoped form of PANI is only soluble in a small number of high-boiling organic solvents, such as *N*-methyl pyrrolidone (NMP). To form a conductive film, the undoped form of PANI is frequently cast from NMP and the film is subsequently exposed to a proton source. Evaporation of NMP is tediously slow, and uniform doping of these films is difficult due to uneven exposure and acid uptake. Efforts have been made to functionalize aniline monomers with acidic side groups to generate self-doping and/or soluble PANI [23–25]; however, they have generally been met with suppressed conductivities and miniscule gains in solubility.

Recent studies have sought to use polymer acids as dopants [26,27]. This approach has two primary advantages. The large polyanions ionically associate with PANI, thereby eliminating issues of dopant volatility and providing long-term conductivity. Excess acid groups along the polyanion can hydrogen bond with water, rendering the complex water dispersible and processable from aqueous media. Development is currently underway on polymer-acid-doped PANI variants analogous to PEDOT–PSS. The most promising system to date utilizes poly(2-acrylamido-2-methyl-1-propane sulfonic acid), or PAAMPSA, as both a template during synthesis and as the proton source to yield an aqueous PANI–PAAMPSA dispersion [28–30]. The conductivity of films cast from these dispersions can reach 2.5 S cm^{-1}, and can be elevated by another order of magnitude upon secondary doping [31].

4.2 Chemistry of Water-Dispersible Conducting Polymers

4.2.1 Synthesis and Redox Chemistry

One of the attractive qualities of conducting polymers is the degree of synthetic control offered prior to and during polymerization. Proper functionalization of the monomer, for instance, can alter the solubility, stability, and reactivity of both the initial monomer and final polymer. By manipulating the reaction temperature, reactant or oxidizing agent concentration, or any other of the array of reaction conditions under which the monomer is polymerized, the electrical and mechanical properties of the resulting polymer can be changed significantly. Although optimization of a desired synthesis can be a daunting task for polymer scientists, it ultimately allows materials to be tailored toward specific pursuits.

The synthesis of conducting polymers can be performed through electropolymerization or via traditional wet chemistry routes. As an exploratory technique, electrochemical synthesis holds several advantages. Provided a clean system without competing reactions, potentiometric experiments allow the strict management of charge flowing across the surface of the electrode and thus permit precise control over the amount of material reacted. In this manner, nanoscopically thin films of polymer can be produced. Electrochemical polymerization methods can also reveal the mechanistic details of the synthesis process. As an industrial technique, however, electrochemical synthesis suffers from scale-up problems. To match the output of conventional solution-phase chemistries, electrodes possessing large surface areas would be required; removal of the electropolymerized product from these surfaces would be excessively laborious and inefficient. As such, traditional chemical synthesis is still the dominant method of producing conducting polymers.

4.2.1.1 PEDOT synthesis and redox chemistry

The oxidative polymerization of EDOT (see **Scheme 2**) is initiated when an oxidizing agent abstracts an electron from EDOT's thiophene moiety to form a radical cation. The mechanism behind the ensuing propagation, however, is not well studied. It has been suggested that initiation is the rate-limiting step in PEDOT synthesis [32]. The radical cation has a weaker electron affinity than the initiator. Rather than propagating directly, two EDOT cations thus prefer to dimerize. Due to resonant stabilization effects, abstracting an electron from these dimers is less difficult than from a single monomer; this subsequent reaction therefore occurs at a faster rate. Propagation consequently is thought to proceed via oxidation and dimerization of oligomers [32]. This synthesis yields PEDOT in its neutral, dark blue, and insulating state. Conductivity is conferred by further oxidizing the neutral polymer into its colorless polycationic state; its chemical structure is shown in **Scheme 3**. The positive charges induced by

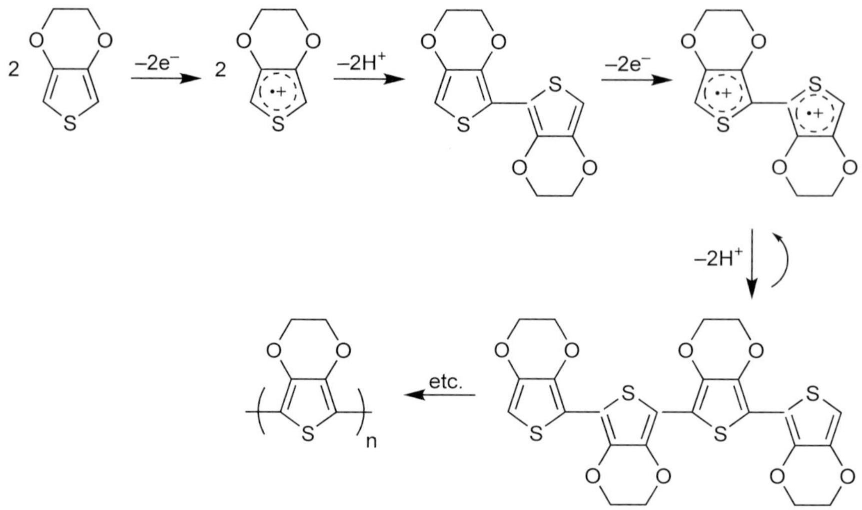

Scheme 2 Mechanism for the polymerization of EDOT to PEDOT in the neutral, undoped state. Adapted with permission from Kirchmeyer S and Reuter K (2005) Scientific importance, properties and growing applications of poly(3,4-ethylenedioxythiophene). *Journal of Materials Chemistry* 15: 2077–2088.

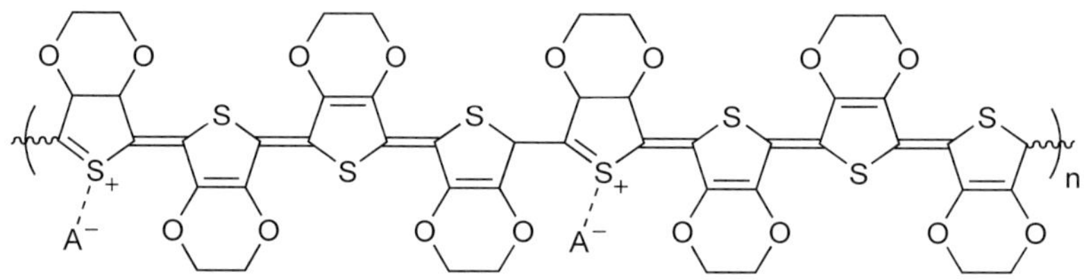

Scheme 3 Two repeat units of PEDOT in its doped form where A^- represents the charge-balancing counter-ion. Each repeat unit consists of approximately four mers. Adapted with permission from Kirchmeyer S and Reuter K (2005) Scientific importance, properties and growing applications of poly(3,4-ethylenedioxythiophene). *Journal of Materials Chemistry* 15: 2077–2088.

oxidation must be balanced by anions; appropriate choice of these counter-ions significantly contributes to the properties of the conductive complex. The use of PSS, for instance, confers aqueous dispersibility, while the use of styrene sulfonate (the monomeric form of PSS) along with tetramethacrylate-terminated PEDOT chains confers dispersibility in many noncorrosive organic solvents [33].

4.2.1.2 PANI synthesis and redox chemistry

The mechanism underlying polymerization of PANI has received much attention and is more thoroughly studied than that of PEDOT [34–37]. Synthesis is initiated by the oxidation of aniline monomers to form resonantly stabilized anilinium radical cations. In highly acidic media ($pH \leq 3$), these cations preferentially couple with other anilinium monomers in a head-to-tail fashion at the *para* position [38]. This newly formed dimer ejects two electrons and two protons to make room for the next anilinium radical. The growing chain propagates by repeating the two-proton/two-electron/anilinium-addition process, as shown in **Scheme 4** [39].

Due to the acidic conditions in which linear PANI chains are formed, synthesis yields PANI in its proton-doped state [35]. Similar to PEDOT and polyacetylene, the conductive form of PANI is a polycationic molecule that requires charge-balancing anions. This metallic-green complex is known as the 'emeraldine salt' form of PANI, or ES, and is the protonated version of PANI's intermediate oxidation state. Exposing ES to a strong base de-dopes the complex and reveals the neutral intermediate oxidation state. This intermediate oxidation state of PANI is blue and is known as

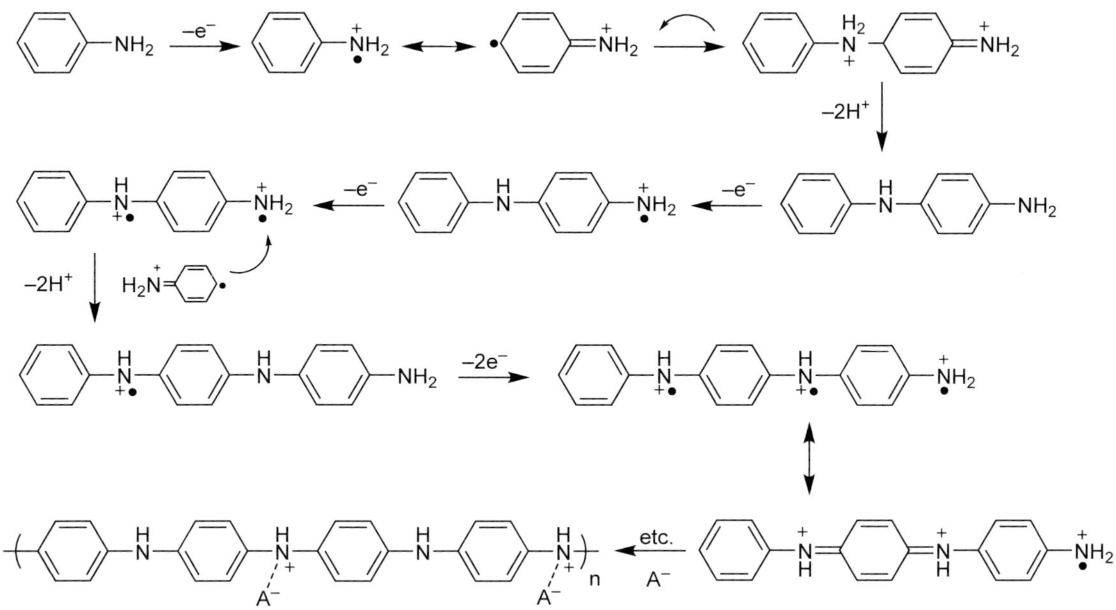

Scheme 4 Mechanism for the polymerization of aniline to PANI in the doped-ES state, where A^- represents a charge-balancing counter-ion. Adapted with permission from Geniès EM, *et al.* (1990) Polyaniline: A historical survey. *Synthetic Metals* 36: 139–182.

'emeraldine base' (EB). Subsequent oxidation or reduction of EB can further modify PANI (**Scheme 5**) [40]. In its fully oxidized form, the polymer is known as 'pernigraniline base' (PB) and appears violet, while in its fully reduced state, the polymer becomes colorless, and is known as 'leucoemeraldine base' (LB). It should be noted that, aside from ES, all forms of PANI are electrically insulating.

4.2.2 Synthetic Control

In general, PEDOT is synthesized from its monomer, EDOT, by oxidative polymerization in the presence of either small molecules or polymers bearing charge-balancing counter-ions. Oxidative polymerization of EDOT in the presence of small-molecule counter-ions results in the precipitation of PEDOT as a brittle and highly crystalline polymer with limited potential for further processing [41,42]. In a manner similar to PEDOT synthesis, aniline is oxidatively polymerized in the presence of small-molecule or polymer acids. Early research on PANI focused on small-molecule acid dopants, such as hydrochloric acid [43], camphorsulfonic acid [44], or dodecylbenzene sulfonic acid [21]. PANI produced through electrochemical or chemical synthesis in the presence of such small-molecule acids exhibited very interesting properties, most notably electrical conductivities in the order of 10^1–$10^2\ \mathrm{S\,cm^{-1}}$ [22,45]. A major drawback of small-molecule-doped PANI, much like small-molecule-balanced PEDOT, stems from its negligible solubility.

The solubility problem of these conducting polymers was subsequently mitigated by using water-soluble polyelectrolytes as the charge-balancing dopants during polymerization. Excess water-soluble pendant groups on such polyelectrolytes can impart water-dispersibility to the conducting polymers. The resulting polymers are therefore not only electrically conductive, but can also be processed directly from aqueous dispersions.

4.2.2.1 Synthetic control of PEDOT

The discovery of PEDOT in 1988 was quickly followed by Bayer's [46] development of the first polyelectrolyte-templated polymerization suitable for practical use in 1991. This technique involves polymerization of EDOT in an aqueous solution containing PSS with sodium persulfate as the oxidizing agent. PSS is a commercially available water-soluble polyanion and thus serves as an appropriate dispersant for aqueous PEDOT–PSS. Performing this reaction at room temperature yields a dark-blue aqueous PEDOT–PSS dispersion. This PEDOT–PSS dispersion is composed of electrostatically stabilized

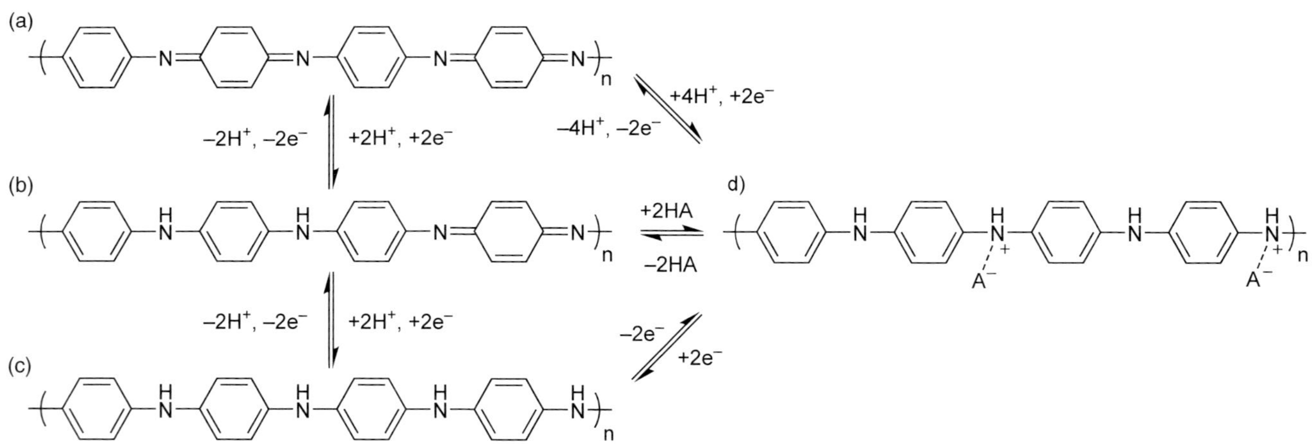

Scheme 5 Redox transitions in PANI are depicted in the vertical reaction sequence: (a) PB, (b) EB, and (c) LB. The horizontal reaction reflects proton doping of EB to (d) ES, where A^- represents charge-balancing counter-ions. Adapted with permission from Tarver J, *et al.* (2009) Polymer acid doped polyaniline is electrochemically stable beyond pH 9. *Chemistry of Materials* 21, 280–286.

particles formed by high-molecular-weight PSS chains interlinked by shorter PEDOT segments [32]. PEDOT–PSS films comprising coalesced particles cast from this dispersion are conductive, translucent (if sufficiently thin), mechanically durable, and insoluble in many common organic solvents. As such, their applications are wide ranging for light-emitting diodes [47], solar cells [48], and ECDs [49]. PEDOT has also been synthesized in the presence of other polyelectrolytes, such as poly(2-acrylamido-2-methyl-1-propanesulfonic acid) or PAAMPSA [50]. PEDOT–PAAMPSA, however, has yet to find practical use due to conductivities that are 3 orders of magnitude lower than those of comparable PEDOT–PSS systems.

4.2.2.2 Synthetic control of PANI

The influence of polymer acids on PANI – both as dopants and polymerization templates – was first explored in 1989 by Hwang and Yang [51]. Later studies by Angelopoulos *et al.* [27] established polymer acid template synthesis as a novel method of inducing water-dispersibility in PANI systems. A large variety of polymer acid dopants, for instance, poly(amic acid) [52], poly(acrylic acid) [26], poly(styrene sulfonic acid) [53], and poly(ethenesulfonic acid) [51] have been used to dope PANI. These polymer acid-doped PANI dispersions are also composed of electrostatically stabilized particles similar to those observed in PEDOT–PSS dispersions. In many cases, the structure of the PANI/polymer-acid complex is locked by strong ionic interactions between PANI and its polymer-acid template [52]. Consequently, the structure and properties of the polymer acid have a strong influence over those of the resulting conducting polymer complex [28,54]. The conductivities of these polymer-acid-doped PANI systems, however, have generally been low (10^{-5}–10^{-2} S cm^{-1}) [26,54]. Numerous efforts have been made to improve the electrical conductivity of PANI via synthetic manipulation of the polymer acid itself [29,55]. For example, Yoo *et al.* [28,29] tuned the molecular weight and molecular weight distribution of PAAMPSA during the template synthesis of PANI. The electrical conductivity of the resulting PANI–PAAMPSA systems systematically improved with decreasing PAAMPSA molecular weight and molecular weight distribution up to 2.5 S cm^{-1}, the highest conductivity among those previously reported for polymer-acid-doped PANI systems [29].

4.2.3 Secondary Doping Effects

The study of conducting polymers received a major wave of momentum in the 1990s when Heeger *et al.* discovered secondary-doping effects in PANI systems [56,57]. Their studies showed that exposure of camphor sulfonic-acid-doped PANI to *m*-cresol can increase the conductivity of the original polymer from 1–2 S cm^{-1} to nearly 400 S cm^{-1} [21]. This dramatic rise in conductivity was said to be induced by structural rearrangement from a compact-coil form of PANI to a more expanded conformation; this structural change persisted even after complete removal of *m*-cresol [22]. Given the ability to dramatically improve PANI's conductivity, these solvents were commonly termed as 'secondary dopants' for PANI.

Recently, it has been discovered that the conductivities of water-dispersible polymer-acid-doped conducting polymers can also be enhanced more than 100-fold via exposure to secondary dopants. For example, the conductivity of PEDOT–PSS films can be enhanced by more than 2 orders of magnitude when ethylene glycol (EG), dimethyl sulfoxide (DMSO), *N,N*-dimethylformamide (DMF), or sorbitol is added to the aqueous dispersion prior to film formation [58,59]. Despite several studies, however, the mechanism responsible for this conductivity enhancement remains unclear.

The conductivity of polymer-acid-doped PANI can also be enhanced by more than 2 orders of magnitude, through a postprocessing solvent annealing process rather than preprocessing solvent modification [31]. Solvents which are good plasticizers for the employed polymer acid dopant and possess lower acid dissociation constant (pK_a) than the polymer acid dopant can improve the conductivity of polymer-acid-doped PANI [31]. During solvent annealing, the solvent plasticizes the conducting polymer films, and unlocks the strong ionic interactions responsible for arresting the structure during polymerization. This process allows structural rearrangement to a conformationally favorable structure that enhances charge transport, and thus the conductivity is increased dramatically [31]. When these highly conductive, solvent-treated materials are incorporated into OTFTs as source and drain electrodes, or in OSCs and OLEDs as anodes, the devices show improved performances due to a significant reduction in the bulk resistances relative to analogous devices utilizing untreated conducting polymers [31].

4.3 Conductivity Relationships – Characterization of Optical, Electrochemical, and Structural

Successful synthesis of a conducting polymer is the first step toward device implementation; the second step involves thorough characterization of the molecular structure and its influence on material properties. The parameter of primary importance to many of the applications proposed for conducting polymers is electrical conductivity. Understanding the nature of conduction or charge transport has thus been a major focus of research on water-dispersible conducting polymers.

Efforts to systematically improve the electrical properties of conducting polymers have focused on optical, electrochemical, and structural characterization in the hopes of identifying specific attributes unique to highly conductive systems. Characterization of the structure–property relationships in PEDOT systems, however, has received minimal attention due to the commercial availability of highly conductive forms, like PEDOT–PSS. As such, the majority of studies on structure–property relationships in conductive polymers has been performed with PANI systems rather than PEDOT.

The most common optical technique used in the characterization of conducting polymer systems is ultraviolet-visible near-infrared (UV-Vis/NIR) spectroscopy. UV-Vis/NIR absorption can be measured in solution and in the solid state. Spectral characterization is most often used to quickly assess the oxidation state of conductive polymer systems. UV-Vis/NIR can also be used to determine the optical band gap of materials.

Electrochemical methods can be used to characterize the redox properties of solution-cast and electropolymerized thin films. The most common electrochemical technique, cyclic voltammetry (CV), is often used to determine the reversibility and potential dependence of redox transitions in conductive polymers. This information can subsequently be used to estimate the energy levels of the conductive polymers. By utilizing an optically transparent electrode, such as indium tin oxide (ITO), electrochemical and spectral characterization can be combined to simultaneously assess the optical and electronic properties of conducting polymers as they undergo redox transitions.

Methods of structural characterization to assess polymer conformation in dilute solutions include dynamic light scattering (DLS). The solid-state structure of films cast from these polymer dispersions can be further assessed using a variety of imaging techniques, such as atomic force, optical, and electron microscopy. Other techniques, such as solid-state nuclear magnetic resonance (NMR) and X-ray diffraction (XRD), are well suited for investigating the ensemble-average properties of the bulk, such as the presence of specific chemical environments and the extent of crystallinity, respectively.

4.3.1 Electrical Conductivity of Conducting Polymers

Water-dispersible PEDOT–PSS is commercially produced by Bayer AG under the trade name Clevios P. Various types of dispersion are commercially available and cover a wide range in conductivities, spanning from 10^{-5} to $300\,\mathrm{S\,cm^{-1}}$. The difference in conductivity between these PEDOT–PSS systems is obtained by changing the PEDOT to PSS mass ratio between 1:20 and 1:2.5, or by adding a secondary dopant into the PEDOT–PSS dispersion. For example, Clevios P CH 8000 is synthesized at a 1:20 mass ratio of PEDOT to PSS to yield a polymer complex with conductivities of $10^{-5}\,\mathrm{S\,cm^{-1}}$. Clevios P Al 4083 and PH 500 are synthesized at 1:6 and 1:2.5 PEDOT to PSS mass ratios; these polymer complexes exhibit conductivities of 10^{-3} and $1\,\mathrm{S\,cm^{-1}}$, respectively. Recently, a variant of Clevios P PH 500 has been released in which DMSO is added in 5 wt% quantities relative to PEDOT–PSS solids content; conductivities up to $300\,\mathrm{S\,cm^{-1}}$ have been reported for this material.

To determine the nature of macroscopic conduction in PEDOT–PSS, the morphology of PEDOT–PSS films has been widely investigated. Greczynski *et al.* [60]. have shown by X-ray photoelectron spectroscopy (XPS) that an insulating layer of PSS preferentially segregates to the surface of PEDOT–PSS films. This phase segregation has been also examined by Jukes *et al.* using small-angle neutron scattering [61]. Given the vertical phase separation, it is not surprising that PEDOT–PSS films exhibit anisotropic conductivities. Conduction, as expected, is inefficient through the depth of the film and enhanced in the plane of the film (10^{-5} and $10^{-3}\,\mathrm{S\,cm^{-1}}$, respectively) [62]. **Figure 1** shows (a) the surface and (b) the cross-sectional morphologies of PEDOT–PSS films. Together, these images suggest the morphological model shown in **Figure 1(c)**. The images taken on thin-film cross sections show

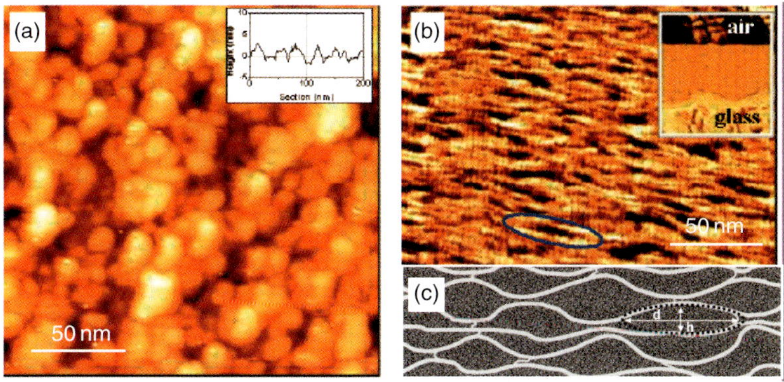

Figure 1 (a) 200 nm × 200 nm scanning tunneling microscopy (STM) image of PEDOT–PSS on indium tin oxide (ITO). Inset shows a line cross section; (b) 200 nm × 200 nm cross-sectional atomic force microscopy (AFM) phase image of cleaved PEDOT–PSS on glass. A pancake-like particle is highlighted by the ellipse; (c) Cross-sectional view of the schematic morphological model for PEDOT–PSS thin films. PEDOT-rich regions (dark) are separated by insulating PSS (light). With permission from Nardes AM, *et al.* (2007) Microscopic understanding of the anisotropic conductivity of PEDOT-PSS thin films. *Advanced Materials* 19: 1196–1200.

horizontal layers of PEDOT-rich regions separated vertically by insulating PSS [62]. This layered morphology plays a large role in governing the conductivity of PEDOT–PSS.

Researchers have identified a number of methods capable of enhancing the conductivity of PEDOT–PSS through structural modification. The degree of phase separation between PEDOT-rich regions and the surrounding PSS, for instance, can be reduced when PEDOT–PSS films are heated [63]. **Figure 2** shows the conductivity of PEDOT–PSS films with annealing temperature. The conductivity of PEDOT–PSS films increases with annealing temperature until 200 °C, at which degradation of the

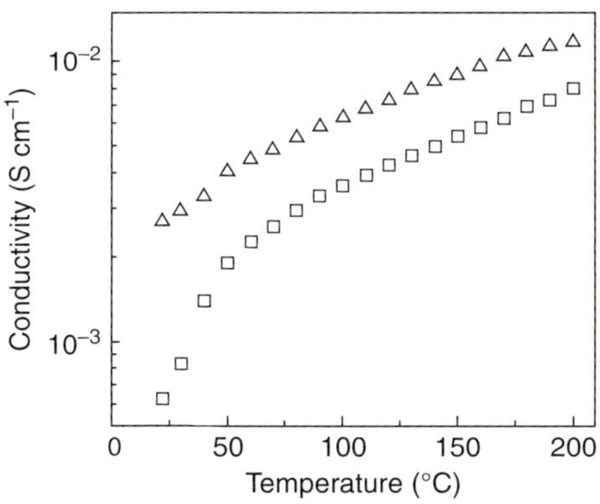

Figure 2 Conductivity enhancement of PEDOT–PSS films with annealing. Triangles represent measurements in N_2 and squares represent measurements in air. With permission from Huang J, *et al.* (2003) Influence of thermal treatment on the conductivity and morphology of PEDOT-PSS films. *Synthetic Metals* 139: 569–572.

conductive polymer takes place [63]. Studies by Kirchmeyer and Reuter [32] utilized ultracentrifugation to fractionate particles of PEDOT–PSS. The conductivity of films cast from the fractionated samples scaled directly with the mean particle size. This phenomenon was speculated to result from the decreased volume fraction of the insulating PSS surface layer as the PEDOT-rich core increased in size.

The addition of water-miscible, high-boiling solvents, such as DMSO, EG, NMP, or DMF, has also been observed to strongly impact conductivity. It is thought that the solvent relaxes the PSS overlayer, allowing structural rearrangement to a conformationally favorable state. Proper solvent annealing thus results in conductivity enhancements of more than 2 orders of magnitude [59,64].

Identifying methods of manipulating structure to increase the conductivity of water-dispersible polymer acid-doped PANI has also been a subject of much research interest. The conformation of PANI complexes is highly dependent on synthetic parameters, such as reaction temperature and molar ratio of reactants, as well as the specific type of oxidant or dopant used during the synthesis. As these parameters have a large influence on the structure of the final product, they consequently have a strong influence on the conductivity of the resulting complex. For example, **Figure 3** shows the conductivity of PANI that is doped with poly(thiophenylenesulfonic acid) (PTPSA), as a function of the molar ratio of sulfonic acid groups in PTPSA to aniline [65]. When polymerized at low acid to aniline-mer molar ratios, the conductivity of the resulting polymer complex is

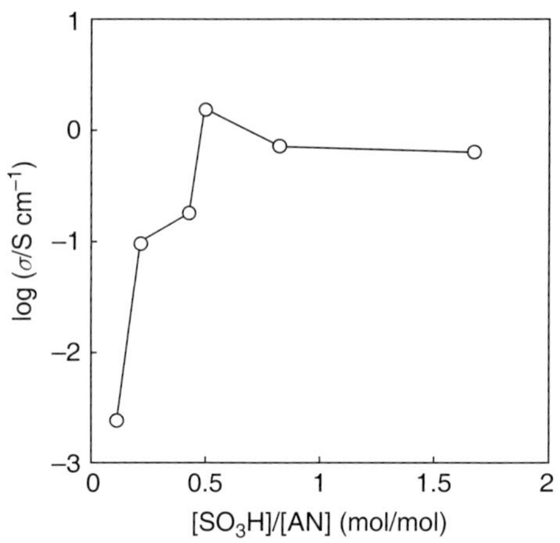

Figure 3 Effect of molar ratio of sulfonic acid group to aniline unit ([SO$_3$H]/[AN]) on the conductivity of PANI–PTPSA. With permission from Miyatake K, *et al.* (1999) Preparation and properties of polyaniline doped with poly(thiophenylenesulfonic acid). *Macromolecular Chemistry and Physics* 200: 2597–2601.

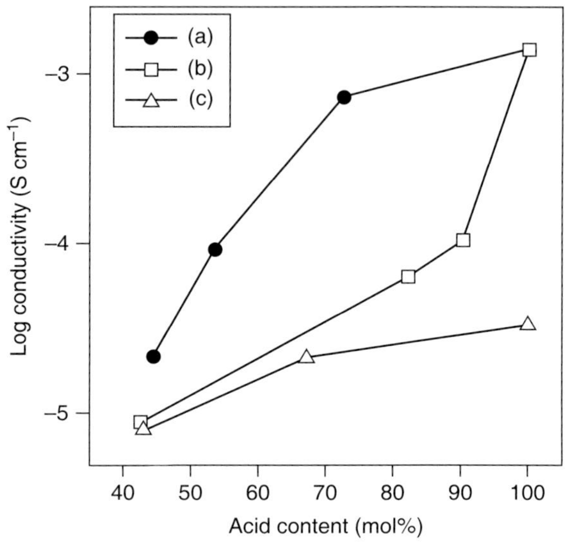

Figure 4 Conductivities of doped PANI vs. copolymer composition: (a) PANI–PMMA$_{1-n}$-co-PSS$_n$; (b) PANI–PS$_{1-n}$-co-PSS$_n$; (c) PANI–PMMA$_{1-n}$-co-PAAMPSA$_n$, where *n* indicates the mole percent of acidic mers in the copolymer.With permission from Moon HS and Park JK (1998) Structural effect of polymeric acid dopants on the chracteristics of doped polyaniline composites: Effect of hydrogen bonding. *Journal of Polymer Science, Part A: Polymer Chemistry* 36: 1431–1439.

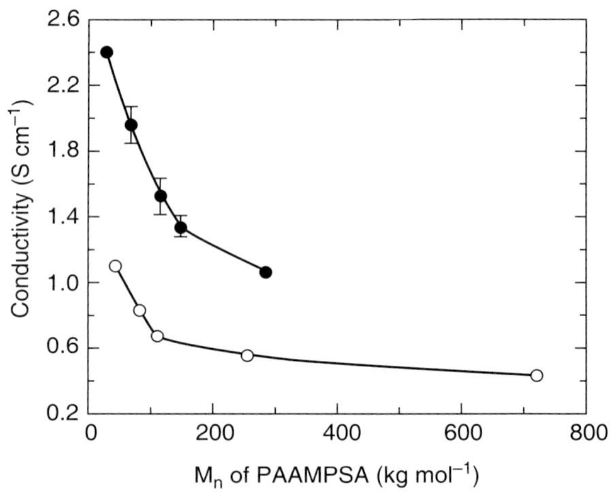

Figure 5 Conductivity of PANI–PAAMPSA as a function of PAAMPSA molecular weight. Broad (open circles) and narrow (close circles) molecular weight distribution PAAMPSA were used to dope PANI. With permission from Yoo JE, *et al.* (2008) Narrowing the size distribution of the polymer acid improves PANI conductivity. *Journal of Materials Chemistry* 18: 3129–3135.

limited by the concentration of dopant. Beyond a 1:2 acid to aniline molar ratio, however, the presence of additional sulfonic acid groups have negligible effects on PANI conductivity [65]. This trend agrees with the doping mechanism presented in **Scheme 5**, in which one acid group is required for a pair of aniline mers to achieve complete doping.

Moon and Park [66] investigated the effects of co-monomer sequence distribution within the polymer acid dopants on the conductivity of PANI; they employed three different copolymers comprising neutral and acidic mers, specifically poly(methyl methacrylate-co-styrenesulfonic acid) (denoted PMMA-co-PSS), poly(styrene-co-styrenesulfonic acid) (denoted PS-co-PSS), and poly(methylmethacrylate-co-2-acrylamido-2-methyl-1-propanesulfonic acid) (denoted PMMA-co-PAAMPSA) for the template synthesis of PANI, and assessed the influence of acid content within the copolymers on the conductivity of PANI (**Figure 4**). PANI conductivity increased with increasing acid content in all cases; PMMA-co-PSS doped PANI consistently exhibited the highest conductivities among the three conductive polymer complexes. The elevated conductivity in this case was attributed to enhanced hydrogen-bonding opportunities provided by the carbonyl groups within the MMA-mers.

Yoo *et al.* [28,29] investigated the influence of the molecular characteristics of PAAMPSA on the electrical properties of PANI (**Figure 5**). The conductivity of PANI–PAAMPSA was found to increase with decreasing PAAMPSA molecular weight and narrowing molecular weight distributions

[28,29]. Further examination revealed that PANI–PAAMPSA films are comprised of particles, the size and size distribution of which scale with PAAMPSA molecular weight and molecular weight distribution [67]. Depth profile measurements of PANI–PAAMPSA by XPS show PANI on the outside of the particles [67], contrary to the structure of PEDOT–PSS particles [68,69]. The macroscopic conduction of PANI–PAAMPSA is consequently along, rather than through, the particles within the cast film. Accordingly, conductivity in PANI–PAAMPSA scales with particle surface area per unit film volume, which increases with decreasing particle size and particle size distribution [67].

To further increase the conductivity of polymer-acid-doped PANI, spun-cast films can be solvent annealed. One solvent used for this conductivity improvement in PANI is dichloroacetic acid (DCA). Annealing in the presence of DCA can enhance the PANI conductivity by more than 2 orders of magnitude [31]. Such solvent annealing allows the conducting polymer to rearrange from the particulate structure arrested during synthesis to a conformationally more favorable structure. This relaxation dramatically enhances charge transport. Preliminary studies suggest that this solvent annealing is general for many polymer-acid-templated systems; the conductivity of PEDOT–PSS can be improved to near 250 S cm^{-1} by this process [31]. Conducting polymer films treated in this manner makes efficient source and drain electrodes for OTFTs, as well as anodes for OLEDs and OPVs [31].

4.3.2 Optical Characterization of Conducting Polymers

Virtually, all conducting polymers are redox active. In many of these polymers, the conductive form is achieved through oxidative doping. Although PANI utilizes a proton-doping mechanism, it has a number of redox states as well. Along with electrical conductivity, optical properties are sensitive to changes in the redox states. This phenomenon serves as the basis for all ECDs [70]. Oxidation and reduction of conductive polymers can induce significant changes in the coloration. One of the simplest methods of characterizing correlations between the oxidation state and optical properties of conducting polymer systems is UV-Vis/NIR spectroscopy.

The UV-Vis/NIR spectra of PEDOT–PSS are shown as a function of increasing electrical potential in **Figure 6** [71]. At reducing potentials (<−0.4 V vs.

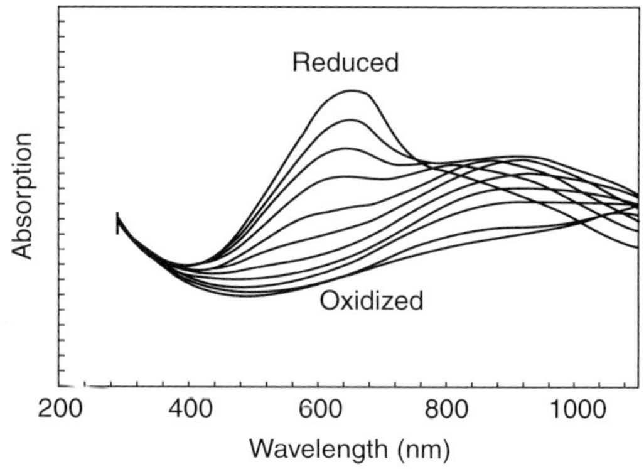

Figure 6 *In situ* UV-Vis/NIR absorption spectra for PEDOT:PSS as a function of increasing applied potential. Adapted with permission from DeLongchamp D and Hammond PT (2001) Layer-by-layer assembly of PEDOT/polyaniline electrochromic devices. *Advanced Materials* 13: 1455–1459.

Ag/AgCl), the polymer exhibits a peak at 600 nm that is characteristic of its neutral de-doped state [72,73]. As the potential is steadily increased, the intensity of this peak declines at the expense of a broad absorption band centered near 800 nm [71–73]. The emergence of this band is associated with the oxidation of PEDOT and the concurrent onset of conductivity [72]. Further oxidation produces bipolarons that give rise to a free carrier tail extending into the NIR region of the spectrum. At highly oxidizing potentials (>0.8 V vs. Ag/AgCl), the free carrier tail grows at the expense of the peak at 800 nm as more bipolarons are generated, but conductivity begins to decline due to disruption of continuous conductive pathways [72]. By tracking changes in absorption at specific wavelengths, namely 600 and 800 nm, UV-Vis/NIR thus provides a valuable method of quickly determining the extent of PEDOT oxidation, and hence, the extent of doping or the macroscopic conductivity. It should be noted that while secondary doping of PEDOT–PSS dramatically increases the electrical conductivity, it has a negligible influence on the system's absorption spectrum, as shown in **Figure 7**.

The UV-Vis/NIR absorption characteristics of PANI are more diverse than those of PEDOT due to PANI's complex redox and proton-doping chemistries. Each of the oxidation and doped states shown in **Scheme 5** possesses unique optical properties, as shown in **Figure 8**. In the fully reduced LB state, PANI exhibits an isolated peak near 330 nm attributed to the π–π^* transition of the benzenoid

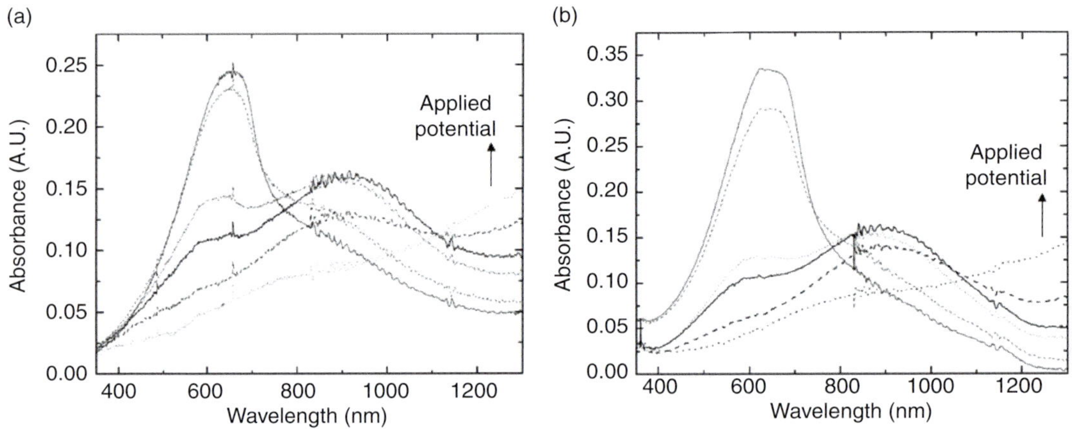

Figure 7 UV-Vis/NIR spectra of PEDOT–PSS as a function of applied potential (a) before and (b) after secondary doping with ethylene glycol. With permission from Ouyang J, *et al.* (2004) On the mechanism of conductivity enhancement in poly(3,4-ethylenedioxythiophene):poly(styrene sulfonate) film through solvent treatment. *Polymer* 45: 8443–8450.

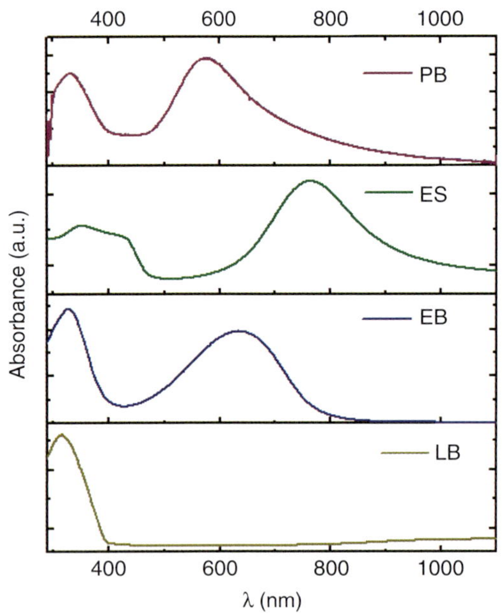

Figure 8 Representative UV-Vis/NIR spectra for PB, ES, EB, and LB forms of PANI. Adapted with permission from Tarver J, *et al.* (2008) Polymer acid doped polyaniline is electrochemically stable beyond pH 9. *Chemistry of Materials* 21, 280–286.

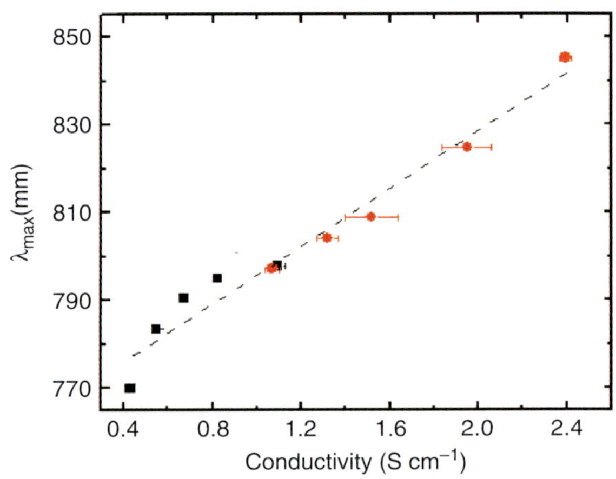

Figure 9 (a) UV-Vis/NIR polaron peak maximum, λ_{max}, of PANI–PAAMPSA as a function of the conductivity of the materials. PANI is doped with broader (■) and narrower (●) molecular weight distribution PAAMPSA.

moiety [74,75]. Upon oxidation to the conductive ES state, the intensity of this peak declines and a shoulder emerges near 420 nm [76,77]. A separate peak near 800 nm corresponding to a polaron interband transition appears, and is accompanied by a free carrier tail extending into the NIR region for highly conductive samples [74,77]. The precise position of this peak (λ_{max}) has been reported by Yoo *et al.* to red shift with increasing conductivity in PANI–PAAMPSA, shown in **Figure 9** [28]. This red shift is suggestive of delocalized charges over larger length scales, and thus implies an increase in the PANI conjugation length, ultimately resulting in more conductive PANI [28,75,77]. Deprotonation of ES yields EB and is marked by the elimination of the shoulder near 420 nm as well as a hypsochromic shift of the ~800 nm peak toward 650 nm [77]. Oxidation to PB further blue shifts this peak toward 550 nm [78]. Unlike PEDOT systems, in which secondary doping has little spectral manifestation, the UV-Vis/NIR spectra of PANI systems exhibit a dramatic increase

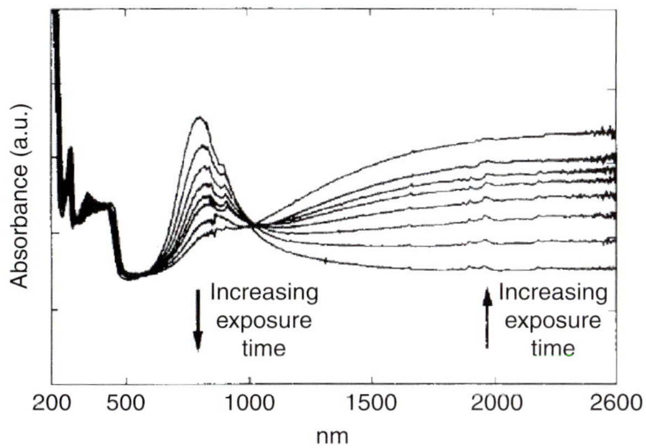

Figure 10 UV-Vis/NIR spectra of camphor sulfonic-acid-doped PANI as a function of time exposed to *m*-cresol. Adapted with permission from MacDiarmid AG and Epstein AJ (1994) The concept of secondary doping as applied to polyaniline. *Synthetic Metals* 65: 103–116.

in the free carrier tail upon exposure to secondary dopants. This observation is consistent with the UV-Vis/NIR spectra obtained on small-molecule acid-doped PANI systems after exposure to secondary dopants, shown in **Figure 10**. Secondary doping is proposed to enhance delocalization of electrons within the polaron band through expansion of the PANI coil [22].

UV-Vis/NIR spectroscopy can also be used to monitor the optical changes that occur during chemical synthesis. Kinetic and mechanistic information can be determined by performing time-resolved spectroscopy on polymerizations carried out under dilute conditions. This technique has been applied to the template synthesis of PANI along both PSS [79] and a variety of poly(amidosulfonic acids) [28,29, 80–82]; the time-resolved spectra obtained during one such polymerization is shown in **Figure 11**. The initial stages of synthesis appear to yield PANI in a mixture of ES and PB states, as indicated by peaks near 550 nm and 750 nm. As the polymerization proceeds, these peaks merge at roughly 600 nm and grow in intensity; during the final stages of synthesis this peak red shifts to 750 nm with a simultaneous emergence of the 420-nm shoulder, indicating the conversion to conductive ES.

4.3.3 Electrochemical Characterization of Conducting Polymers

Like UV-Vis/NIR, electrochemical characterization can be applied to solutions or to thin films. The most common electrochemical technique used to characterize conducting polymer systems is CV. CV experiments probe a system's characteristic redox potentials and assess the electrochemical reversibility between oxidation states. Used in combination with spectroscopic methods, CV can probe concurrently occurring structural and electronic changes in conducting polymer systems.

A typical cyclic voltammogram for PEDOT:PSS films is shown in **Figure 12**. The oxidation and reduction peaks are broad and poorly resolved, suggestive of significant capacitive charging [71,83,84]. Spectroelectrochemical measurements acquired on a similar PEDOT during a CV cycle demonstrate that changes in the electronic structure occur across the entire range of applied potential, as

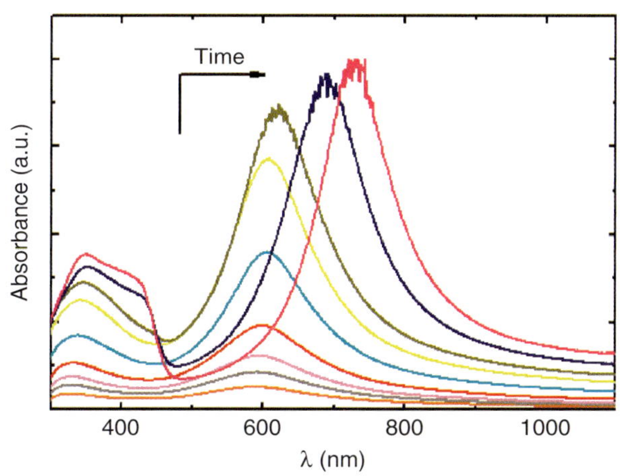

Figure 11 Time-resolved UV-Vis/NIR traces acquired during the chemical synthesis of PANI–PAAMPSA with ammonium persulfate.

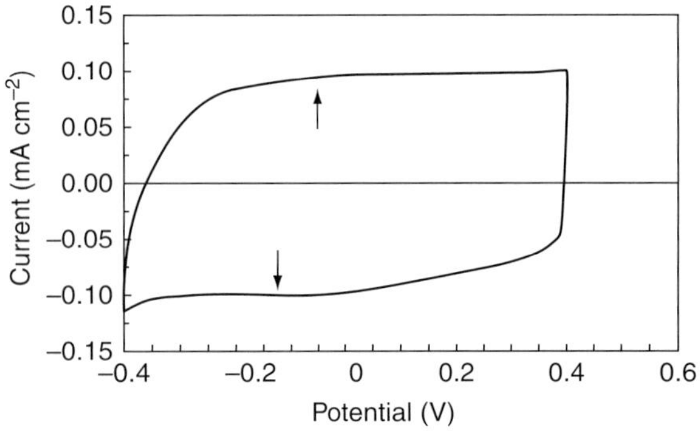

Figure 12 Cyclic voltammogram of PEDOT:PSS. A poorly resolved redox couple appears near −0.1 V. ↑ and ↓ represents the oxidation and reduction reaction potentials, respectively. Adapted with permission from DeLongchamp D and Hammond PT (2001) Layer-by-layer assembly of PEDOT/polyaniline electrochromic devices. *Advanced Materials* 13: 1455–1459.

shown in **Figure 13** [71]. Similar studies performed on electropolymerized PEDOT suggest that potentials beyond 1.4 V result in over-oxidation of the polymer, indicated by an increase in absorbance in the NIR range and an irreversible drop in conductivity [72].

The electrochemistry of PANI is more complex than that of PEDOT. While PEDOT only possesses doped and de-doped states, PANI can dwell in any of its three oxidation states, in addition to its doped ES form. As a consequence, PANI systems typically show at least two redox couples in acidic media: transition between PB and emeraldine and that between emeraldine and LB. Whether the emeraldine intermediate is that of the conductive ES or that of the insulating EB is obscured by the fact that both states share the same number of electrons. EB and ES thus exhibit similar electrochemical properties. The protonic nature of PANI's redox chemistry, however, confers pH dependence to its redox potentials in accordance to the Nernst equation [85]; ES can thus be distinguished from EB by determining the pH dependence of the redox potentials [40].

This analysis has been performed on poly(acrylic acid)- [86], PSS- [87], and PAAMPSA-templated PANI [40]. Poly(acrylic acid)- and PAAMPSA-templated PANI, for instance, show transitions between PB and ES and between ES and LB below pH 4, as detected by spectroelectrochemical experiments at varying pHs. At pH 4 and higher, PANI that is electropolymerized in the presence of poly(acrylic acid) shows only direct LB to PB transitions [86]. Chemically synthesized PANI–PAAMPSA, however, exhibits an ES intermediate state that is stable through pH 7, and direct LB to PB transitions beyond pH 9 (see **Figure 14**) [40]. This difference is said to arise from stronger interactions between PANI and its PAAMPSA template, compared to those with other polymer acids, which provide a higher degree of structural integrity relative to other template polymerized systems.

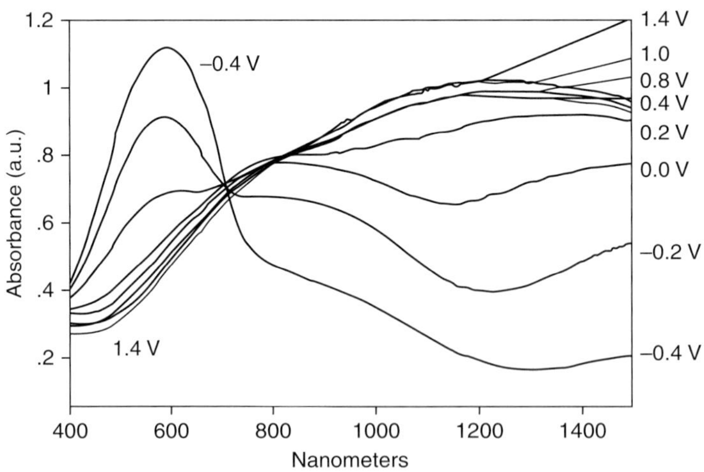

Figure 13 *In situ* UV-Vis/NIR absorption spectra for PEDOT as a function of applied potential (vs. Ag/AgCl). Positive potentials are oxidizing while negative potentials reduce PEDOT. Adapted with permission from Lapkowski M and Pron A (2000) Electrochemical oxidation of poly(3,4-ethylenedioxythiophene) – *in-situ* conductivity and spectroscopic investigations. *Synthetic Metals* 110: 79–83.

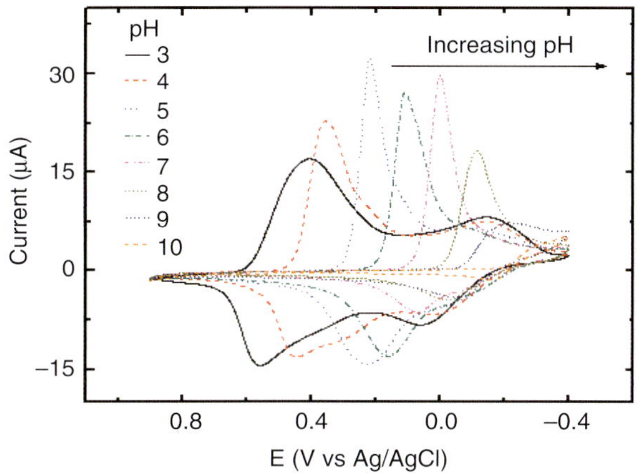

Figure 14 Cyclic voltammograms of PANI–PAAMPSA as a function of pH. The scan rate is held constant at 20 mV s^{-1}. Adapted with permission from Tarver J, *et al.* (2009) Polymer acid doped polyaniline is electrochemically stable beyond pH 9. *Chemistry of Materials* 21, 280–286.

4.3.4 Structural Characterization of Conducting Polymers

4.3.4.1 Solid-state NMR

Solid-state NMR offers a promising way to determine the structure and conformation of many conducting polymer systems. Owing to the delocalization of electrons in the conductive form of PEDOT, however, NMR signals suffer from paramagnetic broadening [88]. This broadening is very strong for both the thiophene nuclei atoms as well as for the oxyethylene bridge in the conducting polymer [88]. Reduction with hydrazine dedopes PEDOT; this elimination of delocalized electrons along the polymer backbone results in higher-resolution NMR spectra [88]. It is difficult, however, to remove the polymer acid template completely when dedoping PEDOT–PSS due to the mass transfer limitations associated with the removal of high-molecular-weight polymers. This complication hinders the quantitative characterization of PEDOT–PSS by solid-state NMR.

Solid-state NMR studies have focused more on PANI than PEDOT, primarily due to the presence of disparate nitrogen environments identifiable via ^{15}N NMR in the former [55,89,90]. In addition to the more conventional ^{1}H NMR and ^{13}C NMR, ^{15}N NMR is a useful technique for studying PANI due to its sensitivity to the chemical environments of charged and uncharged nitrogens in the polymer backbone. The chemical shifts due to differences between charged and uncharged nitrogens in ^{15}N NMR are large compared to those due to differences between carbon environments detected by ^{13}C NMR. ^{15}NMR is, therefore, better suited for characterizing the structural changes that may occur as a result of synthesis. For example, the chemical shift difference between the resonances arising from amine and imine structures is ~270 ppm, whereas the chemical shift difference between the corresponding benzenoid and quinoid moieties is only ~10–20 ppm in ^{13}C NMR. In addition, it is possible to distinguish the extent of charge distribution on various nitrogen species in PANI with ^{15}N NMR.

Figure 15 shows the ^{15}N solid-state NMR spectra of PANI doped with (a) poly(vinylphosphonic acid), PVP [55], and (b) that of PANI doped with PAAMPSA [28]. The conductivities of PANI–PVP and PANI–PAAMPSA are 10^{-2} and 1 S cm^{-1}, respectively, due to differences in their doping levels [28,55]. PANI–PVP is not fully doped, as indicated by the multiple peaks in the ^{15}N NMR spectrum (**Figure 15(a)**). These peaks arise from the presence of multiple nitrogen environments, such as residual unprotonated imine nitrogens (denoted as I in **Scheme 6(a)**), protonated imine nitrogens (denoted as III in **Scheme 6(b)**), and uncharged secondary amine nitrogens (denoted as IV in **Scheme 6(b)**) [55,91]. In contrast, the ^{15}N NMR spectrum of fully doped PANI–PAAMPSA shows only one relatively

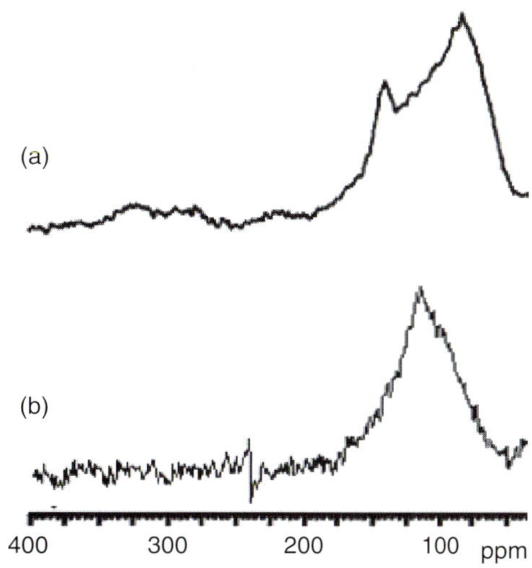

Figure 15 ^{15}N solid state NMR spectra of (a) PANI–PVP and (b) PANI–PAAMPSA. (a) From Sahoo SK, *et al.* (2004) An enzymatically synthesized polyaniline: A solid-state NMR study. *Macromolecules* 37: 4130–4138. (b) From Yoo JE, *et al.* (2007) Improving the electrical conductivity of polymer acid-doped polyaniline by controlling the template molecular weight. *Journal of Materials Chemistry* 17: 1268–1275.

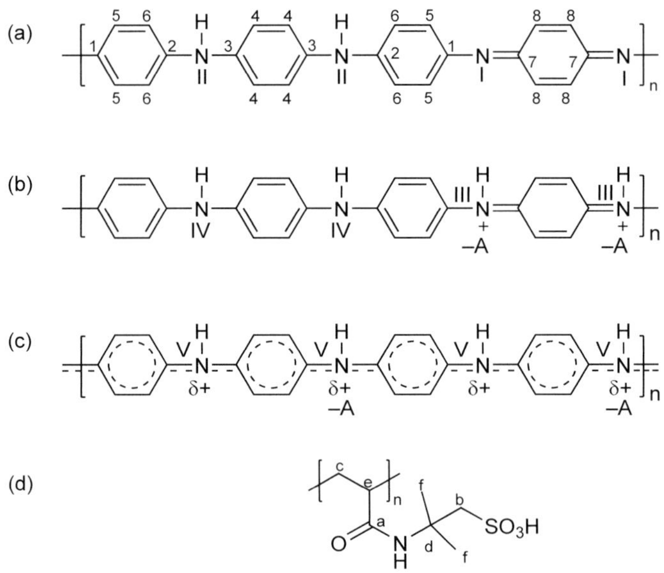

Scheme 6 (a) EB form of PANI is doped to (b, c) ES with (d) PAAMPSA polymer template. Carbon environments of EB are labeled 1–8 while those of PAAMPSA are labeled a–f. Nitrogen environments for PANI are labeled I–V. From Yoo JE, *et al.* (2007) Improving the electrical conductivity of polymer acid-doped polyaniline by controlling the template molecular weight. *Journal of Materials Chemistry* 17: 1268–1275.

narrow peak near 115 ppm (**Figure 15(b)**), which is assigned to the protonated amines whose charges are delocalized along the PANI–PAAMPSA backbone (denoted as V in **Scheme 6(c)**) [55,89,90].

4.3.4.2 X-ray diffraction

Empirical studies have established strong correlations between crystallinity and conductivity in conducting polymer systems [54,92–94]. The crystallinity of conducting polymers is, in turn, highly dependent on the conditions with which they are synthesized, particularly the type of dopant used. While the structures of PEDOT salts, including *p*-toluenesulfonate [92], hexafluorophosphate [93], and perchlorate [94], have been successfully investigated by XRD, structural studies on PEDOT–PSS films using XRD have failed to yield meaningful results [95]. Typically, polymer-acid-doped conducting polymers have very low crystallinity because the high-molecular-weight polymer acid is amorphous. Further, the presence of the high-molecular-weight polymer acid can kinetically suppress the crystallization of the conducting polymer with which it is ionically associated. As such, the XRD patterns for PEDOT–PSS films only show broad amorphous halos and do not reveal any sharp Bragg reflections [95].

Nevertheless, XRD studies on polymer-acid-doped PANI have been carried out to determine the influence of structure on electrical properties, and unlike work with PEDOT–PSS, have yielded informative findings. The relative molar composition of polymer acid and PANI has been observed to influence ordering and chain alignment (**Figure 16**) [54]. In general, the XRD patterns of PANI–PSS only show a weak peak at $2\theta = 9°$. The XRD pattern of PANI–PSS with a 1:1 molar ratio of sulfonic acid to aniline mer, however, shows a clear reflection peak

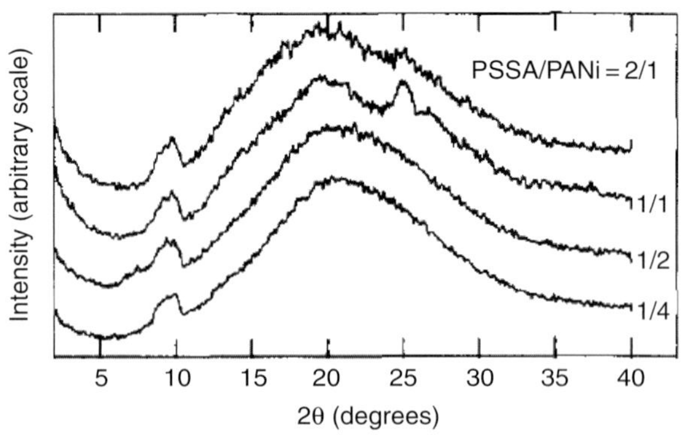

Figure 16 XRD patterns of PANI–PSS with varying molar ratios of PSS (termed PSSA in figure) to PANI. From Moon HS and Park JK (1998) Structural effect of polymeric acid dopants on the characteristics of doped polyaniline composites. *Synthetic Metals* 92: 223–228.

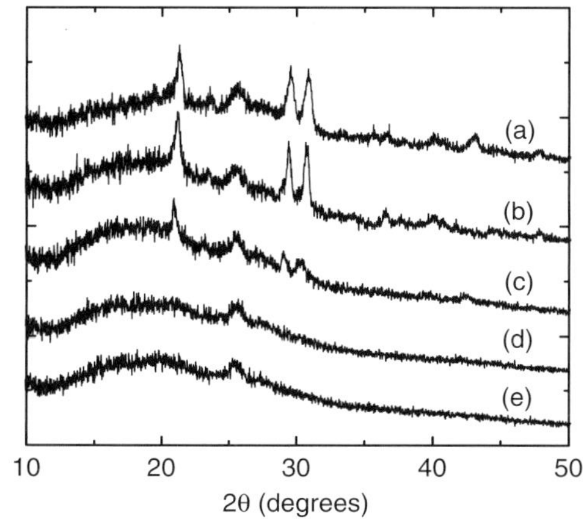

Figure 17 XRD patterns of PANI–PAAMPSA. The molecular weight of PAAMPSA is (a) 45, (b) 83, (c) 106, (d) 255, and (e) 724 kg mol^{-1}. From Yoo JE, *et al.* (2007) Improving the electrical conductivity of polymer acid-doped polyaniline by controlling the template molecular weight. *Journal of Materials Chemistry* 17: 1268–1275.

at $2\theta = 26°$ that is closely associated with π–π stacking in crystalline PANI domains; this PANI–PSS has, in turn, shown the highest conductivity [54]. The crystallinity of PANI can be further increased by properly tuning the molecular characteristics of polymer acid template. **Figure 17** shows the XRD patterns of PANI–PAAMPSA with varying PAAMPSA molecular weight. The level of crystallinity increases, as observed by increases in the Bragg reflection intensities, with decreasing PAAMPSA molecular weight, which is directly correlated with electrical conductivities of PANI–PAAMPSA [28].

4.4 Applications

Electronic devices have historically depended on inorganic materials for both conductor and semiconductor components: copper and silver are the two leading sources for wiring, while silicon is by far the most used semiconductor material for conventional electronics. Although the electrical properties of these materials are perfectly suited for present-day, high-tech electronic applications, their mechanical properties leave much to be desired in terms of flexibility and elasticity in specialty pursuits. In addition, processing techniques for inorganic materials, though well established, frequently involve extreme temperatures and/or greatly reduced pressures, therefore necessitating complex and costly equipment.

Conducting polymers, alternatively, are produced through versatile synthetic routes involving inexpensive monomeric reactants and reaction conditions that only deviate slightly from ambiance. When dispersed in solution, these polymers can be deposited with ease via ink-jet printing or spin coating, obviating the need for elevated temperatures and reduced pressures associated with conventional processing. In addition to these cost-reduction benefits, conductive polymers often possess the mechanical properties associated with plastics, such as light weight, flexibility, elasticity, and moldability. The promise of this unique combination of mechanical and electrical properties in conductive polymers permits their novel implementation in a number of specialty applications.

OLEDs and OTFTs already utilize small molecules as active components in thin-panel televisions and other electronic appliances, resulting in devices that are lower in current relative to inorganics. Much work is underway toward the development of organic photovoltaic devices with the intent of trading efficiency for reduced cost and ease in manufacturing. These first-generation organic devices, however, still rely on metal electrodes and silicon-active layer, and thus necessitate several of the costly processing steps associated with inorganics. Although conductive polymers are not anticipated to replace copper wires, they are perfectly suited substitutes for conducting and semiconducting materials in low-current specialty applications.

Industries previously unassociated with electronics could also incorporate conductive polymers. Clothes, for example, could be manufactured with light-weight heating elements and OLED displays throughout; food and beverage packaging materials could be produced with cheap chemical sensors to detect contamination or decomposition products; multiple newspapers, blogs, and books could be stored on a single sheet of electronic paper with flexibility rivaling its pulp-based counterpart. Electrochromic smart windows used in tandem with OSCs could automatically adjust the amount of light transmitted into or out of a room, thus serving as both tint and curtain. Supplanting the metallic components of a circuit board with conducting polymer analogs, for instance, could be accomplished without compromising the integrity of many low-end products while simultaneously reducing their fabrication costs. In short, conductive polymer-based devices hold much promise for low-current situations in which inorganic materials are overqualified, cost prohibitive, or mechanically deficient.

4.4.1 Organic Thin-Film Transistors

The basic concept of the insulated-gate field-effect transistor (FET) was successfully demonstrated with the invention of the metal–oxide–semiconductor FET (MOSFET) in the early 1960s. The first organic transistors containing conducting polymers were devised in the early 1980s utilizing thin films of polyacetylene as the semiconducting layer, but suffered from poor performance [96]. Significantly improved devices appeared in the late 1980s with the advent of semiconducting polythiophene [97]. Since the early 1990s, remarkable progress has been made in the development of OTFTs utilizing conducting polymer active layers.

Figure 18 shows cross-sectional views of the two types of OTFT architectures. The active part of the device is composed of an organic semiconductor thin film in contact with two electrodes, the source and the drain. The region between the source and the drain is known as the channel. A thin insulating layer, known as the gate dielectric layer, isolates the source and drain electrodes and the organic semiconductor layer from the gate, the third electrode in OTFTs. In bottom-contact devices (**Figure 18(a)**), the source and drain electrodes are prepatterned on the gate dielectric layer, and the organic semiconductor is deposited on top. In top-contact devices (**Figure 18(b)**), the organic semiconductor layer is applied directly to the gate insulator and the source and drain electrodes are deposited on the organic semiconductor subsequently to define the channel region of the OTFT.

Without any gate bias, the conductivity of the organic semiconductor channel is extremely low; the concentration of charge carriers between source and drain electrodes is minimal. The application of a positive (or negative) gate voltage (V_G) gives rise to a large electric field at the organic semiconductor and insulator interface; this field induces electrons (or holes) to accumulate at the interface. These charge carriers can thus flow from the source electrode to the drain electrode when a potential (V_{SD}) is applied across the channel. The channel thus becomes highly conductive; the flow of charge carriers results in a significant source–drain current (I_{SD}). This type of device operation is the most common in OTFTs, and is known as accumulation mode operation. An alternative form of operation, known as depletion mode operation, allows large current flows for $V_G = 0$. In devices operating via this mode, current flow is throttled down by the application of a finite gate voltage.

To evaluate the device characteristics of OTFTs, I_{SD} is measured as a function of V_{SD} under constant V_G. Assuming ohmic contacts between the organic semiconductor channel and the source and drain electrodes, the current can be described in the linear ($I_{SD,lin}$) and saturation ($I_{SD,sat}$) regions by **equations 1** and **2**, respectively:

$$I_{SD,lin} = \frac{W}{L} C_i \mu (V_G - V_T) V_{SD} \tag{1}$$

$$I_{SD,sat} = \frac{W}{2L} C_i \mu (V_G - V_T)^2 \tag{2}$$

where W and L represent the channel width and length, respectively, and C_i represents capacitance per unit area of the gate insulator. The field-effect mobility, μ, measures how efficiently charge carriers travel through the organic semiconductor channel, while threshold voltage, V_T, represents the voltage at which the OTFT turns on. Figures of merit that characterize device operation, including μ and V_T, can be estimated from **equations 1** and **2**, given the geometry and dimensions of the OTFT, as well as the operating voltages. The on–off current ratio (I_{on}/I_{off}) is a measure of the contrast between the on and off states of OTFTs, and is obtained from the ratio of the maximum and minimum source and drain currents in the gate voltage operation range.

In most initial research that investigated the characteristics of OTFTs, noble metals, such as gold and platinum, were used as source and drain electrodes.

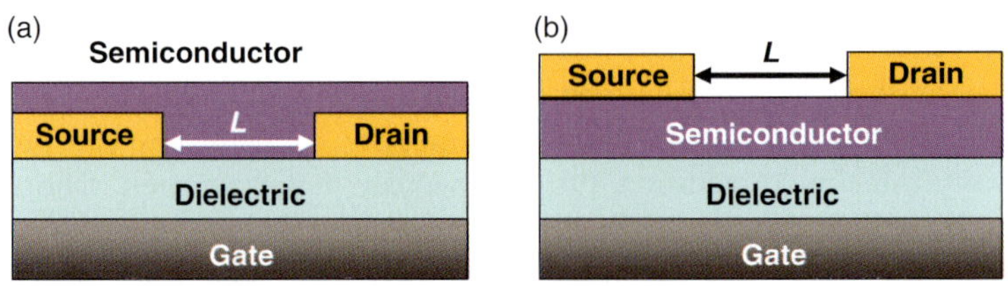

Figure 18 Two device geometries for organic thin-film transistors: (a) bottom-contact and (b) top-contact devices. The distance between source and drain electrodes defines the channel length, L.

These hybrid OTFTs cannot be employed in low-cost, large-area, and flexible displays, as the manufacturing processes and equipment for fabrication are still limited by those needed for depositing metals [98]. Replacing the conventional metal electrodes with water-dispersible conducting polymers has, therefore, been the focus of a significant amount of research. The incorporation of polymer materials permits solution processing, such as spin coating or inkjet printing, to be employed in the inexpensive fabrication of OTFTs at room temperature. Electrically conducting polymers, such as PEDOT and PANI [3,99,100], have been used as materials for both active channels and electrodes.

Many studies on OTFTs have focused on the use of conducting polymers as source and drain electrodes. PEDOT–PSS, in particular, has been heavily exploited in the development of OTFTs due to its commercial availability and facile processability. When PEDOT–PSS is used as source and drain electrodes in bottom-contact OTFTs, the device performance is dramatically improved compared to that of devices with gold electrodes, likely due to reduced contact resistances between the organic semiconductor and the electrodes [101,102]. Indeed, surface potential measurements performed on bottom-contact OTFTs using PANI–PAAMPSA source and drain electrodes show significantly reduced potential drops across the charge injection and extraction interfaces compared to those in devices with gold electrodes (**Figures 19(a)** and **19(b)**) [30]. This reduced contact resistance is directly correlated with morphological differences across the channel/electrode interface between the two types of devices. An abrupt discontinuity in the organic semiconductor layer is observed at the interface of the channel and gold electrodes, whereas this layer is continuous when PANI–PAAMPSA electrodes are employed (**Figures 19(c)** and **19(d)**) [30]. Although the performance of such OTFTs is not limited by contact resistance, it remains limited by the bulk conductivity of the conducting polymer electrodes [30].

To minimize contributions from bulk resistance of electrodes, highly conductive polymers, such as PEDOT–PSS that have been exposed to DMSO or EG, have been introduced as source and drain electrodes in OTFTs [103,104]. OTFTs with DMSO-treated PEDOT–PSS electrodes, for example, showed significantly improved current outputs and on–off ratios relative to devices with unmodified PEDOT–PSS electrodes [103]. **Figure 20** shows

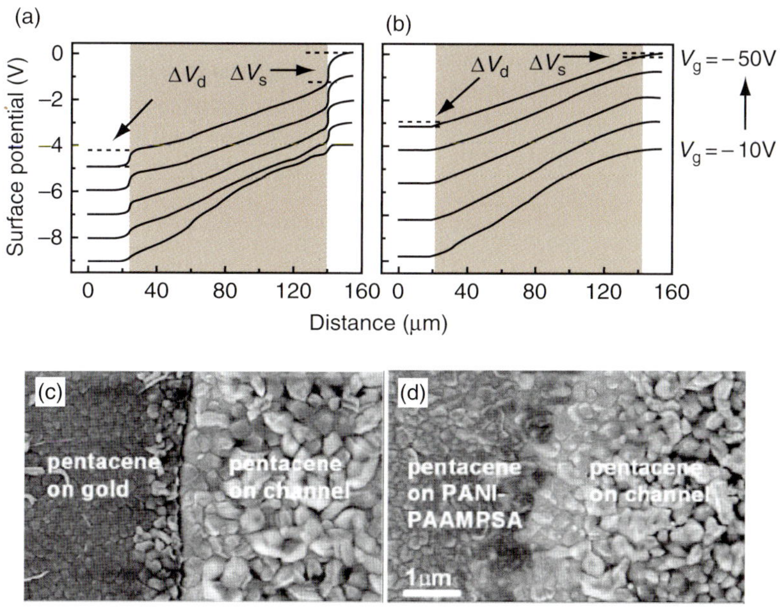

Figure 19 Surface potential profiles of the OTFTs with (a) gold electrodes and (b) PANI–PAAMPSA electrodes. The potential drops at the pentacene or source and pentanece or drain interfaces are labeled ΔV_s and ΔV_d, respectively. The darker area indicates the channel region. SEM images of (c) pentacene or gold interface and (d) pentacene or PANI–PAAMPSA interface. From Lee KS, *et al.* (2006) High-resolution characterization of pentacene–polyaniline interfaces in thin-film transistors. *Advanced Functional Materials* 16: 2409–2414.

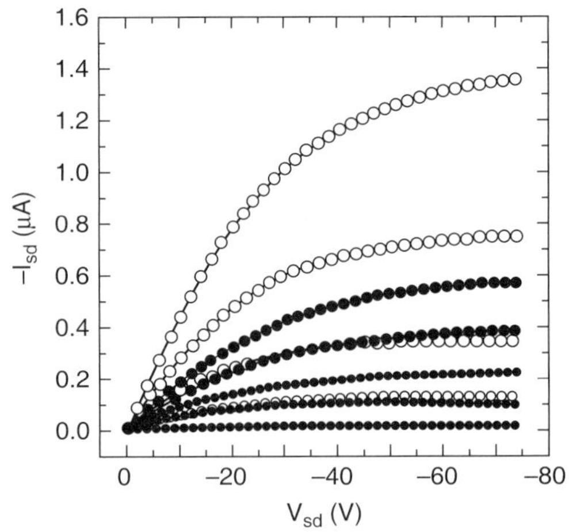

Figure 20 Current-voltage characteristics of poly(3-hexylthiophene) OTFTs with untreated (filled circles) and DMSO-treated (open circles) PEDOT–PSS source and drain electrodes. From Lim JA, *et al.* (2006) Solvent effect of inkjet printed source/drain electrodes on electrical properties of polymer thin-film transistors. *Applied Physics Letters* 88: 082102/1-3.

the current–voltage (*I–V*) characteristics of OTFTs with untreated and DMSO-treated PEDOT–PSS source and drain electrodes [103]. The enhancement in device performance mainly results from the decreased bulk resistivity of the polymer electrodes. The electrical conductivity of PANI–PAAMPSA can also be enhanced by more than 2 orders of magnitude by exposure to secondary dopants, such as DCA. PANI–PAAMPSA modified in this manner has been used as source and drain electrodes in OTFTs. Increases in output current, mobility, and on–off ratio are observed in the OTFTs with DCA-treated PANI–PAAMPSA electrodes relative to those with untreated PANI–PAAMPSA electrodes [31].

Conducting polymers have also been implemented as other device components. For example, blends of PEDOT–PSS with polyelectrolytes, such as poly(ethyleneimine) (PEI), have been used as active layers in OTFTs [105,106]. When a positive bias is applied to the gate, the sulfonic acid groups (SO_3^-) of PSS dissociate from PEDOT and associate with the NH_3^- groups of PEI (**Figure 21(a)**). This process, in turn, reduces the conductivity of the active layer. With increasing gate voltage, the active layer experiences a stronger field and becomes less conductive, giving rise to gate bias-dependent IV characteristics that resemble TFTs operating in depletion mode. (**Figure 21(b)**)

4.4.2 Organic Light-Emitting Diodes

Although the flat panel industry is still dominated by liquid-crystal technology, there has been tremendous interest in, and research on, OLEDs. Unlike liquid crystals, OLEDs are emissive, bright, and have a wide intrinsic viewing angle, thus obviating backlights or filters [107]. These advantages enable the realization of thinner, lighter, and potentially more power-efficient displays. In fact, interest in the OLED display market has been enormous since Sony introduced the first commercial OLED television in 2007 [108]. In the development of OLEDs, conducting polymers

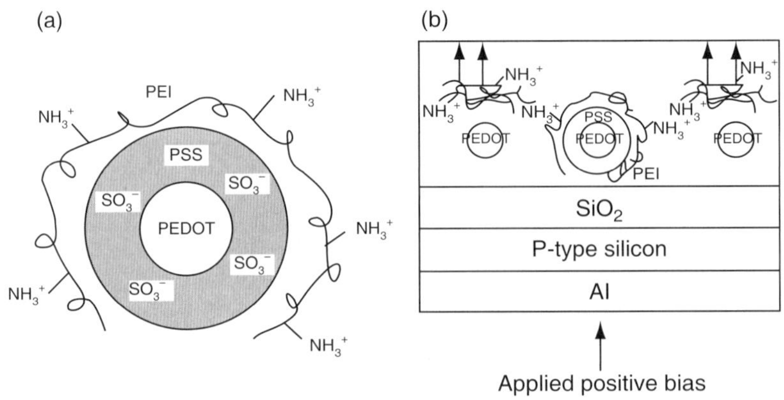

Figure 21 (a) The structure of PEDOT–PSS with PEI. (b) Operating mechanism of a depletion mode OTFT with PEDOT–PSS and PEI as the active layer. From Lin Y-J, *et al.* (2007) Blending of poly(3,4-ethylenedioxythiophene) poly(styrene-sulfonate) with poly(ethyleneimine) as an active layer in depletion-mode organic thin film transistors. *Applied Physics Letters* 91: 253501/1-3.

are among the leading candidates for buffer layers and electrodes.

The process for light emission in OLEDs was first developed by Pope *et al.* [109] in 1963 with single crystals of anthracene sandwiched between two electrodes, although these devices suffered impractical efficiencies and high operating voltages (400 V). Subsequently, Tang and VanSlyke [110] demonstrated more efficient OLEDs in 1987 composed of a double layer of tris(8-hydroxyquinoline)aluminum (AlQ_3) and an aromatic diamine as the emitter and hole transporter, respectively. In 1990, polymer-based light-emitting diodes (PLEDs) were introduced by Burroughes *et al.* using poly(*p*-phenylenevinylene) [111].

An OLED is an organic diode that emits light when a potential bias is applied to the device. **Figure 22(a)** shows the energy level diagram of a single-layer OLED. When a bias is applied to the device, electrons are injected from the cathode into the conduction band (CB) and holes from the anode into the valence band (VB) of the organic layer [112]. When two oppositely charged carriers meet, they form bound electron–hole pairs (excitons) and subsequent recombination of these excitons results in the emission of the light, the color of which is characteristic of the bandgap of organic material.

The performance of an OLED is evaluated by its quantum efficiency η_Q (number of emitted photons per injected electron) using **equation 3**:

$$\eta_Q = \gamma \cdot \eta_{ex} \cdot \eta_{pl} \cdot \eta_{oc} \tag{3}$$

where γ is the ratio of injected electrons and holes, η_{ex} is the fraction of excitons that can decay radiatively, η_{pl} is the efficiency of radiative decay, and η_{oc} is the efficiency of light out coupling (the fraction of photons actually leaving the device). The power efficiency, η_p, of OLEDs is proportional to η_Q and the ratio of photon energy, ε_p, and applied bias, V, as shown in **equation 4**.

$$\eta_p = \eta_Q \cdot \frac{\varepsilon_p}{e \cdot V} \tag{4}$$

Since the charge carriers have to travel within the organic layer after they are injected from the electrodes, the charge-carrier mobility (μ) is an important parameter that largely determines where excitons are generated and where recombination takes place. For many conjugated organic materials, the mobility of one charge carrier (say, holes) is larger than the mobility of its counterpart (electrons), often by several orders of magnitude [113,114]. This difference in mobility is often related to the presence of chemical and structural defects within the organic semiconductor, which can act as preferential traps for one of the charge carriers [113,114]. Successful design of a homogeneous material with comparable hole and electron mobility is difficult. Multilayer structures have, therefore, been proposed for OLEDs to ensure that exciton recombination takes place at a predetermined position within the device structure. **Figure 22(b)** shows the energy level diagram of a multilayer OLED. Such multilayer OLEDs contain an emission layer (EL) sandwiched between a hole-transport layer (HTL) and an electron-transport layer (ETL). The HTL provides a conductive pathway for holes to migrate from the anode into the EL. The ETL aids the transport of electrons from the cathode into the EL. In this manner, exciton generation and recombination is confined within the EL.

Conducting polymers have been considered as favorable materials for several OLED components [115,116]. Specifically, commercially available PEDOT–PSS has been widely used as buffer layers between the anode and the HTL. The benefits of using PEDOT–PSS as a buffer layer between ITO anodes and HTLs in OLEDs have been discussed

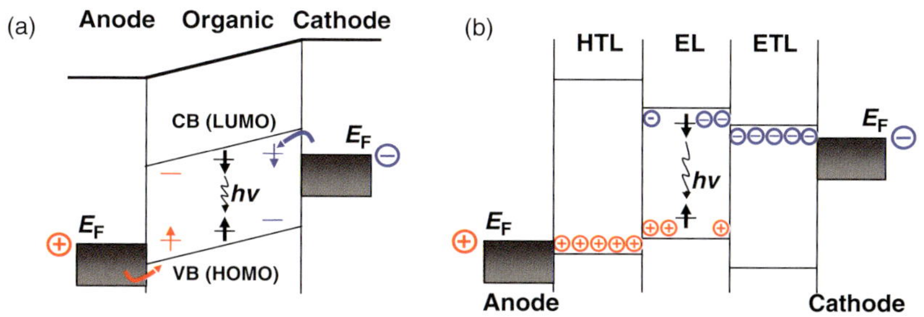

Figure 22 Schematic energy-level diagram of: (a) a single-layer; (b) a multilayer OLED. Adapted with permission from Li Z and Meng H (2007) *Organic Light-Emitting Materials and Devices*. Boca Raton, FL: Taylor and Francis.

extensively. Used in this manner, PEDOT–PSS serves as a hole injection layer (HIL), enhancing charge injection from the anode [117–119]. Similar to HIL, PEDOT–PSS also smoothes out the rough surface of ITO. This smoothing layer reduces the chance of micro-shorts during device operation [120]. The use of PEDOT–PSS between the anode and the HTL also blocks electrons [117,121]. For example, Elschner *et al.* [120] compared the characteristics of OLEDs with and without PEDOT–PSS HILs. **Figures 23(a)** and **23(b)** show the cross section of their OLEDs, as well as the current density ($\mathcal{J}$) and luminance (L) responses of the devices as a function of applied voltage (V), respectively. The OLED with a PEDOT–PSS HIL showed improved device characteristics, with efficiency that is an order of magnitude higher than that of the device without the PEDOT–PSS HIL.

The use of PANI has also been explored in OLEDs, although to a lesser degree [122,123]. While most initial studies have focused on small-molecule acid-doped PANI [116, 124], water-dispersible PANI–PSS has recently been introduced as a candidate for HILs in OLEDs. For example, the incorporation of PANI–PSS HILs can significantly improve the lifetimes of OLEDs [125]. Similar to the use of PEDOT–PSS as a buffer layer on ITO, the incorporation of a PANI–PSS within OLEDs can also effectively reduce surface roughness (**Figures 24(a)** and **24(b)**), thereby eliminating microshorts during device operation [126]. Compared to PEDOT–PSS, however, PANI–PSS has a higher work function; its incorporation within OLEDs thus further enhances the device characteristics compared to devices with PEDOT–PSS HILs. In particular, the use of PANI–PSS in OLEDs leads to a reduction in operation voltage and a simultaneous increase in luminance compared with devices that use PEDOT–PSS (**Figures 24(c)** and **24(d)**).

Conducting polymers have also been used to replace the ITO anode in OLEDs. Although ITO is highly conductive and optically transparent, it is brittle and thus quite fragile. Moreover, the scarcity of indium and high production costs of ITO have encouraged the development of solution-processable and low-cost conducting polymers as replacement electrodes for OLEDs. The low conductivity of polymeric anodes relative to ITO, however, appears to be the current road block; the use of conducting polymers in place of ITO thus often results in devices with reduced light output.

Significant improvements in the conductivity of conducting polymers, specifically PEDOT–PSS, have led to successful attempts in replacing ITO electrodes in OLEDs. Recent studies have shown that the conductivity of PEDOT–PSS can be dramatically increased without any loss in its optical transparency by exposure to small amounts of

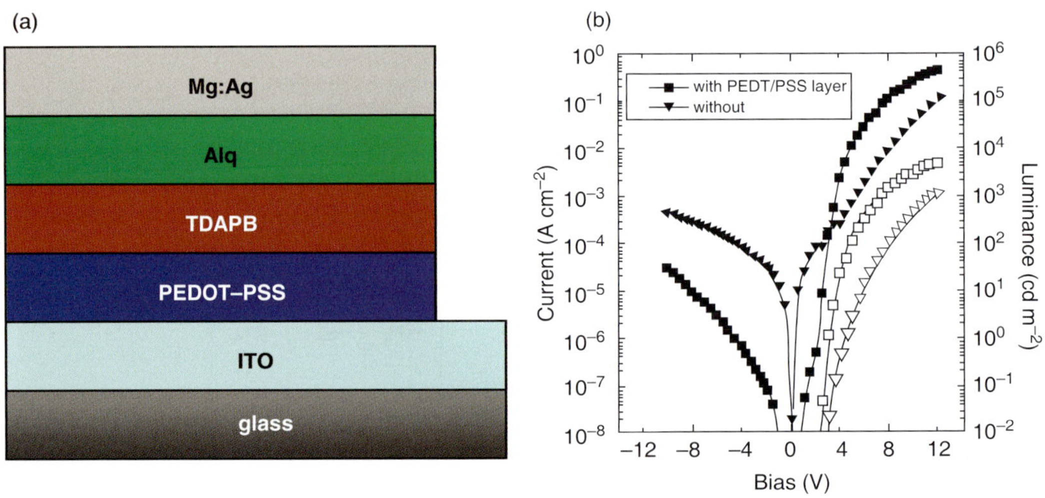

Figure 23 (a) Cross section of a multilayer OLED: ITO is used as the anode, PEDOT–PSS as the hole-injection layer, 1,3,5-tris[4-(diphenyl-amino)phenyl]-benzene (TDAPB) as the hole transport layer, tris-(8-hydroxyquinolinol)aluminium (Alq) as the emissive layer, and alloys of Mg and Ag as the cathode. (b) Accompanying *J–V* (solid symbols) and *L–V* (open symbols) curves of OLEDs with (squares) and without (triangles) PEDOT–PSS HIL. From Elschner A, *et al.* (2000) PEDT/PSS for efficient hole-injection in hybrid organic light-emitting diodes. *Synthetic Metals* 111-112: 139–143.

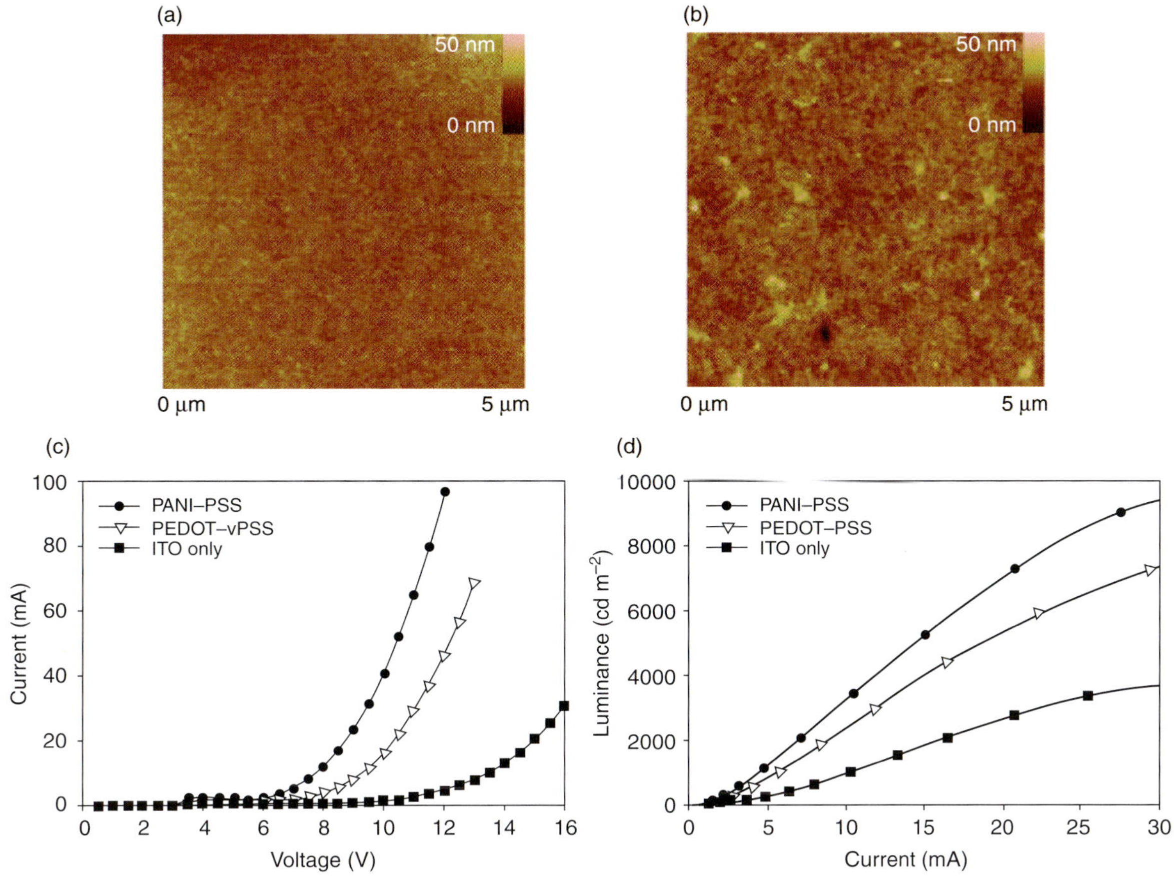

Figure 24 AFM images of (a) PANI–PSS; (b) PEDOT–PSS on ITO. The (c) *I–V* and (d) *L–I* characteristics of OLEDs with PANI–PSS HIL (solid circles), PEDOT–PSS HIL (open triangles), and bare ITO (solid squares). From Jang J, *et al.* (2008) Organic light-emitting diode with polyaniline-poly(styrene sulfonate) as a hole injection layer. *Thin Solid Films* 516: 3152–3156.

glycerol or sorbitol [127,128]. These transparent, yet highly conductive polymers are ideal candidates for use as anode substitutes in OLEDs. For example, Kim *et al.* [129] demonstrated that anodes fabricated from aqueous dispersions of PEDOT–PSS with glycerol (G-PEDOT–PSS) perform well in OLEDs. **Figure 25** shows *J–V* and *L–V* curves for devices fabricated using G-PEDOT–PSS and untreated PEDOT–PSS as anodes. The addition of glycerol dramatically decreases the surface sheet resistance while maintaining the optical transparency of PEDOT–PSS. The performance of devices using this highly conductive polymer anode is greatly improved and compares well with OLEDs using standard ITO anodes.

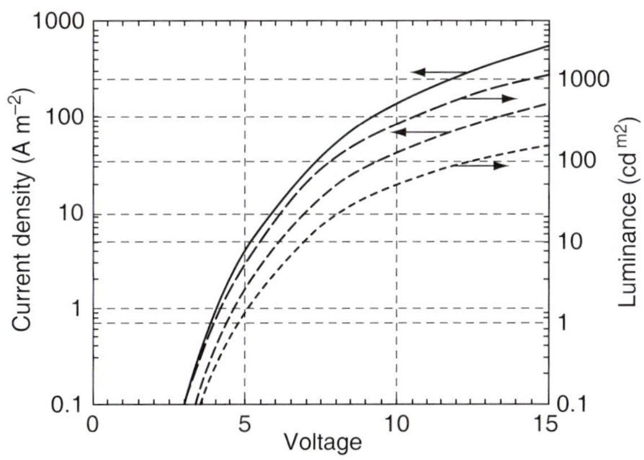

Figure 25 *J–V* and *L–V* characteristics of OLEDs fabricated with G-PEDOT–PSS (solid and dashed dot lines) and untreated PEDOT (dashed and dotted lines) anodes. From Kim WH, *et al.* (2002) Molecular organic light-emitting diodes using highly conducting polymers as anodes. *Applied Physics Letters* 80: 3844–3846.

4.4.3 Organic Solar Cells

A solar cell is a device that converts incident light into electrical energy. The first solar cell was made of gold and selenium in 1883 [130]. About 60 years later, the first silicon-based solar cell was invented [131]. In 1986, Tang [132] reported one of the first OSCs composed of copper phthalocyanine and a perylene tetracarboxylic derivative. The power efficiency of this OSC, however, was below 1 %. Although single crystal inorganic solar cells can exhibit efficiencies as high as 40% [133], they require exotic and expensive materials. In the past 20 years, research has focused on enhancing the power conversion efficiencies of OSCs in hopes of achieving practical performance at a fraction of the cost associated with their inorganic counterparts.

The architecture of OSCs is similar to that of OLEDs. Two organic semiconductor materials, one with a propensity to donate electrons and the other with a propensity to accept electrons, are sandwiched between an anode and a cathode. **Figure 26** shows the four steps by which OSCs are believed to operate; the process ultimately leads to the photogeneration of excitons and collection of charges in a simple bilayer device architecture. In the first step, bound electron–hole pairs are generated within one of the organic layers when light is absorbed. Generation of excitons can occur in the electron-donor layer, as shown in **Figure 26**, or within the electron-acceptor layer, depending on the absorption properties of the respective materials. While there is a finite probability for dissociation of excitons into holes and electrons to occur within the organic layer in which they are generated, exciton dissociation is significantly more efficient at the electron donor–acceptor interface. The photogenerated excitons thus generally have to diffuse toward the organic–organic interface (step 2), at which they dissociate. Dissociation results in the partitioning of the electron to the electron-acceptor layer and the hole to the electron-donor layer. The final step involves the transport of the newly generated charge carriers toward their respective electrodes.

Exciton diffusion lengths in polymeric and other disordered materials are generally in the range of a few nanometers to a few tens of nanometers [134], but can increase to a few hundred nanometers for highly ordered crystalline samples [134]. This limitation implies that only excitons that are generated close to the electron donor–acceptor interface contribute to photocurrent generation; excitons that are generated further away from the interface generally recombine before they make it to the interface. With this in mind, the bilayer structures of OSCs have been supplanted by bulk-heterojunction (BHJ)-type devices, wherein the active layer consists of nanoscale domains of the electron donor and acceptor. Yu *et al.* [135] first described the BHJ architecture for OSCs with an active layer composed of a blend of poly(2-methoxy-5-(2′-ethyl-hexyloxy)-1,4-phenylene vinylene), MEH-PPV, and poly(2,5,2′,5′-tetrahexyloxy-7,8′-dicyanodi-*p*-phenylenevinylene), CN-PPV, as the electron donor and electron acceptor, respectively. The BHJ architecture provides a significantly larger interfacial area between the electron donor and electron acceptor (**Figure 27**) compared to the bilayer architecture, which effectively shortens the average distance excitons must travel before separation, thereby enhancing photocurrent generation. BHJs have therefore been used extensively since their introduction.

The device performance of an OSC can be evaluated by its power conversion efficiency (η_p). η_p is the ratio of electromagnetic radiation input power and electrical output power. For a known irradiance energy (E) over an active device area, *A*, η_p can be calculated from the maximum current (short circuit

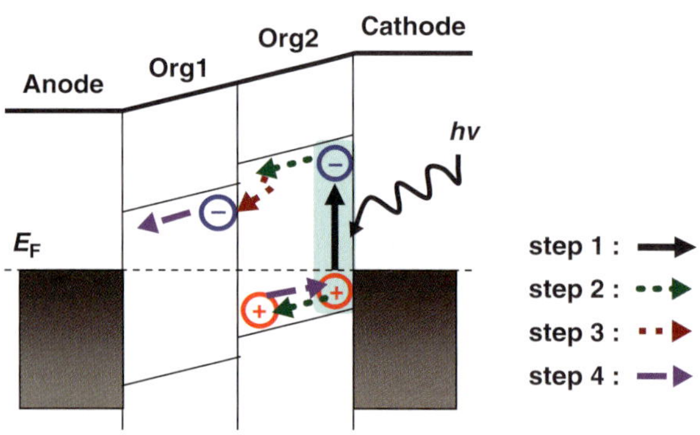

Figure 26 Schematic of the four steps that are involved during the operation of a bilayer OSC.

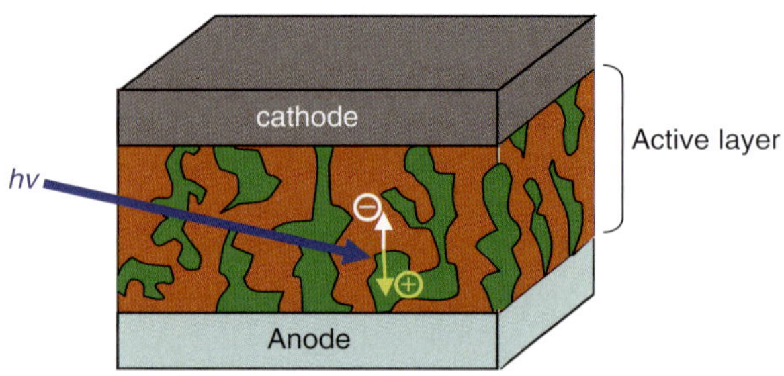

Figure 27 Schematic representation of a bulk-heterojunction solar cell, which comprises enhanced electron donor–acceptor interface area (electron donor–acceptor constituents are depicted in different colors).

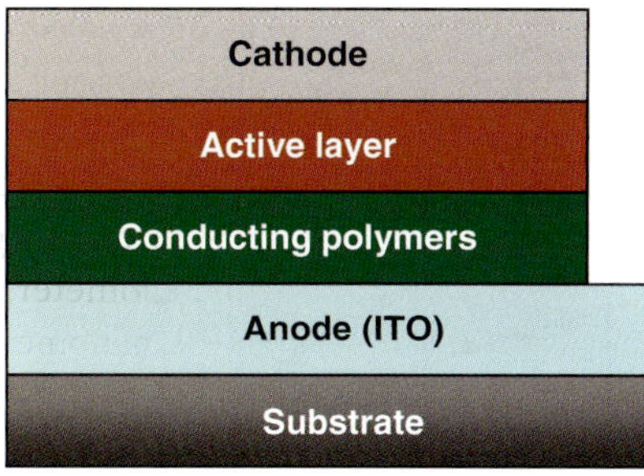

Figure 28 Cross section representation of a typical OSC including the conducting polymer buffer layer.

current I_{sc}), maximum voltage (open-circuit voltage V_{oc}), and the fill factor (FF) using **equation 5**:

$$\eta_p = \frac{FF \cdot I_{SC} \cdot V_{OC}}{E \cdot A} \tag{5}$$

where FF is defined as the maximum power that can be extracted from the actual OSC divided by the product I_{sc} and V_{oc}.

In the last decade, conducting polymers, such as PEDOT or PANI, have become established materials in OSCs research. Like in OLEDs, PEDOT and PANI were initially used as buffer layers between the metal oxide anode and the photoactive layer, as shown in **Figure 28**. In a manner identical to their roles in OLEDs, PEDOT–PSS buffer layers in OSCs smooth the ITO surface. These buffer layers also improve hole extraction to the external anode [136]. Various modifications of PEDOT–PSS have been also discussed as alternative buffer layers in OSCs. Zhang *et al.* [137], for example, have compared various types of PEDOT–PSS as buffer layers between ITO and the organic photoactive layer. Given that PSS preferentially segregates to the surface when PEDOT–PSS is deposited as a thin film, the researchers discovered that they can influence the extent of charge injection by controlling the thickness of the PSS wetting layer. In doing so, they can modulate the V_{oc}, and consequently the efficiency of the OSCs. PEDOT–PSS buffer layers have also been introduced in OSCs, in which a mat of single-walled carbon nanotubes (SWNTs) replaces ITO as printable and flexible anodes [138,139]. **Figure 29** shows the AFM images of the SWNT network (a) before and (b) after PEDOT–PSS deposition; PEDOT–PSS deposition on SWNT film results in decreased roughness. The introduction of PEDOT–PSS in these OSCs significantly improves device performance [138,139].

PEDOT–PSS has been also used as a hole transport material in dye-sensitized solar cells (DSSCs) [140]. In DSSCs, a dye molecule absorbs visible light and, after excitation, injects electrons into the conduction band of TiO_2. The dye is thus oxidized in the process. The dye is rapidly regenerated through reduction by a hole transporting material, usually a redox electrolyte [141,142]. TiO_2 and the hole conductor, therefore, function as selective contacts to the dye, and these contacts transport the separate carriers toward their respective electrodes (see energy diagram in **Figure 30**) [143]. The efficiency of such DSSCs with liquid electrolytes can be as high as 10% [144], but the devices suffer from poor air stability. Further, the liquid electrolytes can often

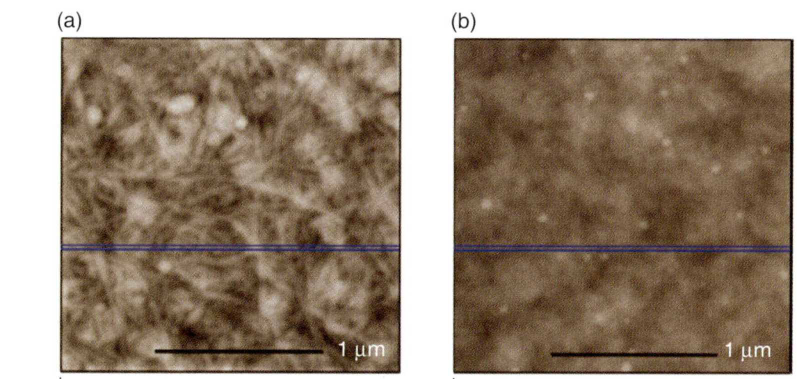

Figure 29 AFM images of SWNT network: (a) before; (b) after PEDOT-PSS deposition. The root mean square (rms) roughness is 7 nm and 3.5 nm before and after the deposition of PEDOT–PSS, respectively. From Rowell MW, *et al.* (2006) Organic solar cells with carbon nanotube network electrodes. *Applied Physics Letters* 88: 233506/1-3.

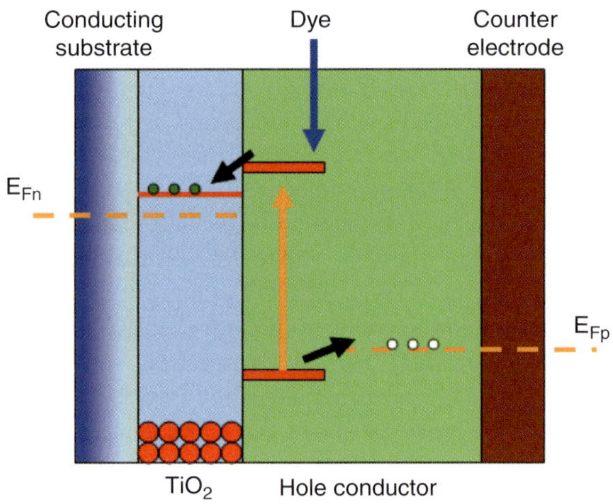

Figure 30 Schematic energy diagram of DSSC. Adapted with permission from Bisquert J, *et al.* (2004) Physical chemical principles of photovoltaic conversion with nanoparticulate, mesoporous dye-sensitized solar cells. *Journal of Physical Chemistry B: Polymers* 108: 8106–8118.

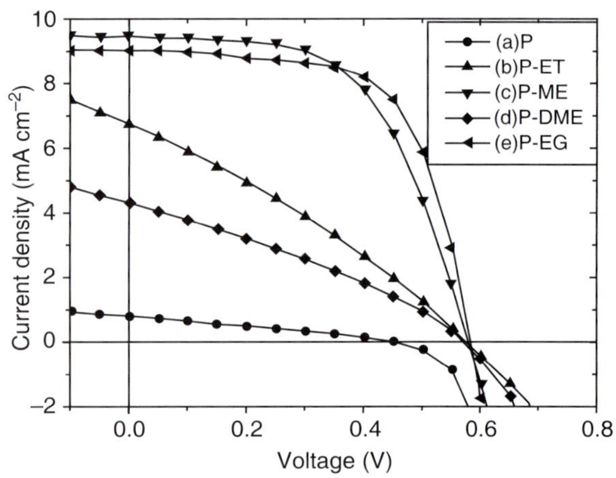

Figure 31 *J–V* characteristics of OSCs incorporating various solvent-modified PEDOT–PSS as anodes: (a) P, (b) P-ET, (c) P-ME, (d) P-DME, and (e) P-EG. All the devices were measured under illumination of AM 1.5 (100 mW c m^{-2}) with active area of 0.01 cm^2. From Hsiao Y-S, *et al.* (2008) High-conductivity poly(3,4-ethylenedioxythiophene): poly(styrene sulfonate) film for use in ITO-free polymer solar cells. *Journal of Materials Chemistry* 18: 5948–5955.

leak out of the cells. Many studies have, therefore, resorted to using solid- or quasi-solid-state hole transport materials in place of the liquid electrolyte. Conducting polymers are of practical interest in replacing liquid electrolytes because they are inexpensive and efficient solid-state hole transporters. For example, Lee *et al.* [145] fabricated DSSC devices using PEDOT–PSS and glycerol-modified PEDOT–PSS (G-PEDOT–PSS) as hole transport materials. They investigated solar cell characteristics as a function of glycerol concentration and reported efficiencies as high as 2.62%. Despite reduced efficiencies relative to liquid-electrolyte devices, these conducting polymer-based devices hold much promise in regard to long-term stability.

Another opportunity for the use of conducting polymers in solar cells is the replacement of metal or metal oxide electrodes. For optoelectronic devices, such as OLEDs and OSCs, at least one of the electrodes needs to be transparent for either light emission or light absorption within the photoactive layer. Although ITO is frequently used as the transparent electrode in flexible OSC and OLED devices, the scarcity of indium makes using ITO a problem in the long run. In addition, ITO is brittle and cracks easily when the underlying substrate is bent. It is, therefore, crucial to develop organic-based electrodes with both high transparency and high conductivity for plastic optoelectronics. Conducting polymers possessing conductivity enhanced through secondary doping have been considered promising materials for this pursuit [64,146]. Devices using such modified conducting polymers as anodes exhibit performances close to those of the devices using ITO as the anode [147]. **Figure 31** summarizes device performances of solar cells incorporating various PEDOT–PSS films [148]. PEDOT–PSS films were modified with ethanol (P-ET), methoxyethanol (P-ME), 1,2-dimethoxyethane (P-DME), and ethylene glycol (P-EG), and used as anodes. For comparison, a control device (P) comprised of unmodified PEDOT–PSS is shown alongside the other devices of interest. In general, the efficiency of OSCs is strongly related to the sheet resistances of the PEDOT–PSS films; the OSCs with the most conductive PEDOT–PSS exhibit the best performance. A similar phenomenon has been observed when PANI is incorporated as anodes in OSCs as well [31]. Recently, the use of PANI–PAAMPSA as the anode in OSCs has been investigated. OSCs with PANI–PAAMPSA modified by secondary doping exhibit higher photocurrent compared to the devices with untreated PANI–PAAMPSA anodes [31].

4.4.4 Electrochromic Devices

Electrochromic materials undergo reversible changes in optical absorbance upon electrically induced oxidation and reduction. Devices comprising these materials only require an initial electrical stimulus sufficient to induce a change in oxidation state, and hence color, rather than a constant stimulus to sustain emission. Unlike the active layers in OLEDs that are responsible for the emission of light, the active materials in ECDs possess transmissive properties and therefore require an external light source. Although poorly suited for displays that require high refresh rates, like those found in televisions [149], this quality is well matched for passive applications, such as smart windows or electronic billboards, in which ambient light is freely available and the intended coloration persists for an extended duration.

The concept of ECDs originated in the 1960s [150,151]. Early research focused on thin films of tungsten oxide capable of electrochemical oscillations between its bleached, oxidized and colored, reduced forms [152,153]. This redox process is reversible, but poor film formability and costly processing steps hindered the incorporation of tungsten-based materials in practical large-area devices [154]. The viability of using small-molecule organic materials, specifically viologen and its derivatives, in ECDs was first examined in 1973 [155]. These devices, however, suffered irreversible bleaching upon exposure to oxygen, greatly reducing their lifetimes [149]. Kaufman *et al.* reported the first construction of ECDs utilizing solution-processable electrochromic polymers in 1979 [156]. In addition to oscillations between the bleached and colored states that tungsten oxide and viologen-based materials possess, many conjugated polymers possess polyelectrochromic properties [157]. As such, they are capable of exhibiting multicolored oxidation states beyond those associated with inorganic and small-molecule materials. A number of other advantages have been attributed to polymer-based ECDs, namely smooth film formation across large areas and systematic tunability of both coloration and efficiency [156].

The basic architecture of an ECD is shown in **Figure 32**. An electrochromic material is located between an electrolytic layer and an electrode, both of which must be optically transparent. The electrolytic layer provides the ionic transport necessary for electrochemical oxidation and reduction of the electrochromic material, and should accordingly possess high ion conductivity. Most ECDs utilize solid-state polymers intercalated with mobile ions, such as poly(ethylene oxide) saturated with lithium chlorate as the electrolytic layer [158]. Solution- or gel-phase electrolytes, such as ionic liquids [159], aqueous

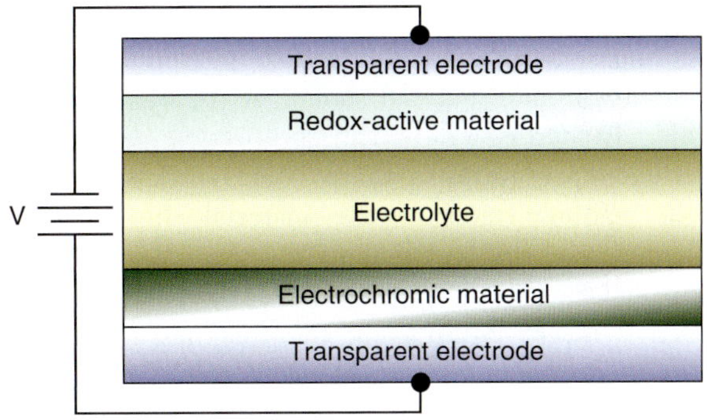

Figure 32 Schematic of an electrochromic device.

solutions, or plasticized poly(methyl methacrylate) loaded with salt [160], can be used, but often present complications related to sealing difficulties and drifts in coloration due to diffusion of the electrolyte into the active material [149]. A secondary redox-active layer, capable of supporting an electrochemical reaction opposing that of the first electrochromic layer, is positioned on the reverse side of the electrolyte and capped with an additional optically transparent electrode. The material comprising this second layer can also be electrochromically active, but if so, should exhibit optical properties complementary to those of the first electrochromic layer. For instance, if one electrochromic layer is colored in its oxidized state, the opposing layer should be transparent when reduced, and vice versa. If the material is not electrochromically active, it should be transparent in both its oxidized and reduced forms.

The most important quantifiable characteristic of an ECD operation is coloration efficiency, η, and is defined as the ratio of optical density contrast (ΔOD) to charge consumed or released per unit area (Q_d) during an electrochromic transition, as shown in **equation 6**.

$$\eta = \frac{\Delta \mathrm{OD}}{Q_d} \tag{6}$$

The charge involved in an electrochemical transition is dictated by the molar quantity of active material. The contrast in optical density, alternatively, is intrinsically related to the specific electrochromic material employed and therefore plays a large role in determining the efficiency of an ECD. As such, proper selection of electrochromic material is an important consideration in any ECD.

Electrochromically active layers in ECDs must possess a number of favorable material properties.

To minimize the power necessary to drive electrochromic transitions, the redox potential of the electrochromic material should be as low as possible. In terms of optical properties, coloration efficiency requires a sharp color contrast among the various states of the material. In addition, large absorption coefficients for the colored states are necessary to minimize the thickness of the active layers. Minimizing the thickness of active layers decreases the amount of material that must be oxidized or reduced during electrochromic transitions, therefore reducing the charge flux and switching times.

To assess the kinetics and reversibility of electrochromic transitions, a square-wave potential step is frequently applied to ECDs, and the current and optical response at the most dynamic region of the visible spectrum is measured as a function of time. **Figure 33** shows typical optical transmittance curves for ECDs utilizing small-molecule acid-doped PANI and water-dispersible PANI–PAAMPSA [161]. By comparing the maximum and minimum optical densities achieved during each cycle, these curves convey the reversibility of repeated cycling. The timescales involved in switching between oxidation states of conducting polymers generally exceed those associated with inorganic or small-molecule based devices due to retarded flow of charge-balancing ions through the tortuous polymer network. Recent work using PANI–PAAMPSA and PEDOT–PSS films modified with secondary-doping solvents, however, has shown significant reductions in bleaching times [162,163]. Although not yet thoroughly investigated, the enhanced response could be attributed to the relaxation of the polymer network, allowing faster diffusion of counter-ions into and out of the film [163].

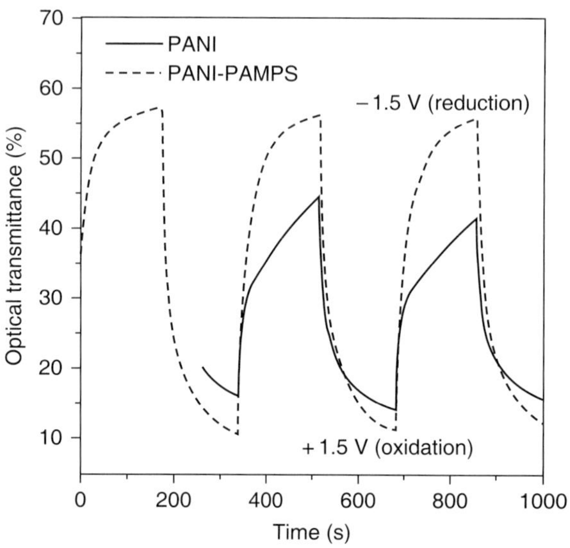

Figure 33 Optical transmittance curves for HCl-doped PANI and PANI–PAAMPSA (denoted PANI--PAMPS in the figure). From Hu HL, *et al.* (2003) Optical and electrical responses of polymeric electrochromic devices: Effect of polyacid incorporation in polyaniline film. *Solid State Ionics* 161: 165–172.

A major advantage of using conducting polymers in ECDs relative to inorganic materials is synthetic tunability. By adding or withdrawing substituent groups to a monomer, the electrochemical and optical properties of the final product can be significantly altered. Work in the mid-1990s by Gustafsson-Carlberg *et al.*, for example, showed that the optical bandgap of neutral polythiophene derivatives could be modified across 400–600 nm [164]. An enormous amount of research on the electrochromic properties of poly(3,4-alkylenedioxythiophenes), or PXDOTs, has subsequently been performed by Reynolds and co-workers [49,160,165–169]. Much like PEDOT, these polymers are conductive and transparent when oxidatively doped, and reveal colored states upon reduction [169]. **Figure 34** shows a photograph of wide-ranging colors exhibited by derivatized PEDOT samples [160], while **Table 1** provides a quantitative summary of the large variations in the material properties of alkylated PEDOT systems [169]. Many of the different PXDOT variants studied can be solubilized by association with a polymer acid in a manner analogous to PEDOT–PSS dispersions, and thus possess enormous potential for use in ECDs [160].

The PXDOT systems created by Reynolds *et al.* have permitted the development of dual-polymer ECDs covering a wide chromic spectrum [160]. Commonly, PANI is employed as the second electrochromic layer. Unlike PXDOTs, PANI reveals colored states when oxidized and becomes transparent only upon reduction. Oxidatively colored PANI therefore provides an ideal complement to PXDOTs in dual-polymer ECDs.

Devices created in such a manner possess efficiencies in excess of 1400 $cm^2\ C^{-1}$ [167]. The efficiencies of inorganic and hybrid PANI–tungsten ECDs, in comparison, have only been reported as high as 115 and 170 $cm^2\ C^{-1}$, respectively [170, 171]. Considering the large-area applications proposed for ECDs in electrochromic windows and billboards, solution-

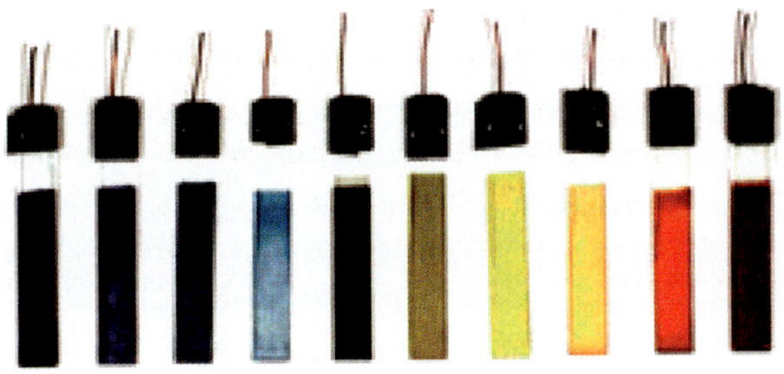

Figure 34 Photograph demonstrating the variation in optical properties among derivatized PEDOTs. From Groenendaal BL, *et al.* (2000) Poly(3,4-ethylenedioxythiophene) and its derivatives: Past, present, and future. *Advanced Materials* 12: 481–494.

Table 1 Variations in redox potential ($E^0_{polym.}$) and conductivity (σ) with varying length of EDOT monomer's alkylation group (-R).

	-R =	$E^0_{polym.}$ *[V]*	σ *[S cm^{-1}]*
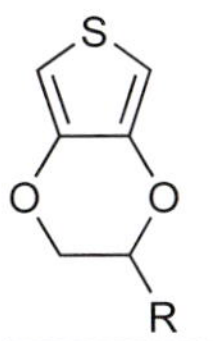	-H	−0.42	650
	$-CH_3$	−0.65	300
	$-C_6H_{13}$	−0.35	200
	$-C_8H_{17}$	−0.32	NA
	$-C_{10}H_{21}$	−0.27	550
	$-C_{14}H_{29}$	−0.06	850

NA, not applicable. Adapted with permission from Groenendaal L, et al. (2003) Electrochemistry of poly(3,4-alkylenedioxythiophene) derivatives. Advanced Materials 15: 855–879.

processable electrochromic polymers offer significant contributions to its development.

4.5 Concluding Remarks

The development of organic electronics has seen an enormous amount of progress in the three decades following the initial discovery of electroactive polymers by Heeger *et al.* Incorporation of polymer-acid counterions has contributed water-dispersibility to many conducting polymers and has greatly enhanced processability. Although initial methods conferred water-dispersibility at the expense of conductivity, the recent development of secondary doping procedures for polymer-acid-doped polymers has returned conductivities to the level of small-molecule doped systems while maintaining facile processibility. These novel materials have already found use in a wide array of applications. Their electrical conductivity and selectivity as a hole transporter make them well suited for use as electrodes and buffer layers in OTFTs, OLEDs, and OSCs, while their optical properties and synthetic tunability make them ideal candidates for active layers in ECDs. The field of organic electronics is now entering its commercial phase, and will soon call for the wide implementation of conducting polymers in a number of applications.

Acknowledgments

The authors acknowledge funding from the National Science Foundation (NSF CAREER DMR-0753148), the Beckman Foundation, the Sloan Foundation, and the W. M. Keck Foundation. Jacob Tarver thanks the NSF for a Graduate Research Fellowship.

References

1. Loo Y-L and McCulloch I (2008) Progress and challenges in commercialization of organic electronics. *MRS Bulletin* 33: 653–662.
2. Loo Y-L (2007) Solution-processable organic semiconductors for thin-film transistors: Opportunities for chemical engineers. *AIChE Journal* 53: 1066–1074.
3. Lee KS, Blanchet GB, Gao F, and Loo Y-L (2005) Direct patterning of conductive water-soluble polyaniline for thin-film organic electronics. *Applied Physics Letters* 86: 074102.

4. Heeger AJ (2001) Nobel Lecture: Semiconducting and metallic polymers: The fourth generation of polymeric materials. *Reviews of Modern Physics* 73: 681–700.
5. MacDiarmid AG (2001) Nobel Lecture: "Synthetic metals": A novel role for organic polymers. *Reviews of Modern Physics* 73: 701–712.
6. Shirakawa H (2001) Nobel Lecture: The discovery of polyacetylene film – the dawning of an era of conducting polymers. *Reviews of Modern Physics* 73: 713–718.
7. Chiang CK, Fincher CR, Jr., Park YW, *et al.* (1977) Electrical conductivity in doped polyacetylene. *Physical Review Letters* 39: 1098–1101.
8. Chiang CK, Park YW, Heeger AJ, *et al.* (1978) Conducting polymers: Halogen doped polyacetylene. *Journal of Chemical Physics* 69: 5098–5104.
9. Yamamoto T, Sanechika K, and Yamamoto A (1980) Preparation of thermostable and electric-conducting poly(2,5-thienylene). *Journal of Polymer Science, Part C: Polymer Letters* 18: 9–12.
10. Lin JWP and Dudek LP (1980) Synthesis and properties of poly(2,5-thienylene). *Journal of Polymer Science, Part A: Polymer Chemistry* 18: 2869–2873.
11. Heywang G and Jonas F (1992) Poly(alkylenedioxythiophoene)s – new, very stable conducting polymers. *Advanced Materials* 4: 116–118.
12. Pei QB, Zuccarello G, Ahlskog M, and Inganas O (1994) Electrochromic and highly stable poly(3,4-ethylenedioxythiophene) switches between opaque blue-black and transparent sky blue. *Polymer* 35: 1347–1351.
13. Bayer AG (1988) Process for the Preparation of Polythiophenes. European Patent 339340.
14. Letheby H (1862) XXIX – On the production of a blue substance by the electrolysis of sulphate of aniline. *Journal of the Chemical Society* 15: 161–163.
15. Green AG and Woodhead AE (1910) CCXLIII – Aniline-black and allied compounds. Part I. *Journal of the Chemical Society, Transactions* 97: 2388–2403.
16. Green AG and Woodhead AE (1912) CXVII – Aniline-black and allied compounds. Part II. *Journal of the Chemical Society, Transactions* 101: 1117–1123.
17. De Surville R, Jozefowicz M, Yu LT, Pepichon J, and Buvet R (1968) Electrochemical chains using protolytic organic semiconductors. *Electrochimica Acta* 13: 1451–1458.
18. MacDiarmid AG, Chiang JC, Huang WS, Humphrey BD, and Somasiri NLD (1985) Polyaniline – protonic acid doping to the metallic regime. *Molecular Crystals and Liquid Crystals* 125: 309–318.
19. MacDiarmid AG, Chiang JC, Halpern M, *et al.* (1985) Polyaniline – interconversion of metallic and insulating forms. *Molecular Crystals and Liquid Crystals* 121: 173–180.
20. Chiang JC and MacDiarmid AG (1986) Polyaniline – protonic acid doping of the emeraldine form to the metallic regime. *Synthetic Metals* 13: 193–205.
21. Cao Y, Smith P, and Heeger AJ (1992) Counter-ion induced processibility of conducting polyaniline and of conducting blends of polyaniline in bulk polymers. *Synthetic Metals* 48: 91–97.
22. MacDiarmid AG and Epstein AJ (1994) The concept of secondary doping as applied to polyaniline. *Synthetic Metals* 65: 103–116.
23. Wei X and Epstein AJ (1995) Synthesis of highly sulfonated polyaniline. *Synthetic Metals* 74: 123–125.
24. Lukachova LV, Shkerin EA, Puganova EA, *et al.* (2003) Electroactivity of chemically synthesized polyaniline in neutral and alkaline aqueous solutions: Role of self-doping and external doping. *Journal of Electroanalytical Chemistry* 544: 59–63.
25. Li C and Mu S (2005) The electrochemical activity of sulfonic acid ring-substituted polyaniline in the wide pH range. *Synthetic Metals* 149: 143–149.
26. Chen SA and Lee HT (1995) Structure and properties of poly(acrylic acid)-doped polyaniline. *Macromolecules* 28: 2858–2866.
27. Angelopoulos M, Patel N, Shaw JM, Labianca NC, and Rishton SA (1993) Water soluble conducting polyanilines: Applications in lithography. *Journal of Vacuum Science and Technology B: Microelectronics and Nanometer Structures* 11: 2794–2797.
28. Yoo JE, Cross JL, Bucholz TL, *et al.* (2007) Improving the electrical conductivity of polymer acid-doped polyaniline by controlling the template molecular weight. *Journal of Materials Chemistry* 17: 1268–1275.
29. Yoo JE, Bucholz TL, Jung S, and Loo Y-L (2008) Narrowing the size distribution of the polymer acid improves PANI conductivity. *Journal of Materials Chemistry* 18: 3129–3135.
30. Lee KS, Smith TJ, Dickey KC, *et al.* (2006) High-resolution characterization of pentacene–polyaniline interfaces in thin-film transistors. *Advanced Functional Materials* 16: 2409–2414.
31. Yoo JE, Lee KS, Gomez ED, *et al.* (2008) unpublished.
32. Kirchmeyer S and Reuter K (2005) Scientific importance, properties and growing applications of poly(3,4-ethylenedioxythiophene). *Journal of Materials Chemistry* 15: 2077–2088.
33. NSF (2004) New molecule heralds breakthrough in electronic plastics. *National Science Foundation News Archives.* Press Release 04-047.
34. Mohilner DM, Argersinger WJ, and Adams RN (1962) Investigation of kinetics and mechanism of anodic oxidation of aniline in aqueous sulfuric acid solution at a platinum electrode. *Journal of the American Chemical Society* 84: 3618–3622.
35. Trivedi DC (1997) Polyanilines. In: Nalwa HS (ed.) *Handbook of Organic Conductive Molecules and Polymers.* vol. 2, pp. 505–572. New York: Wiley.
36. Breitenbach M and Heckner K-H (1971) Studies on kinetics of anodic oxidation of aniline and acetonitrile on rotating platinum electrode. *Journal of Electroanalytical Chemistry* 29: 309–323.
37. Volkov A, Tourillon G, Lacaze PC, and Dubois JE (1980) Electrochemical polymerization of aromatic-amines – IR, XPS and PMT study of thin-film formation on a Pt electrode. *Journal of Electroanalytical Chemistry* 115: 279–291.
38. Ohsaka T, Ohnuki Y, Oyama N, Katagiri G, and Kamisako K (1984) IR absorption spectroscopic identification of electroactive and electroinactive polyaniline films prepared by the electrochemical polymerization of aniline. *Journal of Electroanalytical Chemistry* 161: 399–405.
39. Geniès EM, Boyle A, Lapkowski M, and Tsintavis C (1990) Polyaniline: A historical survey. *Synthetic Metals* 36: 139–182.
40. Tarver J, Yoo JE, Dennes TJ, Schwartz J, and Loo Y-L (2009) Polymer acid doped polyaniline is electrochemically stable beyond pH 9. *Chemistry of Materials* 21: 280–286.
41. Kim TY, Park CM, Kim JE, and Suh KS (2005) Electronic, chemical and structural change induced by organic solvents in tosylate-doped poly(3,4-ethylenedioxythiophene) (PEDOT-OTs). *Synthetic Metals* 149: 169–174.
42. Feng W, Li Y, Wu J, *et al.* (2007) Improved electrical and optical properties of poly(3,4-ethylenedioxythiophene) via ordered microstructure. *Journal of Physics: Condensed Matter* 19: 186220.

43. Pouget JP, Jozefowicz ME, Epstein AJ, Tang X, and MacDiarmid AG (1991) X-ray structure of polyaniline. *Macromolecules* 24: 779–789.
44. Xia YN, Wiesinger JM, MacDiarmid AG, and Epstein AJ (1995) Camphorsulfonic acid fully doped polyaniline emeraldine salt - conformations in different solvents studied by an ultraviolet-visible near-infrared spectroscopic method. *Chemistry of Materials* 7: 443–445.
45. Holland ER, Pomfret SJ, Adams PN, and Monkman AP (1996) Conductivity studies of polyaniline doped with CSA. *Journal of Physics: Condensed Matter* 8: 2991–3002.
46. Bayer AG (1991) New Polythiophene Dispersions, Their Preparation and Their Use. European Patent 440957.
47. Kim WH, Makinen AJ, Nikolov N, *et al.* (2002) Molecular organic light-emitting diodes using highly conducting polymers as anodes. *Applied Physics Letters* 80: 3844–3846.
48. Zhang F, Mammo W, Andersson LM, *et al.* (2006) Low-bandgap alternating fluorene copolymer/methanofullerene heterojunctions in efficient near-infrared polymer solar cells. *Advanced Materials* 18: 2169–2173.
49. Welsh DM, Kumar A, Meijer EW, and Reynolds JR (1999) Enhanced contrast ratios and rapid switching in electrochromics based on poly(3,4-propylenedioxythiophene) derivatives. *Advanced Materials* 11: 1379–1382.
50. Tengstedt C, Crispin A, Hsu C-H, *et al.* (2005) Study and comparison of conducting polymer hole injection layers in light emitting devices. *Organic Electronics* 6: 21–33.
51. Hwang JH and Yang SC (1989) Morphological modification of polyaniline using polyelectrolyte template molecules. *Synthetic Metals* 29: 271–276.
52. Angelopoulos M, Patel N, and Saraf R (1993) Amic acid doping of polyaniline: Characterization and resulting blends. *Synthetic Metals* 55–57: 1552–1557.
53. Jayanty S, Prasad GK, Sreedhar B, and Radhakrishnan TP (2003) Polyelectrolyte templated polyaniline – film morphology and conductivity. *Polymer* 44: 7265–7270.
54. Moon HS and Park JK (1998) Structural effect of polymeric acid dopants on the characteristics of doped polyaniline composites. *Synthetic Metals* 92: 223–228.
55. Sahoo SK, Nagarajan R, Roy S, *et al.* (2004) An enzymatically synthesized polyaniline: A solid-state NMR study. *Macromolecules* 37: 4130–4138.
56. Avlyanov JK, Min YG, MacDiarmid AG, and Epstein AJ (1995) Polyaniline – conformational-changes induced in solution by variation of solvent and doping level. *Synthetic Metals* 72: 65–71.
57. Min Y, Xia Y, MacDiarmid AG, and Epstein AJ (1995) Vapor phase "secondary doping" of polyaniline. *Synthetic Metals* 69: 159–160.
58. Jonsson SKM, Birgerson J, Crispin X, *et al.* (2003) The effects of solvents on the morphology and sheet resistance in poly(3,4-ethylenedioxythiophene)–polystyrenesulfonic acid (PEDOT–PSS) films. *Synthetic Metals* 139: 1–10.
59. Kim JY, Jung JH, Lee DE, and Joo J (2002) Enhancement of electrical conductivity of poly(3,4-ethylenedioxythiophene)/poly(4-styrenesulfonate) by a change of solvents. *Synthetic Metals* 126: 311–316.
60. Greczynski G, Kugler T, and Salaneck WR (1999) Characterization of the PEDOT-PSS system by means of x-ray and ultraviolet photoelectron spectroscopy. *Thin Solid Films* 354: 129–135.
61. Jukes PC, Martin SJ, Higgins AM, *et al.* (2004) Controlling the surface composition of poly(3,4-ethylenedioxythiophene)-poly(styrene sulfonate) blends by heat treatment. *Advanced Materials* 16: 807–811.
62. Nardes AM, Kemerink M, Janssen RAJ, *et al.* (2007) Microscopic understanding of the anisotropic conductivity of PEDOT-PSS thin films. *Advanced Materials* 19: 1196–1200.
63. Huang J, Miller PF, Mello JC, Mello AJ, and Bradley DDC (2003) Influence of thermal treatment on the conductivity and morphology of PEDOT-PSS films. *Synthetic Metals* 139: 569–572.
64. Ouyang J, Xu Q, Chu C-W, *et al.* (2004) On the mechanism of conductivity enhancement in poly(3,4-ethylenedioxythiophene):poly(styrene sulfonate) film through solvent treatment. *Polymer* 45: 8443–8450.
65. Miyatake K, Cho J-S, Takeoka S, and Tsuchida E (1999) Preparation and properties of polyaniline doped with poly(thiophenylenesulfonic acid). *Macromolecular Chemistry and Physics* 200: 2597–2601.
66. Moon HS and Park JK (1998) Structural effect of polymeric acid dopants on the chracteristics of doped polyaniline composites: Effect of hydrogen bonding. *Journal of Polymer Science, Part A: Polymer Chemistry* 36: 1431–1439.
67. Yoo JE, Krekelberg WP, Sun Y, *et al.* (2009) Polymer conductivity through particle connectivity. *Chemistry of Materials* 21: 1948–1954.
68. Hwang J, Amy F, and Kahn A (2006) Spectroscopic study on sputtered PEDOT· PSS: Role of surface PSS layer. *Organic Electronics* 7: 387–396.
69. Xing KZ, Fahlman M, Chen XW, Inganas O, and Salaneck WR (1997) The electronic structure of poly(3,4-ethylene-dioxythiophene): Studied by XPS and UPS. *Synthetic Metals* 89: 161–165.
70. Pernaut J-M and Reynolds JR (2000) Use of conducting electroactive polymers for drug delivery and sensing of bioactive molecules. A redox chemistry approach. *Journal of Physical Chemistry B: Polymers* 104: 4080–4090.
71. DeLongchamp D and Hammond PT (2001) Layer-by-layer assembly of PEDOT/polyaniline electrochromic devices. *Advanced Materials* 13: 1455–1459.
72. Lapkowski M and Pron A (2000) Electrochemical oxidation of poly(3,4-ethylenedioxythiophene) – "*in situ*" conductivity and spectroscopic investigations. *Synthetic Metals* 110: 79–83.
73. Carlberg JC and Inganas O (1998) Fast optical spectroscopy of the electrochemical doping of poly(3,4-ethylenedioxythiophene). *Journal of the Electrochemical Society* 145: 3810–3814.
74. McManus PM, Cushman RJ, and Yang SC (1987) Influence of oxidation and protonation on the electrical conductivity of polyaniline. *Journal of Physical Chemistry* 91: 744–747.
75. Ginder JM and Epstein AJ (1990) Role of ring torsion angle in polyaniline: electronic structure and defect states. *Physical Review B: Condensed Matter* 41: 10674–10685.
76. Huang WS and MacDiarmid AG (1993) Optical properties of polyaniline. *Polymer* 34: 1833–1845.
77. Stafstrom S, Bredas JL, Epstein AJ, *et al.* (1987) Polaron lattice in highly conducting polyaniline: Theoretical and optical studies. *Physical Review Letters* 59: 1464–1467.
78. Sun Y, MacDiarmid AG, and Epstein AJ (1990) Polyaniline - synthesis and characterization of pernigraniline base. *Journal of the Chemical Society, Chemical Communications,* 529–531.
79. Jin Z, Su Y, and Duan Y (2001) A novel method for polyaniline synthesis with the immobilized horseradish peroxidase enzyme. *Synthetic Metals* 122: 237–242.
80. Ivanov VF, Gribkova OL, Cheberyako KV, *et al.* (2004) Template synthesis of polyaniline in the presence of poly-(2-acrylamido-2-methyl-1-propanesulfonic acid). *Russian Journal of Electrochemistry* 40: 299–304.

81. Nekrasov AA, Gribkova OL, Eremina TV, *et al.* (2008) Electrochemical synthesis of polyaniline in the presence of poly(amidosulfonic acid)s with different rigidity of polymer backbone and characterization of the films obtained. *Electrochimica Acta* 53: 3789–3797.
82. Caramyshev AV, Evtushenko EG, Ivanov VF, *et al.* (2005) Synthesis of conducting polyelectrolyte complexes of polyaniline and poly(2-acrylamido-3-methyl-1-propanesulfonic acid) catalyzed by pH-stable palm tree peroxidase. *Biomacromolecules* 6: 1360–1366.
83. Wakizaka D, Fushimi T, Ohkita H, and Ito S (2004) Hole transport in conducting ultrathin films of PEDOT/PSS prepared by layer-by-layer deposition technique. *Polymer* 45: 8561–8565.
84. Bobacka J, Lewenstam A, and Ivaska A (2000) Electrochemical impedance spectroscopy of oxidized poly(3,4-ethylenedioxythiophene) film electrodes in aqueous solutions. *Journal of Electroanalytical Chemistry* 489: 17–27.
85. Bard AJ and Faulkner LR (2000) *Electrochemical Methods,* 2nd edn. New York: Wiley.
86. Cai LT and Chen HY (1997) Preparation and electroactivity of polyaniline/poly (acrylic acid) film electrodes modified by platinum microparticles. *Journal of Applied Electrochemistry* 28: 161–166.
87. Kang Y, Lee M-H, and Rhee SB (1992) Electrochemical properties of polyaniline doped with poly (styrenesulfonic acid). *Synthetic Metals* 52: 319–328.
88. Meng H, Perepichka DF, Bendikov M, *et al.* (2003) Solid-state synthesis of a conducting polythiophene via an unprecedented heterocyclic coupling reaction. *Journal of the American Chemical Society* 125: 15151–15162.
89. Sahoo SK, Nagarajan R, Chakraborty S, *et al.* (2002) Variation in the structure of conducting polyaniline with and without the presence of template during enzymatic polymerization: A solid-state NMR study. *Journal of Macromolecular Science, Part A: Pure and Applied Chemistry* A39: 1223–1240.
90. Sahoo SK, Nagarajan R, Samuelson L, *et al.* (2001) Enzymatically synthesized polyaniline in the presence of a template poly(vinylphosphonic acid): A solid state NMR study. *Journal of Macromolecular Science, Part A: Pure and Applied Chemistry* A38: 1315–1328.
91. Wehrle B, Limbach H-H, Mortensen J, and Heinze J (1989) ^{15}N CPMAS NMR study of the structure of polyaniline. *Angewandte Chemie, International Edition in English* 28: 1741–1743.
92. Aasmundtveit KE, Samuelsen EJ, Pettersson LAA, *et al.* (1999) Structure of thin films of poly(3,4-ethylenedioxythiophene). *Synthetic Metals* 101: 561–564.
93. Niu L, Kvarnstrom C, Froberg K, and Ivaska A (2001) Electrochemically controlled surface morphology and crystallinity in poly(3,4-ethylenedioxythiophene) films. *Synthetic Metals* 122: 425–429.
94. Granström M and Inganas O (1995) Electrically conductive polymer fibres with mesoscopic diameters: 1. Studies of structure and electrical properties. *Polymer* 36: 2867–2872.
95. Brabec C, Scherf U, and Dyakonov V (eds.) (2008) *Organic Photovoltaics*. Weinheim: Wiley-VCH.
96. Ebisawa F, Kurokawa T, and Nara S (1983) Electrical properties of polyacetylene/polysiloxane interface. *Journal of Applied Physics* 54: 3255–3259.
97. Koezuka H, Tsumura A, and Ando T (1987) Field-effect transistor with polythiophene thin film. *Synthetic Metals* 18: 699–704.
98. Dimitrakopoulos CD and Malenfant PR (2002) Organic thin film transistors for large area electronics. *Advanced Materials* 14: 99–117.
99. Drury CJ, Mutsaers CMJ, Hart CM, Matters M, and Leeuw DM (1998) Low-cost all-polymer integrated circuits. *Applied Physics Letters* 73: 108–110.
100. Nilsson D, Chen M, Kugler T, *et al.* (2002) Bi-stable and dynamic current modulation in electrochemical organic transistors. *Advanced Materials* 1: 51–54.
101. Sirringhaus H, Kawase T, Friend RH, *et al.* (2000) High-resolution inkjet printing of all-polymer transistor circuits. *Science* 290: 2123–2126.
102. Blanchet GB, Fincher CR, Lefenfeld M, and Rogers JA (2004) Contact resistance in organic thin film transistors. *Applied Physics Letters* 84: 296–298.
103. Lim JA, Cho JH, Park YD, *et al.* (2006) Solvent effect of inkjet printed source/drain electrodes on electrical properties of polymer thin-film transistors. *Applied Physics Letters* 88: 082102.
104. Xue F, Su Y, and Varahramyan K (2005) Modified PEDOT-PSS conducting polymer as S/D electrodes for device performance enhancement of P3HT TFTs. *IEEE Transactions on Electron Devices* 52: 1982–1987.
105. Epstein AJ, Hsu F-C, Chiou N-R, and Prigodin VN (2003) Doped conducting polymer-based field effect devices. *Synthetic Metals* 137: 859–861.
106. Lin Y-J, Li Y-C, Yeh H-J, *et al.* (2007) Blending of poly(3,4-ethylenedioxythiophene) poly(styrenesulfonate) with poly(ethyleneimine) as an active layer in depletion-mode organic thin film transistors. *Applied Physics Letters* 91: 253501.
107. Prache O (2001) Active matrix molecular OLED microdisplays. *Displays* 22: 49–56.
108. Richer T (2007) The Sony Drive XEL-1 OLED TV: 1,000,000:1 contrast starting December 1st. http://www.engadget.com/2007/10/01/the-sonydrive-xel-1-oled-tv-1-000-000-1-contrast-starting-decem (accessed July 2009).
109. Pope M, Kallmann H, and Magnante P (1963) Electroluminescence in organic crystals. *Journal of Chemical Physics* 38: 2042–2043.
110. Tang CW and VanSlyke SA (1987) Organic electroluminescent diodes. *Applied Physics Letters* 51: 913–915.
111. Burroughes JH, Bradley DDC, Brown AR, *et al.* (1990) Light-emitting diodes based on conjugated polymers. *Nature* 347: 539–541.
112. Li Z and Meng H (2007) *Organic Light-Emitting Materials and Devices*. Boca Raton, FL: Taylor and Francis.
113. Scherf U and List EJW (2002) Semiconducting polyfluorenes – towards reliable structure–property relationships. *Advanced Materials* 14: 477–487.
114. Jurchescu OD, Baas J, and Palstra TTM (2005) Electronic transport properties of pentacene single crystals upon exposure to air. *Applied Physics Letters* 87: 052102.
115. Wang HL, MacDiarmid AG, Wang YJ, Gebler DD, and Epstein AJ (1996) Application of polyaniline (emeraldine base, EB) in polymer light-emitting devices. *Synthetic Metals* 78: 33–37.
116. Yang Y and Heeger AJ (1994) Polyaniline as a transparent electrode for polymer light-emitting diodes: Lower operating voltage and higher efficiency. *Applied Physics Letters* 64: 1245–1247.
117. Seeley AJAB, Friend RH, Kim JS, and Burroughes JH (2004) Trap-assisted hole injection and quantum efficiency enhancement in poly(9,9 dioctylfluorene-alt-benzothiadiazole) polymer light-emitting diodes. *Journal of Applied Physics* 96: 7643–7649.
118. Brewer PJ, Lane PA, Huang J, *et al.* (2005) Role of electron injection in polyfluorene-based light emitting diodes containing PEDOT:PSS. *Physical Review B: Condensed Matter* 71: 205209.

119. Koch N, Elschner A, and Johnson RL (2006) Green polyfluorene-conducting polymer interfaces: Energy level alignment and device performance. *Journal of Applied Physics* 100: 024512.
120. Elschner A, Bruder F, Heuer H-W, *et al.* (2000) PEDT/PSS for efficient hole-injection in hybrid organic light-emitting diodes. *Synthetic Metals* 111–112: 139–143.
121. Poplavskyy D, Nelson J, and Bradley DDC (2003) Ohmic hole injection in poly(9,9-dioctylfluorene) polymer light-emitting diodes. *Applied Physics Letters* 83: 707–709.
122. Gustafsson G, Cao Y, Tracy GM, *et al.* (1992) Flexible light-emitting diodes made from soluble conducting polymers. *Nature* 357: 477–479.
123. Scott JC, Kaufman JH, Brock PJ, *et al.* (1996) Degradation and failure of MEH-PPV light-emitting diodes. *Journal of Applied Physics* 79: 2745–2751.
124. Yang Y, Westerweele E, Zhang C, Smith P, and Heeger AJ (1995) Enhanced performance of polymer light-emitting diodes using high-surface area polyaniline network electrodes. *Journal of Applied Physics* 77: 694–698.
123. Karg S and Angelopoulos M (1996) Improved brightness and lifetime of MEH-PPV light-emitting diodes using polyaniline anodes. In: *38th TMS Electronic Materials Conference.* Santa Barbara, CA, USA, 26–28 June. http://www.tms.org/Meetings/EMC96/wedH1.html (accessed July 2009).
126. Jang J, Ha J, and Kim K (2008) Organic light-emitting diode with polyaniline-poly(styrene sulfonate) as a hole injection layer. *Thin Solid Films* 516: 3152–3156.
127. Granlund T, Pettersson LAA, and Inganäs O (2001) Determination of the emission zone in a single-layer polymer light-emitting diode through optical measurements. *Journal of Applied Physics* 89: 5897–5902.
128. Ghosh S and Inganäs O (2001) Nano-structured conducting polymer network based on PEDOT-PSS. *Synthetic Metals* 121: 1321–1322.
129. Kim WH, Makinen AJ, Nikolov N, *et al.* (2002) Molecular organic light-emitting diodes using highly conducting polymers as anodes. *Applied Physics Letters* 80: 3844–3846.
130. Fritts CE (1883) New form of selenium cell. *American Journal of Science* 26: 465–472.
131. Ohl R (1946) Light Sensitive Device. US Patent 2,402,662.
132. Tang CW (1986) Two-layer organic photovoltaic cell. *Applied Physics Letters* 48: 183–185.
133. Geisz JF, Kurtz S, Wanlass MW, *et al.* (2007) High-efficiency GaInP/GaAs/InGaAs triple-junction solar cells grown inverted with a metamorphic bottom junction. *Applied Physics Letters* 91: 3.
134. Bulovic V and Forrest SR (1995) Excitons in crystalline thin films of 3,4,9,10-perylenetetracarboxylic dianhydride studied by photocurrent response. *Chemical Physics Letters* 238: 88–92.
135. Yu G, Gao J, Hummelen JC, Wudl F, and Heeger AJ (1995) Polymer photovoltaic cells: Enhanced efficiencies via a network of internal donor–acceptor heterojunctions. *Science* 270: 1789–1791.
136. Pichler K (1997) Conjugated polymer electroluminesence: Technical aspects from basic devices to commercial products. *Philosophical Transactions of the Royal Society of London, Series A: Mathematical and Physical Sciences* 335: 829–842.
137. Zhang FL, Gadisa A, Inganaes O, Svensson M, and Andersson MR (2004) Influence of buffer layers on the performance of polymer solar cells. *Applied Physics Letters* 84: 3906–3908.
138. Rowell MW, Topinka MA, McGehee MD, *et al.* (2006) Organic solar cells with carbon nanotube network electrodes. *Applied Physics Letters* 88: 233506.
139. Reale A, Brown TM, Di Carlo A, *et al.* (2006) Nanocomposites for organic and hybrid organic–inorganic solar cells. *Proceedings of SPIE* 6334: 63340Y.
140. Tan S, Zhai J, Xue B, *et al.* (2004) Property influence of polyanilines on photovoltaic behaviors of dye-sensitized solar cells. *Langmuir* 20: 2934–2937.
141. Arango AC, Johnson LR, Bliznyuk VN, *et al.* (2000) Efficient titanium oxide/conjugated polymer photovoltaics for solar energy conversion. *Advanced Materials* 12: 1689–1692.
142. Stathatos E, Lianos P, and Krontiras C (2001) Dye-sensitized photoelectrochemical cell using a nanocomposite SiO_2/poly(ethylene glycol) thin film as electrolyte support. Characterization by time-resolved luminescence and conductivity measurements. *Journal of Physical Chemistry B: Polymers* 105: 3486–3492.
143. Bisquert J, Cahen D, Hodes G, Rühle S, and Zaban A (2004) Physical chemical principles of photovoltaic conversion with nanoparticulate, mesoporous dye-sensitized solar cells. *Journal of Physical Chemistry B: Polymers* 108: 8106–8118.
144. O'Regan B and Gratzel M (1991) A low-cost, high-efficiency solar cell based on dye-sensitized colloidal TiO_2 films. *Nature* 353: 737–740.
145. Lee J-K, Jang S-i, Jung B-H, *et al.* (2008) Photovoltaic properties of dye-sensitized solar cells using glycerol-modified PEDOT-PSS. In: *Korea–Japan Joint Forum.* Chitose, Hokkaido, Japan, 23–25 October. http://www.kjf2008.org/program/Poster25/P25-072.pdf (accessed July 2009).
146. Crispin X, Marciniak S, Osikowicz W, *et al.* (2003) Conductivity, morphology, interfacial chemistry, and stability of poly(3,4-ethylene dioxythiophene)–poly(styrene sulfonate): A photoelectron spectroscopy study. *Journal of Polymer Science, Part B: Polymer Physics* 41: 2561–2583.
147. Ouyang J, Chu C-W, Chen F-C, Xu Q, and Yang Y (2005) High-conductivity poly(3,4-ethylenedioxythiophene):poly(styrene sulfonate) film and its application in polymer optoelectronic devices. *Advanced Functional Materials* 15: 203–208.
148. Hsiao Y-S, Whang W-T, Chen C-P, and Chen Y-C (2008) High-conductivity poly(3,4-ethylenedioxythiophene):poly(styrene sulfonate) film for use in ITO-free polymer solar cells. *Journal of Materials Chemistry* 18: 5948–5955.
149. Lampert CM (1984) Electrochromic materials and devices for energy-efficient windows. *Solar Energy Materials* 11: 1–27.
150. Platt JR (1961) Electrochromism, a possible change of color producible in dyes by an electric field. *Journal of Chemical Physics* 34: 862–863.
151. Deb SK (1969) Novel electrophotographic system. *Applied Optics Supplement* 3: 192–195.
152. Green M, Smith WC, and Weiner JA (1976) A thin film electrochromic display based on the tungsten bronzes. *Thin Solid Films* 38: 89–100.
153. Scrosati B (ed.) (1993) *Applications of Electroactive Polymers*. London: Chapman and Hall.
154. Isamu S, Makoto S, and Eiich I (1979) Solid-state electrochromic device consisting of amorphous WO_3 and Cr_2O_3. *Journal of Applied Physics* 50: 4027–4032.
155. Schoot CJ, Ponjee JJ, van Dam HT, van Doorn RA, and Bolwijn PT (1973) New electrochromic memory display. *Applied Physics Letters* 23: 64–65.
156. Kaufman FB, Schroeder AH, Engler EM, and Patel VV (1980) Polymer-modified electrodes: A new class of

electrochromic materials. *Applied Physics Letters* 36: 422–425.
157. Argun AA, Aubert PH, Thompson BC, *et al.* (2004) Multicolored electrochromism in polymers: Structures and devices. *Chemistry of Materials* 16: 4401–4412.
158. Gustafsson JC, Inganas O, and Andersson AM (1994) Conductive polyheterocycles as electrode materials in solid-state electrochromic devices. *Synthetic Metals* 62: 17–21.
159. Lu W, Fadeev AG, Qi BH, *et al.* (2002) Use of ionic liquids for pi-conjugated polymer electrochemical devices. *Science* 297: 983–987.
160. Groenendaal BL, Jonas F, Freitag D, Pielartzik H, and Reynolds JR (2000) Poly(3,4-ethylenedioxythiophene) and its derivatives: Past, present, and future. *Advanced Materials* 12: 481–494.
161. Hu HL, Hechavarria L, and Campos J (2003) Optical and electrical responses of polymeric electrochromic devices: Effect of polyacid incorporation in polyaniline film. *Solid State Ionics* 161: 165–172.
162. Fabretto M, Hall C, Vaithianathan T, *et al.* (2008) The mechanism of conductivity enhancement in poly(3,4-ethylenedioxythiophene)-poly(styrenesulfonic) acid using linear-diol additives: Its effect on electrochromic performance. *Thin Solid Films* 516: 7828–7835.
163. Tarver J, Yoo JE, and Loo Y-L (submitted) Water-dispersible polyaniline exhibiting stable and reversible switching in the visible extending into the near-IR. *Chemistry of Materials.*
164. Gustafsson-Carlberg JC, Inganäs O, Andersson MR, *et al.* (1995) Tuning the bandgap for polymeric smart windows and displays. *Electrochimica Acta* 40: 2233–2235.
165. Sankaran B and Reynolds JR (1997) High-contrast electrochromic polymers from alkyl-derivatized poly(3,4-ethylenedioxythiophenes). *Macromolecules* 30: 2582–2588.
166. Kumar A, Welsh DM, Morvant MC, *et al.* (1998) Conducting poly(3,4-alkylenedioxythiophene) derivatives as fast electrochromics with high-contrast ratios. *Chemistry of Materials* 10: 896–902.
167. Sapp SA, Sotzing GA, and Reynolds JR (1998) High contrast ratio and fast-switching dual polymer electrochromic devices. *Chemistry of Materials* 10: 2101–2108.
168. Thompson BC, Schottland P, Zong KW, and Reynolds JR (2000) *In situ* colorimetric analysis of electrochromic polymers and devices. *Chemistry of Materials* 12: 1563–1571.
169. Groenendaal L, Zotti G, Aubert PH, Waybright SM, and Reynolds JR (2003) Electrochemistry of poly(3,4-alkylenedioxythiophene) derivatives. *Advanced Materials* 15: 855–879.
170. Faughnan BW, Crandall RS, and Heyman PM (1975) Electrochromism in WO_3 amorphous films. *Radio Corporation of America Review* 36: 177–197.
171. Michalak F and Aldebert P (1996) Flexible electrochromic device based on colloidal tungsten oxide and polyaniline. *Solid State Ionics* 85: 265–272.

5 Electronic Properties of Alkanethiol Molecular Junctions: Conduction Mechanisms, Metal–Molecule Contacts, and Inelastic Transport

G Wang, T-W Kim, and T Lee, Gwangju Institute of Science and Technology, Gwangju, Korea

W Wang, University of Wyoming, Laramie, WY, USA

M A Reed, Yale University, New Haven, CT, USA

5.1 Introduction

Molecular electronics utilizing functional molecules as the ultimate nanoscale electronic components has recently generated considerable interest in both the basic transport physics of molecular systems and potential technological applications in a variety of functional electronic device components for ultra-high density future electronics [1–4].

However, despite the numerous potential advantages of molecular electronics as compared to traditional silicon-based electronics, there are many issues and challenges that need to be overcome to apply molecules to actual electronic circuits. For example, some reports of molecular mechanisms in electronic devices [3a,5,6a,b] have been shown to be premature due to filamentary conduction [3c,7], highlighting the fabrication sensitivity of molecular structures and the need to institute reliable controls and methods to validate true molecular transport [8]. A related problem is the characterization of molecules in the active device structures, including their configuration, bonding, and indeed even their very presence. In addition, metal–molecule contact is important not only for understanding the transport properties of molecular devices but also for realizing reproducible molecular electronic devices due to its role in controlling metal–molecule interfaces [9,10].

Here we present results on well-understood molecular assemblies, which exhibit an understood classical transport behavior, and which can be used as a control for eliminating (or understanding) fabrication variables. Utilizing tunneling spectroscopic methods, we present the unambiguous evidence of the presence of molecules in the junction. Using the statistical analysis on the current–voltage characteristics of molecular junctions, we investigate the effect of metal–molecule contacts and present the contact resistances of the junctions.

A molecular system whose structure and configuration are sufficiently well characterized such that it can serve as a standard is the extensively studied alkanethiol self-assembled monolayer (SAM) [11]. There are two kinds of alkanethiols: alkanemonothiols ($CH_3(CH_2)_{n-1}SH$), where there is one thiol at an end of molecule; and alkanedithiols ($HS(CH_2)_nSH$), where there are thiols at both end of molecule. This molecule system is useful as a control since properly prepared SAMs form single van der Waals crystals [11,12], and presents a simple classical metal–insulator–metal (M-I-M) tunnel junction when fabricated between metallic contacts due to the large gap of approximately 8 eV [13–15] between the highest occupied molecular orbital (HOMO) and the lowest unoccupied molecular orbital (LUMO).

Various surface analytical tools have been utilized to investigate the surface and bulk properties of the alkanethiol SAMs, such as X-ray photoelectron spectroscopy [16], Fourier transform infrared spectroscopy (FTIR) [17], Raman spectroscopy [18], scanning tunneling microscopy (STM) [12], etc. For example, studies have shown that the bonding of the thiolate group to the gold surface is strong with a bonding energy of ~1.7 eV [11]. STM topography examinations revealed that alkanethiols adopt the commensurate crystalline lattice characterized by a $c(4\times2)$ superlattice of a $(\sqrt{3}\times\sqrt{3})R30°$ [12,19]. FTIR investigation showed that the orientation of the alkanethiol SAMs on Au(111) surfaces are tilted ~30° from the surface normal [20].

Electronic transport through alkanethiol SAMs have also been characterized by STM [21,22], conducting atomic force microscopy [23–26], mercury-drop junctions [29–30], cross-wire junctions [31], and electrochemical methods [32–34]. These investigations are exclusively at ambient temperature – clearly useful – but insufficient for an unambiguous claim that the transport mechanism is tunneling (of course expected, assuming that the Fermi levels of the contacts lie within the large HOMO–LUMO gap). However, in the absence of temperature-dependent current–voltage ($I(V,T)$) characteristics, other conduction mechanisms (e.g., thermionic, hopping, or filamentary conduction) cannot be excluded complicate the analysis, and thus such a claim is premature.

Utilizing a solid-state device structure that incorporates alkanethiol SAMs, we demonstrate devices that allow $I(V,T)$ and structure-dependent measurements [35,36] with results that can be compared with accepted theoretical models of M-I-M tunneling. The use of this fabrication approach is not special in any way (other than that we have so far found it to be successful) – indeed we stress that any successful device fabrication method should yield the results described below if one is characterizing the intrinsic molecular transport properties.

The electronic transport is further investigated with the technique of inelastic electron tunneling spectroscopy (IETS) [36]. IETS was developed in the 1960s as a powerful spectroscopic tool to study the vibrational spectrum of organic molecules confined inside metal-oxide–metal junctions [37–41]. In our study, IETS is utilized for the purpose of molecule identification, chemical bonding, and conduction mechanism investigations of the 'control' SAMs. The exclusive presence of well-known vibrational modes of the alkanes used are direct evidence of the molecules in the device structure, something that has to date only been inferred (with good reason, but nonetheless not unambiguously). The vibrational modes, exclusively identified as alkanes (as well as contact modes) are difficult to interpret in any other way other than as components in the active region of the device. The inelastic tunneling spectra also demonstrate that electronic tunneling occurs through the molecules, confirming the conduction mechanism obtained by $I(V,T)$ characterizations. The specific spectral lines also yield intrinsic linewidths that may give insight into molecular conformation, and may prove to be a powerful tool in future molecular device characterization.

We also present the influence of metal–molecule contacts in molecular junctions using a proposed multibarrier tunneling (MBT) model where the metal–molecule–metal junction can be divided into three parts: the molecular-chain body with metal–molecule contacts on either side of molecule [9,10]. The MBT model will help introduce an insight for studying charge transport mechanisms, focused on the metal–molecule contacts in molecular electronic devices or other nanoscale devices.

5.2 Experiment

Electronic transport measurements on alkanethiol SAMs were performed using two different device structures. The first device structure is similar to the nanoscale device structure reported previously, the so-called 'nanopore' devices [3a,35,36,42–44]. Another device structure is vertical metal–molecule–metal device structure with microscale via-holes [9,10,45].

In the nanopore devices, as illustrated in **Figure 1(a)** (not drawn to scale in the relative

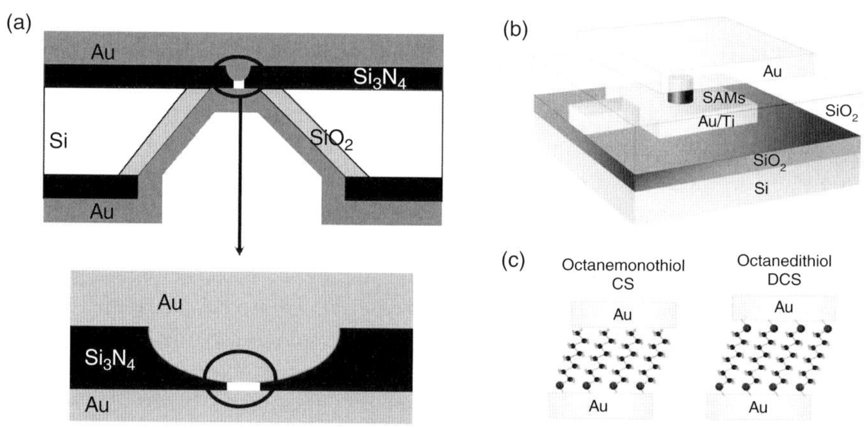

Figure 1 (a) Schematics of a nanometer scale device used in this study (not drawn to scale in the relative thickness). Top schematic is the cross section of a silicon wafer with a nanometer scale pore etched through a suspended silicon nitride membrane. Middle and bottom schematics show a Au/SAM/Au junction formed in the pore area. (b) Schematics of microscale via-hole junctions used in this study. (c) The structures of octanethiol and octanedithiol are shown as examples.

thickness), a number of molecules (several thousands) are sandwiched between two metallic contacts. This technique provides a stable device structure and makes cryogenic measurements possible. The device fabrication starts with a high-resistivity silicon wafer with low-stress Si_3N_4 film deposited on both sides by low-pressure chemical vapor deposition (LPCVD). By standard photolithography processing, a suspended Si_3N_4 membrane (size of 40 μm×40 μm and thickness of ~70 nm) is fabricated on the topside of the wafer. Subsequent e-beam lithography and reactive ion etching creates a single pore with a diameter of tens of nanometers through the membrane. As the next step, 150 nm gold is thermally evaporated onto the topside of the wafer to fill the pore and form one of the metallic contacts.

The device is then transferred into a molecular solution to deposit the SAM layer. For our experiments, a ~5 mM alkanethiol solution is prepared by adding ~10 μl alkanethiols into 10 ml ethanol [46]. The deposition is done in solution for 24 h inside a nitrogen-filled glove box with an oxygen level of less than 100 ppm. Alkanemonothiols and alkanedithiols of different molecular lengths: octanemonothiol ($CH_3(CH_2)_7SH$, C8), dodecanemonothiol ($CH_3(CH_2)_{11}SH$, C12), hexadecanemonothiol ($CH_3(CH_2)_{15}SH$, C16), octanedithiol ($HS(CH_2)_8SH$, DC8), nonanedithiol ($HS(CH_2)_9SH$, DC9), and decanedithol ($HS(CH_2)_{10}SH$, DC10) were used to form the active molecular components in molecular devices [46]. As representative examples, the chemical structures of octanethiol and octanedithiol are shown in **Figure 1(c)**.

In order to statistically determine the pore size in nanopore devices, test patterns (arrays of pores) were created under similar fabrication conditions. This indirect measurement of device size is done since SEM examination of the actual device can cause hydrocarbon contamination of the device and subsequent contamination of the monolayer. From regression analysis of 298 pores, the device sizes of the C8, C12, C16, and C8-dithiol samples are determined as 50 ± 8, 45 ± 2, 45 ± 2, and 51 ± 5 nm in diameters, respectively. A more ideal (less parasitic) C8 sample supercedes that of previous reports [35], and derived parameters from the two data sets agree to within a standard error. We will use these device areas as the effective contact areas. Although one could postulate that the actual area of metal that contacts the molecules may be different, there is little reason to propose that it would be different as a function of length over the range of alkanethiols used, and at most would be a constant systematic error.

The other device structure is shown in **Figure 1(b)**, that is, a vertical metal–molecule–metal junction device structure having a micrscale via-hole in which the molecules are self-assembled.

The sample is then transferred in ambient conditions to an evaporator that has a cooling stage to deposit the opposing Au contact in case of both device structures. During the thermal evaporation (under the pressure of 10^{-7}–10^{-8} Torr), liquid nitrogen is kept flowing through the cooling stage in order to avoid thermal damage to the molecular layer [35,47]. This technique reduces the kinetic energy of evaporated Au atoms at the surface of the monolayer, thus preventing Au atoms from punching through the monolayer. For the same reason, the evaporation rate is kept very low. The deposition rate is typically 0.1–0.5 Å s^{-1}. A total of 50–200 nm gold is deposited to form the contact.

The device is subsequently packaged and loaded into a low-temperature cryostat. The sample temperature is varied from 300 to 4.2 K by flowing cryogen vapor onto the sample (and thermometer) using a closed-loop temperature controller. Two-terminal direct current (DC) $I(V)$ measurements are performed using a semiconductor parameter analyzer. Inelastic electron tunneling spectra are obtained via a standard lock-in second harmonic measurement technique [37,38]. A synthesized function generator is used to provide both the modulation and the lock-in reference signal. The second harmonic signal (proportional to d^2I/dV^2) is directly measured using a lock-in amplifier, which is checked to be consistent with a numerical derivative of the first-harmonic signal (proportional to dI/dV). Various modulation amplitudes and frequencies are utilized to obtain the spectra. The alternating current (AC) modulation is added to a DC bias using operational amplifier-based custom circuitry [48].

5.3 Theoretical Basis

5.3.1 Possible Conduction Mechanisms

In **Table 1**, possible conduction mechanisms are listed with their characteristic current, temperature, and voltage dependencies [49] (we do not discuss filamentary tunneling mechanisms, which are easier to categorize[50a–c]). Based on whether thermal activation is involved, the conduction mechanisms fall into two distinct categories: (i) thermionic or hopping conduction, which has temperature-dependent $I(V)$ behavior and (ii) direct tunneling or

Table 1 Possible conduction mechanisms

Conduction mechanism	*Characteristic Behavior*	*Temperature dependence*	*Voltage dependence*
Direct tunneling[a]	$J \sim V\exp\left(-\frac{2d}{\hbar}\sqrt{2m\Phi}\right)$	None	$J \sim V$
Fowler–Nordheim tunneling	$J \sim V^2\exp\left(-\frac{4d\sqrt{2m}\Phi^{3/2}}{3q\hbar V}\right)$	None	$\ln\left(\frac{J}{V^2}\right) \sim \frac{1}{V}$
Thermiomic emission	$J \sim T^2\exp\left(-\frac{\Phi - q\sqrt{qV/4\pi\varepsilon d}}{kT}\right)$	$\ln\left(\frac{J}{T^2}\right) \sim \frac{1}{T}$	$\ln(J) \sim V^{1/2}$
Hopping conduction	$J \sim V\exp\left(-\frac{\Phi}{kT}\right)$	$\ln\left(\frac{J}{V}\right) \sim \frac{1}{T}$	$J \sim V$

[a] This characteristic of direct tunneling is valid for the low bias regime (see **equation 3a**).
Adapted with permission from Sze SM (1981) *Physics of Semiconductor Devices*. New York: Wiley.

Fowler–Nordheim tunneling, which does not have temperature-dependent $I(V)$ behavior. For example, thermionic and hopping conductions have been observed for 4-thioacetylbiphenyl SAMs[42] and 1,4-phenelyene diisocyanide SAMs [43b]. On the other hand, the conduction mechanism is expected to be tunneling when the Fermi levels of contacts lie within the large HOMO–LUMO gap for short-length molecule, as for the case of alkanethiol molecular system [13–15]. Previous work on Langmuir–Blodgett alkane monolayers[51a,b] exhibited a significant impurity-dominated transport component, complicating the analysis. The $I(V)$ measurements on self-assembled alkanethiol monolayers have also been reported [21–31,52]; however, all of these measurements were performed at fixed temperature (300 K), which is insufficient to prove tunneling as the dominant mechanism.

5.3.2 Tunneling Models

To describe the transport through a molecular system having HOMO and LUMO energy levels, one of the applicable models is the Franz two-band model [53–56]. This model provides a nonparabolic energy–momentum $E(k)$ dispersion relationship by considering the contributions of both the HOMO and LUMO energy levels [53]:

$$k^2 = \frac{2m^*}{\hbar^2}E\left(1 + \frac{E}{E_g}\right) \tag{1}$$

where k is the imaginary part of wave vector of electrons, m^* the electron effective mass, h $(= 2\pi\hbar)$ the Planck's constant, E the electron energy, and E_g the HOMO–LUMO energy gap. From this nonparabolic $E(k)$ relationship, the effective mass of the electron tunneling through the SAM can be deduced by knowing the barrier height of the metal–SAM–metal junction.

When the Fermi level of the metal is aligned close enough to one energy level (either HOMO or LUMO), the effect of the other distant energy level on the tunneling transport is negligible, and the widely used Simmons model [57] is an excellent approximation. Simmons model expressed the tunneling current density through a barrier in the tunneling regime of $V < \Phi_B/e$ as [27,57]

$$\mathcal{J} = \left(\frac{e}{4\pi^2\hbar d^2}\right)\left\{\left(\Phi_B - \frac{eV}{2}\right)\exp\left[-\frac{2(2m)^{1/2}}{\hbar}\alpha\left(\Phi_B - \frac{eV}{2}\right)^{1/2}d\right] - \left(\Phi_B + \frac{eV}{2}\right)\exp\left[-\frac{2(2m)^{1/2}}{\hbar}\alpha\left(\Phi_B + \frac{eV}{2}\right)^{1/2}d\right]\right\} \tag{2}$$

where m is the electron mass, d the barrier width, Φ_B the barrier height, and V the applied bias. For molecular systems, the Simmons model has been modified with a parameter α [27,35]. Here α is a unitless adjustable parameter that is introduced to provide either a way of applying the tunneling model of a rectangular barrier to tunneling through a nonrectangular barrier [27], or an adjustment to account for the effective mass (m^*) of the tunneling electrons through a rectangular barrier [27,35,56,58], or both. Here $\alpha = 1$ corresponds to the case for a rectangular barrier and bare electron mass. By fitting individual $I(V)$ data using equation 2, Φ_B and α values can be obtained.

Equation 2 can be approximated in two limits: low bias and high bias as compared with the barrier height Φ_B. For the low bias range, **equation 2** can be approximated as [57]

$$\mathcal{J} \approx \left(\frac{(2m\Phi_B)^{1/2}e^2\alpha}{h^2 d}\right) V \exp\left[-\frac{2(2m)^{1/2}}{\hbar}\alpha(\Phi_B)^{1/2}d\right] \quad (3a)$$

To determine the high bias limit, we compare the relative magnitudes of the first and second exponential terms in **equation 2**. At high bias, the first term is dominant and thus the current density can be approximated as

$$\mathcal{J} \approx \left(\frac{e}{4\pi^2\hbar d^2}\right)\left(\Phi_B - \frac{eV}{2}\right) \times \exp\left[-\frac{2(2m)^{1/2}}{\hbar}\alpha\left(\Phi_B - \frac{eV}{2}\right)^{1/2} d\right] \quad (3b)$$

The tunneling currents in both bias regimes are exponentially dependent on the barrier width d. In the low-bias regime the tunneling current density is $\mathcal{J} \propto (1/d)\exp(-\beta_0 d)$, where β_0 is bias-independent decay coefficient:

$$\beta_0 = \frac{2(2m)^{1/2}}{\hbar}\alpha(\Phi_B)^{1/2} \quad (4a)$$

while in the high-bias regime, $\mathcal{J} \propto (1/d)^2 \exp(-\beta_V d)$, where β_V is bias-dependent decay coefficient:

$$\beta_V = \frac{2(2m)^{1/2}}{\hbar}\alpha\left(\Phi_B - \frac{eV}{2}\right)^{1/2} = \beta_0\left(1 - \frac{eV}{2\Phi_B}\right)^{1/2} \quad (4b)$$

At high bias, β_V decreases as bias increases, which results from barrier-lowering effect due to the applied bias.

5.4 Results

5.4.1 Tunneling Current–Voltage Characteristics

5.4.1.1 Temperature-variable current–voltage (I(V,T)) measurement

In order to determine the conduction mechanism of self-assembled alkanethiol molecular systems $I(V)$ measurements in a sufficiently wide temperature range (300 to 80 K) and resolution (10 K) were performed. **Figure 2(a)** shows a representative $I(V,T)$ characteristic of dodecanemonothiol (C12) measured with the device structure as shown in **Figure 1(a)**. Positive bias corresponds to electrons injected from the physisorbed Au contact (bottom contact in **Figure 1(a)**) into the molecules. By using the contact area of 45 ± 2 nm in diameter determined from SEM study, a current density of 1500 ± 200 A cm^{-2} at 1.0 V is determined. No significant temperature dependence of the characteristics (from V = 0–1.0 V) is observed over the range from 300 to 80 K. An Arrhenius plot (ln(I) versus $1/T$) of this is shown in **Figure 2(b)**, exhibiting little temperature dependence in the slopes of ln(I) versus $1/T$ at different bias and thus indicating the absence of thermal activation. Therefore, we conclude that the conduction mechanism through alkanethiol is tunneling contingent on demonstrating a correct molecular length dependence. The tunneling through alkanethiol SAMs has been assumed as 'through-bond' tunneling, that is, along the tilted molecular chains between the metal contacts [24,25,34,59a,b]. Based on the applied bias as compared with the barrier height (Φ_B), the tunneling through an SAM layer can be categorized into either direct ($V < \Phi_B/e$) or Fowler–Nordheim ($V > \Phi_B/e$) tunneling. These two tunneling mechanisms can be distinguished due to their distinct voltage dependencies (see **Table 1**). Analysis of ln(I/V^2) versus $1/V$ [in **Figure 2(c)**] shows no significant voltage dependence, indicating no obvious Fowler–Nordheim transport behavior in this bias range (0 to 1.0 V) and thus determining that the barrier height is larger than the applied bias, that is, $\Phi_B > 1.0$ eV. This study is restricted to applied biases ≤ 1.0 V and the transition from direct to Fowler–Nordheim tunneling requires higher bias.

The importance of variable temperature measurements to validate tunneling is demonstrated in **Figure 3**. Here the $I(V)$ of an octanemonothiol (C8) device is shown (**Figure 3(a)**), whose $I(V)$ shape looks very similar to **Figure 2** (i.e., direct tunneling), and indeed can be fit to a Simmons model. However, further $I(V,T)$ measurements display an obvious temperature dependence (**Figure 3(b)**), which can be fit well to a hopping conduction model (**Table 1**) with a well-defined activation energy of 190 meV (**Figure 3(c)**). This and other similar impurity-mediated transport phenomena (such as Coulomb blockade) are observed in a subset of devices and is indicative of the unintentional incorporation of a trap or defect level in those devices. This study instead focuses on devices that do not show any defect-mediated transport and probes the intrinsic behavior of the molecular layer.

Having established tunneling as the conduction mechanism in a device, we will now obtain the barrier height by comparing experimental $I(V)$ data with theoretical calculations from tunneling models.

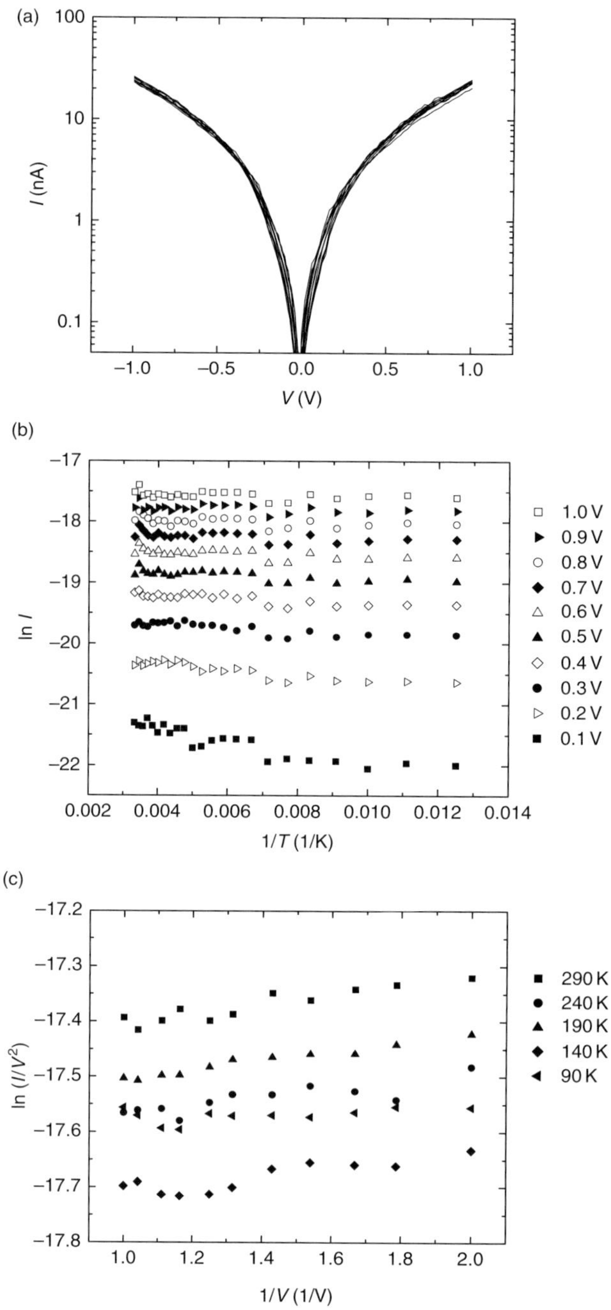

Figure 2 (a) Temperature-dependent *I*(*V*) characteristics of dodecanethiol (C12). *I*(*V*) data at temperatures from 300 to 80 K with 20 K steps are plotted on a log scale. (b) Arrhenius plot generated from the *I*(*V*) data in (a), at voltages from 0.1 to 1.0 V with 0.1 V steps. (c) Plot of $\ln(I/V^2)$ versus 1/*V* at selected temperatures.

5.4.1.2 Tunneling characteristics through alkanethiols

From the modified Simmons model (**equation 2**) by adjusting two parameters Φ_B and α, a nonlinear least-squares fitting can be performed to fit the measured C12 $I(V)$ data (calculation assuming $\alpha = 1$ has been previously shown not to fit $I(V)$ data well for some alkanethiol measurements at fixed temperature (300 K)) [27]. By using a device size of 45 nm in diameter, the best-fitting parameters (minimizing χ^2) for the room temperature C12 $I(V)$ data were found to be $\Phi_B = 1.42 \pm 0.04$ eV and $\alpha = 0.65 \pm 0.01$, where the error ranges of Φ_B and α are dominated by potential device size fluctuations of 2 nm. Likewise, data sets

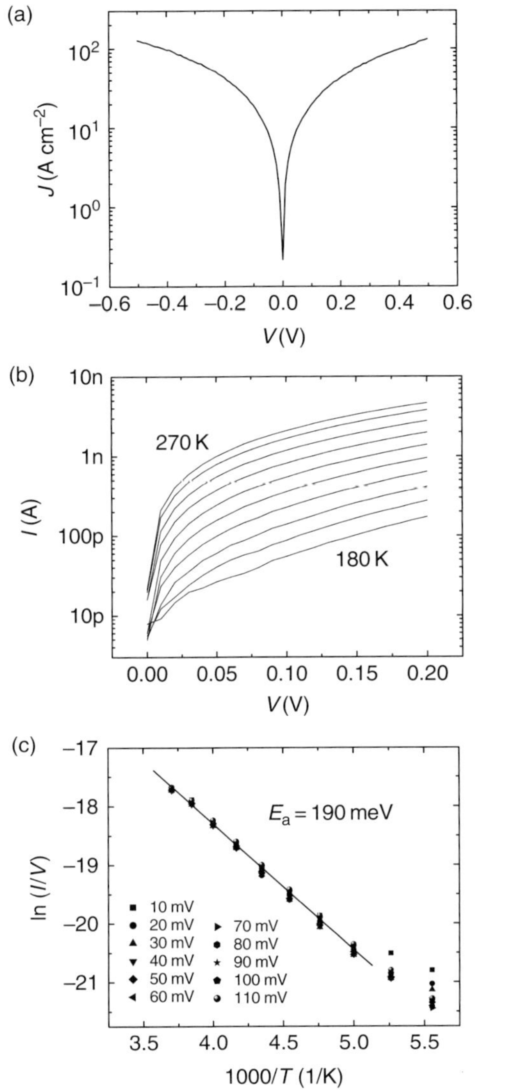

Figure 3 (a) $I(V)$ characteristics of an octanethiol (C8) device at 270 K. (b) Temperature dependence of the device from 270 to 180 K (in 10 K increments). (c) Plot of $\ln(I/V)$ versus $1/T$ at various voltages. The activated behavior is independent of bias voltage; thus, the behavior is hopping (in this device) due to incorporation of a defect of energy 190 meV. This class of devices is not suitable for investigation of the intrinsic transport mechanism in the SAM as it is dominated by a defect.

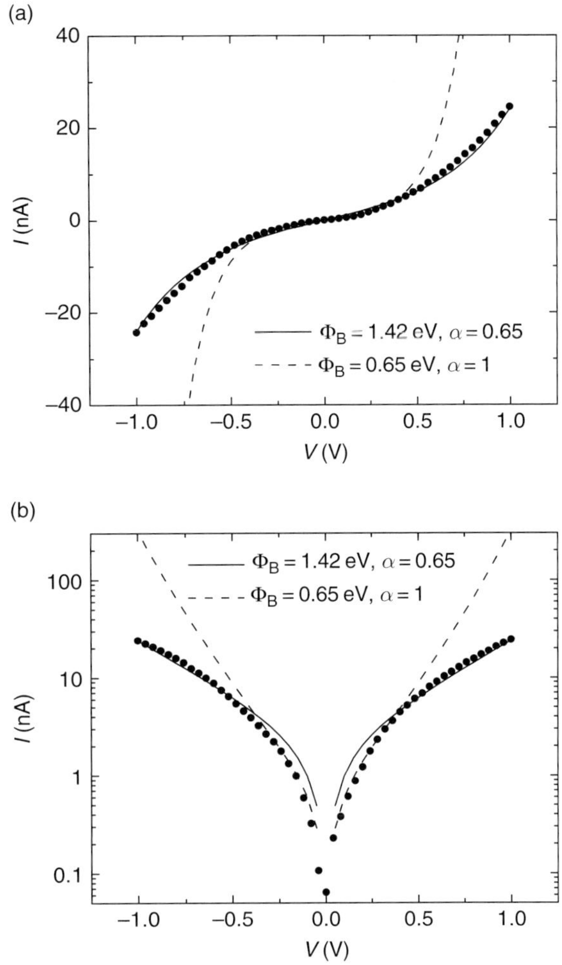

Figure 4 Measured C12 $I(V)$ data (circular symbols) is compared with calculation (solid curve) using the optimum fitting parameters of $\Phi_B = 1.42$ eV and $\alpha = 0.65$. The calculated $I(V)$ from a simple rectangular model ($\alpha = 1$) with $\Phi_B = 0.65$ eV is also shown as the dashed curve. Current is plotted (a) on linear scale and (b) on log scale.

were obtained and fittings were done for octanethiol (C8) and hexadecanethiol (C16), which yielded values ($\Phi_B = 1.83 \pm 0.10$ eV and $\alpha = 0.61 \pm 0.01$) and ($\Phi_B = 1.40 \pm 0.03$ eV, $\alpha = 0.68 \pm 0.01$), respectively.

Using $\Phi_B = 1.42$ eV and $\alpha = 0.65$, a calculated $I(V)$ for C12 is plotted as a solid curve in a linear scale (**Figure 4(a)**) and in a semi-log scale (**Figure 4(b)**). A calculated $I(V)$ for $\alpha = 1$ and $\Phi_B = 0.65$ eV (which gives the best fit at low-bias range) is shown as the dashed curve in the same figure, illustrating that with $\alpha = 1$ only limited regions of the $I(V)$ can be fit (specifically here, for $V < 0.3$ V). For the case of a rectangular barrier, the α parameter fit presented above corresponds to an effective mass m^* $(= \alpha^2 m)$ of $0.42m$.

In order to investigate the dependence of the Simmons model fitting on Φ_B and α, a fitting minimization analysis was undertaken on the individual Φ_B and α values as well as their product form of $\alpha\Phi_B^{1/2}$ in **equation 4a**. The quantity $\Delta(\Phi_B, \alpha) = \left(\sum \left| I_{\exp,v} - I_{cal,v} \right|^2\right)^{1/2}$ was calculated and plotted where $I_{\exp,V}$ is the experimental current–voltage values and $I_{cal,V}$ is calculated using **equation 2**. Seven thousand five hundred different $\{\Phi_B, \alpha\}$ pairs were used in the fittings with Φ_B ranging from 1.0 to 2.5 eV (0.01 eV increment) and α from 0.5 to 1.0 (0.01 increment). **Figure 5(a)** is a representative contour plot of $\Delta(\Phi_B, \alpha)$ versus Φ_B and α values generated for the C12 $I(V)$ data, where darker regions correspond to smaller $\Delta(\Phi_B, \alpha)$ and various shades represent half order of magnitude $\Delta(\Phi_B, \alpha)$ steps. The darker regions represent better fits of **equation 2** to the measured $I(V)$ data. In the inset in **Figure 5(a)** one can see there is a range of possible Φ_B and α values yielding minimum fitting parameters. Although the tunneling parameters determined from the previous Simmons tunneling fitting ($\Phi_B = 1.42$ eV and $\alpha = 0.65$) lie within this minimum region in this figure, there is a distribution of other possible values.

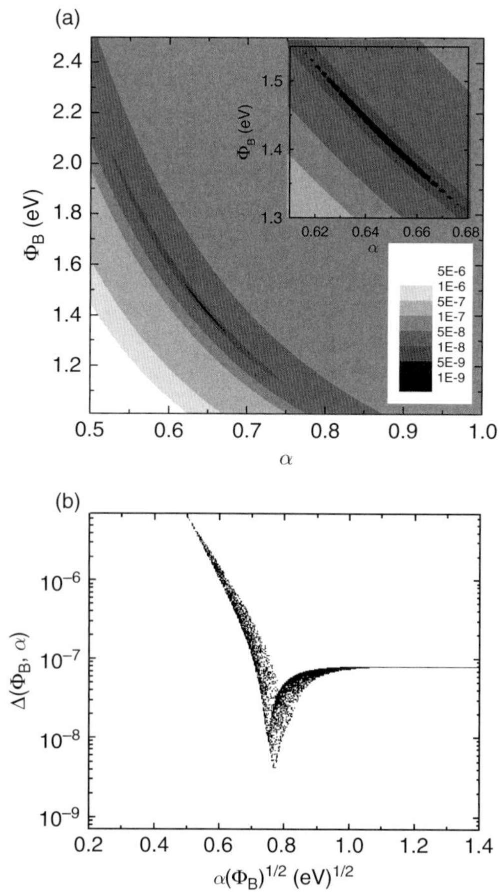

Figure 5 (a) Contour plot of $\Delta(\Phi_B, \alpha)$ values for C12 nanopore device as a function of Φ_B and α, where the darker region corresponds to a better fitting. Inset shows detailed minimization fitting regions. (b) A plot of $\Delta(\Phi_B, \alpha)$ as a function of $\alpha\Phi_B{}^{1/2}$.

A plot of $\Delta(\Phi_B, \alpha)$ versus $\alpha\Phi_B{}^{1/2}$ for the same device reveals a more pronounced dependence, and is shown in **Figure 5(b)**. This plot indicates the fitting to the Simmons model sharply depends on the product of $\alpha\Phi_B{}^{1/2}$. For this plot the $\Delta(\Phi_B, \alpha)$ is minimized at $\alpha\Phi_B{}^{1/2}$ of 0.77 $(eV)^{1/2}$ corresponding to a β_0 value of 0.79 Å^{-1} from **equation 4a**. The C8 and C16 devices showed similar results, indicating the Simmons tunneling model has a strong $\alpha\Phi_B{}^{1/2}$ dependence. For the C8 device, although Φ_B obtained from the fitting is a little larger, combined α and Φ_B gives a similar β_0 value within the error range as the C12 and C16 devices (**Table 2**).

5.4.1.3 Length-dependent tunneling through alkanethiols

Three alkanemonothiols of different molecular length (C8, C12, and C16) were investigated to study the length-dependent tunneling behavior. **Figure 6** is a semi-log plot of tunneling current densities multiplied by molecular length ($\mathcal{J}d$ at low bias and $\mathcal{J}d^2$ at high bias) as a function of the molecular length for these alkanethiols. The molecular lengths used in this plot are 13.3, 18.2, and 23.2 Å for C8, C12, and C16, respectively. Each molecular length was determined by adding an Au-thiol bonding length to the length of molecule [24]. Note that these lengths assume through-bond tunneling [24,25,34,59a,b]. The high- and low-bias regimes are defined somewhat arbitrarily by comparing the relative magnitudes of the first and second exponential terms in **equation 2**. Using $\Phi_B = 1.42$ eV and $\alpha = 0.65$ obtained from nonlinear least-squares fitting of the C12 $I(V)$ data, the second term becomes less than ~10 % of the first term at ~0.5 V that is chosen as the boundary of low- and high-bias ranges.

As seen in **Figure 6**, the tunneling current shows exponential dependence on molecular length, which is consistent with the Simmons tunneling model (**equation 3**). The β values can be determined from the slope at each bias and are plotted in **Figure 7**. The error bar of an individual β value in this plot was obtained by considering both the device size uncertainties and the linear-fitting errors.

The β values determined are almost independent of bias in the low-bias range ($V <$ ~0.5 V), and an average β of 0.77 ± 0.06 Å^{-1} in this region (from 0 to 0.5 V) can be calculated from **Figure 7**. **Table 3** is a summary of previously reported alkanethiol transport parameters obtained by different techniques. The current densities ($\mathcal{J}$) listed in **Table 3** are for C12 monothiol or dithiol devices at 1 V, which are

Table 2 Summary of alkanethiol tunneling parameters in this study

Molecules	*J at 1 V (A cm^{-2})*	Φ_B *(eV)*	α	m^a *(m)*	β_0 *(Å^{-1})^a*
C8	31 000 ± 10 000	1.83 ± 0.10	0.61 ± 0.01	0.37	0.85 ± 0.04
C12	1500 ± 200	1.42 ± 0.04	0.65 ± 0.01	0.42	0.79 ± 0.02
C16	23 ± 2	1.40 ± 0.03	0.68 ± 0.01	0.46	0.82 ± 0.02
C8-dithiol	93 000 ± 18 000	1.20 ± 0.03	0.59 ± 0.01	0.35	0.66 ± 0.02

[a] β_0 values were calculated using **equation 4a**.

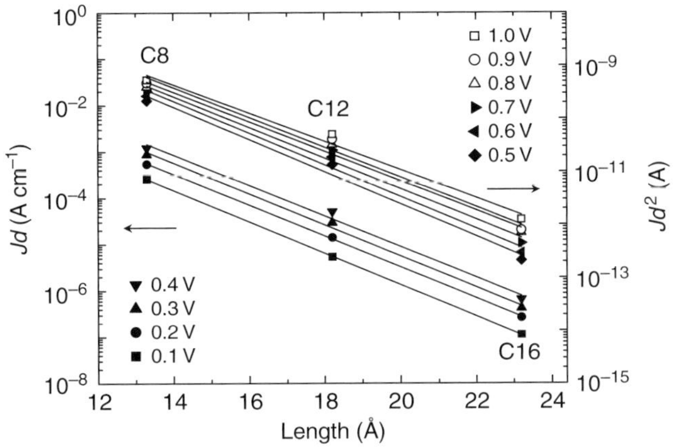

Figure 6 Log plot of tunneling current densities multiplied by molecular length d at low bias and by d^2 at high bias (symbols) versus molecular lengths. The lines through the data points are linear fittings.

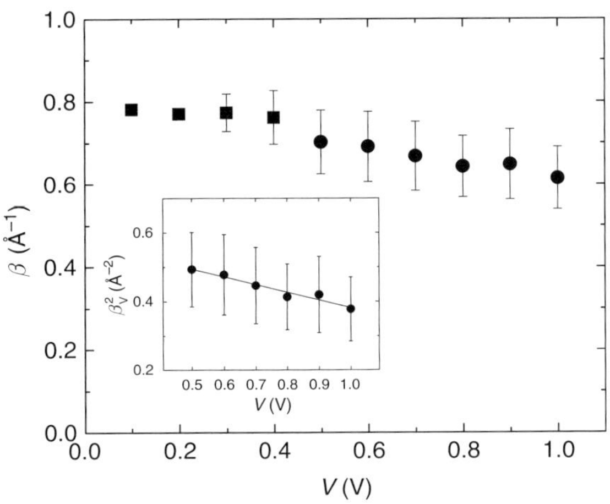

Figure 7 Plot of β versus bias in the low-bias range (square symbols) and high-bias ranges (circular symbols). The inset shows a plot of β_V^2 versus bias with a linear fitting.

extrapolated from published results of other length alkane molecules. The large variation of $\mathcal{J}$ of these reports can be attributed to the uncertainties in device contact geometry and junction area, as well as complicating inelastic or defect contributions. The β value (0.77 ± 0.06 Å^{-1} $\approx$ 0.96 ± 0.08 per methylene) for alkanethiols reported here is comparable to previously reported values as summarized in **Table 3**. This β value agrees with the value of 0.79 Å^{-1} (β_0) calculated via **equation 4a** from fitting individual $I(V)$ characteristic of the C12 device. The calculated β_0 of C8 and C16 devices also have similar values, as summarized in **Table 2**.

According to **equation 4b**, β_V^2 depends on bias V linearly in the high-bias range. The inset in **Figure 7** is a plot of β_V^2 versus V in this range (0.5 to 1.0 V) along with linear fitting of the data. From this fitting, $\Phi_B = 1.35 \pm 0.20$ eV and $\alpha = 0.66 \pm 0.04$ were obtained from the intercept and the slope, respectively, consistent with the values ($\Phi_B = 1.42$ eV and $\alpha = 0.65$) obtained from the nonlinear least-squares fitting in the previous section.

The β values for alkanethiols obtained by various experimental techniques have previously been reported and are summarized in **Table 3** [21–41,52]. In order to compare with these reported β values, we also performed a length-dependent analysis on our experimental data according to the generally used equation [22–30,35]:

$$G = G_0 \exp(-\beta d) \qquad (5)$$

Table 3 Summary of alkanethiol tunneling characteristic parameters

Junction	β *(Å^{-1})*	*J(A cm^{-2}) at 1 V*	Φ_B *(eV)*	*Technique*	*Ref.*
(Bilayer) monothiol	0.87 ± 0.1	25–200^a	2.1^e	Hg-junction	[25]
(Bilayer) monothiol	0.71 ± 0.08	0.7–3.5^a		Hg-junction	[27]
Monothiol	0.79 ± 0.01	1500 ± 200^b	1.4^e	Solid M-I-M	[33]
Monothiol	1.2			STM	[19]
Dithiol	0.8 ± 0.08	$3.7–5 \times 10^{5c}$	5 ± 2^f	STM	[20]
Monothiol	0.73 − 0.95	1100–1900^d	2.2^e	CAFM	[21]
Monothiol	0.64–0.8	10–50^d	2.3^e	CAFM	[23]
Dithiol	0.46 ± 0.02	$3–6 \times 10^{5c}$	1.3–1.5^e	CAFM	[24]
Monothiol	1.37 ± 0.03		1.8^f	Tuning fork AFM	[49]
Monothiol	0.97 ± 0.04			Electrochemical	[30]
Monothiol	0.85			Electrochemical	[31]
Monothiol	0.91 ± 0.08			Electrochemical	[32]
Monothiol	0.76	2×10^4 (at 0.1 V)^c	1.3–3.4^g	Theory	[58]
Monothiol	0.76			Theory	[59]
Monothiol	0.79			Theory	[56]

^{a} The junction areas were estimated by optical microscope.
^{b} The junction areas were estimated by SEM.
^{c} The junction areas were estimated by assuming single molecule.
^{d} The junction areas were estimated by assuming Hertzian contact theory.
^{e} Barrier height Φ_B values were obtained from Simmons equation.
^{f} Barrier height Φ_B values were obtained from bias dependence of β.
^{g} Barrier height Φ_B values were obtained from a theoretical calculation.
Notes: Some decay coefficients β were converted into the unit of Å^{-1} from the unit of per methylene. Current densities (*J*) for C12 monothiol or dithiol at 1 V are extrapolated from published results for other length molecules by using conductance $\propto \exp(-\beta d)$ relationship.

This gives a β value from 0.84 to 0.73 Å^{-1} in the bias range from 0.1 to 1.0 V, which is comparable to results reported previously. For example, Holmlin *et al.* [27] reported a β value of 0.87 Å^{-1} by mercury drop experiments and Wold *et al.* [24] have reported β of 0.94 Å^{-1} and Cui *et al.* [25] reported β of 0.64 Å^{-1} for various alkanethiols by using a conducting atomic force microscope technique. These reported β were treated as bias-independent quantities, contrary to the results reported here and that observed in a slightly different alkane system (ligand-encapsulated nanoparticle/alkane-dithiol molecules) [26]. We also caution against the use of parameters that have not been checked with a temperature-dependent analysis, since small nontunneling components can dramatically affect derived values of β.

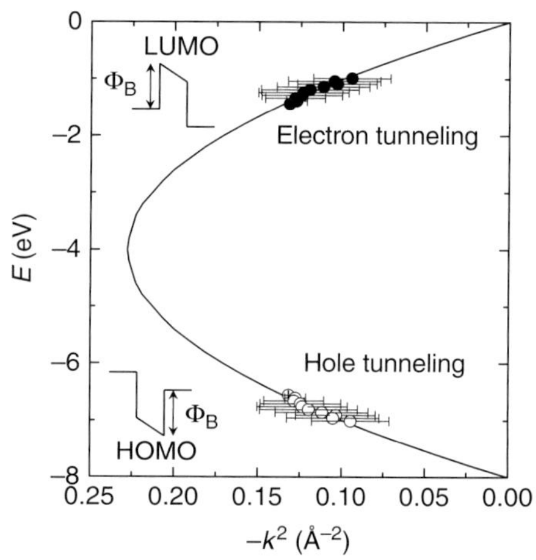

Figure 8 *E*(*k*) relationship (symbols) generated from the length-dependent measurement data for alkanethiols. Solid and open symbols correspond to electron and hole tunneling, respectively. The insets show the corresponding energy-band diagrams. The solid curve is the Franz two-band expression for $m^* = 0.43m$.

5.4.1.4 Franz model

We have analyzed our experimental data using a Franz two-band model [53–56]. Since there are no reliable experimental data on the Fermi level alignment in these metal–SAM–metal systems, Φ_B and m^* are treated as adjustable parameters. We performed a least-squares fit on our data with the Franz nonparabolic *E*(*k*) relationship (**equation 1**) using an alkanethiol HOMO–LUMO gap of 8 eV [14,15]. **Figure 8** shows the resultant *E*(*k*) relationship and the corresponding energy band diagrams. The zero of energy in this plot was chosen as the LUMO energy. The best-fitting parameters obtained by minimizing χ^2 were $\Phi_B = 1.49 \pm 0.51$ eV and $m^* = 0.43 \pm 0.15m$, where the error ranges of Φ_B and m^* are dominated by the error fluctuations of β ($k^2 = -(\beta/2)^2$). Both electron tunneling near the LUMO and hole tunneling near the HOMO can be described by these parameters. The value of $\Phi_B = 1.49$ eV indicates that the Fermi level is aligned close to one energy level in either case; therefore, the Simmons model is a valid approximation. The Φ_B and m^* values obtained here are in reasonable agreement with the previous results obtained from the Simmons model.

5.4.2 Metal–Molecule Contacts for Alkanethiols Junction

5.4.2.1 Statistical analysis of contact properties through alkanethiols

The yield of molecular electronic devices of even these robust alkanethiol molecular systems, however, is very low, mainly because of electrical shorts caused by the penetration of the top electrode through the molecular layer and making contact with the bottom electrode [8,60a,b]. A recent study, with the objective of preventing electrical shorts by using a layer of a highly conducting polymer resulted in a significant improvement in the yield of molecular electronic devices [61]. However, studies on the device yield of simple M-M-M junctions have not been extensive. In particular, systematic studies with the goal of defining 'working' molecular devices, device yield, and even selecting 'representative' devices have not been reported. Furthermore, determining the average transport parameters from a statistically meaningful number of molecular working devices is important, because the statistically averaged transport parameters can provide more accurate and meaningful characteristics of molecular systems. Statistical measurement has been performed, for example, to extract the electrical conductance of single molecules using mechanically controllable break junctions [62]. As mentioned above, the yields of the molecular electronic devices are very low, mainly due to electric short problems [8,60a,b,61]. However, thorough and systematic studies on what 'working' devices are and on the yields of the molecular electronic devices have not been reported. Typically, working devices might be defined as a device showing nonlinear $I(V)$ behavior and not being electrical open and short. Electrical open and short devices can be readily recognized. Open devices are noisy with a current level typically in the pA range and short devices show ohmic $I(V)$ characteristics with a current level larger than a few mA [63]. However, criteria are needed for determining working devices more precisely. Although the choice of such a criterion is not universal, current density can be a good criterion for determining working devices, because $I(V)$ data are major characteristics that are measured initially and the current directly reflects the conductivity of different lengths (or contacts) of alkanethiols or different molecular systems.

We fabricated and characterized a significantly large number of microscale molecular devices shown as **Figure 1(b)** (27 840 devices in total) to statistically analyze the molecular electronic properties of a sufficient number of 'working' molecular electronic devices (427 devices) [9,45].

The working devices displaying molecular properties were determined based on the statistical distribution of the current densities of the fabricated devices. Note that we selected the 99.7% of the devices by using the normal distribution function (Gaussian function) from the overall population, which are included in the interval of the 3σ range between $\mu+3\sigma$ and $\mu-3\sigma$, where μ is the average and σ is the standard deviation. When current densities are within the 3σ range (indicated as dotted lines in **Figure 9**), they are defined as working molecular electronic devices, whereas the others are defined as 'nonworking' devices when the current densities are out of this range. **Figure 9** shows an example of a histogram plot for logarithmic current densities of all C8 candidate devices. Similarly, we are able to define the working device ranges of alkanemonothiols and alkanedithiols devices. Basically, working molecular electronic devices were extracted from devices showing a majority of current densities in the statistical distribution. As summarized in **Table 4**, the numbers of C8, C12, C16,

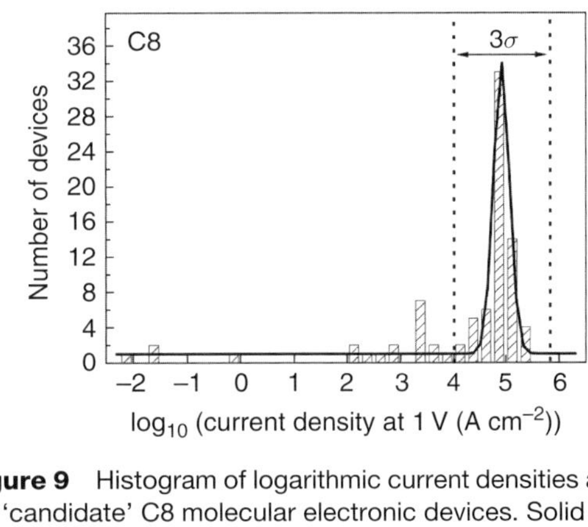

Figure 9 Histogram of logarithmic current densities at 1 V for 'candidate' C8 molecular electronic devices. Solid lines are Gaussian fitting curves. (See text for the definition of candidate and working devices.)

Table 4 Summary of results for the fabricated devices

						Working[a]			
	Number of fabricated devices	*Fab. failure*	*Short*	*Open*	*Nonworking*	*C8/DC8*	*C12/DC9*	*C16/DC10*	*Device yield*
Monothiol	13,440 (100%)	392 (2.9%)	11,744 (87.4%)	1,103 (8.2%)	45 (0.3%)	63(1.41%)	33 (0.69%)	60 (1.44%)	156 (1.2%)
Dithiol	14,400 (100%)	472 (3.28%)	12.340 (85.7%)	1,252 (8.69%)	65 (0.45%)	84 (1.75%)	94 (1.96%)	93 (1.94%)	271 (1.9%)

[a] CS correspond to monothiol and DCs to dithiol.

Note: working and nonworking devices were defined by statistical analysis with Gaussian fitting on the histogram of the logarithmic scale current densities.

DC8, DC9, and DC10 working devices were 63, 33, 60, 84, 94, and 93, respectively, among the total 27 840 fabricated devices. Then, the device yields were found as ~1.2% (156/13 440) for monothiol and ~1.9% (271/14 440) for dithiol devices. Since the device yield (~1.75%) of DC8 dithiol devices is not much different from that of C8 monothiol devices (~1.41%), this result may suggest that device yield is not much affected by metal–molecule contact, but rather affected more by the device structures, fabrication condition, and quality of the self-assembled monolayers. In this study, the use of a statistical approach is very significant, as the analysis of a large number of devices increases the ability to develop more accurate and meaningful characteristics of molecular systems. **Figures 10(a)–10(f)** present the statistical histograms of current densities in logarithmic scale for different lengths of alkanemonothiols (C8, C12, and C16) and alkanedithiols (DC8, DC9, and DC10) at 1.0 V with the mean positions as representative devices indicated with arrows from the fitting results by Gaussian functions. The current densities for these representative devices were found to be ~8.3×10^4, 1.2×10^3, 3.5, 4.9×10^5, 2.0×10^5, and 6.3×10^4 A cm^{-2} at 1.0 V for C8, C12, C16, DC8, DC9, and DC10, respectively. The current densities–voltage (*J–V*) characteristics for these six representative devices are plotted in **Figure 10(g)**. The conductance and *J–V* characteristics are clearly dependent on the molecular length and metal–molecular contacts (i.e., monothiol vs. dithiol). This observation is supported by previous reports of M-M-M junctions that have shown that the current density for alkanedithiol is higher than that for alkanemonothiol due to their different natures of metal–molecule contact properties (chemisorbed vs. physisorbed contact) at Au–molecule contacts [64,65]. The histograms in **Figures 10(a)–10(f)** show the distribution of the logarithmic current densities, indicating the existence of fluctuation factors causing the exponential distribution in the current densities. The variation of junction area may exist, but the area fluctuation does not produce exponential distribution in current, instead fluctuation in the tunneling path is probably responsible for the distribution data of **Figures 10(a)–10(f)**. Some fluctuations in molecular configurations in the self-assembled monolayers in the device junctions are possible, such as fluctuations in molecular configuration or microstructures in metal–molecule contacts [66,67].

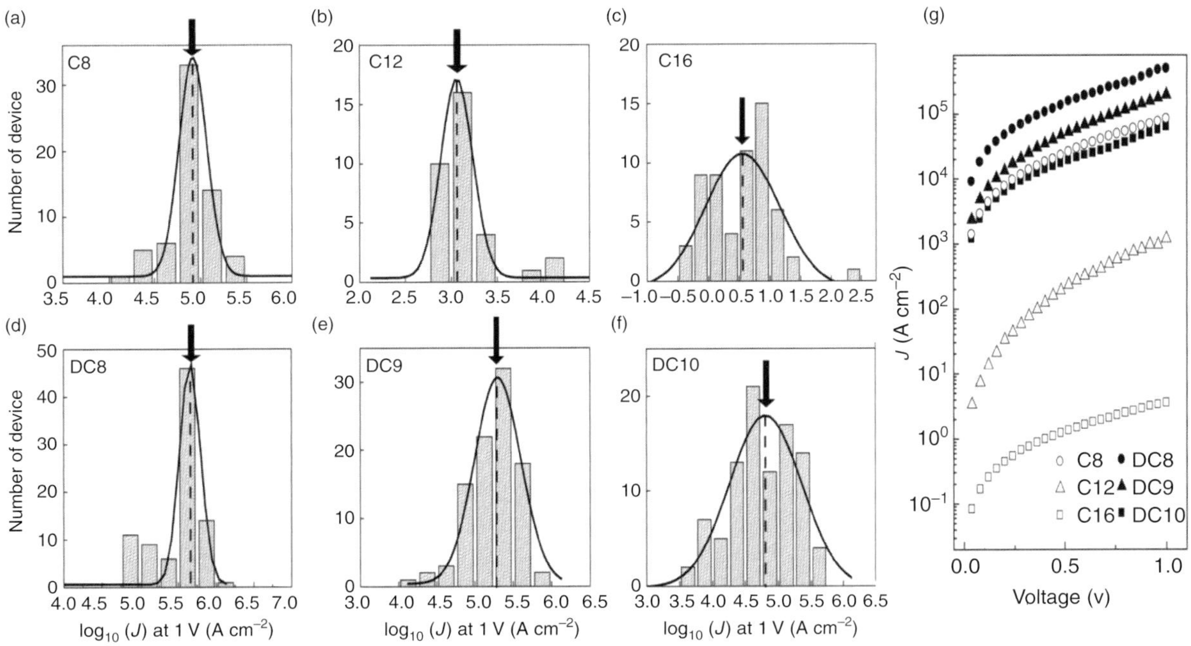

Figure 10 The statistical histograms of log(*J*) measured at 1.0 V for C8 (a), C12 (b), C16 (c), DC8 (d), DC9 (e), and DC10 (f). The line curves are fitting results obtained from the histograms with Gaussian functions, and the mean positions are indicated with arrows. (g) Current density–voltage characteristics of representative devices chosen from the mean positions of the fitted Gaussian functions.

5.4.2.2 *Contact/length-dependent decay coefficients by multibarrier tunneling model*

To investigate the effect of metal–molecule contacts on the electronic transport, we propose a multibarrier tunneling (MBT) model, which generalizes the Simmons tunneling model, a widely-used model for describing a rectangular tunneling barrier [9,10,57]. As compared to the Simmons tunneling model, where the tunneling barrier is represented by a single barrier, the M-M-M junction in MBT model can be divided into three parts: a molecular-chain body and metal–molecule contacts on either side of molecule, represented as three individual conduction barriers, as schematically illustrated in **Figure 11(a)**. In the alkanedithiol M-M-M junction, there is one molecular-chain body barrier [$(CH_2)_n$] (n is the number of carbon units), and two chemisorbed contact barriers [Au–S–C] on either side. Conversely, the alkanemonothiol M-M-M junction with the same a molecular-chain body barrier [$(CH_2)_n$] as the alkanedithiol junction has one chemisorbed contact barrier [Au–S–C] and one physisorbed contact barrier [CH_3/Au]. Please retain square brackets. This approach of separation of metal–molecule contact and molecular body from alkanethiol M-M-M junction is reasonable, since hybridization of the metal–molecule wave function decays rapidly into the junction for alkanethiol devices [68,69]. In the typical nonresonant tunneling case, the resistance is exponentially dependent on the molecular length $d(=d_1 + d_{1(2)} + d_{Body})$. The widths of the barrier for d_1, d_{Body}, and d_2 in alkyl molecular system represent the length of the chemisorbed contact on the molecule [Au–S–C], the molecular body region [$(CH_2)_n$], and the physisorbed contact on the molecule [CH_3/Au], respectively. Here, we assume d_{Body} is the projected length along the molecular body with the incremental length per carbon atom ($\Delta d_{Body}[CH_2]$) of ~1.25 Å and the contact lengths (d_1 and d_2) are the vertical distances between contact sites of molecule and electrode. The length d_{Body} is identical for n-alkanemonothiol and n-alkanedithiol with the same n value. For example, octanemonothiol (C8) and octanedithiol (DC8) have an identical length, $d_{Body}[(CH_2)_8]$, ~8.75 Å. And, d_1 ([Au–S–C]) is ~3.2 Å and d_2([CH_3/Au]) is ~1.4 Å [70].

For small-length molecules with a large HOMO–LUMO energy gap, such as alkyl chain molecules, coherent tunneling is the main conduction mechanism of the electronic charge transport at relatively

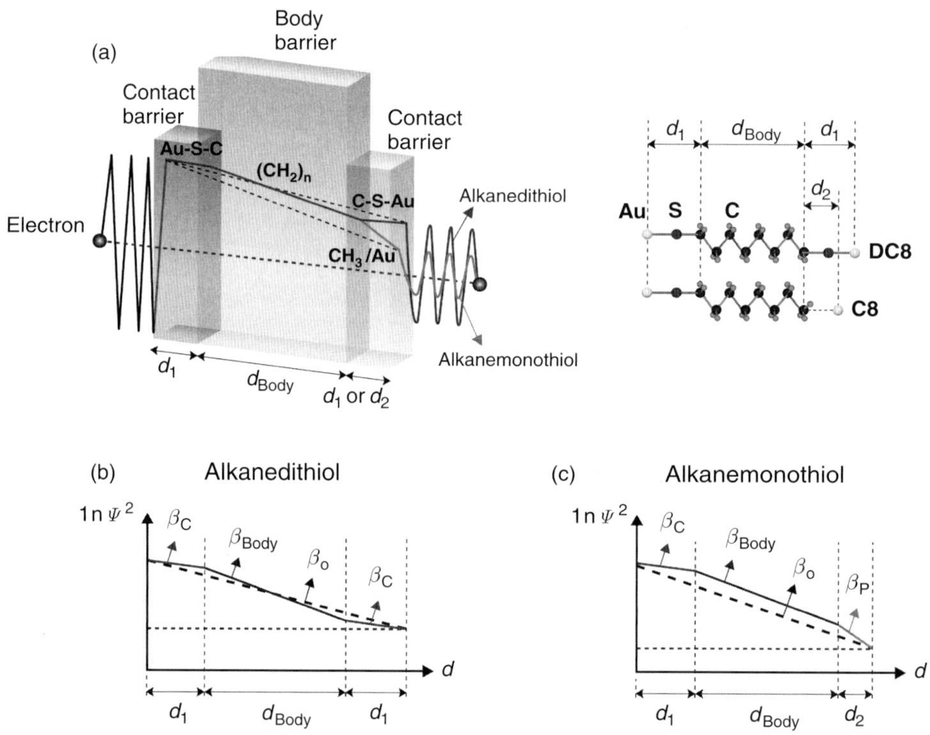

Figure 11 (a) (Left) An illustration of MBT model; (Right) a schematic of barrier widths for C8 and DC8. Schematics of MBT model for an alkanedithiol M-M-M junction (b) and for an alkanemonothiol M-M-M junction (c).

low bias regime [35]. As mentioned above, the tunneling current density in low-bias regime can be approximated as **equation 4a**. From **equation 4a**, β_o is the decay coefficient in a low-bias regime, which reflects the degree of decrease in wave function of the tunneling electron through the molecular tunnel barrier. A higher decay coefficient implies a faster decay of the wave function, that is, lower electron tunneling efficiency.

In MBT model, it is possible to describe the overall slope of wave function decay through the barriers based on the magnitude of the β_o value, and this overall decay can be further decomposed to three individual decays through three individual barriers, as shown in **Figure 11**. The β_o can be expressed as **equation 6** for alkanemonothiol (alkanedithiol) junctions from the consideration of geometric configurations.

$$\beta_o = \frac{\beta_C d_1 + \beta_{Body} d_{Body} + \beta_{C(P)} d_{1(2)}}{d_1 + d_{Body} + d_{1(2)}} \qquad (6)$$

One can see that β_o converges to β_{Body} for a very long molecule. Also, $\alpha(\Phi_B)^{1/2}$ can be expressed as **equation 7** by combining **equations 4a** and **6**:

$$\alpha(\Phi_B)^{1/2} = \frac{\hbar}{2(2m)^{1/2}} \frac{\beta_C d_1 + \beta_{Body} d_{Body} + \beta_{C(P)} d_{1(2)}}{d_1 + d_{Body} + d_{1(2)}} \qquad (7)$$

As mentioned above, because the main conduction mechanism is coherent (elastic) tunneling at low-bias regime (and at room temperature), it is assumed that the energy of electron tunneling through the molecular barriers does not decrease, as expressed by the horizontal dashed line in **Figure 11(a)**. Furthermore, due to the different nature of the metal–molecule contact properties, electron transmission for the chemisorbed contact [Au–S–C] is found to be more efficient than that for the physisorbed contact [CH_3/Au]. As a result, the slope (β_o) for alkanemonothiol junctions is steeper than that for alkanedithiol junctions, as illustrated by the dashed lines in **Figure 11**. In this MBT model, it was possible to define β_C (β_P) as the components of the decay coefficients corresponding to the chemisorbed (physisorbed) contact barrier width d_1 (d_2), as expressed by the solid lines in **Figure 11**. Similarly, β_{Body} is the decay coefficient component for the molecular-chain body barrier (center solid lines).

Figure 12 shows the statistical distribution of β_o values obtained for different-length alkanemonothiol and alkanedithiol M-M-M devices. In this plot, β_o values were determined from fitting the *I–V* data of all the 'statistically defined working' molecular electronic devices (total 427 devices) with the Simmons tunneling model. The values for the mean and standard deviation of β_o are presented as 0.81 ± 0.05, 0.83 ± 0.03, and 0.87 ± 0.05 Å^{-1} for C8, C12, and C16 alkanemonothiols, and 0.55 ± 0.06, 0.57 ± 0.06, and 0.58 ± 0.08 Å^{-1} for DC8, DC9, and DC10 alkanedithiols, respectively. As previously mentioned, the β_o values for alkanemonothiol devices appear to be larger than those for alkanedithiol devices due to the poor tunneling rate of physisorbed contact [CH_3/Au] in alkanemonothiol junctions, as compared to alkanedithiol junctions. Also, a slight increase of β_o values in **Figure 12** can be seen as the molecular length increases, which reflects the different tunneling rates for different lengths of alkanethiols, that is, the wave function of the tunneling electron decays further when it tunnels through longer molecules. The solid lines in **Figure 12** are the results calculated using the estimated β_{Body}, β_C, and β_P values determined from MBT model (**Table 5**). Moreover, **Figure 12** shows that the difference in β_o values between monothiol and dithiol becomes larger as the molecular length decreases. This phenomenon explains that the metal–molecular contact effect becomes relatively more important than the molecular-chain body effect in electronic transport for shorter molecules. On the contrary, if the molecular length increases, the molecular-chain body effect becomes more important and the β_o values of monothiol and dithiol molecular systems become closer and eventually converge to body decay coefficient (β_{Body}), as seen in **Figure 12**.

At low bias, **equations 3a** and **6** can be used to determine the resistance *R* of the ohmic regime as

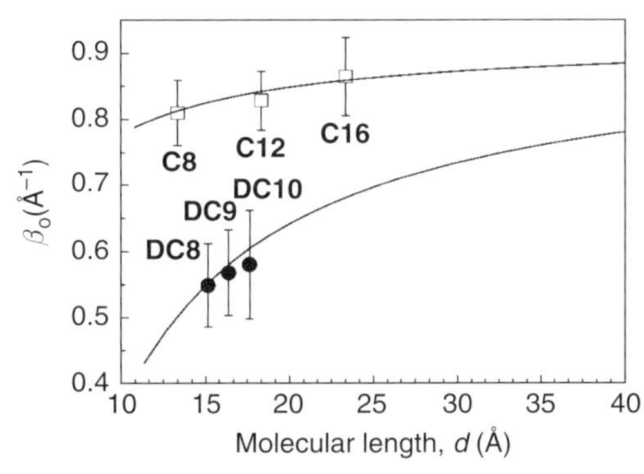

Figure 12 The mean (symbols) and standard deviations (error bars) of β_o versus molecular length *d*. The black solid lines were calculated from MBT model.

Table 5 A summary of the experimental and calculated values for decay coefficients, contact resistances, and specific contact resistances

	β_{Body} (Å^{-1})	β_1(Å^{-1})	β_2(Å^{-1})	R_o(Ω)	R_c(Ω cm^2)
Alkanemonothiol					
Calculated value	0.92	0.05	1.89	0.19	0.58×10^{-8}
Experimental value	0.93 ± 0.03			0.34 ± 0.30	$1.08 \pm 0.94 \times 10^{-8}$
Alkanedithiol					
Calculated value	0.92	0.05		0.03	0.98×10^{-9}
Experimental value	0.92 ± 0.08			0.04 ± 0.03	$1.13 \pm 0.98 \times 10^{-9}$

$$R = \frac{4\pi^2\hbar^2}{A(2m)^{1/2}e^2}\left(\frac{d_1 + d_{Body} + d_{1(2)}}{(\Phi_B)^{1/2}\alpha}\right) \times \exp[\beta_1 d_1 + \beta_{Body} d_{Body} + \beta_{1(2)} d_{1(2)}] \quad (8)$$

where R_o is the contact resistance that can be defined in the limiting case when d_{Body} approaches zero, and expressed as **equation 9** for alkanemonothiol and alkanedithiol, respectively,

$$R_o = \frac{8\pi^2\hbar}{Ae^2}\left(\frac{(d_1 + d_{1(2)})^2}{\beta_1 d_1 + \beta_{1(2)} d_{1(2)}}\right)\exp[\beta_1 d_1 + \beta_{1(2)} d_{1(2)}] \quad (9)$$

Unlike the β_o value that describes the overall decay coefficient, the β_{Body} value is the decay coefficient component only for the molecular-chain body barrier. The molecular-chain body decay coefficient $\beta_{Body} = \Delta \ln R/\Delta d_{Body}$ can be determined from the slopes in the semilog plot of resistance R versus the molecular-chain body length d_{Body}, as shown in **Figure 13(a)**. Here, R is the resistance in the low-bias regime obtained from the linear fit of low-bias I–V data ($0 < V < 0.3$ V) for each device. From the slopes in **Figure 13(a)**, the β_{Body} values were determined to be $\sim 0.93 \pm 0.03$ and $\sim 0.92 \pm 0.08$ Å^{-1} for alkanemonothiol and alkanedithiol, respectively –almost identical values for the two molecular systems. Thus, one should note that the β_{Body} value is the molecular length-independent decay coefficient that is dependent upon molecular structure but not on metal–molecule contacts, whereas the β_o value is the molecular length-dependent overall decay coefficient that depends not only on the molecular structures but also on the form of metal–molecule contact (i.e., chemisorbed or physisorbed). The β_C, β_P, and β_o for the alkyl M-M-M junctions can be calculated from the observed β_{Body} ~0.92 Å^{-1}, β_o values for C8 and DC8, and the widths of barriers (d_1, d_{Body}, and d_2). The contact resistance (R_o) can be considered a method of investigating the metal–molecule contacts. However, since R_o depends on

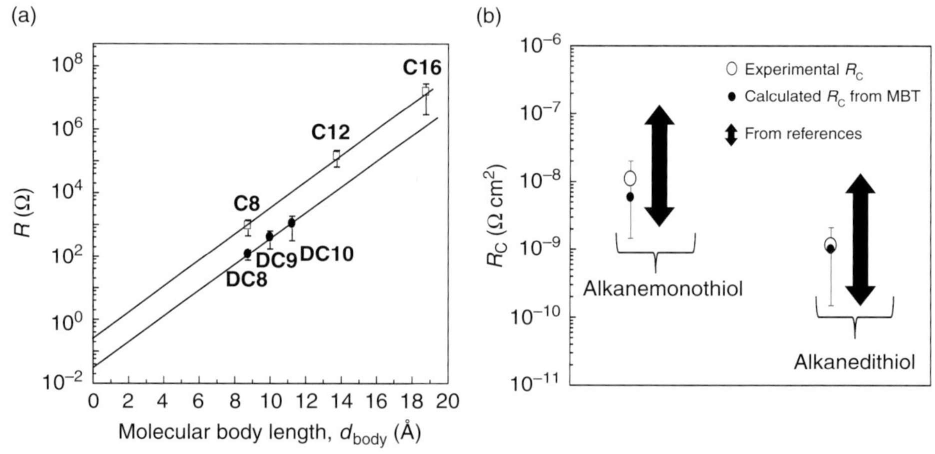

Figure 13 (a) Semilog plot of the resistance R versus the molecular-chain body length d_{Body} for alkanemonothiol and alkanedithiol junctions. The solid lines are exponential fitting results, giving the molecular-chain body decay coefficient β_{Body}. (b) Experimental and calculated specific contact resistance R_C. The blue arrows represent the range of the estimated R_C values from the contact resistances reported in literature [24,64,71a,b].

the junction area, the specific contact resistance (R_C) (a junction-area-compensated quantity) is generally obtained and compared among devices with different junction areas. **Figure 13(b)** presents the experimental and theoretical values for R_C for our alkanemonothiol and alkanedithiol devices at a low-bias regime. Here, the specific contact resistance $R_C (= R_o \times A)$ can be obtained by multiplying the contact resistance (R_o) with the contact junction area (A) ($\sim 3.14 \times 10^{-8}\ cm^2$ for our molecular devices). The contact resistance (R_o) was found by extrapolating the observed resistance (**Figure 13(a)**) to a zero molecular-chain body length ($\sim 0.34 \pm 0.3\ \Omega$ for alkanemonothiol, and $\sim 0.04 \pm 0.03\ \Omega$ for alkanedithiol). Then, R_C was calculated as $(\sim 1.08 \pm 0.94) \times 10^{-8}\ \Omega\ cm^2$ for alkanemonothiol and $(\sim 1.13 \pm 0.98) \times 10^{-9}\ \Omega\ cm^2$ for alkanedithiol. Using MBT model, R_C could also be estimated as $\sim 0.58 \times 10^{-8}\ \Omega\ cm^2$ for alkanemonothiol and $\sim 0.98 \times 10^{-9}\ \Omega\ cm^2$ for alkanedithiol, both of which are in good agreement with the experimental values we obtained. **Table 5** summarizes the experimental and calculated quantities of decay coefficients and contact and specific contact resistances for our measurements.

Note that our analysis with MBT model does not consider the details of the Fermi level alignment and molecular binding sites, which will generally influence the charge transport of molecular devices. Furthermore, the transport properties values obtained from our experimental results with micro-scale molecular junctions are an ensemble average effect with various microstructures of metal–molecule contacts and binding sites, and thus should not be compared with single-molecular measurement results, due to the contribution from the probability amplitude of multiple reflections and the possibility of cooperative effects between individual molecules in ensemble of molecules.

5.4.2.3 *Contact properties through various electrodes by multi-barrier tunneling model*

Similarly, using **equation 4a**, the contact decay coefficients $\beta_{C(P)}$ for Au contacts can be expressed as

$$\beta_{C(P)} = \frac{2(2m)^{1/2}}{\hbar} \alpha_{C(P)} (\Phi_{C(P)})^{1/2} \qquad (10)$$

where $\Phi_{C(P)}$ is the contact barrier height at zero bias and $\alpha_{C(P)}$ the α value through the contact barrier. These two value can be obtained from α and Φ_B in Au–alkyl molecule contacts, which can be deduced from the molecular body decay coefficients (β_{Body}) and widths (d_1, d_2, and d_{Body}) of each barrier part.

Furthermore, the decay coefficients and specific contact resistance can be determined for molecular junctions with various metal contacts other than Au. If molecular monolayers are sandwiched between other metals (Ag and Cu), the contact barrier heights can be expressed as $\Phi_{C(other\ metals)} = \Phi_{C(Au)} + \Delta\Phi_{(Au-other\ metals)}$ and $\Phi_{P(other\ metals)} = \Phi_{P(Au)} + \Delta\Phi_{(Au-other\ metals)}$ for chemisorbed contact and physisorbed contact, respectively, by assuming the tunneling around HOMO levels (i.e., a hole type tunneling). Note that $\Delta\Phi_{(Au-other\ metals)}$ is the difference between the work function of Au and that of the other metal. From **equation 10** with using Φ_C, Φ_P, α_C, and α_P, the contact decay coefficient $\beta_{C(P)(Ag\ and\ Cu)}$ can be calculated as ($\beta_{C(Ag)} = 0.51$ Å^{-1} and $\beta_{C(Cu)} = 0.38$ Å^{-1}) and ($\beta_{P(Ag)} = 2.76$ Å^{-1} and $\beta_{P(Cu)} = 2.28$ Å^{-1}) for chemisorbed contacts and physisorbed contacts, respectively.

Figure 14 shows the R_C values of various metal–molecule contacts (Ag, Cu, and Au) in alkyl M-M-M junctions obtained from experiment, reported studies, and the MBT model, by assuming tunneling around HOMO, that is, a hole type tunneling. In this **Figure 7**, the labels are designed such that, for example, [Au/Ag] refers to [Au–S–$(CH_2)_{n-1}CH_3$/Ag], that is, the chemisorbed contact to the Au electrode and physisorbed contact to the Ag electrode for alkanemonothiol (**Figure 14(a)**) or [Au–S–$(CH_2)_n$–S–Ag], that is, the chemisorbed contacts to both Au and Ag electrodes for alkanedithiol (**Figure 14(b)**). And for different metallic junctions (e.g., [Au/Ag], [Ag/Cu], etc.), the average value of the two individual metal work functions was assigned as the work function. As mentioned earlier, the natures of the chemisorbed and physisorbed contacts are quite different. Because the chemisorbed contacts ([metal–S–C]) can form stronger bondings by molecular overlapping than physisorbed contacts (metal/CH_3 or metal/H), generally the contact decay coefficient for chemisorbed contacts are smaller than that of physisorbed contacts ($\beta_C < \beta_P$), that is, less tunneling electron decay through chemisorbed contacts. In MBT model, the contact decay coefficients (β_C, β_P) in various metallic junctions are dependent on the contact barrier height (Φ_C, Φ_P) and effective mass (α_C and α_P), which can be affected by metal work function, as expressed in **equation 10**. Note that the contact decay coefficient was observed to decrease when metal work function is increased. The R_C values were found to be different for asymmetric

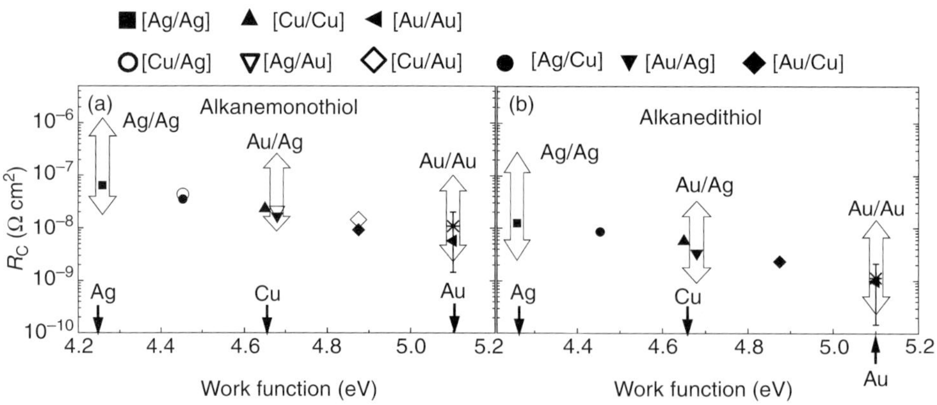

Figure 14 The specific contact resistance R_C for (a) alkanemonothiol and (b) alkanedithiol obtained from the MBT model as a function of metal work function. Open arrows are a range of R_C values estimated from literature [24,64,71a,b]. The black crosses for [Au/Au] are the experimental values obtained in our study.

metal contacts (e.g., [Ag/Au] and [Au/Ag]) for alkanemonothiol because of the different natures of metal–molecule contacts (physisorbed vs. chemisorbed contact side), as shown in **Figure 14(a)**, whereas R_c values for that of alkanedithiol were found to be same for even asymmetric contacts because of the same nature of metal–molecule contact, as shown in **Figure 14(b)**. As a result, it was determined that when the average metal work function increases, R_C decreases due to a reduction of the contact barrier height (or contact decay coefficients). The R_C values calculated by the MBT model are in good agreement with those obtained from reported literatures [24,64,71a,b], as indicated by the arrows, as shown in **Figure 14**.

5.4.3 Inelastic Tunneling

5.4.3.1 *Inelastic electron tunneling spectroscopy*

Electronic transport through alkanethiol SAMs is further investigated with the technique of inelastic electron tunneling spectroscopy [36], for example, by Jaklevic and Lambe, who studied, in 1966, the conductance of tunnel junctions with encased organic molecules [37]. Since then it has become a powerful spectroscopic tool for chemical identification, chemical bonding investigation, and studies in surface chemistry and physics [40]. In an inelastic tunneling process, the electron loses energy to a localized vibrational mode with a frequency ν when the applied bias satisfies the condition of $eV = h\nu$. As a result, an additional tunneling channel is opened for the electron, resulting in an increase in the total current at the applied bias corresponding to the vibrational mode energy [39]. Typically only a small fraction of tunneling electrons are involved in the inelastic tunneling process (determined by the electron–vibronic mode coupling coefficient), resulting in a small conductance change, which is commonly measured in the second harmonics of a phase-sensitive detector that yields the characteristic frequencies of the corresponding vibrational modes as well as other information [38–40].

Measurements of $I(V,T)$ and additional IETS studies have been performed on an octanedithiol (C8-dithiol) SAM using the aforementioned device structure shown in **Figure 1(a)** [36]. **Figure 15(a)** is the $I(V,T)$ data for this device obtained from 300 to 4.2 K. An Arrhenius plot shown in **Figure 15(b)** exhibits mild temperature dependence, verifying that tunneling is the main transport mechanism for C8-dithiol SAM. This result is in good agreement with the tunneling transport characteristics observed previously. **Figure 15(c)** shows the room temperature $I(V)$ measurement result. Using a junction area of 51 ± 5 nm in diameter (obtained from statistical studies of the nanopore size with SEM), a current density of $(9.3 \pm 1.8) \times 10^{-4}$ A cm^{-2} at 1.0 V is calculated. As a comparison, the current density of $(3.1 \pm 1.0) \times 10^{4}$ A cm^{-2} at 1.0 V was observed for C8 monothiol SAM. Using the modified Simmons model (**equation 2**), the transport parameters of $\Phi_B = 1.20 \pm 0.03$ eV and $\alpha = 0.59 \pm 0.01$ ($m^* = 0.34m$) were obtained for this C8-dithiol SAM.

Figure 16 shows the IETS spectrum of the same C8-dithiol SAM device obtained at $T = 4.2$ K. An AC modulation of 8.7 mV (root-mean-square (RMS) value)

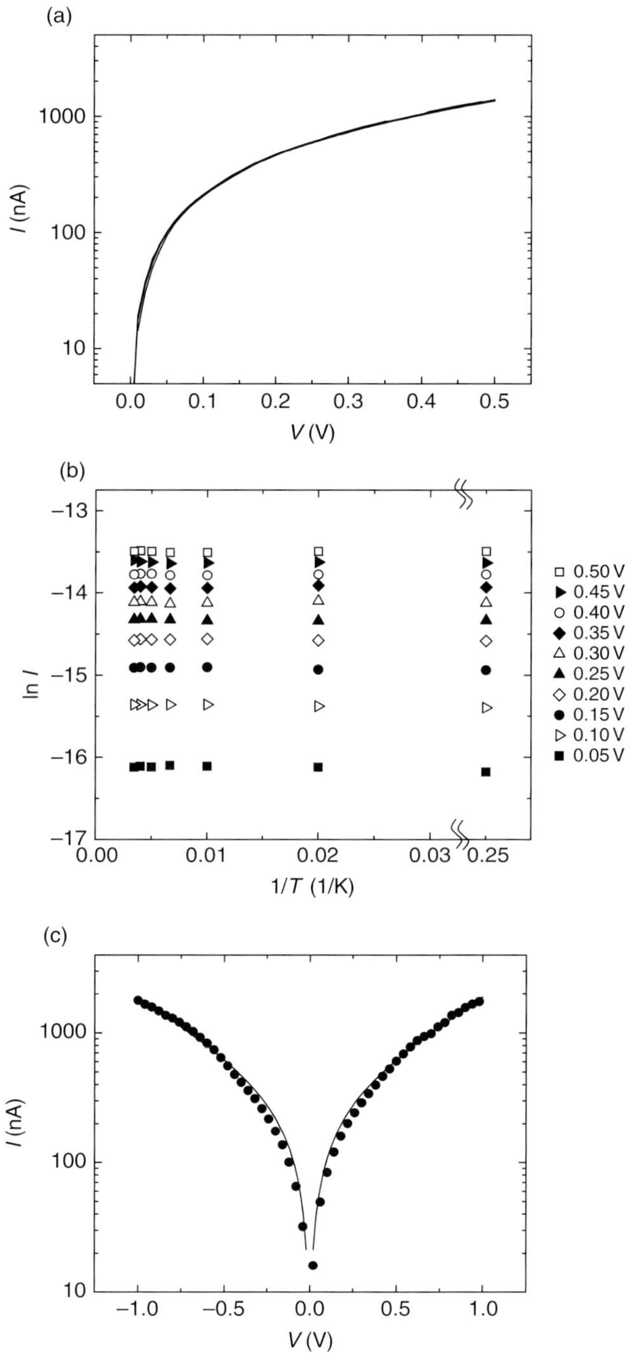

Figure 15 (a) $I(V,T)$ characteristics of C8 dithiol SAM at selected temperatures (4.2, 50, 100, 150, 200, 250, and 290 K). (b) Arrhenius plot generated from the data in (a), at voltages from 0.1 to 0.5 V with 0.05 V steps. (c) Measured C8-dithiol $I(V)$ data at room temperature (circular symbols) is compared with calculation (solid curve) using the optimum fitting parameters of $\Phi_B = 1.20$ eV and $\alpha = 0.59$.

at a frequency of 503 Hz was applied to the sample to acquire the second-harmonic signals. The spectra are stable and repeatable upon successive bias sweeps. The spectrum at 4.2 K is characterized by three pronounced peaks in the 0–200 mV region at 33, 133, and 158 mV. From a comparison with previously reported infrared (IR), Raman, and high-resolution electron energy loss (HREEL) spectra of SAM covered gold surfaces

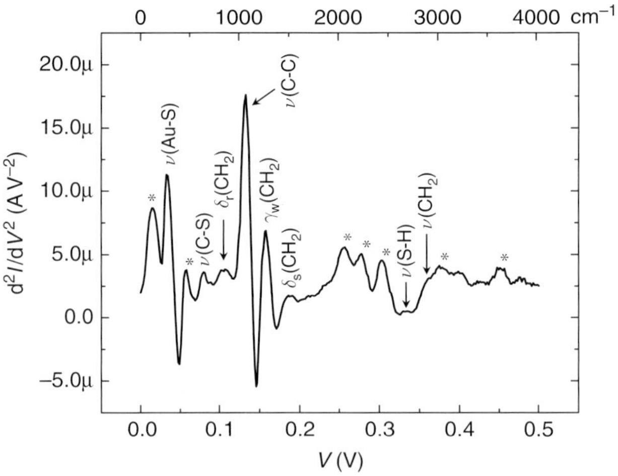

Figure 16 Inelastic electron tunneling spectrum of C8 dithiol SAM obtained from lock-in second-harmonic measurement with an AC modulation of 8.7 mV (RMS value) at a frequency of 503 Hz ($T = 4.2$ K). Peaks labeled * are most probably background due to the encasing Si_3N_4.

Table 6 Summary of the major vibrational modes of alkanethiolates

Modes	*Methods*	*Wavenumber* (cm^{-1})	*(meV)*
ν(Au–S)	HREELS [61]	225	28
ν(C–S)	Raman [60a,b]	641	79
	Raman [60a,b]	706	88
$\delta_r(CH_2)$	HREEL [61]	715	89
	IR [62]	720	89
	IR [62]	766	95
	IR [62]	925	115
ν(C–C)	HREEL [61]	1050	130
	Raman [60a,b]	1064	132
	Raman [60a,b]	1120	139
	IR [62]	1230	152
	HREELS [61]	1265	157
	IR [62]	1283	159
	IR [62]	1330	165
$\delta_s(CH_2)$	HREELS [61]	1455	180
ν(S–H)	Raman [60a,b]	2575	319
$\nu_s(CH_2)$	Raman [60a,b]	2854	354
	HREELS [61]	2860	355
$\nu_{as}(CH_2)$	Raman [60a,b]	2880	357
	Raman [60a,b]	2907	360
	HREELS [61]	2925	363

Note: There is a vast amount of literature with spectroscopic assignments for alkanethiols. The references given are representative for IR, Raman, and HREELS assignments. From Bryant MA and Pemberton JE (1991) *Journal of the American Chemical Society* 113: 8284; Kato HS, Noh J, Hara M, and Kawai M (2002) *Journal of Physical Chemistry B* 106: 9655; Castiglioni C, Gussoni M, and Zerbi GJ (1991) *Chemical Physics* 95: 7144.

(**Table 6**), these three peaks are assigned to ν(Au–S), ν(C–C), and $\gamma_w(CH_2)$ modes of a surface bound alkanethiolate [72–75]. The absence of a strong ν(S–H) signal at ~329 mV suggests that most of the thiol groups have reacted with the gold bottom and top contacts. Peaks are also reproducibly observed at 80, 107, and 186 mV. They correspond to ν(C–S), $\delta_r(CH_2)$, and $\delta_s(CH_2)$ modes. The stretching mode of the CH_2 groups, $\nu(CH_2)$, appears as a shoulder at 357 meV. The peak at 15 mV is due to vibrations from either Si, Au, or δ(C–C–C) [76]. We note that all alkanethiolate peaks without exception or omission occur in the spectra. Peaks at 58, 257, 277, and 302 mV, as well as above 375 mV are likely to originate from Si–H and N–H vibrations related to the silicon nitride membrane [76a,77a,b], which forms the SAM encasement. To the best of our knowledge, alkanethiols have no vibrational signatures in these regions. Measurement of the background spectrum of an 'empty' nanopore device with only gold contacts to obtain background contributions from Si_3N_4 is hampered by either too low (open circuit) or too high (short circuit) currents in such a device. Similar IETS result has also been obtained using a different test structure recently [78].

Although there are no selection rules in IETS as there are in IR and Raman spectroscopy, certain selection preferences have been established. According to the IETS theory [79], molecular vibrations with net dipole moments perpendicular to the interface of the tunneling junction have stronger peak intensities than vibrations with net dipole moments parallel to the interface (for dipoles close to the electrodes). Thus, vibrations perpendicular to the electrode interface, that is, ν(Au–S), ν(C–S), ν(C–C), and $\gamma_w(CH_2)$, dominate the IETS spectrum, while modes parallel to the interface, that is, $\delta_{r,s}(CH_2)$ and $\nu(CH_2)$, are weak, as clearly shown in **Figure 16**.

5.4.3.2 Linewidth study

In order to verify that the observed spectra are indeed valid IETS data, the peak width broadening was examined as a function of temperature and modulation voltage. IETS was performed with different AC modulations at a fixed temperature, and at different temperatures with a fixed AC modulation. **Figure 17(a)** shows the modulation dependence of the IETS spectra obtained at 4.2 K, and **Figure 17(b)** shows the modulation broadening of the C–C

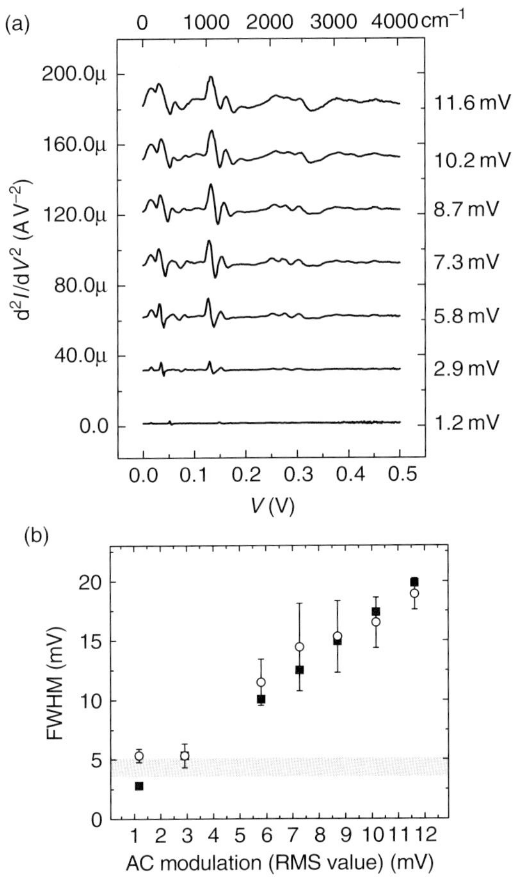

Figure 17 (a) Modulation dependence of IETS spectra obtained at 4.2 K. (b) Line (C–C stretching mode) broadening as a function of AC modulation. The circular symbols are experimental FWHMs and the square symbols are theoretical calculations considering both modulation and thermal contributions. The shaded bar denotes the expected saturation due to the derived intrinsic linewidth (including a $5.4k_BT$ thermal contribution) of 3.73 ± 0.98 meV.

stretching mode at 133 meV. The circular symbols are the full widths at half maximum (FWHMs) of the experimental peak at $T = 4.2$ K with various modulation voltages. A Gaussian distribution function was utilized to obtain an FWHM and the error range [80]. The square symbols are calculated FWHM values ($W_{\mathrm{theoretical}}$) taking into account both a finite-temperature effect ($W_{\mathrm{thermal}} \sim 5.4\,k_BT$)[38] and a finite-voltage modulation effect ($W_{\mathrm{modulation}} \sim 1.7\,V_{\mathrm{ac\_RMS}}$) [81]. These two broadening contributions add as the squares: $W^2_{\mathrm{theoretical}} = W^2_{\mathrm{thermal}} + W^2_{\mathrm{modulation}}$. The agreement is excellent over most of the modulation range, but we note a saturation of the linewidth at low-modulation bias indicating the influence of a non-negligible intrinsic linewidth. Taking into account the known thermal and modulation broadenings, and including the intrinsic linewidth (W_I) [82] as a fitting parameter, the measured peak width (W_{exp}) is given by

$$W_{\mathrm{exp}} = \sqrt{W_I^2 + W^2_{\mathrm{thermal}} + W^2_{\mathrm{modulation}}} \tag{11}$$

W_I can be determined by using a nonlinear least-squares fit to the AC modulation data (**Figure 17**) with **equation 11**, giving an intrinsic linewidth of 3.73 ± 0.98 meV for this line. This is shown (with the error range) in **Figure 17(b)** as a shaded bar, including the thermal contribution.

We can independently check the thermal broadening of the line at fixed modulation. **Figure 18(a)** shows the temperature dependence of the IETS spectra obtained with an AC modulation of 8.7 mV (RMS value). In **Figure 18(b)** the circular symbols (and corresponding error bars) are experimental FWHM values of the C–C stretching mode from **Figure 18(a)**, determined by a Gaussian fit (and error of the fit) to the experimental lineshape. For simplicity, we have considered only Gaussian lineshapes [80], resulting in increased error bars for the lower-temperature range due to an asymmetric lineshape. The square symbols are theoretical calculations considering thermal broadening, modulation broadening, and the intrinsic linewidth determined above. The error ranges of the calculation (due to the intrinsic linewidth error) are approximately the size of the data points. The agreement between theory and experiment is very good, spanning a temperature range from below (×0.5) to above (×10) the thermally broadened intrinsic linewidth. This linewidth should be a sensitive test to compare to theoretical models of transmission probabilities [83].

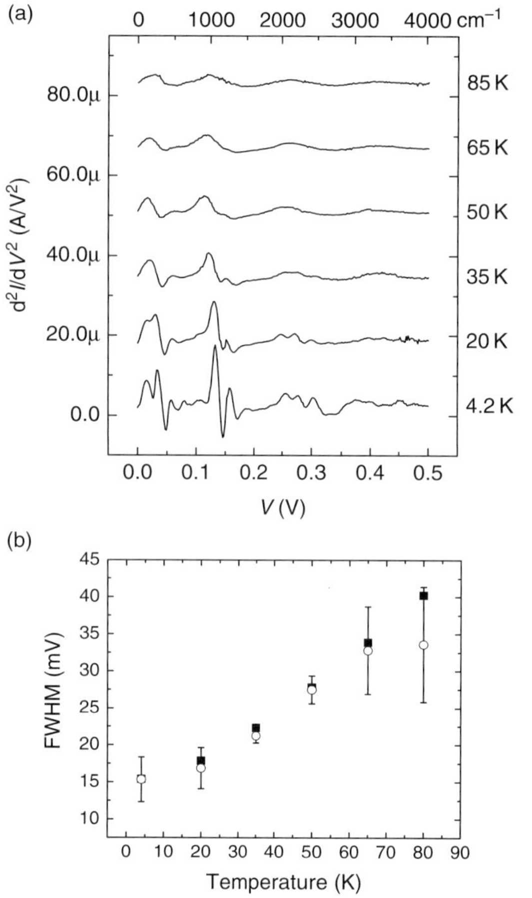

Figure 18 (a) Temperature dependence of IETS spectra obtained at a fixed AC modulation of 8.7 mV (RMS value). (b) Line (C–C stretching mode) broadening as a function of temperature. The circular symbols are experimental FWHMs and the square symbols are theoretical calculations considering thermal broadening, modulation broadening, and the intrinsic linewidth.

5.5 Conclusions

We present here a study of electronic transport properties of alkanethiol SAMs, with the intent that this system can serve as a simple standard for the development of well-characterized molecular junctions. The characteristics are consistent with accepted models of tunneling junctions, as well as presenting a system on which tunneling spectroscopy can be performed. The metal–molecule contact plays a crucial role in the charge transport through the molecular junctions.

The field of 'molecular electronics' is rich in the proposal and promise of numerous device concepts [84,85], but unfortunately reliable data and characterization techniques upon which to test these ideas are not available. It is incumbent upon the experimentalists to carefully institute controls to validate claims of intrinsic molecular behavior. Systematic controls, such as the model system presented here, should assist in guiding further work toward a rational development of the fascinating device structures and systems that the field promises.

Acknowledgments

This work was supported by DARPA/ONR (N00014-01-1-0657), ARO (DAAD19-01-1-0592), AFOSR (F49620-01-1-0358), NSF (DMR-0095215), and NASA (NCC 2-1363). Authors (G.W., T.-W. K., and T. L.) thank financial supports from the National Research Laboratory (NRL) Program of the Korea Science and Engineering Foundation (KOSEF), the Program for Integrated Molecular System at Gwangju Institute of Science and Technology.

References

1. Reed MA and Lee T (2003) *Molecular Nanoelectronics*. Stevenson Ranch, CA: American Scientific Publishers.
2a. Aviram A and Ratner MA (1974) Molecular rectifier. *Chemical Physics Letters* 29: 277.
2b. Chen J, Reed MA, Rawlett AM, and Tour JM (1999) Electron transport in molecular wire junctions. *Science* 286: 1550.
3a. Chen J, Reed MA, Rawlett AM, and Tour JM (1999) Large on-off ratios and negative differential resistance in a molecular electronic device. *Science* 286: 1550.
3b. Chen Y, Jung G-Y, Ohlberg DAA, *et al.* (2003) Nanoscale molecular-switch crossbar circuits. *Nanotechnology* 14: 462.
3c. Stewart DR, Ohlberg DAA, Beck PA, *et al.* (2004) Molecule-independent electrical switching in Pt/organic monolayer/Ti devices. *Nano Letters* 4: 133.
3d. Green JE, Choi JW, Boukai A, *et al.* (2007) A 160,000 bit molecular electronic memory circuit patterned at 1011 bits per square centimeter. *Nature* 445: 414.
4a. Selzer Y, Salomon A, and Cahen D (2002) The importance of chemical bonding to the contact for tunneling through alkyl chains. *Journal of Physical Chemistry B* 106: 10432.
4b. Guisinger NP, Greene ME, Basu R, Baluch AS, and Hersam MC (2004) Room temperature negative differential resistance through individual molecules on silicon surfaces. *Nano Letters* 4: 55.
4c. Xiao X, Xu B, and Tao NJ (2004) Measurement of single molecule conductance: Benzenedithiol and benzenedimethanethiol. *Nano Letters* 4: 267.
4d. Donhauser ZJ, Mantooth BA, Kelly KF, *et al.* (2001) Conductance switching in single molecules through conformational changes. *Science* 292: 2303.
4e. Ramachandran GK, Hopson TJ, Rawlett AM, Nagahara LA, Primak A, and Lindsay SM (2003) A bond-fluctuation mechanism for stochastic switching in wired molecule. *Science* 300: 1413.

5. Hewlett-Packard Quantum Science Research, Beyond Silicon: Breakthroughs in Molecular Electronics. http://www.hpl.hp.com/research/qsr/ (accessed September 2009).
6a. Luo Y, Collier CP, Jeppesen JO, *et al.* (2002) Two-dimensional molecular electronics circuits. *ChemPhysChem* 3: 519.
6b. Collier CP, Mattersteig G, and Wong EW, *et al.* (2000) A [2]catenane-based solid state electronically reconfigurable switch. *Science*. 289: 1172.
7. Lau CN, Stewart DR, Williams RS, and Bockrath M (2004) Direct observation of nanoscale switching centers in metal/molecule/metal structures. *Nano Letters* 4: 569.
8. Lee T, Wang W, Klemic JF, Zhang JJ, Su J, and Reed MA (2004) Comparison of electronic transport characterization methods for alkanethiol self-assembled monolayers. *Journal of Physical Chemistry B* 108: 8742.
9. Wang G, Kim T-W, Lee H, and Lee T (2007) Influence of metal-molecule contacts on decay coefficients and specific contact resistances in molecular junctions. *Physical Review B* 76: 205320.
10. Wang G, Kim T-W, Jang YH, and Lee T (2008) Effects of metal-molecule contact and molecular structure on molecular electronic conduction in nonresonant tunneling regime: Alkyl versus conjugated molecules. *Journal of Physical Chemistry C* 112: 13010.
11. Ulman A (1991) *An Introduction to Ultrathin Organic Filmsfrom Langmuir–Blodgett to Self-Assembly*. Boston, MA: Academic Press.
12. Poirier GE (1997) Characterization of organosulfur molecular monolayers on Au(111) using scanning-tunneling-microscopy. *Chemical Reviews* 97: 1117.
13. Ratner MA, Davis B, Kemp M, Mujica V, Roitberg A, and Yaliraki S (1998) In: Aviram A and Ratner M (eds.) *Molecular Electronics: Science and Technology, the Annals of the New York Academy of Sciences*, vol. 852. New York: New York Academy of Sciences.
14. Although the HOMO–NUMO gap of alkyl chain type molecules has been reported (see Ref. 15), there is no experimental data on the HOMO–LUMO gap for Au/alkanethiol SAM/Au system. 8 eV is commonly used as HOMO–LUMO gap of alkanethiol.
15a. Boulas C, Davidovits JV, Rondelez F, and Vuillaume D (1996) Suppression of charge carrier tunneling through organic self-assembled monolayers. *Physical Review Letters* 76: 4797.
15b. Fujihira M and Inokuchi H (1972) Photoemission from polyethylene. *Chemical Physics Letters* 17: 554.
15c. Lias SG, Bartmess JE, Liebman JF, Holmes JL, Levin RD, and Mallard WG (1998) Gas-phase ion and neutral thermochemistry. *Journal of Physical and Chemistry Reference Data* 17(1): 24.
15d. Yang H-H and McCreery RL (1999) Electron-transfer kinetics and adsorption of methyl viologen and phenothiazine derivatives on glassy carbon electrodes. *Analytical Chemistry* 71: 4081.
16. Walczak MW, Chung C, Stole SM, Widrig CA, and Porter MD (1991) Effects of surface monolayers on the el, Structure and interfacial properties of spontaneously adsorbed *n*-alkanethiolate monolayers on evaporated silver surfaces. *Journal of the American Chemical Society* 113: 2370.
17. Nuzzo RG, Zegarski BR, and Dubois LH (1987) Fundamental studies of the chemisorption of organosulfur compounds on gold(111). Implications for molecular self-assembly on gold surfaces. *Journal of the American Chemical Society* 109: 733.
18. Widrig CA, Chung C, and Porter MD (1991) The electrochemical desorption of *n*-alkanethiol monolayers from polycrystalline Au and Ag electrodes. *Journal of Electroanalytical Chemistry* 310: 335.
19. Poirier GE and Tarlov MJ (1994) The $c(4\times2)$ superlattice of alkanethiol self-assembled monolayers on Au(111). *Langmuir* 10: 2853.
20. Porter MD, Bright TB, Allara DL, and Chidsey CED (1987) Spontaneously organized molecular assemblies. 4. Structural characterization of *n*-alkyl thiol monolayers on gold by optical ellipsometry, infrared spectroscopy, and electrochemistry. *Journal of the American Chemical Society* 109: 3559.
21. Bumm LA, Arnold JJ, Dunbar TD, Allara DL, and Weiss PS (1999) Electron tunneling through organic molecules. *Journal of Physical Chemistry B* 103: 8122.
22. Xu B and Tao NJ (2003) Measurement of single-molecule resistance by repeated formation of molecular junctions. *Science* 301: 1221.
23. Wold DJ and Frisbie CD (2001) Fabrication and characterization of metal–molecule–metal junctions by conducting probe atomic force microscopy. *Journal of the American Chemical Society* 123: 5549.
24. Wold DJ, Haag R, Rampi MA, and Frisbie CD (2002) Distance dependence of electron tunneling through self-assembled monolayers measured by conducting probe atomic force microscopy: Unsaturated versus saturated molecular junctions. *Journal of Physical Chemistry B* 106: 2813.
25. Cui XD, Zarate X, Tomfohr J, *et al.* (2002) Making electrical contacts to molecular monolayers. *Nanotechnology* 13: 5.
26. Cui XD, Primak A, Zarate X, *et al.* (2002) Changes in the electronic properties of a molecule when it is wired into a circuit. *Journal of Physical Chemistry B* 106: 8609.
27. Holmlin R, Haag R, Chabinyc ML, *et al.* (2001) Electron transport through thin organic films in metal–insulator–metal junctions based on self-assembled monolayers. *Journal of the American Chemical Society* 123: 5075.
28. Rampi MA and Whitesides GM (2002) A versatile experimental approach for understanding electron transport through organic materials. *Chemical Physics* 281: 373.
29. Slowinski K, Fong HKY, and Majda M (1999) Mercury–mercury tunneling junctions. 1. Electron tunneling across symmetric and asymmetric alkanethiolate bilayers. *Journal of the American Chemical Society* 121: 7257.
30. York RL, Nguyen PT, and Slowinski K (2003) Long-range electron transfer through monolayers and bilayers of alkanethiols in electrochemically controlled Hg–Hg tunneling junctions. *Journal of the American Chemical Society* 125: 5948.
31. Kushmerick JG, Holt DB, Pollack SK, *et al.* (2002) Effect of bond-length alternation in molecular wires. *Journal of the American Chemical Society* 124: 10654.
32. Smalley JF, Feldberg SW, Chidsey CED, Linford MR, Newton MD, and Liu Y (1995) The kinetics of electron transfer through ferrocene-terminated alkanethiol monolayers on gold. *Journal of Physical Chemistry* 99: 13141.
33. Weber K, Hockett L, and Creager S (1997) Long-range electronic coupling between ferrocene and gold in alkanethiolate-based monolayers on electrodes. *Journal of Physical Chemistry B* 101: 8286.
34. Slowinski K, Chamberlain RV, Miller CJ, and Majda M (1997) Through-bond and chain-to-chain coupling. Two pathways in electron tunneling through liquid alkanethiol monolayers on mercury electrodes. *Journal of the American Chemical Society* 119: 11910.
35. Wang W, Lee T, and Reed MA (2003) Electron conduction mechanism in self-assembled alkanethiol monolayer devices. *Physical Review B* 68: 035416.

36. Wang W, Lee T, Kretzschmar I, and Reed MA (2004) Inelastic electron tunneling spectroscopy of alkanedithiol self-assembled monolayers. *Nano Letters* 4: 643.
37. Jaklevic RC and Lambe J (1966) Molecular vibration spectra by electron tunneling. *Physical Review Letters* 17: 1139.
38. Lambe J and Jaklevic RC (1968) Molecular vibration spectra by inelastic electron tunneling. *Physical Review* 165: 821.
39. Adkins CJ and Phillips WA (1985) Inelastic electron tunnelling spectroscopy. *Journal of Physics C: Solid State Physics* 18: 1313.
40. Hansma PK (ed.) (1982) *Tunneling Spectroscopy: Capabilities, Applications, and New Techniques*. New York: Plenum.
41. Stipe BC, Rezaei MA, and Ho W (1998) Single-molecule vibrational spectroscopy and microscopy. *Science* 280: 1732.
42. Zhou C, Deshpande MR, Reed MA, Jones L, II, and Tour JM (1997) Nanoscale metal/self-assembled monolayer/metal heterostructures. *Applied Physics Letters* 71: 611.
43. Chen J, Calvet LC, ReedMA, Carr DW, Grubisha DS, and Bennett DW (1999) Electronic transport through metal–1,4-phenylene diisocyanide–metal junctions. *Chemical Physics Letters* 313: 741.
44. Ralls KS, Buhrman RA, and Tiberio RC (1989) Fabrication of thin-film metal nanobridges. *Applied Physics Letters* 55: 2459.
45. Kim T-W, Wang G, Lee H, and Lee T (2007) Statistical analysis of electronic properties of alkanethiols in metal–molecule–metal junctions. *Nanotechnology* 18: 315204.
46. Ethanol and alkane molecules were purchased from Sigma-Aldrich
47. Metzger RM, Chen B, Holpfner U, *et al.* (1997) Unimolecular electrical rectification in hexadecylquinolinium tricyanoquinodimethanide. *Journal of the American Chemical Society* 119: 10455.
48. Horowitz P and Hill W (1989) *The Art of Electronics*. New York: Cambridge University Press.
49. Sze SM (1981) *Physics of Semiconductor Devices*. New York: Wiley.
50a. Thurstans RE and Oxley DP (2002) The electroformed metal–insulator–metal structure: A comprehensive model. *Journal of Physics D: Applied Physics* 35: 802.
50b. Simmons JG and Verderber RR (1967) New conduction and reversible memory phenomena in thin insulating films. *Proceedings of the Royal Society of London Series A* 301: 77.
50c. Dearnaley G, Stoneham AM, and Morgan DV (1970) Electrical phenomena in amorphous oxide films. *Reports on Progress in Physics* 33: 1129.
51a. Mann B and Kuhn H (1971) Tunneling through fatty acid salt monolayers. *Journal of Applied Physics* 42: 4398.
51b. Polymeropoulos EE and Sagiv J (1978) Electrical conduction through adsorbed monolayers. *Journal of Chemical Physics* 69: 1836.
52. Fan FF, Yang J, Cai L, *et al.* (2002) Charge transport through self-assembled monolayers of compounds of interest in molecular electronics. *Journal of the American Chemical Society* 124: 5550.
53. Franz W (1956) In: Flugge S (ed.) *Handbuch der Physik* vol. 17, 155p. Berlin: Springer.
54a. Lewicki G and Mead CA (1966) Experimental determination of *E–k* relationship in electron tunneling. *Physical Review Letters* 16: 939.
54b. Stratton R, Lewicki G, and Mead CA (1966) The effect of nonparabolic energy bands on tunneling through thin insulating films. *Journal of Physics and of Chemistry Solids* 27: 1599.
54c. Parker GH and Mead CA (1968) Energy–momentum relationship in InAs. *Physical Review Letters* 21: 605.
55. Brar B, Wilk GD, and Seabaugh AC (1996) Direct extraction of the electron tunneling effective mass in ultrathin SiO_2. *Applied Physics Letters* 69: 2728.
56. Joachim C and Magoga M (2002) The effective mass of an electron when tunneling through a molecular wire. *Chemical Physics* 281: 347.
57. Simmons JG (1963) Generalized formula for the electric tunnel effect between similar electrodes separated by a thin insulating film. *Journal of Applied Physics* 34: 1793.
58a. Simmons JG (1971) Conduction in thin dielectric films. *Journal of Physics D: Applied Physics* 4: 613.
58b. Maserjian J and Petersson GP (1974) Tunneling through thin MOS structures: dependence on energy (E–κ). *Applied Physics Letters* 25: 50.
59a. Yamamoto H and Waldeck DH (2002) Effect of tilt-angle on electron tunneling through organic monolayer films. *Journal of Physical Chemistry B* 106: 7469.
59b. Napper AM, Liu, Haiying, and Waldeck DH (2001) The Nature of electronic coupling through insulating barriers on Au electrodes. The importance of chain composition and interchain coupling. *Journal of Physical Chemistry B* 105: 7699.
60a. Lee JO, Lientschnig G, Wiertz F, *et al.* (2003) Absence of strong gate effects in electrical measurements on phenylene-based conjugated molecules. *Nano Letters* 3: 113.
60b. Haick H, Ghabboun J, and Cahen D (2005) Pd versus Au as evaporated metal contacts to molecules. *Applied Physics Letters* 86: 042113.
61. Akkerman HB, Blom PWM, de Leeuw DM, and de Boer B (2006) Towards molecular electronics with large-area molecular junctions. *Nature* 441: 69.
62. González MT, Wu S, Huber R, van der Molen SJ, Schönenberger C, and Calame M (2006) Electrical conductance of molecular junctions by a robust statistical analysis. *Nano Letters* 6: 2238.
63. Kim T-W, Wang G, Song H, Choi N-J, Lee H, and Lee T (2006) Charge transport of alkanethiol self-assembled monolayers in micro-via hole devices. *Journal of Nanoscience and Nanotechnology* 6: 3487.
64. Engelkes VB, Beebe JM, and Frisbie CD (2004) Length-dependent transport in molecular junctions based on SAMs of alkanethiols and alkanedithiols: Effect of metal work function and applied bias on tunneling efficiency and contact resistance. *Journal of the American Chemical Society* 126: 14287.
65. Cui XD, Primak A, Zarate X, *et al.* (2001) Reproducible measurement of single-molecule conductivity. *Science* 294: 571.
66. Lörtscher E, Weber HB, and Riel H (2007) Statistical approach to investigating transport through single molecules. *Physical Review Letters* 98: 176807.
67. Hu Y, Zhu Y, Gao H, and Guo H (2005) Conductance of an ensemble of molecular wires: A statistical analysis. *Physical Review Letters* 95: 156803.
68. Kushmerick JG (2005) Metal–molecule contacts. *Materials Today* 8: 26.
69. Salomon A, Cahen D, Lindsay S, Tomfohr J, Engelkes VB, and Frisbie CD (2003) Comparison of electronic transport measurements on organic molecules. *Advanced Materials* 15: 1881.
70. All of these lengths were determined by Chem3D software.
71a. Beebe JM, Engelkes VB, Miller LL, and Frisbie CD (2002) Contact resistance in metal–molecule–metal junctions based on aliphatic SAMs: Effects of surface linker and metal work function. *Journal of the American Chemical Society* 124: 11268.

71b. Kaun C-C and Guo H (2003) Resistance of alkanethiol molecular wires. *Nano Letters* 3: 1521.

72. Bryant MA and Pemberton JE (1991) Surface Raman scattering of self-assembled monolayers formed from 1-alkanethiols: Behavior of films at gold and comparison to films at silver. *Journal of the American Chemical Society* 113: 8284.

73. Kato HS, Noh J, Hara M, and Kawai M (2002) An HREELS study of alkanethiol self-assembled monolayers on Au(111). *Journal of Physical Chemistry B* 106: 9655.

74. Castiglioni C, Gussoni M, and Zerbi GJ (1991) Charge mobility in s-bonded molecules: The infrared spectrum of polymethylene chains in the solid and liquid phases. *Chemical Physics* 95: 7144.

75. The symbols, δ, γ, and ν denote in-plane rocking (r) and scissoring (s), out-of-plane wagging (w) and twisting (t), and stretching modes, respectively.

76a. Molinari M, Rinnert H, Vergnat M, and Weisbecker P (2003) Evolution with annealing treatments of the size of silicon nanocrystallites embedded in a SiN_x matrix and correlation with optical properties. *Materials Science and Engineering: B* 101: 186.

76b. Bogdanoff PD, Fultz B, and Rosenkranz S (1999) Vibrational entropy of $L12Cu_3Au$ measured by inelastic neutron scattering. *Physical Review B* 60: 3976.

76c. Mazur U and Hipps KW (1982) Spectroscopic characterization of tricyanovinyl alcoholate and pentacyanopropenide salts in solid, solution, and adsorbed phases: NMR, IR, Raman, and tunneling spectra. *Journal of Physical Chemistry* 86: 2854.

77a. Mazur U and Hipps KW (1981) Characterization of alumina supported O_ySiH by inelastic electron tunneling spectroscopy. *Journal of Physical Chemistry* 85: 2244.

77b. Kurata H, Hirose M, and Osaka Y (1981) Wide optical-gap, photoconductive a-Si_xN_{1-x}:H. *Japanese Journal of Applied Physics* 20: L811.

78. Kushmerick JG, Lazorcik J, Patterson CH, Shashidhar R, Seferos DS, and Bazan GC (2004) Vibronic contributions to charge transport across molecular junctions. *Nano Letters* 4: 639.

79. Kirtley J and Hall JT (1980) Theory of intensities in inelastic-electron tunneling spectroscopy orientation of adsorbed molecules. *Physical Review B* 22: 848.

80. Lauhon IJ and Ho W (1999) Single-molecule vibrational spectroscopy and microscopy: CO on Cu(001) and Cu(110). *Physical Review B* 60: R8525.

81. Klein J, Léger A, Belin M, Défourneau D, and Sangster MJL (1973) Inelastic-electron-tunneling spectroscopy of metal–insulator–metal junctions. *Physical Review B* 7: 2336.

82. Lauhon LJ and Ho W (2001) Effects of temperature and other experimental variables on single molecule vibrational spectroscopy with the scanning tunneling microscope. *Review of Scientific Instruments* 72: 216.

83. Kastner MA (1993) The quantum dot: A journey into the future of microelectronics. *Physics Today* January. 24–31.

84. Aviram A and Ratner M (eds.) (1998) *Molecular Electronics: Science and Technology, The Annals of the New York Academy of Sciences,* vol. 852. New York: New York Academy of Sciences.

85. Reimers J, Picconatto C, Ellenbogen J, and Shashidhar R (2003) *Molecular Electronics III, The Annals of the New York Academy of Sciences*, vol. 1006. New York: New York Academy of Sciences.

6 Nanoscale Transistors

Y Li and C-H Hwang, National Chiao Tung University, Hsinchu 300, Taiwan

6.1 Introduction

This chapter presents an evolution of device structures in modern nanometer scale (nanoscale) field-effect-transistors (FETs). It begins with an introduction of nanoscale transistors from planar device to vertical channel, such as double-gate, triple-gate, and surrounding-gate transistors. Intrinsic parameter fluctuations on electrical characteristics of nanoscale FETs and circuits are then characterized and discussed through device and circuit design viewpoints.

6.1.1 Nanoscale Transistor Architecture and Ultimate

Evolution of complementary metal–oxide–semiconductor (CMOS) technology in the past several years has followed the path of device scaling for achieving density, speed, and power improvements. The 2007 International Roadmap for Semiconductors (ITRS) projected that sub-10-nm-gate length will be launched before 2015 [1]. The most critical issues for continuing the device scaling will be the performance enhancement (short-channel effect (SCE), leakage current, power consumption, etc.) and the yield (intrinsic parameter fluctuations).

In the past several decades, planar metal–oxide–semiconductor FETs (MOSFETs) have been at the core of very large scale integration (VLSI) circuits; however, as gate-length scales decreased, they started to suffer from undesirable SCEs in scaled dimensions. The significant SCEs not only increase standby power dissipation, but also enlarge electrical characteristic fluctuations, such as the deviation of threshold voltage, drive current, mismatch, etc. Various technologies, such as mobility enhancement [2–4], metal gate with high-κ dielectrics (κ is the dielectric constant) [5–8], optimal doping profile design [9–11], lithography [12–14], vertical-channel transistor [15–39], etc., have been proposed to enhance the transistor performance and reduce device variability. Among these advanced technologies, succession of device structure from planar to vertical-channel transistors is the main trend in VLSI technologies. The vertical-channel transistors with multiple-gate surrounded have inherently good suppression of SCEs, high transconductance (g_m), and ideal subthreshold swing (SS), and provide more effective suppression of off-state leakage current and enhanced on-state current. Additionally, the vertical-channel transistors are also more compact than conventional planar transistors, enabling higher transistor density which translates to smaller overall nanoscale electronics. **Figure 1** plots an evolution of transistor architecture from planar MOSFETS to ultrathin-body (UTB), silicon-on-insulator (SOI), double-gate, omega gate, and nanowire fin-type FETs (FinFETs). Sub-10-nm-gate nanowire FinFETs with near-optimal gate coverage have been designed, simulated, and fabricated to show the promising characteristics [15–17]. **Figure 2(a)** shows the fabrication flow and the associated process parameters in fabrication. A tilted scanning electron microscope (SEM) image of the nanowire FinFET before gate hard mask removal is exhibited in **Figure 2(b)**. **Figure 3(a)** shows cross-sectional transmission electron microscopy (TEM) images of the rounded silicon nanowire body with an omega-shaped poly-Si gate. The diameter is controlled to about 10 nm, which is twice the gate length. Poly-Si gate with a 5-nm-physical-gate-length is patterned, as shown in **Figure 3(b)**. The drain current –gate voltage (I_d–V_g) and drain current –drain voltage (I_d–V_d) characteristics of nanowire FinFETs have 5-nm-gate-length p-type nanowire FinFETs and 10-nm-gate-length n-type nanowire FinFETs are shown in **Figure 4**. The accumulation-mode p-type nanowire FinFETs have a drive current of 115 μA μm^{-1} at an off-state leakage of 0.5 nA μm^{-1}, giving an on-/off-current ratio of 2.3×10^5. The extremely low leakage is consistent with the nature of depletion region expansion into the source or drain in the off-state ($V_g = 0$ V; $V_d = |1|$ V). Hence, the off-state effective channel length is longer than the physical gate length for accumulation-mode transistors. Excellent short-channel characteristics are obtained with drain-induced barrier lowering (DIBL) of 14 mV V^{-1} and SS of 63 mV dec^{-1}, even at gate length of 5 nm. Conversely, inversion-mode 10-nm-gate-length n-type nanowire FinFETs show a drive current of 522 μA μm^{-1}; adequate threshold voltage was also achieved. The better SCE control

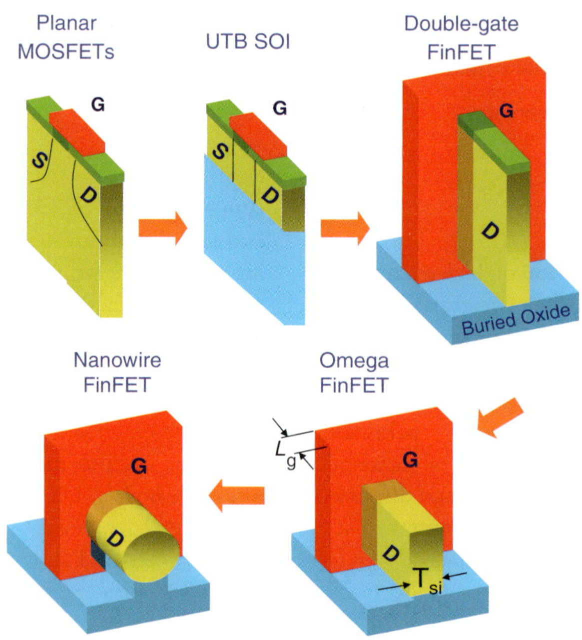

Figure 1 The evolution of transistor architecture from planar MOSFETs to ultrathin-body silicon-on-insulator, double-gate, omega gate, and nanowire FinFETs.

(a)

- Si fin patterning, T_{si} trimmed down upon etching
- Nanowire process for Si fin (H_2 anneal at 900 ± 50 °C)
- V_{th} implantation
- Gate oxide growth (physical thickness 19Å)
- Gate deposition (*in situ* doped N^+ poly silicon)
- Poly-Si CMP
- Gate hard mask deposited and patterning (Hard mask trimmed down upon etching)
- Poly-gate etching
- Pocket implantation (N-FET exclusively)
- Oxide/nitride combo spacer formation
- Source/drain implant
- Activation with spike anneal (>1050 °C)
- Contact formation
- Copper interconnect

(b)

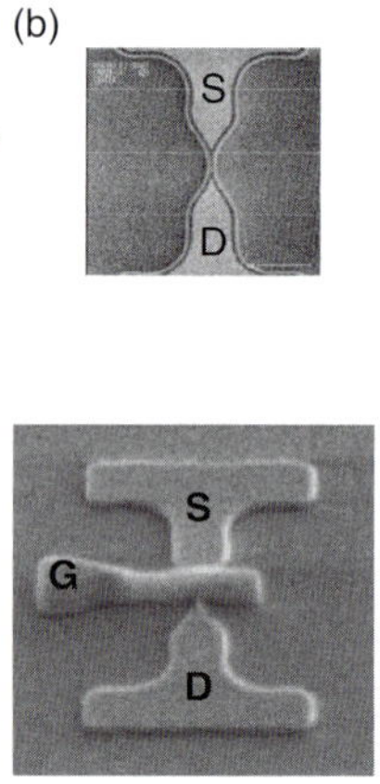

Figure 2 (a) The fabrication flow and the critical process parameters in fabrication. (b) A tilted SEM image of the nanowire FinFET before gate hard mask removal.

in the 5-nm-gate-length p-type nanowire FinFETs than that in the 10-nm-gate-length n-type nanowire FinFETs suggests that the accumulation-mode p-type nanowire FinFETs have an effective gate length that is larger than the physical gate length. This could be an advantage of accumulation-mode over inversion-mode designs for very aggressive gate length scaling in nanowire devices. Accumulation-mode p-type nanowire FinFETs and inversion-mode n-type nanowire FinFETs with physical $L_g = 5$ and 10 nm, respectively, are introduced, n-type nanowire FinFETs gate delay (CV/I) of 0.22 ps and p-type nanowire FinFETs gate delay of 0.48 ps with excellent subthreshold characteristics are achieved, both with very low off-leakage current of less than 10 nA μm^{-1}. The new nanowire FinFETs are promising and compatible with current manufacturing process.

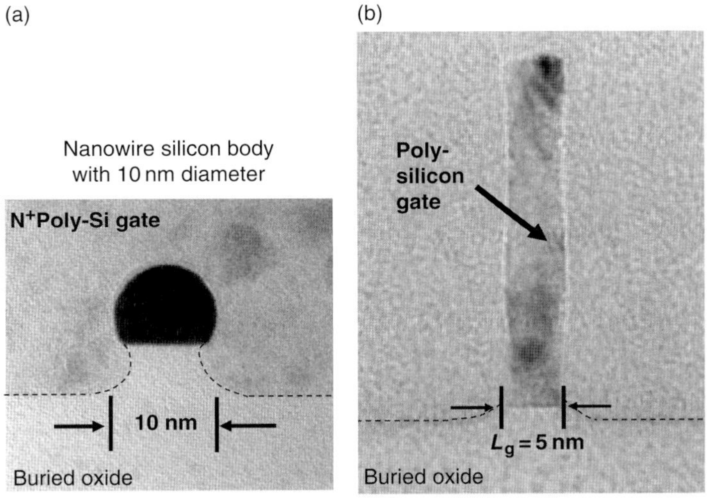

Figure 3 (a) Cross-sectional TEM images of the rounded silicon nanowire body with an omega-shaped poly-Si gate. The diameter is controlled to about 10 nm, which is twice the gate length. (b) Poly-silicon gate with 5 nm physical gate length is patterned.

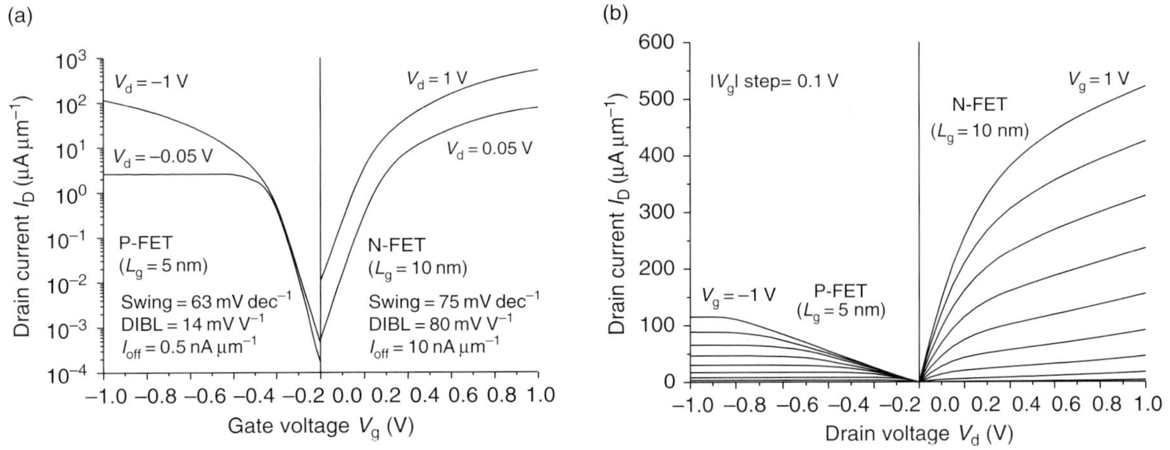

Figure 4 The (a) I_d–V_g and (b) I_d–V_d characteristics of nanowire FinFETs have 5-nm L_g p-type nanowire FinFETs and 10-nm L_g n-type nanowire FinFETs.

6.1.2 Toward Variability of Nanoscale Transistor Era

Gate lengths of scaled MOSFETs are under 30 nm in 45-nm node high-performance circuits [5, 40–43]. Devices with sub-10-nm-gate lengths have recently been studied. Transistor scaling down to sub-20 nm and even beyond has widely encountered many more challenges with short-channel control than previous generations. The worsened short-channel control enlarges electrical characteristic fluctuations, such as the deviation of threshold voltage, drive current, mismatch, etc. Yield analysis and optimization, which take into account the manufacturing tolerances, model uncertainties, variations in the process parameters, etc., are known as indispensable components of the circuit design methodology [44–48]. Additionally, the fluctuation is intrinsically increased with the scaling of transistor feature sizes, even when not considering the worsened short-channel control [15, 20–27].

Figure 5 reviews the threshold voltage (V_{th}) fluctuation versus technology node [15]. At the 32-nm node, thinner gate dielectric thickness could be achieved by the introduction of metal gate (for eliminating poly-depletion) and high-κ materials (for thinner

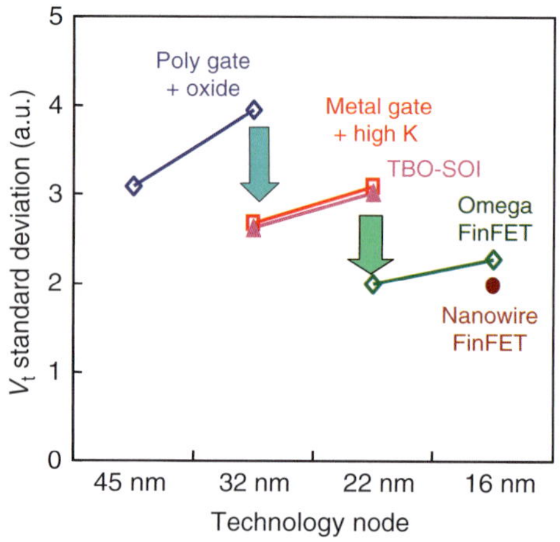

Figure 5 The threshold voltage fluctuation vs. technology node.

equivalent-oxide thickness without increasing the gate leakage). Another method is not scaling the gate dielectric thickness but enhancing the short-channel control with SOI substrate. Special attention will be paid to thin-buried-oxide SOI [49]. At the sub-22-nm node, planar SOI will no longer be a good device option, because a very thin body (less than 10 nm) is necessary for SCE control; however, in the meanwhile the thin body will have degraded drive current due to quantum confinement [50]. As a result, nonplanar transistors such as FinFETs may emerge instead [51–56]. Their V_{th} fluctuation characteristics will be addressed, including the effect of body thickness variation. **Figure 6** shows the major sources of intrinsic parameter fluctuations, the gate length deviation [15, 27, 36, 57–61], line edge roughness (LER) [15, 27, 36, 57–61], and random dopant fluctuation [21–27, 62–89]. The fluctuations caused by granularity of the poly-Si gate [90–93] and silicon film thickness variation [94–97] are also important factors in intrinsic device fluctuation, which depends on the device structure. **Figure 7** shows the gate length deviation, LER, and random dopant fluctuation-induced V_{th} fluctuation versus the equivalent oxide thickness (EOT). The EOT reduction can be achieved by the introduction of metal-gate and high-κ dielectric for low-standby-power devices. All of the comparisons are based on the same off-state (leakage) current. The random dopant fluctuation dominates the intrinsic parameter fluctuation, and therefore is the focus here. In principle, the standard deviation (SD) of V_{th} induced by the three aforementioned sources can all be reduced with decreased gate dielectric thickness, due to less surface potential perturbation under the enhanced gate controllability. However, this improvement will depend on timely implementation of reliable high-κ gate dielectrics and a work-function modulated metal gate. Notably, the dominating factor of V_{th} fluctuation depends on the application and design of nanoscale transistor, which can be verified by various experimental data.

The remainder of this chapter is organized as follows. Section 6.2 introduces the characterization technique for studying the effect of intrinsic parameter fluctuations. Sections 6.3 and 6.4 detail the characteristic fluctuations for planar and vertical-channel bulk and SOI transistors. Then, the impact of intrinsic device variability on digital, analog or high-frequency, and memory circuits is examined in Section 6.5. Fluctuation-suppression techniques proposed from device and circuit design viewpoints are presented and examined in Section 6.6. In Section 6.7, conclusions are drawn and future work is suggested.

6.2 Characterization of Variability in Nanoscale Transistors

This section presents the characterization of the intrinsic parameter fluctuation for the process-variation-induced effect (PVE) and randomness effect. The characterization approaches are examined with

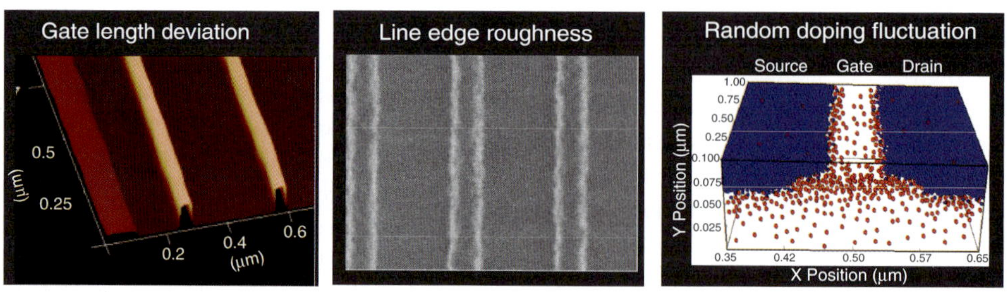

Figure 6 The major sources of intrinsic parameter fluctuations: the gate-length deviation, line edge roughness, and random dopant fluctuation.

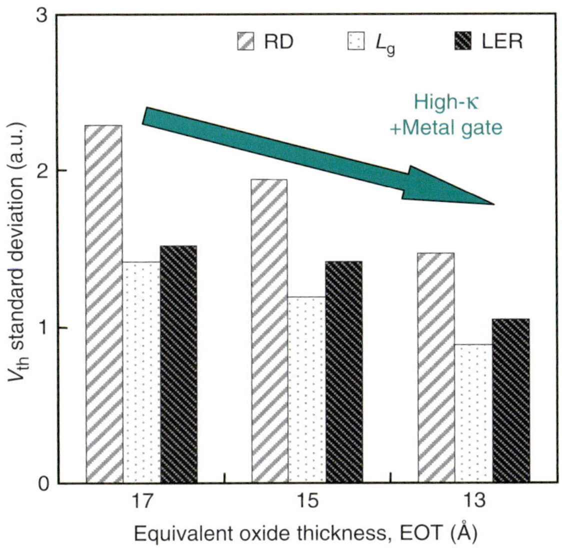

Figure 7 The gate-length deviation, line edge roughness, and random dopant-fluctuation-induced V_{th} fluctuation vs. equivalent oxide thickness.

experimental data and full quantum mechanical simulation. Then, the methodology is extended to study the circuit characteristic fluctuations.

6.2.1 Process Variation Effect

During the manufacturing process, the inevitable variations of processing condition cause variations on device geometry, such as nonuniform gate oxide and material interface facial layer thickness, gate length deviation, LER, etc. [95]. All of these fluctuations worsen with gate length scaling. Among these process-variation-induced fluctuations, the gate length deviation and the LER are the dominating factors; therefore, we focus on the fluctuations induced by these two effects.

Threshold voltage is one of the key device parameters in the characteristics of nanoscale MOSFETs. Base upon the independence of random variables, the total threshold voltage fluctuation, $\sigma V_{\mathrm{th,total}}$, is expressed as follows:

$$\sigma^2 V_{\mathrm{th,total}} = \sigma^2 V_{\mathrm{th,RD}} + \sigma^2 V_{\mathrm{th,PVE}} \quad (1)$$

where $\sigma V_{\mathrm{th,RD}}$ is the randomness fluctuation, $\sigma V_{\mathrm{th},Lg/\mathrm{LER}}$ is the fluctuation caused by the PVE . **Figure 8** shows an atomic force microscope (AFM) scan of poly-gate profile with an LER of 4 nm (3σ) after poly-etching. LER arises not only from the resolution limit of lithography, but also from the grainy nature of the photoresist and poly-gate. The

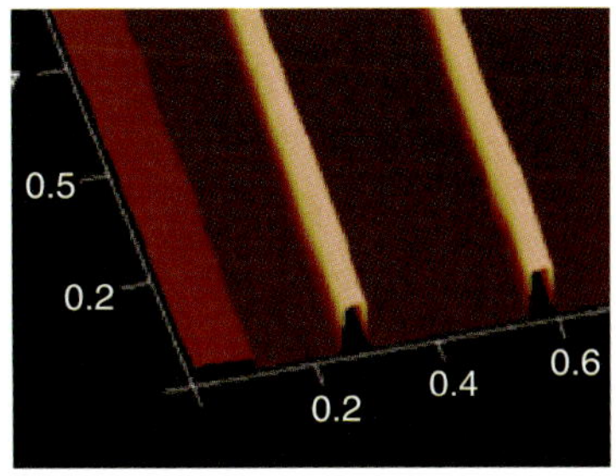

Figure 8 An AFM scan of poly-gate profile with a line edge roughness of 4 nm (3σ) after poly-etching.

effect of line width roughness is random and cannot be corrected by the optical proximity correction. A statistical approach is applied herein to evaluate the effect of process-variation-induced V_{th} fluctuation, $\sigma_{V\mathrm{th},Lg/\mathrm{LER}}$. The magnitudes of the gate-length deviation and the LER are extracted from the projections of ITRS for different technology nodes. A look-up table of the threshold voltage versus gate length is established, as shown in **Figure 9**. It enables us to evaluate the threshold voltage with respect to the deviation of gate length and LER. Following the roadmap of ITRS,, $3\sigma_{Lg} = 0.9$ nm and $3\sigma_{\mathrm{LER}} = 1.2$ nm for the 22-nm node and $3\sigma_{Lg} = 0.7$ nm and $3\sigma_{\mathrm{LER}} = 0.8$ nm for the 16-nm technology node, respectively, as the inset table of Figure 9 illustrates. Using the V_{th} roll-off relation, $\sigma V_{\mathrm{th},Lg/\mathrm{LER}} = \mathrm{d}V_{th} \,/\, \mathrm{d}L_g \times \sigma L_g$, $\sigma V_{\mathrm{th},Lg/\mathrm{LER}}$ can be extracted with the experimental data of V_{th} roll-off

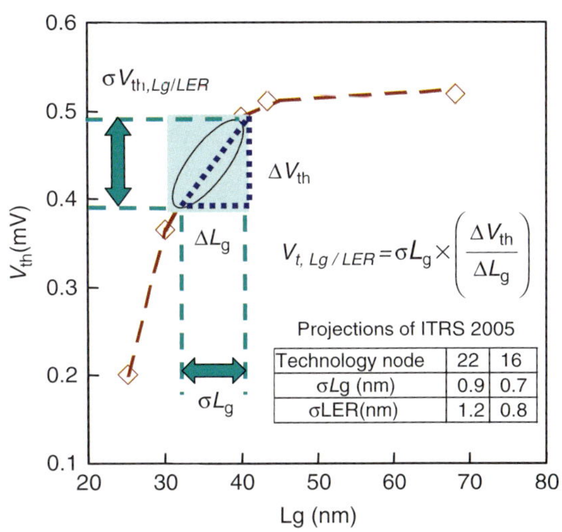

Technology node	22	16
σLg (nm)	0.9	0.7
σLER(nm)	1.2	0.8

Figure 9 A look-up table of the threshold voltage vs. gate length. Using the V_{th} roll-off relation, $\sigma V_{\mathrm{th},Lg/\mathrm{LER}}$ can be extracted with the experimental data of V_{th} roll-off and standard gate-length deviation, σL_g.

and standard gate-length deviation, σL_g. Thus, we can calculate SD of threshold voltage resulting from the deviation of gate length and LER. We note that data for σL_g can be obtained from SEM critical-dimension measurements, as shown in **Figure 6**.

6.2.2 Randomness Effect

The random dopants come from the manufacturing process, such as ion implantation and thermal annealing [98]. The random nature of discrete dopant distribution results in significant random fluctuations, such as the deviation of threshold voltage, drive current mismatch, etc. The fluctuation budget has to be controlled ever tighter due to the doubly increased transistor number along with technology node advancements. Unfortunately, the fluctuation is intrinsically increased with the scaling of transistor feature size, without even considering the worsened short-channel control.

Figure 10 shows the equivalent number of dopants and channeling probability versus the size of channel. Under the same equivalent channel doping concentration, the equivalent number of dopants within the channel decreases significantly from 78 to 17 as the channel size scales from (25 nm)^3 to (15 nm)^3. The dopant distributions are illustrated in the inset of **Figure 10**. For larger device dimension, the surface potential is controlled by a cloud of remote charges. However, as the device size scales, the surface potential is dominated by only a few dopants nearby. The impact of each discrete dopant is large enough to be distinct. A channeling probability is defined by the probability of existence of a channel with (7 nm)^2 cross-section between the source (S) and the drain (D) within the channel surface. The high carrier channeling probability indicates the increasing importance of dopants in scaled channels. Carrier transport in the (7 nm)^2 cross-section behaves as if the carriers are in an undoped channel. As the planar CMOS devices advance to sub-20-nm-gate lengths, double-digit channel dopants make the transistor behavior more complicated to be characterized with conventional continuum modeling because every discrete dopant has its significant weight impacting the resulting transistor performance. The discrete-dopant model is especially validated for quantifying carrier channeling effects between the source and the drain, which cannot be done by conventional continuum modeling.

Various random dopant effects have recently been studied in both experimental and theoretical approaches [21–27, 62–89]. These studies have shown that the fluctuation of electrical characteristics is not purely a result of the variation in average doping density associated with a fluctuation in the number of dopants, but also the particular random distribution of dopants in the channel region. To characterize random dopant-induced standard V_{th} deviation, $\sigma V_{th,RD}$, the doping profile is first analytically approximated for the device measured. Then, a statistically sound atomistic simulation approach is described below, as shown in **Figure 11**, to generate

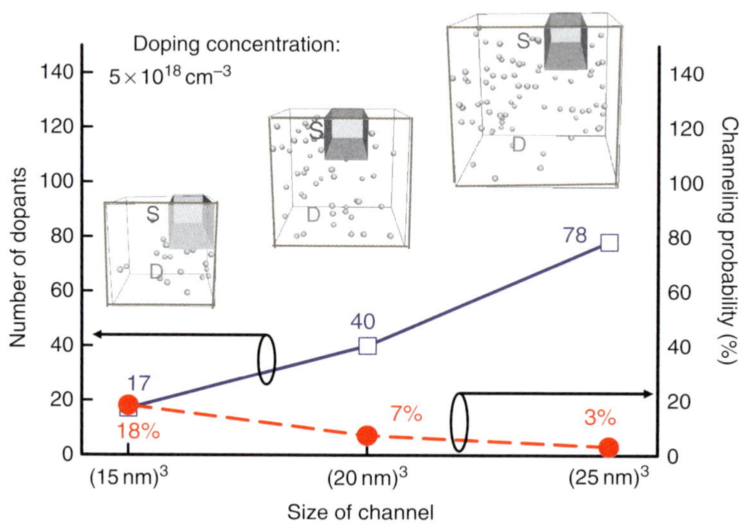

Figure 10 The equivalent number of dopants and channeling probability vs. the size of channel.

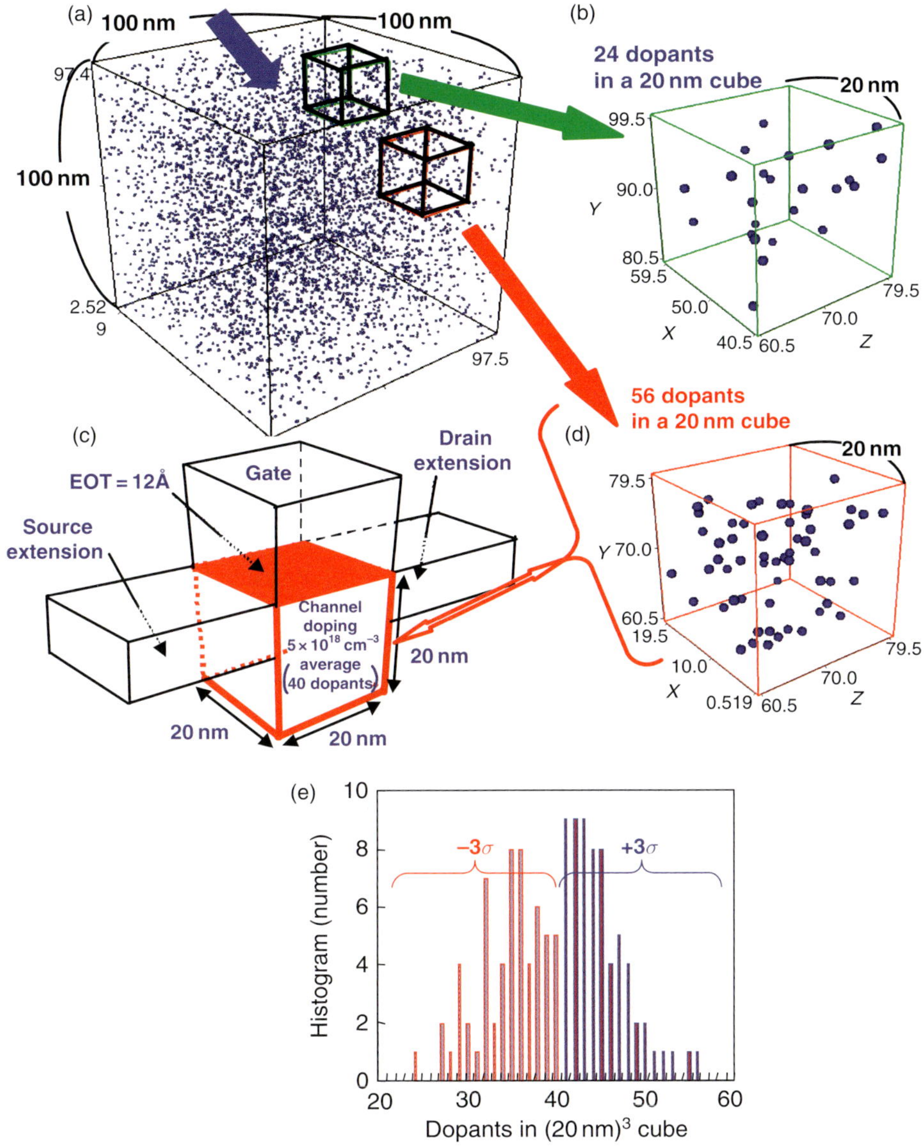

Figure 11 (a) Discrete dopants randomly distributed in $(100\,\text{nm})^3$ cube with an average concentration of $5 \times 10^{18}\,\text{cm}^{-3}$. There will be 5000 dopants within the $(100\,\text{nm})^3$ cube, but dopants vary from 24 to 56 (the average number is 40 and the standard deviation is 6.3) within its 125 subcubes of $(20\,\text{nm})^3$, as shown in (b), (c), and (e), respectively. These 125 subcubes are then equivalently mapped into channel region for dopant-position- and dopant-number-sensitive simulation, as shown in (d).

discrete dopants in the channel region. **Figure 11** briefly illustrates how to generate discrete dopant channels for the aforementioned simulation, concurrently capturing randomness of dopant number and dopant position. **Figure 11 (a)** shows the discrete dopants randomly distributed in the cube of volume $(100\,\text{nm})^3$ with an average concentration of $5 \times 18\,\text{cm}^{-3}$, which is the same as the fabricated device. There will be 5000 dopants within the $(100\,\text{nm})^3$ cube, but dopants vary from 24 to 56 (the average number is 40 and the SD is 6.3) within its 125 subcubes of $(20\,\text{nm})^3$, as shown in **Figures 11 (b), 11(c), and 11(e)**, respectively. These 125 subcubes are then equivalently mapped into the channel region for the discrete dopant simulation, as shown in **Figure 11 (d)**. In principle, three-dimensional (3D) device simulation with the 125 channel structures almost covers $\pm 3\sigma$ cases, shown in

Figure 11(e), and thus will be fairly meaningful to reflect statistical randomness of dopant number and dopant position in channel region.

The device is simulated by solving a set of 3D Poisson equations and electron–hole current continuity equations with quantum corrections [99–108]. A step function, N_A, is used to define the concentration and positions of dopants:

$$N_A = \sum_{i=0}^{k} N_A^{\text{dopant}} \cdot \begin{bmatrix} H(x-x_l, y-y_l, z-z_l) \\ -H(x-x_u, y-y_u, z-z_u) \end{bmatrix} \quad (2)$$

where

$$H(x,y,z) = \begin{cases} 1, x \geq 0, y \geq 0, z \geq 0 \\ 0, \quad \text{otherwise} \end{cases} \quad (3)$$

(x_l, y_l, z_l) and (x_u, y_u, z_u) are the lower and upper coordinates of a discrete dopant; k is the number of dopants in the device channel; N_A^{dopant} is the associated doping concentration for a dopant within a box. Then, N_A is substituted into the source of the Poisson equation and solved, with the electron–hole current continuity equations and density-gradient quantum correction equations, simultaneously for device characteristics. Notably, in atomistic device simulation, the resolution of individual charges within a conventional drift–diffusion simulation using a fine mesh creates problems associated with singularities in the Coulomb potential [81–83]. The potential becomes too steep with fine mesh and therefore the majority carriers are unphysically trapped by ionized impurities and the mobile carrier density is reduced [81–83]. Thus, the density gradient approximation is used to handle the discrete charges by properly introducing the related quantum mechanical effects [99–103]. All statistically generated discrete dopants, as shown in **Figure 11**, are advanced and incorporated into the 3D device simulation under our parallel computing system [109–112]. Such a large-scale computer simulation approach allows us to explore the electrical characteristic fluctuations induced by randomness of dopant number and position in the channel region concurrently. The mobility model used in the device simulation, according to Mathiessen's rule [113, 114], can be expressed as

$$\frac{1}{\mu} = \frac{D}{\mu_{\text{surf\_aps}}} + \frac{D}{\mu_{\text{surf\_rs}}} + \frac{1}{\mu_{\text{bulk}}} \quad (4)$$

where $D = \exp(x/l_{crit})$, x is the distance from the interface, and l_{crit} is a fitting parameter. The mobility consists of three parts: (1) the surface contribution due to acoustic phonon scattering

$$\mu_{\text{surf\_aps}} = \frac{B}{E} + \frac{C(N_i/N_0)^{\tau}}{E^{1/3}(T/T_0)^{K}},$$

where $N_i = N_A + N_D$, $T_0 = 300\,\text{K}$, $\boldsymbol{E}$ is the transverse electric field normal to the interface of the semiconductor and insulator, B and C are parameters which are based on physically derived quantities, N_0 and τ are fitting parameters, T is lattice temperature, and K determines the temperature dependence of the probability of surface phonon scattering; (2) the contribution attributed to surface roughness scattering is

$$\mu_{\text{surf\_rs}} = \left(\frac{(E/E_{\text{ref}})^{\Xi}}{\delta} + \frac{E^3}{\eta} \right)^{-1}$$

where

$$\Xi = A + \frac{\alpha \cdot (n+p) N_{\text{ref}}^{v}}{(N_i + N_1)^{v}}$$

$E_{\text{ref}} = 1\,\text{V cm}^{-1}$ is a reference electric field to ensure a unitless numerator in $\mu_{\text{surf\_rs}}$, $N_{\text{ref}} = 1\,\text{cm}^{-3}$ is a reference doping concentration to cancel the unit of the term raised to the power v in the denominator of Ξ, δ is a constant that depends on the details of the technology, such as oxide growth conditions, $N_1 = 1\,\text{cm}^{-3}$, A, α, and η are fitting parameters; and (3) the bulk mobility is $\mu_{\text{bulk}} = \mu_L \left(\frac{T}{T_0} \right)^{-\xi}$, where μ_L is the mobility due to bulk phonon scattering and ξ is a fitting parameter. The mobility model is quantified with device measurements for the best accuracy.

The aforementioned methodology is one of the approaches in the examination of random dopant-induced V_{th} fluctuation. To extract the experimental data of $\sigma V_{\text{th,RD}}$, $\sigma V_{\text{th,total}}$ and $\sigma V_{\text{th,}Lg\text{/LER}}$ are first measured directly from experimental data since the mean gate-length deviation, LER, and random dopant distribution are the major sources of variation of threshold voltage. The $\sigma V_{\text{th,RD}}$ thus can be extracted from **equation 1**. Notably, the equation implies two important insights: $\sigma V_{\text{th,total}}$ reduction relies substantially on the dominant factor improvement and the second-order factors do not impact the derived value of $\sigma V_{\text{th,RD}}$. For example, assuming that $\sigma V_{\text{th,RD}}$ and $\sigma V_{\text{th,}Lg\text{/LER}}$ are 40 and 30 mV, respectively, $\sigma V_{\text{th,total}}$ is 50 mV. The

16% reduction of $\sigma V_{th,RD}$ achieves 16% reduction of $\sigma V_{th,total}$; however, it requires a 31% reduction of $\sigma V_{th,Lg/LER}$ to obtain the same $\sigma V_{th,total}$. Moreover, even with a new addition fluctuation sources, for example: source/drain dopant fluctuation with 20 mV, is introduced. The value of $\sigma V_{th,total}$ is around 52 mV, which indicates that the second-order factors do not significantly impact the derived value of $\sigma V_{th,RD}$.

Figures 12(a) and **12(b)** illustrate the mesh size dependence of the classical and quantum mechanical potentials for a single discrete dopant within the silicon channel. In the atomistic simulation, the key point to study the random impurities-induced fluctuation relies on how microscopic nonuniformity of localized impurity distributions is introduced in the device. In the conventional drift–diffusion approach for a large device size, the number of impurities included in each mesh exceeds one and the equivalent doping concentration does not change abruptly at every mesh node. Also, the dopant density at each mesh node changes gradually and the nonuniformity of impurity arrangement is averaged, as shown in **Figure 13(a)**. However, for the nanoscale transistor, the corresponding number of impurities is significantly reduced. Most meshes contain no dopant or, at most, one dopant. The dopant density at each mesh node changes its order of magnitude and behaves like a δ-function, as shown in **Figure 13(b)**. The resolution of individual impurities for the conventional drift–diffusion simulation using a fine mesh creates problems of singularities in the Coulomb potential, as shown in **Figure 12(a)**. The sharp Coulomb potential wells may unphysically trap majority carriers, reduce the mobile electron concentration, modify the depletion region, and alter the threshold voltage. Therefore, the density-gradient quantum correction [99–103] is used to handle the discrete dopant effect by properly introducing the related quantum mechanical effects, as plotted in **Figure 12(b)**. The quantum mechanical potential shows less sensitivity to the mesh size and is quite constant for mesh spacing below 0.5 nm. We notice that the potential barrier of the Coulomb well is about 45 mV, which roughly corresponds to the ground state of a hydrogenic model of an impurity in silicon.

6.2.3 Device Variability in Circuit

For the state-of-the-art nanoscale VLSI circuits and systems, the local device variation and uncertainty of signal propagation time have become crucial in the variation of system timing and the high-frequency characteristics. Yield analysis and optimization, which take into account the manufacturing tolerances, model uncertainties, variations in the process parameters, and other factors are known as indispensable components of the circuit design procedure [44–48].

Diverse approaches have recently been employed to investigate the fluctuation-related issues in transistors [21–27, 62–83] and circuits [61–67, 84–89]. However, most of them rely on the use of compact models (i.e., equivalent circuits of transistor). Although the extraction of a compact model provides an efficient way to estimate circuit characteristics fluctuation, the random dopant-induced nonlinear device gate capacitance fluctuations make the device characteristics difficult to be modeled using present

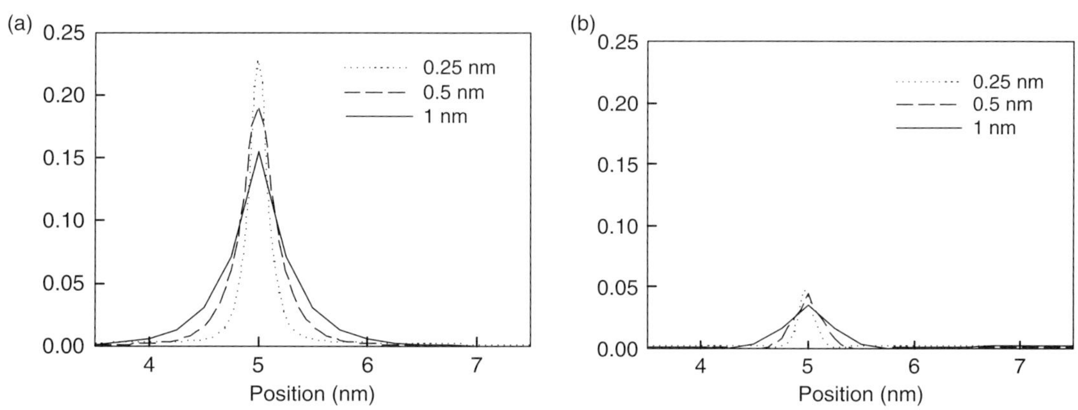

Figure 12 Potential profiles for (a) classical and (b) quantum potential with different mesh size.

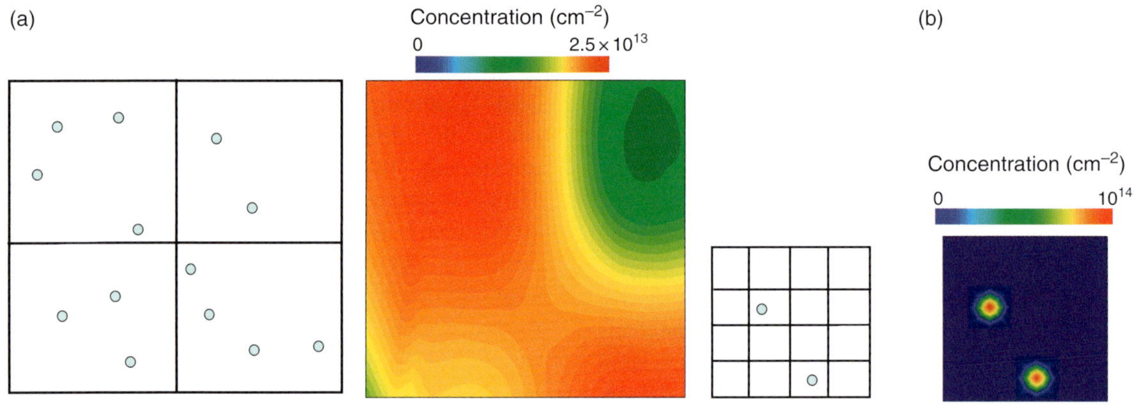

Figure 13 Meshing and dopant distribution for (a) long channel and (b) nanoscale transistors. The fine mesh in nanoscale transistor creates problems of singularities in the Coulomb potential and unphysically trap majority carriers.

compact models [64–67, 77]. Moreover, well-established compact models of ultrasmall nanoscale devices may not be available yet. In estimating circuit characteristics, since no well-established compact model of ultrasmall nanoscale devices is available, to capture the fluctuations induced by discrete dopant position, a coupled device-circuit simulation approach [64–67, 115–118] is then proposed. Characteristics of the devices of the circuit are first estimated by solving the device transport equations. The obtained result is then used as an initial guess in the coupled device-circuit simulation. The nodal equations of the test circuit are formulated, and then directly coupled to the device transport equations (in the form of a large matrix that contains both circuit and device equations), which are solved simultaneously to obtain the circuit characteristics. The device characteristics, such as distributions of potential and current density, obtained by device simulation are input in the circuit simulation through device contacts. The effect of discrete dopants in the transistor on circuit characteristics is thus properly estimated. The time-domain coupled device-circuit simulation chart is shown in **Figure 14**, where a CMOS inverter circuit is shown as an example. For the coupled device-circuit simulation, the time-dependent device transport equations are shown below:

$$\Delta\phi = \frac{q}{\varepsilon_s}(n - p + N_D - N_A) \quad (5)$$

$$\frac{\partial n}{\partial t} + \nabla\cdot\left(\mu_n n \nabla\left(\phi - \frac{KT}{q}\log\left(\frac{n}{n_i}\right) + \gamma_n\right)\right) + R = 0 \quad (6)$$

$$\frac{\partial p}{\partial t} - \nabla\cdot\left(\mu_p p \nabla\left(\phi - \frac{KT}{q}\log\left(\frac{p}{n_i}\right) + \gamma_p\right)\right) + R = 0 \quad (7)$$

and

$$\gamma_n = 2b_n\frac{\nabla^2\sqrt{n}}{\sqrt{n}};\, b_n = \frac{\hbar^2}{12qm_n^*};\, \gamma_p = 2b_p\frac{\nabla^2\sqrt{p}}{\sqrt{p}};\, b_p = \frac{\hbar^2}{12qm_p^*} \quad (8)$$

Equations 5–8 represent the Poisson equation, current continuity equation for electrons, current continuity equation for holes, and quantum-corrected equation for electrons and holes. ϕ, n, p, t, and T are the electrostatic potential, electron densities, hole densities, time, and temperature to be solved, respectively. The silicon permittivity, ε_s, is $1.05 \times 10^{-10}\,\mathrm{F\,m^{-1}}$; q is the elementary charge, 1.06×10^{-19} C; k_B is the Boltzmann constant, $8.6174 \times 10^{-15}\,\mathrm{eV\,K^{-1}}$; R, μ_n, and μ_n are, respectively, the net electron–hole recombination rate and electron and hole mobilities; N_D and N_A are the number of ionized donors and acceptors, respectively; γ_n and γ_p are the quantum potential for electrons and holes; b_n and b_p are density-gradient coefficient for electrons and holes; $\hbar$ is the reduced Planck constant; and m_n^* and m_p^* are the effective mass for the electrons and holes. **Equations 9–11** are the circuit nodal equations of inverter:

$$\text{Node 1}: V_1 = V_G \quad (9)$$

$$\text{Node 2}: V_2 = V_{DD} \quad (10)$$

$$\text{Node 3}: I_{d,\mathrm{PMOS}} = I_{d,\mathrm{NMOS}} \quad (11)$$

$$\text{Node 4}: V_4 = 0 \quad (12)$$

After solving the device transport equation, the device and circuit equations from **equations 5–12**

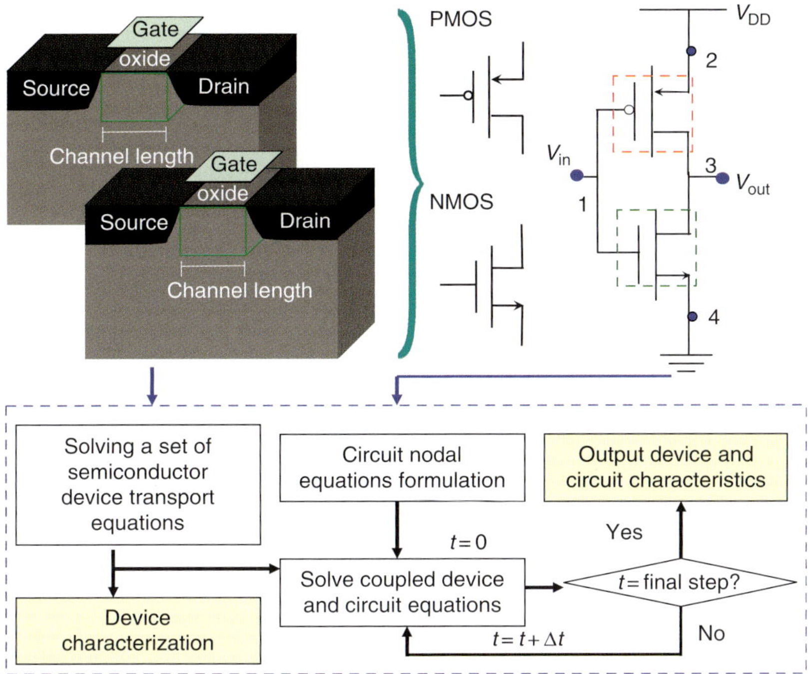

Figure 14 Time domain-coupled device circuit simulation flow, in which an inverter circuit is used as an example.

are coupled and solved simultaneously to obtain circuit characteristics. The coupled device–circuit simulation approach directly transfers the device characteristics to circuit and connects the device physics with circuit characteristics.

6.3 Intrinsic Parameter Fluctuation in Bulk MOSFETs Transistors

We first experimentally quantified the random dopant distribution-induced threshold voltage SD up to 40 mV for 20-nm-gate planar CMOS devices. As the gate length further scales down to 15 nm, the newly developed discrete dopant scheme features an effective solution to suppress the 3σ-edge single-digit dopant-induced V_{th} fluctuation by the gate work-function modulation. Additionally, we explore the fluctuations of threshold voltage roll-off in nanoscale bulk FinFETs. The explored devices are of three different dimensions: $(16\,nm)^3$, $(22\,nm)^3$, and $(30\,nm)^3$. The trend and behavior of discrete impurities inside channel region are illustrated.

6.3.1 Experimentally Validated 20-nm-Gate Planar Bulk MOSFETs

Figures 15 and **16** show the experimental V_{th} fluctuation and the on- and off-state currents characteristics (I_{on}–I_{off}) of the n-typed MOSFETs (NMOSFETs) down to 20-nm gates. The gate-length values, as shown in **Figure 15**, are estimated from the gate capacitances in analysis data, and we presume the widths of all samples are 200 and 20 nm for **Figures 15(a)** and **15(b)**, respectively. As expected, the V_{th} roll-off characteristics of 20-nm-wide devices are much more scattered than that of 200-nm-wide devices. The random dopant-induced V_{th} fluctuation, $\sigma V_{th,RD}$, has then been experimentally extracted, as shown in **Figure 17(a)**. Discrete dopant simulation for L_g = width (W) = 20 nm (data represented with symbol *, as shown in Figure **17(a)**) is in good agreement with the experimental data, which confirms the channel doping is randomly distributed as statistically modeled. As shown in **Figure 11**, more than 100 cases are required for a set of values of L_g and width; note that each 3D simulation case may take about ~3–7 days for a final convergent result. Without loss of generality (WLOG), due to the huge computing resource, we

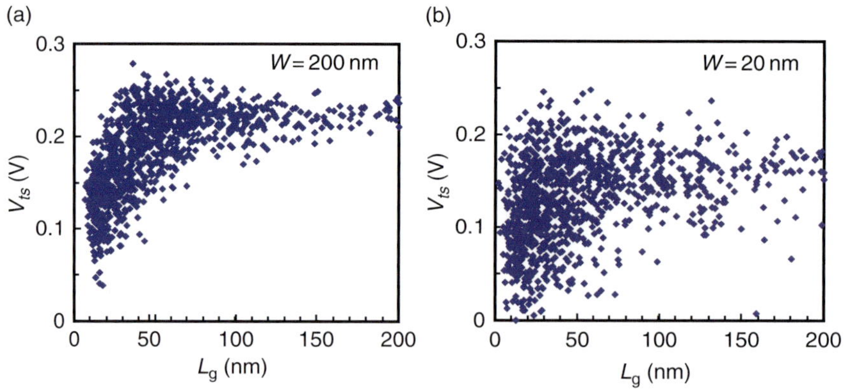

Figure 15 Experimental saturation threshold voltage, V_{th}, of NMOSFETs with L_g down to 20 nm for the (a) width = 200 nm and (b) width = 20 nm at $V_d = 1.0$ V.

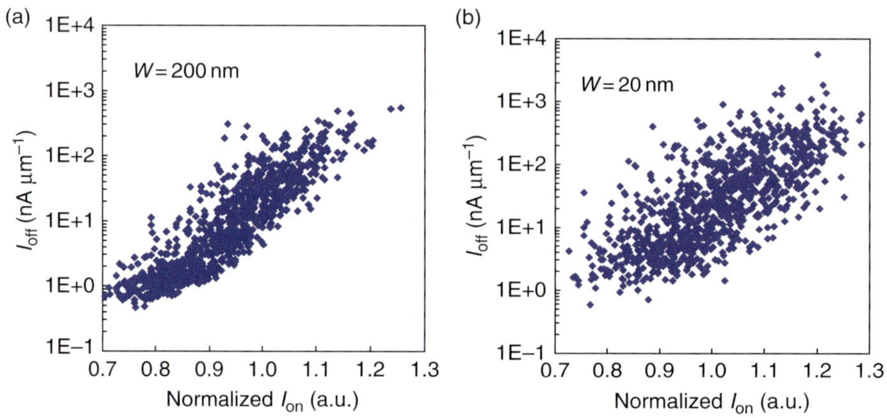

Figure 16 Experimental I_{on}–I_{off} characteristics of NMOSFETs with L_g down to 20 nm for the (a) width = 200 nm and (b) width = 20 nm at $V_d = 1.0$ V. The I_{on} was normalized against the on-current of nominal L_g case, that is, the 20-nm L_g case.

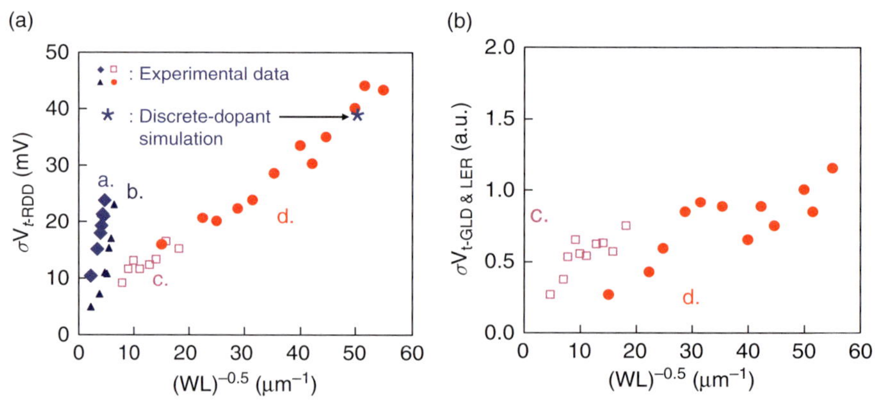

Figure 17 (a) Experimentally extracted $\sigma V_{th,RD}$, and discrete dopant simulation (*, $L_g = W = 20$ nm, EOT = 1.2 nm) for various devices with nominal L_g from 55 nm down to 20 nm. The width is fixed and the length is varying to give the range of values of $(WL)^{-0.5}$. The sample size for each data point of σV_{th} is around 100 points. (b) The extracted $\sigma V_{th,Lg/LER}$ for c and d conditions. The value was normalized against the $\sigma V_{th,Lg/LER}$ of nominal L_g case in d condition.

Table 1 Summary of parameters corresponding to all cases in **Figure 17**. It presents the trend of σV_{th} for technology scaling. The nominal L_g cases are nominal gate lengths for each technology node. The equivalent oxide thickness for a, b, c and d are 2.4, 1.8, 1.2, and 1.2nm, respectively

	Channel doping (cm^{-3})	*Width (nm)*	*Nominal L_g (nm)*
a	$\sim 1 \times 10^{18}$	1000	55
b	$\sim 3 \times 10^{18}$	1000	35
c	$\sim 5 \times 10^{18}$	200	20
d	$\sim 5 \times 10^{18}$	20	20

select the most critical case (i.e., length = width = 20 nm) for comparison between simulation and measurement. **Figure 17(b)** shows the extracted $\sigma V_{th,Lg/LER}$ of (c) and (d) conditions. The deviation $\sigma V_{th,Lg/LER}$ contains contributions from the mean gate-length deviation and the LER. In our experimental data, the $\sigma V_{th,Lg/LER}$ increases as the $(WL)^{-0.5}$ increased, and it has a similar trend, compared with $\sigma V_{th,RD}$. **Table 1** summaries the corresponding parameters for all cases, as shown in **Figure 17**. **Figure 18(a)** shows the extracted mobility versus the doping concentration from samples of the cases (a) and (b), as shown in **Figure 17(a)**. The mobility model used here can generate mobility that is in good agreement with the extracted mobility, as shown in **Figure 18(a)**. The low-field electron mobility at 0.3 MV cm^{-1} is greatly reduced with increasing doping concentration. That is why we limit our channel doping concentration around 5×18 cm^{-3}. Less channel doping concentration may reduce $\sigma V_{th,RD}$, but channel dopants will quickly approach single-digit numbers, as shown in **Figure 18(b)**.

Figures 19(a)–19(c) show the I_{on}, I_{off}, and V_{th} distributions as a function of channel dopants. From the random dopant-number point of view, the equivalent channel doping concentration increases when the dopant number increases, this substantially alters the threshold voltage and the on- and off-state currents, as shown in **Figures 19(a)–19(c)**, respectively. The random dopant-position induced different fluctuation of characteristics in spite of the same number of dopants. Furthermore, the magnitude of the spread characteristics increases as the number of dopants increases. **Figure 20(a)** shows the I_{on}–I_{off} characteristics of the 125 cases from **Figure 11**. **Figures 20(b)–20(d)** disclose three different discrete dopant channels which have similar values of I_{on} or I_{off} but with various dopant distributions. Their corresponding cross-sectional off-state electrostatic potential and on-state current density at 1 nm below the gate oxide are also presented, shown in **Figures 20(b′)–20(d′), and 20(b″)–20(d″)**, which clearly show that distributions of the electrostatic potential and current density are closely

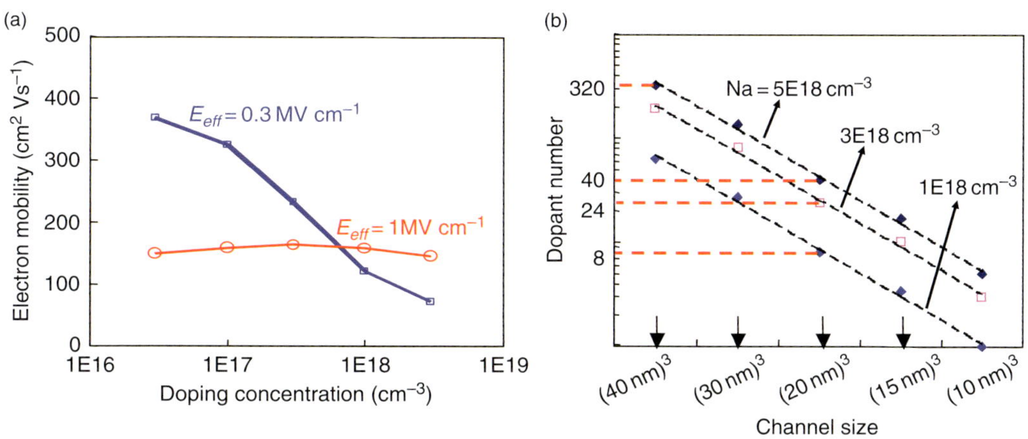

Figure 18 (a) Extracted nonstrain mobility vs. doping concentration at 0.3 and 1 MV cm^{-1} vertical field; (b) scaling of average channel dopant numbers vs. channel size.

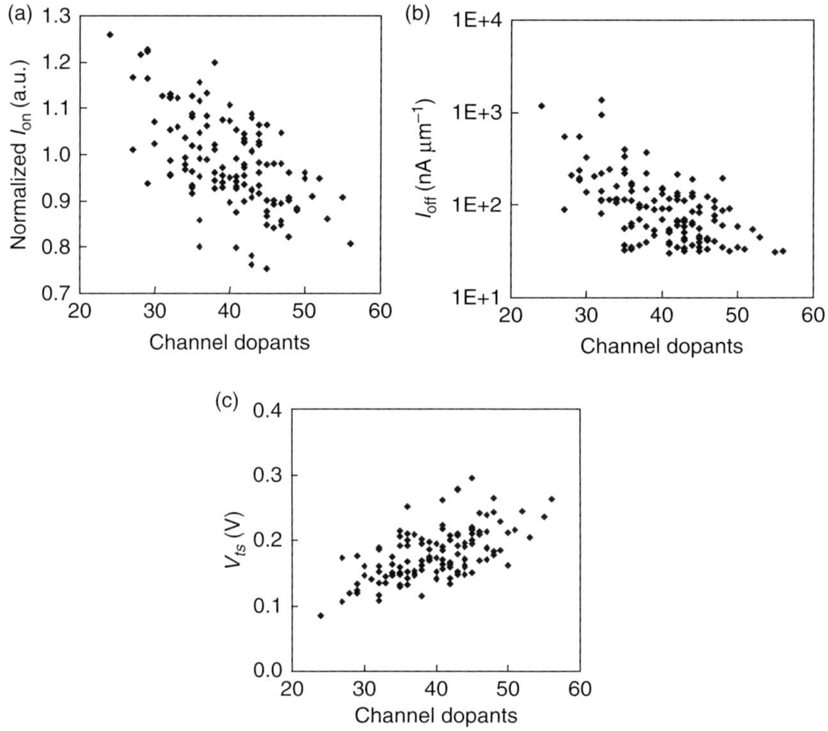

Figure 19 Distributions of (a) I_{on}, (b) I_{off}, and (c) V_{th} vs. channel dopants for the 125 discrete dopant 20-nm-gate transistors ($L_g = W = 20$ nm) shown in **Figure 11(e)**.

related to the dopant arrangements within the cross-sectional area beside source side, as shown in **Figures 20(b)–20(d)**.

Based on the experimental data and discrete modeling of 20-nm gate with 1.2-nm EOT, 0.8-nm EOT seems an effective way for 15-nm-gate CMOS to mitigate the increase of the random dopant-induced V_{th} variation. With the same approach, as shown in **Figure 11**, for generating discrete dopant channels, **Figure 21(a)** shows 343 subcubes of $(15\ \text{nm})^3$ derived from $(105\ \text{nm})^3$ cube with $5 \times 10^{18}\ \text{cm}^{-3}$ doping. **Figure 22** shows the I_{on}–I_{off} characteristics and V_{th} distribution of these cases with 1.2- and 0.8-nm EOT. The case of 0.8-nm EOT shows tighter V_{th} scattering. Furthermore, as channel dopants could be only 7 at 3σ edge, shown in **Figure 21(c)**, we herein propose using a higher work-function gate to increase its intrinsic electrostatic potential barrier height, as shown in **Figure 23(c)**, to prevent source-to-drain punch through at off-state, as shown in **Figure 23(b)**. Thus $\sigma V_{th,RD}$ can be maintained while L_g scaling down to 15 nm from 20 nm, as summarized in **Table 2**. It has been known that the σV_{th} is proportional to the oxide thickness [21, 22, 76]; that is,

$$\sigma V_{th} = \frac{q t_{ox}}{\varepsilon_{ox}} \sqrt{\frac{N_A W_d}{4LW}}$$

where ε_{ox} is the permittivity of the gate oxide and W_d is the width of the depletion layer under the gate. In the examination for 15-nm-gate CMOS, when the EOT changed from 0.8- to 1.2-nm EOT (1.5 times smaller), the $\sigma V_{th,RD}$ was reduced from 54 mV to 41 mV (1.32 times smaller), which conformed to the expression of σV_{th}. Although the advanced devices, such as double-gate or surrounding-gate structures [21–27, 72], or epitaxial channels [22, 75, 76] can reduce the fluctuation, these approaches are much more complicated and still require EOT's improvement to some degree for well suppression of σV_{th}.

6.3.2 Outlook for Sub-20-nm Planar Bulk MOSFETs

The ITRS roadmap has forecasted the change of device for the gate length between 20 and 16 nm. Therefore, the study of characteristic fluctuations in 16 nm is crucial. The characteristic fluctuations of a 16-nm-gate device with nominal channel doping

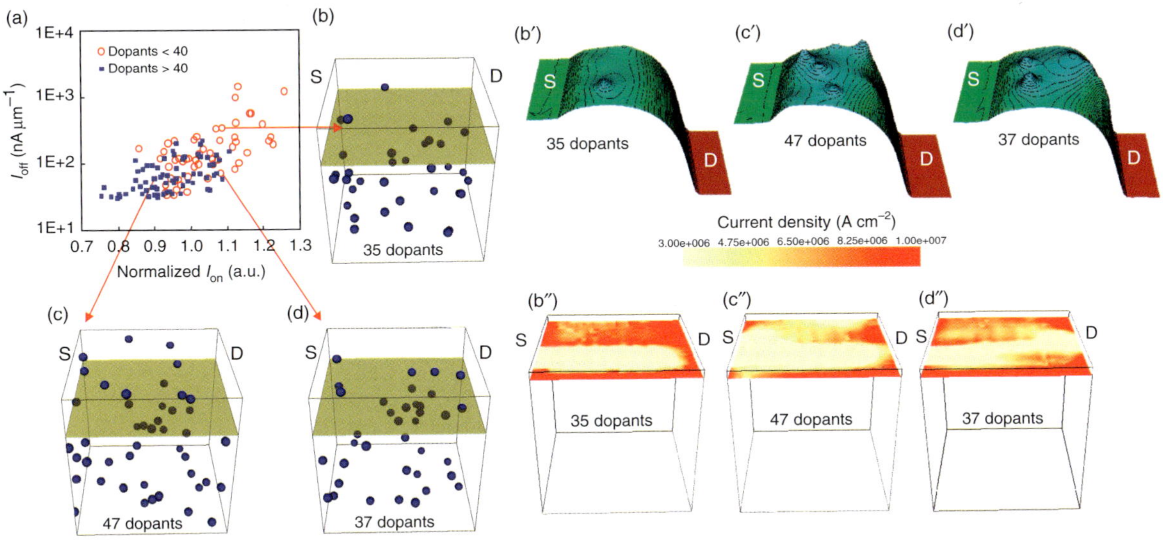

Figure 20 (a) I_{on}–I_{off} characteristics of the 125 discrete dopant 20-nm-gate transistors ($L_g = W = 20$ nm). (b) and (d) represent two cases of channel doping with similar I_{on} but different I_{off}; (c) and (d) represent two cases of channel doping with similar I_{off} but different I_{on}. The corresponding off-state ($V_d = 1.0$ V, $V_g = 0.0$ V) potential contours and on-state ($V_d = 1.0$ V, $V_g = 1.0$ V) current density of (b), (c), and (d) are shown in the **Figures 20(b′), 20(c′), 20(d′)** and **20(b″), 20(c″), 20(d″)**, respectively. All cross-sectional figures of off-state potential contours and on-state current density distributions are extracted at 1 nm below 1.2-nm EOT gate oxide.

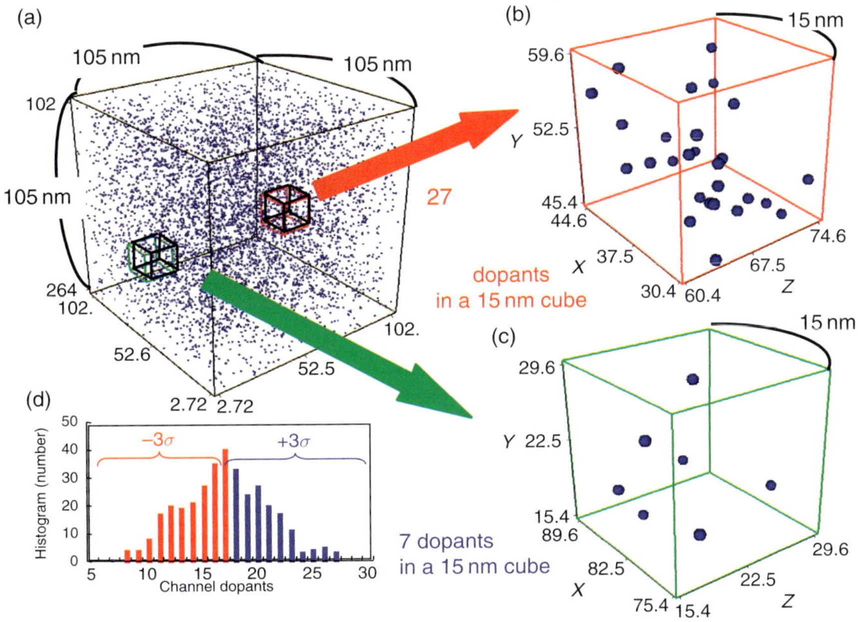

Figure 21 The 5×10^{18} cm^{-3}-doped (105 nm)^3 cube (a) with 343 subcubes of (15 nm)^3. Dopants inside the subcubes range from 7 (c) to 27 (b) with the average number of 17 and one standard deviation of 4 (d).

concentration of 1.48×10^{18} cm^{-3} is thus examined. These devices have a gate-oxide thickness of 1.2 nm with a mid-bandgap metal gate material. The source or drain and background doping concentrations are 1×10^{20} cm^{-3} and 1×10^{15} cm^{-3}, respectively. WLOG, 758 dopants are randomly generated in an

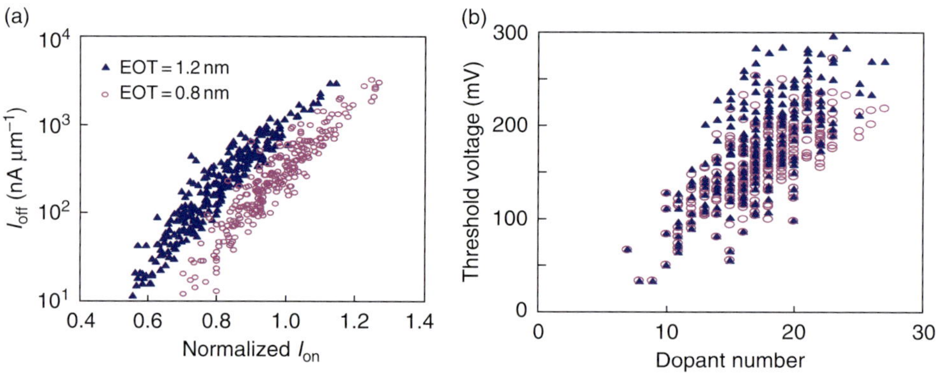

Figure 22 (a) I_{on}–I_{off} characteristics and (b) V_{th} vs. channel dopants of discrete doped 15-nm-gates with EOT = 1.2 nm (solid triangle) and 0.8 nm (open circle). Gate work function is 4.22 eV in the simulation.

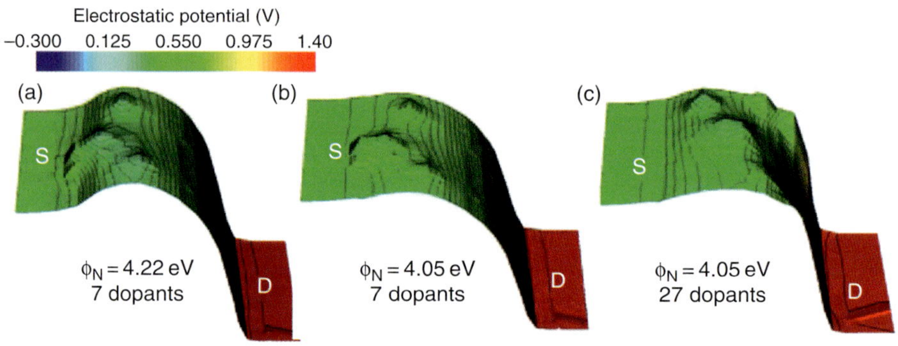

Figure 23 Cross-sectional off-state electrostatic potential contours of two extreme cases in 15-nm channels (W = 15 nm and EOT = 0.8 nm) with (a) 27 dopants/Φ_N = 4.05 eV, (b) 7 dopants /Φ_N = 4.05 eV, and (c) 7 dopants/Φ_N = 4.22 eV; all at 1 nm below the gate oxide.

Table 2 Summary of discrete dopant-fluctuated 20-nm- and 15-nm-gate planar CMOS transistors. The channel doping concentrations of the transistors are 5×10^{18} cm^{-3}

L_g(nm)	Width (nm)	Data source	EOT (nm)	$\sigma V_{th,RD}$ (mV)
20	200	Measurement	1.2	17
	20	Measurement	1.2	40
	20	Simulation	1.2	39
15	15	Simulation	1.2	54
	15	Simulation	0.8	41

(80 nm)^3 cube, yielding an equivalent doping concentration of 1.48×10^{18} cm^{-3}. The (80 nm)^3 cube is then partitioned into 125 subcubes of (16 nm)^3. The number of dopants varies from 0 to 14, and the average number is six. These 125 subcubes are then mapped into the channel region of the device for the 3D atomistic device simulation including discrete dopants. **Figure 24(a)** plots the I_d–V_g characteristics of 16-nm-gate planar NMOSFETs with discrete dopant fluctuations; the solid line represents the nominal case (continuously doped channel with a doping concentration of 1.48×10^{18} cm^{-3}) and the dashed lines are random dopant-fluctuated devices. The on-state currents, off-state currents, and threshold voltage are plotted in **Figures 24(b)–24(d)**, respectively. Each line and symbol in **Figures 24(a)–24(d)** indicates its DC characteristic for each device. The threshold voltage is determined from a

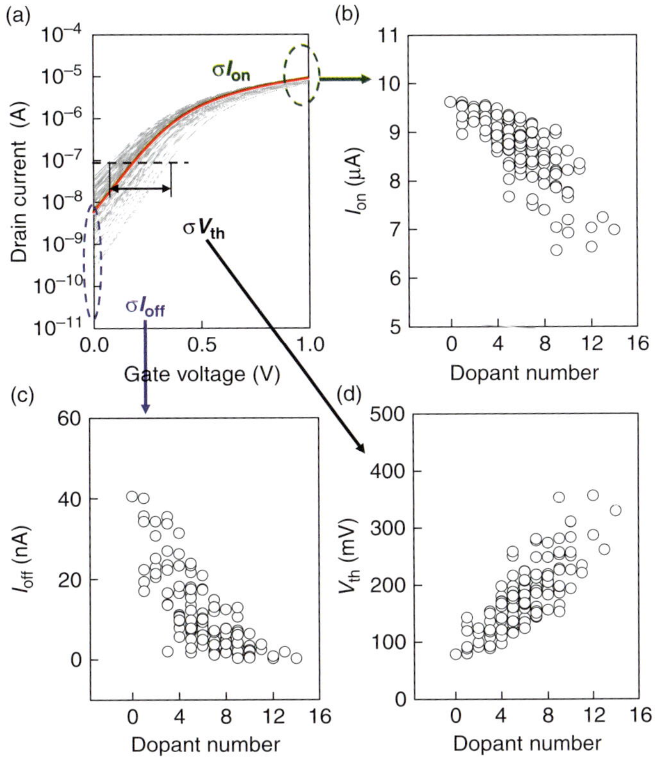

Figure 24 DC characteristic fluctuations of (a) I_d–V_g characteristics, (b) I_{on}, (c) I_{off}, and (d) V_{th} of the discrete dopant-fluctuated 16-nm-gate planar MOSFET. The solid line in I_d–V_g curves shows the capacitance of the nominal case and the dashed lines are random dopant-fluctuated devices.

current criterion that the drain current larger than $10^{-7}(W/L)$ ampere. From the random dopant-number point of view, the equivalent channel doping concentration increases when the dopant number increases, this substantially alters the threshold voltage and the on- and off-state currents. Additionally, the random dopant-position induces different fluctuation of characteristics in spite of the same number of dopants, as marked in the inset of **Figure 24(d)**. Furthermore, the magnitude of the spread characteristics increases as the number of dopants increases. Notably, the maximum and minimum V_{th} are achieved for this specific set of 125 randomized channels and would be different (larger range), if a larger number of samples were taken. The detailed physical mechanism is described in previous Section 6.3.1.

Figure 25(a) plots the capacitance–voltage (C–V) characteristics of 16-nm-planar NMOSFETs with discrete dopant fluctuations. The solid and dot lines are the cases with and without random dopant-position effect, respectively. The cases without random dopant-position effect are simulated by changing their channel doping concentration continuously from 1.0×10^{15} to $3.4 \times 10^{18}\,\mathrm{cm}^{-3}$. **Figure 25(b)** plots the slope of the C–V curves, in which the gate capacitance is fixed to 5×10^{-18} F. The dashed line indicates the cases without random dopant-position effect, and the symbols are the cases with random dopant-position effect. The slope of the C–V curve for the cases without random dopant-position effect is nearly independent of doping concentration, which implies the lateral shift of the C–V curves. The lateral shift of gate capacitance results from the variation of V_{th}, and can be described by the corresponding parameters in compact models. However, the slopes of C–V curves are substantially altered as the random dopant-position effect is taken into consideration, as shown in **Figure 25(b)**. The variation of the slopes of C–V curves indicates the change of shape of C–V curves, as illustrated in **Figure 25(a)**. The variation of C–V curves results from the randomness of dopant position in the depletion region of channel and therefore is hard to describe in current compact models [64–67, 77]. To the best of our knowledge, the fluctuation in gate capacitance has not yet been modeled and a coupled device-circuit simulation must be

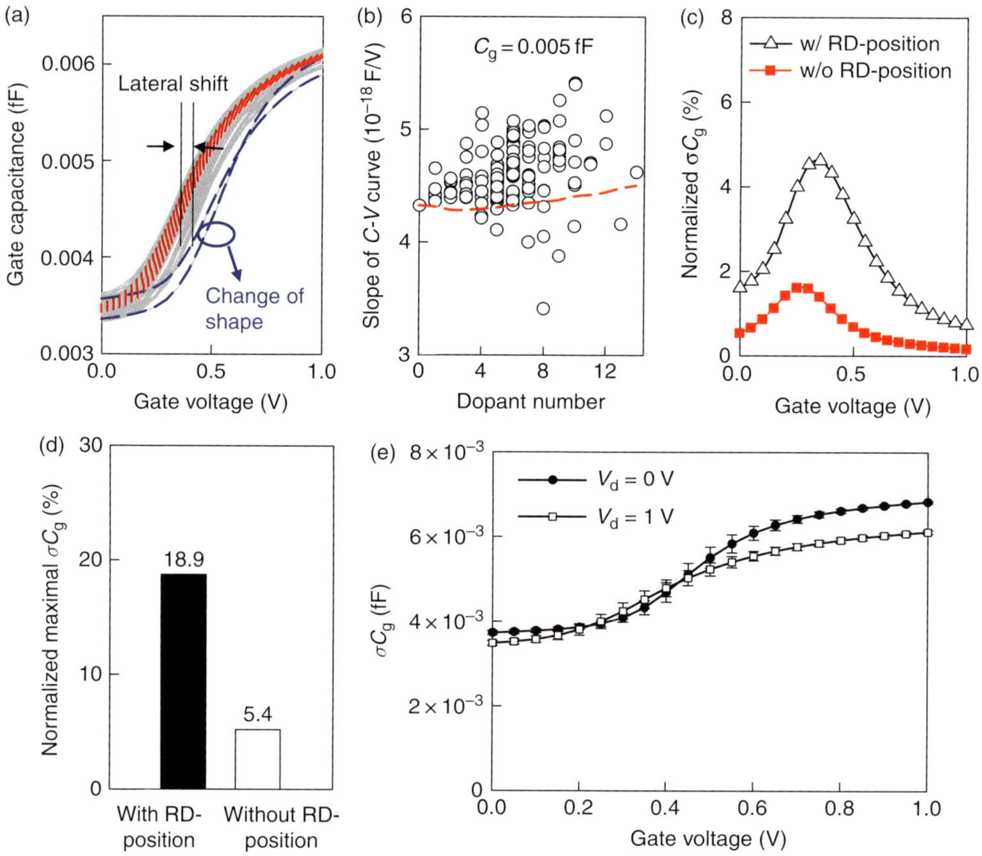

Figure 25 (a) *C–V* curves; (b) the slope of the *C–V* curves for cases with and without taking random dopant-position effect into consideration, where the solid line shows the nominal case and the dashed lines are random dopant-fluctuated devices. The solid and dot lines in (a) are the cases with and without random dopant-position effect, respectively. The dash line in (b) indicates the cases without random dopant-position effect, and the symbols are the cases with random dopant-position effect. The (c) normalized gate capacitance fluctuation and (d) maximum gate capacitance fluctuation are calculated. (e) The C_g fluctuation with different drain bias.

performed to estimate its impact on circuit characteristics. **Figure 25(c)** plots the normalized C_g fluctuation as a function of gate bias. The C_g fluctuations are normalized by the nominal C_g. The result implies the importance of random dopant-position effect. Moreover, the device under subthreshold operation suffers from the largest fluctuation. For devices with high V_g, the screening effect of the inversion layer of the device screens the variation of electrostatic potential and decreases the fluctuation of gate capacitance [77]. The normalized maximum variations of $C_g$are summarized in **Figure 25(d)**, in which the normalized maximum variation of C_g is about 18.9%. The neglect of the random dopant-position effect may underestimate the C_g fluctuation by a factor of five. **Figure 25(e)** shows the C_g fluctuation with different drain bias. The device with high drain bias has less gate capacitance fluctuation due to the pinch-off effect and smaller gate-to-drain capacitance at a high drain bias. The intrinsic cut-off frequency and intrinsic gate delay of the studied device are shown in **Figures 26(a)** and **26(b)**, respectively. The insets give the definition of these characteristics. As the number of dopants in the device channel is increased, the depletion width is decreased, which then increases the gate capacitance. With the decreasing transconductance and increasing gate capacitance, the intrinsic cutoff frequency of the device decreases with increasing dopant number. The intrinsic gate delay is increased as the dopant number increases due to the decreasing on-state current and increasing gate capacitance. The fluctuations of cutoff frequency and intrinsic gate delay are 8.2 GHz and 0.069 ps, respectively, and the magnitude of fluctuation is increased as dopant number increases. The

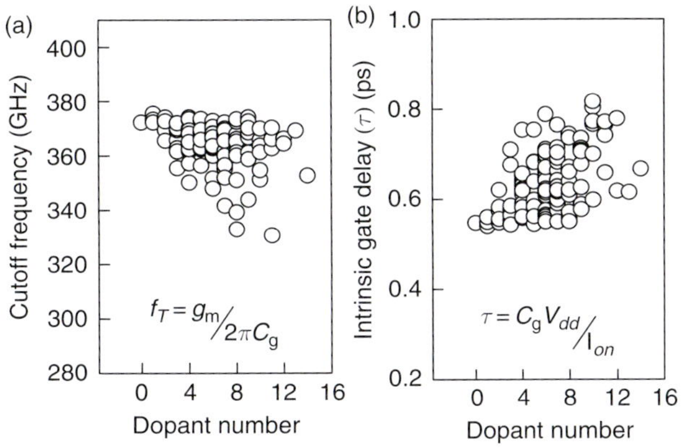

Figure 26 (a) Cutoff frequency; (b) intrinsic gate delay of transistor for the discrete dopant-fluctuated 16-nm-gate planar MOSFETs.

discrete dopant effect not only causes fluctuations in V_{th} and current, but also affects the gate capacitance of the transistor.

6.3.3 Fin-Type Field Effect Transistors

Nanoscale devices with vertical-channel structures, such as bulk FinFETs, are of great interest due to the great control of SCE [51–56]. Here, we explore the fluctuations of threshold voltage roll-off in nanoscale bulk FinFETs by a 3D statistically full-scale atomistic device simulation technique. The explored devices are of three different dimensions: $(16\,\text{nm})^3$, $(22\,\text{nm})^3$, and $(30\,\text{nm})^3$.

In fabricating the nanoscale FinFETs, the minimum dimension of the device has transferred from the gate length to the fin thickness. Top of the fin is formed to a round shape naturally and the fin bottom is not actually rectangular for the lithography and silicon-etching processes, as shown in **Figure 27(a)**. These nonideal processes will result in a wider fin bottom with respect to the fin top; thus it leads to a slanted edge of the channel fin [51–56]. Such a geometric derivation degrades device performance and raises serious SCEs, such as a large SS, low ratio of the on- and off-state currents, and large DIBL. Electrical characteristics of 25-nm-gate round-top-gate FinFETs on silicon wafers are numerically explored first, as shown in **Figure 27(b)**, where the oxide thickness, the fin width, and the fin height are fixed at 1.2, 20, and 50 nm, respectively. The employed physical model has been calibrated with experiment data. For

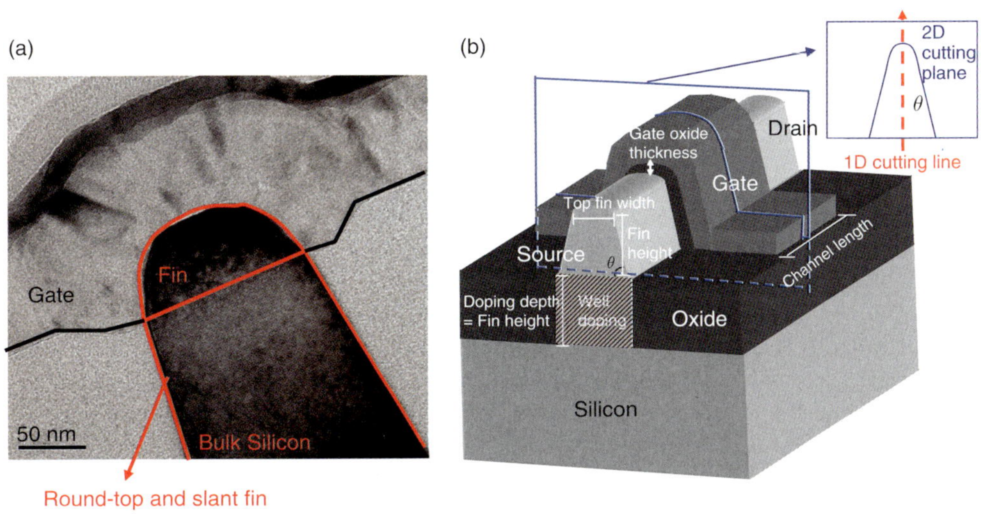

Figure 27 3D schematic plots for the bulk FinFETs. The symbol θ is the fin angle and the inset shows a 2D cutting-plane extracting from the center of channel.

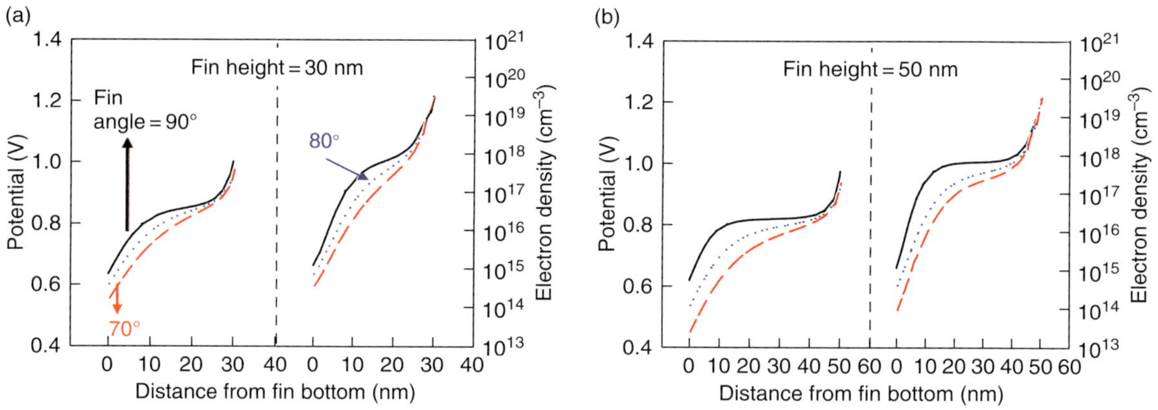

Figure 28 Plots of 1D cutting lines of the potential and electron density at the center of channel. The device is with the fin height of (a) 30 nm; (b) 50 nm, where the solid line is the result of 90°, the dotted line is for the 80°, and the dashed line is for the 70°. The circled windows indicate the regimes where nearly constant potential and electron density are occurred in device channel.

$V_d = V_g = 1.0$ V, the distributions of potential and electron density for device with the fin height = 30 nm (see **Figure 28(a)**) and 50 nm (see **Figure 28(b)**) are extracted according to the 1D cutting line, as shown in the inset of **Figure 27**. The plots are for devices with three different angles of fin-taper: 70°, 80°, and 90°. Nonuniform distributions of the potential and electron density along the longitudinal direction are observed. This is a direct result of donor's and bias' impacts on the channel along the direction of the fin top to bottom. The potential attains the minimum value at the fin bottom, which increases rapidly toward the fin top. For the same distance, the larger potential associated with the larger fin taper angle is perceived from the fin bottom. The electron density is reduced when the distance (from the top fin) is broadened; similarly, it has the minimum value at the fin bottom. For the device with the fin height = 30 nm, shown in the left plot of **Figure 28(a)**, the case of 90° (i.e., the solid lines) exhibits the most uniform distribution of the potential in the center of channel. The potential decreases more quickly with a more slanted fin angle, which leads to a stronger longitudinal electric field and may degrade the device performance. Similar results can be observed for the cases of fin height 50 nm. Moreover, it is found that the variation of potential remains nearly constant within the longest distance inside the channel region. The potential variation is significant when the fin angle is decreased, and the case of 70° (the dashed lines) possesses the largest potential variation among three angles. The variation is even more appreciable when the fin height is increased. For the different fin angles and fin heights, the right plots of **Figures 28(a)** and **28(b)** disclose the variation of the electron density versus the distance from the fin bottom. The variation of this physical quantity predicts the same trend as depicted above. It is found that the device with an approximately ideal fin angle (i.e., the fin angle approaches 90°) behaves with the most uniform distribution of the examined physical quantities within the device channel. Plots of the drain current versus the gate voltage for the three fin angles and two fin heights are shown in **Figure 29(a)**. The case for the fin height of 30 nm, as shown in the left plot of **Figure 29(a)**, allows a larger variation of the fin angle, compared with the result of the fin height = 50 nm. It means that the fin height of 30 nm maintains a highest ratio of the on- and off-state currents and implies better performance. The parameters of SCE versus the fin angle are calculated accordingly, as shown in **Figure 29(b)**. A larger taper angle is necessary for fabrication of nanoscale bulk FinFETs to obtain robust electrical characteristics. Besides, **Figure 29(b)** implies that a smaller fin height is essential for the device of 25-nm-gate length. The calculated SS suggests that an increase of the fin angle will result in better SS; in particular, for those silicon fins with higher fin height, which improve the SCE remarkably. Similarly, we also find the evidence for the DIBL. Consequently, for the cases with a small fin angle, the device may not have an acceptable performance. In addition, for the bulk FinFET with a high fin height, the situation gets worse simultaneously. The fin height and the fin angle are the two critical limiting factors when the dimension of the bulk FinFET is continuously scaled down. **Figures 30(a)** and **30(b)** plot the SS

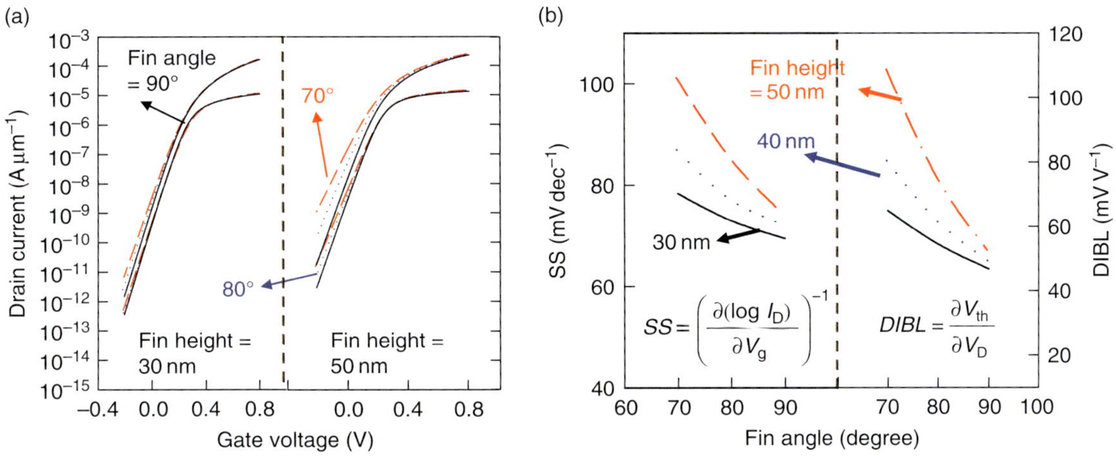

Figure 29 (a) The characteristics of the I_d– V_g curves for the device with respect to different fin angles and heights. The solid lines are the result of 90°, the dotted lines are for the 80°, and the dashed lines are for the 70°; (b) Results of SS (the left plot) and DIBL (the right one) vs. the fin angle, where the solid lines are the result of 30 nm, the dotted lines are for the 40 nm, and the dashed lines are for the 50 nm.

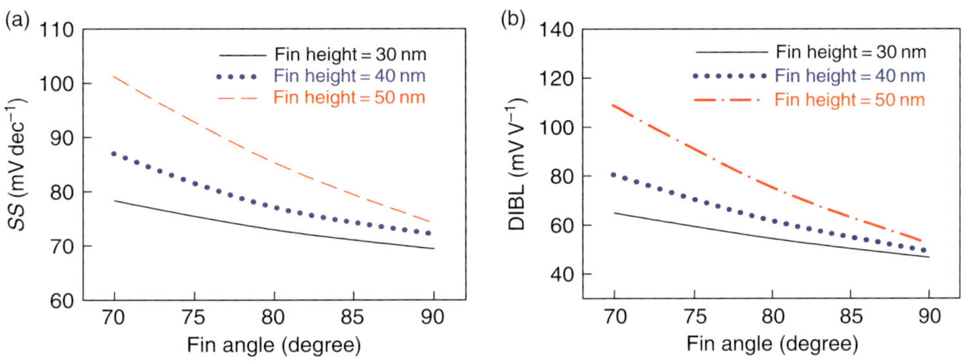

Figure 30 (a) SS; (b) DIBL as a function of fin angle, in which the fin height varies from 30 to 50 nm.

and DIBL as a function of fin angle, in which the fin height varies from 30 to 50 nm. The critical angles for the case of SS <75 mV dec^{-1} and DIBL <75 mV V^{-1} are then estimated. It is found that the critical angles when fin height or top fin width = 1.5, 2, and 2.5 nm, the critical angles are 71.9°, 79.3°, and 87.4°, respectively. For devices with a large ratio between the fin height and the top fin width, the controllability of manufacturing the fin taper angle is exceptionally important and much more efforts on processing should be made. A nearly rectangular-shaped fin is only crucial for the device with a higher fin height. The following discussion of characteristics of FinFETs is based on an assumption that a FinFET has a fin aspect ratio equal to one.

Figure 31(a) shows the large-scale statistically computed results of $I_d$as a function of V_g for the bulk FinFET device with L_g of 16 nm, where the red solid line presents the nominal case (continuous doping profile: 1.48×10^{18} cm^{-3}) and the dotted lines show the discrete dopant fluctuated cases. The spread of the I_d–V_g curves indicates the magnitude of the effect of discrete dopants. The characteristics of fluctuation are further explored with respect to the I_{on}, I_{off}, and V_{th}, as shown in **Figures 31(b)–31(d)**, respectively. From the random dopant-number point of view, the equivalent channel doping concentration increases as the dopant number increases; this substantially alters the V_{th}, I_{on}, and I_{off}. Moreover, it is observed that even for devices with the same numbers of dopants inside the channel, the effect of random dopant position induces different fluctuations of characteristics in spite of there being the same number of dopants. Furthermore, the

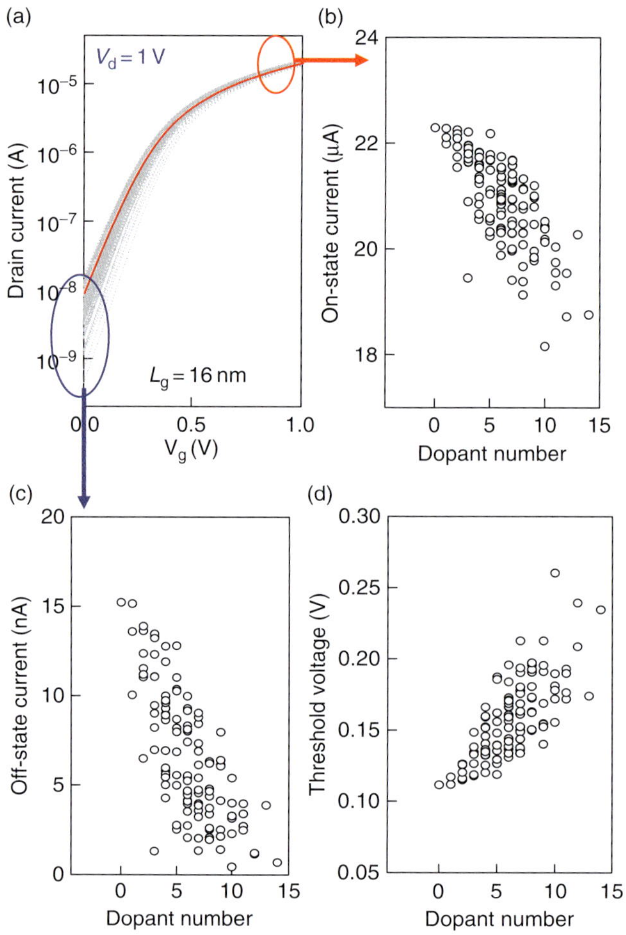

Figure 31 (a) The characteristics of $I_d - V_g$, where the solid line indicates the continuous (i.e., the nominal) case. The fluctuations of (b) on-state current, (c) off-state current, and (c) threshold voltage as a function of the number of dopants.

magnitude of the spread characteristics increases as the number of dopants increases. To explore the random dopant-position-induced V_{th} fluctuation, the on-state potential distributions of the 16-nm-gate bulk FinFETs with six discrete dopants inside the channel (average concentration: $1.48 \times 10^{18}\,cm^{-3}$), as shown by the rectangle in **Figure 31(d)**, and the nominal case (continuous doping profile: $1.48 \times 10^{18}\,cm^{-3}$) are investigated and shown in **Figure 32**, where the potential distributions are 2 nm below the top and lateral gates. We note that the V_{th} is determined from the current criterion that $I_d \geq 10^{-7}(W/L)$ A. Because of the lack of discrete dopants located near the channel surface, the device with minimal V_{th}, as shown in **Figure 32(a)**, exhibits a larger potential distribution than that in the nominal case, as shown in **Figure 32(b)**. Conversely, in a device with maximal V_{th}, as shown in **Figure 32(c)**, there are four dopants located near the surface of the channel. The corresponding potential distributions near these dopants are significantly decreased, which changes the electron-conducting path. The random dopant position induces rather different fluctuations of characteristics in spite of there being the same number of dopants. The results for the studied devices with the same doping concentration of $1.48 \times 10^{18}\,cm^{-3}$ are summarized in **Table 3**. The maximum differences in V_{th} induced by the discrete dopant-position effect for 30-, 22-, and 16-nm-gate lengths are then normalized with respect to their nominal V_{th}, as plotted in **Figure 33**. The result shows a significant increase in discrete dopant-position-induced V_{th} fluctuation in nanoscale transistors.

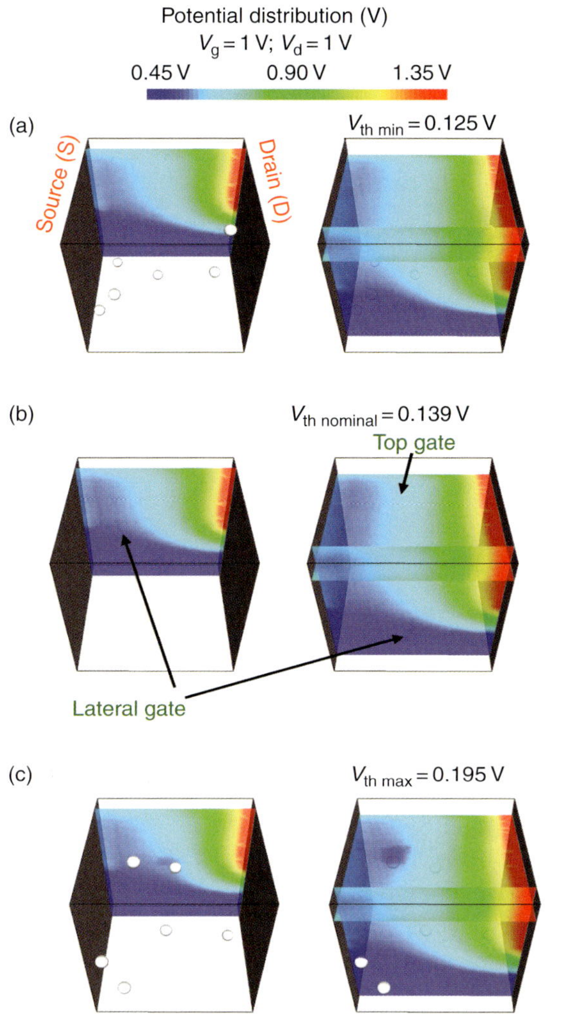

Figure 32 The on-state potential distributions of the 16-nm-gate bulk FinFETs with six discrete dopants inside the channel (average concentration: $1.48 \times 10^{18}\,\mathrm{cm}^{-3}$), as shown in the rectangle of **Figure 31(d)**, and the nominal case (continuous doping profile: $1.48 \times 10^{18}\,\mathrm{cm}^{-3}$). The potential distributions are 2 nm below the top and lateral gates. The potential distributions for the device with (a) minimal threshold voltage, (b) nominal threshold voltage, and (c) maximal threshold voltage.

Table 3 Threshold voltage of the devices studied with equivalent doping concentration of $1.48 \times 10^{18}\,\mathrm{cm}^{-3}$

Channel Size (nm³)	*Nominal V_{th} (V)*	*Maximum V_{th} (V)*	*Minimum V_{th} (V)*
30 × 30 × 30	0.226	0.268	0.211
22 × 22 × 22	0.192	0.225	0.168
16 × 16 × 16	0.139	0.195	0.125

Figure 34 shows the fluctuations of V_{th} for the 30-, 22-, and 16-nm-gate bulk FinFETs. All V_{ths} of the studied devices are normalized by their corresponding nominal V_{th}. As bulk FinFETs shrink in size, the V_{th} scales with an increasing fluctuation of V_{th}. In addition to the fluctuation of V_{th}, the fluctuation of current is also significant. **Figure 35** shows the I_{on}–I_{off} characteristic fluctuations. For cases with similar I_{on}, the maximum fluctuation of I_{off} increases significantly as the gate length is scaled down. The plot of V_{th} roll-off and fluctuation of V_{th} against the gate length of bulk FinFETs is shown in **Figure 36**. The V_{th} fluctuations of the 22- and 16-nm-gate bulk FinFETs are 1.28 and 1.61 times larger than that of the 30-nm-gate device. Fluctuation of V_{th} for bulk FinFETs and planar MOSFETs are presented in **Figure 37**. For the gate length of 16 nm, the V_{th} fluctuation of FinFETs is only half that of planar MOSFETs. Moreover, as gate length is reduced from 30 to 16 nm, the fluctuation of V_{th} is increased by 11 mV, which is better than the increase in the case of planar MOSFETs, 27.3 mV. With varying variations of W and L, the fluctuation of V_{th}, $\sigma_{v_{th}}$, is given as

$$\sigma_{v_{th}} \approx (WL)^{-\gamma} \qquad (13)$$

where γ is 0.5 for the planar MOSFETs [21, 22, 76]. The dependence factor of threshold voltage, $(WL)^{-\gamma}$, is then plotted as a function of γ, as shown in **Figure 38(a)**. As γ decreases, the difference between 60 and 16 nm decreases. The difference is greater than one order for $\gamma = 1$ and less than one order for $\gamma = 0.5$. The dependence on device dimension is eliminated at $\gamma = 0$, which means that the V_{th} fluctuation will be constant regardless of the size of the device. Assume the fluctuation of V_{th} for the 60-nm-gate device to be 1 mV. **Figure 38(b)** shows the increase in V_{th} fluctuation as the device becomes smaller. The fluctuation of V_{th} decreases as γ decreases. Based on this relationship and the aforementioned results, the γ factor of the bulk FinFETs on device dimension is then fitted to be 0.25. For the device with 16-nm-gate length, the fluctuation of device with $\gamma = 0.5$, as shown in **Figure 38(b)**, is about two times larger than that of a device with $\gamma = 0.25$, which is similar to the results shown in **Figure 37**.

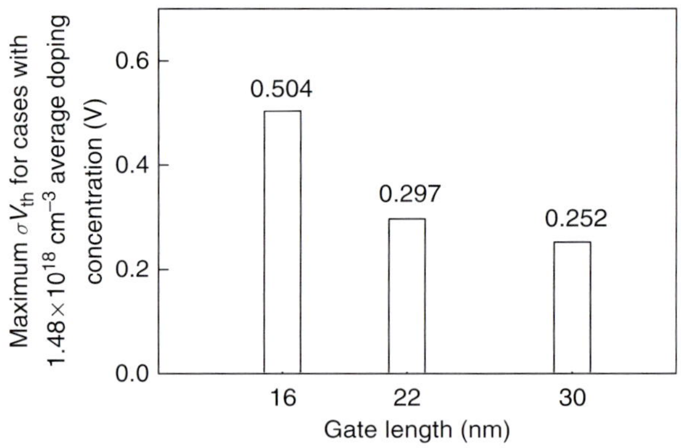

Figure 33 For cases with $1.48 \times 10^{18}\,cm^{-3}$ average channel doping concentration, the maximum differences in threshold voltage induced by discrete dopant-position effect for 30-, 22-, and 16-nm-gate lengths, which are normalized by the corresponding nominal threshold voltage.

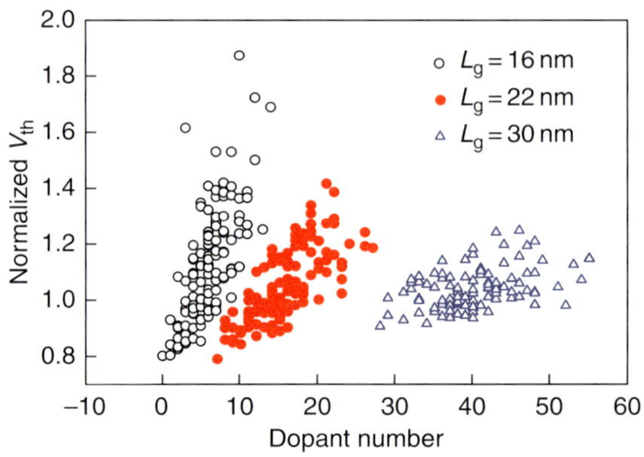

Figure 34 Fluctuations of threshold voltage for the 30-, 22-, and 16-nm-gate bulk FinFETs. The threshold voltages of the studied devices are normalized by the corresponding nominal threshold voltage.

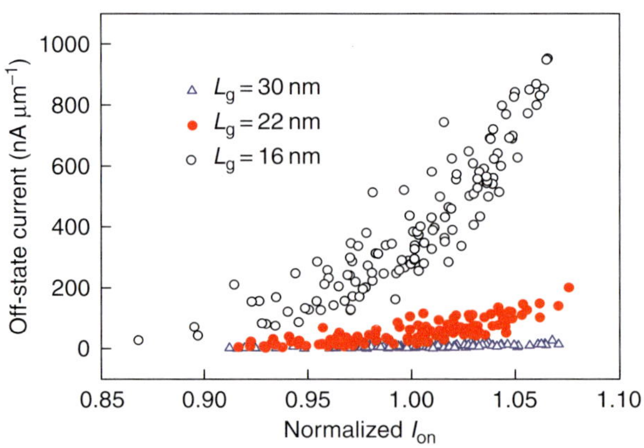

Figure 35 On-off state current characteristics of the 30-, 22-, and 16-nm-gate bulk FinFET devices.

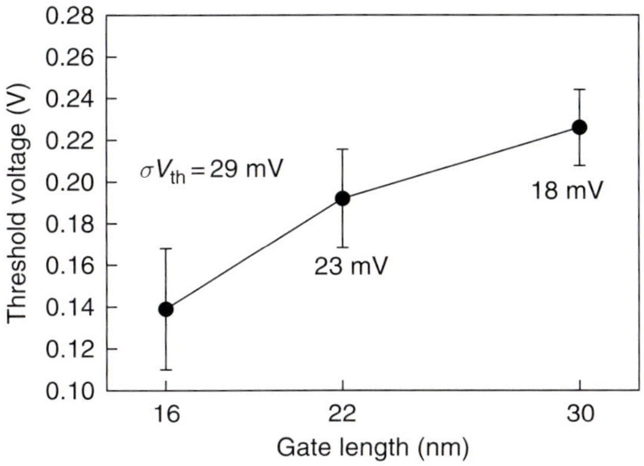

Figure 36 Plot of threshold voltage roll-off for the 30-, 22-, and 16-nm-gate bulk FinFETs, where the bars indicate the fluctuation of threshold voltage.

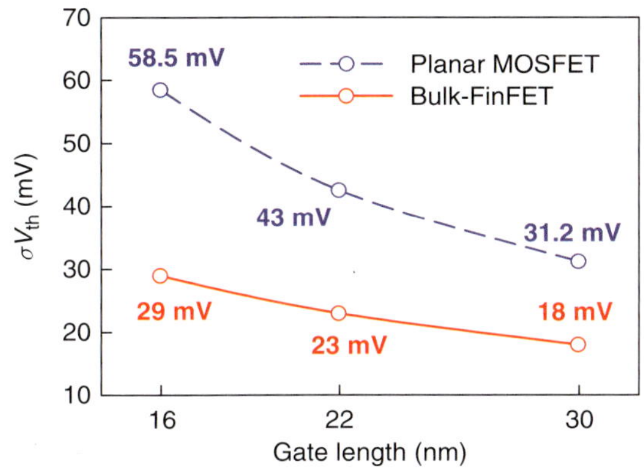

Figure 37 Fluctuations of threshold voltage roll-off (i.e., vs. the L_g) of bulk FinFETs and planar MOSFETs.

Figure 39(a) shows the characteristics of I_{on}–I_{off}. For cases with similar I_{on}, the maximum fluctuation of I_{off} is within 5 times. For a 16-nm-gate conventional planar MOSFET, our calculation shows that the fluctuation of I_{off} is 10 times larger than that of bulk FinFET. **Figures 39(b)–39(d)**, **39(b′)–39(d′)**, and **39(b″)–39(d″)** disclose three different discrete dopant channels, having very similar values of I_{on} or I_{off} but with various dopant positions. The cross-sectional views (both the top- and lateral view) on-state current density and off-state electrostatic potential at 1 and 5 nm below the gate oxide are examined. **Figures 39(b)–39(d)** and **39(b')–39(d')** show the top and lateral views of the distribution of the on-state current. For the top gate, the three different cases of the dopant, shown in **Figures 39(b)–39(d)**, imply that bulk FinFET has very similar current-conducting paths. However, the lateral-view plots, shown in **Figures 39(b′)–39(d′)**, exhibit different current distribution due to an impurity appearing on the surface of the lateral gate, as shown in **Figure 39(c′)**. The different current-conducting paths that locate at the lateral gate result in different on-state currents even if we have very similar off-state currents, as shown in **Figure 39(a)**. For the device having very similar on-state (or off-state) current with different off-state (or on-state) situations, **Figures 39(b″)–39(d″)** are plots of the off-state potential distribution at 5 nm below gate oxide. As shown in **Figure 39(a)**, the device possesses very similar on-state current (the vertical ellipse circled), but with different off-state current (>3 times) resulting from the different

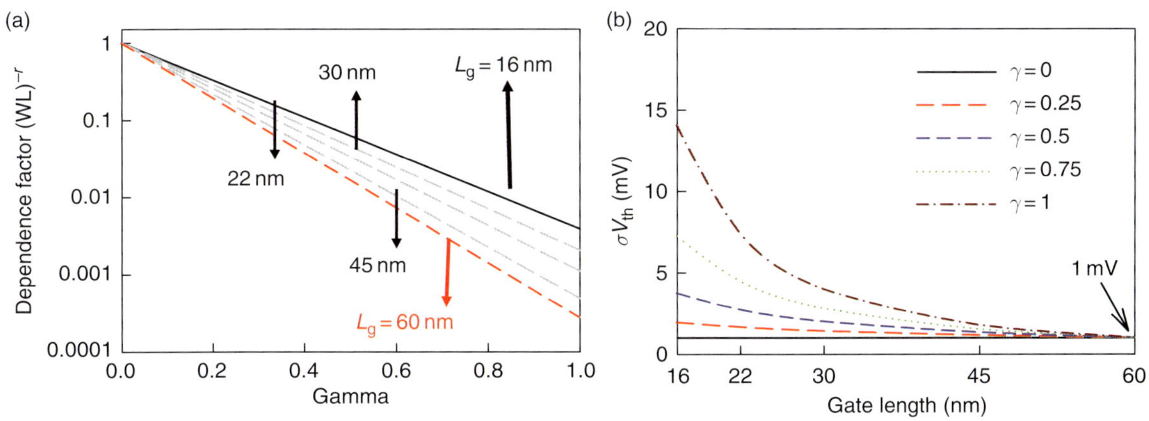

Figure 38 Threshold voltage fluctuation as a function of device dimension. (a) Plots of dependence factor on γ for various device dimensions. (b) Fluctuation of threshold voltage for the 60-nm-gate device was assumed to be 1 mV.

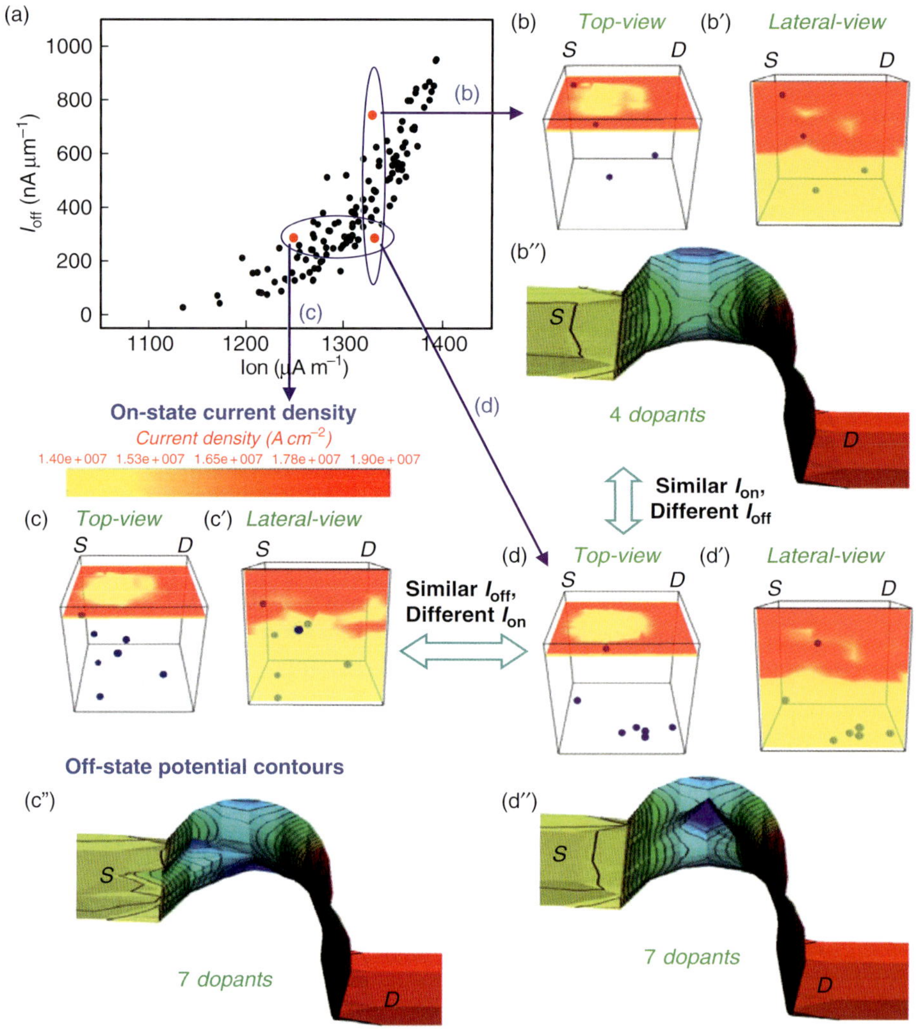

Figure 39 (a) The characteristics of I_{on}–I_{off} of the 125 discrete dopant 16-nm-gate bulk FinFETs. Three cases are selected to evaluate similar I_{on} but different I_{off} (plots of (b) and (c)) and similar I_{off} but different I_{on} (plots of (c) and (d). Plots of (b)–(d) and (b′)–(d′) show the corresponding top and lateral views of the on-state current density, which are extracted at the 1 nm below the gate oxide, respectively. Plots of (b″)–(d″) show the off-state potential contours.

randomness of the dopant number and position, shown in **Figures 39(b″)** and **39(d″)**. Furthermore, the magnitude of the spread characteristics increases as the number of dopants increases.

The dependence of the threshold voltage fluctuation of bulk FinFETs with device dimension was empirically fitted as $\sigma_{v_{th}} \approx (WL)^{-0.25}$, which is superior to that of planar MOSFETs, $\sigma_{v_{th}} \approx (WL)^{-0.5}$. The preliminary results indicate that the bulk FinFETs can not only provide better immunity against discrete dopant-induced threshold voltage fluctuations, but also show less sensitivity to device size than planar MOSFETs. The superior immunity against fluctuation and the more stable fluctuation of threshold voltage roll-off indicate that the bulk FinFET to be a promising device for the sub-16-nm era.

6.4 Silicon-on-Insulator Multiple-Gate Field Effect Transistor

SOI devices have recently been of great interest and relatively present better scaling-down properties than that of conventional bulk MOSFETs. Double-gate SOI devices have been proposed and are found to have more attractive electrical characteristics than that of single-gate SOI devices [15–39]. They inherently have good suppression of SCEs, high transconductance, and ideal SS. To further reduce the SCEs, promising device structures – such as FinFETs – and surrounding-gate nanowire FETs have been proposed, fabricated, and demonstrated with fascinating device characteristics. Although the undoped (the channel doping concentration below 1×10^{15} cm^{-3}) SOI devices can prevent the random dopant-induced fluctuation, they may suffer from more significant SCEs. Thus, the channel doping must be employed to alter the threshold voltage and provide the promising electrical characteristics in today's semiconductor manufacturing processes. Consequently, characteristic fluctuation that is induced by discrete dopants is important in these nanoscale MOSFETs, which are with fascinating structures.

6.4.1 Multiple-Gate MOSFETs from Single to Quadruple

Here, we initially discuss the physical fluctuations that are induced by discrete dopants, and then proceed to study the characteristic fluctuations and mechanism of immunity against fluctuations of multiple-gate SOI devices. **Figure 40** displays the explored 16-nm-gate single-, double-, triple-, and surrounding-gate SOI devices. Discrete dopant-induced fluctuations, caused by local potential spikes, are determined by the corresponding dopants within the device channel. The potential spike alters not only the electric field and current-conducting path, but also the electron velocity, carrier mobility, and electron temperature distribution.

Figure 41 depicts, for example, discrete dopant-induced fluctuations of a triple-gate SOI transistor. All cross-sectional plots of the distributions of the on-state current density and the contours of the off-state potential are extracted 1 nm below the top-gate oxide. **Figures 41(a)** and **41(a′)** refer to the nominal case (continuous doping is assumed) and the discretely doped case, respectively. The on-state potential contour and current density distribution of the discretely doped case, shown in **Figures 41(b′)** and **41(c′)**, are investigated to elucidate the effect of discrete dopant on the potential and current distribution of the device. The potential spikes in **Figure 41(b′)** are associated with the corresponding dopants (spikes A–C) in **Figure 41(a′)**. The potential distribution near spike C does not vary significantly with structure of the triple-gate. Comparison of **Figures 41(c)** and **41(c′)** reveals that the current-conducting path is disturbed and impeded by dopants in the channel. Changing the number and position of discrete dopants in the channel causes a significant potential variation and change in current. The current may not pass through the high-potential barrier region but pass through the valley between two potential spikes (such as spikes A and B). Additionally, the local field distribution, shown in **Figure 41(d′)**, is also perturbed significantly by discrete dopants. In the on-state condition (the gate voltage $V_g = 1$ V; the drain voltage $V_d = 1$ V), since the potential distribution near the dopant is relatively negative in the channel, the dopant acts as a center of a whirlpool-like electric field to repel electrons. As electrons drift from source to drain, some of them encounter a negative electric

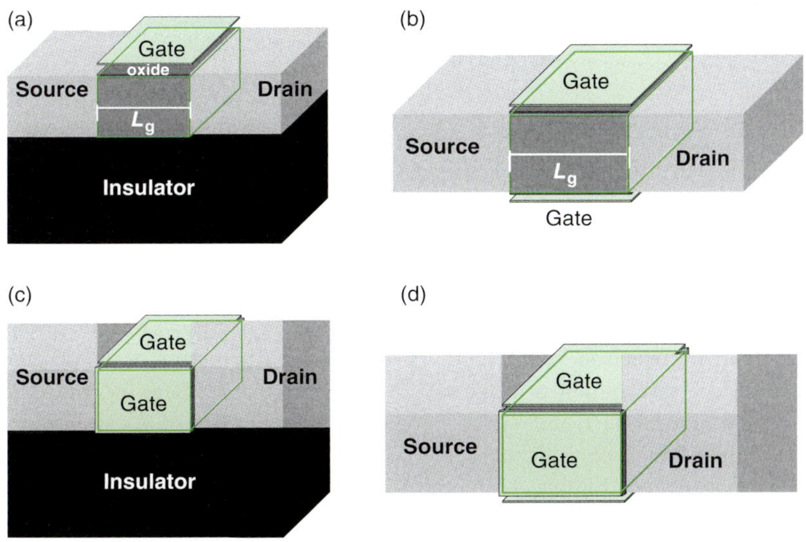

Figure 40 (a) single-gate; (b) double-gate; (c) triple-gate; and (d) quadruple surrounding-gate for the dopant position/number-sensitive device simulation.

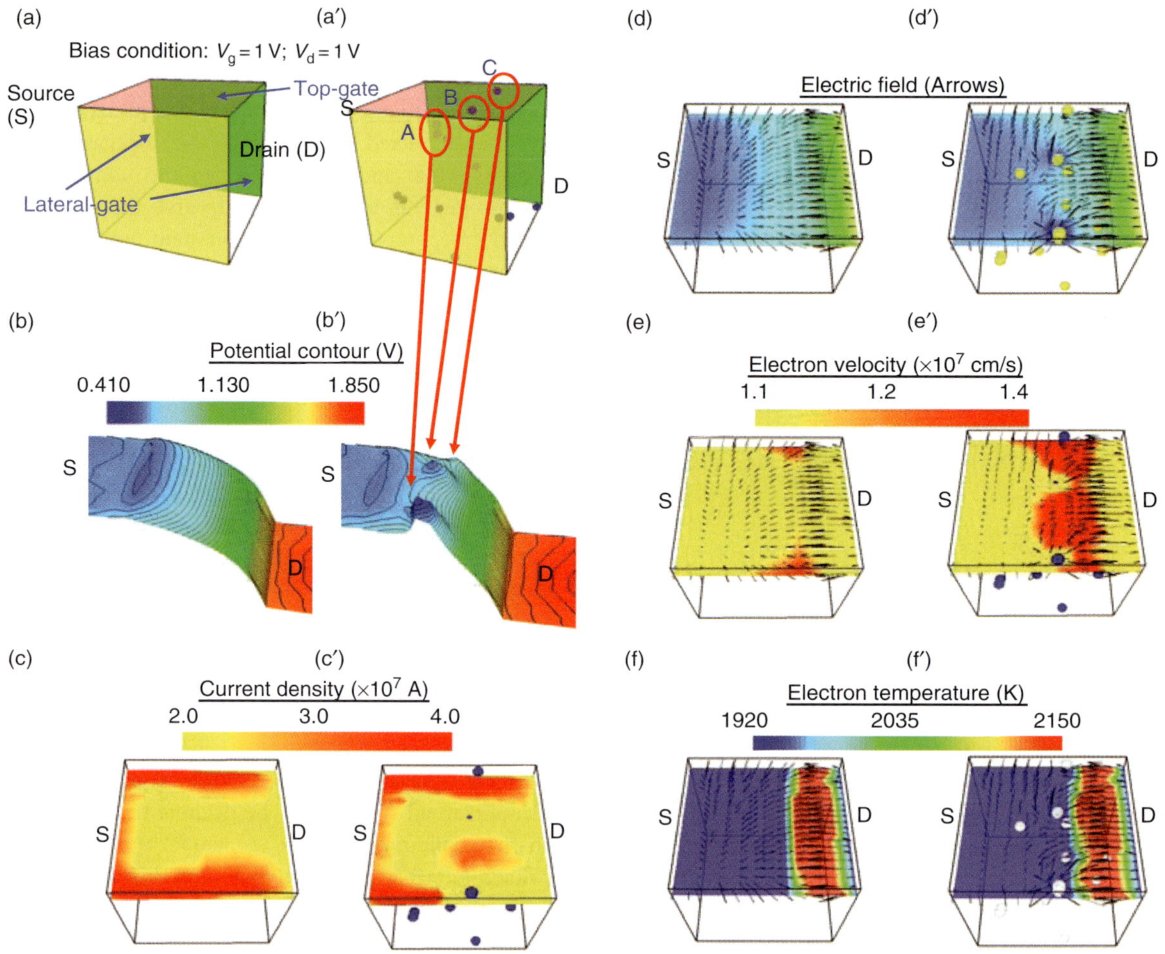

Figure 41 Comparison of the on-state potential contours ((b) and (b′)), the current density distributions ((c) and (c′)), the electric field ((d) and (d′)), the electron velocity ((e) and (e′)), and the electron temperature ((f) and (f′)) of the (a) the nominal case and (a′) discretely doped cases. The potential spikes (marked as A, B, and C) in (b′) are induced by corresponding dopants in (a′)). All cross-sectional plots of the on-state current density distributions and off-state potential contours are extracted at 1 nm below the top-gate oxide.

field that is induced by the dopant. The lateral electric field between the source and the drain, combined with the repulsion of the dopants, twists the electron field and increases the electron velocity near the dopant, as shown in **Figures 41**(e) and **41**(e′). This phenomenon explains why the distribution of the electron velocity remains the same to the left of the dopants at device's source. The fluctuation in the electron velocity also indicates a fluctuation of the distribution of electron temperatures, as plotted in **Figures 41**(f) and **41**(f′).

The characteristic fluctuations of multiple-gate SOI devices are also investigated. **Figure 42** plots fluctuations of I_D–V_G for single- and multiple-gate SOI devices. The solid lines represent the nominal case and the dashed lines represent the discretely doped cases. The spread of the I_D–V_G voltage curves shows the magnitude of the current fluctuation that is induced by discrete dopants. The result shows that the single-gate device exhibits much larger current fluctuations than the multiple-gate devices. **Figure 43** plots the corresponding on- and off-state current characteristics. For cases with similar on-state currents, the maximum fluctuation of the off-state current declines as the number of gates increases; moreover, the maximum fluctuation of I_{off} in multiple-gate devices is within 20 nA μm^{-1}, whereas the planar device (with a single gate) exhibits a much larger fluctuation of I_{off} (>2000 nA μm^{-1}). **Figure 44**(**a**) statistically compares the threshold voltage fluctuations of the aforementioned devices. As expected, the multiple-gate device

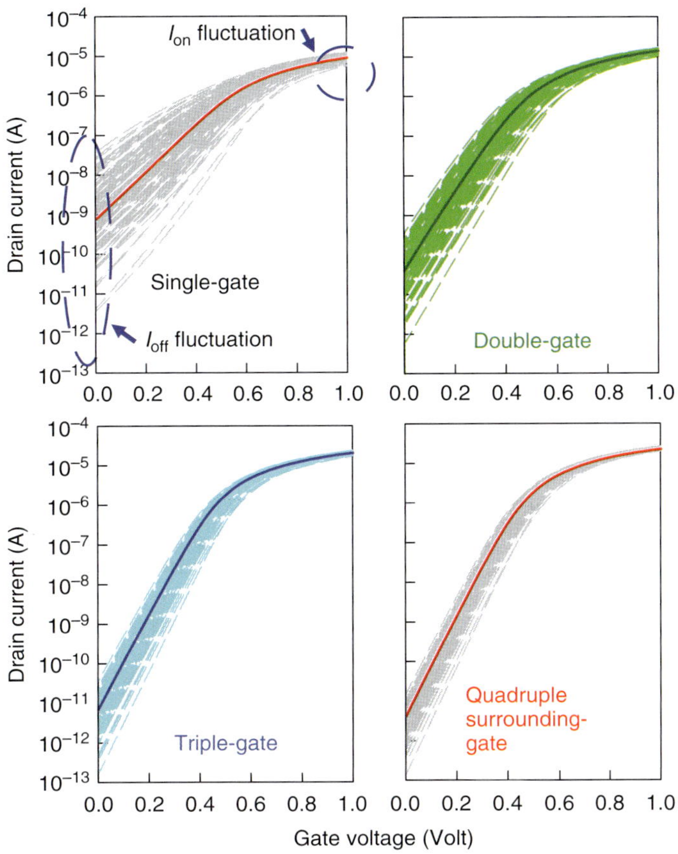

Figure 42 The I_d–V_g characteristics of the 500 discrete dopant-fluctuated 16-nm-gate single- and multiple-gate SOI devices, where the solid lines indicate the nominal case and the dash lines indicate the discretely doped cases.

exhibits better immunity against fluctuation and the magnitude of the V_{th} fluctuation (both SD and difference between maximum and minimum V_{th} declines as the number of gates increases, as plotted in **Figure 44(b)**. The SDs of single-, double-, triple-, and surrounding-gate devices are 102, 46.2, 30.9, and 25.5 mV, respectively. Calculations demonstrate that the fluctuations of V_{th} of the double-, triple-, and quadruple surrounding-gate devices are 2.2, 3.3, and 4 times smaller, respectively, than that of the planar SOI device. The equivalent channel doping concentration increases with the dopant number, substantially altering V_{th}, as shown in **Figure 44(a)**, and the on- and off-state currents, as plotted in **Figure 42**. The magnitude of spread increases as the number of dopants increases. The discrete dopant-position affects the fluctuation of characteristics in a different manner for a fixed number of dopants.

The mechanism of immunity of single- and multiple-gate devices against fluctuation is studied with reference to the extracted physical quantities. **Figures 45(b)–45(e)** plot the on-state potential along the center of the device's channel, shown in **Figure 45(a)**, and display 3D electric fields of single-, double-, triple-, and quadruple surrounding-gate devices, respectively. The potential and electric field of multiple-gate devices are more uniform than those of the planar device because the channel controllability is better than that of a planar device for a large gate-coverage ratio. **Figure 46** compares the on-state current density, plotted in **Figures 46(a)–46(d)**, and the contour of the off-state potential, shown in **Figures 46(a′)–46(d′)**, for the 16-nm-gate single-, double-, triple-, and quadruple surrounding-gate SOI devices, respectively. All cross-sectional plots of the on-state current density and the off-state potential are at 1 nm below the top-gate oxide. As the gate-coverage ratio is increased, the on-state current density is increased and the heights of off-state potential spikes (A–C in **Figure 46(a)**) are reduced. The lateral-gate structure of triple- and quadruple surrounding-gate devices effectively suppresses the potential (spike C), and

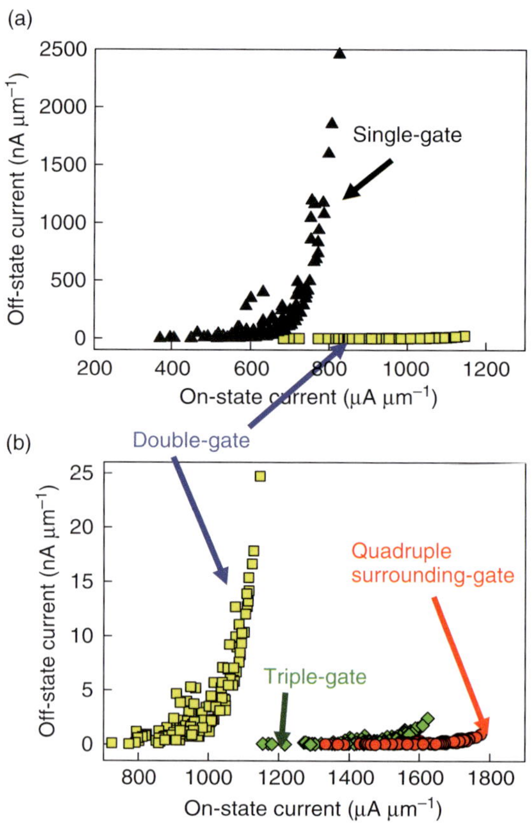

Figure 43 The on-off state current characteristics of the 500 discrete dopant-fluctuated 16-nm-gate single- and multiple-gate SOI devices. (a) Comparison between the single- and double-gate devices. (b) Comparison among the double-, triple-, and quadruple surrounding-gate devices.

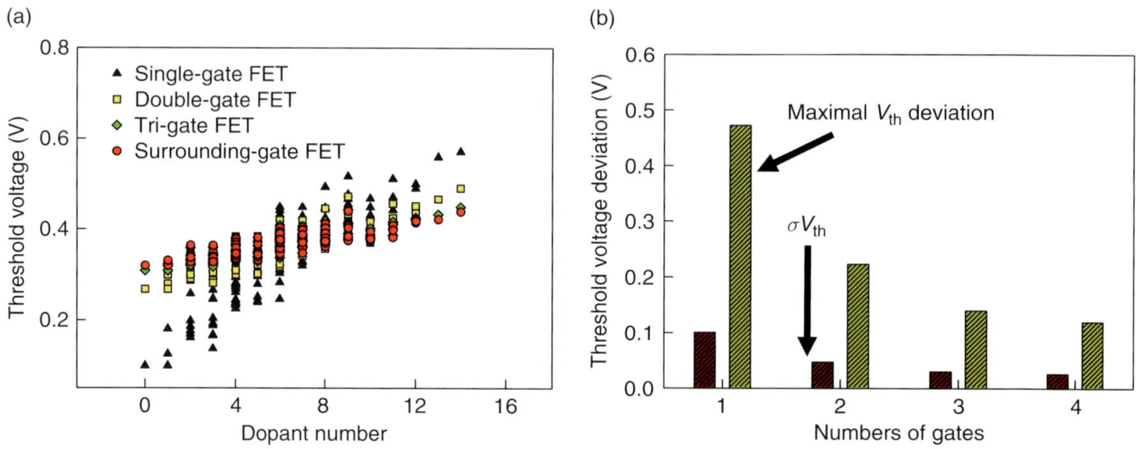

Figure 44 (a) Comparison of V_{th} fluctuation of the 16-nm-gate single- and multiple-gate SOI MOSFETs; (b) Plots of the standard deviation (SD) and the maximum difference of V_{th} with respect to different gate number.

then increases the on-state current at the lower side of the channel, shown in **Figures 46(c)** and **46(d)**. The lateral gate of multiple gate intrinsically enhances the controllability of dopant-induced fluctuations near the sidewall of the channel surface. It also accounts for why spike C in **Figure 2(a′)** has a weaker

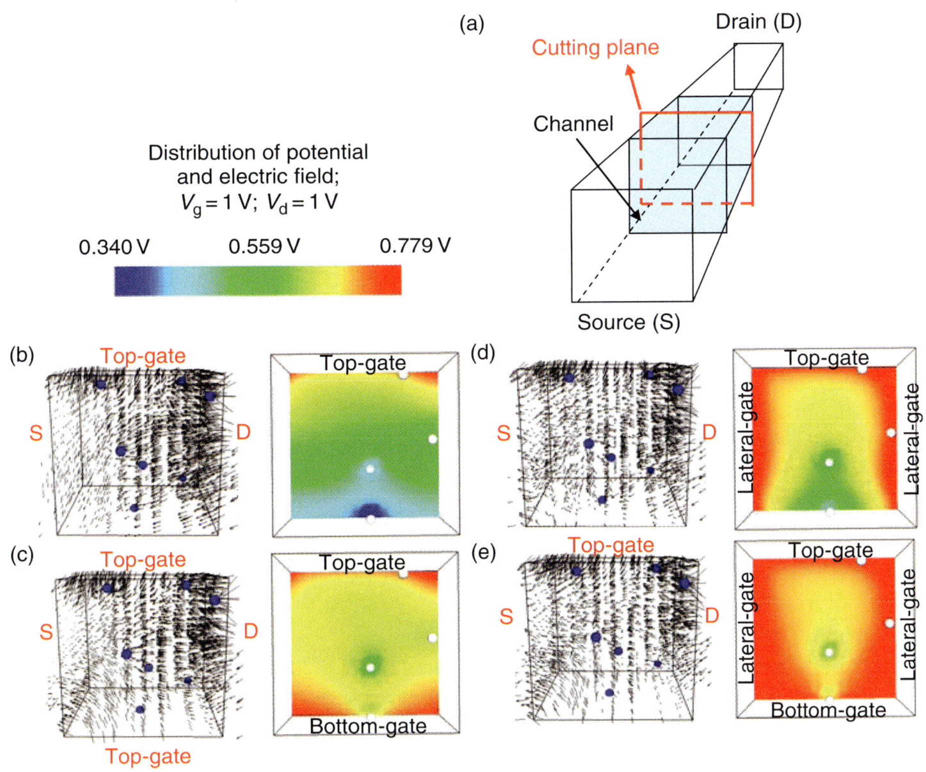

Figure 45 Plots of the on-state potential and 3D electric field distribution of the 16-nm-gate SOI MOSFET with the (b) single- (c) double-, (d) triple- and (e) quadruple surrounding-gate structure. The cross-sectional plots of the on-state potential distributions are extracted along the center of the device's channel, as shown in (a).

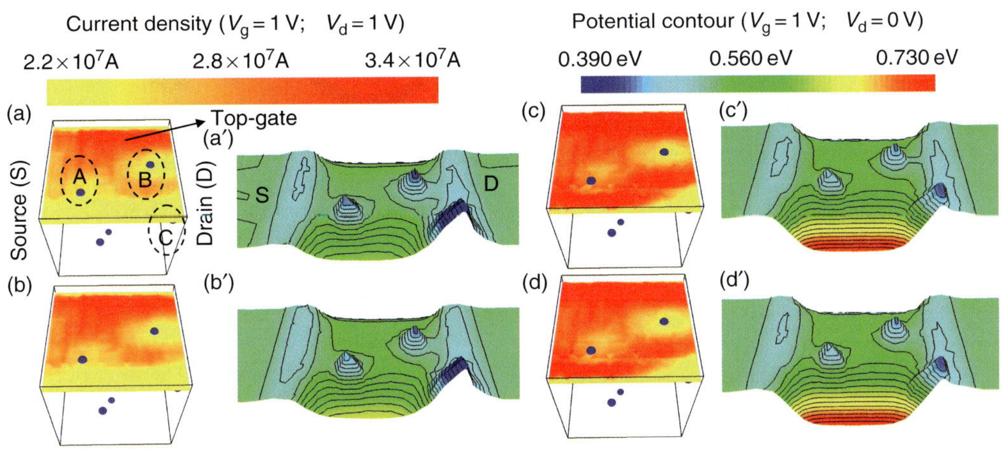

Figure 46 Plots of the on-state current density distributions ((a)–(d)) and the off-state potential contours ((a′)–(d′)) of the 16-nm-gate SOI MOSFETs ((a) and (a′) are for the single-gate; double-gate: (b) and (b′); triple-gate: (c) and (c′); quadruple surrounding-gate: (d) and (d′)). The potential spikes in (a′)–(d′) are induced by corresponding dopants in channel (spikes A–C shown in (a)). The height of potential spikes is decreased as gate number is increased. All cross-sectional plots of the on-state current density distributions and off-state potential contours are extracted at 1 nm below the top-gate oxide.

influence on the potential distribution than spike B. **Figure 47** shows the lateral- and bottom-gate on-state current densities of the structures. Similarly, all cross-sectional plots are obtained at 1 nm below the lateral- and bottom-gate oxides. Once the current paths are impeded by discrete dopants on parts of the channel

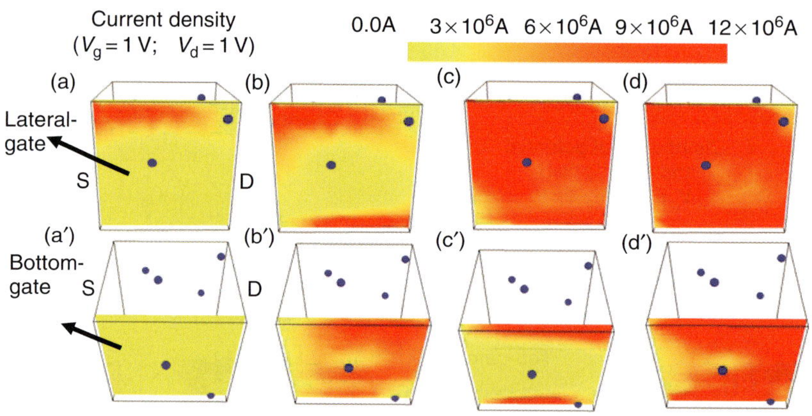

Figure 47 Plots of the lateral-side (a–d) and bottom-gate (a′–d′) on-state current density distributions of the 16-nm-gate devices ((a) single-gate; (b) double-gate; (c) triple-gate; (d) quadruple surrounding-gate). All cross-sectional figures are extracted at 1 nm below the lateral-side and bottom-gate oxides.

surface, the well-gate-controlled multiple-gate structure bridges the alternative conducting paths to prevent a significant fluctuation of the conduction current. Thus, the benefit of the superior vertical-channel structure is that multiple-gate devices suppress potential fluctuation and provide a more stable conduction current than the planar device. As a result, the immunity against fluctuation in multiple-gate SOI devices results mainly from the uniform potential distribution and the fact that the current conduction area (multiple paths) is larger than that of a planar SOI device.

6.4.2 Nanowire Transistor

Nanowire FinFETs are ultimate structures and potential candidates for next generation nanoelectronic devices [16–18, 29–39]. Due to the limitation of manufacturability, nanowire transistors with a perfect gate structure (i.e., a surrounding gate with 100% gate-coverage ratio) theoretically are not always guaranteed. Impact of the discrete dopants on device performance is crucial in determining the behavior of nanoscale semiconductor devices. The immunity of nanowire transistor against random discrete dopant-induced fluctuation may suffer from the variation of gate-coverage-ratio. Here, we first discuss the fluctuations of potential and current density due to discrete dopants; and then proceed to study the impact of nonideal nanowire gate-coverage ratio on immunity against discrete dopant-induced fluctuations.

Figures 48(a)–48(c) illustrates the 16-nm-gate silicon nanowire transistor with structures of surrounding-gate (i.e., 100% coverage), the omega-gate

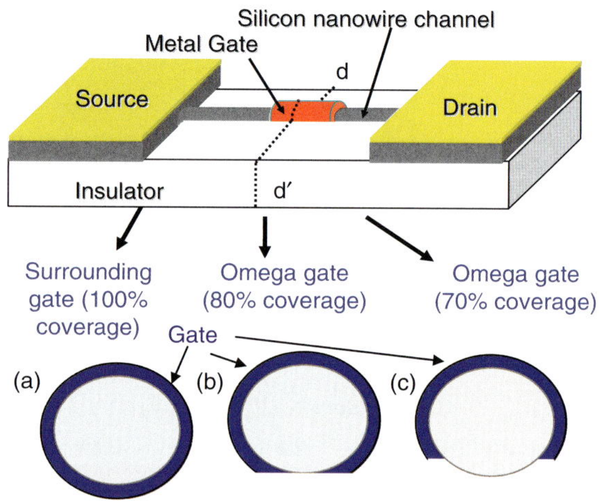

Figure 48 The 16-nm-gate silicon nanowire transistor with structures of: (a) surrounding-gate (i.e., 100% coverage); (b) the omega-gate with 80% coverage-ratio; and (c) the omega-gate with 70% coverage-ratio.

with 80% coverage-ratio, and the omega-gate with 70% coverage-ratio, respectively. The used physical models have been calibrated with experimental data. **Figure 49** shows the I_d–V_g and I_d–V_d characteristics of 5-nm-gate nanowire transistor, in which the symbols are the measurement data and the solid lines are the simulation results. The simulation result shows an agreement with measured data. The experimentally validated models are then used for discrete dopant simulation. The on-state (the device is with $V_d = V_g = 1$ V) potential distribution of the discretely doped case, shown in **Figures 50(a)** and **50(b)**, are investigated to elucidate the effect of discrete dopant on the potential and current distributions of the device.

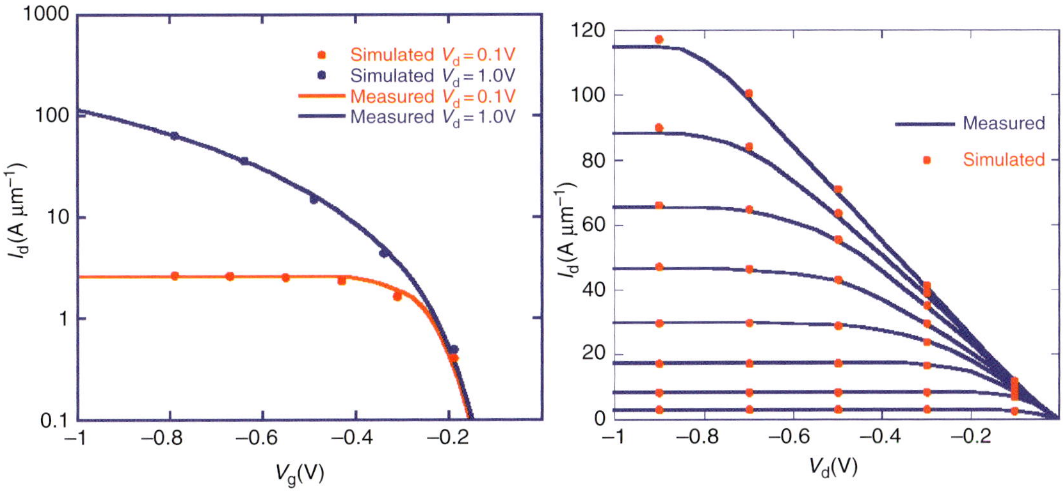

Figure 49 (a) I_d–V_g; (b) I_d–V_d characteristics have been well modeled using 5-nm-gate nanowire FinFET as an example.

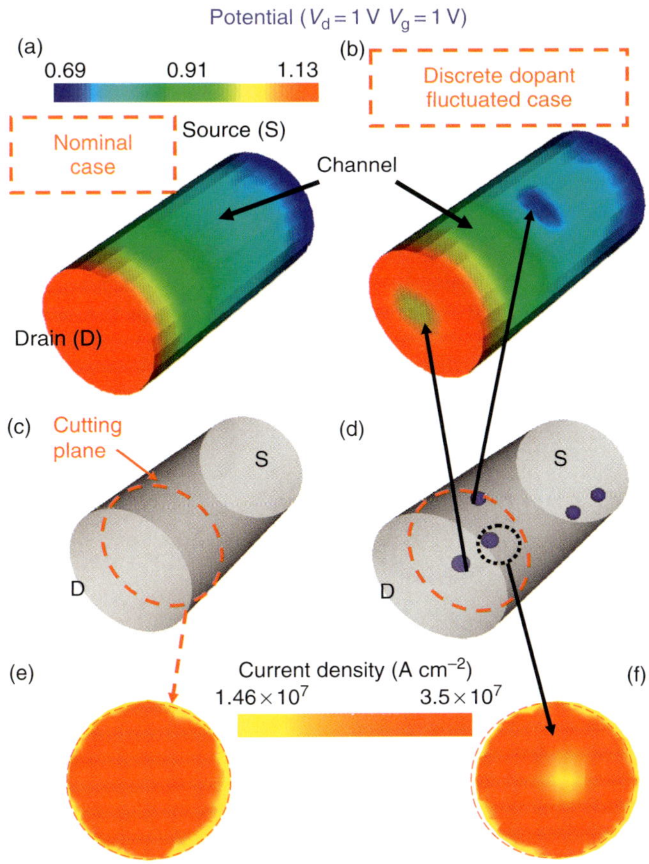

Figure 50 Comparison of the on-state potential (a, b), the current density distributions (e, f) of the (c) the nominal case, and (d) discretely doped cases. The potential fluctuations are induced by corresponding dopants in (d)).

The potential fluctuations, shown in **Figure 50(b)**, are associated with the corresponding dopants in **Figure 50(d)**. Comparison between **Figures 50(e)** and **50(f)** reveals that the current-conducting path is disturbed and impeded by the discrete dopants locating at the channel. Since the potential distribution near the dopant is relatively negative in the channel, the dopant acts as a center of a whirlpool-like electric field to repel electrons. As electrons transport from the end of the source to the drain side, some of them will encounter a negative electric field that is induced by the discrete dopants. The lateral electric field between the source and the drain, combined with the repulsion of the dopants, twists the electric field and increases the electron velocity near the dopants. The potential fluctuations not only alter the electric field and current-conducting path, but also the electron velocity and carrier mobility.

Figure 51 shows the comparisons of potential and current density distributions, respectively, for the nanowire transistors with the surrounding-gate (i.e., 100% gate-coverage-ratio) and the omega-gate (i.e., 80% and 70% gate-coverage-ratios) structures. The discrete dopants positioned in the channel induce a potential fluctuation and substantially disturb the current density distribution and the corresponding conduction path. The magnitude of potential and current density fluctuations is increased as the gate-coverage-ratio decreases. The nonideal gate-coverage disturbs the channel controllability of the explored nanowire transistors and thus decreases the immunity against discrete dopant-induced fluctuation. In our study, for a 16-nm-gate silicon nanowire transistor, the threshold voltage fluctuations of the omega-gate devices with 80% and 70% gate-coverage are 1.04 and 1.19 times larger than that of the surrounding-gate structure, as shown in **Figure 52**. The ratio of current fluctuations is 1.75 and 4.5 times larger than that of the surrounding-gate one, as shown in **Figure 53**. The results confirm the influence of nonideal gate-coverage on the disturbances of the channel controllability of nanowire transistors. The immunity against discrete dopant-induced fluctuation is thus decreased. **Figure 54** shows the characteristics of the on-state and off-state currents (I_{on} $-I_{off}$). For those cases with similar I_{on}, the maximum fluctuation of I_{off} is within 0.05 nA μm^{-1}. This figure discloses three different discrete dopant channels, having similar values of I_{on} or I_{off} but with various dopant positions. The cross-sectional on-state current density and off-state potential distributions extracted from the center of the channel are examined. Due to the difference of discrete dopant position, the different conduction paths of devices result in different on-state currents even if we have very similar off-state currents, as shown in **Figure 54**. For the device having very similar on-state current with different off-state situations, **Figure 55** shows the off-state potential distribution at device's channel. However, due to the effect of discrete dopant position, there is no potential barrier located in the channel region.

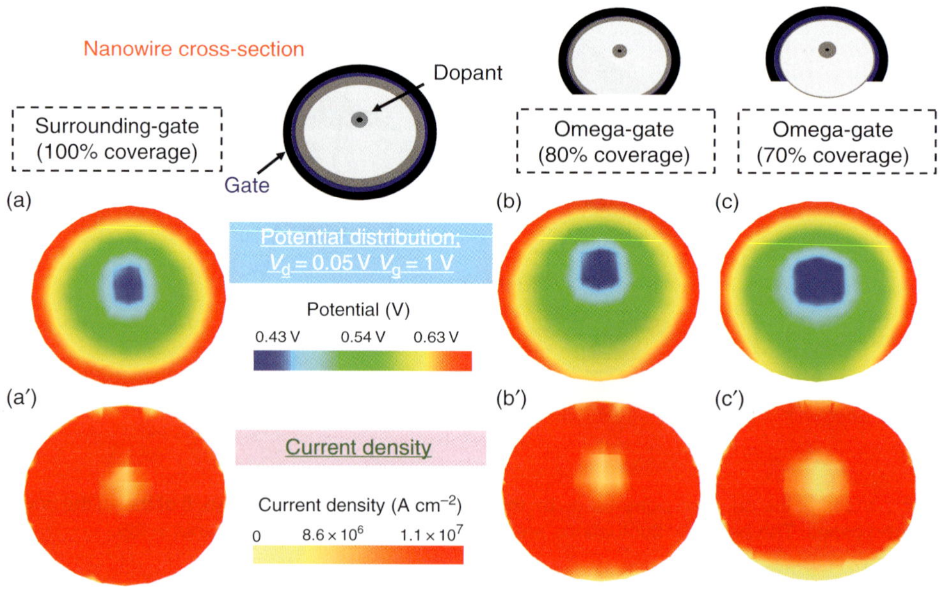

Figure 51 Comparison of potential (plots of (a′), (b′), and (c′)) and current density distribution (plots of (a″), (b″), and (c″)) in nanowire transistors with surrounding-gate (100% gate-coverage-ratio) and omega-gate (80% and 70% gate-coverage-ratio) structures.

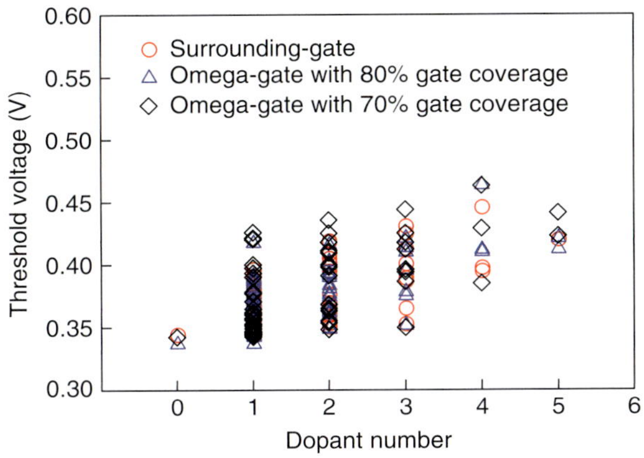

Figure 52 Comparison of the threshold voltage fluctuation of the 16-nm-gate silicon nanowire FET with surrounding-gate, omega-gate with 80%, and 70% gate-coverage-ratio.

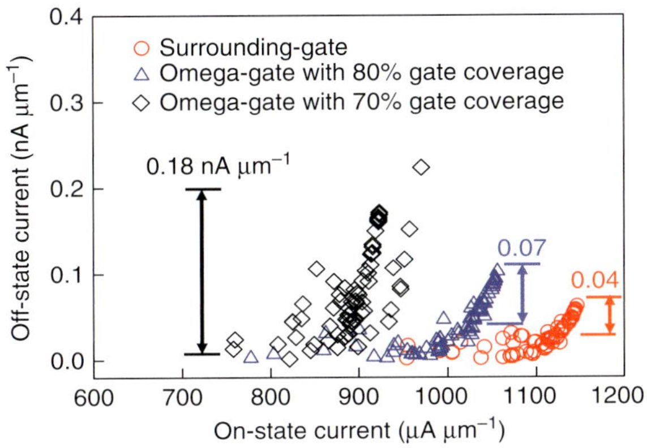

Figure 53 Comparison of the on-off state current fluctuation of the 16-nm-gate silicon nanowire FET with surrounding-gate, omega-gate with 80%, and 70% gate-coverage-ratio.

So far we have discussed the fluctuations of V_{th} of double-, triple-, and quadruple surrounding-gate SOI MOSFETs and the influence of nonideal gate coverage disturbs the channel controllability of nanowire FinFETs. An insight into the intrinsic fluctuation and the mechanism of immunity against fluctuation in multiple-gate SOI devices has been provided.

6.5 Device Variability in Nanoscale Transistor Circuits

Modeling of device variability is crucial for the accuracy of timing in circuits and systems, and the stability of high-frequency applications [64–67]. Unfortunately, due to the randomness of dopant position in devices, the fluctuation of device gate capacitance is nonlinear and hard to be modeled in current compact models. Here, the variations of transition time of digital circuit (inverter, NAND, and NOR gates) and high-frequency characteristic of common-source amplifier are estimated by a large-scale statistically sound atomistic coupled device-circuit simulation approach.

6.5.1 Analogy/High-Frequency Characteristics

For the radio-frequency or mixed-signal applications, a cut-off frequency higher than 200 GHz has been reported [119, 120]. However, attention is seldom drawn to the existence of high-frequency characteristic fluctuations of active devices due to random dopant placement. Section 6.5.1 quantitatively introduces the discrete dopant-induced characteristic fluctuation of 16-nm-gate planar bulk MOSFETs circuit under

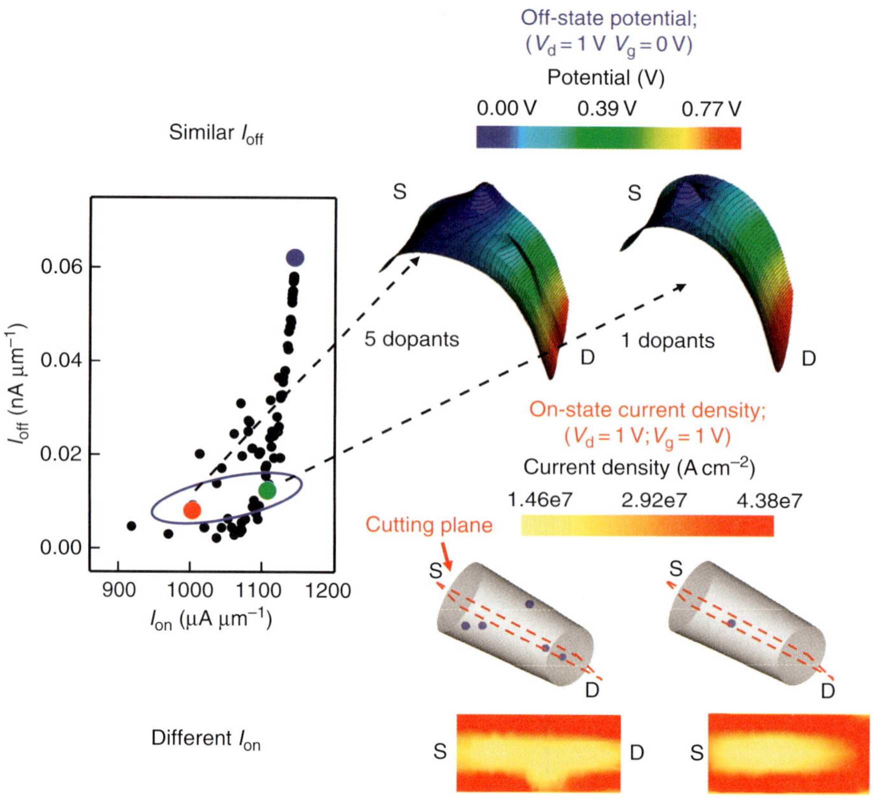

Figure 54 Effect of discrete dopant position in silicon nanowire FET, where the devices are with different I_{on} but similar I_{off}.

high-frequency regime. The circuit gain, the 3 dB bandwidth, and the unity-gain bandwidth of the tested nanoscale transistor circuit are calculated concurrently capturing the discrete dopant-number- and discrete dopant-position-induced fluctuations.

The common-source circuit with sinusoid input wave (offset is equal to 0.5 V), as shown in **Figure 56**, is used as a tested circuit to explore the fluctuation of high-frequency characteristics. The time-domain simulation results are simultaneously used for the calculation of the property of the frequency response, where the frequency is swept from 1×10^8 to 1×10^{13} Hz. **Figures 57(a)** and **57(b)** show the maximum transconductance ($g_{m,\ max}$), and output resistance of transistor (r_o) versus dopant number, where the maximum transconductance and the output resistance of transistor are shown in insets. Since the threshold voltage is increased with increasing channel doping concentration, according to the definition in insets of **Figures 57(a)** and **57(b)**, $g_{m,\ max}$ is decreased and r_o is increased as the number of dopants is increased. The position of random dopants induced different fluctuation of characteristics in spite of the same number of dopants. Furthermore, the magnitude of the spread characteristics increases as the number of dopants increases. **Figures 58(a)** and **58(b)** explore the voltage-transfer-curve (VTC) and corresponding output voltage (V_{out}) of the discrete dopant-fluctuated 16-nm-gate planar MOSFET circuits, respectively. The derived voltage gain and its fluctuation of the studied circuit are shown in **Figures 58(c)** and **58(d)**. The V_{out} is increased as the number of dopants in the device channel is increased due to the decreasing output current and voltage drop across R_1. The variation of V_{out} will influence the capacitance of the studied MOSFETs and alters the high-frequency response. Moreover, as expected, the fluctuations of V_{out} and voltage gain are increased as dopant number is increased. Beside the fluctuations of the nodal voltage of circuits, the output signal is also fluctuated. WLOG, **Figure 59** plots the output signal of the discrete dopant-fluctuation circuits at 10^8 Hz. The solid line is the nominal case and the dashed lines are transistors with discrete dopants. The output signal corresponds to the input signal plotted in **Figure 56**. The discrete dopant fluctuation not only impacts the operation points of circuits, but also influences the magnitude of output signal.

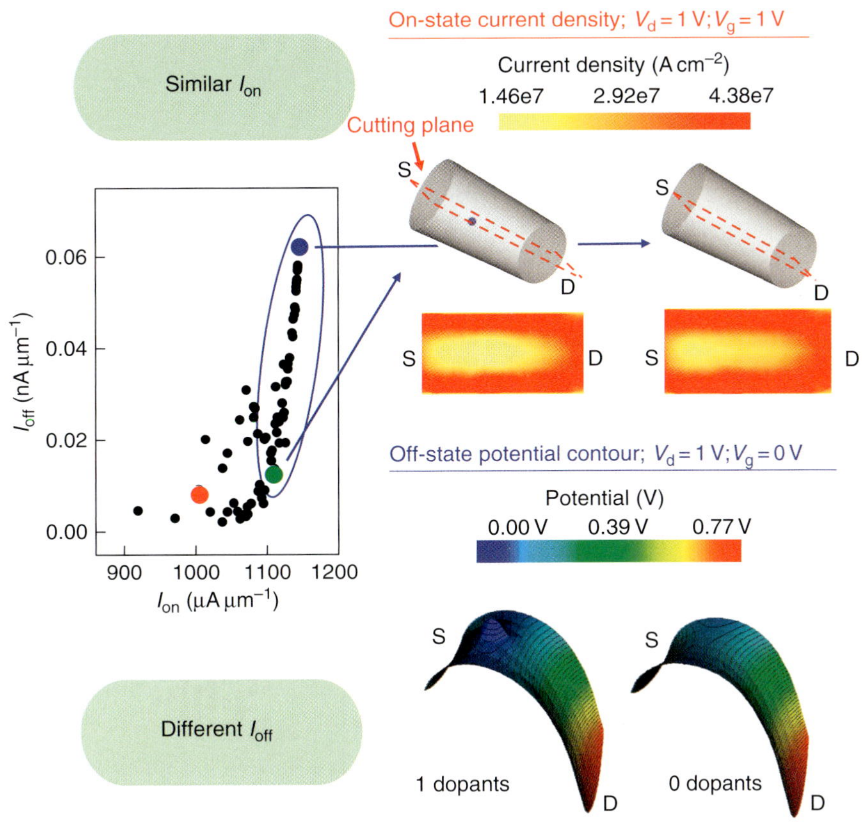

Figure 55 Effect of discrete dopant position in silicon nanowire FET, where the devices are with similar I_{on} but different I_{off}.

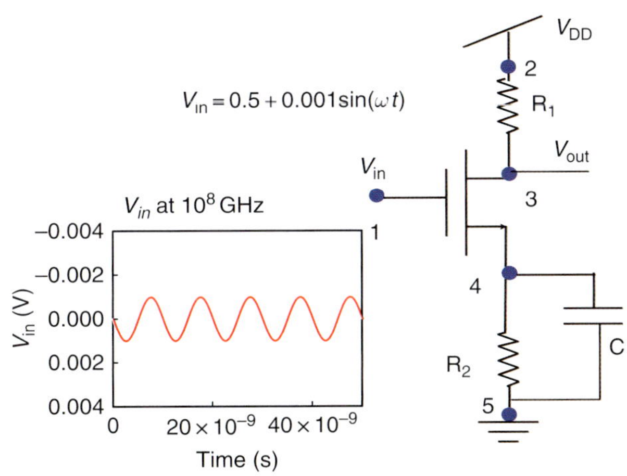

Figure 56 The common-source circuit is used as a tested circuit to explore the fluctuation of high-frequency characteristics. The input signal is a sinusoid input wave with 0.5 V offset. The frequency is sweep from 1×10^8 Hz to 1×10^{13} Hz.

Figure 60(a) shows the circuit gain versus operation frequency for all fluctuated cases, where the solid line shows the nominal case, whose channel doping profile is continuously doped with 1.48×10^{18} cm^{-3}. The circuit gain, 3 dB bandwidth, and unity-gain bandwidth of the nominal case are

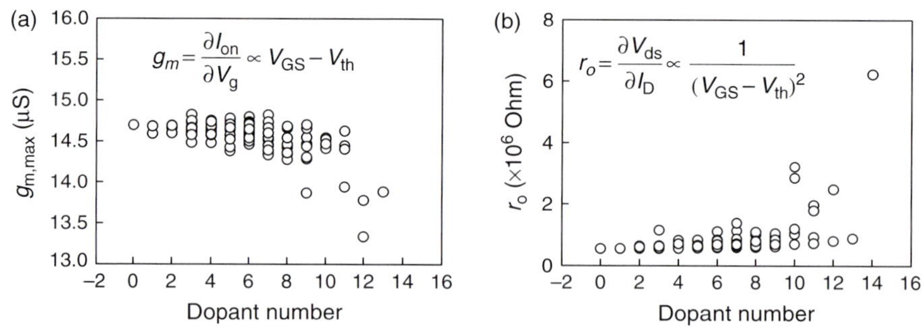

Figure 57 DC characteristic fluctuations of (a) $g_{m,\ max}$, and (b) r_g of the 125 discrete dopant-fluctuated 16-nm-gate planar MOSFET. The definitions are shown in the insets.

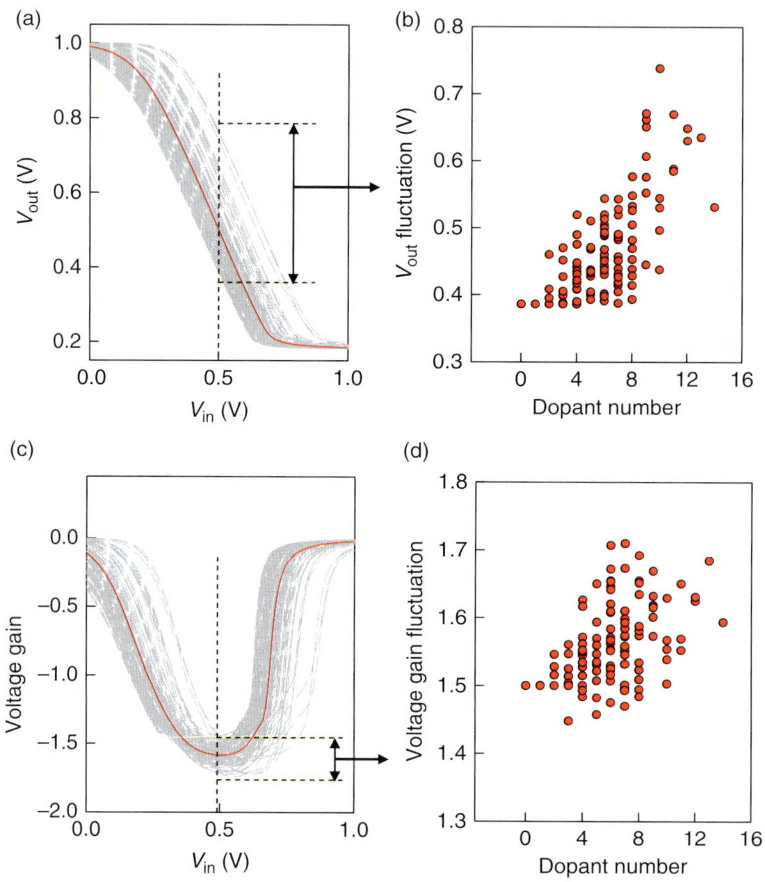

Figure 58 (a) voltage-transfer-curve; (b) the output voltage; (c) the voltage gain; (d) the fluctuation of voltage gain for the studied discrete dopant-fluctuated 16-nm-gate common-source circuits. The solid line shows the capacitance of the nominal case (continuously doped channel with $1.48 \times 10^{18}\ \text{cm}^{-3}$ doping concentration) and the dashed lines are the random dopant-fluctuated devices.

8.14 dB, 68 GHz, and 281 GHz, respectively. The corresponding high-frequency characteristic fluctuations for the explore circuit are explored, as shown in **Figures 60(b)–60(d)**, where the insets show the trend of circuit gain, 3 dB bandwidth, and unity-gain bandwidth as a function of device characteristic and circuit element. The gain of the studied circuit is proportional to transconductance multiplied by output resistance of circuit. The circuit output resistance, R_{out}, is given by

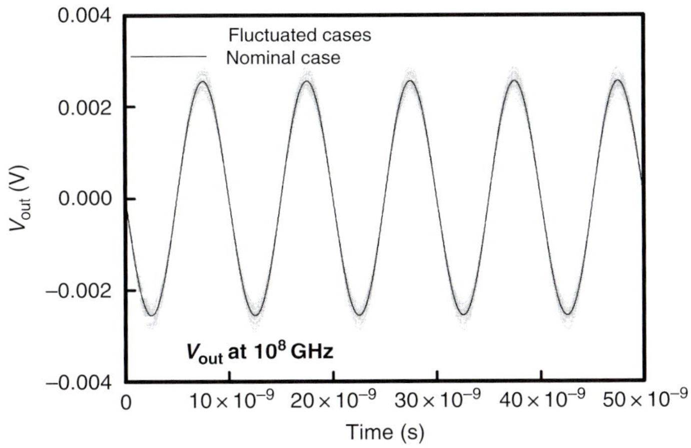

Figure 59 The output signal fluctuation of the discrete dopant-fluctuated circuits at 10^8 Hz. The solid line is the nominal case and the dashed lines are transistors with discrete dopants.

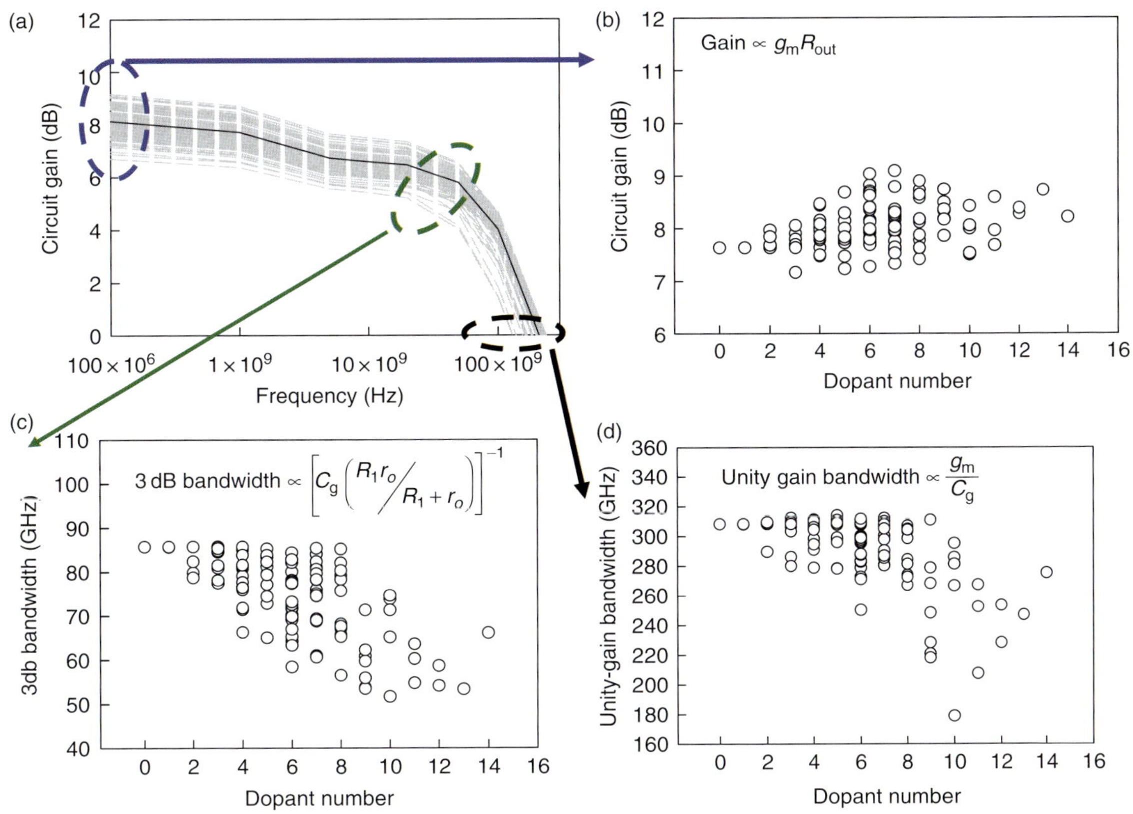

Figure 60 (a) frequency response; (b) high-frequency circuit gain; (c) 3 dB bandwidth, (d) and unity-gain bandwidth fluctuations of the studied discrete dopant-fluctuated 16-nm-gate common-source circuits.

$$R_{out} = \left(\frac{I_{out}}{V_{out}} + \frac{1}{r_o}\right)^{-1} \tag{14}$$

where I_{out} and V_{out} are the output current and voltage, respectively of the studied circuit. As the number of dopants in the device channel is increased, the on-state current of the transistor, which is associated with the output current of the test circuit, is decreased and thus increases the output voltage of the circuit. Additionally, the output resistance of transistor, r_o, is increased with increasing threshold voltage. Therefore, the circuit output resistance, R_{out}, is increased as threshold voltage is increased. We notice that although the dependence of R_{out} and g_m on threshold voltage is opposed, the trend of circuit gain fluctuation is dominated by the output resistance due to the square dependence of r_o on $V_{gs} - V_{th}$. Therefore, the trend of circuit gain fluctuation is dominated by the output resistance and increased as the number of dopant is increased, as shown in **Figure 60(b)**. **Figures 60(c)** and **60(d)** show the fluctuation of 3 dB bandwidth and the unity-gain bandwidth of the nanoMOSFET circuit, which indicates the variations of switching speed nanoMOSFET circuit resulted from random discrete dopants. The insets of **Figures 60(c)** and **60(d)** show the main sources of variations contributed from device characteristics fluctuations, g_m, r_o, and C_g, as shown in **Figures 57** and **25**. As the number of dopants in the device channel is increased, the depletion width is decreased, but gate capacitance is increased.. The fluctuation of $C_g$accompanied with increasing output resistance and decreasing g_m result in a decrement of 3 dB bandwidth and the unity-gain bandwidth on increasing dopant number. Similar to the DC characteristic of a device, the high-frequency characteristic fluctuation of the nanoscale MOSFET circuit is much more scattered as the number of dopants is increased. The SDs of the gain, 3 dB bandwidth, unity-gain bandwidth, and gain-bandwidth product are 0.46 dB, 9.63 GHz, 29.3 GHz, and 64.4 GHz, respectively. The high-frequency characteristic fluctuations of the tested circuit are summarized in **Table 4**. Both the discrete dopant-induced DC and high-frequency characteristics fluctuations of 16-nm-gate MOSFET circuit have been studied using a 3D atomistic simulation technique. Upon using the experimentally calibrated analyzing technique, the results have shown that the discrete dopant fluctuated 16-nm-gate MOSFET circuit exhibits 5.7% variation of the circuit gain, 14.1% variation of the 3 dB bandwidth, and 10.4% variation of the unity-gain bandwidth.

Table 4 Summarized high-frequency characteristic fluctuations of the nano-MOSFET circuit

	Gain (dB)	*3 dB bandwidth (Hz)*	*Unity-gain bandwidth (Hz)*
Nominal	8.138	6.84×10^{10}	2.81×10^{11}
Fluctuation	0.465	9.63×10^{9}	2.93×10^{10}
Normalized Fluctuation	5.7%	14.1%	10.4%

6.5.2 Timing of Digital Circuit

Silicon-based devices are scaled down continually to increase density and speed. In nanoscale VLSI circuits and systems, the local device variation and uncertainty of signal propagation time have become crucial in the variation of system timing and the determination of clock speed [64, 65]. Here, we introduce the timing characteristic fluctuation in the state-of-the-art nanoscale circuits and systems. The function- and circuit-topology-dependent characteristic fluctuations, caused by random nature of the discrete dopants, are indispensable components of the circuit design procedure.

Figures 61(a)– **61(c)** plot the explored 16-nm-gate CMOS digital integrated circuits, inverter, NAND, and NOR, respectively. The characteristics of the inverter are first drawn. **Figure 62(a)** shows the VTCs for the 16-nm-gate CMOS inverters with discrete dopants. Two points on the VTC determine the noise margins of the inverter. These are the maximum permitted logic 0 at the input, V_{IL}, and the minimum permitted logic 1 at the input, V_{IH}. The two points on the VTC are defined as those values of V_{in} where the incremental gain is unity; the slope is $-1\,\mathrm{V\,V^{-1}}$. The nominal value and fluctuations of the V_{IL} and V_{IH} are shown in the insets of **Figure 62(a)**. σV_{IL} exceeds V_{IH} because the σV_{th} of NMOSFETs exceeds that of PMOSFETs. The maximum slope of the VTC indicates the maximum voltage gain of the inverter. The 7% of the normalized voltage gain fluctuation of the inverter is therefore estimated, as shown in the inset of **Figure 62(a)**. **Figure 62(b)** plots the noise margins for the logic 0 and 1, NM_H and NM_L, as a function of the dopant number. Both the NM_H and NM_L are defined in insets. The NM_L is increased with the increasing dopant number in the NMOSFET due to the increased V_{th} of device. For the NM_H, as the number of dopants in the PMOSFET increases, the increased V_{th} of the device may decrease the V_{IH} of VTC and thus increase the NM_H. We notice that even for cases with the same number of dopants within device channel, their noise margins are still quite different due to the different distribution of random dopants. The noise margins of the inverter circuit increases as dopant number increases; however, the fluctuations of the noise margins are also increased due to the greater sources of fluctuation in the device channel region.

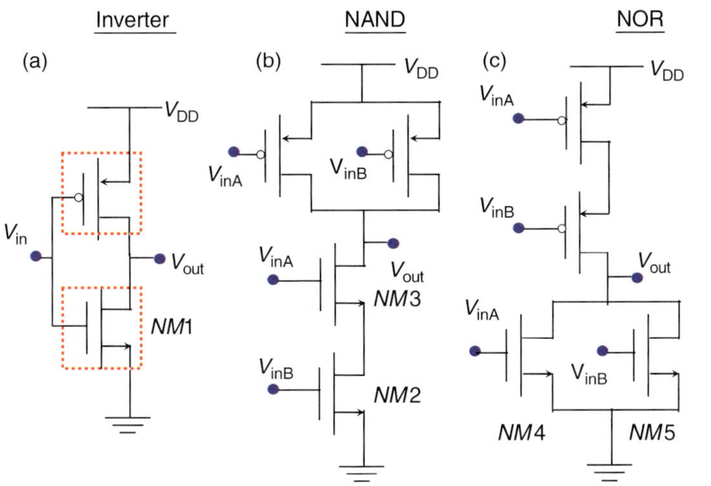

Figure 61 The explored 16-nm-gate CMOS digital integrated circuits, (a) inverter, (b) NAND, and (c) NOR.

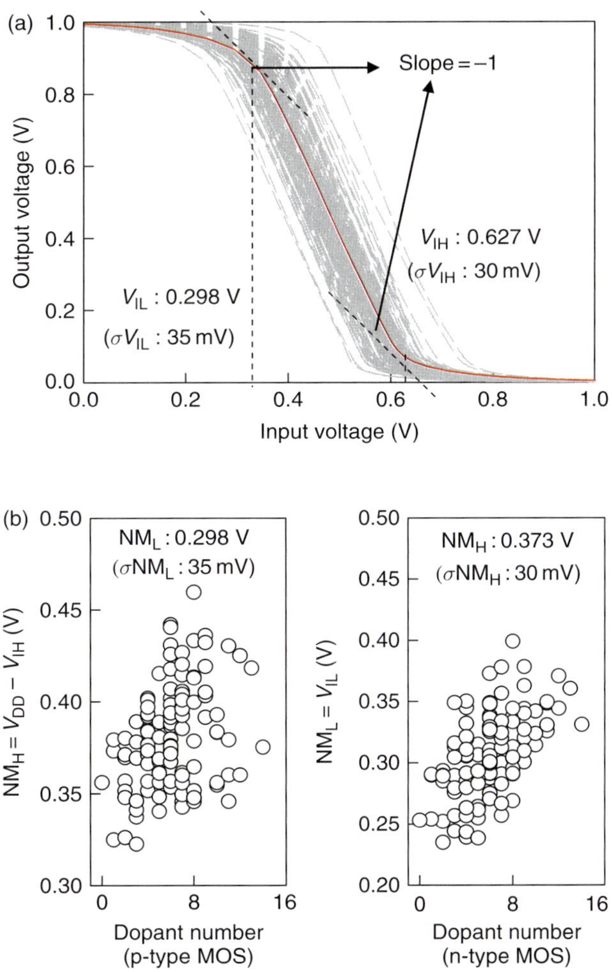

Figure 62 (a) The voltage-transfer curves for the studied 16-nm-gate planar MOSFET circuit; (b) The noise margins, NM_L and NM_H, as a function of the dopant number in the NMOSFET and PMOSFET.

Figure 63(a) presents the input and output signals; the solid line represents the nominal case (continuously doped channel with a channel doping concentration of $1.48 \times 10^{18}\,\text{cm}^{-3}$) and the dashed lines represent cases with discrete dopant fluctuations. The rise time (t_r), fall time (t_f), and hold time of the input signal are 2, 2, and 30 ps, respectively. **Figures 63(b)** and **63(c)** display the zoom-in plots of the falling and rising transitions. The term t_r is the time required for the output voltage (V_{out}) to rise from 10% of the logic 1 level to 90% of the logic 1, and the t_f denotes the time required for the output voltage to fall from 90% of the logic 1 level to 10% of the logic 1 level. The low-to-high delay time and high-to-low delay time are defined as the difference between the times of the 50% points of the input and output signals during the rising and falling of the output signal, respectively. For the high-to-low transition, the NMOSFET is on and starts to discharge load capacitance, causing the output signal to transit from logic 1 to logic 0. Similarly, for the low-to-high transition characteristics, the PMOSFET is turned on and starts to charge the load capacitance, causing the output voltage to transit from logic 0 to logic 1. The 90% ($t_{90\%}$) and 10% ($t_{10\%}$) of the logic 1 level are defined as starting points for the high-to-low and low to-high transition and plotted in **Figures 64(a)** and **64(b)**, respectively. During the high-to-low signal transition, the output signal falls as the NMOSFETs is turned on. Therefore, the fluctuation of the starting points for high-to-low signal transition is determined by the σV_{th} of the NMOSFET. With the increasing number of dopants in NMOSFET, the increased V_{th} delays the starting point of signal transition ($t_{90\%}$) and increases the high-to-low delay time, as shown in **Figure 65(a)**. Similarly, the starting point of low-to-high transition (the time of 10% of the logic 1 level) is influenced by V_{th} of PMOSFET and increased as number of dopants in PMOSFET, as shown in **Figure 64(b)**. **Figures 65(a)** and **65(b)** plot the high-to-low delay time and low-to-high delay time for the inverter circuits with discrete dopants, respectively. Since the delay time is dependent on the start of the signal transition, the t_{HL} and t_{LH} are increased as channel dopant number increases. Notably, even with the same dopant number inside the channel, the delay time can still vary significantly. Take the cases of six dopants inside the NMOSFETs as an example; the maximum t_{HL} difference is about 0.3 ps, where the nominal t_{HL} is 0.59 ps. We refer to this effect as discrete dopant-position-induced fluctuation. The magnitude of discrete dopant-position-induced fluctuations increases as dopant number increases because of the increasing number of fluctuation sources (dopants). **Table 5** summarized the normalized timing characteristic fluctuations (the SD/nominal value × 100%). For the 16-nm-gate CMOS inverter, as the number of discrete dopants varies from 0 to 14, fluctuations of t_r, t_f, t_{LH}, and t_{HL}, of 0.036, 0.021, 0.105, and 0.108 ps respectively, may occur. The normalized fluctuation for t_r, t_f, t_{LH}, and t_{HL} are 3.5%, 2.4%, 13.2%, and 18.3%, respectively. The delay time fluctuations dominate the timing characteristics. The normalized maximum fluctuations (the maximum variation of time / nominal value × 100%) of the low-to-high and high-to-low delay times are about 73.5% and 101.8%, respectively. Notably, the maximum and minimum delays associated with this specific set of 125 randomized channels would vary such that their range would increase, as the number of samples increased. For the high-to-low signal transition of the output signal, the delay time is dominated by the starting points of the signal transition and then controlled by the on/off state of the NMOSFETs in the inverter circuit. Therefore, the fluctuation of the threshold voltage of NMOSFETs substantially affects the high-to-low delay time characteristic. Similarly, the low-to-high delay time fluctuation is strongly influenced by the σV_{th} of PMOSFETs. σt_{HL} exceeds σt_{LH} because the σV_{th} of NMOSFETs exceeds that of PMOSFETs. The rise or fall time fluctuations depend on the charge or discharge capability of the PMOSFETs/NMOSFETs. Therefore, σt_r exceeds the σt_f because the driving capability of PMOSFETs is weaker than that of NMOSFETs in the given device dimensions scenario. The device with larger driving capability requires less time to charge and discharge a given load capacitance and so exhibits a less fall time fluctuations. **Figures 66(a)** and **66(b)** show the fall time and rise time as a function of the threshold voltage for NMOSFETs and PMOSFETs, respectively. The fall time is dependent on the discharge capability of the NMOSFETs, and the rise time is dependent on the charge capability of the PMOSFETs. As the threshold voltages of the NMOSFETs and PMOSFETs are increased, the discharge and charge ability for given values of t_f and t_r decrease. Therefore, the time required for the fall and rise transitions increases. The trend for t_f is not clear because herein only the transistors' gate capacitance was used as the load capacitance. The small load capacitance and strong driving capability of NMOSFETs make the trend of t_f fluctuation insignificant. Notably, the rise and fall

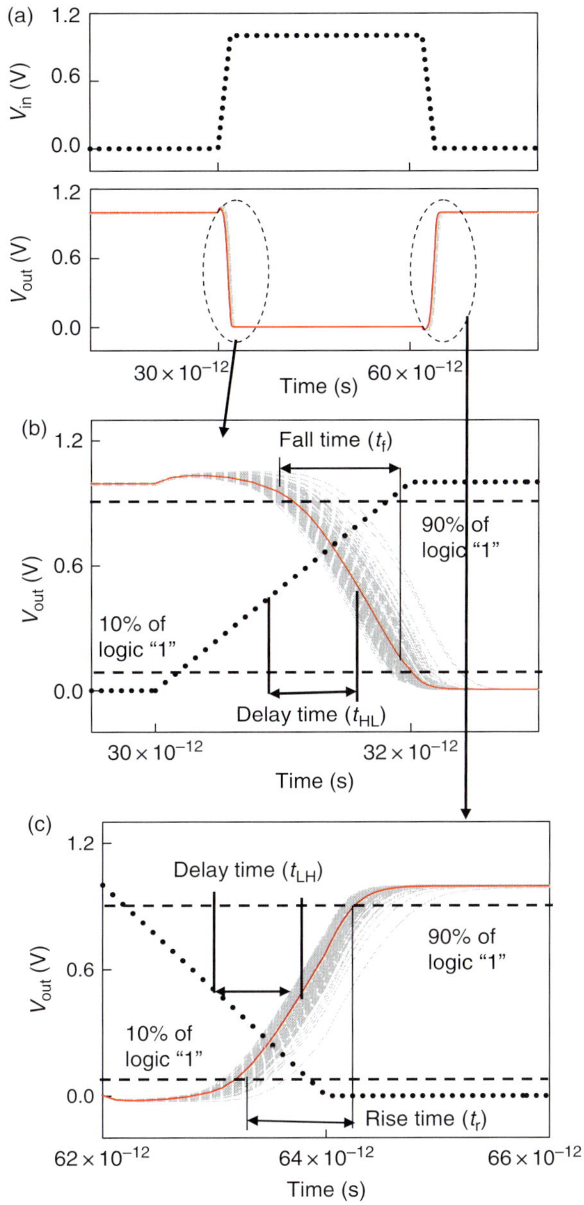

Figure 63 (a) The input and output signals for the discrete dopant-fluctuated 16-nm-gate inverter circuit. The magnified plots show (b) the fall and (c) the rise transitions, where the rise time, fall time, high-to-low delay time, and low-to-high delay time are defined.

time fluctuations generally may not be as important as the delay time fluctuation in circuit timing; however, their maximum variations can exceed 0.237 and 0.110 ps, respectively, which exceed the delay time fluctuation and should therefore, be considered in statistical timing analysis in circuit and system design. Moreover, fluctuations in the rise and fall time can be added to the delay time, and increasing the delay time fluctuations.

Figures 67(a) and **67(b)** plot the input and output transition characteristics for the two-input NAND, and the two-input-NOR circuits, respectively. The solid line represents the nominal case (continuously doped channel with a channel doping concentration of $1.48 \times 10^{18}\,\text{cm}^{-3}$) and the dashed lines represent cases with discrete dopant fluctuations. The timing characteristics are summarized in **Table 6**. As expected, the timing fluctuations of NAND and NOR circuits are

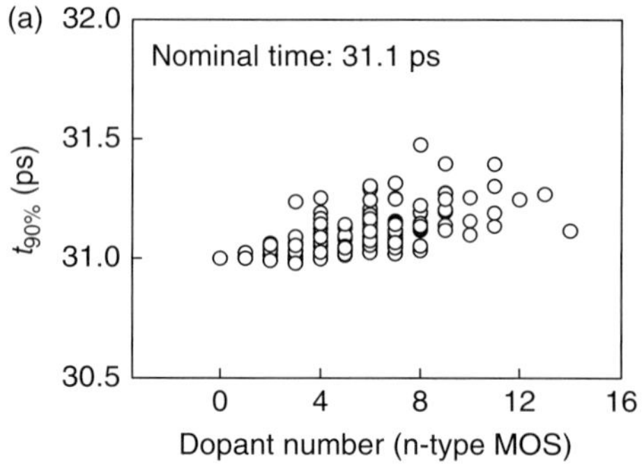

Figure 64 The fluctuations of (a) fall and (b) rise signal transition points as a function of dopant number in n-type and p-type MOSFETs for the discrete dopant-fluctuated inverter circuits.

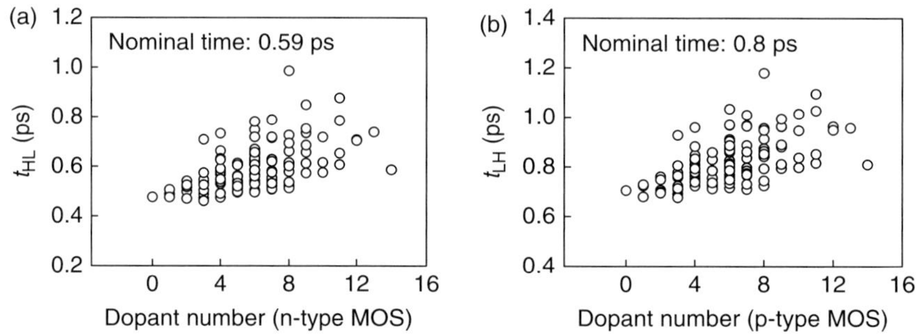

Figure 65 The fluctuations of (a) high-to-low and (b) low-to-high delay time as a function of dopant number in n-type and p-type MOSFETs for the discrete dopant-fluctuated inverter circuits.

Table 5 Transition time variation for the 16-nm-gate inverter circuits

	t_r (ps)	t_f (ps)	t_{LH} (ps)	t_{HL} (ps)
Nominal	1.021	0.897	0.800	0.590
Fluctuation	0.036	0.021	0.105	0.108
Normalized fluctuation[a]	3.5 %	2.4%	13.2%	18.3%
Normalized maximum fluctuation[a]	23.2%	12.3%	73.5%	101.8%

[a] Normalized by the nominal value.

increased as the number of transistors increased. The fluctuation of timing characteristics may suffer from the increase of fluctuation sources in circuit. It is thus reasonable to infer that the NAND and NOR circuits may exhibit larger timing fluctuations than the inverter circuit. Also, they may have similar timing fluctuations due to the same number of transistors in the circuit. However, the obtained results show that the σt_r and σt_f for the NAND circuit are 0.53 and 1.87 times larger than that of the NOR circuit. The different characteristics of timing fluctuation result from the different operation status of device with different input signal. For the fall time transition characteristics of NAND and NOR circuits, their output signal transitions are determined by the devices NM2 and NM4 in **Figures 61(b)** and **61(c)**, respectively. The NM2 and NM4 devices are operated in the linear saturation region, respectively. As aforementioned, the device operated in the saturation region exhibits a better discharge capability to reduce the timing fluctuations. Therefore, the σt_f of the NAND circuit exceeds that of the NOR due to the different operation status of transistors. Moreover, for the NMOSFETs in series (NM2 and NM3 in the NAND circuit), the NM2 may inherit the fluctuations resulted from NM3, such as the current fluctuation and intrinsic resistance fluctuation to influence the operation of NM2. Similarly, we can infer that σt_r of the NOR circuit is larger than that of the NAND circuit. The fluctuation of timing characteristics is influenced by the different input signals in the circuit. Also, the transistors in series are found to accumulate fluctuations and increase

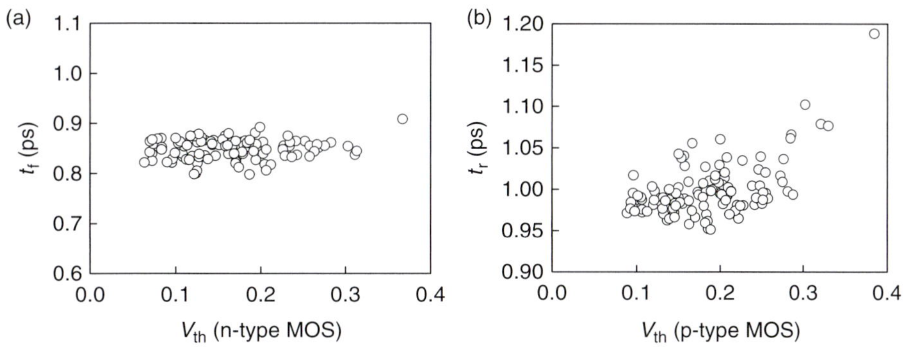

Figure 66 The fluctuations of (a) the fall and (b) rise time as a function of the threshold voltage in the n-type and p-type MOSFETs for the discrete dopant-fluctuated CMOS inverters.

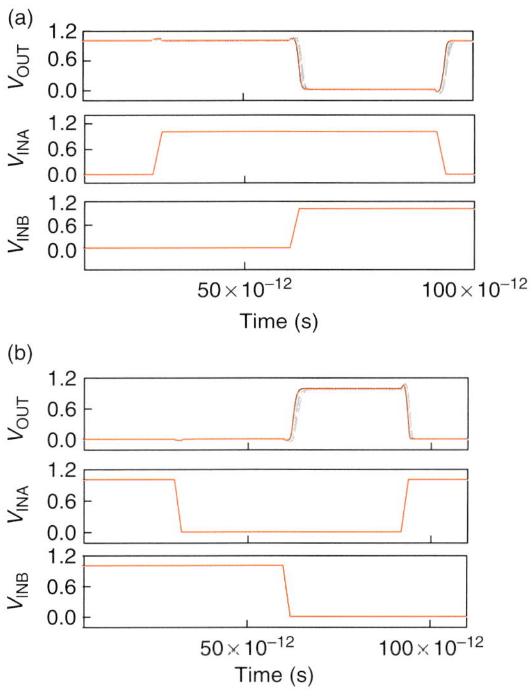

Figure 67 The input and output signal for the studied discrete dopant-fluctuated 16-nm-gate (a) two-input NAND and (b) two-input NOR circuits.

Table 6 Summarized transition time fluctuation for 16-nm-gate NAND and NOR circuits

	t_r (ps)	t_f (ps)	t_{LH} (ps)	t_{HL} (ps)
NAND				
Nominal	1.248	1.197	0.987	1.254
Fluctuation	0.070	0.056	0.107	0.129
NOR				
Nominal	1.762	0.993	1.616	0.761
Fluctuation	0.133	0.030	0.169	0.111

the fluctuations of circuits. Additionally, we can expect that the timing fluctuation can be reduced by the use of shunt transistors. The effect of shunt transistor is similar to increase the device width of a most fluctuation-sensitive element in circuits. The function- and circuit-topology-dependent characteristic fluctuations resulted from random nature of discrete dopants is briefly discussed and worth to be explored for future digital circuit applications in nanoCMOS era.

The experimentally calibrated simulation technique predicted that the discrete dopant-fluctuated 16-nm-gate CMOS inverter circuit may exhibit 0.036, 0.021, 0.105, and 0.108 ps fluctuations in the rise time, fall time, high-to-low delay time, and low-to-high delay time. The delay time fluctuations dominate the timing characteristics. The normalized maximum fluctuations of the high-to-low and low-to-high delay times are about 101.8% and 73.5%. However, the maximum variations of rise and fall time fluctuations can exceed 0.237 and 0.110 ps, respectively, which exceed the delay time fluctuation and should therefore, be considered in statistical timing analysis in circuit and system design. For the NAND and NOR circuits, the timing fluctuation is further increased with increasing number of transistors in circuit. Generally, circuits with the same transistors may have a similar variation of characteristics due to the same fluctuation sources. However, due to the device operation status and circuit topology of circuits, the characteristic fluctuations in NAND and NOR are different. The function- and circuit-topology-dependent characteristic fluctuations resulted from random nature of discrete dopants is discussed. The transistors in series may accumulate fluctuations and increase the fluctuations of circuits. It is considered that links should be established between circuit design and fundamental device technology to allow circuits and systems to accommodate the individual behavior of every transistor on a silicon chip. This study provides an insight into random dopant-induced timing characteristic fluctuations, which may benefit the development of state-of-the-art digital circuits with robust timing characteristics.

6.5.3 Static Random Access Memory

Static random access memory (SRAM) scaling is a major driver in microprocessor development. However, the intrinsic parameter fluctuations limit the performance and functionality of SRAM due to the significant component mismatch in circuits. The stability of an SRAM cell is often related to the static noise margin (SNM) in the read operation [121], defined as the maximum DC noise voltage tolerance to avoid the cell state being flipped. Here, we discuss the dependence of SNM and its fluctuation on transistor's channel size from $(65\,\text{nm})^3$ to $(16\,\text{nm})^3$. Both the PVE and random dopant-fluctuation (RDF)-induced SNM fluctuations are explored. The circuit topology of the SRAM circuit is shown in **Figure 68**.

Figures 69(a) and **69(b)** show the RDF and PVE-induced threshold voltage fluctuations for the 16–65-nm-gate n-type planar MOSFETs, respectively. The nominal value of threshold voltage is rolled-off as the gate length decreased, where the nominal $V_{th}$for 16-, 32-, and 65-nm devices are 140, 220, and 280 mV, respectively. As the gate length scales from 65 to 16 nm, the V_{th} fluctuations increase significantly from 16 to 64 mV. The V_{th} fluctuation of 16-nm-gate MOSFET is four times larger than that of 65 nm, which follows the trend of analytical formula [21, 22, 76]. The RDF dominates the total V_{th} fluctuation in the explored device dimensions. The RDF induces 5 and 3.5 times larger V_{th} fluctuation than PVE for 65-nm- and 16-nm-gate device, respectively. The increasing fluctuations of such extreme small components may cause critical issues with stability. **Figure 70** shows the static transfer characteristics of 65-nm-gate planar SRAM cells, including PVE and RDF. The dashed lines represent the cases with PVE and RDF fluctuation; the solid line stands for the nominal case with continuous channel doping profile of $1.48 \times 10^{18}\,\text{cm}^{-3}$. The nominal SNM for the 65-nm-gate planar MOSFET SRAM is 138 mV. The RDF and PVE induce 2.1% and 4.3% normalized SNM fluctuation, respectively. Since the RDF induces larger device variability in threshold voltage, the SNM fluctuation of SRAM is dominated by random doping fluctuation. Similarly, we can obtain the RDF and PVE-induced SNM fluctuation for the 16- and 32-nm-gate planar SRAM, as shown in **Figures 71(a)** and **71(b)**, respectively. As gate length scales from 65 to 16 nm, the SNM decreases significantly from 138 to 20 mV. Moreover, the RDF dominates the SNM fluctuation. **Figure 72** summarizes the normalized SNM fluctuations. The SNM fluctuation induced by RDF is two to four times larger than the PVE-induced fluctuation in different gate sizes. For 16-nm-gate SRAM, the RDF and PVE induce as much as 80% and 30%, which is much higher than that in 32 and 65 nm technologies. The result shows that the fluctuation of SRAM is highly dependent on the gate size of transistors. Moreover, the RDF becomes more significant in scaled SRAM circuits. The small SNM of scaled SRAM may cause the malfunction of SRAM. The shrinking of the device size can increase the density of memory; however, the decreased SNM and increased SNM fluctuation are crucial issues and may limit the usage of such a small device.

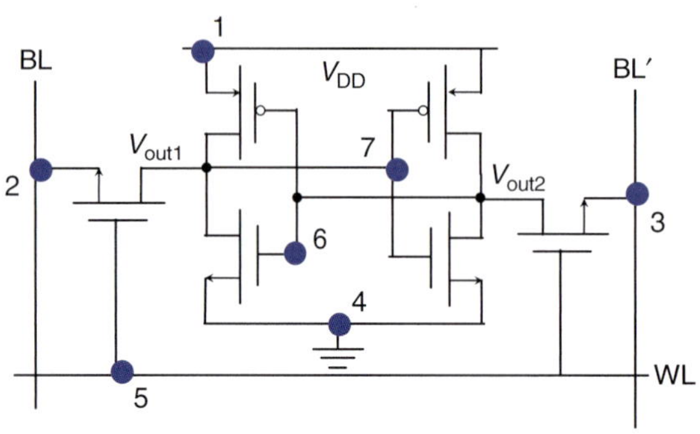

Figure 68 Circuit topology of the explored static random access memory. BL and BL′ are the bit lines; WL is the word line.

6.6 Fluctuation-Suppression Techniques

Random dopants may cause significant timing fluctuations in devices and circuits. Suppression approaches that are based on the circuit and device design viewpoints are implemented to examine the associated characteristic fluctuations [11, 15, 21–27, 63, 64, 67, 68, 75, 76]. The use of shunted transistors in the circuit provides similar suppression to the use of a device with doubled width. It is considered that links should be established between circuit design and fundamental device technology to allow circuits

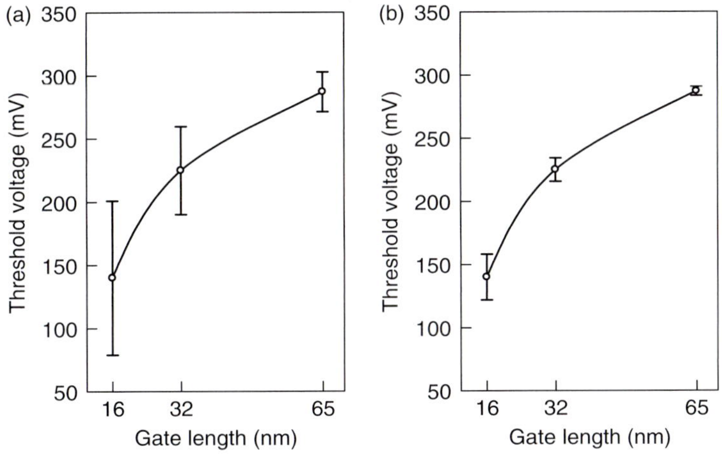

Figure 69 V_{th} fluctuation induced by (a) RDF and (b) PVE for different gate lengths of n-type planar MOSFETs.

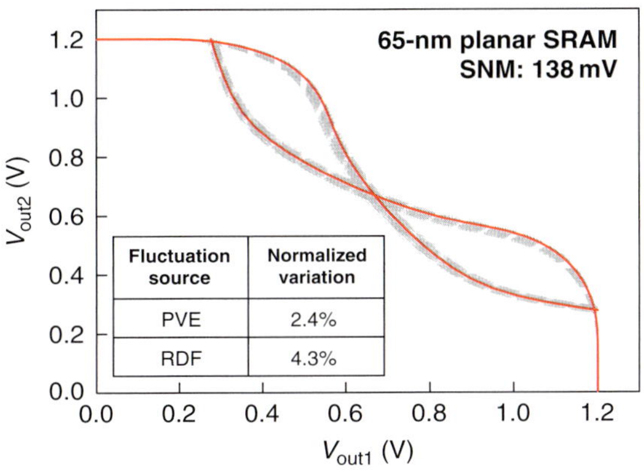

Figure 70 Nominal and fluctuated static transfer characteristics of 65-nm planar SRAM cells.

and systems to accommodate the individual behavior of every transistor on a silicon chip.

6.6.1 Device-Level Suppression

Various approaches have been proposed to suppress random dopant fluctuations, such as scaled gate oxide [27, 63], high-κ gate dielectrics [27, 63], epitaxial layer [22, 75, 76], channel profile engineering [11, 64, 66, 67], change of device architecture [15, 21–27, 68], etc. We now introduce four fluctuation-suppression techniques and examine their characteristics.

6.6.1.1 High-κ dielectrics and vertical-channel transistors

Gate oxide scaling has become the key in scaling silicon CMOS technology. The metal gate and high-κ dielectric are very attractive to maintain low gate leakage and control SCEs [27, 63]. Here, we discuss and compare the dependency of PVE and random dopant-induced V_{th} fluctuation on the gate oxide thickness scaling in 16-nm-gate MOSFETs. The threshold voltage fluctuation for the planar MOSFETs with EOTs from 1.2 to 0.2 nm (e.g., SiO_2 for the 1.2 and 0.8-nm EOTs, Al_2O_3 for the 0.4-nm EOT, and HfO_2 for the 0.2-nm EOT) are compared with the results of 16-nm-gate bulk FinFETs.

Gate capacitance is one of the most important indexes for channel controllability. **Figure 73** plots the gate capacitance as a function of the EOT, where the solid line shows the planar MOSFETs with various EOT and the square symbol indicates the bulk FinFET device with 1.2-nm EOT. The planar MOSFET with 0.4-nm EOT, where Al_2O_3 is used for gate dielectric, exhibits a similar gate capacitance with the bulk FinFET device with 1.2-nm EOT because the gate size of the explored bulk FinFETs is about three times larger than the planar MOSFETs. Since the process variation is due to the enhancement of the SCE, the bulk FinFETs with 1.2-nm EOT is expected to have similar immunity against process-variation-induced fluctuation with the planar MOSFETs with 0.4-nm EOT. The $\sigma V_{th,Lg/LER}$ of the planar MOSFETs and bulk FinFETs presented in **Figure 74** confirmed this viewpoint, in which the V_{th} roll-off characteristics in **Figure 75** are used for the $\sigma V_{th,Lg/LER}$ estimation. The bulk FinFET device with 1.2-nm EOT exhibits a similar immunity against process-variation-induced fluctuation with the planar MOSFET with 0.4-nm EOT.

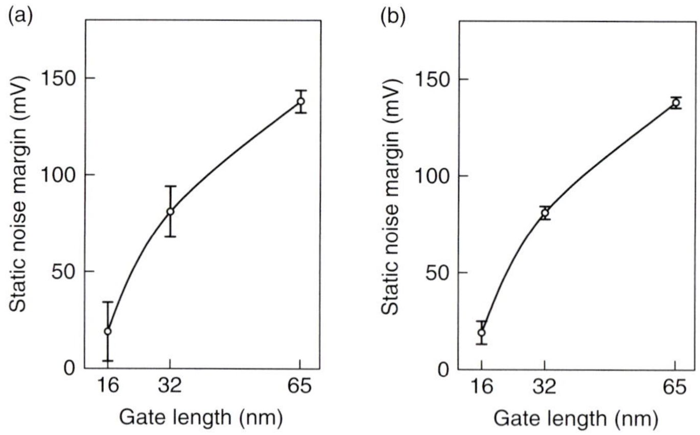

Figure 71 Static noise margin fluctuations induced by (a) RDF and (b) PVE for different gate lengths of SRAM cells.

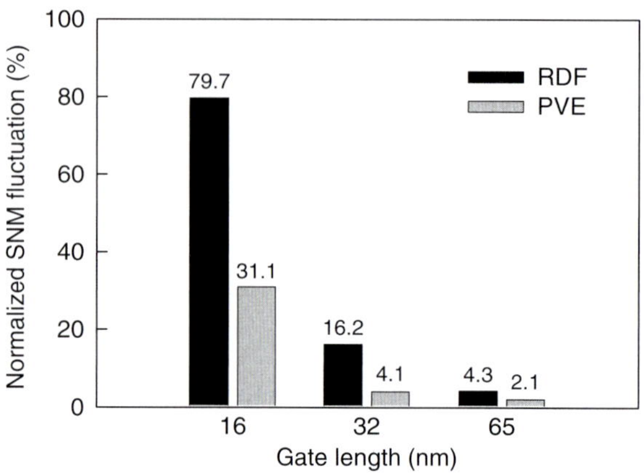

Figure 72 Normalized PVE and RDF-induced SNM variation for 16–65 nm planar SRAM.

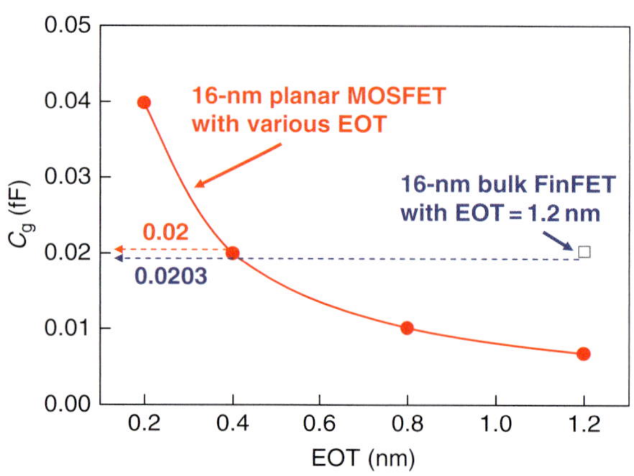

Figure 73 The gate capacitance as a function of the EOT, where the solid line shows the planar MOSFETs with various EOT and the square symbol indicates the bulk FinFET device with 1.2-nm EOT.

However, this trend is still valid in the random dopant-induced fluctuation.

Figure 76 shows the random dopant-induced V_{th} fluctuation, $\sigma V_{\mathrm{th,RD}}$, of the studied devices. The $\sigma V_{\mathrm{th,RD}}$ decreases significantly as the EOT is scaled down. However, the expectation between gate capacitance and $\sigma V_{\mathrm{th,RD}}$ is invalid. The $\sigma V_{\mathrm{th,RD}}$ of bulk FinFETs approximates the planar MOSFETs with 0.2-nm EOT. To further investigate the mechanism of bulk FinFETs in fluctuation suppression, the potential distributions ($V_{\mathrm{g}} = 1$ V; $V_{\mathrm{d}} = 0$ V) extracted 1 nm below the top gate of the channel are examined. **Figure 77(a)** shows the dopant distribution and **Figure 77(b)–77(f)** are the explored devices with various EOT. The potential barriers are induced by the corresponding dopants at positions: A–C, respectively, as shown in **Figure 77(a)**. The potential barrier is largest at C because two discrete dopants are located close to each other there. For planar MOSFETs with various EOTs, the sizes of the potential barriers are suppressed as the equivalent gate-oxide thickness is reduced. The results for planar MOSFETs, as displayed in **Figures 77(b)–77(e)** are then compared with that of a bulk FinFET device, as shown in **Figure 77(f)**, indicating that the potential barriers of bulk FinFET are smaller than those of planar MOSFETs, especially at position A. The potential barrier induced by corresponding dopant in A is significantly reduced in bulk FinFET devices because of the strong electric field around the corners to suppress the potential barrier induced by dopants. The strengthened electric field is due to the changed gate structure, which causes the different suppression mechanisms between the planar MOSFETs and bulk FinFETs. **Figure 78** shows the

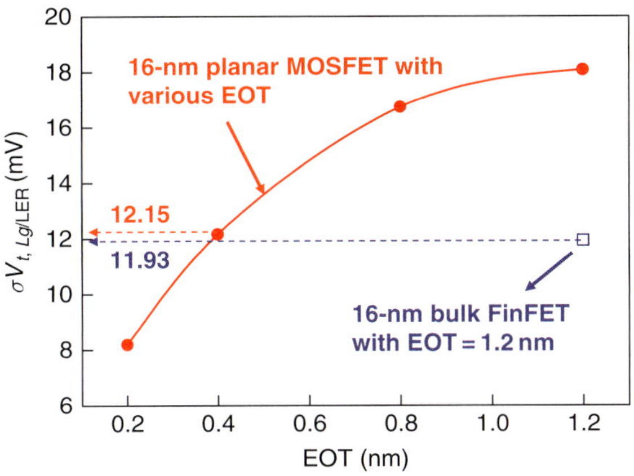

Figure 74 The process-variation-induced threshold voltage fluctuation for the studied devices.

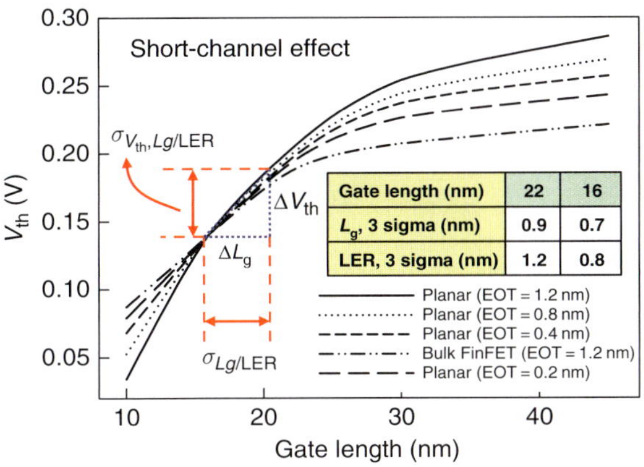

Figure 75 The threshold voltage roll-off of the studied devices, where the variation follows the projection of ITRS 2005 roadmap. These results are used to estimate the V_{th} fluctuation resulting from the gate-length deviation and the LER.

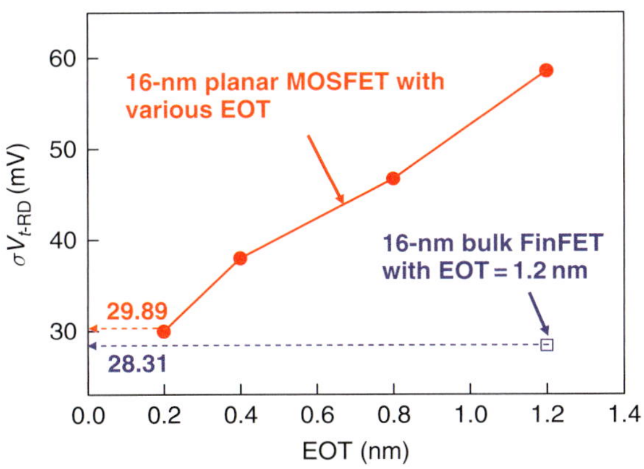

Figure 76 The random dopant-induced threshold voltage fluctuation for the studied devices.

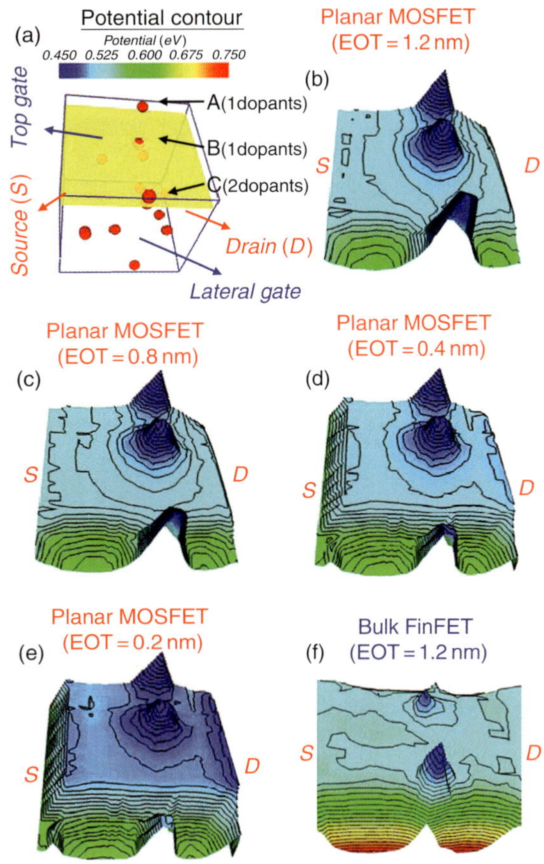

Figure 77 Top-gate potential contours of the planar MOSFETs with various EOT ((b) EOT = 1.2 nm, (c) EOT = 0.8 nm, (d) EOT = 0.4 nm, (e) EOT = 0.2 nm), and (f) bulk FinFETs with 1.2-nm EOT. The distributions of potential barriers are induced by the corresponding dopants location (i.e., A, B, and C). The corresponding distribution of discrete dopants is shown in (a) and all the plots are extracted 1 nm below the gate oxide.

lateral side potential distributions of the studied devices, in which **Figure 78(a)** plots the dopant distribution. **Figures 78(b)** and **78(c)** show the potential contours of the nominal (continuous channel doping concentration: $1.48 \times 10^{18}\ \text{cm}^{-3}$) and discrete dopant fluctuated cases of bulk FinFETs with 1.2-nm EOT. **Figures 78(d)–78(h)** show the nominal and discrete dopant-fluctuated cases of planar MOSFETs with EOT scaling. The potential fluctuation is mitigated as the equivalent gate-oxide thickness is scaled down, as shown in **Figures 78(f)–78(h)**. For the discrete dopant-fluctuated bulk FinFET device in **Figure 78(c)**, although the potential distribution is disturbed by a dopant that is located on lateral side of the channel, the overall potential distribution in the case of fluctuation is still quite similar to that in

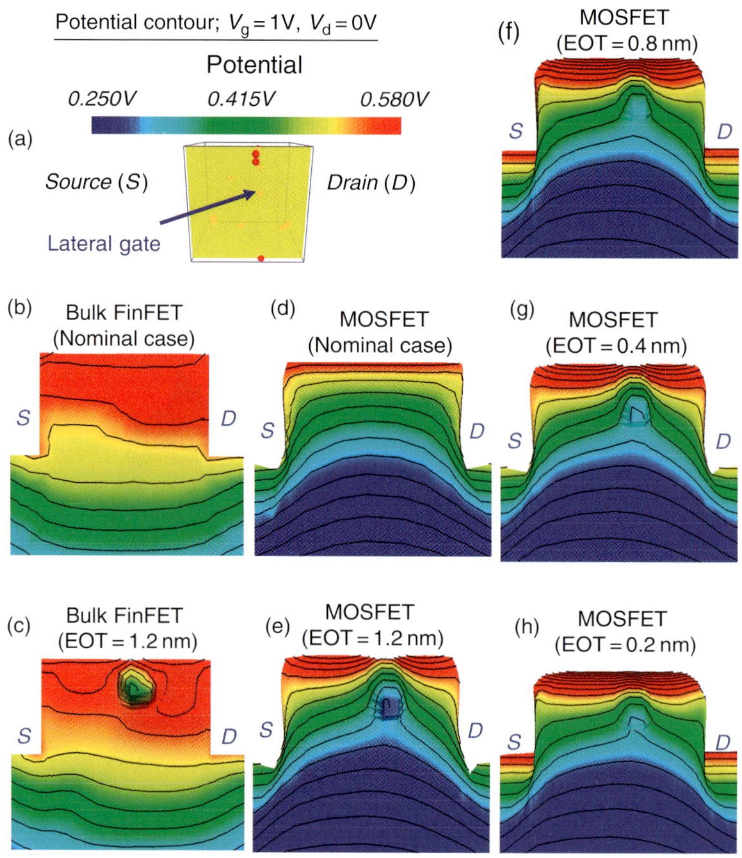

Figure 78 Lateral-gate off-state potential contours for the studied planar MOSFET and bulk FinFET devices, where (b) and (c) show the nominal (continuously doped) and discrete dopant-fluctuated cases of bulk FinFETs. The nominal and discrete dopant-fluctuated cases of planar MOSFETs with different EOTs are shown in (d)–(h).

the nominal case. However, for MOSFETs in **Figure 78(e)**, due to the lack of lateral gate structure, the potential fluctuation is still significant. This result reconfirms the effect of the lateral gate in bulk FinFET devices in suppressing potential fluctuations.

Figure 79 plots the top and lateral views of the on-state current density ($V_g = 1$ V; $V_d = 1$ V) of planar and bulk FinFET devices with 1.2-nm EOT. All cross-sectional plots are from 1 nm below the top and lateral side of channel surface. The top views of the channel, as presented in **Figures 79(b)** and **79(e)**, reveal that bulk FinFET device provides a larger and more uniform current distribution than the planar MOSFET due to the smaller fluctuation of potential. The lateral views of channel, **Figures 79(c)** and **79(d)**, show that the current-conducting paths of planar MOSFETs are easily disturbed by discrete dopants. However, in the bulk FinFETs, **Figures 79(f)** and **79(g)**, even current-conducting paths are retarded in parts of channel surface; the tri-gate structure of bulk FinFETs provides more alternative conducting paths that prevent a significant fluctuation of conduction current. Thus, benefiting from the superiority of the vertical-channel structure, the bulk FinFET device suppresses potential fluctuations and maintains a more stable conduction current than the planar MOSFET. **Figure 80** plots the on-/off- state current characteristics of the studied devices. For devices with similar on-state current, the maximum difference of off-state current is declined from approximately 2000 to 800 nA μm^{-1} as the EOT is scaled from 1.2 to 0.2 nm. Comparing the results for planar MOSFETs with those of bulk FinFETs, even though the planar device with 0.4- and 0.2-nm EOTs has a better on-off state characteristic, the bulk FinFET device exhibits a smaller current fluctuation (about 600 nA μm^{-1}). The bulk FinFET device can provide a more uniform potential distribution and a more stable current flow than that of the planar MOSFET. The additional

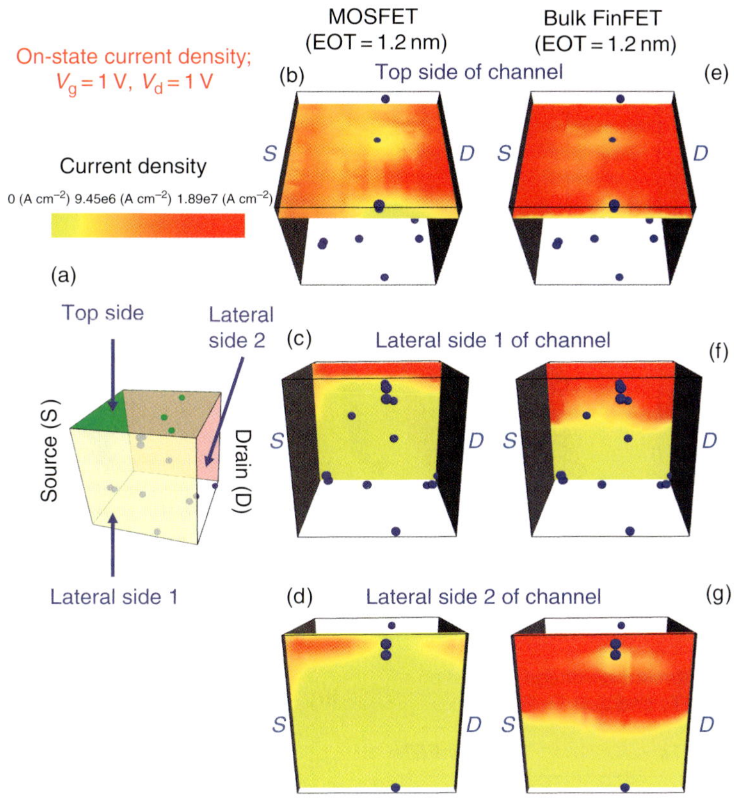

Figure 79 Cross-sectional views of on-state current density distribution in channel of device, where (b)–(d) show the planar MOSFET device and (e)–(g) show the bulk FinFET device. The corresponding distribution of discrete dopants is shown in (a) and all the cross-sectional plots are extracted 1 nm below the channel surface.

structural improvement of bulk FinFET devices enhances the immunity of device against random dopant-induced fluctuation, which cannot be evaluated from the trend of gate capacitance. The PVE and RDF-induced V_{th} fluctuations are summarized in **Table 7** and **Figure 81**. The random dopant effect is the dominating factor in this scenario. The planar MOSFETs with 0.2-nm EOT and the bulk FinFETs possess the best fluctuation immunity. The V_{th} fluctuations are suppressed with the gate-oxide thickness scaling. However, the immunity of the planar MOSFETs suffers from the nature of structural limitations and the use of vertical-channel transistors, such as FinFETs, can alleviate this problem.

WLOG, the high-frequency characteristics for 16-nm-gate planar MOSFETs with 1.2-nm EOT and FinFETs are compared as follows. **Figures 82(a)** and **82(b)** plot the cut-off frequency (F_T) of planar MOSFETs and F_T fluctuation versus V_g for PVE and RDF, respectively. The solid line shows the nominal case with $1.48 \times 10^{18}\,\mathrm{cm}^{-3}$ channel doping; the dashed lines are fluctuated cases; the symbol line shows averaged result. The F_T $(=v_{sat}/2\pi L_g)$ saturates at high V_g due to the velocity saturation of carriers. The PVE impacts the gate length of transistor and therefore influences the F_T characteristics. For the RDF, the discrete impurity impacts the gate capacitance and the carrier-impurity scattering alters the saturation velocity (v_{sat}). Notably, for PVE-induced fluctuation, the nominal and averaged values of F_T are similar. However, the deviation of the nominal and the averaged F_T of RDF increase as V_g increases due to the randomness of carrier-impurity scattering events and carrier velocity variations. The F_T fluctuations are summarized in **Figure 82(c)**. For the RDF, the F_T fluctuation decreases as V_g increases due to the formation of inversion layer of electrons shading the influence of fluctuation. However, the phenomenon is not obvious as in the PVE. Thus, the RDF dominates the F_T fluctuation with V_g smaller than 0.6 V and the PVE dominate F_T fluctuation at high electric field. The F_T characteristics for 16-nm-gate FinFETs are shown in **Figures 82(d)–82(f)**. The F_T of FinFETs is about 1.6 times

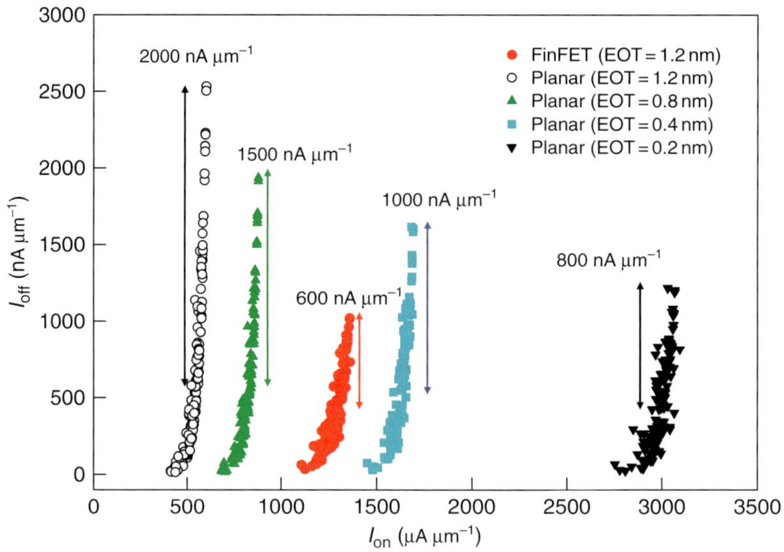

Figure 80 I_{on}–I_{off} current characteristics of the studied 16-nm-gate planar MOSFET and bulk FinFET devices.

Table 7 Summary of the threshold voltage fluctuation of the explored planar MOSFETs and bulk FinFETs

	Planar MOSFETs				*Bulk FinFETs*
EOT (nm)	1.2	0.8	0.4	0.2	1.2
L_g/LER	18.1	16.7	12.2	8.18	12.6
RD	58.5	46.6	37.9	29.9	28.3
Total	61.2	49.5	39.8	31.0	30.7

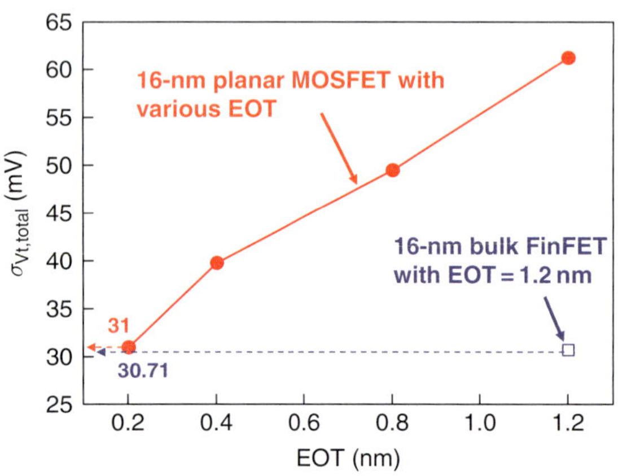

Figure 81 Total threshold voltage fluctuation for the studied devices.

smaller than the planar MOSFETs due to the large gate capacitance of FinFETs. The F_T of FinFETs decreases significantly at high field region because the significant carrier scattering reduces their mobility. Notably, the PVE-induced fluctuation is significantly suppressed in bulk FinFETs; therefore, the RDF dominates the F_T fluctuations. The high-frequency characteristics of planar MOSFETs and FinFETs in common-source amplifiers are calculated and shown in **Figures 83** and **84**. **Figure 83** shows the output power (P_{out}), the gain, and the power-added-efficiency (PAE) of the explored devices versus the input power (P_{in}). The FinFETs possess a higher power and power efficiency characteristics than the planar MOSFETs. The P_{out}, gain, and PAE of FinFETs are 1.6, 1.6, 1.43 times better than planar MOSFETs. The fluctuation of the high-frequency response, gain, 3 dB bandwidth, and unity gain bandwidth, are calculated and shown in **Figures 84(a)–84(d)**. The offset of input voltage is 0.5 V and therefore the RDF dominates the high-frequency characteristics in both planar MOSFETs and FinFETs. Notably, the fluctuation of FinFET is significantly suppressed; therefore, we conclude that the trade-off design between fluctuation and performance can be further improved as the transistor architecture changes. The FinFET's characteristic feature of high power efficiency with small fluctuation is an asset for FinFET as a next-generation nanoscale transistor.

6.6.1.2 Channel profile engineering

Channel engineering technique has been known to be an effective way to suppress the random dopant-

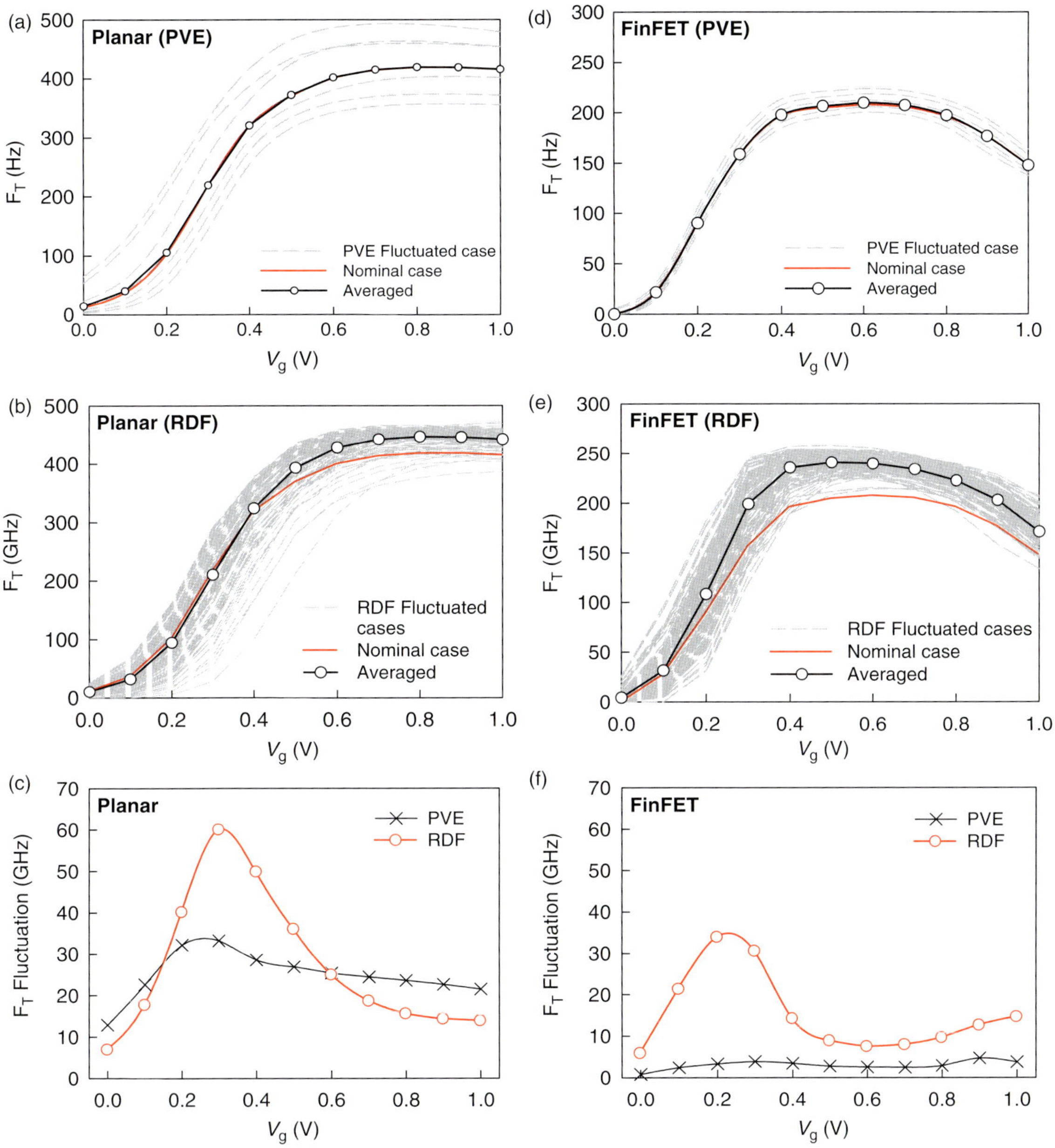

Figure 82 The F_T characteristics induced by process-variation-effect (a–c), and random dopant-fluctuation (d–f) for the planar MOSFETs and FinFETs.

induced characteristic fluctuation [11, 64, 66, 67]. Here, we introduce the vertical- and lateral-direction doping profile engineering to examine their effectiveness in DC and high-frequency fluctuation suppression. The mechanism of the use of epitaxial silicon layer and delta doping [22, 75, 76] is similar to the vertical doping profile engineering and therefore it can be inferred that the channel engineering is a trade-off between fluctuation and performance.

1. *Vertical doping profile engineering.* Discrete dopants located near the channel surface induce potential barriers and disturb the current-conducting path. Therefore, the idea of vertical doping profile engineering is to leave dopants away from the channel surface and reduce the fluctuation. **Figure 85(a)** illustrates the vertical doping profile engineering, where the doping profile along the longitudinal diffusion direction from surface to

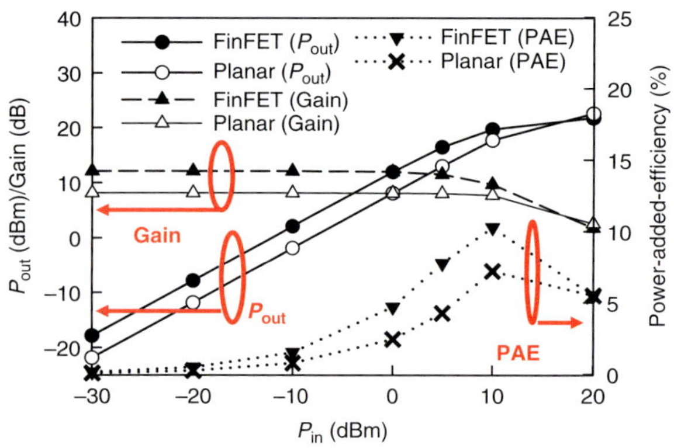

Figure 83 P_{out}, gain, and PAE of the planar MOSFETs and FinFETs as a function of P_{in}.

substrate follows the normal distribution. Both the mean position and the three sigma of this distribution are eight nanometers, which can be controlled by the manufacturing processes of ion implementation and thermal annealing. To compare the vertical-channel engineering with the original cases on clearly, the nominal threshold voltage of the vertical-channel engineering is adjusted to 140 mV. First, 758 dopants are randomly generated in a large rectangular solid ($x \times y \times z = 16 \times 2000 \times 16$ nm), in which the equivalent doping concentration is 1.48×10^{18} cm^{-3}. The inset of **Figure 85(a)** shows the nominal case of the vertical doping profile engineering, where the darker region indicates the higher doping concentration. The large rectangular solid is then partitioned into 125 subcubes of $(16\,\text{nm})^3$ cube and mapped into device channel for discrete dopant and coupled device-circuit simulation. Similar to the original doping profile, the number of dopants may vary from 0 to 14, and the average number is six, as shown in **Figure 85(b)**. The longitudinal dopant distribution of the improved and the original doping profile are shown in **Figures 85(c)** and **85(d)**, respectively. Since the position of discrete dopants of the original is random in each direction, the distribution of dopant number in channel depth is uniform. Additionally, for the vertical doping profile engineering, numbers of dopant appearing near the channel surface is significantly less than that of the original doping profile and thus may induce less surface potential fluctuation than the other.

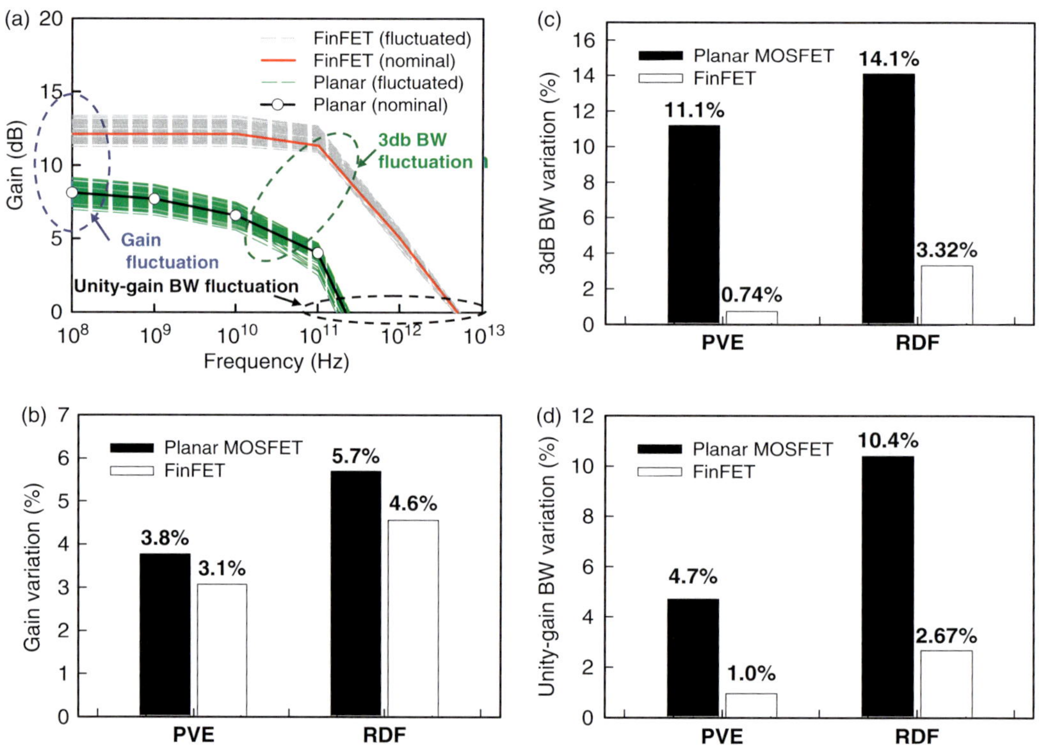

Figure 84 (a) The frequency response of the planar MOSFETs and FinFETs common-source amplifiers, in which (b) circuit gain, (c) 3dB bandwidth, and (d) unity-gain bandwidth are extracted.

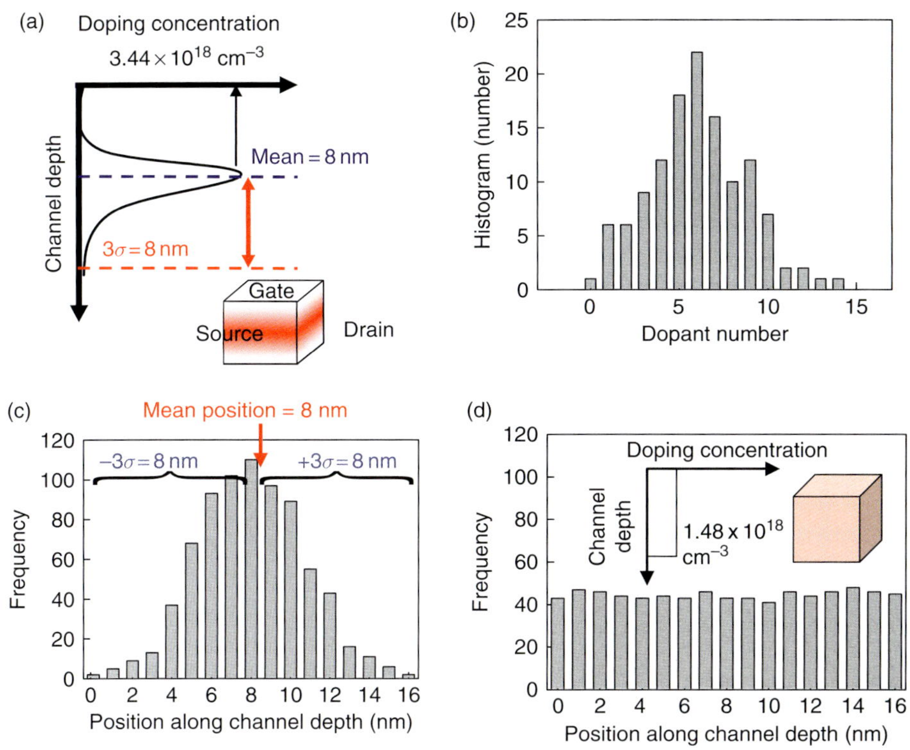

Figure 85 There will be 758 dopants which are within a large rectangular solid, in which the equivalent doping concentration is $1.48 \times 10^{18}\,cm^{-3}$. The dopant distribution in the direction of channel depth follows the normal distribution (a). Similarly, dopants within the $(16\,nm)^3$ cubes may vary from 0 to 14 (the average number is six) (b). The vertical dopant distribution of the vertical doping profile engineering and the original doping profile are shown in (c) and (d), respectively.

Figure 86 shows the comparison of DC characteristic fluctuations for the original and the vertical doping profile engineering. **Figure 86(c)** shows the $I_d - V_g$ characteristics of the discrete dopant-fluctuated device generated from the improved (dashed lines) and the original (solid lines) doping profiles. The spreading range of the vertical doping profile engineering is reduced and indicates a suppression of fluctuation on DC characteristics of device. The fluctuation of the threshold voltage is shown in **Figure 86(a)**, where the square symbol indicates the cases of the vertical doping profile engineering and the x symbols shows the cases of original doping profile. The scattering range of the threshold voltage is significantly suppressed, and the fluctuation of threshold voltage is reduced from 58.5 to 50.2 mV. The fluctuation of the on-state current, off-state current, and maximum transconductance are shown in **Figures 86(a), 86(b), 86(d),** and **86(e)**, respectively. The fluctuation of on-state current and maximum transconductance is effectively suppressed due to the lesser amount of dopant located near the channel surface. Notably, the suppression of I_{off} fluctuation is not significant due to the similar numbers of dopant locating near the path of leakage current (about 4 nm below the gate oxide). The proper selection of the dopant distribution near channel surface is a critical issue. **Figure 87(a)** shows the averaged gate capacitances of the 250 discrete dopant-fluctuated 16-nm-gate planar MOSFET. The total gate capacitance ($C_{g,\ total}$), is composed of gate-drain capacitance (C_{gd}), gate-source capacitance (C_{gs}), and gate-bulk capacitance (C_{gb}), and the shape and the value of the averaged C–V curves for the two doping profiles are quite similar. **Figure 87(b)** shows the C–V characteristics of the discrete dopant-fluctuated device generated from the original (dashed lines) and the improved (solid lines) doping profiles. Although the averaged C–V curves of the two doping profiles are similar, the devices from vertical doping profile engineering show a more condensed distribution and thus a smaller capacitance fluctuation than the others. The fluctuations of C_{gd} and C_{gs} are shown in **Figures 87(c)–87(f)**. The capacitance fluctuation can be reduced effectively by optimization of doping profile. At a drain voltage of 0 V, the fluctuations of $C_{g,\ total}$ are mainly contributed from C_{gd} and C_{gs}

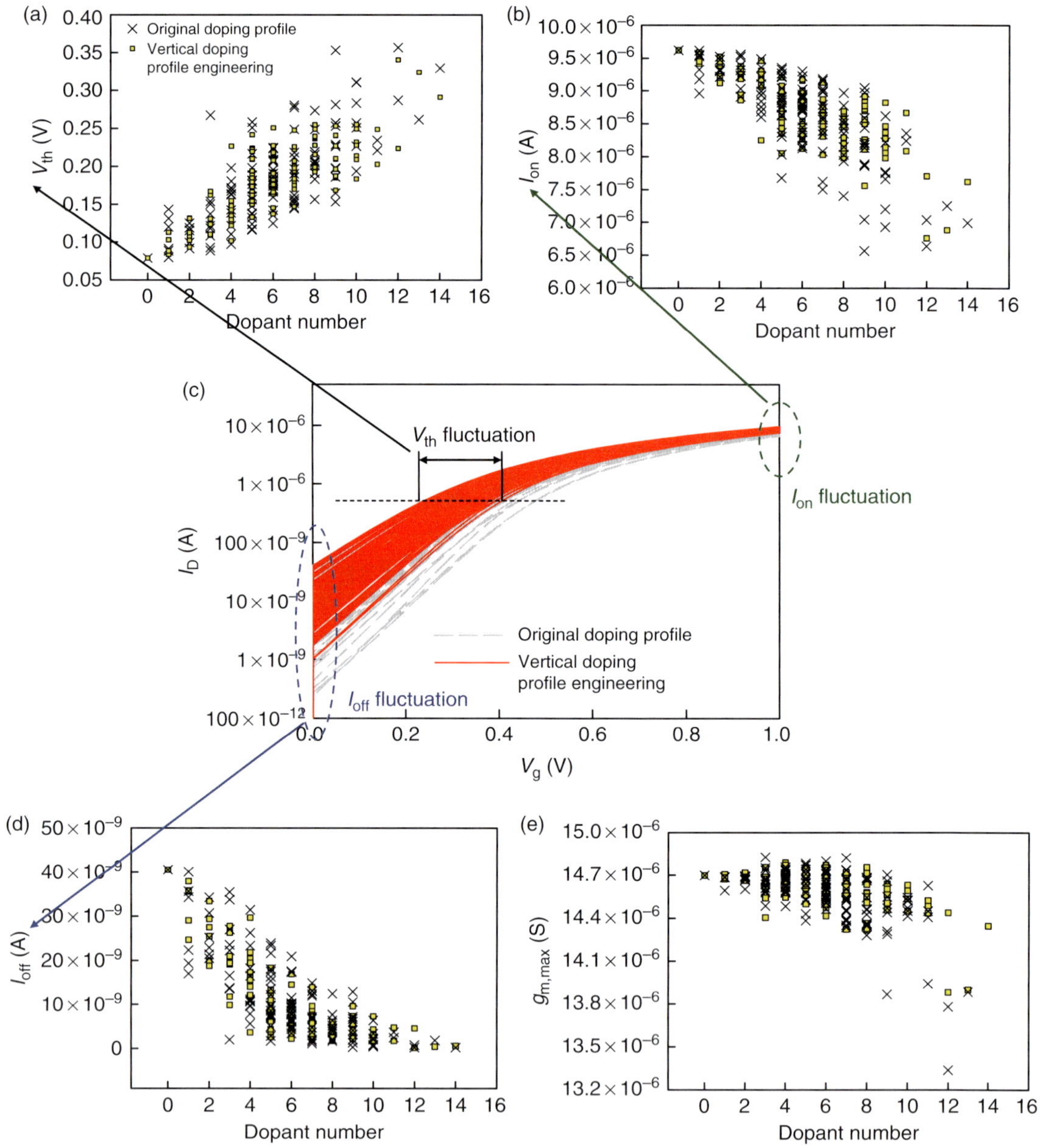

Figure 86 DC characteristic fluctuations of the 250 discrete dopant-fluctuated 16-nm-gate planar MOSFET from the original and the improved doping profile. The studied fluctuations of (a) threshold voltage, (b) on-state current, (c) I_d–V_g curves, (d) off-state current, (e) and maximum transconductance.

fluctuation. The fluctuations of C_{gd} and C_{gs} are similar, as displayed in **Figures 87(c)** and **87(d)**. Conversely, at a drain voltage of 1 V, the fluctuation of gate capacitance is dominated by C_{gs}, as shown in **Figures 87(e)** and **87(f)**.

The suppressed fluctuation of gate capacitance and DC characteristics of device implies the effectiveness of the fluctuation-suppression technique on the high-frequency characteristics. The high-frequency response of the nanoscale MOSFET circuit is shown in **Figure 88(a)**. The spreading range of the improved doping profile is reduced through the improvement of doping profile. The fluctuation of the high-frequency characteristics, the circuit gain, the 3 dB bandwidth, and the unity-gain bandwidth of the nanoscale MOSFET circuit are calculated, as shown in **Figures 88(b)** and **88(c)**, the square and 'x' symbols indicate the cases of the improved and original doping distributions. The effectiveness of fluctuation-suppression technique in nanoscale

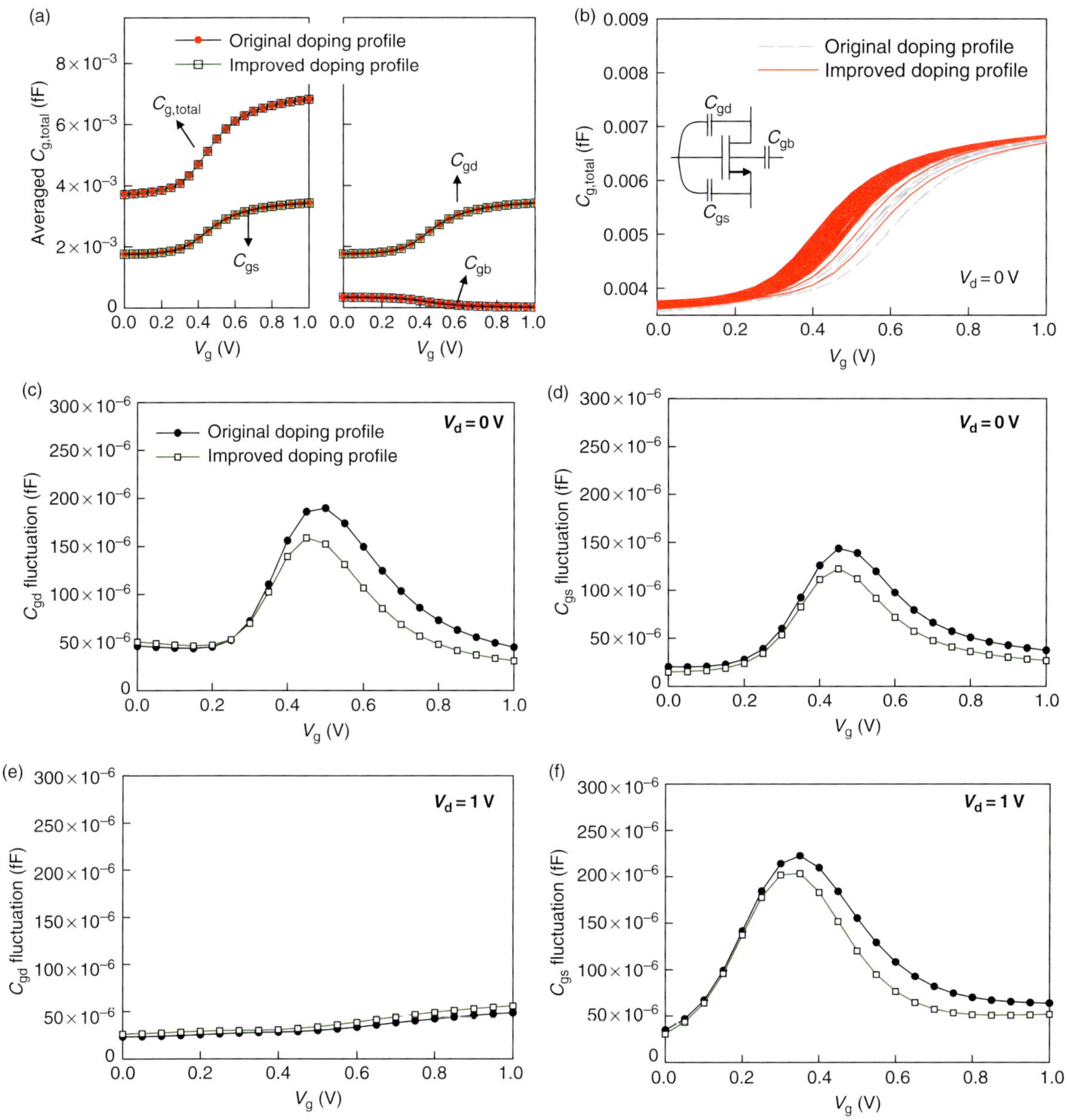

Figure 87 (a) The averaged gate capacitance of the studied devices. ($C_{g,total} = C_{gd} + C_{gs} + C_{gb}$). (b) The C–V characteristics of the discrete dopant-fluctuated devices generated from the original (dashed lines) and the improved (solid lines) doping profiles. (c, d) Fluctuation of C_{gd} and C_{gs} with $V_d = 0$ V. (e, f) Fluctuation of C_{gd} and C_{gs} with $V_d = 1$ V.

device and circuit is summarized in **Figure 89**. The fluctuation suppressions of the characteristics of the nanoscale device circuit are more pronounced than that of device due to the second-order nonlinear effect of circuit characteristics. The fluctuation suppression of the high-frequency characteristics results from the lower DC characteristics and gate capacitance fluctuation of the device by introducing a doping profile of Gaussian distribution along the longitudinal direction from surface to substrate. The result confirms the effectiveness of the fluctuation-suppression technique on a nanoscale device circuit that shows high-frequency characteristic fluctuation.. The comparison of characteristic fluctuations for devices without and with doping profile improvement is summarized in **Table 8**.

Without degrading the device and circuit performance, by considering a doping profile with

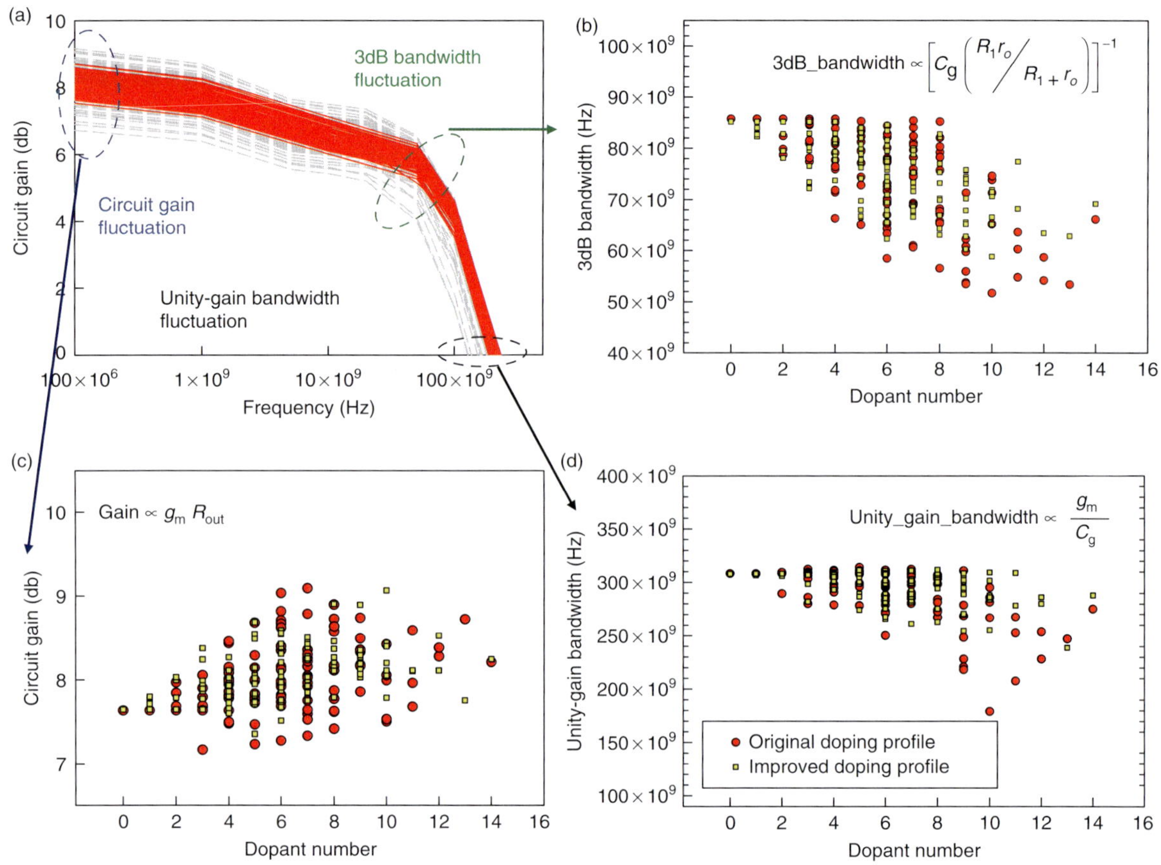

Figure 88 (a) Comparison of frequency response fluctuations of the discrete dopant-fluctuated 16-nm-gate cases generated from the improved (dashed line) and original (solid line) doping distributions. Comparison of (a) gain, (b) 3dB bandwidth, and (c) unity-gain bandwidth fluctuations of the 125 discrete dopant-fluctuated 16-nm-gate planar MOSFET circuit with the improved (squares) and original (x) doping profiles.

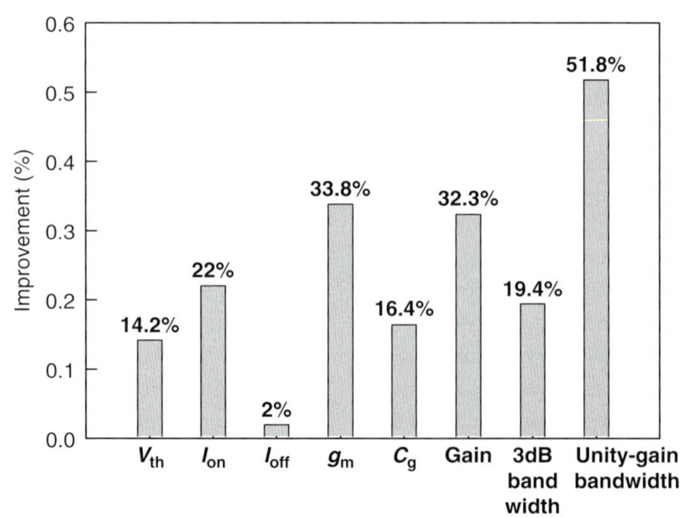

Figure 89 The effectiveness of fluctuation-suppression technique in both the nanoscale device and circuit.

Table 8 Comparison of high-frequency characteristic fluctuations for devices with original and improved vertical doping profile engineering

	Original	*Improved*
V_{th} (mV)	58.5	50.2
I_{on}(A)	6.7×10^{-7}	5.2×10^{-7}
I_{off} (A)	9.5×10^{-9}	9.3×10^{-9}
$g_{m,max}$ (S)	2.1×10^{-7}	1.4×10^{-7}
C_g ($\times 10^{-3}$fF)	0.24	0.20
Gain (dB)	0.465	0.314
3 dB bandwidth (Hz)	9.63×10^{9}	7.76×10^{9}
Unity-gain bandwidth (Hz)	2.93×10^{10}	1.14×10^{10}

less dopants located near the surface of the channel, the suppression technique on high-frequency characteristic fluctuations of the nanoscale transistor circuit have shown its effectiveness. The vertical doping profile engineering significantly reduces the intrinsic characteristic fluctuation in the device level, and thus suppressed fluctuations of the tested circuit gain, the 3 dB bandwidth and the unity-gain bandwidth, and the reductions are 32.3%, 19.4%, and 51.8%, respectively.

2. *Inverse lateral asymmetry doping profile.* In nanoscale FETs, the lateral asymmetric channel (LAC) devices have shown better control of the SCE [122, 123]. The high concentration channel doping near the source-end reduces SCE while low doping concentration near the drain-end ensures high mobility. However, less attention is paid regarding the existence of random doping effects for the lateral asymmetry channel doping profile [64].

Figure 90(a) shows the fluctuation of $C_{g,\ total}$ with different drain bias, V_d, for all discrete dopant fluctuated devices. The $C_{g,\ total}$ is composed of gate-drain capacitance, gate-source capacitance, and gate-bulk capacitance. **Figure 90(b)–90(d)** shows the components of gate capacitance, C_{gd}, C_{gs}, and C_{gb}. As the drain voltage increases from 0 to 1 V, the C_{gd} and fluctuation of C_{gd} are decreased due to the widening of depletion layer near the drain junction and channel pinch-off, as displayed in **Figure 90(b)**. Further, the C_{gs} and its fluctuation are increased, as shown in **Figure 90(c)**. As for the gate-bulk capacitance, the fluctuation is small enough to be neglected, as shown in **Figure 90(d)**. The drain bias-induced variations in C_{gd} and C_{gs} are observed, and the dependence of gate capacitance on different drain bias motivates the design of lateral asymmetry doping profile in the source–drain direction. To verify the asymmetry sketch of RDF, 500 discrete dopant-fluctuated transistors are generated and classified into two groups according to the random distribution of discrete dopants: near-source-end cases and near-drain-end cases, as shown in **Figure 91**. The near-source-end or near-drain-end cases are the devices with more dopants near the source or drain end of the channel. The inset of **Figure 91** describes the selection criteria. To strengthen and make the phenomenon clear, the difference of dopant numbers near the source and near the drain ends must be larger than two. For the near-source-end cases, the selection criterion is that a device should have more than three dopants in near-source end of the channel but with zero or one dopant located near the drain end. The near-source-end cases are then selected accordingly. There are 105 and 122 cases chosen as near-source-end and near-drain-end cases, respectively, which are fairly meaningful to reflect statistical randomness of dopant number and dopant position in the channel region.

Figure 92 plots the average capacitance for the two groups. Both the selected groups show quite similar C–V characteristics with $V_d = 0$ V and $V_d = 1$ V. The $C_{g,\ total}$ fluctuation with $V_d = 0$ V of the studied devices is shown in **Figure 93(a)**. The two groups also exhibit similar $C_{g,\ total}$ fluctuation. **Figure 93(b)** show the corresponding $C_{gd}$and C_{gs} fluctuations. For the group with dopants located near the source-end of the channel, the fluctuation of C_{gs} is more pronounced. Similarly, the C_{gd} fluctuation dominates the $C_{g,\ total}$ fluctuation for the near-drain-end cases. The random dopants located near source or drain side in device channel exhibits different C_{gd} and C_{gs} characteristics in spite of the similar $C_{g,\ total}$ fluctuation. More interestingly, the $C_{g,\ total}$ gate drain capacitance, and C_{gs} fluctuations with $V_d = 1$ V are shown in **Figures 94(a)–94(b)**, respectively. The $C_{g,\ total}$ fluctuation of near-drain-end cases is significantly smaller than that of near-

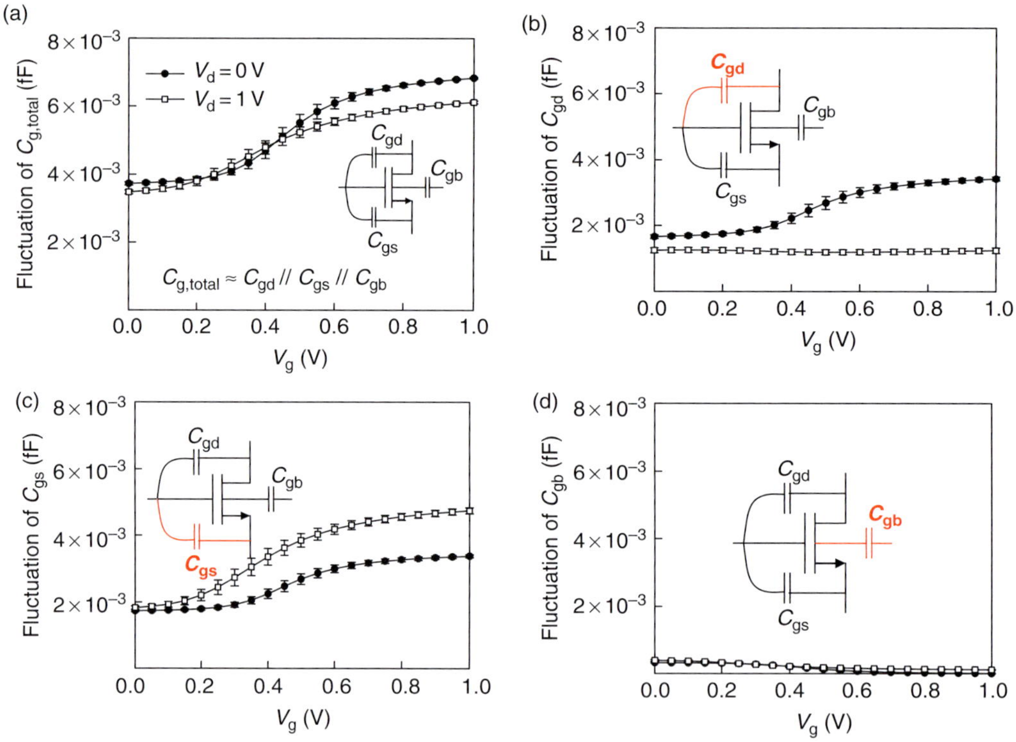

Figure 90 Fluctuations of: (a) the total gate capacitance ($C_{g,\ total} = C_{gd} // C_{gs} // C_{gb}$); (b) C_{gd}; (c) C_{gs}; (d) C_{gb} of all discrete dopant-fluctuated 16-nm-gate planar MOSFETs.

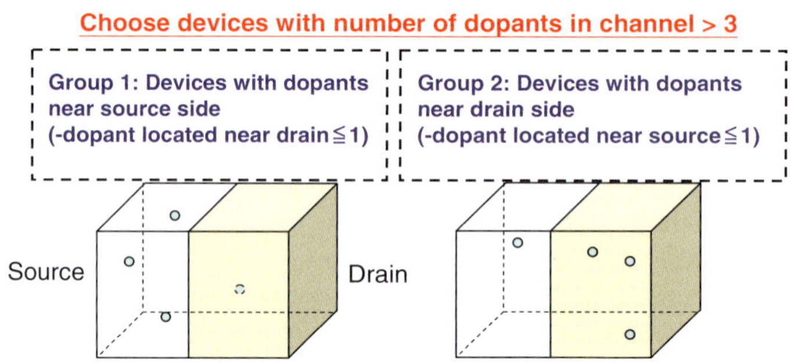

Figure 91 The discrete dopant-fluctuated devices are then classified into two groups: devices with dopants near source side and devices with dopants near drain side.

source-end cases. With the increased V_d, the C_{gs} fluctuations are increased and the C_{gd} fluctuation is significant. The C_{gd} fluctuation is below 10^{-4} fF. For the devices with dopants-near-drain side, since the major source of fluctuation is contributed by C_{gd}, the $C_{g,\ total}$ fluctuation is well controlled. However, for the devices with dopants-near-drain side, the major source of fluctuation, C_{gs}, is not decreased. The non-balance of the C_{gd} and C_{gs} characteristics and drain bias-dependent gate capacitance fluctuation motivate the design of asymmetry channel doping profile for suppression of random doping fluctuation.

The high concentration channel doping near the source-end of the conventional LAC device reduces SCE and ensures high mobility near the drain-end. However, such design contradicts the low fluctuation transistor design. An inverse lateral asymmetry doping profile is thus proposed with respect to the conventional LAC device to examine its resistance against RDF. **Figure 95(a)** illustrates the doping

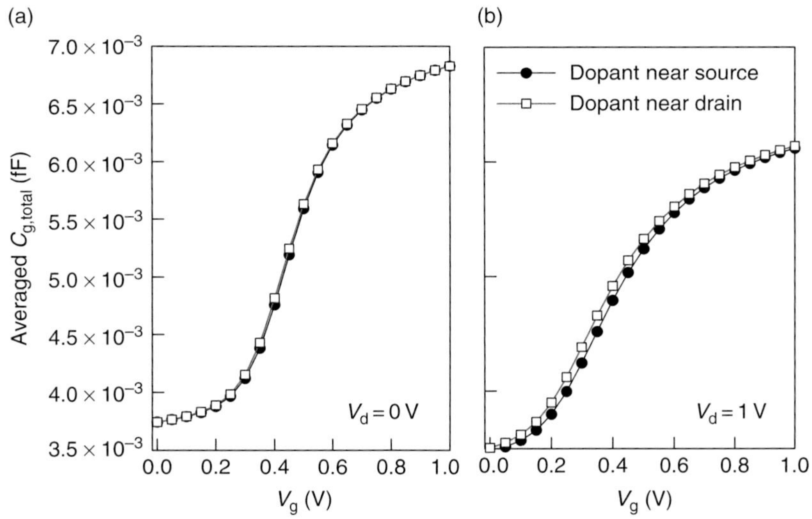

Figure 92 (a) Average of total gate capacitance at (a) $V_d = 0$ V and (b) $V_d = 1$ V.

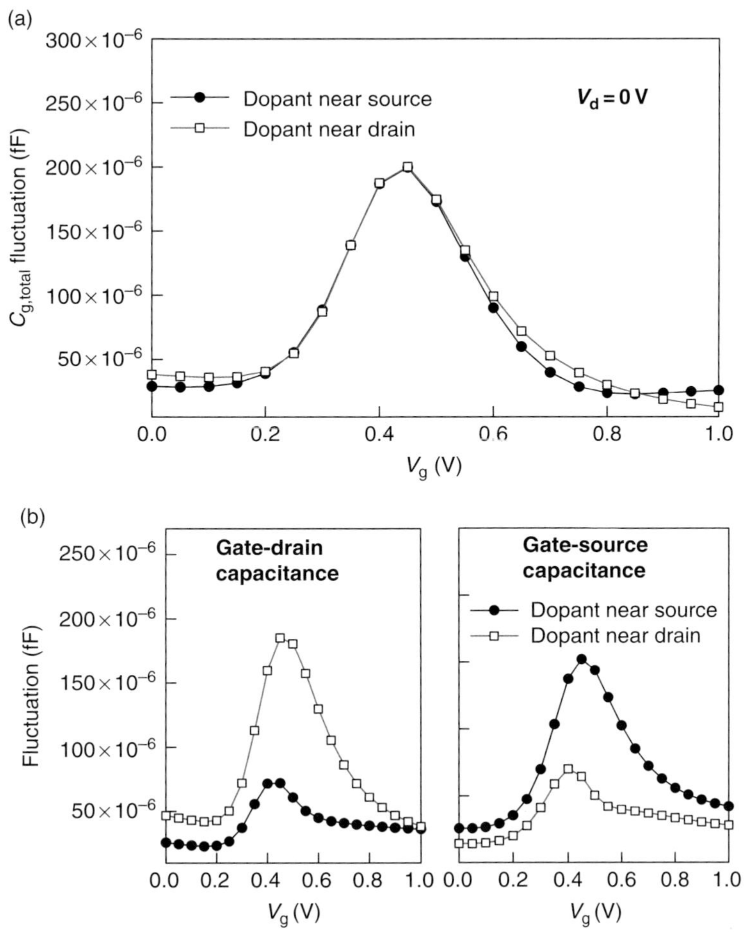

Figure 93 (a) Fluctuations of total gate capacitance at $V_d = 0$ V. (b) Fluctuations of C_{gd} and C_{gs} at $V_d = 0$ V. The major sources of fluctuation of dopant-near-drain and dopant-near-source devices are contributed from C_{gd} and C_{gs}, respectively.

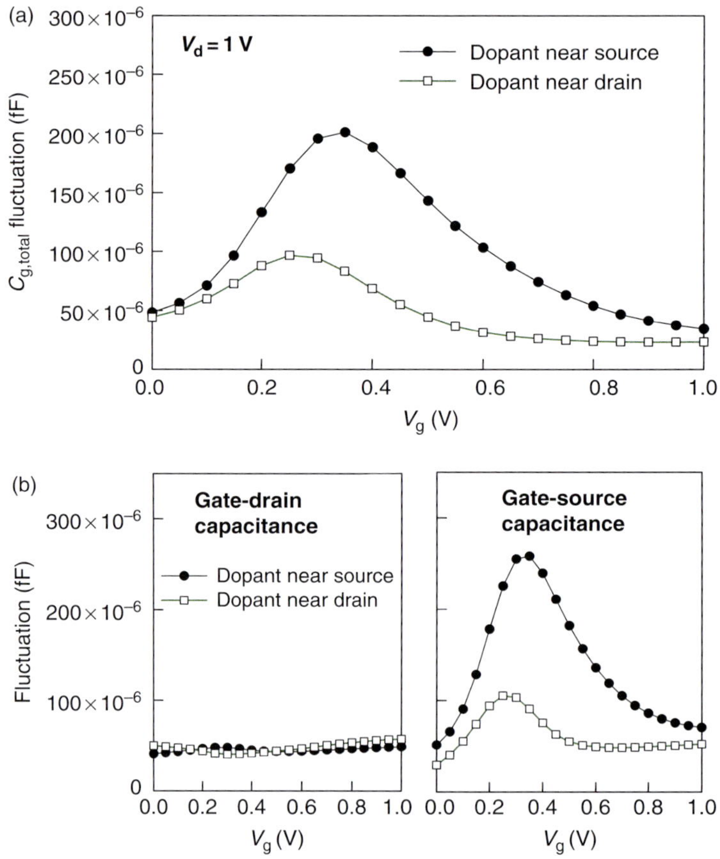

Figure 94 (a) Fluctuations of total gate capacitance at $V_d = 1$ V. (b) Fluctuations of C_{gd} and C_{gs} at $V_d = 1$ V, where the C_{gd} fluctuations are well controlled.

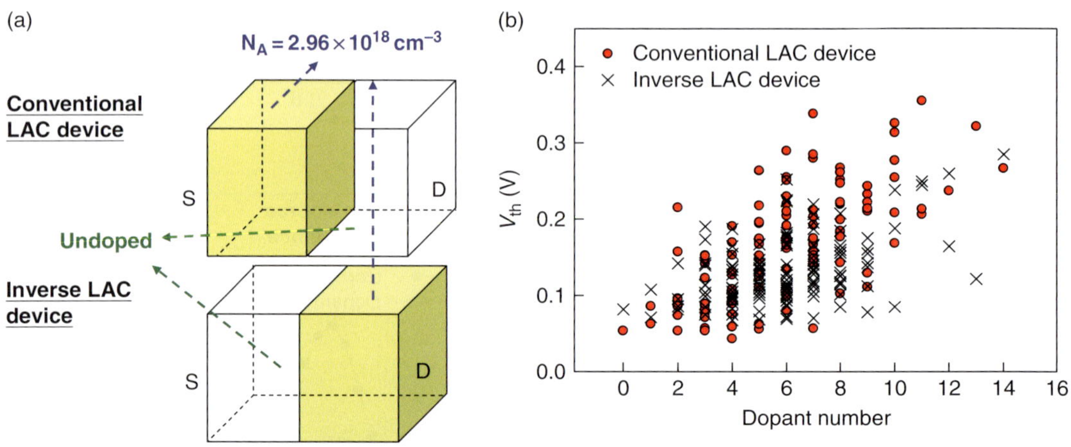

Figure 95 (a) Illustration of conventional and proposed lateral asymmetry doping profile. (b) Comparison of the V_{th} fluctuations for the conventional and proposed lateral asymmetry doping profile.

profile for the conventional LAC profile and inverse lateral asymmetry doping profile structures. The threshold voltage fluctuations are then compared in **Figure 95(b)**, where the x and o indicate the cases of the conventional LAC doping profile and inverse LAC doping profile. The V_{th} fluctuation for the proposed inverse lateral asymmetry doping profile device is 1.6 times smaller than the conventional LAC device because of the smaller injection velocity and current fluctuation for dopants located near the drain-end of the channel [123]. The DC and high-frequency characteristics of proposed inverse lateral asymmetry doping profile are then compared with the original doping profile and shown in **Figures 96(a)** and **96(b)**. The spreading range of the I_d–V_g and high-frequency response characteristics is significantly suppressed. **Figure 97** shows that the threshold voltage, on-state current, gate capacitance, circuit gain, 3 dB bandwidth, and the unity-gain bandwidth are reduced by 21.8%, 32.3%, 62.6%, 22.2%, 63.1%, and 41.4%, respectively. Note that the 3 dB bandwidth is strongly dependent on the gate capacitance and therefore exhibits similar improvement to gate capacitance fluctuation. However, unfortunately, the off-state current fluctuation increases by 45.2% due to the strengthened SCE. In the lateral asymmetry doping profile, unlike the profiles of conventional lateral asymmetry devices with higher channel doping concentrations nearer the source-end of the channel region, the

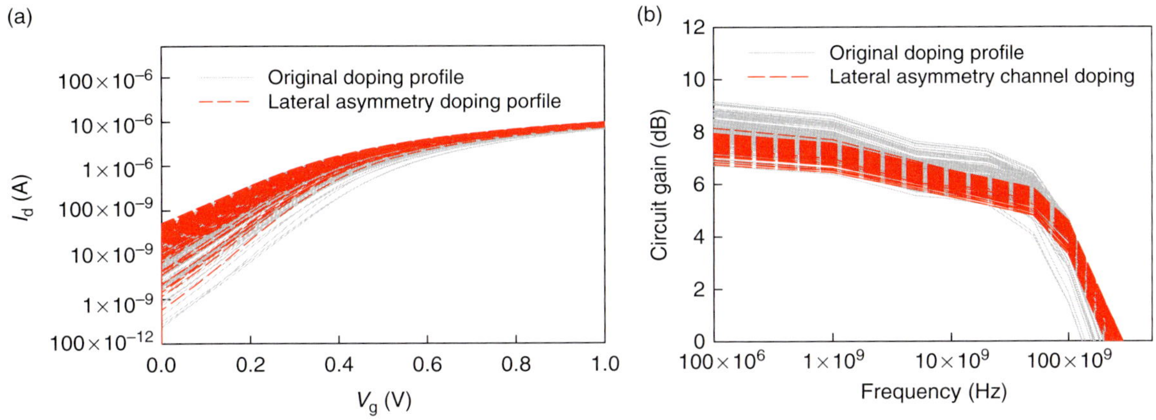

Figure 96 The explored (a) I_d–V_g and (b) high-frequency response characteristics for the proposed inverse lateral asymmetry doping profile and original doping profile.

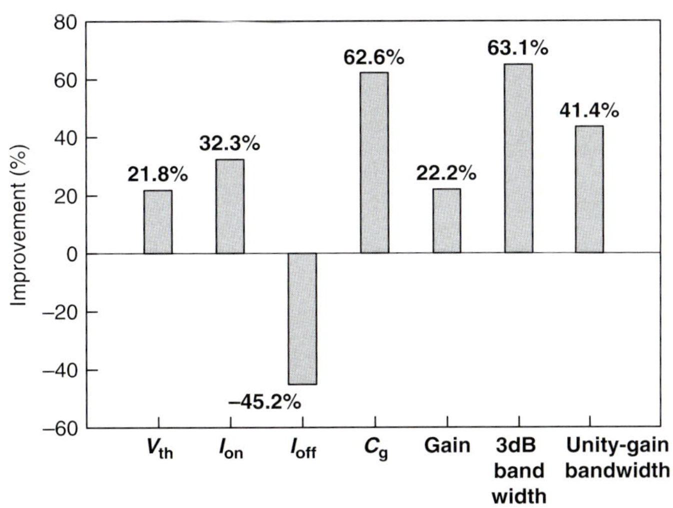

Figure 97 The effectiveness of the fluctuation suppression in both the nanoscale device and circuit, compared with the conventional device's results.

Table 9 Effectiveness of vertical doping profile engineering and lateral asymmetry doping profile with respect to the original doping profile

	Vertical doping profile engineering (%)	*Lateral asymmetry doping profile (%)*
V_{th}	14.2	21.8
I_{on}	22.0	32.2
I_{off}	2.0	−45.2
C_g	16.4	62.6
Gain	32.3	22.2
3 dB bandwidth	19.4	63.1
Unity-gain bandwidth	51.8	41.4

channel doping concentration is higher nearer the drain-end of the channel region. Dopants at the source-end of the channel may induce larger current fluctuation than at the drain-end of the channel.

The effectiveness of vertical and lateral doping profile engineering is summarized in **Table 9**. The lateral asymmetry doping profile exhibits promising fluctuation suppression on threshold voltage, on-state current, gate capacitance, and 3 dB bandwidth characteristics. However, the leakage current fluctuation is increased by 45.2% due to the stronger SCE. For future low-power applications, the vertical doping profile engineering is more suitable. Notably, the channel engineering is a trade-off between fluctuation and performance. The vertical-and lateral-channel profile co-design is an important issue to address in future nanoscale technology.

6.6.2 Circuit-Level Suppression

Here, we examine fluctuation-suppression techniques from the viewpoints of circuit design to reduce timing characteristic fluctuations. Since the transistor in series may inherit fluctuation at previous stage, an additional transistor is shunted on the most fluctuation-sensitive component to reduce circuit characteristic fluctuations [64]. An inverter is used as an example. The σt_{HL} is a dominating factor in timing fluctuations and the PMOSFET is used as the control, the fluctuation-suppression techniques are applied only to NMOSFETs.

Figures 98(a) compare t_r and t_f for the original inverter and the shunted NMOSFETs inverter circuits, in which the circuit topology of the inverter with shunted NMOSFETs is shown in inset. Since the number of NMOSFETs used to discharge the load capacitance is increased with respect to the original inverter, the nominal value and the fluctuations of the fall time are reduced. However, the increased number of transistors increases the load capacitance and the number of fluctuation sources, affecting the rise time and its fluctuations. **Figure 98(b)** compares the t_{LH}, t_{HL}, and their fluctuations. The t_{HL}, σt_{HL}, and σt_{HL} are reduced in the

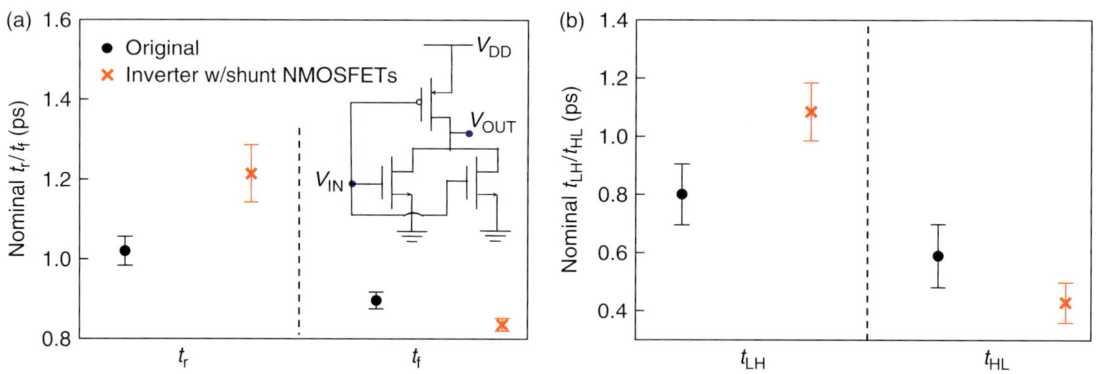

Figure 98 (a) The rise or fall time; (b) low-to-high and high-to-low delay time for the studied discrete dopant-fluctuated 16-nm-gate inverter circuit, where the inset shows the inverter with shunted NMOSFETs for mitigating the timing fluctuations induced by NMOSFETs.

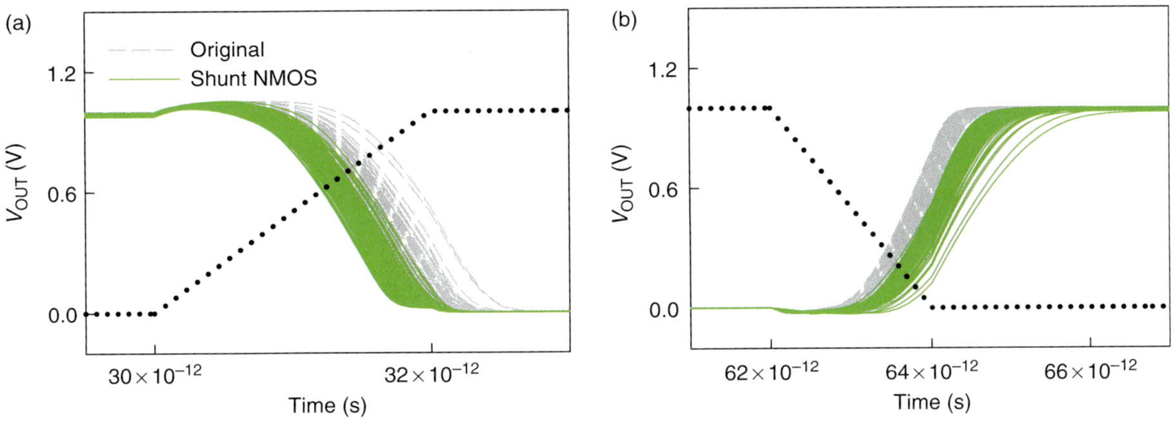

Figure 99 Comparison of (a) high-to-low and (b) low-to-high delay times for the original inverter circuit (dashed lines) and the inverter circuit with shunted NMOSFETs (solid lines).

shunted NMOSFETs inverter circuits. **Figures 99(a)** and **99(b)** compare the high-to-low and low-to-high transitions for the original inverter and the shunted NMOSFETs inverter circuits, respectively. Since the high-to-low transition begins when one of the NMOSFETs is turned on, t_{HL} and σt_{HL} of the shunted NMOSFETs inverter are smaller than the original inverter because the shunted NMOSFETs inverter has a higher probability to have a low V_{th}NMOSFET than the original inverter. The t_{HL} of inverter is possibly to be reduced with a low V_{th} NMOSFET shunted. The range of spread of t_{HL} is therefore reduced. Similarly, in a low-to-high transition, since the driving capability of PMOSFETs is lower than that of NMOSFETs, the rise transition starts when both of the shunted NMOSFETs are turned off. The t_{LH} of inverter is possibly to be increased with a high V_{th} NMOSFET shunted and therefore the range of spread of t_{LH} is reduced. However, since the probability that the inverters with low V_{th} NMOSFETs is reduced, the t_{LH} is increased significantly, as shown in **Figure 98(b)**. As shown in analytical models [21, 22, 76], from the device design viewpoint, the inverter with the shunted transistors can be implemented by designing NMOSFETs with a double device width. To compare the device characteristics clearly, the nominal threshold voltages for devices with doubled width are calibrated to 140 mV. **Figures 100(a)** and **100(b)** display the normalized rise or fall time and delay time for the (A) the original inverter, (B) the inverter with the shunted NMOSFETs, and (C) the inverter with doubled NMOSFET width. As expected, the timing characteristics of the inverter with doubled NMOSFET width are similar to that of the inverter with the shunted NMOSFETs. The high-to-low timing characteristics of the inverter with doubled NMOSFET width are improved at the cost of a worse low-to-high signal transition. Notably, both approaches ((B) and (C)) may increase

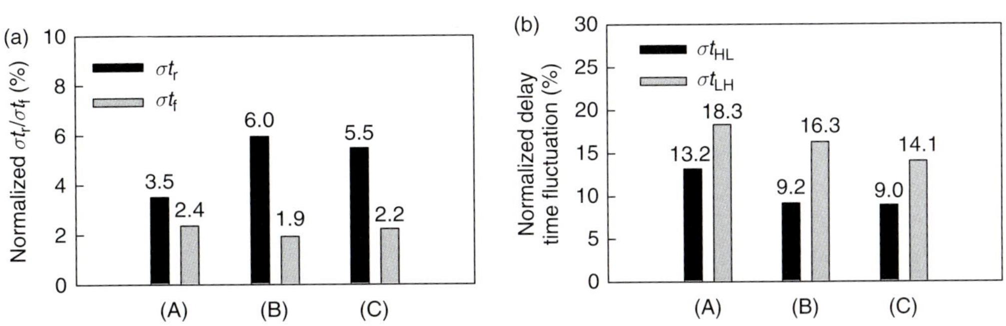

Figure 100 The normalized (a) rise or fall time (b) and delay time for the (A) the original inverter, (B) the inverter with the shunted NMOSFETs, and (C) the inverter with doubled NMOSFET width.

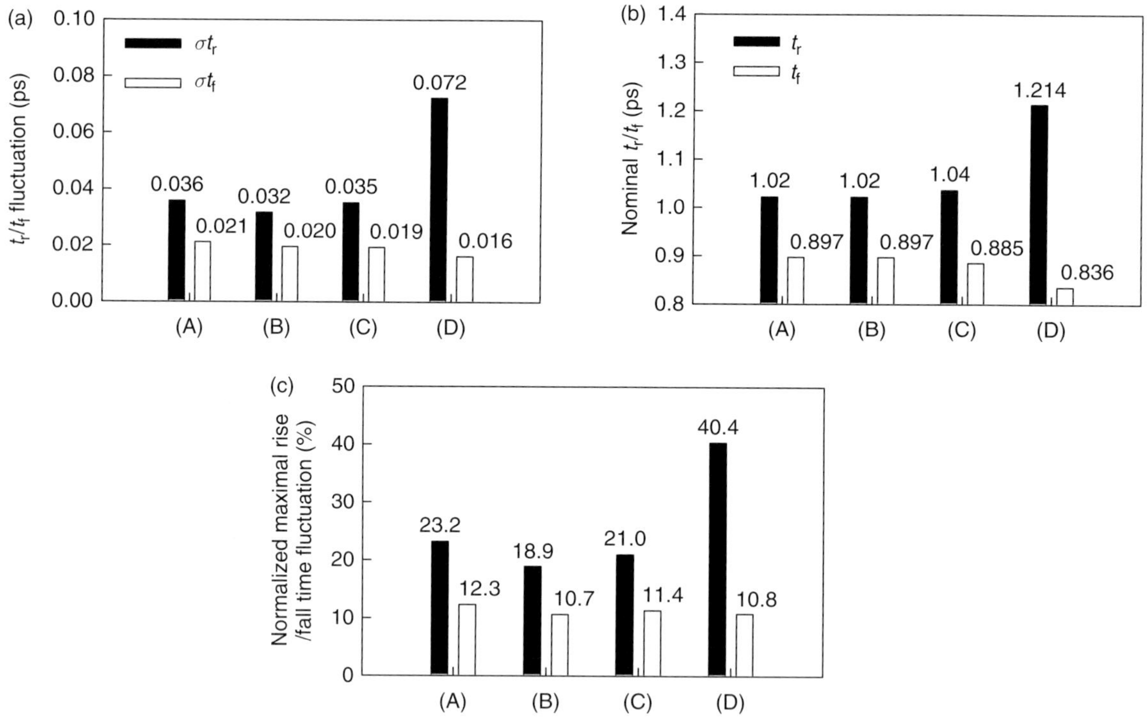

Figure 101 (a) The rise or fall time fluctuation; (b) the nominal rise or fall time, and (c) the normalized maximal rise or fall time fluctuation for (A) the original doping profile, (B) the vertical doping profile engineering, (C) the lateral asymmetry doping profile, and (D) the inverter with shunt n-MOSFETs.

the short circuit leakage power and chip area, potentially limiting the use of these fluctuation-suppression techniques.

WLOG, the channel engineering approaches mentioned in previous sections are implemented and compared with the inverter with shunt NMOSFETs. **Figures 101(a)** and **101(b)** respectively show $\sigma t_r/\sigma t_f$, and the nominal t_r/t_f for (A) the original doping profile devices, (B) the vertical doping profile engineering devices, (C) the lateral asymmetry doping profile devices, and (D) the inverters with shunt n-MOSFETs, respectively. The t_r and σt_r, of the original, vertical, and lateral asymmetry doping profile cases are similar due to the similar characteristics of PMOSFETs and the use of the same nominal V_{th} for n-MOSFETs (140 mV). The nominal t_r for the inverter with shunt n-MOSFETs scenario is increased due to the increased number of transistors, resulting in greater load capacitance in circuit. Moreover, due to the increased number of transistors, the number of fluctuation sources in the circuit is increased, resulting in increased σt_r. For the fall time characteristics, the values of σt_f for the vertical and lateral asymmetry doping profile devices are decreased due to the smaller C_g and I_{on} fluctuations. The reduction in fluctuation is not significant due to the small load capacitance in this simulation scenario. For the CMOS inverters with shunt n-MOSFETs, due to the increased number of n-MOSFETs to discharge the load capacitance, the nominal t_f and σt_f are significantly decreased. **Figure 101(c)** summarizes the normalized maximal rise and fall time fluctuations, which are defined by

$$\max\left(t_{r/f,\text{fluctuation}} - t_{\text{nominal}}\right)/t_{\text{nominal}}$$

The vertical doping profile engineering device shows a slightly greater improvement of fluctuation in rise time and fall time characteristics owing to the smaller SCE and the I_{off} fluctuation than the lateral asymmetry doping profile device. Note that although the CMOS inverters with shunt NMOSFETs do not show the smallest normalized rise and fall fluctuations, its nominal t_f and σt_f are smallest at the cost of worse t_r and σt_r. **Figures 102(a)** and **102(b)** show the delay time fluctuation and the nominal delay time, respectively and **Figure 102(c)** shows the normalized delay time fluctuation.

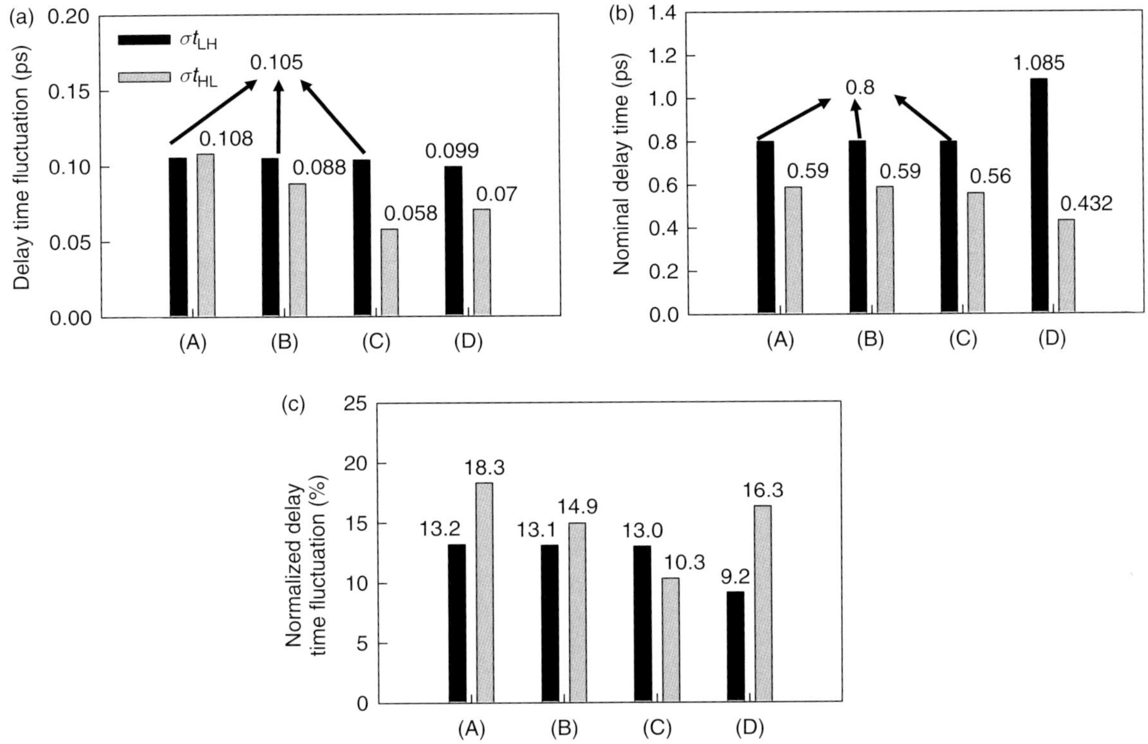

Figure 102 (a) The low-to-high/high-to-low delay time fluctuation; (b) the nominal low-to-high/high-to-low delay time; and (c) the normalized low-to-high/high-to-low delay time fluctuation for (A) the original doping profile, (B) the vertical doping profile engineering, (C) the lateral asymmetry doping profile, and (D) the inverter with shunt n-MOSFETs.

Because of the similar threshold voltage fluctuation of PMOSFETs, t_{LH} and σt_{LH} for (A) the original, (B) the vertical, and (C) the lateral asymmetry doping profiles devices are similar. The lateral asymmetry doping profile devices exhibit the smallest σt_{HL} owing to the smallest threshold voltage fluctuation of n-MOSFETs. For σt_{LH}, (D) the inverters with shunted n-MOSFETs exhibit the smallest fluctuation at the cost of a larger t_{LH}. **Table 10** summarizes the improvement of the timing characteristic fluctuations and nominal timing characteristics associated with the fluctuation-

Table 10 Comparison of the timing characteristics for fluctuation and nominal value among suppression techniques (fluctuation = (($t_{original}$−$t_{suppressed}$)/$t_{original}$) × 100%)

		Device design viewpoint		
	Circuit design viewpoint (A) (%)	*(B) (%)*	*(C) (%)*	*(D) (%)*
Increment/decrement of the nominal timing characteristic				
t_f	−6.86	−4.56	~0.0	−1.29
t_r	+18.84	+20.25	~0.0	+1.42
t_{HL}	−26.66	−20.70	~0.0	−5.23
t_{LH}	+35.53	+37.23	~0.0	−0.36
Timing characteristic fluctuations				
σt_{f*}	23.61	9.88	7.74	8.47
σt_r	−101.97	−88.36	11.32	1.53
σt_{HL}	34.76	39.17	18.42	46.59
σt_{LH}	5.79	6.76	0.52	1.71
Area cost	25	~20	0	0

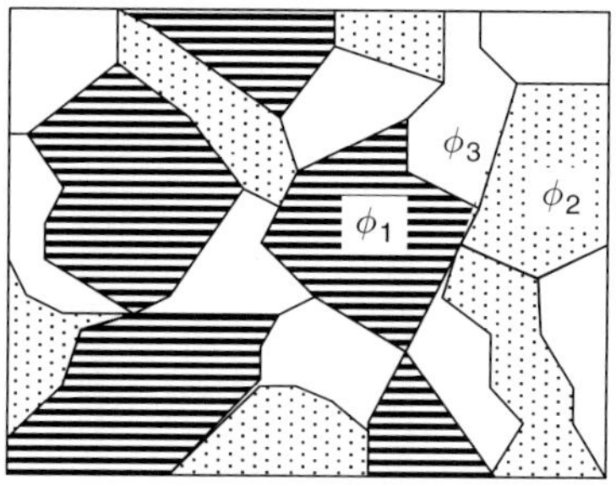

Material	Orienta-tion	Probability	WK
TiN	<200>	60%	4.6 eV
	<111>	40%	4.4 eV
MoN	<110>	60%	5.0 eV
	<112>	40%	4.4 eV

Figure 103 Illustration of grain orientation of metal gate and their distribution.

suppression techniques from the circuit (A: the inverter with shunted NMOSFETs) and the device design viewpoints (B: the inverter with doubled NMOSFET width, C: the inverter with vertical doping profile engineering, and D: the inverter with lateral asymmetry doping profile). The + and − signs in the nominal timing characteristics indicate the increase and decrease of the corresponding timing characteristics, respectively. The − sign in the timing characteristic fluctuations represents the degradation of the corresponding timing characteristics. To improve high-to-low transition characteristics and reduce high-to-low timing fluctuations, (A) the use of shunted NMOSFETs and (B) an increase in NMOSFET width can be considered, at the cost of worse low-to-high transition characteristics and increased chip area and power. In reducing delay time fluctuations, the (D) lateral asymmetry doping profile is the most effective. However, a large SCE of the device and the consumption of leakage power should be considered. Without scarifying the chip area and device performance, (c) vertical doping profile engineering is effective for low-power applications.

6.7 Conclusions and Future Work

In this chapter, the transition of device architecture and performance variability has been guided by experimentally validated planar MOSFETs, bulk FinFETs, and nanowire transistors. The variability of performance and yield in various nanoscale transistors are of great interest and are crucial for circuit design. This chapter introduces the characterization techniques for device variability and its impact on circuits. Moreover, various fluctuation-suppression approaches from device and circuit design viewpoints are examined. Consequently, the links should be established between circuit design and fundamental device technology to allow circuits and systems to accommodate the individual behavior of every transistor on a silicon chip.

It is expected that the nanoscale transistor will face several critical issues in the near future. The first one is the new fluctuation sources in advanced technology, such as the metal-gate material. In device characteristic fluctuations, metal gate is the technology to enhance performance and reduce fluctuations. However, the use of metal as a gate material introduces a new source of random variation due to the dependence of the work function on the orientation of metal grains, as shown in **Figure 103**. The fluctuation induced by work-function variability depends on the gate area and grain size of metal-gate material, which may result in an higher fluctuation than the random dopant effect [124]. Second, to realize the nanoscale device and circuit design, it is imperative to derive a well-established compact model consisting of the randomness effect. The coupled device-circuit simulation may provide an effective way to obtain circuit characteristics without the use of a compact model. However, there is still a lot of room to improve the numerical stability and numerical method in the coupled device-circuit simulation. Besides the above issues, the manufacturing of nanoscale transistor and development of related computer-aided-design software, such as layout and routing tools are additional issues to contend with nowadays. Nanoscale technology is a great challenge, although it promises ample scope for new semiconductor applications.

Acknowledgments

This work was supported in part by Taiwan National Science Council (NSC) under Contract NSC-97-2221-E-009-154-MY2, NSC-96-2221-E-009-210,

NSC-95-2221-E-009-336 and Contract NSC-95-2752-E-009-003-PAE, and by the Taiwan Semiconductor Manufacturing Company, Hsinchu, Taiwan under a 2007-2009 grant.

References

1. International Roadmap for Semiconductors.http://www.itrs.net (accessed July 2009).
2. Lee M (2001) Strained Ge channel p-type metal–oxide–semiconductor field effect transistor grown on Si_xGe_{1-x}/Si virtual substrates. *Applied Physics Letters* 79: 3344.
3. Shang H (2002) High mobility p-channel germanium MOSFETs with a thin Ge oxynitride gate dielectric. *In: Technical Digest of International Electron Devices Meeting,* pp. 441–444.
4. Li Y, Chen HM, Yu SM, *et al.* (2008) Strained CMOS devices with shallow-trench-isolation stress buffer layers. *IEEE Transactions on Electron Devices* 15: 1085–1089.
5. Zhu S, Yu HY, Whang SJ, *et al.* (2004) Schottky-barrier S/D MOSFETs with high-k gate dielectrics and metal-gate electrode. *IEEE Electron Device Letters* 25: 268–270.
6. Chiu C (2002) A sub-400 degree C germanium MOSFET technology with high-κ dielectric and metal gate. *In: Technical Digest of International Electron Devices Meeting,* pp. 437–440.
7. Wilk GD, Wallace RM, and Anthony JM (2001) High-k gate dielectrics: Current status and materials properties considerations. *Journal of Applied Physics* 89: 5243.
8. Frank DJ and Wong HS (2000) Analysis of the design space available for high-k gate dielectrics in nanoscale MOSFETs. *Superlattices and Microstructure* 28: 485–491.
9. Li Y and Yu SM (2007) A coupled-simulation-and-optimization approach to nanodevice fabrication with minimization of electrical characteristics fluctuation. *IEEE Transaction on Semiconductor Manufacturing* 20: 432–438.
10. Juang MH, Ou-Yang CI, and Jang SL (2002) A design consideration of channel doping profile for sub-0.12 μm partially depleted SOI n-MOSFET's. *Solid-State Electronics* 46: 1117–1121.
11. Joshi S, Boyd S, and Dutton RW (2005) Optimal doping profiles via geometric programming. *IEEE Transactions on Electron Devices* 52: 2660–2675.
12. Peng Y, Weifeng Q, and Pan DZ (2007) Fast lithography image simulation by exploiting symmetries in lithography systems. *IEEE Transaction on Semiconductor Manufacturing* 20: 432–438.
13. Zhang F, Zhang L, and Claus RO (2008) Investigation of the photo resist pattern profile contrast improvement in interference lithography technique using 488-nm Laser. *IEEE Transaction on Semiconductor Manufacturing* 21: 464–468.
14. Dobisz EA, Bandic ZZ, Wu TW, *et al.* (2008) Patterned media: Nanofabrication challenges of future disk drives. *Proceedings of the IEEE* 96: 1836–1846.
15. Yang FL, Hwang JR, and Li Y (2006) Electrical characteristic fluctuations in sub-45nm CMOS devices. *In: IEEE Custom Integrated Circuits Conference*, pp. 691–694.
16. Yang FY, Lee DH, Chen HY, *et al.* (2004) 5nm-gate nanowire FinFET. *In: Symposium on VLSI Technology, Digest of Technical Papers*, pp. 196–197.
17. Yang FL, Chen HY, Chen FC, *et al.* (2002) 25 nm CMOS omega FETs. *In: Technical Digest of International Electron Devices Meeting*, pp. 255–258.
18. Li Y, Chou HM, and Lee JW (2005) Investigation of electrical characteristics on surrounding-gate and omega-shaped-gate nanowire FinFETs. *IEEE Transactions on Nanotechnology* 4: 510–516.
19. Li Y and Chou HM (2005) A comparative study of electrical characteristic on sub-10 nm double gate MOSFETs. *IEEE Transactions on Nanotechnology* 4: 645–647.
20. Li Y and Hwang CH (2007) Effect of fin angle on electrical characteristics of nanoscale round-top-gate bulk FinFETs. *IEEE Transactions on Electron Devices* 54: 3426–3429.
21. Li Y and Yu SM (2006) Comparison of random-dopant-induced threshold voltage fluctuation in nanoscale single-, double-, and surrounding-gate field-effect transistors. *Japanese Journal of Applied Physics* 45: 6860–6865.
22. Li Y and Yu SM (2006) A study of threshold voltage fluctuations of nanoscale double gate metal–oxide–semiconductor field effect transistors using quantum correction simulation. *Journal of Computational Electronics* 5: 125–129.
23. Li Y and Hwang CH (2007) Electrical characteristic fluctuations in 16 nm bulk-FinFET devices. *Microelectronics Engineering* 84: 2093–2096.
24. Li Y and Hwang CH (2007) Discrete-dopant-induced characteristic fluctuations in 16 nm multiple-gate silicon-on-insulator devices. *Journal of Applied Physics* 102: 084509.
25. Li Y, Hwang CH, and Huang HM (2008) Large-scale atomistic approach to discrete-dopant-induced characteristic fluctuations in silicon nanowire transistors. *Physica Status Solidi (A)* 205: 1505–1510.
26. Li Y and Hwang CH (2008) Discrete-dopant-fluctuated threshold voltage roll-off in sub-16 nm bulk FinFETs. *Japanese Journal of Applied Physics* 47: 2580–2584.
27. Li Y, Hwang CH, and Cheng HW (2009) Process-variation-and random-dopants-induced threshold voltage fluctuations in nanoscale planar MOSFET and bulk FinFET devices. *Microelectronic Engineering* 86: 277–282.
28. Park T, Choi S, Lee DH, *et al.* (2003) Fabrication of body-tied FinFET's (omega MOSFET's) using bulk Si wafers. *In: Symposium on VLSI Technology, Digest of Technical Papers*, pp. 135–136.
29. Gnani E, Reggiani S, Rudan, M, *et al.* (2006) Design considerations and comparative investigation of ultra-thin SOI, double-gate and cylindrical nanowire FETs. *In: Proceeding of European Solid-State Device Research Conference*, pp. 371–374.
30. Jing W and Lundstrom M (2005) Channel material optimization for the ultimate planar and nanowire MOSFETs: A theoretical exploration. *In: Device Research Conference*, pp. 241–242.
31. Monfray S, Skotnicki T, Morand Y, *et al.* (2002) 50 nm-gate all around (GAA)-silicon on nothing (SON)-devices: A simple way to co-integration of GAA transistors within bulk MOSFET process. *In: Symposium on VLSI Technology, Digest of Technical Papers*, pp. 108–109.
32. Colinge JP, Gao MH, Romano-Rodriguez A, *et al.* (1990) Silicon-on-insulator 'gate-all-around device'. *In: Technical Digest of International Electron Devices Meeting*, pp. 595–598.
33. Auth CP and Plummer JD (1997) Scaling theory for cylindrical, fully-depleted, surrounding-gate MOSFET's. *IEEE Electron Device Letters* 18: 74–76.
34. Chau R, Doyle B, Doczy M, *et al.* (2003) Silicon nano-transistors and breaking the 10 nm physical gate length barrier. *In: Device Research Conference*, pp. 123–126.
35. Balaban SN, Pokatilov EP, Fomin VM, *et al.* (2002) Quantum transport in a cylindrical sub-01 μm silicon-based MOSFET. *Solid State Electronics* 46: 435–444.

36. Wang R, Zhuge J, Huang R, *et al.* (2007) Analog/RF performance of Si nanowire MOSFETs and the impact of process variation. *IEEE Transactions on Electron Devices* 54: 1288–1294.
37. Reza S, Bosman G, SIslam M, *et al.* (2006) Noise in silicon nanowires. *IEEE Transactions on Electron Devices* 5: 523–529.
38. Chau R, Datta S, Doczy M, *et al.* (2005) Benchmarking nanotechnology for high-performance and low-power logic transistor applications. *IEEE Transaction on Nanotechnology* 4: 153–158.
39. Cohen GM, Solomon PM, Laux SE, *et al.* (2007) An n-FET with a Si nanowire channel and doped epitaxially-thickened source and drain regions. *In: Device Research Conference*, pp. 175–176.
40. Fried DM, Hergenrother JM, Topol AW, *et al.* (2004) Aggressively scaled (0143 μm^2) 6T-SRAM cell for the 32 nm node and beyond. *In: Technical Digest of International Electron Devices Meeting*, pp. 261–264.
41. Wakabayashi H, Ezaki T, Sakamoto T, *et al.* (2006) Characteristics and modeling of sub-10-nm planar bulk CMOS devices fabricated by lateral source/drain junction control. *IEEE Transactions on Electron Devices* 53: 715–720.
42. Doris B, Ieong M, Kanarsky T, *et al.* (2002) Extreme scaling with ultra-thin Si channel MOSFETs. *In: Technical Digest of International Electron Devices Meeting*, pp. 267–270.
43. Lee H, Yu LE, Ryu SW, *et al.* (2006) Sub-5nm all-around gate FinFET for ultimate scaling. *In: Symposium on VLSI Technology, Digest of Technical Papers*, pp. 58–59.
44. Bandler JW, Biemacki RM, Cai Q, *et al.* (1992) Integrated physics-oriented statistical modeling, simulation and optimization. *IEEE Transaction on Microwave Theory Technology* 40: 1374–1400.
45. Stoneking DE, Bilbro GL, Gilmore PA, *et al.* (1992) Yield optimization using a GaAs process simulator coupled to a physical device model. *IEEE Transaction on Microwave Theory Technology* 40: 1353–1363.
46. Zaabab AH, Zhang QJ, and Nakhla M (Senior Member, IEEE) (1995) A neural network modeling approach to circuit optimization and statistical design. *IEEE Transaction on Microwave Theory Technology* 43: 1349–1358.
47. Purviance J and Meehan M (1991) CAD for statistical analysis and design of microwave circuits. *International Journal of Microwave and Millimeter-Wave Computer-Aided Engineering* I: 59–76.
48. Li Q, Zhang J, Li W, *et al.* (2001) RF circuit performance degradation due to soft breakdown and hot-carrier effect in deep-submicrometer CMOS technology. *IEEE Transaction on Microwave Theory Technology* 49: 1546–1551.
49. Ohtou T, Sugii N, and Hiramoto T (2007) Impact of parameter variations and random dopant fluctuations on short-channel fully depleted SOI MOSFETs with extremely thin BOX. *IEEE Electron Devices Letters* 28: 740–742.
50. Majima H, Ishikuro H, and Hiramoto T (2000) Experimental evidence for quantum mechanical narrow channel effect in ultra-narrow MOSFETs. *IEEE Electron Devices Letters* 21: 396–398.
51. Choi JA, Lee K, Jin YS, *et al.* (2004) Large scale integration and reliability consideration of triple gate transistors. *In: Technical Digest of International Electron Devices Meeting*, pp. 647–650.
52. Park T, Choi S, Lee DH, *et al.* (2003) Fabrication of body-tied FinFETs (omega MOSFETs) using bulk Si wafers. *In: Symposium on VLSI Technology, Digest of Technical Papers*, pp. 135–136.
53. Tsui BY and Lin CP (2004) A novel 25nm modified Schottky barrier FinFET with high performance. *IEEE Electron Device Letter* 25: 430–433.
54. Kedzierski J, Ieong M, Kanarsky T, *et al.* (2004) Fabrication of metal gated FinFETs through complete gate silicidation with Ni. *IEEE Transactions on Electron Devices* 51: 2115–2120.
55. Wu X, Chan PCH, and Chan M (2005) Impacts of nonrectangular fin cross section on the electrical characteristics of FinFET. *IEEE Transactions on Electron Devices* 52: 63–68.
56. Hsu TH, Lue HY, King YC, *et al.* (2005) A high-performance body-tied FinFET bandgap engineered SONOS (BE-SONOS) for nand-type flash memory. *IEEE Transactions on Electron Devices* 28: 443–445.
57. Asenov A, Kaya S, and Brown AR (2003) Intrinsic parameter fluctuations in decananometer MOSFETs introduced by gate line edge roughness. *IEEE Transactions on Electron Devices* 50: 1254–1260.
58. Fukutome H, Momiyama Y, Kubo T, *et al.* (2006) Direct evaluation of gate line edge roughness impact on extension profiles in sub-50-nm n-MOSFETs. *IEEE Transactions on Electron Devices* 53: 2755–2763.
59. Shiying X and Bokor J (2004) A simulation study of gate line edge roughness effects on doping profiles of short-channel MOSFET devices. *IEEE Transactions on Electron Devices* 51: 228–232.
60. Diaz CH, Tao HJ, Ku YC, *et al.* (2001) An experimentally validated analytical model for gate line-edge roughness (LER) effects on technology scaling. *IEEE Electron Device Letters* 22: 287–289.
61. Leunissen LHA, Jonckheere R, Ronse K, *et al.* (2003) Influence of gate patterning on line edge roughness. *Journal of Vacuum Science and Technology B* 21: 3140–3143.
62. Taur Y and Ning TH (1998) *Fundamentals of Modern VLSI Devices*. Cambridge: Cambridge University Press.
63. Li Y, Yu SM, Hwang JR, *et al.* (2008) Discrete dopant fluctuated 20nm/15nm-gate planar CMOS. *IEEE Transactions on Electron Devices* 55: 1449–1455.
64. Li Y, Hwang CH, and Cheng HW (2009) Discrete-dopant-fluctuated transient behavior and variability suppression in 16-nm-gate complementary metal–oxide–semiconductor field-effect transistors. *Japanese Journal of Applied Physics* 48: 04C051.
65. Li Y, Hwang CH, Yeh TC, *et al.* (2008) Large-scale atomistic approach to random-dopant-induced characteristic variability in nanoscale CMOS digital and high-frequency integrated circuits. *In: Proceeding of International Conference on Computer-Aided Design (ICCAD)*, pp. 278–285.
66. Li Y and Hwang CH (2008) High-frequency characteristic fluctuations of nano-MOSFET circuit induced by random dopants. *IEEE Transaction on Microwave Theory and Techniques* 56: 2726–2733.
67. Li Y, Hwang CH, Yeh TC, *et al.* (2008) Reduction of discrete-dopant-induced high-frequency characteristic fluctuations in nanoscale CMOS circuit. *In: International Conference Simulation Semiconductor Processes and Devices*, pp. 209–212.
68. Li Y, Yu SM, and Chen HM (2007) Process-variation- and random-dopants-induced threshold voltage fluctuations in nanoscale CMOS and SOI devices. *Microelectronics Engineering* 84: 2117–2120.
69. Millar C, Reid D, Roy G, *et al.* (2008) Accurate statistical description of random dopant induced threshold voltage variability. *IEEE Electron Device Letters* 29: 946–948.

70. Wong HS, Taur Y, and Frank DJ (1999) Discrete random dopant distribution effects in nanometer-scale MOSFETs. *Microelectronics Reliability* 38: 1447–1456.
71. Keyes RW (1975) Effect of randomness in distribution of impurity atoms on FET thresholds. *Applied Physics* 8: 251–259.
72. Francis P, Terao A, and Flandre D (1994) Modeling of ultrathin double-gate NMOS/SOI transistors. *IEEE Transactions on Electron Devices* 41: 715–720.
73. Tang XH, De VK, and Meindl JD (1997) Intrinsic MOSFET parameter fluctuations due to random dopant placement. *IEEE Transaction on VLSI Systems* 5: 369–376.
74. Stolk PA, Widdershoven FP, and Klaassen DBM (1998) Modeling statistical dopant fluctuations in MOS transistors. *IEEE Transactions on Electron Devices* 45: 1960–1971.
75. Noda K, Tatsumi T, Uchida T, *et al.* (1998) A 0.1-μm delta doped MOSFET fabricated with post-low-energy implanting selective epitaxy. *IEEE Transactions on Electron Devices* 45: 809–813.
76. Asenov A and Saini S (1999) Suppression of random dopant-induced threshold voltage fluctuations in sub-0.1-μm MOSFET's with epitaxial and δ-doped channels. *IEEE Transactions on Electron Devices* 46: 1718–1724.
77. Brown A and Asenov A (2008) Capacitance fluctuations in bulk MOSFETs due to random discrete dopants. *Journal of Computational Electronics* 7: 115–118.
78. Andrei P and Oniciuc L (2008) Suppressing random dopant-induced fluctuations of threshold voltages in semiconductor devices. *Journal of Applied Physics* 104: 104508.
79. Zhou JR and Ferry DK (1995) 3D simulation of deep-submicron devices. How impurity atoms affect conductance. *IEEE Computational Science and Engineering* 2: 30–37.
80. Vasileska D, Gross WJ, and Ferry DK (1998) Modeling of deep-submicrometer MOSFETs: Random impurity effects, threshold voltage shifts and gate capacitance attenuation. *In: Extended Abstracts of International Workshop on Computational Electronics*, pp. 259–262.
81. Sano N, Matsuzawa K, Mukai M, *et al.* (2000) Role of long-range and short-range coulomb potentials in threshold characteristics under discrete dopants in sub-0.1 μm Si-MOSFETs. *In: Technical Digest of International Electron Devices Meeting*, pp. 275–278.
82. Sano N, Matsuzawa K, Mukai M, *et al.* (2002) On discrete random dopant modeling in drift-diffusion simulations: Physical meaning of 'atomistic' dopants. *Microelectronics Reliability* 42: 189–199.
83. Sano N and Tomizawa M (2007) Random dopant model for three-dimensional drift-diffusion simulations in metal–oxide–semiconductor field-effect-transistors. *Applied Physics Letters* 79: 2267.
84. Tanabe R, Ashizawa Y, and Oka H (2006) Investigation of SNM with random dopant fluctuations for FD SGSOI and FinFET 6T SOI SRAM cell by three-dimensional device simulation. *In: Simulation of Semiconductor Processes and Device Conference*, pp. 103–106.
85. Cheng B, Roy S, Roy G, *et al.* (2006) Impact of intrinsic parameter fluctuations on SRAM cell design. *In: International Conference on Solid-State and Integrated Circuit Technology*, pp. 1290–1292.
86. Mahmoodi H, Mukhopadhyay S, and Roy K (2005) Estimation of delay variations due to random-dopant fluctuations in nanoscale CMOS circuits. *IEEE Journal of Solid-State Circuits* 40: 1787–1796.
87. Tang X, Bowman KA, Eble JC, *et al.* (1999) Impact of random dopant placement on CMOS delay and power dissipation. *In: Proceeding of European Solid-State Device Research Conference*, pp. 184–187.
88. Springer SK, Lee S, Lu N, *et al.* (2006) Modeling of variation in submicrometer CMOS ULSI technologies. *IEEE Transactions on Electron Devices* 53: 2168–2178.
89. Balasubramanian A, Fleming PR, Bhuva BL, *et al.* (2003) Implications of dopant-fluctuation-induced Vt variations on the radiation hardness of deep submicrometer CMOS SRAMs. *IEEE Transaction on Device and Materials Reliability* 8: 135–144.
90. Yamagiwa H, Abo S, Wakaya FM, *et al.* (2006) Local resistance measurement across grain boundaries in low-temperature polycrystalline silicon layer of thin film transistor using scanning spreading resistance microscopy. *Applied Physics Letters* 89: 062101.
91. Arora ND, Rios R, and Huang CL (1995) Modeling the polysilicon depletion effect and its impact on submicrometer CMOS circuit performance. *IEEE Transactions on Electron Devices* 42: 935–943.
92. Tuinhout HP, Montree HA, Schmitz J, *et al.* (1997) Effects of gate depletion and boron penetration on matching of deep submicron CMOS transistor. *In: Technical Digest of International Electron Devices Meeting*, pp. 631–634.
93. Brown AR, Roy G, and Asenov A (2007) Poly-Si gate related variability in decananometre mosfets with conventional architecture. *IEEE Transactions on Electron Devices* 54: 3056–3063.
94. Zhang M, Knoch J, Zhang SL, *et al.* (2008) Threshold voltage variation in soi schottky-barrier MOSFETs. *IEEE Transactions on Electron Devices* 55: 858–865.
95. Knoch J and Appenzeller J (2002) Impact of the channel thickness on the performance of SB-MOSFETs. *Applied Physics Letters* 81: 3082–3084.
96. Ananthan H and Roy K (2006) A compact physical model for yield under gate length and body thickness variations in nanoscale double-gate CMOS. *IEEE Transactions on Electron Devices* 53: 2151–2159.
97. Samsudin K, Cheng B, Brown AR, *et al.* (2005) Impact of body thickness fluctuation in nanometre scale UTB SOI MOSFETs on SRAM cell functionality. *In: Proceeding of European Conference on Ultimate Integration of Silicon*, pp. 45–48.
98. Chang CY and Sze SM (1996) *ULSI Technology*. New York: Mcgraw-Hill College.
99. Ancona MG and Tiersten HF (1987) Macroscopic physics of the silicon inversion layer. *Physical Review B* 35: 7959–7965.
100. Ancona MG and Svizhenko A (2008) Density-gradient theory of tunneling: Physics and verification in one dimension. *Journal of Applied Physics* 104: 073726.
101. Odanaka S (2004) Multidimensional discretization of the stationary quantum drift-diffusion model for ultrasmall MOSFET structures. *IEEE Transaction on Computer-Aided Design Integrated Circuits and Systems* 23: 837–842.
102. Tang TW, Wang X, and Li Y (2002) Discretization scheme for the density-gradient equation and effect of boundary conditions. *Journal of Computational Electronics* 1: 389–393.
103. Roy G, Brown AR, Asenov A, *et al.* (2003) Quantum aspects of resolving discrete charges in 'atomistic' device simulations. *Journal of Computational Electronics* 2: 323–327.
104. Iafrate GJ, Grubin HL, and Ferry DK (1981) Utilization of quantum distribution functions for ultra-submicron device transport. *Le Journal de Physique Colloques* 42: C7–C307.
105. Zhou JR and Ferry DK (1992) Ballistic phenomena in GaAs MESFET: Modeling with quantum moment equations. *Semiconductor Science and Technology* 7: B546.
106. Zhou JR and Ferry DK (1992) Simulation of ultra-small GaAs MESFET's using quantum moment equations II

velocity overshoot. *IEEE Transactions on Electron Devices* 39: 1793–1796.
107. Zhou JR and Ferry DK (1993) Modeling of quantum effects in ultrasmall HEMT devices. *IEEE Transactions on Electron Devices* 40: 421–427.
108. Li Y and Yu SM (2004) A unified quantum correction model for nanoscale single- and double-gate MOSFETs under inversion conditions. *Nanotechnology* 15: 1009–1016.
109. Li Y, Yu SM, and Li YL (2008) Parallel solution of large-scale eigenvalue problem for master equation in protein folding dynamics. *Journal of Parallel and Distributed Computing* 68: 678–685.
110. Li Y and Yu SM (2005) A parallel adaptive finite volume method for nanoscale double-gate MOSFETs simulation. *Journal Computational and Applied Mathematics* 175: 87–99.
111. Li Y, Lu HM, Tang TW, *et al.* (2003) A novel parallel adaptive Monte Carlo method for nonlinear Poisson equation in semiconductor devices. *Mathematics and Computers in Simulation* 62: 413–420.
112. Li Y, Sze SM, and Chao TS (2002) A practical implementation of parallel dynamic load balancing for adaptive computing in VLSI device simulation. *Engineering with Computers* 18: 124–137.
113. Darwish MN, Lentz JL, Pinto MR, *et al.* (1997) An improved electron and hole mobility model for general purpose device simulation. *IEEE Transactions on Electron Devices* 44: 1529–1538.
114. Selberherr S (1984) *Analysis and Simulation of Semiconductor Devices*. New York: Springer.
115. Grasser T and Selberherr S (2000) Mixed-mode device simulation. *Microelectronics Journal* 31: 873–881.
116. Li Y (2007) A two-dimensional thin-film transistor simulation using adaptive computing technique. *Applied Mathematics and Computation* 184: 73–85.
117. Huang KY, Li Y, and Lee CP (2003) A time-domain approach to simulation and characterization of RF HBT two-tone intermodulation distortion. *IEEE Transactions on Microwave Theory and Techniques* 51: 2055–2062.
118. Li Y, Huang JY, and Lee BS (2008) Effect of single grain boundary position on surrounding-gate polysilicon thin film transistors. *Semiconductor Science and Technology* 23: 015019.
119. Dimitrov V, Heng JB, Timp K, *et al.* (2005) High performance, sub-50nm MOSFETS for mixed signal applications. *In: Technical Digest of International Electron Devices Meeting*, pp. 213–216.
120. Post I, Akbar M, Curello G, *et al.* (2006) A 65 nm CMOS SOC technology featuring strained silicon transistors for RF applications. *In: Technical Digest of International Electron Devices Meeting*, pp. 1–3.
121. Seevinck E, List FJ, and Lohstroh J (1987) Static-noise margin analysis of MOS SRAM cell. *IEEE Journal of Solid State Circuits* 22: 748–754.
122. Cheng B, Rao VR, and Woo JCS (1998) Sub-0.18 μm SOI MOSFETs using lateral asymmetric channel profile and Ge pre-amorphization salicide technology. *In: IEEE International SOI Conference*, pp. 113–114.
123. Dollfus P, Bournel A, Galdin S, *et al.* (2004) Effect of discrete impurities on electron transport in ultrashort MOSFET using 3D MC simulation. *IEEE Transactions on Electron Devices* 51: 749–756.
124. Dadgour H, Vivek D, and Banerjee K (2008) Statistical modeling of metal-gate work-function variability in emerging device technologies and implications for circuit design. *In: Proceeding of International Conference on Computer-Aided Design*, pp. 270–277.

7 Spin-Based Data Storage

O Ozatay, San Jose Research Center, San Jose, CA, USA

P G Mather, Everspin Technologies, Chandler, AZ, USA

J-U Thiele, Seagate Recording Media Operations, Fremont, CA, USA

T Hauet and P M Braganca, Hitachi Global Storage Technologies, San Jose Research Center, San Jose, CA, USA

7.1 Introduction

The novel magnetic phenomena that emerge as magnetic structures are geometrically confined to submicron-size scale are the subject of nanomagnetism. Such studies lead to the realization of nanomagnetic devices offering a variety of potential applications in future technologies. Some of the most important goals of the nanomagnetism research include [1] building energy efficient ultrastrong magnets, spin-transistors with gain, spin-based qubits for quantum computing, bio-magnetic sensors, on-chip microwave oscillators, magnetic logic devices, and room temperature magnetic semiconductors for the integration of magneto-electronic devices with the existing semiconductor technology. Some of these applications have already been discussed in 00137.

In addition to all these fascinating device ideas, a major class of studies in magnetism research is targeted towards achieving ultrahigh-density data storage with nanoscale magnets while maintaining thermal stability. The aim of this chapter is to present the potential benefits of utilizing the spin degrees of freedom in storage devices when compared to conventional charge-based electronics approach, through a survey of emerging spin-based storage technologies. The most notable advantages include nonvolatility, more efficient power consumption, and improved data processing-speed (approximately nanosecond to picosecond range) together with high integration density. Coupling all such benefits with low-cost and long-term reliability makes spin-based data-storage approaches highly desirable from a technology perspective. However, the successful implementation of these devices relies upon in-detail physical understanding and utilization of spin-dependent transport and spin dynamic properties in a controllable fashion. More specifically, one needs to know how to change the state of a spin system (write operation), how to maintain (retention), transport, and detect spin information (read operation) coherently. Furthermore, size scales required demand significant developments of nanopatterning techniques compatible with the magnetic materials of choice [2].

The working principles behind a magnetic memory device stem from a few basic physical interactions between a magnetic bit (a spin system) and its environment. **Figure 1** summarizes these fundamental coupling mechanisms in the simplest possible form together with their corresponding device applications. If we consider a magnetic bit as a macrospin, where all the spins within the spin system are strongly coupled, then such a macrospin is subject to three different sources of torque through interactions with its physical environment. These are the spin torque, the damping torque, and the precessional torque that arise due to the effect of an electrical current, the crystal lattice itself, and the magnetic field from a variety of sources, respectively (**Figure 1**).

Before addressing the individual torque effects, it is important to grasp the fundamental spin-dependent conduction properties across magnetic superlattices. The electrical transport through a ferromagnet (FM) results in a net spin polarization of the conduction electrons due to an asymmetry in the scattering probabilities of electrons with different spin orientations. Such nonequilibrium spin population of the conduction electrons when incident on another ferromagnetic electrode is also subject to spin-dependent scattering such that the electrical resistance is determined by the magnetization direction of the FM. This leads to the giant magnetoresistance (GMR) [3] and (in the case of spin-dependent tunneling due to spin-polarized density of states of the FM electrodes) tunneling magnetoresistance (TMR) effects in a spin-valve structure (FM/normal metal (NM)/FM) and a

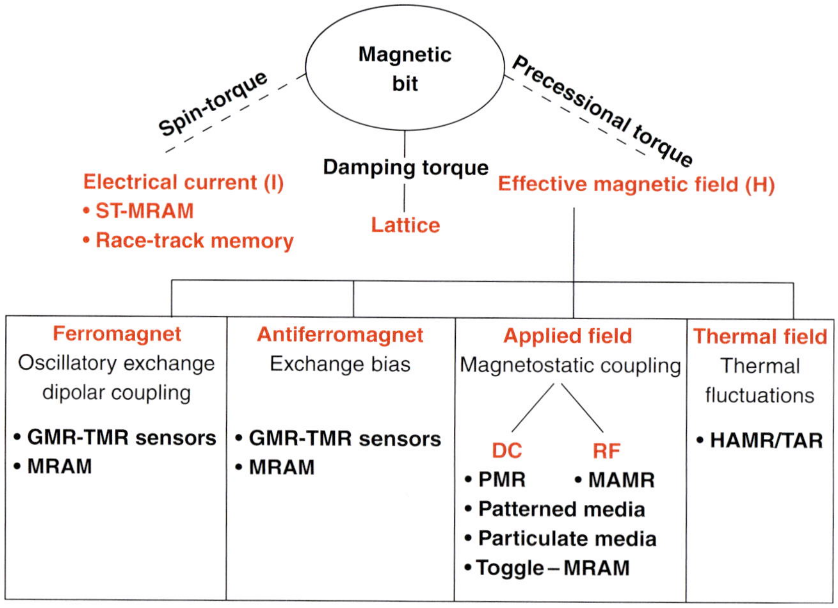

Figure 1 Basic physical coupling mechanisms between a magnetic bit and its environment leading to magnetic memory device applications.

magnetic tunnel junction (MTJ) (FM/tunnel barrier/FM), respectively. The antiparallel alignment of individual magnetic layers in such structures are achieved through thin-film interactions such as dipolar coupling using layers with different coercive fields, oscillatory exchange coupling, or by virtue of exchange bias with an antiferromagnet . GMR and TMR effects are utilized in magnetic sensing (read operation) as well as data storage in magnetic random access memory (MRAM).

As much as the electrical transport itself is influenced by the spin-dependent scattering, the local magnetization also experiences a torque from a spin-polarized current due to the requirement of conservation of transverse spin angular momentum [4]. The resulting spin torque, if strong enough, can overcome the damping torque (the torque that acts to damp out magnetization oscillations) from interactions with the lattice (mostly spin–orbit coupling), causing instabilities that may result in magnetization reversal or sustained radio frequency (RF) oscillations. Such spin transfer torque (STT)-induced reversal is to be exploited in the next-generation MRAM device architectures called spin-torque magnetic random access memory (ST-MRAM). The most notable advantages of using a current-induced switching approach over conventional magnetic field-induced switching include better scalability, elimination of cross-talk, and half-select problems between neighboring cells as well as more efficient power consumption. Another important outcome of the interaction of the spin-polarized current with the local moment manifests itself in the current-induced translation of domain walls between pinning sites in a ferromagnetic nanowire. The race-track memory (a 3D magnetic memory being developed by IBM) [5] utilizes such current-induced domain-wall motion in a vertical magnetic nanowire to increase the data-storage capacity.

While a spin-polarized current results in a net spin torque, the influence of an effective magnetic field is to apply a precessional torque on the local moment around the field direction. Such oscillations are quickly damped out by the damping torque making the moment lineup with the field direction which minimizes the magnetostatic energy of the spin system. The effective magnetic field is a combination of several external and internal fields. The internal contributions include demagnetization fields and anisotropy fields (shape, magnetocrystalline, magnetoelastic, etc.) Externally applied fields can overcome these internal contributions to manipulate the magnetic moment for data-storage applications. A static applied field (e.g., the oersted field from a current-carrying wire) exceeding a critical field (coercivity) can flip the magnetization orientation of the recording medium. Such simple write schemes are used in current perpendicular magnetic recording (PMR)

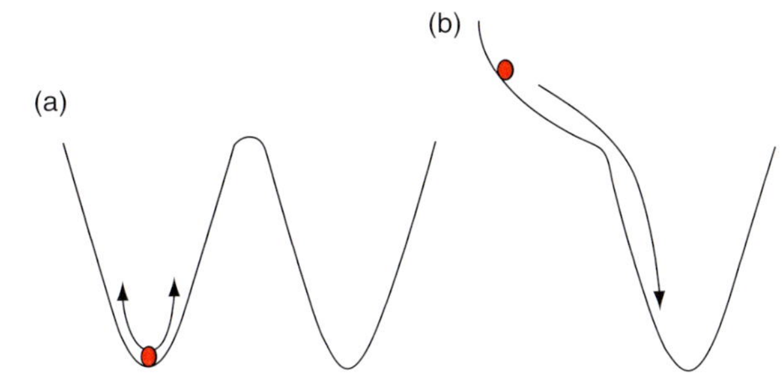

Figure 2 One simple schematic model of the magnetization reversal process assuming uniform magnetization (a) with no external magnetic fields applied, the magnetic bit is stable in one orientation but is still subject to thermal fluctuations; (b) an applied magnetic field larger than the coercivity of the bit destabilizes one magnetic orientation leading to trapping in the other state.

[6], longitudinal tape recording (particulate media), toggle-MRAM, and are also likely to be used in future patterned media recording.

The reversal process (write operation) in these devices is actually of probabilistic nature so that the reversal of the orientation of a magnetic bit can be simply modeled as the excitation of a particle trapped in a double-well potential that leads to jumping from one stable state to the other as shown in **Figure 2**. The probability of reversal is determined by the barrier height (anisotropy energy) which is the product of bit volume (V) and anisotropy (K) (the direction dependence of the preferred magnetization orientation), the temperature of the system (through both thermal fluctuations of the bit and temperature dependence of anisotropy), and also the externally applied fields. When the storage element size (V) is reduced for high-density recording, instability due to thermal fluctuations appears (superparamagnetic limit). High-anisotropy materials (such as FePt or CoPt alloys) are the natural choice to push the thermal stability limitations further down. However, even though these materials in nanometer scale will be able to maintain spin information for 10 years or more, it should also be possible to change their magnetic orientation easily to store new data. For this purpose radio frequency (RF) magnetic fields or heating (effectively a randomly fluctuating field) can be used to resonantly or thermally lower the coercivity (the threshold switching field) of the recording medium for a short period of time (microwave-assisted magnetic recording (MAMR) and thermally assisted recording (TAR)/heat-assisted magnetic recording (HAMR) (TAR/HAMR)).

Even though conceptually all spin-based storage approaches introduced above look simple on paper, the actual realization of nanoscale magnetic data-storage devices, incorporating exciting developments in both sensor and media technologies, is a great challenge and requires the development and integration of advanced deposition and fabrication techniques. The most promising patterning techniques of nanoimprint lithography, electron-beam (e-beam) lithography, focused-ion-beam (FIB) patterning and block-copolymer self-assembly will have to be combined with chemically and physically non-destructive etch methods using ion milling and plasma etching. Furthermore, the magnetic properties of the media materials are highly sensitive to the deposition methods and conditions used. Overcoming the fabrication challenges for nanoscale spin-based storage devices will open the path to storage densities beyond Tb/in^2.

7.2 Spin-Dependent Transport in Magnetic Nanostructures

Unusual conduction properties of nanoscale ferromagnetic materials play a key role in determining the performance of spin-based storage devices. The electrical transport in these devices is modified such that one spin sub-band (called majority spins) is favored over the other (called minority spins) due to the exchange-split band structure of the FM. The spin-polarized current that results from such spin filtering can be transmitted either by metallic conduction through a long spin diffusion length material such as Cu or by tunneling through a thin insulating barrier such as MgO. However, magnetoresistance (MR) effects arise when the electrons have to scatter into different momentum states when they impinge upon another FM whose magnetization is not parallel to the direction of spin polarization. The scattering center itself can experience a reaction torque from the electrons during this interaction via deposition of their transverse spin component on the local moment (in the form of spin torque). All these spin-dependent transport properties are exploited in spintronics devices [7,8] and therefore are discussed in detail in 00137. Here, we only give a brief background to help understand the operation principles of magnetic storage devices.

7.2.1 Spin Filtering

Unlike normal metals with a density of states (DOS) that is symmetric with respect to spin orientation, the DOS of an FM is exchange split so that the number of available spin-momentum states for conduction electrons to scatter into at the Fermi level is unequal as shown in **Figure 3**. At the interface between a FM and a normal metal, a spin-dependent band mismatch occurs, leading to the generation of a spin-polarized current. The spin polarization [9] P is defined as the ratio of the spin current I_s to the charge current I, so that

$$P = \frac{I_s e}{I \mu_B} \quad (1)$$

where e is the electron charge and μ_B is the Bohr magneton.

For a normal metal with equal density of states for both spin sub-bands, P is 0, whereas for a perfect half-metal that has a band gap at the Fermi level for one spin subband, P is 1. The most intuitive description of spin polarization is given in terms of the imbalance in the density of states at the Fermi level for majority and minority electrons. However, such a description is not very accurate to describe spin-dependent transport properties due to the complex band

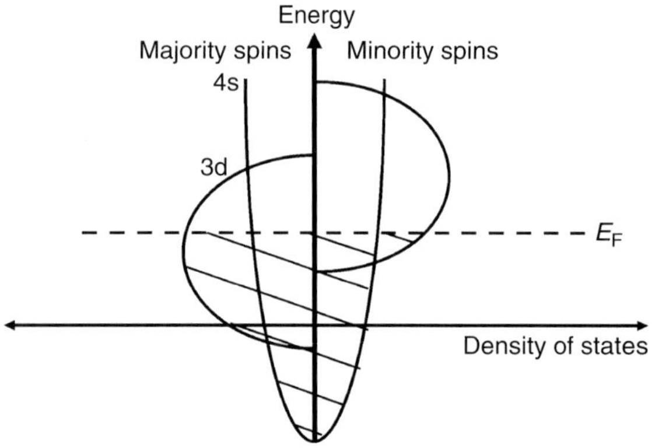

Figure 3 Schematic representation of exchange-split density of states as a function of energy of a FM.

structure of FMs that yield differences in the mobilities of electrons in different bands that cross the Fermi level. For this reason the definition of spin polarization is usually modified according to the transport regime of interest. For instance, tunneling current polarization is defined as the imbalance between the densities of states for spin sub-bands weighted by their transmission probabilities.

When the spin-polarized current is applied to a normal metal through an ohmic contact, the nonequilibrium spin accumulation at the interface diffuses away being subject to an exponential attenuation with a characteristic length scale [10] (known as spin diffusion length) until it completely depolarizes by spin-flip scattering events. Several scattering events can lead to spin flip (loss of spin polarization). Spin–orbit interactions, scattering from magnetic impurities in the normal metal, and scattering from spin waves in the FM are just a few examples. Alternatively, spin information can be transported via tunneling through a thin insulating barrier to minimize spin-flip scattering, taking advantage of the spin-polarized density of states of the emitter and detector electrodes together with spin-dependent transmission properties of the tunnel barrier.

7.2.2 Giant Magnetoresistance

One of the most important outcomes of the interaction of the current-carrying electrons with the local moment of a ferromagnetic material is the GMR effect which was originally discovered in Fe/Cr multilayers in 1988 [11]. With the introduction of GMR read-head sensors by IBM in 1997, the effect has gained a lot of commercial interest in the hard-disk drive (HDD) and sensor industry since it enables sensing much smaller amounts of magnetic flux (provided by ever decreasing sizes of recorded magnetic transitions) as compared to the previous sensing technologies based on inductive and anisotropic magnetoresistance (AMR) heads. The AMR effect is based on the deformation of the electron cloud around each nucleus as the magnetization rotates to line up with an applied field such that the current-carrying electrons experience different amounts of scattering during their transit through the lattice. This effect results in a few percent change of the device resistance usually much smaller than that of the GMR devices.

The GMR effect can be described as the electrical resistance change that is dictated by the relative magnetization orientation of two magnetic electrodes separated by a metallic spacer. This configuration is known as a spin-valve structure. When the magnetizations of the two electrodes are antiparallel, the device resistance can be reduced by applying an external field that triggers magnetization reversal of one of the electrodes (called the free layer) to reach parallel alignment with the other electrode (the fixed or reference layer). There are several ways to ensure antiparallel alignment between the magnetic electrodes in the absence of magnetic fields [12]. In the early GMR experiments (e.g., Fe/Cr and Co/Cu multilayers), it was discovered that due to a quantum mechanical indirect exchange coupling that oscillates with the spacer normal metal thickness, it is possible to achieve antiparallel alignment at some critical thickness values [13].

Another method involves using materials with different magnetic switching fields so that the magnetization of one electrode is less sensitive to an external magnetic field (hard layer) as opposed to the magnetically soft, free layer. One way to generate a magnetically hard electrode is to pin it by forming an interface with an antiferromagnetic (AF) layer such as IrMn. With the proper deposition and annealing treatments, a strong exchange coupling (exchange bias) [14] develops at the ferromagnet or antiferromagnet interface that effectively enhances the coercivity of that electrode. The coupling strength can be improved further by using a thin interlayer metal such as Ru between a pair of magnetic electrodes where one of the electrodes is also pinned with an antiferromagnet. The resulting structure is called a synthetic antiferromagnet (SAF) and is very commonly used in read-sensors due to its excellent magnetic stability. If instead no such pinning strategy is employed but the spin valve consists of the same magnetic materials with different thicknesses, then the thinner layer behaves as a magnetically soft layer due to its lower demagnetizing field when patterned into small areas. The resulting device is called a pseudo-spin-valve.

The spin-dependent transport in the GMR stacks described above can be simply thought of as consisting of two parallel conduction channels for the spin-up and spin-down sub-bands as long as the spin-flip scattering events are rare (two-channel model) [15]. The device resistance contains contributions both from scattering in the bulk of the ferromagnet and at the interface with the normal metal spacer. In the so-called current-perpendicular-to-plane (CPP) geometry (where the current flows perpendicular to the multilayer interfaces), the interfacial spin-dependent scattering dominates the GMR signal ($\%\Delta R/R$-percentage change in resistance with field), whereas in the current-in-plane (CIP) geometry (where the current flows parallel to the multilayer interfaces), it is harder to make such a distinction.

A typical CPP-GMR read-head sensor [16] is shown in **Figure 4**. As the read sensor is exposed to the magnetic field induced by a magnetized object or medium, the magnetization of the free layer rotates accordingly such that the angle between the magnetizations of the free layer and the reference layer changes. Therefore the device resistance varies due to the GMR effect making it possible to detect the recorded magnetic transitions on the medium.

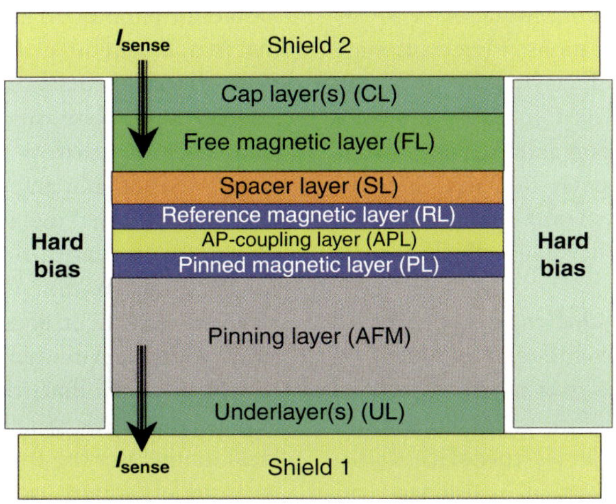

Figure 4 A schematic drawing of a CPP-GMR read head sensor. Modified from Childress JR and Fontana RE (2005) Magnetic recording read head sensor technology *Comptes Rendus Physique* 6: 997.

A constant magnetic field (hard bias) perpendicular to the reference layer magnetization is applied to the sensor to clean up any domain-wall nucleation during reversal that could potentially decrease the signal to noise ratio for clean sensing operation.

7.2.3 Tunneling Magnetoresistance

When the nonmagnetic metal spacer in a spin valve is replaced by a thin insulator, the resulting device structure is called a magnetic tunnel junction (MTJ). The spin-dependent tunneling effect due to the spin-polarized density of the states on the opposite sides of the tunnel barrier leads to the TMR effect. When the emitter and detector electrodes have their magnetizations oriented to be parallel or antiparallel to each other since the majority spin population sees a large or small number of accessible states in the detector electrode, this results in a low resistance or high resistance state. A simple formula for the optimistic TMR ratio is given by the tunneling version of the two-channel model (known as the Julliere model) [17]. In terms of the polarizations of the emitter (P_1) and detector (P_2) electrodes,

$$\mathrm{TMR} = \frac{R_{\mathrm{AP}} - R_{\mathrm{P}}}{R_{\mathrm{P}}} = \frac{2P_1P_2}{1 - P_1P_2} \tag{2}$$

where R_{AP} and R_{P} are the antiparallel state and parallel state resistances, respectively.

Nanoscale MgO-based MTJs providing large TMR ratios at room temperature [18] (>100%) are ideal candidates both for nonvolatile memory applications and sensor applications. Such high TMR ratios are usually accompanied by large device resistances (>1 kΩ) which may be desirable for complementary metal-oxide semiconductor (CMOS) integration in memory applications. However, such large resistance–area (RA) products limit the operating frequencies and introduce more Johnson and shot noise. For this reason, the exponential scaling of tunneling resistance with the barrier thickness suggests that ultrathin barriers (below 1 nm) should be used, especially for read-head applications, which require low-noise operation.

This presents important challenges in terms of device fabrication due to the difficulty of obtaining pinhole-free, uniform barriers over macroscopic distances on a Si wafer. The failure to obtain smooth interfaces [19] causes large device-to-device variations both in terms of electrical resistance and magnetic properties. Roughness-induced magnetic coupling (Orange–Peel coupling), metallic shorts through pinholes and variations of the chemical composition of the barrier through temperature-dependent interfacial reactions with the magnetic electrodes are just a few examples that can adversely affect the performance of TMR-based devices. Nevertheless, the high MR signal available makes it worthwhile to engineer smooth metallic underlayers and magnetic materials with good thermal stability characteristics for the successful implementation of TMR-based sensors and MRAM devices.

7.2.4 Spin Transfer Switching

The effect of the magnetization orientation of a ferromagnet on a spin-polarized current can be viewed as selectively transmitting or reflecting spins depending on their relative orientation with respect to the magnetization direction (acting as an analyzer). This is basically what causes the MR effects that we have discussed so far. When the GMR spin-valve or TMR MTJ structure is patterned into nanoscale dimensions, the constrained geometry allows high enough current densities to observe the inverse-MR effect. That is the impact of the spin-polarized current on the local magnetization of the analyzer FM. In this case, the mutual torque that develops at the interface with the spacer can lead to the current-induced magnetization reversal of the free layer analogous to the magnetic field-induced reversal in GMR structures [4,20].

Figure 5 demonstrates the spin-transfer switching effect in a pseudo-spin-valve structure. When the electrons flow from the fixed layer to the free layer starting from antiparallel alignment (**Figure 5(a)**) the transmitted electrons get spin-polarized so that the majority current carriers have their magnetic moments parallel to that of the fixed layer. Since the two magnetic electrodes can never have their magnetizations perfectly collinear due to thermal fluctuations and/or dipolar coupling, the spin-polarized current carriers always carry a transverse component of spin angular momentum. If the thickness of the normal metal spacer is much less than the spin diffusion length then to a good approximation, the spin polarization is conserved as the electrons traverse the spacer layer to reach the interface with the free layer. A mutual spin torque develops at the interface such that the spin-polarized electrons deposit their transverse angular momentum on the local moment of the free layer. If the resulting torque is strong enough to overcome the intrinsic damping of the material, then the magnetization orientation can be instantaneously destabilized to cause reversal.

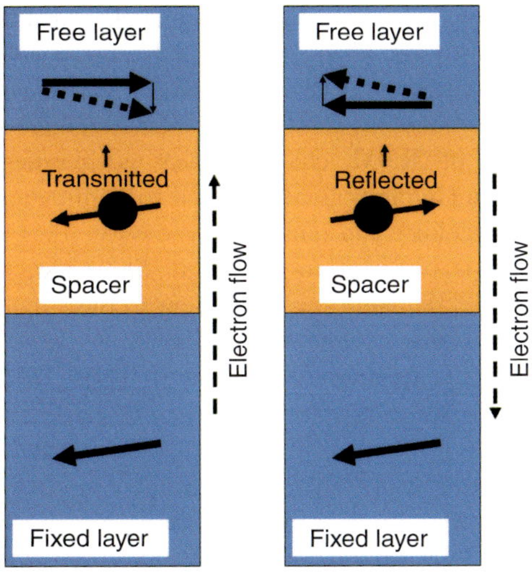

Figure 5 Schematic demonstration of spin-transfer switching in a pseudo-spin-valve structure: (a) spin-transfer switching from the antiparallel to the parallel state; (b) spin transfer switching from the parallel to the antiparallel state.

For the case of electron flow from the free to the fixed layer in the parallel state (**Figure 5(b)**), the electrons that are reflected at the fixed layer–spacer interface apply a torque back on the free layer that can destabilize the parallel alignment. In all cases there is naturally a spin torque that develops at the fixed layer–spacer interface as well due to spin-polarized electrons that travel from the free to the fixed layer. However, the larger magnetic volume of the fixed layer ensures that it is less susceptible to current-induced manipulation, and a stable magnetic configuration is reached by the switching of the free layer instead.

It is important to note that STT can also be utilized to stabilize microwave frequency precession of the magnetization of a nanomagnet in the presence of a magnetic field. The realization of such a nanoscale microwave oscillator is the topic of intensive research efforts [21] since it could be beneficial for high-speed electronic devices. Such spin-torque-induced microwave oscillations actually act as a noise source in read-head sensors [22]. Therefore, for the purpose of improved signal-to-noise operation, it is desirable to suppress spin-torque effects by using materials with high damping (strong, spin–orbit coupling). However, the focus here is on only the manipulation of a nanomagnet's magnetization that leads to reversal for spin-based storage applications. Spin transfer-induced switching is indeed a promising approach for the next-generation MRAM devices.

7.3 Magnetic Random Access Memory

MRAM possesses a unique combination of attributes that provide considerable benefits over those available with conventional memory technology. Here we discuss the state-of-the-art MRAM, from parts currently in production to development efforts underway. A promising future for MRAM is emerging as new write techniques for even higher-density and lower-power modes of operation are developed and optimized. While high density is highly desirable in any data-storage medium, MRAM differentiates itself from the magnetic data storage discussed in this chapter, as the primary focus is operation as a high-speed RAM with fast read and write times and extremely reliable performance coupled with nonvolatility that the magnetic nature of the storage element provides. MRAM provides a nonvolatile high-speed memory array that can be paired directly with a microprocessor, rather than a system for storing extensive volumes of information.

Contemporary MRAM harks back to the early days of computer hardware when ferrite core memories [23] were the primary memory array available for data storage. These arrays consisted of hand-wound toroidal elements with two perpendicular write lines that provided a cross point architecture for writing an individual element. When the superposition of fields provided by a simultaneous set of current pulses along each line was sufficient to switch the magnetization direction of the element, it was written into a new state. A destructive read was enabled through monitoring the inductive signal on a separate read line. To read the bit, an attempt to write it into a known state was made. The lack of an induced voltage pulse revealed that the bit was in that known state, while an inductive pulse revealed the other state. Magnetic memory has since evolved along with the discovery and utilization of new physical processes for sensing magnetic states, analogous to the evolution of sensing techniques for HDDs.

The inductive read was replaced in AMR-based MRAM, which suffered from a lack of signal as compared with the base cell resistance [24]. After the discovery of GMR in 1988 revolutionized the hard-disk industry, utilization of the phenomenon was attempted for MRAM, but the intrinsically low resistance and improved yet still low signal of these elements made meshing with conventional CMOS circuits a significant challenge. While the low resistance could be mitigated with large GMR elements, this created a large disconnect with the density of contemporary memories [25]. The physical phenomenon for the return of a commercially competitive MRAM, the MTJ with high MR at room temperature, was first developed in 1995 [26,27] and has since become the fundamental building block for MRAM technology. The first commercial MRAM product appeared approximately 10 years later. The family of parts currently produced by Everspin Technologies can be traced back to the first MRAM product: the 4 Mb MR2A16A, initially produced by Freescale Semiconductor.

7.3.1 Comparison to Conventional Memories

Settling on the MTJ as the fundamental construct for MRAM technology allows a detailed comparison with the conventional RAM technologies which utilize the manipulation of electronic charge for the storage of the memory state. The heart of a MTJ consists of two ferromagnetic layers placed on either side of an insulating tunnel barrier layer. Through the TMR effect, the resistance is determined by the relative orientation of the magnetization of these two ferromagnetic layers, and can take one of two stable values or memory states. As no power is required to maintain the relative magnetic orientation and the intrinsic speed for magnetic response is in the GHz range, the basis for a high-speed nonvolatile memory system is present. No known write-induced wear-out mechanisms are present because the switching of the magnetic memory does not involve the physical movement of electrons or atoms, simply a reorientation of the magnetic moment in one of the ferromagnetic layers. Magnetic charge cannot leak from the storage layer and thus there is no need for a constant refresh as in dynamic RAM (DRAM).

It then becomes clear that an advantageous combination of RAM characteristics is manifest in MRAM technology, lending itself to applications in both standalone and embedded memory. This combination is outlined in **Table 1**. The slow, high-power, and damaging write present in flash memory (10^5–10^7 write cycles) is avoided in a nonvolatile high-speed memory. While static RAM (SRAM) has fast read and write speeds, its volatile nature and relatively large cell size becomes expensive in large arrays, especially for embedded applications. DRAM's requirement for a constant refresh can be problematic for device power consumption. MRAM provides an additional advantage: the division of labor between fast volatile and slow nonvolatile memory types does not have to be predetermined at manufacturing and multiple memories need not be employed with the associated inter-memory data transfers to work around the limitations of conventional memories. Memory roles may be dynamically adjusted if desired for improved performance in a product containing MRAM.

7.3.2 Magnetic Tunnel Junctions for MRAM Read

The read mechanism is similar for all MTJ-based MRAM technologies. **Figure 6(a)** shows the resistance versus applied field curve for a 0.6 μm × 1.2 μm MTJ. The magnetic field is applied along the bit easy (long) axis. When the field exceeds the switching field of the bit, the bit-free layer reverses state. The magnetization direction of the fixed reference layer is held through AF coupling to a pinned layer and

Table 1 Comparison of MRAM with conventional memory technologies. MRAM offers a combination of benefits not found in other technologies

	Toggle MRAM (180 nm)	*Toggle MRAM (65 nm)[a]*	*ST MRAM (65 nm)[a]*	*FLASH (65 nm)[b]*	*DRAM (65 nm)[b]*	*SRAM (65 nm)[b]*
Read time	35 ns	10 ns	10 ns	10–50 ns	10 ns	1 ns
Program time	5 ns	5 ns	5 ns	0.1–100 ms	10 ns	1 ns
Endurance	$>10^{15}$	$>10^{15}$	$>10^{15}$	$>10^{15}$ read $>10^5$ write	$>10^{15}$	$>10^{15}$
Nonvolatility	Yes	Yes	Yes	Yes	No	No

[a] 65 nm MRAM values are projected.
[b] These values are from the ITRS roadmap.

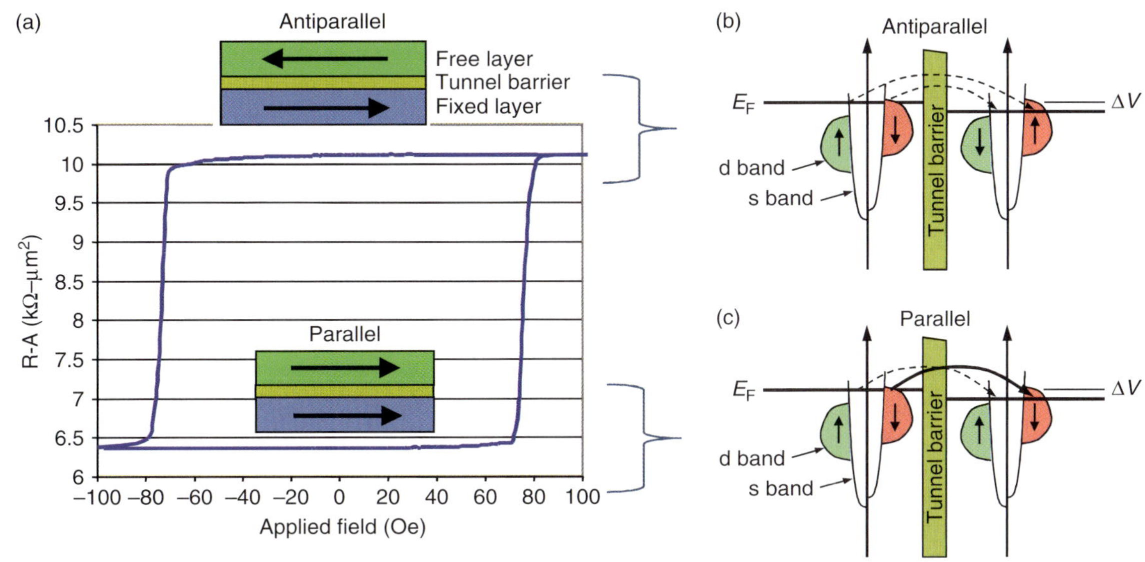

Figure 6 Tunneling magnetoresistance hysteresis loop. A low-resistance state arises when magnetization is parallel in both electrodes, leading to matching of densities of states on either side of the tunnel barrier.

cannot respond. Due to the magnetic hysteresis, the free layer will stay in its current state in the absence of an applied field. The data state is stored until a magnetic field greater than the switching field is applied in the opposite direction, and hence the bit requires no external power source for data retention. It is this magnetic hysteresis that forms the basis for the nonvolatility of MRAM. The resistance difference stems from the fact that a spin-polarized electrical current originating from one FM is tunnel-injected into the other. Appealing to the two-band model for ferromagnetic polarization (**Figure 3**), when both ferromagnetic electrodes are aligned parallel to one another (**Figure 6(c)**), there is a better overlap for electron tunneling: spin-up (down) electrons in the source electrode tunnel into available states in the spin-up (down) band in the destination electrode, and a high current or low resistance results. The transmission probability for tunneling systems is related to the product of the DOS available for electron occupation for source and destination electrodes. Hence, in the parallel case shown in **Figure 6(c)**, the high conductance spin-down to spin-down tunneling channel (bold arrow) results in a low resistance as the large DOS contributed by the d-bands dominates the tunneling current. When both ferromagnetic electrodes are aligned antiparallel, (**Figure 6(b)**) spin-down states in one electrode must tunnel into spin-up states in the other. The transmission probability is then related to the product of a small number (spin-up DOS) and large number (spin-down DOS), and hence a high resistance results. The TMR is thus maximal for strongly polarized materials which have a vanishingly small DOS for one spin channel.

The situation described above is a simplification of the tunneling process, as surface states, impurities in the tunnel barrier and electrodes, details of the electrode or barrier interface, and the electronic structure of the barrier itself [28] all work in concert to determine both the sign and magnitude of the MR. This is clearly evident in the high MR present in magnesium oxide tunnel barriers, theoretically predicted in 2001 [29,30] and then experimentally demonstrated in 2004 [31,32]. Here, the electronic structure of MgO serves as a spin filter when coupled with the appropriate ferromagnetic electrodes, and this exciting system has yielded MR values as high as 500% [33]. However, the crystalline MgO tunneling barrier results in wider resistance distributions than amorphous barriers, and the system requires further development. Another important manifestation of these real properties is the decrease of the MR with increasing voltage bias across the tunnel junction. As a circuit will necessarily apply a finite bias across the tunnel junction, the actual MR that may be utilized for MRAM design will be less than the values from the ideal cases usually mentioned in literature. Additionally, high-temperature operation will further reduce the available MR as the polarization of the ferromagnetic electrodes is reduced as their Curie points are approached.

Moving beyond the experimental demonstration of the MTJ stack into a commercial memory product requires a detailed consideration of the distributions of the individual bit resistances at bias and operation temperature. For integration with CMOS elements, an $\sim$10 kΩ bit resistance is desired, requiring an extremely thin AlO_x tunnel barrier of $\sim$15 Å. The tunnel barrier resistance is exponentially dependent upon its thickness, and exquisite control of barrier uniformity is needed. Despite this challenge, it is possible to achieve good uniformity and repeatability in resistance, with a distribution width (σ) of 5% and MR (σ: 1%) across a 200 mm wafer [34]. The tunnel barrier is formed through the plasma oxidation of a sputtered metal film. This level of uniformity was achieved using an Al thickness uniformity of 0.5% and meticulous matching of the oxidation process uniformity with that of the Al metal.

While the wafer-level uniformity is important, even more crucial to MRAM operation is the uniformity across the array. In operation, reference cells are utilized to determine the midpoint between the high and low resistance states of the bits, and variation across the array can cause an erroneous bit state determination. This is best understood by referring to **Figure 7(a)**. The midpoint generated by the reference cell is compared to the resistance of the queried bit, and bits that fall on the low end of the high-resistance (AP state) distribution or the high end of the P-state resistance distribution may be read out incorrectly if insufficient separation is present between the two distributions. A separation of 6σ from the reference or 12σ between states is the minimum required for operation of a 4 Mb memory. Additional separation creates more margin between the 6σ tails and the reference (ΔR_{use}), allowing higher speed and operation temperature.

As the resistance of the bit is proportional to its area, variations in etch and photolithography both factor into the final bit resistance distributions. Process damage or veils creating a low resistance shorting path will contribute as well. However, it is the short-range resistance variations that play the largest role in determining the distribution width. The extremely thin dielectric layer utilized may have variations in thickness on a fine size scale, due both to the roughness from underlayers propagating upward, and to the intrinsic barrier roughness. Minimizing this barrier roughness is important, because it produces local thin regions where tunneling may preferentially take place, that is, local tunneling hot spots [35]. If the size scale for variation of these hot spots is comparable with the size of the bit, statistical variations in both the strength and the number of hot spots each bit encloses will contribute to the resistance distribution width. The deposition of an Al metal layer followed by oxidation helps with intrinsic barrier roughness, and underlayer roughness is largely mitigated through addition of texturing layers. **Figure 7(b)** shows the resistance distributions for an optimized material stack and patterning process in a 4 Mb toggle MRAM array. The separation between distributions is 23σ.

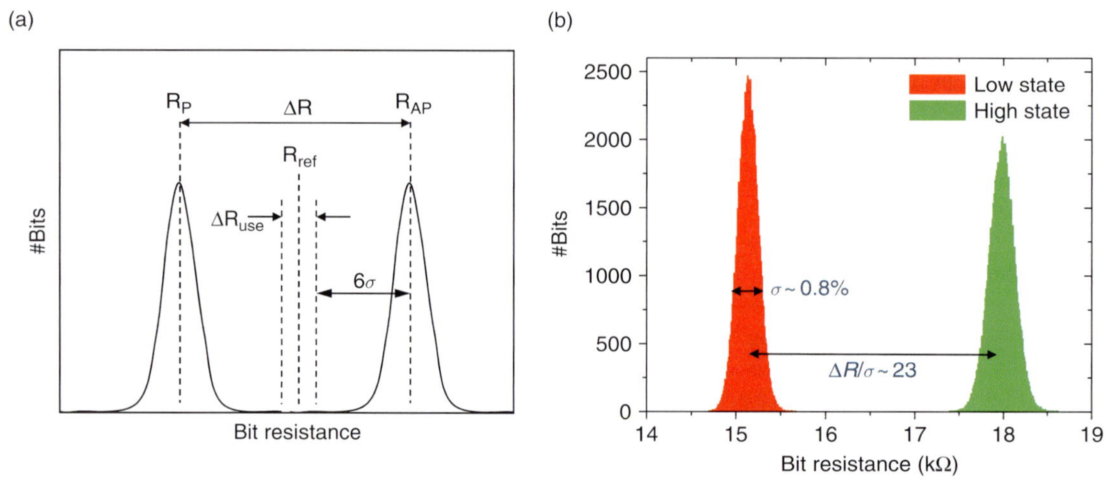

Figure 7 (a) Hypothetical resistance distributions for high- and low-resistance states. The mean for each state must be at least 6σ from the reference (12σ separation) for error-free read out; (b) Optimized tunnel barrier material showing 23σ separation.

In addition to contributing to the resistance distributions, if one considers the propensity of current to flow through predominately through these hot spots, it is clear that they determine the device reliability. Over time, continued exposure to the voltages utilized for the read operation and exposure to high temperatures lead to two main failure mechanisms: time-dependent dielectric breakdown (TDDB) and drift of the bit resistance. Dielectric breakdown is a sudden, dramatic decrease in the bit resistance, a manifestation of a short in the tunnel barrier layer. Resistance drift causes a gradual reduction of the junction resistance over time and leads to a decrease in the separation between the high and low states. Studies [36,37] utilizing temperature and stress voltage to accelerate both mechanisms have shown that MRAM product reliability is well beyond industrial and automotive requirements [38].

As MRAM technology is scaled to future technology nodes, the smaller size and denser packing of transistors and MTJ bits results in less precise transistor matching and increased parasitic contributions to read signal. A smaller MTJ bit results in a reduction in the number of hot spots enclosed and hence increase in the relative variation from bit to bit. In general, a widening of measured resistance distributions will be observed. The clearest avenue to overcome the detrimental effect on the distribution width is to increase the MR, resulting in improved resistance separation between the two distributions. The high MR values available with MgO tunnel barriers offer a very encouraging path. Despite an increase in resistance distribution width (σ:1.5%), the increase in MR yields an increase in separation from 28σ to 49σ for an MgO tunneling barrier with a CoFeB free layer as evaluated in an Everspin 4 Mb CMOS circuit. Read access time is improved to the 17 ns limitation of the underlying CMOS circuitry [39]. An 8 kbit array integrated with 90 nm CMOS has been demonstrated [40]. However, MgO-based MRAM is still under development. The wider intrinsic resistance distributions stemming from the crystalline nature of the MgO tunnel barrier gave a foreshadowing of the increase in hot-spot variation and larger observed distributions of extrinsic low breakdown bits [41].

7.3.3 Magnetic Random Access Memory Types

There are three basic types of MRAM, differentiated by write technique. Conventional MRAM operates in a fashion similar to that outlined for ferrite core memory: the superposition of two current pulses creates a field which is larger than the coercive field (H_c) of the bit causing a reversal of its magnetic orientation. However, each of the bits along the digit line experience a reduced field (half-select) that may be sufficient to reverse bits that lie in the tails of H_c distributions of the memory array. Hence, careful control of the bit properties and details of the architecture are required. The half-select problem may be eliminated through the write scheme employed in Savtchenko, or toggle MRAM [42]. This technique is in use in commercial MRAM today, and employs a free layer that consists of several magnetic layers, and a precise pulse timing sequence for the two current lines. In this case, the bit is still switched through the time-dependent superposition of two magnetic fields, but the half-selected bits are stabilized against reversal. Clear scaling paths exist, in each of the field-switched cases above, but the cell size and power consumption may be further reduced by exploiting the recently discovered phenomenon of spin transfer [43–45]. ST-MRAM is still under development by number of companies and shows strong promise as a memory technology. The key benefits of field-switched MRAM are supplemented with the low-power, high-density architecture of ST-MRAM. Elimination of the bulky write lines required for field-induced switching reduces the process complexity and cell size, opening opportunities to significantly reduce cost and enabling dramatic increases in memory density. The highly efficient switching mode through direct transfer of angular momentum considerably lowers the power consumption as compared with the indirect field induced-switching mechanism employed in toggle MRAM.

7.3.4 Conventional Field-Switched MRAM

The fundamental cell for a field-switched MTJ is shown in **Figure 8**, consisting of an isolation transistor, current lines passing above and below the bit, and the MTJ itself. The isolation transistor is off for a current pulse to switch the bit, and open to read out the bit state. The MTJ consists of a fixed reference layer, generally SAF layer exchange coupled to a pinned layer, for which the magnetization is pinned in a particular direction, the tunnel barrier layer, and the free layer which can be switched between one of its two stable states in the presence of an applied field. These stable states are determined by the combination of shape anisotropy of the patterned bit and the intrinsic anisotropy of the ferromagnetic free

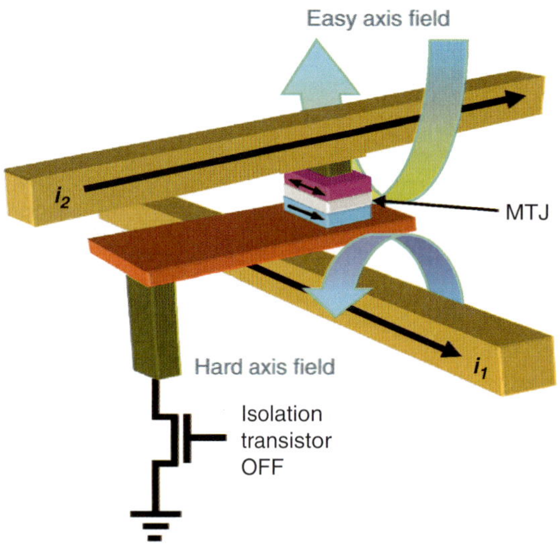

Figure 8 MRAM bit cell. Current applied along the bit and digit line with the isolation transistor off cause the free layer (double-headed arrow) in the MTJ stack to reverse, storing a 1 or 0 state. The isolation transistor is turned on and a current is passed from the top line through the transistor to read the state.

layer. As an electrical current is applied through the bit and digit lines, the resultant field at the selected bit causes the free layer to reverse state and the bit to switch between the high- and low-resistance states. An important development to reduce the current required for switching was the introduction of a layer of highly permeable magnetic material along three sides of the bit and digit lines [46]. The resultant magnetic flux changes from a circular pattern enveloping the write line to a focused region around the bit, reducing the current required for switching by at least a factor of 2.

Early MRAM used a field generated along the bit hard axis to lower the field threshold for switching, simultaneously with a field applied along the easy axis to switch the bit into the desired state. Referring to **Figure 9**, this process can be understood by looking at the influences of the individual fields on the bit-free layer. **Figure 9(a)** shows the bit in one of its stable states in the absence of any applied field. In **Figure 9(b)**, a field is applied along the hard (short) axis of the bit, which causes the magnetization as represented by the arrow to tilt along this axis. In **Figure 9(c)**, a field is additionally applied along the easy axis of the bit, pointing to the left. In **Figure 9(d)**, both fields are removed, and the bit is oriented in its reversed stable state. In practice, both fields are applied simultaneously, but the effect of the hard axis field to lower the barrier for magnetization reversal is the same. The border between switching and nonswitching behavior can be represented in a plot of the hard axis field against the easy axis field, known as a switching astroid. This is shown in **Figure 9(e)** for high-speed switching pulses for a $0.6 \times 1.2\text{-}\mu m^2$ bit. The operating region to switch a single bit is then the area outside of the astroid.

In the cross point architecture, (**Figure 10**) as current is applied down each of the digit and bit lines, each bit along the lines is exposed to one component of the field utilized to switch the selected device. Similar to the resistance distributions mentioned earlier, there are bits on the high end of the switching distribution as well as the low end. While in an ideal environment bits with only the hard axis field applied (half-selected bits) will not switch, thermal noise can cause a low number of such bits to reverse. For a bit lying on the low end of the switching distribution, the easy axis field combined with thermal fluctuations may be sufficient to cause reversal. Therefore, the field must be high enough to switch the bits with the highest switching fields, yet low enough that those half-selected bits do not undergo unintended reversal. Practically, this means

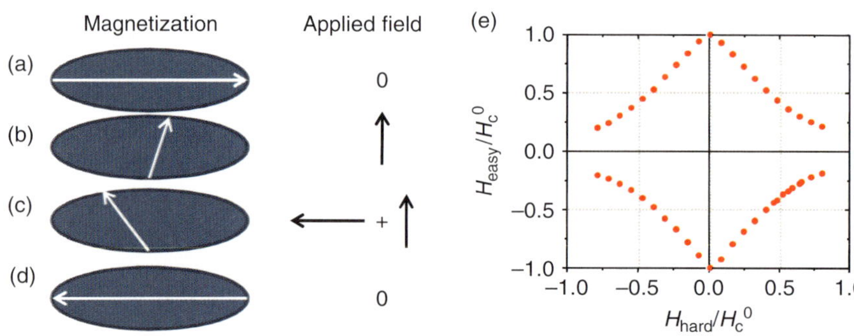

Figure 9 Illustration of conventional MRAM switching procedure and switching astroid for simultaneous application of hard and easy axis fields.

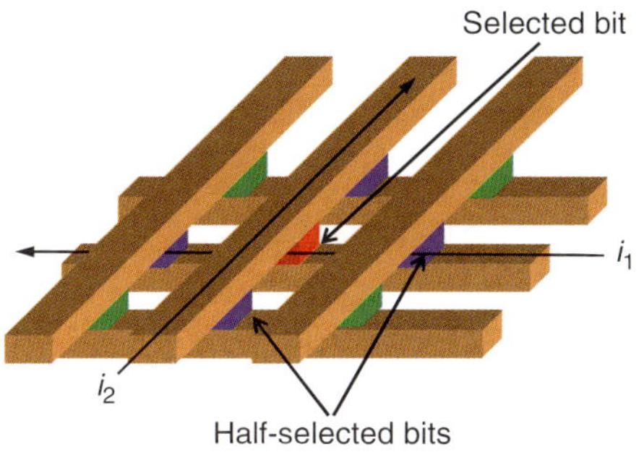

Figure 10 Bits along the bit and digit line experience a partial half-select field, which may be sufficient to disturb bits low in the switching distribution.

that the switching current must be 6σ above the mean, and the half-select must be 6σ below the switching mean. Again, reducing the tunnel barrier roughness plays an important role in reducing switching distributions through reduced coupling. **Figure 11** shows a section of the switching astroid for a 256 kb array with cross point architecture, demonstrating the significant challenge involved. The small operating window for a single array allows for diminishing process variation and continues to shrink as the array size is increased. The small window for bit and digit line currents would require careful die-by-die tuning of the programming currents. An adaptation of the cross point architecture is to utilize a segmented write line [47], whereby a reduced set of bits are exposed to a very strong hard axis field, has been recently shown to be scalable to Mb-size memory arrays [48]. A small easy axis field is then required to write all of the half-selected bits at once into their desired state. Since the easy axis field is small, no half-select problem is present for this longer bit line. However, this approach reduces the array efficiency over that of cross point architecture, and is subject to programming errors from external fields.

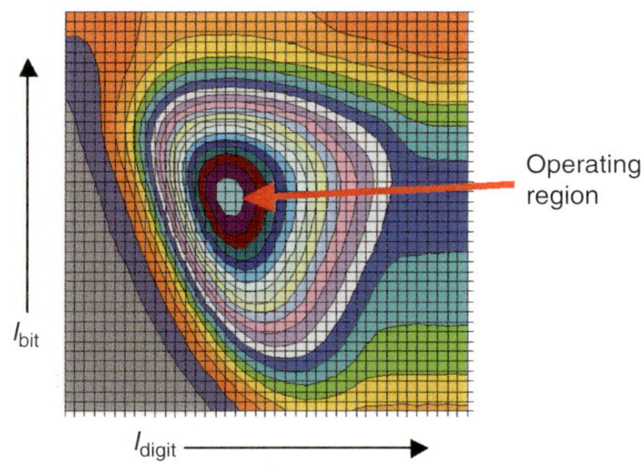

Figure 11 The switching astroid for a 256 kb conventional MRAM array. Due to half-select disturbs, very small operating region is present, requiring precise tuning of the bit and digit line currents on a die by die basis.

7.3.5 Toggle-Switched Magnetic Random Access Memory

A unique switching mode invented by Savtchenko [29] relying on specific pulse sequence coupled with particular free-layer structure and bit orientation allows the density of the cross point architecture while eliminating the half-select problem inherent in conventional MRAM, enabling the successful commercialization of the MRAM product. The basis for toggle switching originates from the introduction of a moment balanced SAF free layer. In the absence of an applied field, both moments will be coupled antiparallel to one another, with each moment lying along the long axis of the bit (**Figure 12(a)**). Here, the blue arrow represents the magnetization of the layer nearest to the tunnel barrier. It is this magnetic layer which determines the bit cell resistance; due to the nature of tunneling, the interfacial polarization governs the bit-cell resistance. In the presence of an applied field greater than the spin flop field (H_{sw}) (**Figure 12(b)**) both moments will rotate orthogonal to the field (flop), each layer scissoring in the direction of the field [49]. The bit is oriented at 45° to each current line, and hence application of a single pulse stabilizes the bit against reversal.

In operation, the pulse sequence proceeds on to **Figure 12(c)**, when the field induced by both current pulses is incident upon the selected bit. The scissored magnetization rotates about the applied field (now the sum of the two pulses), and then first pulse is turned off, allowing the magnetization to further rotate along so that the magnetization is now scissored around the remaining field from the second pulse. After the removal of this pulse, the bit relaxes into its new stable state, reversed from the previous state. Another application of the same phase relationship will rotate the bit with the same chirality, reversing the state again. This reversal symmetry requires a decision write whereby each bit must be pre-read to determine its state and then written if required. This reduces power consumption, as only about half of the time is a write sequence required, and requires only unipolar transistors, desirable for CMOS integration. An additional advantage of this

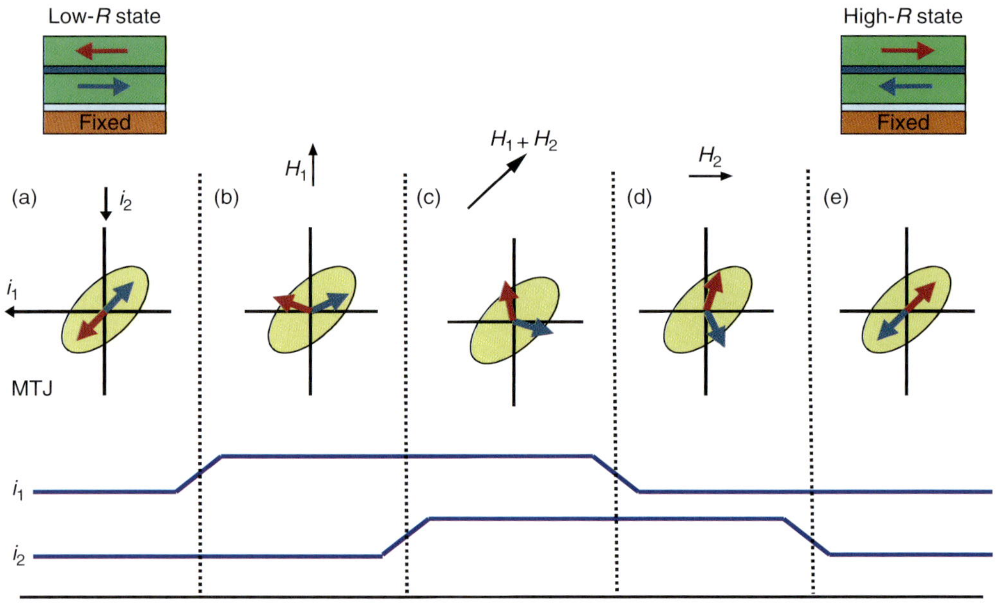

Figure 12 Toggle switching sequence for SAF free layer. The initial current pulse causes the SAF free-layer magnetizations to scissor around the resultant field. They then rotate around the changing total field direction which is determined by the phase of the current pulses along the two current lines. As the resistance is determined by tunneling from the layer closet to the tunnel barrier, the bit resistance is changed as the SAF free layer is reversed.

innovative approach is that the balanced SAF layer which forms the basis for this toggle reversal is largely immune to interfering fields, as for fields below the flop field, the low magnetic moment results only in weak coupling of external fields into the magnetic state of the bit. Maintaining the precise moment balance between the two magnetic layers in the free layer again requires careful control of the deposition process to avoid moment imbalances that will interfere with the switching symmetry.

Figure 13 shows the large operating window for a 4 Mb toggle MRAM. The yellow region is the 0% switching region, where the field is either too weak or in the incorrect direction for toggling. The purple region is the operating region above the transition to toggling for all the bits, where 100% of the bits are toggling properly. The transition between the two regions shows the switching field variation of the switching field for individual bits within the array. The dramatic improvement of the operating region over conventional MRAM is evident.

For MRAM to maintain its status as a competitive memory in future technology nodes, scalability of the essential parameters is crucial. The main factors influencing write scalability are the increased switching field required for small bit sizes, its generation, and thermal stability of the memory bits. As the bit

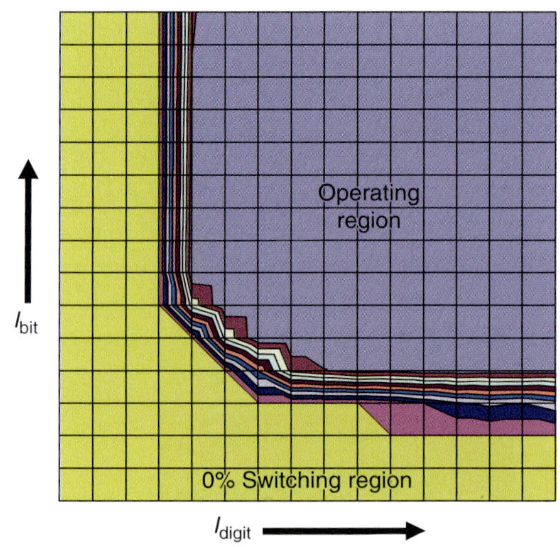

Figure 13 Switching astroid for 4 Mb toggle MRAM array. A large operating region is present once the bit and digit line currents exceed the threshold values. The disturb mechanism has been eliminated.

width decreases, maintaining sufficient stability against thermal disturbances during the 20-year operation lifetime requires that the thickness of the ferromagnetic layers be increased and hence a switching field increase. However, as the feature size decreases, so do the bit and digit line widths and the interlayer dielectric separating them from the bit. The final result is that the field generated at the bit increases at the same rate as the required switching field, while sufficient thermal stability is generated for operation down to the 45 nm node. Additionally, alternative toggle configurations have been proposed to decrease the mean switching field for use at these nodes [50,51]. Scaling has also been demonstrated in MRAM utilizing and MgO tunneling barrier integrated with a 90 nm front end CMOS in 8 kb array, exhibiting read and toggle switching operations [45].

7.3.6 Spin-Torque Magnetic Random Access Memory

Spin transfer switching – discussed in the previous section – first observed in GMR structures [30], and later in MTJS [52,53], provides an alternative method for switching the bit, and thereby a unique approach that mitigates some of the scaling challenges in MRAM for 45 nm and below. In ST-MRAM, bit reversal requires only passing a current directly through the tunnel barrier, and thus the lower current line and via for contact to the underlying CMOS are no longer necessary (see **Figure 8**). Cladding of the bit and digit line for field amplification is not required. The currents required for switching are substantially reduced, and both the read and write current paths share the same transistor. The spin transfer effect becomes more important as the bit decreases [30], and can form the dominant interaction for a spin-polarized tunneling current at bit sizes of ~100 nm and below, leading to a favorable scaling. The result of these attributes is a low-power, dense MRAM with substantial simplification in the bit cell and fabrication process.

However, ST-MRAM is not without its challenges, the primary of which is reducing the currents required for switching the bits. Approximating the magnetization in the free layer as a single direction (the macrospin model), the critical current to initiate spin transfer switching in zero temperature and field (relevant for high-speed reversal) is given by

$$I_c = \left(\frac{2e}{\hbar}\right)\left(\frac{\alpha M_s V}{g}\right)(H_K + 2\pi M_s), \tag{3}$$

with spin torque efficiency g, and free-layer parameters: α (Gilbert damping), M_s (saturation magnetization), V (Volume), and H_K (anisotropy). For typical material parameters, the resultant critical current density is 1–10 × 10^6 A cm^{-2}, for a current of ~0.5 mA for conventional devices of 100 nm dimensions. Independent of CMOS node, a minimum-sized transistor can source ~0.6 mA per micron of transistor width, and as each MTJ requires a transistor to apply switching current, the cell size will be determined by the transistor width unless the critical current can be reduced. If the critical current can be reduced to <100 μA while maintaining thermal stability, a high-density ST-MRAM array becomes viable.

Forming a tunnel barrier with sufficient transparency such that the application of a reasonable bias voltage (~1 V) across the memory element induces sufficient current density for ST switching requires ultrathin tunneling barriers with very low resistance–area (RA) products (RA < ~20 Ω μm^2). The application of the write voltage across the bit may cause a tunnel barrier failure in weaker bits, and read and write share the same path creating a mechanism for read disturbs. A battle must be waged on three fronts. The separation of the means and the widths of the distributions of the read, write, and breakdown voltages, all, must be carefully controlled with 12σ separation between each distribution (**Figure 14**).

The distribution of required write voltages stems from differences in the magnetic volume or shape of the memory element to be switched, from variances in the patterning process, as well as differing amounts of magnetic disorder. The distribution in the required read voltages to successfully determine the state of the bit comes primarily from resistance and

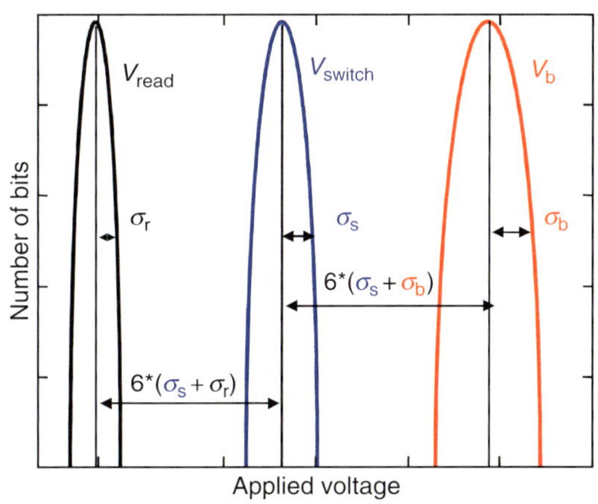

Figure 14 Critical separations for ST-MRAM. Read distribution must be separated by 12σ from write and write distribution must be separated by 12σ from the breakdown distribution for successful ST-MRAM demonstration.

MR variations between bits, as well as differences in the resistances of the pass transistors. The distribution of breakdown voltages arises from defects that may be present at the interfaces or within the tunnel barrier, tunneling hot spots, and size differences of the memory elements. Apart from the clear need to maximize the breakdown voltages and minimize the write voltages (or critical currents), it is necessary that each of the distributions must be controlled. Additionally, the necessity for a low read voltage and low RA barrier requires an MTJ with very high MR for enough read separation to cope with the low read signal and resistance distributions inherent in the small patterned bits. The recent discovery of MgO as a tunneling barrier provides not only higher MR for improved read signal and spin torque efficiency, but also lower barrier height, allowing a thicker, more pinhole-free barrier to be fabricated than that of that is, AlO_x for a given RA.

Equation 3 shows that there are a number of approaches to reduce the critical current. The second term in parenthesis draws attention to thin, low moment, low-damping alloys with high polarization (for high-spin torque efficiency). However, there are a limited number of ferromagnetic materials that offer the simultaneous combination of some of these traits, the most widely utilized of which is CoFeB, especially when coupled with an MgO tunneling barrier. While CoFeB alloys can be made with low moment, the spin polarization is reduced with the moment. Similarly, there are limits in how low the volume may be reduced before thermal stability problems begin to creep in. There are a number of alternative alloys with low disorder and hence low damping, but finding a combination with low M_s, and high polarization has proved elusive. Active research in this area is ongoing.

The spin torque efficiency is closely related to the polarization of the tunnel current impinging upon or reflected back toward the free layer, with a maximum efficiency of 1 for 100% polarization of incident electrons. Dramatic increases in polarization have been developed with the introduction of MgO tunnel barriers, but only another ~10% I_c reduction is available through these means [54]. Additionally, while increasing read signal, perfect spin polarization may not be desirable, as this would correspond to a high resistance state that is effectively infinite. In such a case, the voltage bias that must be applied across the tunnel barrier becomes unreasonably large, and tunnel barrier breakdown will result before I_c can be reached.

In the macrospin model, spin torque reversal involves precessional motion of the magnetization with an out-of-plane component, due to the direction with which spin torque is applied. The final term in **equation 3** contains the in-plane (H_K) and out-of-plane ($2\pi M_s$) anisotropies of the bit that must be overcome for reversal. For thin films, the out-of-plane anisotropy, which comes from the demagnetization field, is much larger than the shape anisotropy and largely determines I_c. This enables a distinct approach for reduction: the introduction of either partial or complete perpendicular anisotropy in the free layer. This can be induced either through using multilayers of thin films to take advantage of the surface anisotropies (e.g., Co/Ni and Co/Cu) [55,20], or through the use of ordered alloys [56]. The induced anisotropy then subtracts from the $2\pi M_s$ term. With a careful optimization, the magnetization may lie in-plane but the demagnetization contribution can be nearly removed. The difficulty in this approach is that two relatively large terms are subtracted to leave a small difference, which will likely exhibit a significant switching current distribution.

When the induced out-of-plane anisotropy is larger than $2\pi M_s$, as in Ref. [57], the magnetization of the free layer lies entirely out-of-plane. The shape anisotropy of the bit is then no longer important for preservation of the thermal stability and very small circular bits can be patterned. The lithographic challenges associated with defining a high aspect ratio bits at 100 nm size scales are then no longer relevant. While active research is currently underway, it has been a formidable challenge to develop high polarization, low-damping materials with perpendicular anisotropy that can be successfully grown on tunnel barrier layers with good magnetic response. ST-MRAM is currently under development in a number of industrial research laboratories, with reports of initial CMOS results in 4 kb arrays [57], design of a 2 Mb chip [58], and several recent CMOS demos with discussion of read and write separations in 16 [59] and 4 kb[45] arrays. Additionally, performance of the individual devices with perpendicular anisotropy integrated with 1 kb arrays has been reported [57].

7.3.7 Conclusion

The unique combination of performance characteristics present in MRAM technology provide a high-speed nonvolatile memory with superior reliability characteristics, the ideal combination for numerous applications. Toggle MRAM is in volume production at Everspin, with parts currently available at up to 4 Mb-memory densities and numerous temperature ranges. A number of companies have reported demonstrations of a 16- [60] and 32-[61]Mb chip, as well as demonstration circuits with the segmented write line approach [48]. The promise of extension into an ultradense, low-power regime with next generation ST-MRAM has placed it at the center of an active development effort.

7.4 Magnetic Recording

While high-speed information flow requirements may potentially be met using MRAM, the need to store large quantities of data will require intensive efforts to improve old magnetic recording methods such as hard disk or magnetic tape drives. Information or data can be recorded using virtually any form of energy, from manual muscle power in handwriting, to acoustic vibrations in phonographic recording, to electromagnetic energy modulating magnetic tape and optical discs. The capacity of stored information is usually measured in bits, where one bit corresponds to a 1 or a 0 in binary language. Today, the most commonly used storage systems involve magnetic recording, where data is stored on a magnetic medium, providing nonvolatility and unlimited rewriting capability. The first practical application of this type of recording was in magnetic tape recorders demonstrated in 1935 by BASF in Germany. In 1956, IBM introduced the 305 RAMAC computer system with the first magnetic disk drive, the IBM 350, which had a storage capacity of 4.4 Mb, was the size of two large refrigerators, and weighed two tons. By comparison, today's laptop computers commonly have disk drives that can store in excess of 100 Gb and are the size of a pack of cards. This represents an increase in areal density (number of bits per square inch of disk surface) of more than eight orders of magnitude from 0.002 Mb in^{-2} in 1956 to 400 Gb in^{-2} in today's state-of-the-art drives. Through the first 5 years, IBM sold a total of about 1000 magnetic disk drives, with a total storage capacity of about 32 Gb (or 32×10^9 bits), some of which are still operational to this day in museums. For 2008, it is estimated that 250 Eb (2×10^{21} bits) of data were created and replicated worldwide and that, by 2010, that figure will grow fourfold to almost 1000 Eb [62]. As of 2009, magnetic storage media, primarily HDD or tapes, are widely used to store computer data as well as audio and video signals. Magnetic tape has been used when storing large amounts of data mostly for backups and archives, since tape can be substantially less expensive than HDD or other data-storage options. However, the highest achieved areal density demonstrations are only 6, 7 Gb in^{-2}; about 100 times lower than in HDD. Since development of magnetic tape recording follows pretty much the same path as HDD recording, and since the rapid improvements in disk-storage density and pricing have reduced the market share of tape storage, the following discussion focuses only on HDD technology.

The present overview describes recent improvements and challenges concerning the two main magnetic components of the HDD, head and media. There are several issues that current HDD technology faces in terms of performance, scalability, cost, and complexity of manufacturing. The two prevalent approaches to overcome or delay these limits include bit-patterned media (BPM) or discrete track media (DTM) and energy-assisted magnetic recording. However, each approach has its own challenges both in terms of the recording physics and material system requirements to achieve potentially good and scalable performance in addition to nanofabrication challenges for the actual realization (cost and complexity of manufacturing). Overcoming these challenges will enable reaching recording densities beyond 1 Tb in^{-2}.

7.4.1 Current Technology – Continuous Perpendicular Recording

7.4.1.1 Components of a hard-disk drive

All HDDs share a basic structure and contain the same main elements. A schematic of a magnetic recording system is shown in **Figure 15**. One of the main components of an HDD is the disk, made from glass or Aluminum, which is coated on both sides with a magnetic recording layer formed by sputter deposition. This recording layer is made up of nanoscale magnetic grains, and the signal-to-noise ratio (SNR) needed for high-density recording is determined by statistically averaging the contributions from a large number of such weakly interacting magnetic grains per bit. The granular structure limits the magnetic correlations and allows information to be written on a finer scale than is possible in a homogeneous magnetic film. As sketched in **Figure 15(a)**, the transitions generally follow the grain boundaries, and thus the storage density of the data is ultimately limited by the grain size. Scaling of magnetic media

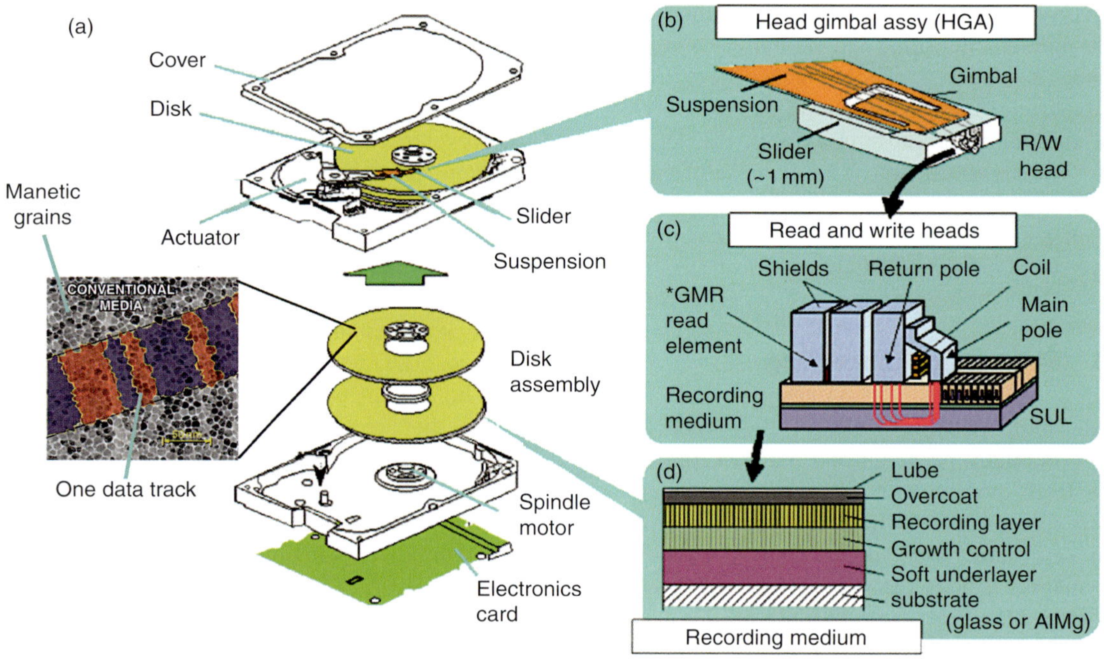

Figure 15 The components of a hard-disk drive. The basic design involves a rotating disk plus a rotary actuator that provides the access mechanism for the read/write heads (a). The heads are fabricated onto a ceramic slider that glides just a few nanometers above the surface of the spinning disk (b). A perpendicular recording configuration is illustrated (c), and recording medium constituents (d) are detailed in Ref. [62].

involves reducing the grain diameter and reducing the media's magnetic thickness $M_r t$ (where M_r is the remanent magnetization of the media and t is the media thickness). The original disk-recording medium consisted of iron oxide particles in an organic binder that was spun onto the disk. This technology was in use until the 1980s; when the inability to scale particle size and particle-size distributions was in part responsible for the introduction of sputtered thin-film CoCr alloy media. For these granular CoCr films, the Cr diffuses toward the grain boundaries during growth, resulting in grains with a magnetic core and nonmagnetic or weakly magnetic grain boundaries. These films have evolved to the current media using CoPtCr alloys with the so-called segregates, materials like B or oxides such as SiO_2 or TiO_2 that segregate at the grain boundaries, and help define and control the exchange coupling between adjacent grains. Alloy developments, along with considerable improvements in underlayers, allowed the grain size and media thickness to scale appropriately to grain diameters in the order of 8 nm (**Figure 15(d)**). However, media scaling is limited by thermal instabilities when the grain volume V decreases to the point where the magnetic energy per particle $K_U V$ (where K_U is the magnetic anisotropy energy density) becomes comparable with thermal energies. The reversal of the magnetization of individual media grains due to thermal activation and the resulting loss of data is called the superparamagnetic effect. The minimum energy needed to maintain stability for 10 years is $K_U V \approx 45\, k_B T$, where k_B is the Boltzmann constant and T is absolute temperature. Reductions in V can be compensated by increasing K_U. However, increases in K_U are limited by available write fields needed to overcome the media's coercive field. The combination of readability requirements (SNR), write-field limitations, and thermal activation of small particles is often referred to as the trilemma of magnetic recording and sets the basic requirements for future HDD technologies.

In the current state-of-the-art HDD products, the magnetic data are written in the medium on ~100-nm-wide concentric tracks. Along these tracks, a succession of approximately 15-nm long zones, having either a magnetization pointing up or down relative to the disk plane, define the magnetic bits (**Figure 15(c)**). At these dimensions, approximately 400×10^9 bits/in^2 or 400 Gb in^{-2} can be written on the magnetic media surface. Note that each bit is not square but rectangular with a bit aspect ratio (BAR = width/length) of 5. The recording layer is generally covered with a 3–4-nm-thick carbon-based overcoat layer that protects the magnetic grain from oxidation and mechanical damage, and a 1-nm-thick lubrication layer made from long organic molecules that allows the write/read head to fly stably across the disk surface (**Figure 15**). In a current HDD, typically one to five disks are mounted to a rotating spindle that turns all the disks in unison. The spindle motor is mounted right below the spindle and spins the disks at a constant rotational speed ranging from 5000 to 15 000 rotations per minute.

The read/write head is the device that reads and writes data in the medium while the disk is rotating. This head contains various functional elements. The write element consists of a high-permeability magnetic pole with a few turns of a conducting coil wound around it. When an electrical current flows through the coil, magnetic flux flows through the writing pole, emanates from the pole, and penetrates the magnetic medium (**Figure 15(c)**), causing the bits to be magnetized up or down. In recent HDDs using media with a perpendicular orientation of the magnetization, a CoFe-based so-called soft underlayer (SUL) is introduced between the disk substrate and the magnetic medium (**Figures 15(c)** and **15(d)**). This SUL conducts the flux from the writing pole of the head and the medium to the return pole and helps increasing and localizing the effective write field of the recording head. Binary data is encoded via the direction (up or down) of the bit magnetization, in coincidence with a clock, which is synchronized with the disk. The second magnetic element of the recording head, the so-called read head or sensor, is used to sense the magnetic flux emanating from the recorded bits in the magnetized medium during the read back process. Then, sophisticated signal processing electronics are used to convert the waveforms of the signal measured by the read head back into digital data. The read element is protected from any spurious magnetic fields by two large soft-magnetic shields (**Figure 15(c)**).

Since the magnetic field coming from the bits decreases rapidly with distance from the magnetic medium, the space between the head and medium must be reduced as much as possible. As of 2009, the separation between the recording head and the media surface is only about 5 nm. When in motion, this spacing is maintained by an aerodynamic air-bearing surface patterned lithographically onto the bottom of the slider carrying the magnetic head elements (**Figure 15(b)**), and even though the head–disk separation is less than the typical mean free path of an air molecule, the head is said to fly or glide over the disk. (As a scale comparison, with the same spacing constraint, a 70 m- long Boeing 747 Jumbo Jet would have to fly 1.5 mm from the ground.) If the head hits the disk's surface or any defects like dust at the disk surface, the aerodynamic stability of the head–disk interface can be disturbed and the head can crash into the disk surface and damage it, which may in turn result in data loss and an inoperable device. This necessitates the hard drive assembly to be done in a clean room environment in which the concentration of dust in the air is strongly reduced.

Many HDDs have multiple disks and heads, but typically all heads are attached to a common head actuator so that they all move in unison. However, only one of the heads is active at a time, that is, reading or writing data. The actuator controls the movement of the head arm, also called head gimbal assembly. The guidance system that keeps the head on a track of data is called a servo. Outside of the head disk assembly is the electronic card that controls the movements of the internal parts and controls the movement of data into and out of the drive (**Figure 15**).

7.4.1.2 Areal density increases – past and future

Throughout the first half-century of disk-drive development, the basic recording principle – longitudinal magnetic recording (LMR) – had remained the same, and the increase in HDD recording density was achieved primarily by scaling the components of the disk drive to ever smaller dimensions. In the write/read head, as shown in **Figure 16**, tremendous progress in the ability to fabricate the magnetic elements at ever smaller dimensions and the invention of new sensor schemes with increasing sensitivity have enabled the rapid progression of the areal density. Over the years, the read head sensor has evolved from inductive heads, that is, sensing coils wound around magnetic poles, to inductive thin-film heads, where the poles and coils were deposited by thin-film processing, and then starting in the early 1990s a rapid succession of MR effects to probe the field emerging from the medium, including AMR and GMR. Starting in 2005, the newest generations of heads have been using a related technology based on the so-called TMR effect. In both these cases, the sensor is based on a trilayer stack in which a

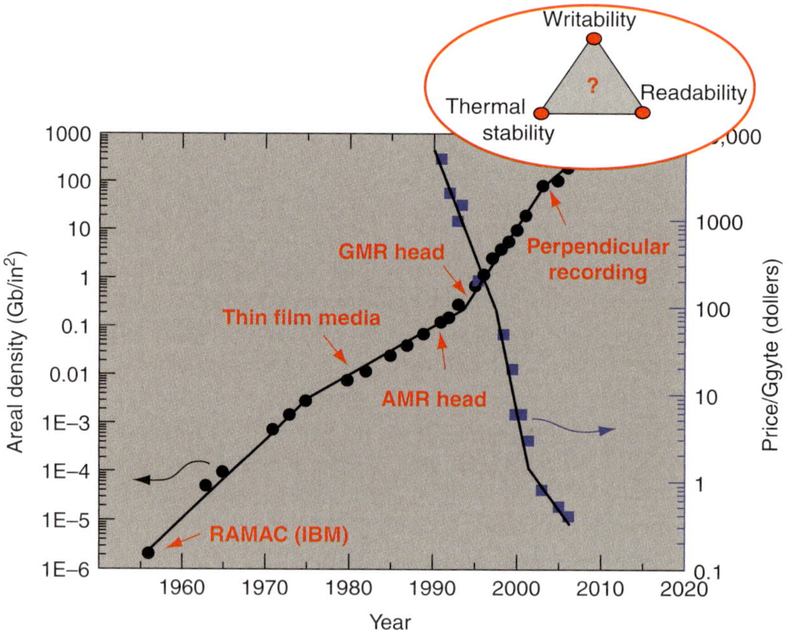

Figure 16 Evolution of areal density and price per Gigabyte (1 byte = 8 bits) of magnetic HDD with time since 1956 (RAMAC). After 2010, the recording trilemma, that is, the requested equilibrium between writability, readability, and thermal stability, leads to uncertainties on the areal density growth trend Ref. [63].

nonmagnetic spacer layer magnetically decouples two ferromagnetic layers. The fields originating from successive bits orient the magnetization of the probe layer relative to the second reference layer whose magnetization is kept fixed. During the reading process, an electrical current is sent to the sensor stack whose resistance varies with the relative orientation of the magnetizations (see Section 7.2.2). These technologies have led to dramatic increases in sensitivity and SNR of the read sensors and allowed to continuously decrease the area allocated to one bit (as well as the number of grains per bit). Moreover, as a result of scaling, the head flies closer to the disk and the effective write field from the head increases as do the write and read resolution.

All these advances in head technology have been accompanied by similar improvements in media technology enabled by many years of intense research on magnetic thin-film growth. The first media were based on grains with randomly orientated anisotropy from iron oxide embedded in epoxy binder materials. In the 1980s, in-plane magnetized thin-film media made by vacuum deposition were introduced, which enabled a reduced media grain size and anisotropy distributions by improved process control and well-defined crystallographic orientation of the media grains, with the magnetization of the grains aligned in the plane of the medium. A few years ago the industry has transitioned to a technology called PMR. That has allowed further increase in areal density enabled by increases in available head write field due to more efficient pole designs and the introduction of the so-called SUL that guides the magnetic flux in the recording media. On the media side, continued improvements in process control and technology have led to reductions in media grain size and anisotropy distributions.

However, the continuation of the seemingly unstoppable growth in HDD areal storage density can no longer be taken for granted. For the first time, the challenges faced by the HDD industry are no longer merely engineering challenges, that is, technical challenges related to scaling all the HDD components as the hard drives becomes smaller and smaller. Write head design, reader-sensitivity, head–media spacing, track-following accuracy, and signal processing are some of the limiting factors in this category. And for the first time over the last decade the challenge faced by the industry has been of a more fundamental, physical nature, namely maintaining of magnetic information in volumes so small that the total magnetic energy related to preserving the direction of the magnetization ($K_U V$) becomes comparable to the energy of thermal fluctuations at room temperature (superparamagnetic limit), commonly expressed as the product of Boltzmann's constant, k_B, and the temperature, T. As the areal density increases, the grain volume V has to be reduced. It is then obvious to conclude that the solution to maintain the magnetic stability of the media while decreasing the grain size is to increase the magnetic anisotropy of the grain. But stability is not the only important parameter for an HDD to perform; equally important are writability and readability. The need of tuning these three correlated features leads to the so-called recording trilemma sketched in **Figure 16**. First, granular perpendicular recording readability (i.e., high SNR) is insured by a large number of magnetic grains per bit. Therefore to increase the bit areal density one has to reduce the size of the each grain to reduce the overall size of the bits. However, the thermal stability of a grain is given by the product $K_U V$ defined earlier. So, as the grain volume V is reduced, the anisotropy K_U has to be increased to maintain sufficient thermal stability. Finally, the ability to switch the grain magnetization with the head field is in first approximation proportional to K_U. Since, the field currently produced by the head is close to its maximum, writability places an upper limit in the value of K_U for the medium.

7.4.1.3 Beyond the limits of single media structures – exchange-coupled-continuous media

While generally writability and stability for conventional recording media are correlated through the magneto-crystalline anisotropy, it has been demonstrated in recent years that advanced multilayered media structures can be designed in such a way that an improved writability to stability ratio can be achieved. Using such multilayered structures first allowed the extension of LMR technology beyond 100 Gb in^{-2} and more recently has led to advances in PMR technology to beyond 600 Gb in^{-2}, more than twice the areal density that was thought possible just 3 or 4 years ago. In perpendicular recording the media technology used to achieve these areal densities is called exchange-coupled-continuous (ECC) media. Within this new multilayer system, the usual granular magnetic media is coupled to an assist layer with a lower anisotropy that allows fundamental changes in the switching mechanism of the media magnetization. Indeed, in single granular media the magnetization switches via the so-called coherent reversal mechanism (also often referred to as Stoner–Wohlfarth switching), that is, the magnetic moments inside one grain switch all at once and can be described by the response of a single spin, the so called macrospin, to an external field. In contrast, the reversal mechanism in an ECC media can best be described by a domain-wall motion through the depth of the media stack: at first a domain wall, that is, a continuous twist of the magnetization is forced into the depth of the assist layer, and is pinned at the interface between the assist layer and the granular media. At a certain field, the domain wall depins from the interface and propagates through the media layer, switching its magnetization in the process. For optimized composition of this bilayer system, the field at which the domain wall propagates irreversibly into the media layer is lower than the switching field of the media layer by itself [64–66]. Therefore, the addition of the assist layer allows reducing the required write field for a given media without affecting its thermal stability, or, conversely, allows the use of higher K_U materials for a given write field and a further reduction in grain size and resulting increase in areal density.

Using advanced media schemes such as the ECC structures mentioned above, experts estimate that PMR technology using the current set of materials will scale to recording densities of 750 Gb in^{-2} or slightly beyond by utilizing incremental improvements and continued optimization before reaching thermal stability limitations. Although, to meet the growing consumer demands and continue the annual areal density growth of 30–40%, new technologies to extend magnetic recording into the terabit/in^2 regime will have to be developed in the next few years. Quite generally, the choices are twofold. On the one hand, the approach called patterned media proposes to lithographically define magnetic islands, each one corresponding to one bit, whose volume can be increased as compared with the current grain size that rules the thermal stability. SNR would then be maintained through reduction in grain-size distribution and position jitter. One possible stepping stone toward this technology is the so-called DTM, which only use lithography to define the tracks on the recording media. On the other hand, the approach called energy-assisted recording consists of increasing the magnetic anisotropy of the media and overcome the resulting limitation in writability by using some sort of assist mechanism to (temporarily) lower the required switching field of the media during the writing process.

7.4.2 Patterned Media

In this part, the status of two proposals for extending the storage areal density by patterning the storage medium, namely DTM and BPM (**Figure 17**), is reviewed.

7.4.2.1 Discrete track media

In conventional magnetic recording, both the bit length, along the data track, and the bit width, across the track, are defined by the write field from the recording head. The track locations are defined using servo marks which enable the head to accurately go to a certain track and stay on-track over millions of disk rotations. The servo marks consist of multiple radius lines of information written perpendicularly to the data track in a regular manner over the entire disk. Achieving high-density recording with the current perpendicular media will require increasing the track density. For example, to achieve a density of 1 Tb in^{-2} with a BAR of 6 would require a track width of less than 60 nm. As a consequence of such narrow track width, the read head and write head elements will have to be shrunk drastically as well, and both readability and writability will pose great challenges. First, as the track width decreases, the read head starts sensing more than one track at a time which induces noise or may lead to errors in the reading process. To avoid such edge-noise problems, the sensor element that is already around 50 nm wide has to be further reduced, constituting a real nanofabrication challenge. Moreover, as the track width is reduced, the write pole size of the head has to be decreased to avoid writing more than one track at a time. But this reduced write pole width without a corresponding reduction in head-media spacing, in turn, leads to a decrease in delivered write-field intensity. Therefore, errors during the write process can occur when the head field amplitude will be reduced below the field value requested to switch the bits' magnetization.

The edge-noise errors and adjacent track overwrite can be reduced by physically defining the track locations on the disk and thus no longer relying solely on the mechanical positioning capability of the HDD head. In DTM technology, the tracks' widths and locations are lithographically defined [67,68]. From a recording perspective, the main impact of DTM is at the head design level. Indeed, by using discrete magnetic tracks, the write and read elements can be made wider than the patterned track. The requirements on head tolerances may be eased, thus improving manufacturing yield. As far as magnetic media is concerned, while the track width is defined by lithography, the bit transitions along the track are still defined by the field from the write head as in conventional magnetic recording. The transition between successive bits along the track is still determined by the reversal of magnetization direction of multiple grains. Hence, the linear recording density still depends on the size and uniformity of the grains and the material requirements on DTM are thus very similar to those of continuous media.

Besides the gains on the head design, efforts to develop new lithography methods for DTM will help to introduce patterned servos. In current HDD, a certain amount of eccentricity is inherent in the rotation of the disk. Moreover, the head actuator position in the drive is not controlled very precisely. So, before the first data recording, servo marks are written for measuring the exact head actuator position and enable the head to repeatedly read/write the data bits. In the current PMR media, the servo writing process consists of flying the head over the entire disk surface and recording sequences of data called servo marks. This step has to be performed on all HDD disks, which is both slow and expensive. Patterning the servo marks onto the disk surface poses a lithography challenge due to various duty cycles, but ultimately offers the potential for significant cost savings. Indeed, with prepatterned bits or tracks, the servo information is no longer detected by the head via a specific magnetic sequence but due to a specific shape of the media patterns that are uniformly magnetized. A simple saturation of the magnetization under a DC magnetic field on each disk would thus be enough to set the servo marks. The same servo-pattern technology will be implemented for BPM.

In conclusion, DTM offers reduced adjacent track interference and edge noise, resulting in higher track density and better SNR. Moreover, the development of nanofabrication methods for mass production will help introduce low-cost patterned servo fabrication. Nevertheless, there are still enormous difficulties from the standpoint of practical application such as eliminating patterning-induced edge-noise effects, developing planarization process, and improving magnetic edge corrosion resistance. Moreover, production cost is another issue in manufacturing DTM. It could take several years to further develop nanoprocessing technology and establish a cost-effective infrastructure to produce double-sided disks with DTM. Finally, in terms of scalability and areal

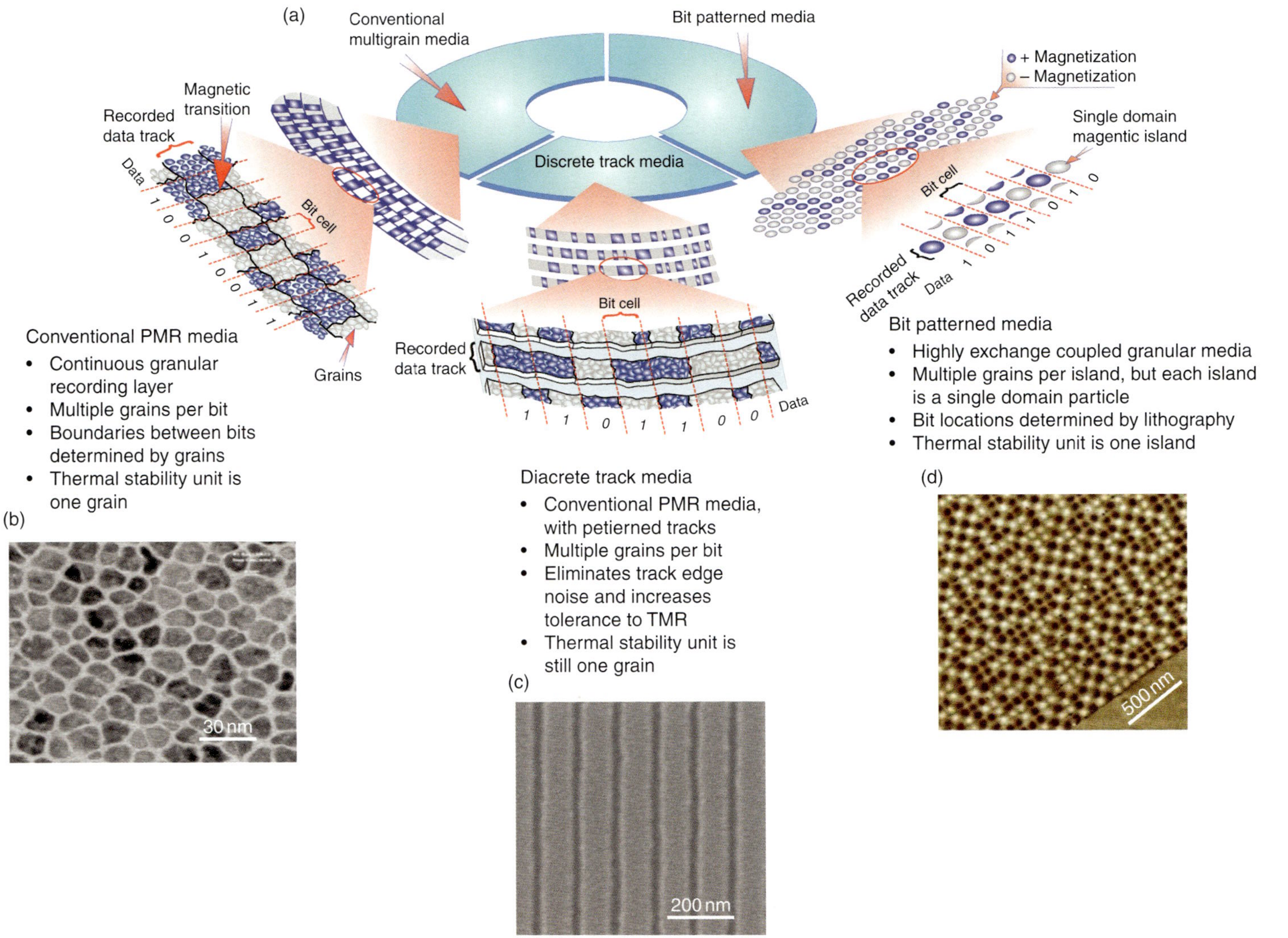

Figure 17 (a) Sketches and main features of current of continuous granular media, discrete track media, and bit patterned media. (b) Transmission electron microscopy image of a conventional granular media. (c) Scanning electron microscopy image of patterned tracks and grooves of a DTM media. (d) Magnetic force microscopy image of bit-patterned media in a demagnetized state. Courtesy of Hitachi GST.

density increase, the fact that DTM is based on the current granular media is one of the limitations of this new concept. Indeed, DTM does not solve the problem of superparamagnetism in the media as one still has to shrink the grain size to increase the areal density.

7.4.2.2 Bit patterned media

Contrary to DTM technology, BPM technology is expected to address the increasingly difficult challenges of reducing grain size in continuous media, by using lithographically patterned media where each magnetic island (or dot) corresponds to a single bit (**Figure 17**). As an example, 1 Tb in^{-2}recording in conventional PMR media would correspond to about 30 grains with a diameter of about 4 nm. In the case of a patterned recording medium, it is assumed that magnetic dots or entities of 12.5 nm diameter would be required for storing the same amount of information.. For the same thickness and anisotropy, the thermal stability, defined as K_uV, of a 12.5-nm-diameter dot is up to nine times higher than the stability of each 4-nm-diameter grain. Thus, patterned media recording offers a strong advantage in terms of thermal stability compared to current granular media. Many investigations have been conducted exploring both BPM manufacturing processes and recording simulations of BPM [66,69]; however, there are also bigger challenges than in the case of DTM. Fabrication methods and tolerances (BAR and patterning quality), magnetic bit switching field distributions (SFDs), write-synchronization jitter, as well as the gradient of the write-field profile, are just some of the main parameters that need to be addressed in designing bit pattern recording systems.

1. *Bit patterning methods.* One of the most difficult challenges is to define a method of forming 10–16-nm-diameter islands with extremely tight tolerances on the width, length, and positional fluctuation of dots. To achieve 1 Tb in^{-2}, one has to develop a lithography process to create billions of individual identical nanoislands that must be equally spaced within the circular symmetry of the rotating disk, while meeting an economically acceptable cost – the cost of manufacturing a current PMR disk (i.e., disk substrate and medium) is of the order of a few US dollars. Furthermore, according to the 2008 International Technology Roadmap for Semiconductors (ITRS), a DRAM gate length of 15 nm would not be reached until about 2020. Nevertheless, although these dimensional requirements for patterned media may seem daunting when compared with the current state-of-the-art silicon (Si) lithography, there are several important differences between the two. Patterned media is likely a one-step mask process, while the fabrication of memory cells and logic gates requires many lithographic steps with precise mask alignments. Moreover the pattern does not even need to be well centered on the disk, as the disk will not be centered to better than approximately 10 μm on the drive spindle.

There are many ways to obtain isolated magnetic nanoobjects (see Section 7.2.3). For BPM, two classes of pattern transfer processes can be distinguished according to where in the process flow the lithography step occurs. For the prepatterning process, SUL and seed layers are first deposited on the substrate and patterned to create a surface of pillars and trenches, or with holes, over which the magnetic medium is sputter deposited (**Figure 18(a)**). In that case, the final magnetic bits stand either on the dots (**Figure 18(a)**) or in the holes [69–71]. Conversely, in the processes referred to as postpatterning, SUL, seed layer, and magnetic film are first deposited; then subtractive lithography steps are applied to create an array of magnetic dots (**Figure 18(b)**). The relative advantages and disadvantages of these two approaches are still under debate in view of their integration into real HDD recording. In the case of the prepatterning processes, the main problem arises from the magnetic noise created by the magnetic materials deposited on both sidewall and trenches (**Figure 18(a)**). In the case of the postpattern process, the magnetic material subtraction step may potentially damage the magnetic layer in the bits.

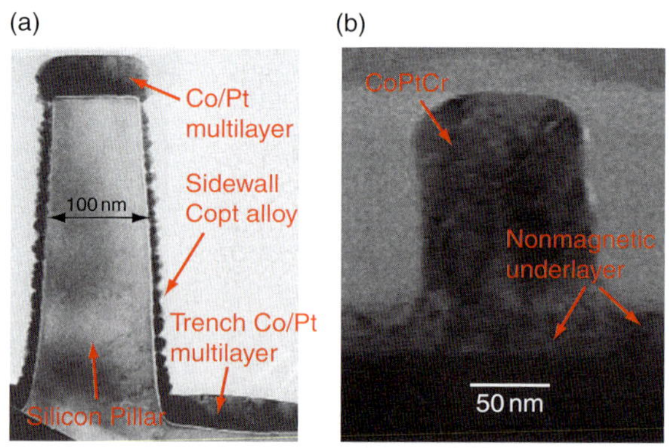

Figure 18 (a) Cross sectional TEM view of a 200 nm height prepatterned dot prepared by e-beam lithography followed by silicon RIE and covered by Co/Pt multilayer [69]. (b) Cross sectional TEM image of CoCrPt bit postpatterned media obtained by etching from a CoPtCr continuous film Ref. [70].

In both postpatterning and prepatterning approaches, the empty space between the dots has to be filled to protect the magnetic material (mainly against oxidation as in regular granular media) and support read/write head flyability. As already mentioned, in a HDD, the head–disk spacing has to be minimized down to 5 nm or less to optimize the writing and reading quality. This requires an extremely smooth surface, free of any resist residue or other contamination from the patterning process. To avoid the problems of planarization, studies have been conducted to create magnetic elements in a nonmagnetic matrix without physically separating between magnetic elements of patterned media (**Figure 19(a)**). In these cases, a magnetic film with perpendicular anisotropy (e.g., Co/Pt multilayer) is locally irradiated with low-energy ions such as Ar^+, He^+, and Ga^+. The bombarding induces local loss of magnetic anisotropy due, for example, to the intermixing of Co and Pt at the interfaces in the case of Co/Pt multilayer (**Figure 19(b)**). Multiple experimental demonstrations have been reported using FIB irradiation [71]. However, by using ion irradiation, two successive nonmodified magnetic regions, that is, two successive bits, are not fully magnetically decoupled and the low anisotropy region in between strongly affects the magnetization reversal mechanism of bits. Therefore, the main advantage of the ion irradiation method could be time and cost related but a careful investigation has to be performed to confirm whether or not this patterning technique can provide magnetically stable bits with a reliable switching mechanism from bit-to-bit.

Finally, self-assembled arrays of high anisotropy magnetic nanoparticles could be an ultimate patterned medium with densities higher than 10 Tbit in^{-2}. During the past two decades, significant research efforts has shown that chemical synthesis provides a powerful and rapid way to obtain large quantities of organized nanometric particles in a single reaction (**Figure 19(a)**). The magnetic nanostructures obtained through colloidal chemistry could achieve better crystallinity, and are typically much smaller in size than what is typically achieved using either pre- or postpatterning methods. FePt nanoparticle assemblies with particle diameters as small as 4 nm and an extremely narrow distribution of sizes (<5%) was demonstrated for the first time in 2000 [73]. However, to use FePt nanoparticles as a recording medium, significant challenges have to be overcome. One of those originates from the coating approach that is currently used to grow the particles on the substrate that is very different from the currently used sputtering technique [74]. Another current limitation is the extremely large distribution in the magnetic anisotropy direction of the particles which far exceeds the tolerances of BPM.

2. *Media design.* As described in Ref. [69], the important magnetic features of the future bit pattern medium include the following: (1) the medium has a well-defined uniaxial anisotropy axis, whose orientation is constant from one dot to another, (2) the discrete island must always maintain a single-domain remanent state,(3) the coercivity and hence anisotropy need to match the available write field while conserving good thermal stability, (4) the SFD must be sufficiently narrow so that the write head field addresses only the island intended to be written, and (5) the saturation magnetization should be tuned to optimize write and read back processes.

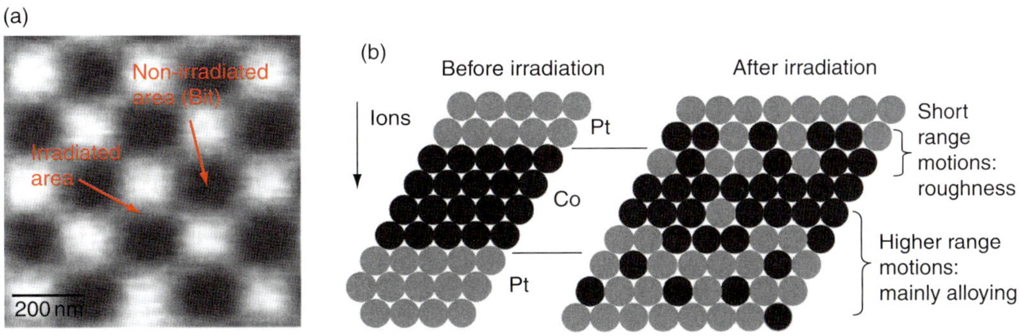

Figure 19 (a) Magnetic force microscopy image of a 2 × 2 μm² FIB patterned region. The black and white areas correspond to regions where the magnetization is pointing up and down. The grey regions in between are nonmagnetized areas. (b) Side view sketch of the irradiation induced intermixing of Co and Pt atoms in the depth of a Co/Pt multilayer film Ref.[72].
(a) Courtesy of B.D. Terris.

For BPM, it is likely that magnetic materials with perpendicular anisotropy will be chosen. The main reason for this choice is again related to the difficulty in inducing an in-plane magnetic anisotropy with a radial symmetry as required for the HDD. In addition to the patterning dot-to-dot reproducibility requirements for BPM, the magnetic properties of each island have to be closely controlled since there is no longer averaging over multiple grains. Perpendicular media for first BPM generations will most probably be multilayers or alloys based on transition metals (Co, Fe, Ni) and 5d metals (Pt, Pd), or media similar to those proposed for conventional perpendicular recording, that is, based on CoPtCr alloys grown with the *c*-axis normal to the substrate.

Single-domain behavior at remanence is expected when the island size is lower than the size of a magnetic domain wall. A domain-wall extension, that is, the zone of magnetization rotation between domains of different (at maximum opposite) magnetization direction, is given in first approximation by the square root of the exchange ($\mathcal{J}$) divided by the anisotropy (Ku). Therefore, to achieve single-domain islands, it will be desirable to have high exchange coupling within the individual islands, rather than the low exchange between grains as required for current magnetic recording. Generally, for the multilayers or alloys cited above with K_U up to $5.10^6\,\mathrm{J\,m^{-3}}$, magnetic dots with diameter lower than 50 nm cannot stabilize a domain wall at remanence [75]. At $1\,\mathrm{Tb\,in^{-2}}$, BPM would require dots with diameter less than 15 nm. Therefore, as far as 15-nm dots can be patterned, each dot will conserve a single magnetic domain at remanence.

As emphasized above, stability is the main advantage of BPM compared with current continuous media technology. Considering Co/Pt multilayer dots with $K_U = 2.10^6\,\mathrm{J\,m^{-3}}$ and a bit thickness of 10 nm, and assuming that each bit behaves as a macrospin, the thermal stability of each dot at $1\,\mathrm{Tb\,in^{-2}}$ (defined earlier as K_UV) would be much higher than 100 times the thermal energy, k_BT, at 350 K. Moreover for the same system, the switching field value that is needed for reversing the magnetization can be easily tuned to be equal to the field provided by the write head (i.e., a static coercivity around 6 kOe) similar to the field provided by the write head. As a consequence, writability is currently not an issue for BPM up to $1\,\mathrm{Tb\,in^{-2}}$. The use of very high anisotropy materials, like chemically ordered crystallographic $L1_0$ phase FePt or CoPt, will not be needed until at least $2\,\mathrm{Tbit\,in^{-2}}$. At that time, novel write mechanisms, such as ECC technology or energy-assisted recording mixed with the patterned technology may be required.

Reproducibility in the structural and magnetic properties from bit-to-bit is another main issue that BPM technology has to solve before introduction at a manufacturing level. Indeed, bit-to-bit variations in the structural and/or magnetic properties induce a variation of switching field. In other words, the field required to reverse the magnetization of one single dot varies over the entire disk. During the recording process, such a distribution of switching field can result in nonwriting errors in the case of higher reversal field dots or overwritten neighboring bits with a lower reversal field. There are multiple origins of SFD. The first one, the so-called dipolar field-induced SFD, is inherent to the bit pattern medium and defines the minimum SFD that can be achieved. Indeed, in an array of single magnetic dots, the magnetic field generated by each individual dot affects the reversal of all the other dots. As a consequence, a magnetic bit with all its neighbors pointing in the down direction would have a different switching field than another bit with all its neighbors pointing in the up direction. In addition to this dipolar field-induced SFD, the difficulty in controlling all the intrinsic and extrinsic parameters that influence the magnetization reversal mechanism causes various SFD as described in **Figure 20**. Intrinsic usually stands for the properties that are not induced by the patterning process including anisotropy distribution or crystal boundaries which would be present in the full film (**Figure 20**). The extrinsic parameters, that is, patterning induced, can be edge defects or interface mixing and strongly depend on the patterning method (prepatterning or postpatterning).

Finally, the optimum value for the saturation magnetization (M_s) of the medium is determined by considering the competing requirements of the write and read processes. On the one hand, the magnetic flux coming from the patterned bits being proportional to the bit volume and the M_s value, this M_sV needs to be large enough for the bit data to be probed by the reader element of the head. On the other, as discussed in the previous paragraph, the magnetic field generated by each individual bit affects the reversal of all the other bits. And the resulting SFD increases the probability of an unintended reversal or overwriting of adjacent dots during the write process. This argument would then favor a lower value for M_s. Incorporating these requirements into a

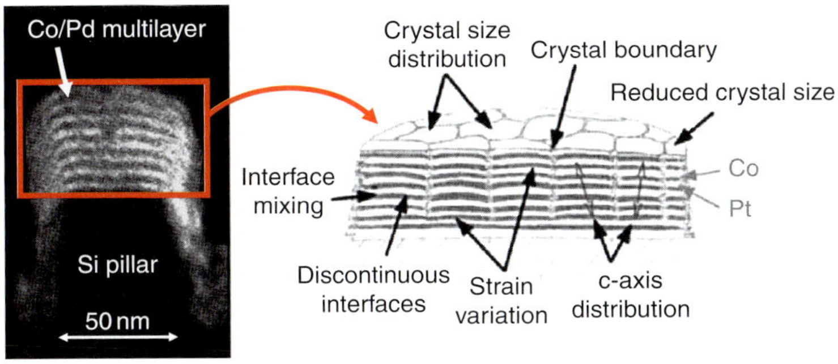

Figure 20 Transmission electron microscopy image (left) and sketch (right) of a Co/Pt multilayer-based deposited on top of a prepatterned silicon substrate with 50-nm-diameter dots. Some of the possible switching field distribution origins are pointed out on the sketch. Courtesy of Hitachi GST.

recording model, current theoretical designs for 1 Tb in^{-2} use a medium with M_s of 500 emu cm^{-3} [76].

3. *Head design and recording constrains for BPM.* Unlike discrete track recording, BPM technology needs to cope with many novel challenges in terms of the recording physics. Detailed theoretical analyses have been performed in the last 5 years concerning the geometrical and magnetic tolerances of BPM at densities up to about 10 Tb in^{-2} [75]. The parameter space for head/media design for BPM recording is extremely large. In the following, only the leading order terms are highlighted: BAR value, head size, write/read synchronization, field gradient, SFD, and thermal stability for fixed values of the fabrication tolerances.

To achieve the large density gains expected from using BPM, the BAR, currently around 5–6 for continuous perpendicular media, would have to be reduced to 1 or 2. Of course, the final dimensions of the bit will be restricted by the smallest achievable lithographic feature. A BAR equal to 1 at 1 Tb in^{-2} would correspond to a bit pitch (average space between the centers of two successive dots) of 25 nm. As a consequence, research efforts to develop recording heads with lateral dimensions small enough to avoid adjacent track erasure and adjacent track noise, respectively during the writing and reading process are required in parallel to developments in media patterning processes. As explained in the case of DTM and as sketched in **Figure 21(a)**, if the write element within the recording head is not centered on one bit or if it is too wide, the magnetic field emerging from the write pole affects more than one bit. This can lead to a misregistration of the information coupled to an erasure of a neighboring bit. Similarly, during the reading process, interferences between magnetic signals coming from multiple bits could drastically affect the SNR of a wide read sensor.

In addition to head size constraints, synchronized writing is another important issue in the implementation of BPM technology. The frequency and the phase (i.e., position relative to the dots center) of the record field pulse produced by the head needs to be accurately synchronized with the location of successive bits along the data track (**Figure 21(a)**). Even if the head pole is very small, variation of the head position along the track can lead to an unintended bit writing. In current granular media, the bit locations along the track are not specified before the first writing process since the magnetic grains cover the entire surface of the hard disk. On the contrary, in the case of BPM, each bit has a physically defined location. Depending on the tolerance in positioning the write head over the desired bit, simulations shows that switching of the neighboring bits can be avoided by adjusting the magnetic exchange and anisotropy, as well as the BAR of the dots [78]. This behavior can be used to enlarge the window where the switching of one bit occurs without reversing the neighboring bits even if they are partially covered by the write field.

If head field synchronization can be tightly controlled, the areal bit density, and particularly the bit pitch along the data track, will then be limited by the bit thickness and the write head field gradient. This last parameter corresponds to the lateral dispersion in the medium plane of the magnetic flux with respect to the write pole position. Indeed, contrary to the perfect sketch in **Figure 21(a)**, the magnetic flux emitted by the write head spreads over a surface larger than the dimensions of the write head. Therefore, the head field gradient has to be very

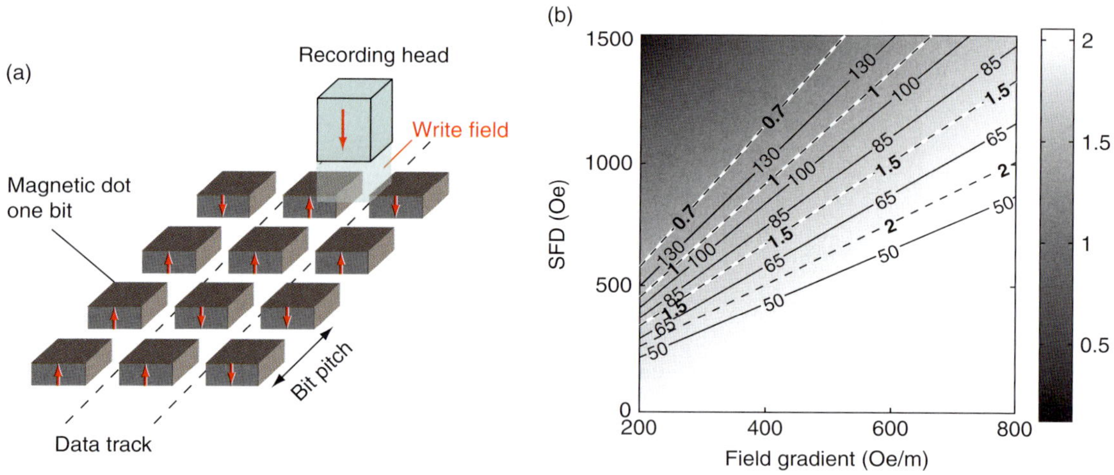

Figure 21 (a) Schematic view of the bit patterned medium and read/write head during the data writing process. (b) Areal density achievability as a function of field gradient and the Switching field distribution (SFD) Ref. [75]. A synchronization error of 1 nm, a bit-aspect-ratio of 1, a saturation magnetization $M_s = 430$ emu cm^{-3}, an anisotropy $K_U = 3.10^6$ erg cm^{-3} are considered. The gray scale encodes the areal density in Tb in^{-2}. The dashed contours with labels in bold font are areal-density contours in Tb in^{-2}. Other contours are thermal energy-barrier contours of the islands at the underlying areal density and temperature $T = 350$ K.

sharp. On the other hand, for a given field gradient value, the taller the bits, the greater the head field that a neighboring bit experiences, and hence the more likely it is to be inadvertently written. Thus, media with a thickness of the order of the bit spacing will be required.

Figure 21(b) gives a summary of theoretical predictions of areal densities that can be achieved as a function of field gradient and the SFD, as well as the corresponding thermal stability [77]. With a synchronization error of 1 nm, a BAR of 1, a thickness equal to the lateral bit dimension, a saturation magnetization $M_s = 430$ emu cm^{-3} and an anisotropy $K_U = 3.10^6$ erg cm^{-3}, and densities of 1–2 Tb in^{-2} can be obtained within an SFD ranging from 500 to 1000 Oe and within a field gradient ranging from 400 to 800 Oe m^{-1}. These values define the current objectives of HDD industry research laboratories. However, at this stage there is more interest in evaluating the feasibility of patterned media rather than improving the fabrication process. It is expected that the approach of current PMR to its limit, dictated by the superparamagnetic effect, will encourage more researchers to investigate the nanofabrication, magnetic switching, and recording aspects of BPM systems. There are many technical challenges as well with regard to the characterization and analysis of the different types of media discussed here. Some of them involve the accurate estimation of the SFD, analysis of the reversal modes, and switching properties of media grains. As the volume of the bits shrinks further into the nanoscale regime, the measurement problem will get accentuated and will require novel tools and techniques.

7.4.3 Energy-Assisted Magnetic Recording

Besides patterning the magnetic medium into discrete bits, there are two alternative approaches currently under consideration to overcome the limitations imposed by the superparamagnetic effect by using granular media with significantly higher magneto-crystalline anisotropy (K_U) and thus stability against thermal magnetization reversal at reduced grain size. A number of materials with sufficiently high K_U are known and have been studied experimentally for many years, among them artificially layered systems such as Co/Pt and Co/Pd multilayers, and the face-centered tetragonal (fct) phases of binary alloys like CoPt, FePd and FePt, also referred to as CuAu(I) or $L1_0$ phases [79]. Especially the latter have attracted great interest in recent years. The $L1_0$ phase consists of a natural superlattice structure based on alternating monoatomic layer of two different elements. When grown with the monatomic layers parallel to the film plane, that is, with the *c*-axis of the fct unit cell in the film normal direction, this structure results in perpendicular magnetic anisotropy. The magneto-crystalline anisotropy of the $L1_0$ phase of FePt and CoPt alloys are among the highest reported in the literature, and estimated to offer a 5–10 times benefit over conventional CoPtCr media materials, making them attractive base materials for recording media for HAMR.

However, as the same high K_U that provides thermal stability at small grain sizes also results in a very high switching field, significantly higher than can be obtained from a conventional write head. The simplest approach to use such high-K_U media would be to employ the ECC mechanism discussed in Section 7.4.1.3. However, while in principle large reductions in switching field should be feasible, it is currently not fully understood whether this effect can indeed be sufficiently efficient for a material such as FePt within the boundary conditions of a HDD recording system [80]. Alternatively, HAMR (also often referred to as TAR or hybrid recording) is an approach that aims to enable the use of media materials with very high K_U by temporarily heating the magnetic media by means of an optical heat source and therefore lowering its coercivity during the magnetic write process. A schematic drawing of such a system and the recording process is depicted in **Figure 22**.

Initially, most of the experimental studies on HAMR were using diffraction limited laser spots of 0.4–1 μm to heat the magnetic recording media [81]. As first pointed out by Ruigrook *et al.* [82], however, any approach using a heat spot significantly larger than the track width will be limited by thermal erasure due to repeated heat exposure of adjacent data tracks during writing. Based on a simple geometric model, a maximum increase in areal density of about a factor of 1.5–2 for HAMR using a large heat spot over conventional magnetic recording can be estimated. Consequently, ultrahigh recording densities will require a sufficiently small and intense optical heat source that avoids heating of neighboring tracks, as depicted in **Figure 23**. However, it is a well-known limitation in classical far-field optics that the efficiency of light transmitted through a subwavelength aperture rapidly falls off with the size of the aperture [83]. At the Tb in^{-2} areal densities at which HAMR is thought to become competitive with conventional recording technology, aperture sizes of 50 nm and below will be required. At such aperture sizes the transmission efficiencies of simple apertures are below 0.01% at wavelength in the visible range of the spectrum, too low to induce significant heating of the media with any reasonably sized laser and sufficiently short pulses or exposure times. Consequently,

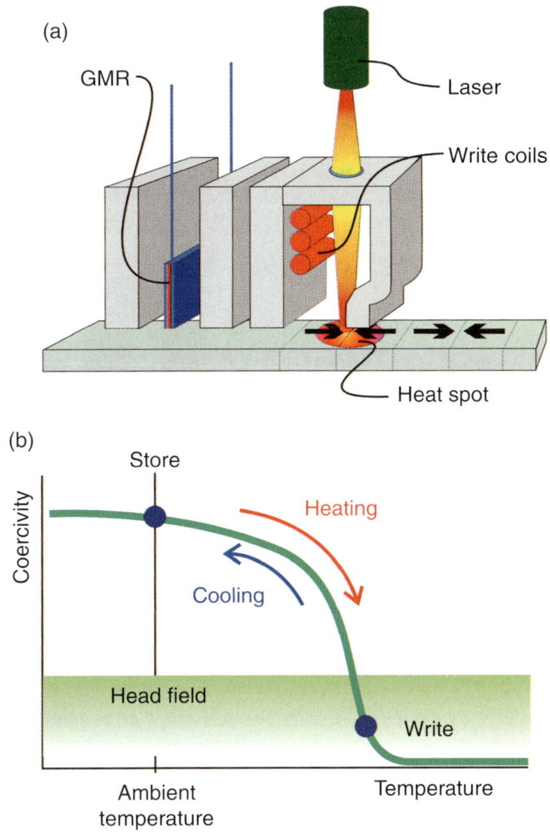

Figure 22 (a) Schematic drawing of an HAMR recording system with the magnetic write and read elements and an optical heat source. (b) The HAMR recording process uses the temperature dependence of the anisotropy and magnetization of the media material to temporarily lower the coercivity of the media and consequently the required write field provided by the recording head. After writing the bit the media is cooled back and at ambient temperature provides sufficient thermal stability for long-time data retention even at very-small grain sizes and high areal-recording densities.

until about 10 years ago any optical recording technology at such densities was considered fundamentally possible but impractical due to the resulting low data rates for writing (and reading in all-optical systems) [84].

Since then a number of interesting discoveries have been made in the field of resonant optical structures and the so-called plasmonic devices, and some of these very small apertures, solid-immersion lenses, dimple lenses and antennas lend themselves to fabrication directly onto the emitting surfaces of a laser diode or an optical waveguide [85,86]. In recent experimental and theoretical studies of optical sources with subwavelength apertures in the shape of tiny bow-ties and the so-called C-apertures very

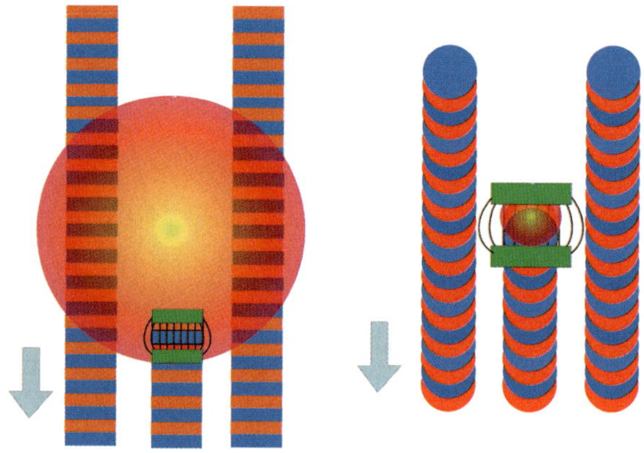

Figure 23 Comparison of large heat spot and small heat spot approaches to HAMR. As first pointed out by Ruigrook *et al.* [82] large heat spots lead to thermal exposure and accelerated magnetization reversal in neighboring data tracks and limit the maximum areal density gain of HAMR to 1.5–2× relative to conventional recording. Small spot HAMR employing optical near-field transducers will be required to achieve ultimate areal density gains.

high transmission efficiencies of >10% were demonstrated, a 1000-fold increase over classical diffraction limited apertures, as shown in **Figure 24** [87].

Such power densities raise the prospect of an HAMR system for ultrahigh densities at reasonable data rates using readily available laser technology such as the red laser diodes commonly used in CD and DVD players. First demonstrations of optical and magnetooptical recording using such near-field apertures at realistic spot-sizes, laser powers, and recording rates have recently been published [88–90]. Beyond the demonstration of sufficient transmission efficiency of these nanoapertures, still any reduction in the total power required to heat the media as well as in the recording temperature itself can be expected to greatly enhance the power efficiency and reliability of an HAMR system, as it also reduces the collateral heat load, especially on the sensitive magnetic parts of the recording head and the tribological interface between the recording head and the media.

Before discussing the details of and media technology for HAMR, let us take a look at the implications of using a combination of light and magnetic field for the recording process. In conventional magnetic recording a field profile such as the one shown in **Figure 25(a)** is swept over the recording media. The optimum point for data recording is near the point of the steepest field gradient at the trailing edge of the field profile, as this will result in the sharpest bit transition. In an optimally balanced recording system the switching field of the media will be matched to the available head field at that point, as depicted in **Figure 25(a)**. By contrast, in an HAMR recording system the recording process becomes much more complicated as here the final width of the transition written in the media is determined by the convolution of the media temperature profile, the change in media switching field with temperature, the isothermal media SFD, and the write field profile. For the thermal contribution we can consider the product of the media switching field temperature dependence, dH_K/dT, and the temperature gradient of the media, dT/dx, at the write temperature, T_W, to be an effective thermal field gradient, dH_t/dx. For a given temperature profile and magnetic head the total field gradient, $dH_t/dx + dH_W/dx$, will then depend on the thermo-magnetic response of the media, as pointed out by Thiele *et al.* [91] and in more detail by Rausch *et al.* [92]. **Figure 25(b)** shows this scenario, where the temperature profile at a fixed position for a simple media structure is convolved with a typical temperature dependence of H_K for a simple high-K_U magnetic material such as the one shown in **Figure 22**, resulting in a switching field versus time profile as shown by the blue line. For this

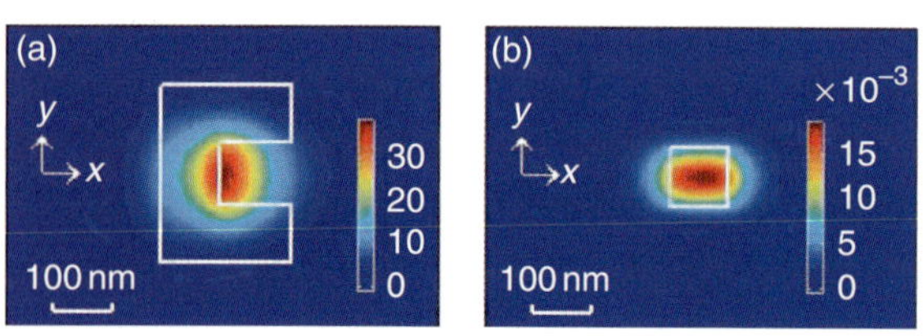

Figure 24 Power transmission of (a) *C*-aperture vs. (b) square aperture from Ref. [87] – note the difference in intensity scale by a factor of 2000.

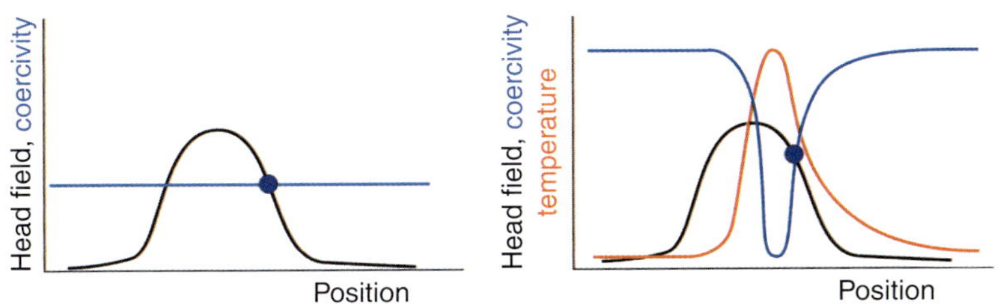

Figure 25 Schematic drawing of the recording process in conventional recording and HAMR. In both cases the bit transition ideally occurs at the location of the steepest magnetic field gradient in the media, but while in the case of conventional recording this is given by the head field gradient only, in the case of HAMR it is the convolution of the head field profile, the thermal profile in the media and the temperature dependence of the magnetic properties of the media that define the location of the bit transition.

plot the peak temperature has arbitrarily been adjusted to the Curie temperature of the material.

Some features of this profile are noteworthy: At room temperature the media switching field H_K is significantly higher than the maximum available head write field, H_W. As the media passes the laser spot, the switching field drops to below H_W. At the trailing edge of the temperature profile H_K rapidly rises again, and quickly reaches H_W, at which point it becomes too high to be written by the magnetic head field. To understand the implications for the recording process, consider first conventional magnetic recording: the center of the transition between two magnetic bits occurs at the point where the head field is equal to the average switching field of the magnetic media. In simplistic terms, the initial write transition width is the quotient of the media SFD width (in Oe) and the write field gradient of the head (in Oe/nm) where the latter is evaluated at the average media switching field. Field gradients of a few 100 Oe/nm can be achieved in current recording heads.

In HAMR, combining, for example, media with a temperature coefficient of the coercivity of about $20\,\mathrm{Oe\,K^{-1}}$ found in modern CoPtCr-based alloys with a realistic thermal gradient of $10\,\mathrm{K\,nm^{-1}}$ at the trailing edge of the temperature results in a value of the effective thermal field gradient, dH_t/dx, of about $200\,\mathrm{Oe\,nm^{-1}}$. This is comparable to the magnetic field gradient of the head. For optimum recording performance on such a media material it will therefore be essential to align the laser beam and the magnetic write elements such that the maximum media temperature occurs at or near the point of maximum magnetic field. Consequently, the sharpest transition will be written at the point of the steepest gradients at the trailing edges of both the magnetic field and the temperature contours. Such behavior was indeed observed experimentally in a study by Alex *et al.* [81]. However, this scenario also imposes very challenging constraints on the design of the recording head as it requires very close proximity and accurate alignment of the magnetic pole with the optical near-field transducer.

The situation changes somewhat for media with higher K_U and lower Curie temperature, T_C: here, temperature coefficients of the coercivity of several $100\,\mathrm{Oe\,K^{-1}}$ have been observed at temperatures near the Curie point. Accordingly, values of dH_t/dx ranging from several hundred Oe/nm to almost $10\,\mathrm{kOe\,nm^{-1}}$ can be obtained for such materials. It is interesting to note that for such media materials dH_t/dx is significantly higher than even the highest conceivable magnetic write field gradients. Consequently, in this case the bit transition written into the magnetic media will be dominated by the thermal profile, relieving some of the constraints on the head design. Recording heads with integrated magnetic write poles and optical wave guides using a solid immersion lens to focus light down to approximately 100 nm spot size, such as the one shown schematically in **Figure 26**, have recently been demonstrated [86]. **Figure 27(a)** shows the light emanating from an aperture fabricated at the bottom of such a recording head, and **Figure 27(b)** shows the intensity profile mapped using a near-field scanning optical microscope (SNOM). **Figure 28** shows a comparison of conventional recording and HAMR on high-K_U media. The reversal field of this media is too high for the conventional recording head to record any bit transitions, however, with the additional heat assist the HAMR head can write clear data tracks.

One important aspect of an HAMR recording system is the location of the laser diode - proposals range from putting the laser on the arm that holds the read-write head and project it through air onto a

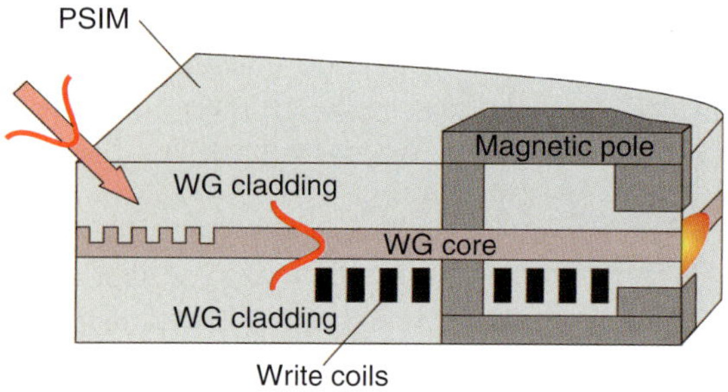

Figure 26 Schematic drawing of integrated HAMR head with light delivery through free space onto a grating coupler, planar waveguide, and planar solid immersion lens (PSIM) focusing element. The near-field optical aperture would be fabricated onto the bottom (right end) of the waveguide, but is not shown here. Drawing based on Ref. [93].

grating coupler on the read/write head and from there through a planar waveguide to the near-field aperture [94], using a waveguide to guide the light along the arm, and then around the top edge of the slider to the planar waveguide, mount the laser directly onto the slider or even use the laser body as a slider. Each one of these approaches has considerable technical and economical challenges, and which solution will ultimately prevail is far from clear at this point.

To fully realize the potential of very narrow optical spots it is also very important to design media that can sustain very steep thermal gradients and very fast heating and cooling curves. In a simple picture this requires the use of very efficient heat sink materials underneath the magnetic layers [92]. However, efficient heat sinks have the disadvantage of increasing the optical power required to achieve a given desired recording temperature. Unfortunately, thermal conductivity in inhomogeneous, directional materials on a nanoscale is probably one of the least developed areas of modern solid-state physics, in particular due to the difficulties of investigating interface resistances and the large surface-to-volume ratio in granular media at relevant time- and length-scales [95]. Consequently, the technological limits of the thermal properties of the recording media are not fully understood and presumably offer considerable room for improvement.

Significant progress has recently been made in the process technology required to fabricate magnetic recording media based on FePt with regular microstructures, very high K_U of about 3×10^7, approximately half that achieved in epitaxial continuous films, and grain sizes down to about 10 nm [96]. However, significant challenges remain in growing media with all required magnetic and microstructural properties and grain sizes <5 nm as required for HAMR at multiple-Tb in^{-2} recording densities. Furthermore, it is fundamentally at present not completely understood how at these very small length scales the change in surface-to-volume ratio of the magnetic grains may influence the thermodynamic properties responsible for the chemical ordering process of the FePt alloys and the resulting magnetic properties.

In addition, the simple single-layer high-K_U media considered so far face a dilemma in that both K_U and the saturation magnetization, M_s of a ferromagnetic material decrease with increasing temperature such that the write field $H_0 = K_U/M_s$ does not decrease as fast with temperature as the anisotropy. Consequently, to sufficiently reduce the coercivity of these media to enable writing often requires heating to temperatures close to their Curie temperature, T_C. At such high temperatures (e.g., around 500 °C for FePt) significant problems with the head-disk interface of the recording system such as lubricant degradation and large thermal stress can be expected. A number of schemes have been proposed to use more complex magnetic multilayer structures showing, for example, a first-order magnetic phase

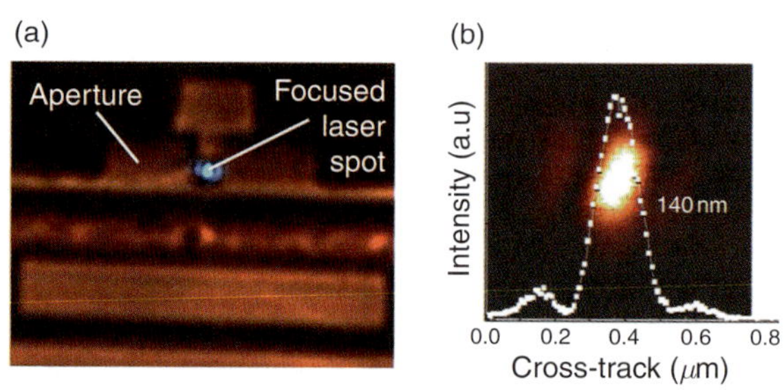

Figure 27 (a) ABS view optical image of an HAMR head with an aperture and with the light coupled into the waveguide. (b) Scanning near-field optical image of the ABS of an HAMR head with the magnetic write poles and with an aperture from Ref. [93].

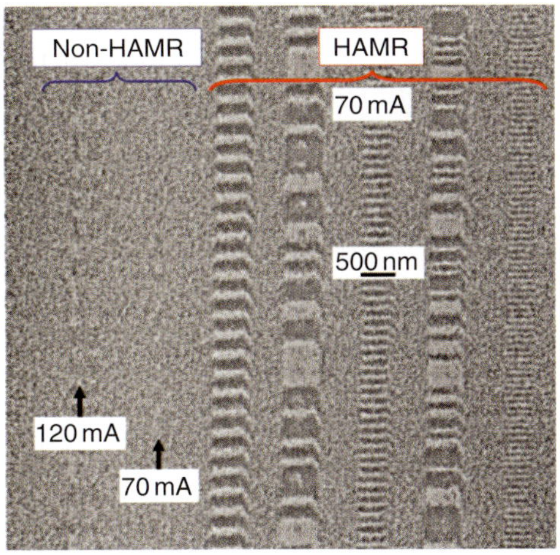

Figure 28 Magnetic force microscopy image of high coercivity media being written without heat assist and with heat assist, clearly demonstrating the writability advantage of HAMR with this particular media from Ref. [93].

transition from an AF to a ferromagnetic phase in alloys like FeRh, to significantly lower the recording temperatures [97], but it remains to be seen whether these much more complex materials and layer structures can be brought to sufficient technical maturity for real-world applications.

Even with a number of uncertainties surrounding the ultimate limits of HAMR technology, it should be emphasized that tremendous progress has been made in the last 10 years, and HAMR recording demonstrations at reasonable areal densities such as the one shown in **Figure 28** have been published in recent years.

In summary, compared to 10 years ago, when companies first started to look at HAMR as an alternative technology for HDDs, many of the fundamental problems of the technology, in particular the limitations of power transmission through nanometer scale apertures have been resolved. However, many other aspects of a real-life HAMR system remain very challenging. While many of these can be categorized as mere engineering problems, they may nevertheless pose considerable hurdles to a product introduction of HAMR systems. Among these are the practical difficulty of manufacturing the high-K_U media with appropriate effort and cost, the integration of the laser and the light delivery system with the magnetic write head structure, and a reliable tribological interface between the head and the media.

While magneto-optical recording systems have been used for many years, and ultrahigh density schemes such as HAMR discussed above have been under intense investigation development for a number of years, another relatively recent addition to energy-assisted recording schemes is MAMR. This is a scheme where the coercivity of the magnetic media is temporarily lowered by using a microwave field to excite ferromagnetic resonance, that is, precession of the magnetization in the media. In principle, the recording physics of the scheme is very similar to HAMR, however, a near-field source of microwaves that can be confined to spot sizes of <50 nm is not readily available. One idea that is currently being investigated is to use spin-torque oscillators built from GMR- or MTJ-type sensor structures as a highly localized microwave source, as shown in **Figure 29** [98].

One of the key aspects of this technology is the control of the frequency characteristics of these oscillators, which in devices fabricated to date typically varies over a wide range. This is important in that sufficient efficiency of the microwave transmission into the media can only be achieved for very narrow resonances. Furthermore, granular recording media tend to have grain-to-grain variations of their magnetic properties – these are already an important limiting factor in conventional recording, in patterned media and in HAMR, but in MAMR they become even more important as here distributions in the magnetic properties will directly result in a broadening of the resonance peak and influence the coupling efficiency of the microwave field into the media. MAMR has only been worked on for a relatively short period of time, and many fundamental questions are still open, so it is probably too early to give a definitive answer about the prospects of the technical approach, but compared to patterned media and even HAMR the fundamental challenges seem daunting.

7.4.4 Summary and Conclusions

HDD technology offers a reliable and low-cost solution to storing large amounts of data but future improvements in the development of HDDs introduce a variety of challenges. In the current granular media, one bit of information is stored within multiple magnetic grains and the thermal stability of the data is defined by the thermal stability of each grain. Since 2000, amazing improvements have been brought about by the introduction of CoCrPt:oxide-based PMR together with very

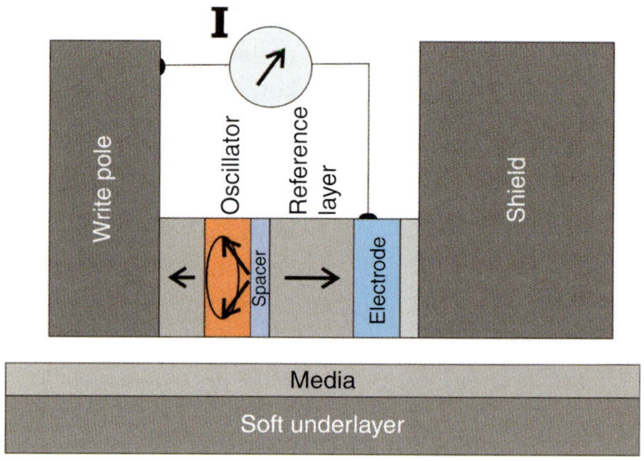

Figure 29 Schematic illustration of the microwave field-assisted perpendicular head design. The ac field generator drawing at the top is rotated 90° with respect to the drawing below. From Hardie C (2008) Challenges of integrating heat assist magnetic recording. *Presentation at the 2008 ISOM Conference*, 13–17 July 2008, Hawaii. Based on Ref. [96].

sensitive GMR and TMR sensors. However, fundamental limits of the physics underlying the HDD technology make the extension of this trend into the next decade very challenging. In particular, the design of media having good thermal stability, writability, and SNR (readability) is quickly approaching fundamental limits. Finding more suitable materials could help to obtain fine and uniform grains and then achieve good signal-to-noise ratio. Looking for materials or methods to obtain high magnetic anisotropy in recording media could help to improve thermal stability. Significant work is also carried out on new multilayer concepts like exchange coupled continuous (ECC). However, how far the areal density can ultimately be pushed with continuous granular media is an open question of great technological and economical consequences. Most experts currently expect it to reach close to $1\,\text{Tb in}^{-2}$, which at current annual growth rates could be reached within the next 3–5 years.

Therefore, prospects for data areal densities above $1\,\text{Tb in}^{-2}$ involve novel advanced media schemes either using a patterned media or energy-assisted recording (**Figure 30**). On the one hand, the patterned media approach proposes either to physically defined track within current granular media (DTM) or to lithographically defined magnetic islands (BPM). In the extreme case of bit pattern media, one magnetic island corresponds to one bit, whose volume can be increased compared with the grain size in conventional media, resulting in increased thermal stability. On the other hand, energy-assisted magnetic recording uses heat, or a high-frequency magnetic field to temporarily lower the reversal field of granular high-K_U media while maintaining high stability at rest or during the reading process. For each one of these alternative approaches several critical issues remain to be solved, and which one will provide the most promising and extensible entry into the Tb in^{-2} era of magnetic recording is very much an open question at this point. However, the challenges at $10\,\text{Tb in}^{-2}$ will most probably require the two discussed approaches to be combined, that is, BP media with various forms of energy-assisted writing. There are finally many technical challenges as well with regard to the characterization and analysis of the different types of media discussed here. Some of them involve the accurate estimation of the SFD, analysis of the reversal modes, and switching properties of media grains or nanodots. As the volume of the medium storing a bit of information shrinks further into the nanoscale regime, these measurement challenges require novel tools and techniques. Consequently, magnetic recording technology even 110 years after its invention still is evolving at a rapid pace and offers a rich field for basic research and technological invention.

7.5 Three-Dimensional Magnetic Memory

The advanced memory technologies we have discussed so far aim to push the data-storage density limitations due to thermally induced instabilities of nanoscale magnetic bits down to $10\,\text{Tb/in}^2$ and beyond. Undoubtedly, these clever approaches will draw the roadmap for magnetic storage in the next

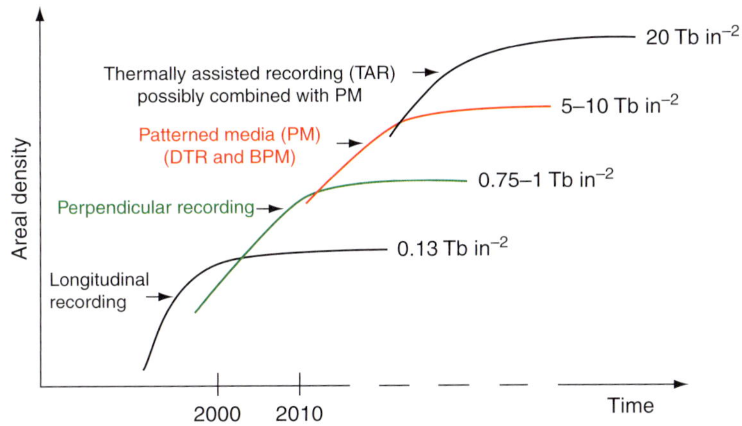

Figure 30 One possible roadmap of the areal density of HDD products vs. year. Patterned media (DTM and BPM) and thermally assisted recording, must be introduced to solve the fundamental trilemma that forces trade-offs between writability, signal-to-noise, and thermal stability.

10–20 years. However, it naturally looks very attractive to explore the possibility of utilizing the vertical dimension to boost the data-storage capacities since a 3D memory solution may potentially be the ultimate universal storage device by providing orders of magnitude improvement in the storage-density as the planar confinement restriction is removed.

Magnetic race-track memory [99] (based on the controlled motion of magnetic bits in a vertical shift register) is considered as one such promising avenue for future 3D magnetic recording with the possible advantages of combining the low cost of HDDs with the high performance of solid-state drives (SSDs) in a single device. Despite the several important challenges that need to be overcome to realize this technology, recent improvements in the fundamental understanding of the response of magnetic domains with nanoscale confinement to applied magnetic fields and electrical currents have led scientists to believe that these devices may indeed prove to have superior performance compared to other leading memory technologies.

7.5.1 Magnetic Field-Induced Domain-Wall Motion

A magnetic domain wall (the boundary-region between two magnetic domains with different magnetization orientations) with nanoscale confinement can be considered as a particle trapped in a pinning potential well. Its dynamic response to external factors such as applied magnetic fields or electrical currents depend heavily on the structure of the domain wall. For instance, in permalloy nanowires three distinct domain-wall structures have been identified [100] as transverse, vortex, and antivortex types as shown in **Figure 31**. Transverse and vortex wall structures are further classified according to their chiralities (the sense of spin rotation from left to right) to be clockwise and anticlockwise (with this definition the chiralities of the transverse and vortex domain walls shown in **Figure 31** are both anticlockwise). Since the geometrical constraints such as the width and the length of the nanowire lead to a shape anisotropy due to the effective demagnetizing fields, they also control which wall structures are favorable in a given device configuration.

The behavior of the magnetic domain walls with an applied magnetic field can be qualitatively described in three different field regimes [101]. For very low applied fields, the domain-wall motion is thermally activated and consists of discrete jumps from one pinning potential to another (creep-like propagation). When the applied field exceeds the pinning field, a viscous propagation regime where the domain-wall velocity scales linearly with the applied field starts as shown in **Figure 32(a)**. When

Figure 31 Three domain-wall structures observed in permalloy nanowires. Modified from Hayashi M, *et al*. (2008) Real time observation of the field driven periodic transformation of domain walls in permalloy nanowires at the Larmor frequency and its first harmonic. *Applied Physics Letters* 92: 112510.

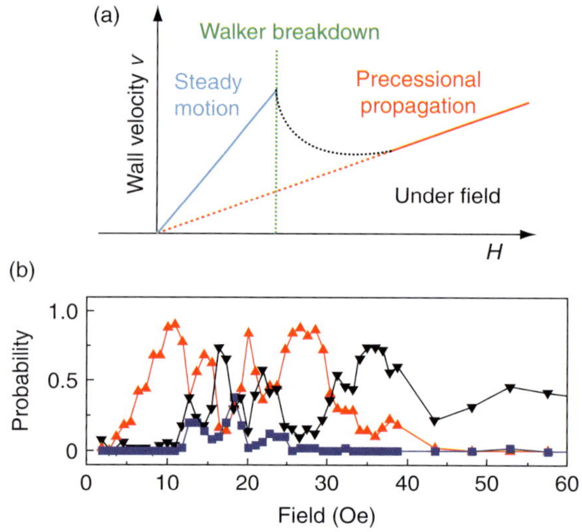

Figure 32 (a) Sketch of domain-wall velocity vs field for a 180° Bloch wall. (b) The probability of trapping a domain wall in a 200-nm-wide permalloy nanowire at a triangular notch for a clockwise vortex wall (blue squares), a clockwise transverse wall (red triangles), and an anticlockwise transverse wall (black triangles) as a function of applied magnetic field along the nanowire. (a) Modified from Mougin A, *et al.* (2007) Domain wall mobility, stability and Walker breakdown in magnetic nanowires. *European Physics Letters* 78: 57007. (b) From Hayashi M, *et al.* (2007) Direct observation of the coherent precession of magnetic domain walls propagating along permalloy nanowires. *Nature Physics* 3: 21–25.

the precessional torque from the applied field overcomes the torque due to demagnetizing field within the wall and the damping torque, a net azimuthal torque develops marking the onset of precession (Walker breakdown) [102]. For higher fields, the motion of the domain wall is oscillatory with a nonzero forward displacement.

The motion of the domain walls can be controlled by trapping them with engineered pinning sites during their propagation along the nanowire. However, the trapping probability is highly sensitive to the structure of the wall [103] as shown in **Figure 32(b)**. For instance, in the case of permalloy nanowires the trapping probability of a vortex wall in a triangular notch can be quite small compared to transverse walls. Furthermore in the precessional propagation regime the domain-wall state evolves periodically by changing the chirality through transformations to other intermediate states. Therefore, the trapping probability of a domain wall is also sensitive to the location of the pinning site with respect to the domain-wall injection site.

7.5.2 Spin-Torque-Induced Domain-Wall Motion

Apart from an external magnetic field, the domain-wall motion can also be driven by an applied electrical current. The effect of the bias current is threefold. First, it increases the local temperature via Joule heating making thermally activated propagation mechanisms more effective. The heating effect, if sufficiently high, may also cause structural transformations and variations of the domain-wall width due to reduced anisotropy. Second, the current-carrying electrons apply a pressure on the wall due to a linear momentum transfer to the local spins during reflections from the wall [104]. The third and the most dominant effect is the dragging of the domain wall due to spin-torque [105]. Since a current carrying electron has to accommodate the local change in magnetization while traversing the domain wall it has to deposit its transverse spin component on the domain wall dragging it in the direction of its drift velocity. This spin-torque effect on a domain wall is demonstrated in **Figure 33**.

The domain-wall dynamics due to the combined effects of an effective magnetic field H_{eff} and electrical current density $\mathcal{J}$ in the x-direction is governed by the Landau–Liftshitz–Gilbert equation with the spin-torque terms (LLGS),

$$\frac{\partial \vec{M}}{\partial t} = \gamma \vec{H}_{\mathrm{eff}} \times \vec{M} + \frac{\alpha}{M_s} \vec{M} \times \frac{\partial \vec{M}}{\partial t} - u\Delta_x \vec{M} + \frac{\beta u}{M_s} \vec{M} \times \Delta_x \vec{M} \quad (4)$$

where γ is the gyromagnetic ration, α is the Gilbert damping parameter, β is the phenomenological nonadiabatic spin-torque parameter and u is the current-dependent velocity term given by $u = g\mathcal{J}\mu_B P/2eM_s$ (g is the Landé factor, μ_B is the Bohr magneton, P is the current polarization, e is the electron charge, and M_s is the saturation magnetization). **Equation 4** basically states that the magnetization in a domain wall evolves due to the precessional torque from an effective field, the damping torque (mostly spin–orbit coupling), the adiabatic spin-torque which acts as an effective damping and nonadiabatic spin torque, which may also be considered as a precessional torque term due to an effective field along the magnetization gradient parallel to the current flow direction.

Figure 34 shows the additional effect of electrical current on the domain-wall velocity in the presence of an applied field for a 180° Bloch wall. It can be seen that the domain-wall velocity can be enhanced

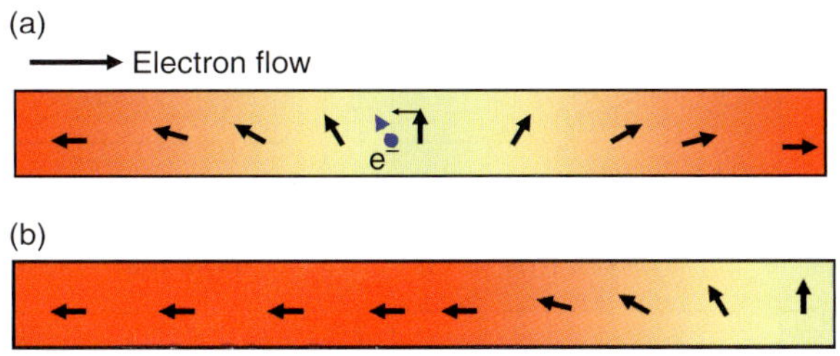

Figure 33 Schematic demonstration of spin-torque effect on a tail-to-tail domain wall. The color scale represents the *y*-component of the magnetization with red and yellow corresponding to 0 and maximum values. The thick arrows represent the local moment. (a) The blue arrow is the spin of the current carrying electron and the thin arrow marks the spin-torque direction. (b) The domain-wall position after the current is turned off.

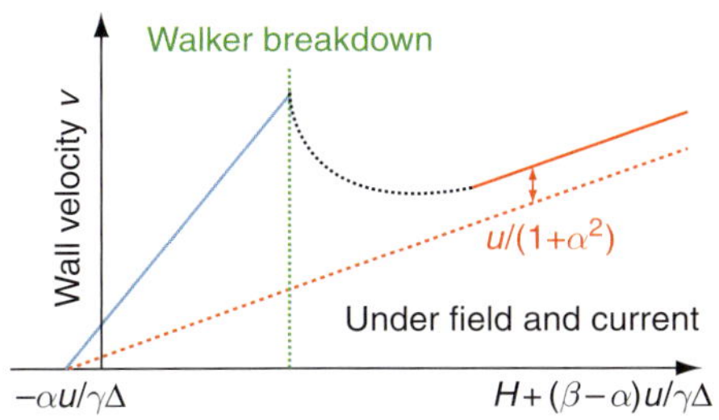

Figure 34 The effect of electrical current on the velocity of a 180° domain wall. Modified from Mougin A, *et al*. (2007) Domain wall mobility, stability and Walker breakdown in magnetic nanowires. *European Physics Letters* 78: 57007.

due to the additional effective field from the nonadiabatic spin-torque term of magnitude $\beta u/\gamma\Delta$, where Δ is the domain-wall width. Furthermore the adiabatic spin-torque term introduces an additional offset of $u/(1+\alpha^2)$ far above the Walker breakdown limit. It is important to note that domain walls can also be moved even with no applied field provided that the applied current exceeds the critical depinning current (to overcome the pinning effects due to defects and roughness). In permalloy nanowires with typical 5–10-Oe depinning fields, the critical depinning current densities have been measured [106] to be on the order of $10^8\,\mathrm{A\,cm^{-2}}$. After depinning, the maximum attainable domain-wall velocity at zero field is about $110\,\mathrm{m\,s^{-1}}$ with a $1.5\times10^8\,\mathrm{A\,cm^{-2}}$ current density.

For memory device applications, it is desirable to have periodic pinning sites to gain control over the domain-wall location and also to stabilize the domain-wall position against thermal fluctuations so as to maintain the stored information. In this case, the domain wall's oscillatory motion in the potential well can be resonantly amplified [107] using a series of current pulses synchronized with the domain-wall oscillations so as to effectively reduce the critical current density required to achieve depinning. **Figure 35** demonstrates the idea of resonant depinning of a domain wall. Under the application of a small dc current the domain wall displays low amplitude oscillations in the potential well. Such oscillations are quickly damped out when the current is turned off leaving the domain wall in its initial equilibrium position. However, applying a series of pulses with the same amplitude as the dc current level, the induced domain-wall precession can lead to depinning provided that the pulse length is approximately half-integral multiple of the precession period, as shown in **Figure 35**(d).

7.5.3 Magnetic Race-Track Memory

The idea of utilizing domain-wall motion in a nonvolatile memory device dates back to 1970s with the magnetic bubble memory concept. This was considered as a potential breakthrough in memory technologies since such a device did not need any moving parts and small bubble sizes enabled very high density data storage. However, the bubbles had to be shifted using magnetic fields for the stored information to be read, adding complexity and additional cost for reliable device operation. With the discovery of spin-torque effect, it is now possible to manipulate domain walls without the need for cumbersome magnetic fields. Therefore, it has become worthwhile to revisit the idea of controllably moving domain walls for nonvolatile data-storage applications, this time using spin-polarized currents in a magnetic race track.

The operation of the magnetic race-track memory is summarized in **Figure 36**. Data is encoded as a bit pattern in a ferromagnetic nanowire. A series of spin-

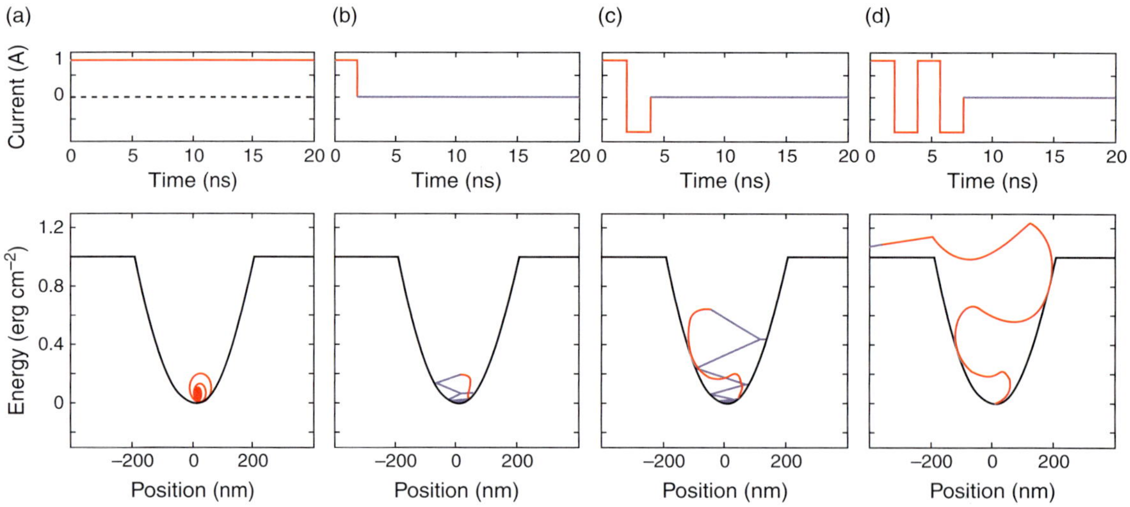

Figure 35 Demonstration of the resonant amplification of domain-wall oscillations in a deep potential well. The domain-wall trajectories are shown for a dc current (a), a single current pulse at resonance (b), a bipolar current pulse at resonance (c), and two bipolar pulses at resonance (d). Modified from Parkin SSP (2008) Magnetic domain-wall racetrack memory. *Science* 320: 190–194.

polarized current pulses can shift the whole pattern from one side of the track to the other lining up the region of the data pattern of interest with the reading element. A MTJ senses the presence or absence of a domain wall through the TMR effect. Data can be written to the track using the magnetic field from a write line. This could, for example, be the fringing field from another domain wall stored in the write line. Ultimately, building a large array of such race-tracks is necessary to maximize the data-storage density. Even though the vertical version of the race-track memory carries the additional benefit of using the third dimension, the fabrication of such a device is considerably more difficult compared to the horizontal version. Additional challenges include engineering very well-defined pinning sites with tight pinning field distributions to avoid possible domain-wall collisions while moving a set of domain walls simultaneously.

Based on an analysis of the technical specifications of products or prototypes (for emerging technologies) various memory technologies are compared to race-track memory [108] in **Table 2**, assuming a critical current density of $5 \times 10^7\,\mathrm{A\,cm^{-2}}$, domain-wall velocity of $100\,\mathrm{m\,s^{-1}}$ and a TMR of 200% using 65-nm-node CMOS technology. It can be seen that the magnetic race-track memory can be potentially advantageous in several aspects. It does not suffer from any wear-out mechanisms due to cycling making it very reliable for long-term operation. It is very efficient in terms of power consumption and the vertical race track can boost the information density by approximately a factor of 100.

However, as in most magnetic storage technologies currently under consideration new advancements in nanofabrication techniques are needed to overcome the challenges in building such structures with a reasonable cost.

7.6 The Role of Fabrication Development in Magnetic Storage

For the past 50 years, magnetic storage has seen a steady increase in areal densities accomplished mainly through novel and innovative strategies for fabricating the sensor and media components involved in HDD operation. The original RAMAC hard drive first introduced by IBM in 1956 stood 68″ high, 60″ long, and 29″ deep, with 50 24″ disks comprising its approximate 5 MB storage capacity – quite a departure from what is available today for use in personal computers and laptops! Over time, the magnetic oxides used in the media gave way to smoother, more robust thin-film sputter-grown Co alloys with smaller grain sizes, leading to significantly higher areal densities due to smaller track widths and head/media separations. These achievements in media growth would not have been as significant for HDD operation if not for similar advances in

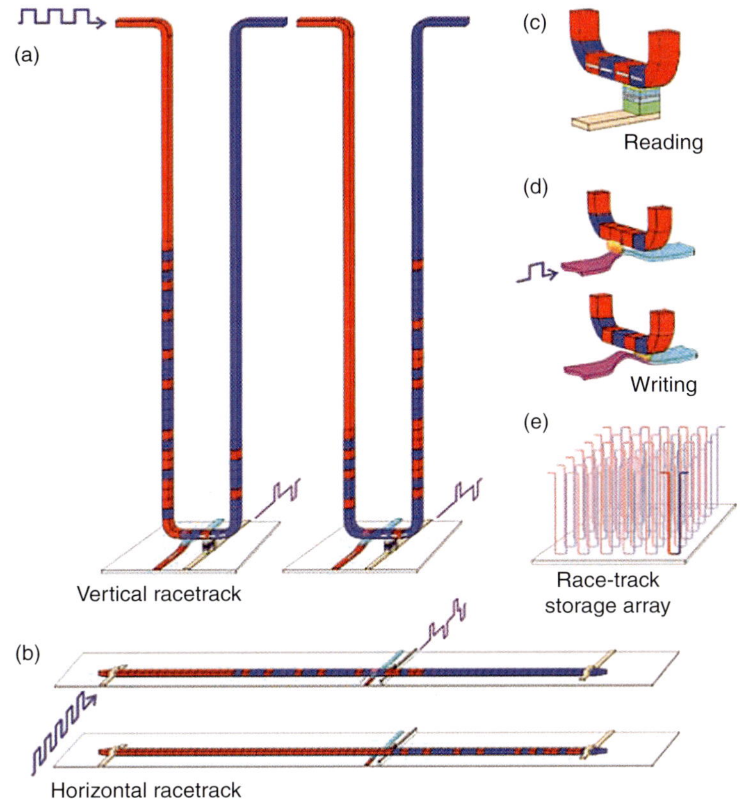

Figure 36 The operation of the magnetic race-track memory: (a) vertical race-track configuration; (b) horizontal race-track configuration; (c) read operation; (d) write operation; and (e) race-track storage array. Parkin SSP, Hayashi M, and Phomas L (2008) Magnetic domain-wall racetrack memory. *Science* 320: 190–194.

read/write head technologies, as head dimensions were required to scale down with track sizes. Smaller thin-film inductive read/write heads replaced the original ferrite core heads, and eventually were split into separate modules utilizing forms of MR discussed in this chapter (AMR, GMR, or TMR) for reading and thin-film inductive elements for writing. Developing the capabilities to build these components has been crucial in the approximate 10^8 increase in hard-drive areal densities, which has shaped modern technology by providing convenient, inexpensive storage for multiple forms of electronics.

A variety of issues remain to be addressed as areal densities are pushed to 1 Tb in^{-2} and beyond. Some involve exploring the physics governing the properties of magnetic elements on the order of 10 nm, giving rise to material considerations discussed earlier in this chapter. However, these considerations are purely theoretical without the means to reliably pattern media and heads to these extreme dimensions, an entirely nontrivial undertaking. In addition, implementation of any such fabrication process requires high throughput with relatively low cost to be industrially feasible. There are several potential strategies currently being explored to lithographically define hard-drive components required for future HDD generations. However, each approach has its own issues with successful incorporation into product lines.

Similar issues exist in the development of magnetic memory technologies, which must compete with more established Si-based products. Although MRAM [45] offers several advantages over convention memory, such as nonvolatility, low power consumption, and fast read/write speeds, it must prove to be scalable to warrant consideration as more than a niche technology. As with Si-based memories, increased densities require decreasing critical dimensions of the magnetic elements so that various lithographic strategies must be explored. Different architectures, using the spin

Table 2 Comparison of the product/prototype device specifications from various memory technologies with race-track memory

	Cost			*Performance*				*Power*	*Reliability*		
	Cell size F^2 *[effective cell size* F^2*]*	*Multibits (bits per cell)*	*Embedded masks*	*Random read access (ns)*	*Random write time (ns)*	*Erase time (ms)*	*Read data rate (Mb/s per access point)*	*Write energy/bit (pJ)*	*Endurance – read*	*Endurance – write*	*Retention*
SRAM	120	1	0	2–100	2–100	NA	500	2	10^{15}	10^{15}	(with power)
DRAM	6–12	1	4	6–40	6–40	NA	800	2	10^{15}	10^{15}	~64 ms
NOR FLASH	9.5 [4.8]	2	8–11	70	2.5×10^4	100	≤80	160	10^{15}	10^5	10 years
NAND FLASH	4.5 [2.3]	2	NA	10^4	10^5	2	≤30	65	10^{15}	10^5	10 years
HDD	NA	NA	NA	$>2 \times 10^6$	$>2 \times 10^6$	NA	1000	NA	NA	NA	10 years
Field MRAM	25–40	1	≥3	3–30	3–30	NA	≤200	50	10^{15}	10^{15}	10 years
Spin MRAM	8–25	1	≥3	10–20	3–30	NA	≤100	2	10^{15}	10^{15}	10 years
PCRAM	5.8	>1	2–4	50	100	NA	50	100	10^{15}	10^{12}	10 years
FeRAM	15	1	2–4	50	50	NA	20	2	10^8–10^{14}	10^8–10^{14}	10 years
Horizontal racetrack[(1)] (no reservoir)	2–4 [<1-2]	>1	6–8	20–32	20–32	NA	250–670	2	10^{15}	10^{15}	10 years
Vertical racetrack[(1)] (no reservoir)	[1/2–1/8]	up to 128	7–8	20–32	20–32	NA	250–670	2	10^{15}	10^{15}	10 years
Horizontal racetrack[(1)] (w/reservoir)	4–8 [<2-4]	>1	6–8	9.5	9.5	NA	400–670	2	10^{15}	10^{15}	10 years
Vertical racetrack[(1)] (w/reservoir)	[1–1/4]	up to 128	7–8	9.5	9.5	NA	400–670	2	10^{15}	10^{15}	10 years

NA, not available.
From Parkin SSP, Hayashi M, and Thomas L (2008) Magnetic domain wall race-track memory. Supporting online material. *Science* 320: 190–194.

transfer effect [58,59], or building in 3D as with race-track memory [5], offer attractive scaling paths for MRAM.

7.6.1 Trending Conventional Lithography to Smaller Length Scales

A common thread linking all magnetic storage media is the continued need for patterning large arrays of increasing smaller magnetic elements. As such, it is important to review some of the more common lithography techniques currently being examined for defining these features in the next generations of high-density storage devices and discuss the advantages and limitations of each. Although extensive literature exists on basic lithography processes [109–111](also discussed in several chapters of this book), a brief review will help to better illustrate more advanced techniques. The workpiece, an unpatterned film or substrate, is first spin-coated with a radiation-sensitive polymer known as resist, with typical thicknesses on the order of a micron. Selected areas of the resist are then exposed to a radiation source, often through a mask (**Figure 37**). Upon sufficient exposure, the polymer chains in the resist are either broken (positive resist), or become cross-linked (negative resist, poorer resolving power). Immersing the workpiece in a solvent typically called a developer preferentially dissolves part of the resist, resulting in either a positive or negative image of the mask, depending on the resist type. Usually, there is no impact on the dissolution rate up to a certain dose, at which point it changes logarithmically with the dose until the resist is completely dissolved. Pattern transfer can then be realized in two general processes: from the resist to an unpatterned film by wet or dry etching; or post-deposition onto patterned resist by lift-off and/or electrodeposition as shown in **Figure 37**.

Lithography systems typically consist of four fundamental elements: (1) an illumination system complete with an energy source; (2) a mask containing the patterns corresponding to the structures to be fabricated; (3) an exposure system to generate an aerial image of the mask pattern; and (4) the resist for recording the image generated by the exposure system. The illumination system uniformly illuminates the patterns on the mask using a radiation source, such as a mercury lamp operating in the broadband regime (300–450 nm wavelengths), or at more specific wavelengths such as 436 nm (g-line) and 365 nm (i-line). More recently, optical sources for lithography have moved into the deep ultraviolet (DUV) regime using excimer lasers with krypton fluoride (248 nm KrF) and argon fluoride (193 nm ArF) [112]. In an optical projection lithography system, the smallest features

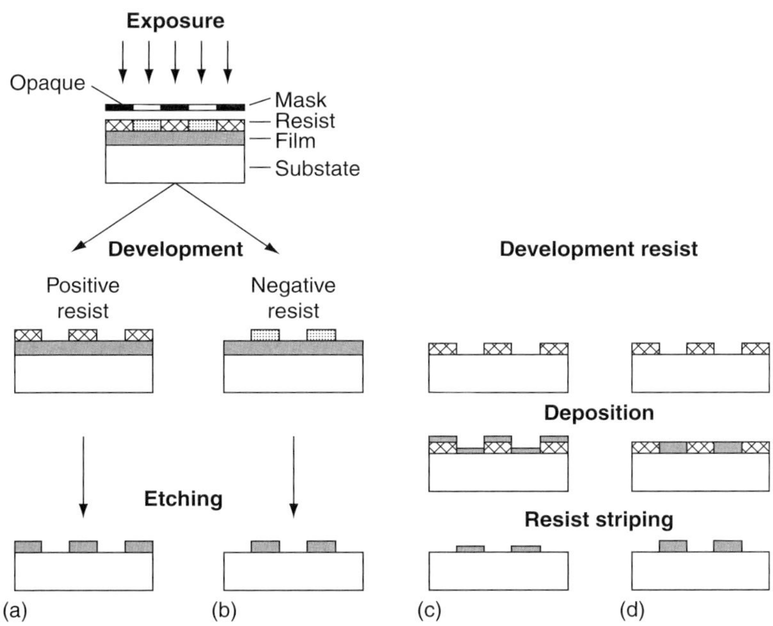

Figure 37 Schematics of lithography processes for (a) positive and (b) negative resists in conjunction with (a), (b) etching, (c) lift-off, and (d) electrodeposition. Reprinted from Martın JI, Nogues J, Liu K, Vicent JL, and Schuller IK (2003) *Journal of Magnetism and Magnetic Materials* 256: 449–501.

that can be printed R, that is, the resolution, are given by the relation [113]

$$R = K_1 \frac{\lambda}{NA} \quad (5)$$

where K_1 is a process-dependent parameter usually ~0.6 for conventional lithography, with a theoretical limit of 0.25, λ is the wavelength of the exposure tool, and NA is the numerical aperture. From **equation 5**, the minimum feature size achievable scales down with decreasing radiation wavelength, making it preferable to continue developing shorter wavelength sources, although this will require significant development of masks, resists, and tool optics.

The mask is typically a glass blank containing the layout information of the structures to be patterned on the wafer in the form of physical or optical coding. Conventional masks, known as chrome-on-glass (COG), are made with chrome (Cr) patterns on a blank, which must be made of a material transparent to the wavelength of the exposing radiation. Addition types of masks control the phase of the imaging radiation to form high-resolution images.

The exposure system is used to project an optical image of the mask patterns onto the workpiece. Several types of these systems exist, including contact, proximity, and projection tools. In contact exposure systems, the mask remains in hard contact with the wafers and transfers an image of the mask to the wafer in a single exposure. While contact lithography offers high-resolution patterning capabilities, it suffers from mask damage due to hard contact between the mask and the resist wafer, which ultimately leads to the generation of defects on wafers. Proximity lithography reduces mask damage by keeping the mask a few (10–25) μm away from the wafer. However, this results in poorer resolution caused by diffraction effects introduced by having a gap between the mask and wafer. By far, the most common exposure system used today is the projection lithography system, which projects a demagnified (4–10×) mask image onto the workpiece through a system of lenses [114]. Because of the reduced field size, the wafer is exposed using multiple shots, for which the exposure dose and focus can be controlled independently.

Several resolution-enhancement techniques have been developed and integrated with DUV lithography to enable significant improvements in pattern resolution without the development of new lithography tools. One such method involves the use of phase shift masks (PSMs), which work by combining the phase and amplitude information of the exposing radiation to increase pattern resolution using the principle of destructive interference. Variations of these masks include attenuated and alternating PSMs. Attenuated PSMs are similar to conventional masks, except with the opaque chrome replaced by a partially transmitting layer [115,116] such as molybdenum silicide, with a 180° phase shift with respect to the clear regions. Alternating PSMs are somewhat different, in that the exposing radiation passing through clear areas on either side of a line are phase shifted 180° with respect to one another. In either case, the diffracted light from either side of the chrome line cancel each other due to phase-cancellation effects and keep the line dark and narrow. To illustrate these effects, a schematic comparison of attenuated and alternating PSMs with conventional binary mask is shown in **Figure 38**. **Figures 38(a)** and **38(b)** show a 3D top view and cross sectional view of the mask, respectively; the grey region of the attenuated PSM indicates partial transmission. In **Figure 38(c)**, a sketch of the electric field transmission function of the mask is shown, with the π phase shown as a negative field. Interestingly, the net step height in the electric field function is increased for attenuated PSM compared with the binary mask, resulting in higher image contrast upon imaging through the projection system. In physical terms, π-phase background transmission strengthens the higher diffraction orders at the cost of 0th orders as sketched in **Figure 38(d)**. Apart from these magnitude changes, there is no distinction between an attenuated PSM and binary imaging. **Figure 38(e)** shows the increase in image contrast due to the reduced background and strengthened higher diffraction orders, which increase further with the increase in background transmission. However, the position of diffraction orders remains unaltered; so, there is no gain in the theoretical resolution limit using an attenuated PSM.

In an alternating PSM, the adjacent apertures are made 180° out of phase by modulating the mask blank thickness between apertures (**Figure 38(f)**). The electric field transmission function of the mask, where the negative field corresponds to π-phase transmission, is shown in **Figure 38(c)**. Interestingly, the period of the electric field transmission function is doubled in comparison with the binary mask. The diffraction orders in an alternating PSM then become separated by a spatial frequency of $1/2p$ instead of $1/p$ in the binary case as shown in **Figure 38(d)**. The reduction in separation of

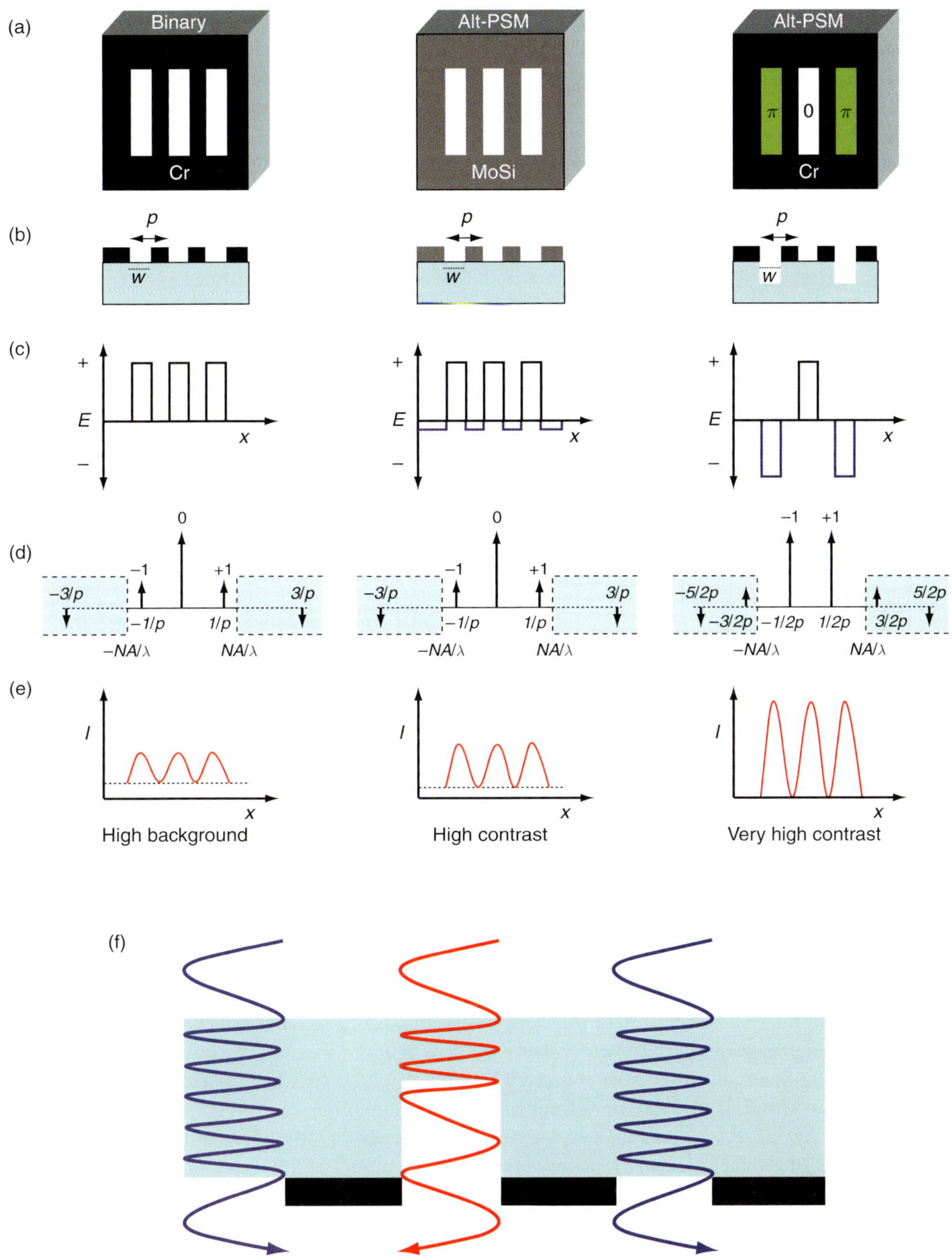

Figure 38 (a)–(e) Schematic comparison of a binary COG mask, an attenuated PSM, and an alternating PSM; *w* and *p* are the space width and the pitch, respectively. Lens apertures are drawn as semitransparent blocks rejecting the higher diffraction orders. $\pm NA/\lambda$ is the cut-off frequency. (f) 180° phase modulation for adjacent apertures in alternating PSMs. Phase differences caused by mask thickness differences between apertures. Reprinted from Adeyeye AO and Singh N (2008) *Journal of Physics D: Applied Physics* 41: 153001.

diffraction orders to half doubles the cut-off spatial frequency and thus increases the resolution by a factor of 2. Furthermore, the alternating PSM does not have 0th diffraction order and thus imaging occurs using only the higher diffraction orders, which lie symmetrically inside the lens pupil. The complete elimination of the 0th order results in a very high image contrast as there is no background intensity, shown in **Figure 38(e)**. The alternating PSM was introduced in 1982; however, the full potential of this technology for resolution enhancement has yet to be realized due to a number of problems. Besides implementation of phases on complex circuits [117], phase error control in mask fabrication is an obvious problem. The fabrication of an alternating PSM is challenging due to the precise etching requirements of the mask blank to create π-phase apertures. In addition, the biggest problem with the phase shifting method may be the restrictions placed on pattern layout, as this method can be easily applied to periodic patterns, but is difficult to apply to random patterns.

Since 2002, 193 nm immersion lithography has taken over as the frontrunner technology for the 45-nm node on the ITRS [118]. Unlike dry lithography, this strategy involves immersing a wafer in a fluid with refractive index greater than 1. The key benefit in doing so is the possibility of constructing projection optics with $NA > 1$, decreasing the minimum feature size as shown in **equation 5**. This improves resolution without changing the total lithography platform (resist type, lens material, and light source), and can yield a resolution gain equivalent to two generations of advancement [119]. A commonly used immersion medium is water, which has low cost, low toxicity, and a refractive index of 1.44. Modern immersion tools using highly purified water have demonstrated that features with linewidth and pitch dimension down to 37 nm [120] can be achieved using a chemically amplified resist. However, there are several key problems to be addressed in using this technology. Some of these issues involve the use of an immersion fluid, such as the formation of micro- or nanosize bubbles, which scatter the incoming radiation and affect the imaging process, contamination of immersion media by the resist, damage to optical components from contaminated immersion media, as well as fluid heating changing the refractive index of the media during exposure [121,122]. Further development of resist materials for immersion lithography is also required, which will have to deal with the obvious issues related to exposing the imaging stack to aqueous or other immersion media. These issues can generally be classified in the areas of penetration of immersion media into the imaging stack, effects of immersion media uptake on photoresist chemistry, and the leaching of photoresist components into immersion media. One possible solution may be to use a top barrier coating [123], isolating the resist layer from the immersion medium, and thus making it possible to use existing resist formulations for dry lithography with minimal formulation optimization. Although the challenges facing 193-nm immersion technology are not insurmountable, there are still critical needs that must continue to be addressed at this stage of development.

Lithography tools with illumination sources capable of providing radiation at wavelengths below 193 nm have been explored for next generation lithography systems. These include F_2 line (157 nm), extreme ultraviolet lithography or EUV (~50–5 nm), and X-ray lithography or XRL (<5 nm). While the benefits of reducing the radiation wavelengths are clear from **equation 5**, significant issues must be resolved in both the design and cost of such systems before introduction into mainstream use can occur. Obtaining sufficient quantities of high-quality CaF_2, the sole lens material for 157 nm lithography, is difficult at an affordable price. The photoresists have too much absorption, let alone satisfying the other requirements, such as etch resistance, sensitivity, line edge roughness, postexposure bake sensitivity, and cost [124]. Photomask materials are unclear as conventional fused silica does not have acceptable transmission at these wavelengths, and materials such as CaF_2 and BaF_2 with high transmissivity also have large thermal expansion coefficients, which act to increase pattern distortion during mask definition [125] In addition, 157-nm technology would require an oxygen- and water-vapor-free atmosphere [124], increasing costs associated with the system. The inherent problems discussed above, coupled with the extension of 193-nm technology through resolution enhancements such as PSMs and immersion mediums, have all but eliminated development of 157-nm systems. EUV and XRL offer greater potential for replacing 193-nm lithography when it ultimately reaches its limit, but they suffer from many of the same issues as 157-nm lithography, namely the design and materials used for the optics, materials for resists and photomasks, and contamination issues, most likely requiring the wafer stage to be placed in high vacuum [126]. Reliable and cost-effective radiation sources must also be found for these lithography systems. Originally, synchrotron sources were the best option for delivering the required radiation for both EUV [127,128] and XRL [129,130], which has large cost and space requirements limiting its practical use in production tools. More recently, plasma-produced sources have been developed [131,132] which depend on high-temperature emitting plasmas with several 100 000 K [133], introducing a better option for production tools. However, further improvement in both the power and durability of these sources is required.

Electron beam lithography (EBL) has rapidly become a common high-resolution method for patterning magnetic nanostructures down to the nanometer length scale which is attractive for rapid prototyping and process development. Here, a focused beam of electrons is scanned over a thin layer of resist to directly write the desired features. The advantage in using electrons comes from the fact that electron beams are not diffraction limited (due to the small diffraction lengths for electrons), but rather aberration limited, allowing beam widths on the order of nanometers. Because of this, much higher resolution can be achieved than with optical sources – in fact, masks for many of the optical lithography systems described above are defined using EBL. However, there are limitations to the use of EBL in fabricating large arrays of magnetic nanostructures. The writing process in EBL is serial and very slow – taking up to orders of magnitude longer than with optical lithography depending on the total area of the exposed pattern, and thus making large-area fabrication extremely difficult, although it can be used in the preparation of masks for both optical and nanoimprint lithography. Several strategies have been proposed to substantially improve the throughput of EBL but have yet to be adopted into mainstream use, including using multiple coordinated electron beams [134] or projecting a wide area beam of electrons through a mask and onto the substrate, a strategy known as PREVAIL [135]. In addition, it is also very difficult to fabricate closely packed high aspect ratio nanostructure arrays due to proximity effects arising from electron scattering within the resist layers and the substrate. Eventually, EBL may prove to be acceptable as a production process for fabricating hard-drive read/write heads [136]; however, it is not seriously being considered for direct fabrication of discrete- or bit- patterned media and magnetic memories, although it may play a role.

7.6.2 Challenges for Bit-Patterned Media

Over the past 50 years, areal densities for HDD media have increased by an astonishing factor of 70 million. This remarkable achievement has been made possible by scaling the grain size in the media with the bit size, thus keeping the number of grains per bit approximately constant and maintaining a sufficient SNR. However, if the average grain volume is continually reduced without innovations in the media design, grains will become susceptible to superparamagnetic effect [137,138]. This effect results in information loss, and will be a limiting factor for areal densities in conventional media. One approach to delaying the onset of the superparamagnetic effect is to use patterned bit media, wherein each bit is a single magnetic switching volume and is produced using a lithographic process. Two forms of such media have already been discussed in this chapter, DTM and BPM. In DTM, the tracks are separated physically and magnetically by grooves defined through lithography (with the most likely candidates being EBL or nanoimprint) [139,140]. While the track widths are lithographically defined through this process (see **Figure 39**), the along track bit transitions are still defined by the field gradient from the write head as in conventional magnetic recording. Regions between tracks should be either nonmagnetic or recessed from the head such that the read element senses little or no magnetic response from them. This approach reduces the problem of poorly written and partially erased information at the edges of the track, which is a significant source of noise. Thus, by using discrete tracks, the read head can be made wider and comparable to the write head width, as the so-called side reading is no longer an issue. By eliminating this noise, somewhat higher areal densities [141] may be accomplished. In addition, with the track widths physically defined in the media, the requirements on head tolerances may be eased, thus improving manufacturing yield. Unfortunately, the bit transition is still determined by the reversal of magnetization direction in adjacent grains, making DTM susceptible to many of the same areal density limitations as continuous media.

For BPM, grains in the media material are strongly coupled, unlike with continuous media. Magnetic isolation between bits is controlled by lithographically patterning the media into the isolated islands. Due to the strong coupling, grains in each bit reverse together as one unit in a Stoner–Wohlfarth [142] fashion. In the past, the HDD industry has had the

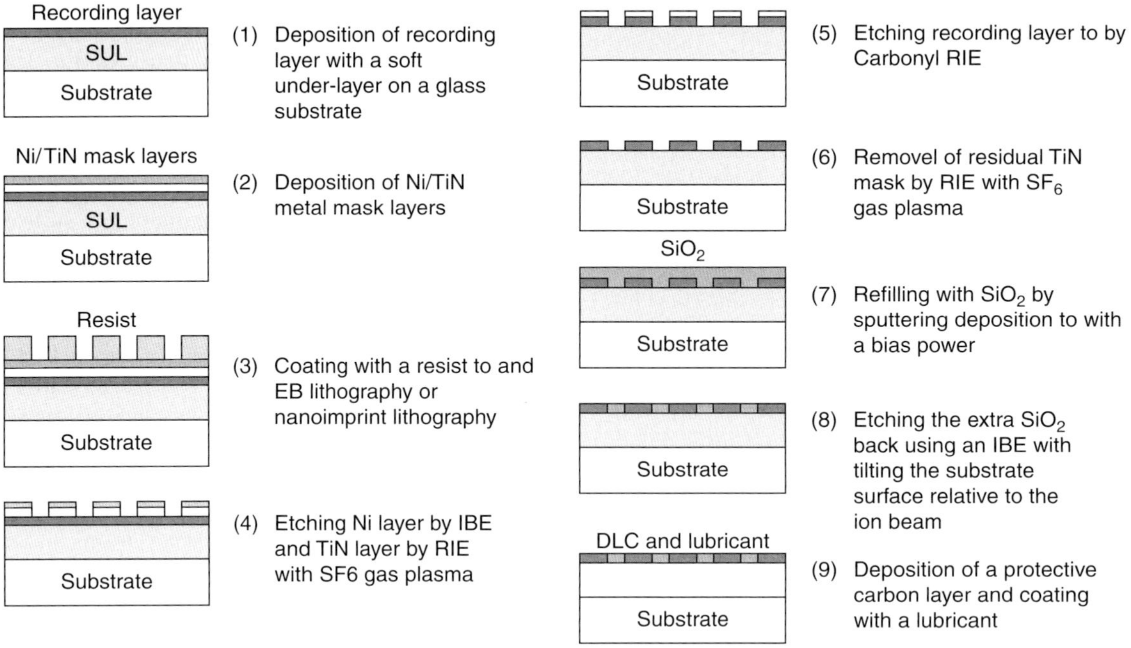

Figure 39 Typical process for DTM fabrication. Reprinted from Hattori K, Ito K, Soeno Y, Takai M, and Matsuzaki M (2004) *IEEE Transactions on Magnetics* 4: 2510–2515.

luxury of utilizing the mature lithographic tool and technology infrastructure already developed by the semiconductor industry for fabricating these kinds of structures. However, the resolution requirements placed on BPM negates these tool infrastructure advantages and will require the development of entirely new strategies for fabricating $\geq$1 Tb in^{-2} islands on disks in a very high-volume manufacturing process. **Table 3** shows projected scaling over the next 5 years of lithographic patterns with a 1:2 BAR [143] for patterned media at different data-storage densities and compares it to line and space periods on the ITRS timeline. In addition, pattern placement accuracy must meet strict specifications so that a recording head can follow bits arranged on circular tracks, as it is desired to continue using a rotating disk and flying head. For 1 Tb in^{-2} patterned media, there can be no deviation in position of the bits off the track larger than 3 nm [144].

Although EBL has been shown capable of achieving patterned arrays with densities of 4.5 Tb in^{-2} [145], the stitching errors found in the best x–y raster scan machines of 15 nm are too great to achieve the tolerances required, and thus rotary stage electron

Table 3 Roadmap of disk-drive storage density

Year	*Disk bit density (Gbin^{-2})*	*Track pitch (nm)*	*Down track bit pitch (nm)*	*FLASH line-center-to-center (nm)[19]*
2009	300	63	31.5	80
2011	500	50	25	64
2012	750	41	20.5	56
2013	1000	35	17.5	50
	2500	21	10.5	

Islands have a 2:1 aspect ratio, with the shorter direction being in the downtrack direction and the island area filling ~60% of spacing.
Reproduced with permission from Dobisz EA, Bandic ZZ, Wu TW, and Albrecht T (2008) *Proceedings of IEEE* 96: 1836–1846.

beam writers will be required. Even with the development of these high-resolution rotary stage tools, estimates for wafer throughput (ranging from 1 day and above for a single 2.5″ disk [144,146]) fall woefully short of required production values.

Of all the possible options, nanoimprint lithography (NIL) appears most likely to meet the lithographic specifications for resolution, placement accuracy, and throughput. While there are many imprint lithography techniques, all are based on the basic premise that a template or mold with a prefabricated topography is pressed into a displaceable material (i.e., resist). That material takes on the shape of the master pattern defined in the template, and the shaped material is then cured into a solid. A variation of this process known as 'step and flash' [147] is shown in **Figure 40**. One possible strategy for patterned media disk manufacturing may be to write a master disk pattern with high-resolution EBL on a rotary stage tool, a procedure currently used in HD-DVD and Blu-Ray optical disk manufacturing. This expensive master mold, either rigid or flexible, would be replicated into subsequent daughter and granddaughter molds, which could then imprint hundreds of disks per hour and eliminate the time penalty associated with EBL [148,149]. Following imprinting of the disk pattern into the resist on the disk, the pattern is transferred into the disk using physical and/or chemical etching processes. In practice, a substantial amount of characterization is required to optimize conditions for pattern transfer. The imprinted resist must adhere to the substrate and also completely wet the mold pattern, making the proper choices of low-viscosity resist and surface

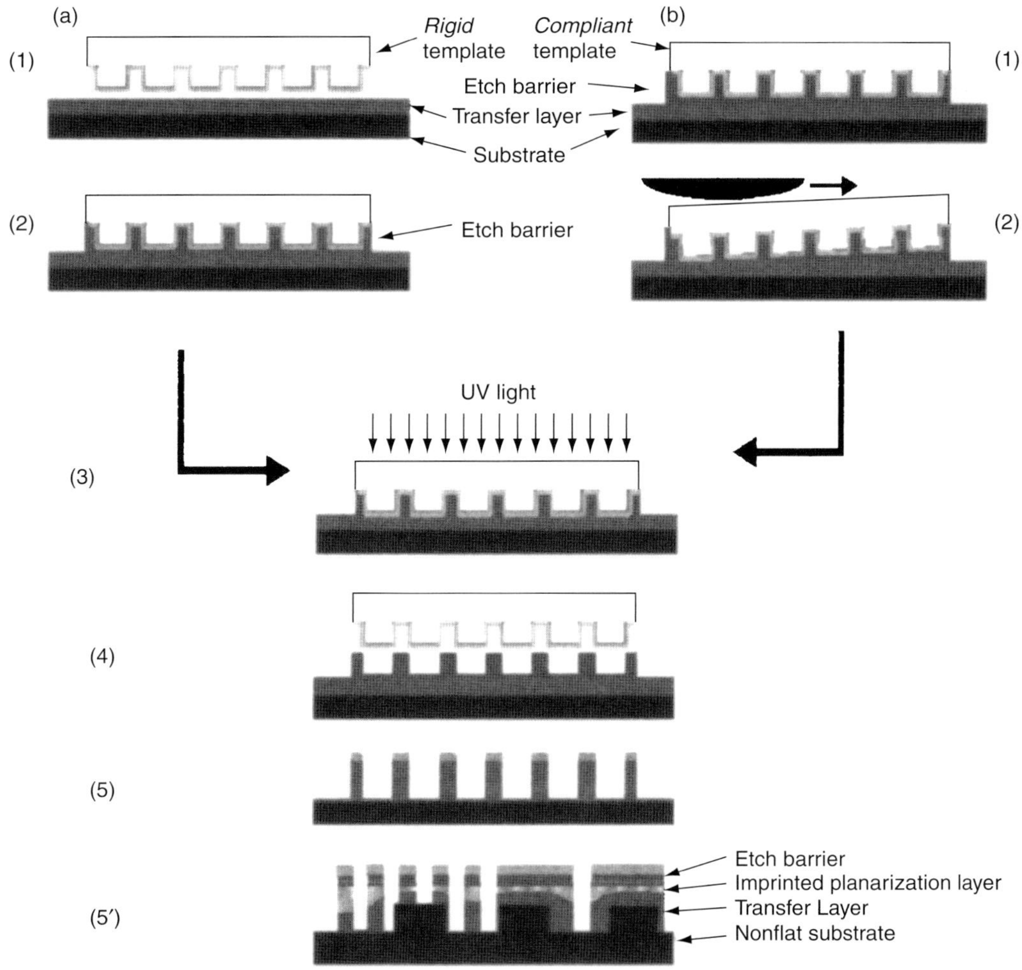

Figure 40 Step and flash imprint lithography process using: (a) a rigid template and (b) a compliant template. Reproduced with permission from Colburn M, Grot A, Choi BJ, *et al.* (2001) *Journal of Vacuum Science and Technology B* 19: 2162.

treatments of the mold essential for high reproduction fidelity. After the resist cures, it must then fully release from the mold or risk tearing of the pattern. Following its release, the imprinted resist pattern on a disk must have uniform thickness and be as thin as possible so that base layer topography variations will not be replicated into the etched disk.

Figure 41 shows an example of the replication of a 300 Gb in^{-2} pattern, with (1) a 300 Gb in^{-2} pillar pattern in a quartz master; (2) its replica mold; and (3) the imprinted disk pattern.

Separate issues remain in transferring patterns from the resist to the media itself. Following the base layer etch, the remaining resist pattern on the disk must retain sufficient thickness to properly mask off the substrate. Two generic approaches to this pattern transfer have been suggested: (1) prepatterning a substrate and subsequent deposition of a magnetic film; and (2) depositing a magnetic film and then etching or ion milling it through a mask to define the islands. The approach of prepatterning a substrate, as shown in **Figure 42**[150], has probably been the front running approach to date as (1) no good reactive ion etching (RIE) chemistry has been found for the magnetic materials of interest; (2) ion milling tends to result in redeposition; (3) it would be necessary to clean all resist and redeposition to produce a clean, flyable disk; and (4) prepatterning the substrate puts the patterning process earlier in the manufacturing process, thereby minimizing the impact of lower yields. This approach has progressed significantly, with published patterned densities of 300 Gb in^{-2}[151], although potential issues with magnetic noise from magnetic material in the trenches have been identified [152, 153] and must be addressed. Regardless of the eventual choice for

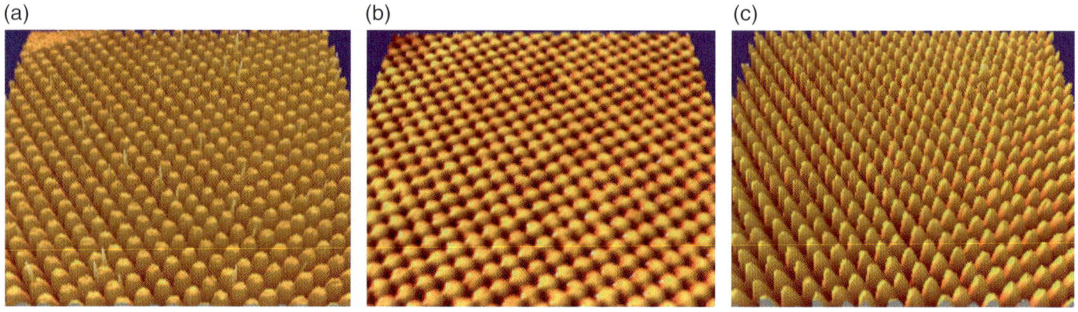

Figure 41 AFM micrographs of etched 45-nm-period pattern in: (a) quartz master; (b) daughter mold; and (c) substrate imprint (318 Gb). Reproduced with permission from Dobisz EA, Bandic ZZ, Wu TW, and Albrecht T (2008) *Proceedings of IEEE* 96: 1836–1846.

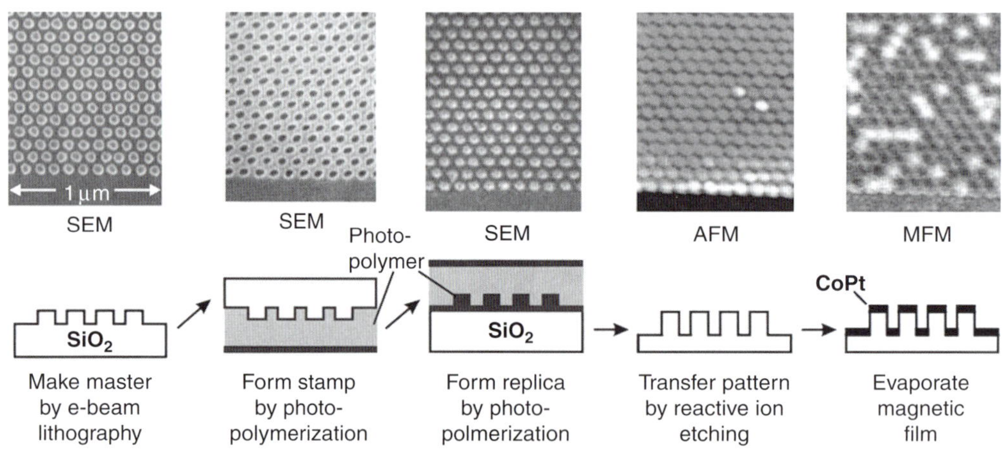

Figure 42 Steps in the fabrication of patterned magnetic media. The final panel displays a magnetic force microscope image, showing quantized up and down magnetization of isolated domains. Reproduced with permission from McClelland GM, Hart MW, Rettner CT, Best ME, Carter KR, and Terris BD (2002) *Applied Physics Letters* 81: 1483.

patterning the films, it is clear that significant progress has already been made and must continue in tooling and processing protocols.

It is important to note that ultimately NIL is a 1× contact process, and so the ultimate resolution is only as good as the imprint mask used. The use of diblock copolymers has been explored in conjunction with NIL for the purpose of improving the fabrication of BPM [154]. As opposed to the more traditional top-down fabrication approaches discussed in this section, diblock copolymers represent a bottom-up approach, using the uniform nano- and supermolecular structures that result from organized macromolecule packing. Block copolymers tend to self-assemble into a variety of well-ordered nanostructures with almost continuously tunable resolution from several to hundreds of nanometers because of the chemical immiscibility of the covalently linked segmental chains [155–157]. Through various templating processes with these block copolymer microdomain masks, high-resolution functional nanostructures can be generated, such as spheres, cylinders, and lamella.

The thermodynamically stable microdomain patterns formed in the bulk state are governed by the positive mixing enthalpy and low mixing entropy of the component segments. Because of the covalent bonding between the segments, the system can not macroscopically phase separate, and so it minimizes the interfacial energy by adopting well-defined microdomain patterns with constant interfacial curvature and stretched interfacial chain configurations [156,157]. However, intrinsic obstacles remain for large-scale nanofabrication with block copolymer patterns: (1) the inherent small domain size or, in other words, short-range lateral order; and (2) unfavorable domain orientation. Fortunately, it has been shown that both substrate topography [158–160] and chemical modification of the substrate surface [161,162] can generate periodic thickness or surface or interface energy profiles which guide the overlaying block copolymer nanopattern into long-range order. More recent work [155] has shown that pattern density multiplication is possible, enhancing patterning resolution while reducing EBL exposure time.

The eventual solution for lithographically defining and manufacturing patterned media for hard drive applications will most likely be some combination of the strategies mentioned above. Patterned media has become a clear example of how the storage device fabrication has aggressively entered the nanoscale regime, requiring an evolution in the tools and techniques used. The storage industry is developing leading edge capabilities in nanolithography, nanofabrication, metrology, and nanoscale physics, and is also pioneering plans for the implementation of nanofabrication in high volume. These exciting new capabilities open up possibilities for new device applications as well as exploring the basic physics of the nanoscale world.

7.6.3 Fabrication of 3D Memory Structures

Magnetic memory technologies such as MRAM represent a possible universal memory with the potential to replace the various Si memories used in a variety of applications. However, serious consideration must be given to the ability for 3–4 generation of device scaling due to the significant costs involved with changing tooling and fabrication practices from more traditional Si fabs. Three-dimensional memory architectures, such as the race-track memory architecture mentioned in this chapter, offer the potential for this scaling. The development of 3D or vertical race-track memory is likely to benefit from techniques that have been and are continuing to be developed for building 3D microelectronic devices. One approach to building vertical race-track memory may be to take advantage of the transistor industry's knowledge in fabricating vias (i.e., holes) in Si and silicon oxide (SiO_2), which can then be filled with metals using techniques such as electroplating. For example, arrays of vias can be formed by anodization of aluminum metal [163,164] or by etching nuclear tracks in polymers [165,166], after which the very long, high-aspect-ratio vias can readily be filled using electrodeposition [167–170]. Another technique for the fabrication of vias is the use of steerable proton beams [171,172], although this would require considerable development to be accurate, fast, and cheap enough for manufacturing real devices. The use of self-assembly, for example, using diblock copolymer templates [173] might also be a possibility in creating vias.

A method for building the vertical racetrack shown in **Figure 36** is described in Ref. [174]. Here, a via is constructed in a substrate which is formed from a multilayer laminate composed of two different alternating dielectric materials. By choosing dielectric materials with different etch rates, and then using a combination of selective and nonselective etchants, a via can be constructed with corrugated sides on the nanometer scale. When the via is filled by a magnetic material, then the corrugated sides provide the pinning sites for the domain-wall memory elements.

Other approaches for building vertical racetrack do not require the fabrication of very narrow and tall vias but can use more conventional approaches. For example, one can fabricate racetracks on the sides and the bottom of a deep trench whose width is many race-tracks wide [175]. Conventional patterning and etching can form the racetracks on the sides of the trench which are connected by the material deposited on the bottom of the trench to form the U-shaped racetrack shown in **Figure 36**. The integration of magnetoresistive sensing devices for reading data in the racetrack can readily be accomplished using the techniques developed in recent years to form high-quality deep submicron-sized MTJ elements. While there are many significant fabrication challenges for building these 3D architectures, a clear path has been outlined for fabricating and manufacturing these devices.

7.6.4 Conclusion

The demand for faster, higher-density magnetic storage in both hard drives and magnetic memory has pushed the limits of cutting-edge nanofabrication techniques. For the future technologies discussed in this chapter, a significant amount of interdisciplinary research and development will be required to establish the tooling and fabrication procedures necessary for manufacturing future generations of magnetic storage. Achieving these goals will continue to involve interplay between several fields of science and engineering, as well as an understanding of magnetism on the relevant length scales. The challenges in developing these techniques are not insurmountable, and will lead to increased understanding of physical properties, nanofabrication techniques, and manufacturing know-how.

7.7 Conclusions

Spin-based memory devices offer a unique combination of low cost, low power, high speed and high-density solutions to the current technologies' data storage needs. However, the demand to increase the storage density and to enable information flow at much higher speeds is driving an exhaustive study of alternative storage schemes for the future generations of magnetic memory devices. For a nonvolatile memory with fast read and write access times, spin-transport properties are utilized in MRAM. ST-MRAM promises an even lower power, higher-density solution. The needs for large amounts of data storage are to be met with improvements in the data-storage density in HDDs through patterned media and/or energy-assisted magnetic recording schemes. These clever approaches will enable an extension of the physical limitations on storage densities due to thermal stability issues. While not yet realized, there will be a limit to how much information can be recorded on a 2D medium. The ultimate solution will be a magnetic memory device with a 3D architecture, such as race-track memory. All these developments require parallel research and advancements in the fabrication area, to ensure low-cost and reliable manufacturing schemes are feasible for each technology. Although there are many challenges for the implementation of these devices, a combination of existing lithography technologies and promising new approaches such as nanoimprint lithography and block-copolymer self assembly is expected to provide a pathway to storage densities Tb/in^2 and beyond.

References

1. Bader SD (2006) Colloquium: Opportunities in nanomagnetism. *Reviews of Modern Physics* 78: 1–15.
2. Comstock RL (2002) Review: Modern magnetic materials in data storage. *Journal of Materials Science: Materials in Electronics* 13: 509–523.
3. Thompson SM (2008) Topical review: The discovery, development and future of GMR: The Nobel Prize 2007. *Journal of Physics D: Applied Physics* 41: 1–20.
4. Ralph DC and Stiles MD (2008) Current perspectives: Spin transfer torques. *Journal of Magnetism and Magnetic Materials* 320: 1190–1216.
5. Parkin SSP (2008) Magnetic domain-wall racetrack memory. *Science* 320: 190–194.
6. Richter HJ (2007) Topical review: The transition from longitudinal to perpendicular magnetic recording. *Journal of Physics D: Applied Physics* 40: R149–R177.
7. Ziese M and Thornton MJ (eds.) (2001) *Spin Electronics*. Berlin: Springer
8. Awschalom DD, Buhrman RA, Daughton JM, Von Molnar S, and Roukes ML (eds.) (2004) *Spin Electronics*. Dordrecht: Kluwer Academic Publishers.
9. Chalsani P, Upadhyay SK, Ozatay O, and Buhrman RA (2007) Andreev reflection measurements of spin polarization. *Physical Review B* 75: 411–426.
10. Dennis CL, *et al.* (2002) The defining length scales of mesomagnetism: A review. *Journal of Physics D: Applied Physics* 14: R1175–R1262.
11. Baibich MN, *et al.* (1988) Giant magnetoresistance of (001)Fe/(001)Cr magnetic superlattices. *Physical Review Letters* 61: 2472–2475.
12. Daughton JM, *et al.* (1999) Applications of spin dependent transport materials. *Journal of Physics D: Applied Physics* 32: R169–R177.

13. Parkin SSP, *et al.* (1990) Oscillations in exchange coupling and magnetoresistance in metallic superlattice structures: Co/Ru,Co/Cr and Fe/Cr. *Physical Review Letters* 64: 2304–2307.
14. Stamps RL (2000) Mechanisms for exchange bias. *Journal of Physics D: Applied Physics* 33: R247–R268.
15. Valet T and Fert A (1996) Theory of perpendicular magnetoresistance in magnetic multilayers. *Physical Review B* 48: 7099–7113.
16. Childress JR and Fontana RE (2005) Magnetic recording read head sensor technology. *Comptes Rendus Physique* 6: 997–1012.
17. Julliere M (1975) Tunneling between ferromagnetic films. *Physics Letters A* 54: 225–226.
18. Parkin SSP, *et al.* (2004) Giant tunneling magnetoresistance at room temperature with MgO (100) tunnel barriers. *Nature Materials* 3: 862–867.
19. Ozatay O, *et al.* (2006) Analytical electron microscopy study of growth mechanism for smoothing of metallic multilayer thin films. *Applied Physics Letters* 89: 162509.
20. Katine J and Fullerton E (2008) Device implications of spin-transfer torques. *Journal of Magnetism and Magnetic Materials* 320: 1217–1226.
21. Kiselev S, *et al.* (2003) Microwave oscillations of a nanomagnet driven by a spin-polarized current. *Nature* 425: 380–383.
22. Smith N, *et al.* (2006) Thermal and spin-torque noise in CPP (TMR and/or GMR) read sensors. *IEEE Transactions on Magnetics* 42: 114–119.
23. Russell L, Whalen R, and Leilich H (1968) Ferrite memory systems. *IEEE Transactions on Magnetics* 4: 134–145.
24. Hurst A and Granley G (1996) Projected applications, status, and plans for honeywell high density, high performance nonvolatile memory. *Proceedings of the 1996 Nonvolatile Memory Technology Conference*, Albuquerque, NM.
25. Chen EY, Tehrani S, Zhu T, Durlam M, and Goronkin H (1997) Submicron spin valve magnetoresistive random access memory cell. *Journal of Applied Physics* 81: 3992.
26. Moodera JS, Kinder LK, Wong TM, and Meservey R (1995) Large magnetoresistance at room temperature in ferromagnetic thin film tunnel junctions. *Physical Review Letters* 74: 3273–3276.
27. Miyazaki T and Tezuka N (1995) Giant magnetic tunneling effect in Fe/Al2O3/Fe junction. *Journal of Magnetism and Magnetic Materials* 139: L231–L234.
28. de Teresa JM, Barthelemy A, Fert A, *et al.* (1999) Inverse tunnel magnetoresistance in Co/SrTiO3/La0.7/Sr0.3/MnO3: New ideas on spin-polarized tunneling. *Physical Review Letters* 82: 4288–4291.
29. Butler WH, Zhang XG, Schulthess TC, and MacLaren JM (2001) Spin-dependent tunneling conductance of FelMgOlFe sandwiches. *Physical Review B* 63: 054416.
30. Mathon J and Umerski A (2001) Theory of tunneling magnetoresistance of an epitaxial Fe/MgO/Fe(001) junction. *Physical Review B* 63: 220403.
31. Parkin SSP, Kaiser C, Panchula A, *et al.* (2004) Giant tunneling magnetoresistance at room temperature with MgO (100) tunnel barriers. *Nature Materials* 3: 862–867.
32. Yuasa S, Nagahama T, Fukushima A, *et al.* (2004) Giant room-temperature magnetoresistance in single-crystal Fe/MgO/Fe magnetic tunnel junction. *Nature Materials* 3: 868–871.
33. Lee YM, Hayakawa J, Ikeda S, Matsukura F, and Ohno H (2007) Effect of electrode composition on the tunnel magnetoresistance of pseudo-spin-valve magnetic tunnel junction with a MgO tunnel barrier. *Applied Physics Letters* 90: 212507.
34. Durlam M, Addie D, Åkerman J, *et al.* (2003) A 0.18um 4Mb toggling MRAM. *IEEE International Electron Devices Meeting* 34: 1–3.
35. Dimopoulos T, Da Costa V, Tiusan C, Ounadjela K, and Van den Berg HAM (2001) Interfacial phenomena related to the fabrication of thin Al oxide tunnel barriers and their thermal evolution. *Applied Physics Letters* 79: 3110.
36. Akerman J, Brown P, Gajewski D, *et al.* (2005) Reliability of 4Mbit MRAM. *Reliability Physics Symposium*, 163–167.
37. Pon SK and Chung Y (2006) Resistance drift of aluminum oxide magnetic tunnel junction devices. *IEEE International Reliability Physics Symposium*, 437–441.
38. Tehrani S (2006) Status and outlook of MRAM memory technology (invited). *Electron Devices Meeting*, pp. 1–4; 11–13.
39. Dave RW, Steiner G, Slaughter JM, *et al.* (2006) MgO-based tunnel junction material for high-speed toggle MRAM. *IEEE Transactions on Magnetics* 42: 1935–1939.
40. Slaughter JM, Dave RW, Durlam M, *et al.* (2005) High speed toggle MRAM with MgO-based tunnel junctions. *IEDM Technical Digest, IEEE International*, 893–896.
41. Beach R, Min T, Horng C, *et al.* (2008) A statistical study of magnetic tunnel junctions for high-density spin torque transfer-MRAM (STT-MRAM). *Electron Devices Meeting, IEDM '08.* International, pp. 305–308.
42. Savtchenko L, Engel BN, Rizzo ND, Deherrera MF, and Janesky JA (2003) Method of writing to scalable magnetoresistance random access memory element, US Patent 6,545,906 B1.
43. Katine JA, Albert FJ, Buhrman RA, Myers EB, and Ralph DC (2000) Current-driven magnetization reversal and spin-wave excitations in Co/Cu/Co pillars. *Physical Review Letters* 84: 3149–3152.
44. Slonczewski JC (1996) Current-driven excitation of magnetic multilayers. *Journal of Magnetism and Magnetic Materials* 159: L1–L7.
45. Berger L (1996) Emission of spin waves by a magnetic multilayer traversed by a current. *Physical Review B* 54: 9353–9358.
46. Tehrani S, Slaughter JM, DeHerrera M, *et al.* (2003) Magnetoresistive random access memory using magnetic tunnel junctions. *Proceedings of IEEE* 91(5): 703.
47. Reohr WR and Scheuerlein RE (2002) Segmented Write Line Architecture for Writing Magnetic Random Access Memories. US Patent 6,335,890, 1 Jan.
48. Min T, Heim D, Chen Q, and Wang P (2008) A scalable field switching MRAM design. *Proceedings MMM DC-02*, 225.
49. Zhu J (1999) Spin valve and dual spin valve heads with synthetic antiferromagnets. *IEEE Transactions on Magnetics* 35: 655–660.
50. Abraham DW and Worledge DC (2006) Low power scaling using parallel coupling for toggle magnetic random access memory. *Applied Physics Letters* 88: 262505.
51. Fukumoto Y, Suzuki T, and Tahara S (2006) Low writing field with large writing margin in toggle magnetic random access memories using synthetic antiferromagnet ferromagnetically coupled with soft magnetic layers. *Applied Physics Letters* 89: 061909.
52. Huai Y, Albert F, Nguyen P, Pakala M, and Valet T (2004) Observation of spin-transfer switching in deep submicron-sized and low-resistance magnetic tunnel junctions. *Applied Physics Letters* 84: 3118–3120.
53. Fuchs GD, Emley NC, Krivorotov IN, *et al.* (2004) Spin-transfer effects in nanoscale magnetic tunnel junctions. *Applied Physics Letters* 85: 1205–1207.
54. Sun JZ and Ralph DC (2008) Magnetoresistance and spin-transfer torque in magnetic tunnel junctions. *Journal of Magnetism and Magnetic Materials* 320: 1227–1237.

55. Mangin S, Ravelosona D, Katine JA, Carey MJ, Terris BD, and Fullerton EE (2006) Current-induced magnetization reversal in nanopillars with perpendicular anisotropy. *Nature Materials* 5: 210.
56. Kishi T, Yoda H, Kai T, *et al.* (2008) Lower-current and fast switching of a perpendicular TMR for high speed and high density spin-transfer-torque MRAM. *Digest of International Electron Devices Meeting, IEDM*.
57. Hosomi M, Yamagishi H, Yamamoto T, *et al.* (2005) A novel nonvolatile memory with spin torque transfer magnetization switching: Spin-RAM. *IEDM 2005*, p. 459 (IEEE Cat. 05CH37703C).
58. Kawahara T, Takemura R, Miura K, *et al.* (2007) 2Mb spin-transfer torque RAM (SPRAM) with bit-by-bit bidirectional current write and parallelizing-direction current read. *2007 IEEE International Solid-State Circuits Conference*, p. 480.
59. Slaughter JM, Rizzo ND, Mancoff FB, *et al.* (2009) Recent results on spin-torque MRAM arrays and technology outlook. *IEEE Intermag 2009 Technical Digest*, CF-02.
60. Worledge D, Abraham DW, Brown S, *et al.* (2008) Magnetic reliability of toggle MRAM. *Proceedings from 53rd Annual Conference of Magnetism and Magnetic Material*, DC-04.
61. Nebashi R, Sakimura N, Honjo H, *et al.* (2009) A 90nm 12ns 32Mb 2T1MTJ MRAM. *Proceedings from IEEE ISSCC*, 27.4.
62. Wood R (2009) Future hard disk drive systems. *Journal of Magnetism and Magnetic Materials* 321: 555–561.
63. Sbiaa R and Piramanayagam SN (2007) Patterned media towards nano-bit magnetic recording: fabrication and challenges. *Recent Patents on Nanotechnology* 1: 29.
64. Victora R and Shen X (2005) Exchange coupled composite media for perpendicular magnetic recording. *IEEE Transactions on Magnetics* 41: 2828–2833.
65. Suess D, Schrefl T, Fahler S, *et al.* (2005) Exchange spring media for perpendicular recording . *Applied Physics Letters* 87(1): 012504.
66. Berger A, Supper N, Ikeda Y, *et al.* (2008) Improved media performance in optimally coupled exchange spring layer media. *Applied Physics Letters* 93: 122502.
67. Soeno Y, Moriya M, Ito K, *et al.* (2003) Feasibility of discrete track perpendicular media for high track density recording. *IEEE Transactions on Magnetics* 39: 1967–1971.
68. Soeno Y, Moriya M, Kaizu A, and Takai M (2005) Performance evaluation of discrete track perpendicular media for-high recording density. *IEEE Transactions on Magnetics* 41: 3220–3222.
69. Terris BD and Thomson T (2005) Nanofabricated and self-assembled magnetic structures as data storage media. *Journal of Physics D: Applied Physics* 38: R199–R222.
70. Moritz J (2003) Enregistrement ultra-haute densité sur réseau de plots magnétiques nanométriques à aimantation perpendiculaire au plan. PhD Thesis, Universite Joseph Fourier, Grenoble.
71. Kikitsu A (2009) Prospects for bit patterned media for high-density magnetic recording. *Journal of Magnetism and Magnetic Materials* 321(6): 526–530.
72. Rettner CT, Best ME, and Terris BD (2001) Patterning of granular magnetic media with a focused ion beam to produce single-domain islands at >140 Gbit/in(2). *IEEE Transactions on Magnetics* 37: 1649–1651.
73. Fassbender J and McCord J (2008) Magnetic patterning by means of iron irradiation and implantation. *Journal of Magnetism and Magnetic Materials* 320: 579–596.
74. Sun S, Murray CB, Weller D, Folks L, and Moser A (2000) Monodisperse FePt nanoparticles and ferromagnetic FePt nanocrystal superlattices. *Science* 287: 1989–1992.
75. Lina X-M and Samia ACS (2006) Synthesis, assembly and physical properties of magnetic nanoparticles. *Journal of Magnetism and Magnetic Materials* 305: 100–109.
76. Thomson T, Hu G, and Terris BD (2006) Intrinsic distribution of magnetic anisotropy in thin films probed by patterned nanostructures. *Physical Review Letters* 96: 257204.
77. Schabes M (2008) Micromagnetic simulations for terabit/in(2) head/media systems. *Journal of Magnetism and Magnetic Materials* 320(22): 2880–2884.
78. Sbiaaa R and Piramanayagam SN (2008) High frequency switching in bit-patterned media: A method to overcome synchronization issue. *Applied Physics Letters* 92: 012510.
79. Weller D, Moser A, Folks L, *et al.* (2000) High K-u materials approach to 100 Gbits/in(2). *IEEE Transactions on Magnetics* 36(1): 10–15.
80. Suess D (2007) Micromagnetics of exchange spring media: Optimization and limits. *Journal of Magnetism and Magnetic Materials* 308(2): 183–197.
81. Alex M, Tselikov A, McDaniel T, *et al.* (2001) Characteristics of thermally assisted magnetic recording. *IEEE Transactions on Magnetics* 37(4): 1244–1249.
82. Ruigrok JJM, Coehoorn R, Cumpson SR, *et al.* (2000) Disk recording beyond 100 Gb/in.(2): Hybrid recording? (invited). *Journal of Applied Physics* 87(9): 5398–5403.
83. Bethe H (1944) *Physical Review* 66: 163–182.
84. Betzig E and Trautman JK (1992) Near-field optics – microscopy, spectroscopy, and surface modification beyond the diffraction limit. *Science* 257(5067): 189–195.
85. Ebbesen TW, Lezec HJ, Ghaemi HF, *et al.* (1998) Extraordinary optical transmission through sub-wavelength hole arrays. *Nature* 391(6668): 667–669.
86. Thio T, Pellerin KM, Linke RA, *et al.* (2001) Enhanced light transmission through a single subwavelength aperture. *Optics Letters* 26(24): 1972–1974.
87. Shi XL and Hesselink L (2002) Ultra high light transmission through a C-shaped nanoaperture. *Japanese Journal of Applied Physics* 41: 1632.
88. Rottmayer RE, Batra S, Buechel D, *et al.* (2006) Heat-assisted magnetic recording. *IEEE Transactions on Magnetics* 42(10): 2417–2421.
89. Matsumoto T, Anzai Y, Shintani T, *et al.* (2006) Writing 40 nm marks by using a beaked metallic plate near-field optical probe. *Optics Letters* 31(2): 259–261.
90. Stipe B (2008) Magneto-optic writing using the ridge waveguide. *Presentation at the MMM Conference 2008.* Austin/TX.
91. Thiele J-U, Coffey KR, Toney MF, *et al.* (2002) Temperature dependent magnetic properties of highly chemically ordered Fe(55-x)Ni(x)Pt(45)LI(0)films. *Journal of Applied Physics* 91(10): 6595–6600.
92. Rausch T, Bain JA, Stancil DD, *et al.* (2004) Thermal Williams-Comstock model for predicting transition lengths in a heat-assisted magnetic recording system. *IEEE Transactions on Magnetics* 40(1): 137–147.
93. Seigler MA, Challener WA, Gage E, *et al.* (2008) Integrated heat assisted magnetic recording head: Design and recording demonstration. *IEEE Transactions on Magnetics* 44: 119–124.
94. Rausch T, Mihalcea C, Pelhos K, *et al.* (2006) Near field heat assisted magnetic recording with a planar solid immersion lens. *Japanese Journal of Applied Physics Part 1-Regular Papers Brief Communications and Review Papers* 45(2B): 1314–1320.

95. Cahill DG, Ford WK, Goodson KE, *et al.* (2003) Nanoscale thermal transport. *Journal of Applied Physics* 93(2): 793–818.
96. Hardie C (2008) Challenges of integrating heat assist magnetic recording. *Presentation at the 2008 ISOM Conference*, 13–17 July 2008, Hawaii.
97. Thiele JU, Maat S, and Fullerton EE (2003) FeRh/FePt exchange spring films for thermally assisted magnetic recording media. *Applied Physics Letters* 82(17): 2859–2861.
98. Zhu JG (2008) Microwave assisted magnetic recording. *IEEE Transactions on Magnetics* 44: 125–131.
99. Parkin SSP (2004–2007) US Patents 6,834,005, 6,898,132 6,920,062, 7,031,178, and 7,236,386.
100. Hayashi M, *et al.* (2008) Real time observation of the field driven periodic transformation of domain walls in permalloy nanowires at the Larmor frequency and its first harmonic. *Applied Physics Letters* 92: 112510.
101. Mougin A, *et al.* (2007) Domain wall mobility, stability and Walker breakdown in magnetic nanowires. *European Physics Letters* 78: 57007.
102. Schryer NL and Walker LR (1974) The motion of 180° domain walls in uniform dc magnetic fields. *Journal of Applied Physics* 45: 5406–5421.
103. Hayashi M, *et al.* (2007) Direct observation of the coherent precession of magnetic domain walls propagating along permalloy nanowires. *Nature Physics* 3: 21–25.
104. Tatara G and Kohno H (2004) Thoery of current-driven domain wall motion: Spin transfer versus momentum transfer. *Physical Review Letters* 92: 086601.
105. Berger L (1984) Exchange interaction between ferromagnetic domain-wall and electric-current in very thin metallic-films. *Journal of Applied Physics* 55: 1954–1956.
106. Hayashi M, *et al.* (2007) Current driven domain wall velocities exceeding the spin angular momentum transfer rate in permalloy nanowrires. *Physical Review Letters* 98: 037204.
107. Thomas L, *et al.* (2007) Resonant amplification of domain-wall motion by a train of current pulses. *Science* 315: 1553–1556.
108. Parkin SSP, Hayashi M, and Thomas L (2008) Magnetic domain-wall race track memory. Supporting online material. *Science* 320: 190–194.
109. Thompson LF, Wilson CG, and Bowden M (eds.) (1994) *Introduction to Microlithography,* 2nd edn. Washington, DC: American Chemical Society.
110. Elliott D (1986) *Microlithography: Process Technology for IC Fabrication*. New York: McGraw-Hill.
111. Sheats JR and Smith BW (1998) *Microlithography Science and Technology*. New York: Marcel Dekker.
112. Bruning JH (1997) *Proceedindgs of SPIE* 3049: 14–27.
113. Adeyeye AO and Singh N (2008) *Journal of Physics D: Applied Physics* 41: 153001.
114. Ulrich W, Beiersdorfer S, and Mann HJ (2000) *Proceedings of SPIE* 4146: 13–24.
115. Saito Y, Kawada S, Yamamoto T, Hayashi A, Isao A, and Tokoro Y (1994) *Proceedings of SPIE* 2254: 60–63.
116. Smith BW, Alam Z, Butt S, Kurinec S, Lane RL, and Arthur G (1997) *Microelectronic Engineering* 35: 201–204.
117. Pierrat C, Cote M, and Patterson K (2002) *Proceedings of SPIE* 4691: 325–335.
118. International Technology Roadmap for Semiconductors (ITRS) (2007) http://www.itrs.net/Links/2007ITRS/Home2007.htm (accessed July 2009).
119. Lin BJ (2004) *Proceedings of SPIE* 5377: 46–67.
120. Wagner C, Finders J, de Klerk J, and Kool R (2007) *Solid State Technology* 50: 51.
121. Allen RD, Brock PJ, Sundberg L, *et al.* (2005) *Journal of Photopolymer Science and Technology* 18: 615–619.
122. Bratton D, Yang D, Dai J, and Ober CK (2006) *Polymer Advanced Technology* 17: 94–103.
123. Dammel RR, Houlihan FM, Sakamuri R, Rentkiewicz D, and Romano A (2004) *Journal of Photopolymer Science and Technology* 17: 587–602.
124. Lin BJ (2006) *Microelectronic Engineering* 83: 604–613.
125. Bates AK, Rothschild M, Bloomstein TM, *et al.* (2001) *IBM Journal of Research and Development* 45: 605–614.
126. Wu B and Kumar A (2007) *Journal of Vacuum Science and Technology B* 25: 1743–1761.
127. Kinoshita H, Kuirhara K, Ishii Y, and Torii Y (1989) *Journal of Vacuum Science and Technology B* 7: 1648.
128. Bjorkholm JE, Bokor J, Eichner L, and Freeman RR (1990) *Journal of Vacuum Science and Technology B* 8: 1509.
129. Xiao J, Cerrina F, and Rippstein R (1994) *Journal of Vacuum Science and Technology B* 12: 4018.
130. Cerrena F (2000) *Journal of Physics D: Applied Physics* 33: R103–R116.
131. Bllard WP, Tichenor KA, O'Connell DJ, and Bernardez LJ (2003) *Proceedings of SPIE* 5037: 47.
132. Miura T, Murakami K, Suzuki K, Kohama Y, Ohkubo Y, and Asami T (2006) *Proceedings of SPIE* 6151: 615105.
133. Hansson BAM and Hertz H (2004) *Journal of Physics D* 37: 3233.
134. Chang THP, Mankos M, Lee KY, and Muray LP (2001) *Microelectronic Engineering* 57–58: 117–135.
135. Pfeiffer HC (2000) PREVAIL—IBM's E-beam technology for next generation lithography. *In: Proceedings of SPIE, Emerging Lithographic Technologies—IV,* vol. 3997, pp. 206–213.
136. Fontana RE, Katine J, Rooks M, *et al.* (2002) *IEEE Transactions on Magnetics* 38: 95–100.
137. Moser A, Takano K, Margulies DT, *et al.* (2002) *Journal of Physics D: Applied Physics* 35: R157–R167.
138. Weller D and Moser A (1999) *IEEE Transactions on Magnetics* 35: 4423.
139. Hattori K, Ito K, Soeno Y, Takai M, and Matsuzaki M (2004) *IEEE Transactions on Magnetics* 4: 2510–2515.
140. Soeno Y, Moriya M, Ito K, *et al.* (2003) *IEEE Transactions on Magnetics* 4: 1967–1971.
141. Che XD, Tang Y, Lee HJ, *et al.* (2007) *IEEE Transactions on Magnetics* 43: 2292.
142. Stoner EC and Wohlfarth EP (1948) *Transactions of the Royal Society* 240: 599.
143. Richter HJ, *et al.* (2006) *IEEE Transactions on Magnetics* 42: 2255.
144. Dobisz EA, Bandic ZZ, Wu TW, and Albrecht T (2008) *Proceedings of IEEE* 96: 1836–1846.
145. Yang X, Xiao S, Wu W, *et al.* (2007) *Journal of Vacuum Science and Technology B* 25: 2202.
146. Terris BD (2008) *Journal of Magnetism and Magnetic Materials* doi:10.1016/j.jmmm.2008.05.046.
147. Colburn M, Grot A, Choi BJ, *et al.* (2001) *Journal of Vacuum Science and Technology* B19: 2162.
148. Heidari B, Moller T, Palm R, Bolmsjo E, and Beck M (2005) *Proceedings of IEEE International Conference on Microprocesses and Nanotechnology Conference*, 25–28 Oct. 2005, pp. 144–145.
149. Bogdanov A, Holmqvist T, Jedrasik P, and Nilsson B (2003) *Microelectronic Engineering* 67–68: 381–389.
150. McClelland GM, Hart MW, Rettner CT, Best ME, Carter KR, and Terris BD (2002) *Applied Physics Letters* 81: 1483.
151. Terris BD and Thomson T (2005) *Journal of Physics D: Applied Physics* 38: R199–R222.
152. Moser A, Hellwig O, Kercher D, and Dobisz E (2007) *Applied Physics Letters* 91: 162502.

153. Moser A, Olson T, Hellwig O, Kercher D, and Dobisz E (2007) *Paper DB-09 at Intermag/MMM 2007*.
154. Ruiz R, Detcheverry KMFA, Dobisz E, *et al.* (2008) *Science* 321: 936–939.
155. Hamley IW, *et al.* (1998) *The Physics of Block Copolymers*. New York: Oxford Science Publications.
156. Leibler L (1980) *Macromolecules* 13: 1602.
157. Bates FS and Fredrickson GH (1990) *Annual Review of Physical Chemistry* 41: 525.
158. Fasolka MJ, *et al.* (2000) *Macromolecules* 33: 5702.
159. Asakawa K, *et al.* (2002) *Journal of Photopolymer Science and Technology* 15: 465.
160. Segalman RA, *et al.* (2001) *Advanced Materials* 13: 1152.
161. Kim SO, *et al.* (2003) *Nature* 424: 411.
162. Heier J, *et al.* (1997) *Macromolecules* 30: 6610.
163. Masuda H and Fukuda K (1995) *Science* 268: 1466.
164. Singh GK, Golovin AA, Aranson IS, and Vinokur VM (2005) *Europhysics Letters* 70: 836.
165. Huczko A (2000) *Applied Physics A* 70: 365.
166. Whitney TM, Jiang JS, Searson PC, and Chien CL (1993) *Science* 261: 1316.
167. Dubois S, *et al.* (1997) *Applied Physics Letters* 70: 396.
168. Schönenberger C, *et al.* (1997) *Journal of Physical Chemistry B* 101: 5497.
169. Lee JH, *et al.* (2007) *Angewandte Chemie International Edition* 46: 3663.
170. Martin CR (1994) *Science* 266: 1961.
171. Teo EJ, *et al.* (2004) *Applied Physics Letters* 84: 3202.
172. van Kan JA, Bettiol AA, and Watt F (2006) *Nano Letters* 6: 579.
173. Thurn-Albrecht T, *et al.* (2000) *Science* 290: 2126.
174. Chen T-C and Parkin SSP (2005) Method of fabricating data tracks for use in a magnetic shift register memory device. US Patent #6,955,926.
175. Chen T-C and Parkin SSP (2006) Method of fabricating a shiftable magnetic shift register. US Patent #7,108,797.

8 Optical Holographic Data Storage

K Curtis, L Dhar, and W L Wilson, InPhase Technologies, Longmont, CO, USA

8.1 Introduction

The use of digital data permeates almost every aspect of our lives today. Applications such as the distribution of content, digital video, interactive multimedia, small personal data storage devices, archiving of valuable digital assets, and downloading over high-speed networks, continue to grow. Of current technologies, optical data storage, such as the compact disk (CD), digital versatile disk (DVD), and recently introduced Blu-ray disk (BD), have found wide utility due to their ability to provide random access to data, rapid replication of content, and removable, inexpensive, recording media. However, with the growth and spread of digital data, these technologies find themselves constantly challenged to provide ever-increasing improvements in performance.

Traditional optical storage technologies are based on surface recording and the serial writing and reading of data. In these approaches, individual bits of data are recorded as changes in the reflectivity of small marks on the media surface. To increase the density in optical storage, the representative size of these marks is decreased. CDs yield a storage density of ~0.66 Gbits per square inch and BDs currently reside at a density of ~17 Gbits per square inch. With this evolution, the size of the marks has been reduced to the practical limit of far-field optical recording limiting further improvements. In addition, because data are recorded and recovered serially (as single bits), the transfer rates are limited with the practical limit of BD thought to be 15–20 Mbytes s^{-1}. These challenges have triggered a search for the next generation of optical technologies.

Holographic data storage (HDS) breaks through the density limitations of conventional storage technologies by going beyond two dimensions to write data in three dimensions. HDS has demonstrated the highest storage densities (663 Gbits per square inch) of any removable technology [1] and has a theoretical achievable density of ~40 Tbits per square inch . Unlike serial technologies, which record one data bit at a time, holography records and reads over a million bits of data with a single flash of light enabling transfer rates significantly higher than traditional optical storage devices. High storage densities, fast transfer rates, and random access, combined with durable, reliable, low-cost media, make holography a compelling choice for next-generation storage and content distribution applications.

In this article we will review the fundamentals of holographic storage, highlight several important milestones in its development, and discuss the current state of the technology.

8.2 Fundamentals of Holographic Storage

8.2.1 Essentials of Holography

The principles of holography or 'two-step optical imagery' were proposed by Dennis Gabor in 1948 [2]. As a method to record optical wave fronts, holography enables the storage of the amplitude and phase profiles of optical fields. **Figure 1** illustrates the formation of a simple off-axis hologram. Two planes waves, optical beams of wavelength λ that originate from the same laser source and are coherent with each other intersect within a recording medium (**Figure 1(a)**). The resulting optical interference pattern is recorded into the media volume as a modulation of the dielectric constant of the material. This image (**Figure 1(b)**) is the hologram and extends throughout the volume of the media wherever the optical interference pattern existed. To reconstruct the hologram, one of the beams used during the recording process diffracts off the hologram reconstructing the other beam (**Figure 1(c)**).

Holograms are recorded via linear changes of the real and/or imaginary parts of the dielectric constant (absorption or index of refraction, respectively) of a material. The lossy nature of absorptive gratings (pure amplitude) limits their maximum diffraction efficiency to less than 4%. Refractive index gratings (pure phase gratings), on the other hand, can achieve diffraction efficiencies of 100%. If the thick recording media are used, the holograms exhibit Bragg selectivity allowing multiplexing or multiple holograms to be recorded through the same volume of material.

More complex object waves can be recorded. For example, the diffracted field of any imaged object can be described as a Fourier sum of plane waves of

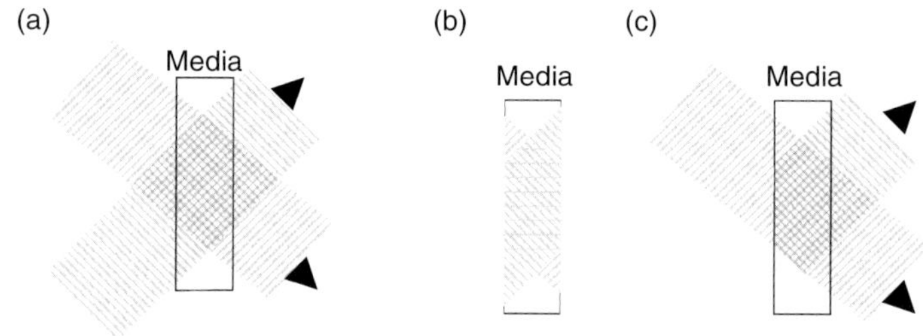

Figure 1 Diagram of the formation of a hologram and reading out of a hologram. (a) The intersection of two beams creates an interference pattern of light and dark regions. The media records this interference pattern. (b) The record of this pattern is a 'hologram'. (c) Reconstruction of the other beam by using one of the beams during the recording process (reconstruction of the hologram).

varying angles or spatial frequencies. The interference pattern resulting from the overlapping of a complex object wave and a plane wave generates a 'grating spectrum' in the material's volume. Diffraction of the plane wave from the resulting hologram reconstructs the complex field.

8.2.2 Volume Holograms and the Bragg Effect

One of the enabling phenomena for data storage is the ability to multiplex holograms. As discussed above, this capability depends on using thick media. In order to determine whether a medium satisfies the thickness criteria, Klein and Cook [3,4] defined a grating strength parameter, Q. The parameter Q is given in **equation 1**, where λ_0 is the wavelength of the recording light, d is the thickness of the material, n_0 is the bulk refractive index of the material, and Λ is the period of the hologram grating.

$$Q = 2\pi\lambda_0 d / n_0 \Lambda^2 \tag{1}$$

If $Q > 1$, the gratings are considered to be thick, volume gratings that can support Bragg diffraction. In essence, Q provides a metric of how many grating planes the light encounters while traveling through the recording medium.

Thick media enable multiplexing via the Bragg effect. A hologram recorded in thick media is fully reconstructed through readout with light of the same angle of incidence and wavelength as used during recording. (Deviations in the angle of incidence or wavelength of the readout light lead to drops in the power of the diffracted light.) This specificity in readout allows the recording of multiple holograms in the same volume. By 'detuning' the angle of incidence or wavelength of the recording light from that used previously, holograms can be superimposed.

The behavior of the diffracted light can be understood through Kogelnik's coupled wave treatment of plane wave holograms [5]. For a pure refractive index grating in transmission geometry, the diffraction efficiency is given by **equation 2**:

$$\eta = \sin^2\left(v + \xi^2\right) / \left(1 + \frac{\xi^2}{v^2}\right) \tag{2}$$

where the parameters are given by **equations 3** and **4**:

$$v = \pi n_1 d / \lambda \left(\cos^2\theta - \frac{K\cos\theta\cos\phi}{\beta}\right)^{1/2} \tag{3}$$

and

$$\xi = \left(\Delta\theta \cdot K\sin(\varphi - \theta_0) - \Delta\lambda K^2 / 4\pi n\right) d / 2\left(\cos\theta - \frac{K\cos\phi}{\beta}\right) \tag{4}$$

In the above equations, η is the diffraction efficiency, K the grating $(2\pi/\Lambda)$, where Λ is the grating period, d the material thickness, θ the angle of the reference beam (outside the media) on the readout measured from the media normal, n the bulk index of the material, n_1 the index perturbation, ϕ the grating slant angle inside the material measured from normal, λ the wavelength of light, and β is $2\pi(\varepsilon_0)^{1/2}/\lambda$, where ε_0 is the bulk dielectric constant. The parameters $\Delta\lambda$ and $\Delta\theta$ are the deviations in wavelength and angle, respectively, relative to the Bragg condition.

To illustrate the selectivity of holograms in thick media, the diffraction efficiency of a transmission grating as a function of angular detuning $\Delta\theta$ only $(\Delta\lambda = 0)$ is plotted in **Figure 2**. This Bragg selectivity curve was calculated for a medium of thickness 1.5 mm and bulk refractive index of 1.5, a 50° angle between the recording beams, and a recording wavelength of 405 nm. The rapid drop-off of diffraction efficiency as the readout angle is

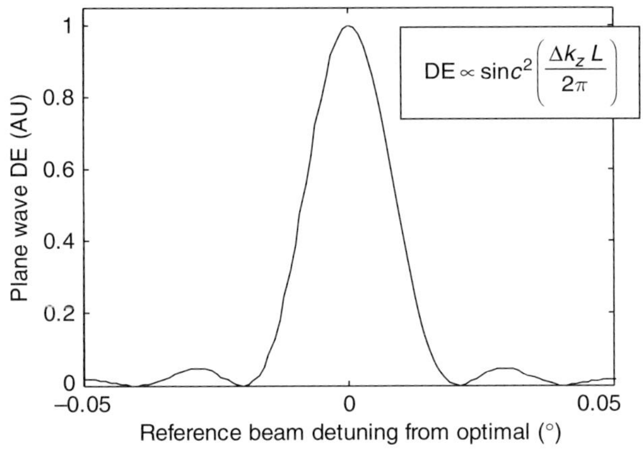

Figure 2 Example of a Bragg selectivity curve.

tuned away from the recording value demonstrates how holograms can be multiplexed.

8.2.3 Applications to Data Storage

In 1963, Van Heerden [6] was the first to formulate the application of holography to data storage. He recognized that two important aspects of holography could be exploited to provide significant enhancements of data transfer rates and storage densities. Rather than recording bits of data one by one, a collection of bits could be stored as a hologram and recovered in single events enabling rapid transfer rates. In addition, because the physics of holography allows multiplexing, ultrahigh storage densities can be attained. (Theoretical calculations predict a limit of ~1 bit per λ^3 of material.)

In general, in a HDS system, light from a source laser beam is split into two arms, signal and reference beams. The digital data to be stored are arranged as 'pages' or a series of matrices of the bits. The pages are encoded onto the signal beam via a spatial light modulator (SLM), which represents the bits (the 0's and 1's) as pixels that are either opaque or transmitting to light. Traveling through the SLM, the signal beam becomes an optical checkerboard pattern. The signal and reference beam intersect in a light-sensitive medium, storing the bits of the page in parallel. Multiple holograms are stored within the same volume of the medium (creating 'books') by changing some aspect (e.g., the angle of incidence, wavelength, phase profile) of the incident reference beam. **Figure 3** is a schematic of the write process.

To recover each data page, the reference beam with the same characteristics as used during recording diffracts from the stored hologram reconstructing the information. The diffracted pattern is imaged onto a detector that reads the bits of the page in parallel. **Figure 4** illustrates the reading process.

8.2.4 Approaches to Multiplexing

A series of strategies to multiplex holograms have been developed. **Figure 5** captures a summary of the various options. The strategies fall into three categories: Bragg-, momentum-, and correlation-based methods. (Some methods, such as peristrophic multiplexing, can be dominated by either Bragg effects or conservation of momentum, depending on the exact geometry and system configuration used.) In addition, many of the multiplexing methods can be combined.

8.2.4.1 Bragg-based techniques

Using the Bragg effect, multiple holograms can be superimposed in the same volume of the recording media by exploiting the destructive interference of Bragg mismatch to suppress undesired reconstructions. Of the multiplexing techniques based on the Bragg effect, angle multiplexing is the most common [7–9], but wavelength [10,11], phase code [12–15], peristrophic [16], shift [17,18] and correlation multiplexing methods [19–21] use the same principles. Angle multiplexing, as described above, uses a set of reference beams with each a distinct plane wave with different angles of incidence. Each data page is stored by interfering the signal beam encoded with the data page with one of the plane wave reference beams. The data page is then changed and stored using a reference beam at a different angle of incidence.

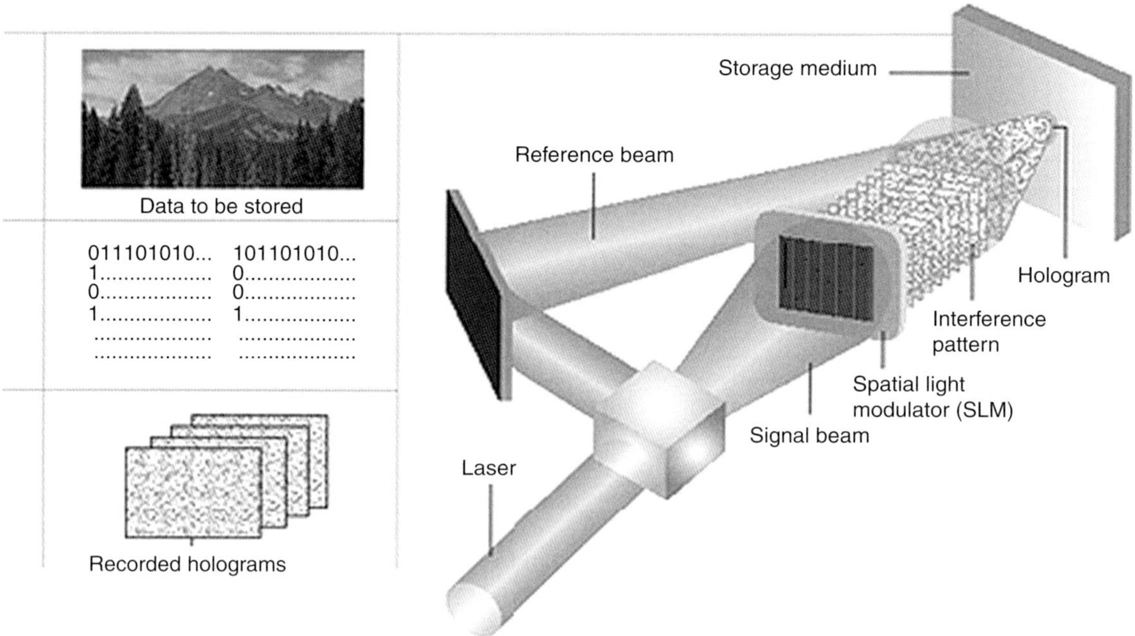

Figure 3 Recording data in a holographic storage system.

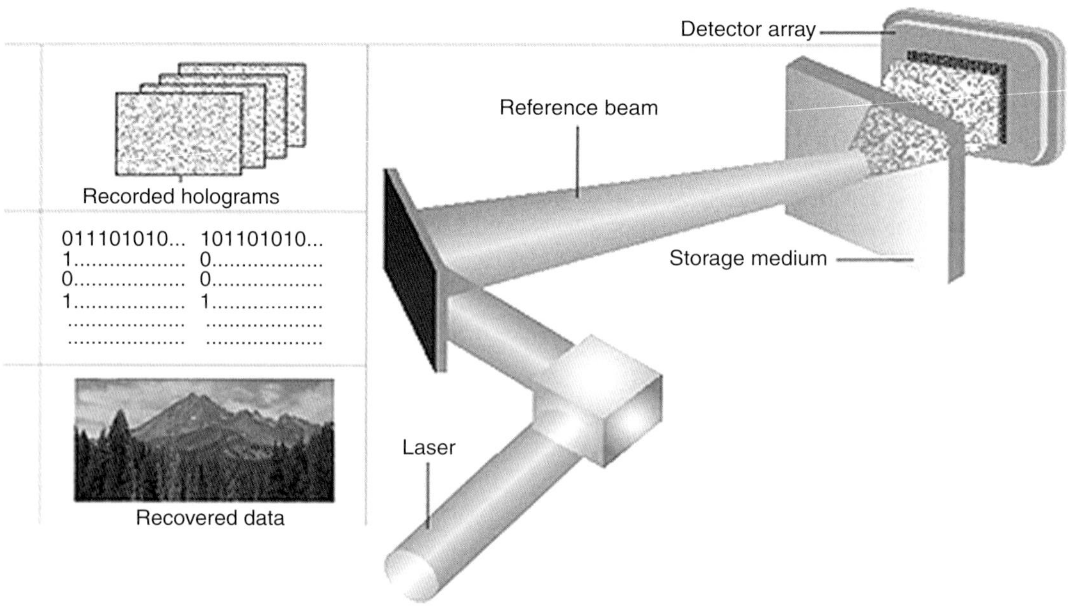

Figure 4 Reading out data in a holographic storage system.

The angular selectivity, that is, the spacing between the diffraction maximum and the first null is given by **equation 5**:

$$\Delta\theta = \frac{\lambda}{L} \frac{\cos\theta_s}{\sin(\theta_r + \theta_s)} \tag{5}$$

where L is the media thickness, θ_r and θ_s are reference and signal angles, respectively, and λ is of course the wavelength of light. Data pages are placed at the 'null' positions of previously recorded holograms allowing data recovery with a high signal-to-noise ratio (SNR).

Bragg based	Momentum based	Correlation based
• Angle • Wavelength • Phase code • Peristrophic • Shift • Correlation	• Fractal • Peristrophic • Aperture • Polytopic	• Correlation • Correlation distance

Figure 5 Summary of multiplexing options.

Wavelength multiplexing works in a similar manner – the Bragg mismatch is defined by a change in wavelength needed to reach the first null of the selectivity curve. The wavelength selectivity for a reflection hologram is given in **equation 6**:

$$(\Delta\lambda)_{Bragg} = \frac{\lambda^2 \cos\theta_s}{2L\sin^2\frac{1}{2}(\theta_f + \theta_s)} \qquad (6)$$

Phase-code multiplexing utilizes a set of reference beams with unique phases and varying reference angles spaced by Bragg selectivity. Accurate readout requires simultaneously replicating the phases and the Bragg conditions for all of the plane wave components. The number of holograms that can be stored is the same as an angle-multiplexed system with the same media thickness, wavelength, and geometry. The performance is determined by the orthogonality of the set of phases and, most importantly, how well the phase modulation is repeated from recording to readout. Theoretically, phase code multiplexing should have excellent cross-talk performance, but practically, the difficulty of accurately phase modulating the optical beams makes this technique difficult to implement.

Shift multiplexing uses a focused beam, such as a spherical or cylindrical wave, made up of a spectrum of plane waves. By moving the focal point relative to the media, the media locally sees a different part of the reference wave front or different part of the plane wave spectrum. This change in the angles of incidence will have the same Bragg selectivity as angle multiplexing; however, the overall selectivity also has an additional term, namely the uncertainty in the focus spot's location. Thus, the shift selectivity of a plane wave signal beam and a spherical reference beam in the Bragg direction (the plane of intersection defined by the center rays of the reference and data beams) is given in **equation 7**:

$$\Delta x = \frac{\lambda z_0}{L\tan\theta_s} + \frac{\lambda}{2(N.A.)} \qquad (7)$$

where L is the thickness of the media, λ the wavelength, z_o the distance from the center of the material to the focus point of the reference beam, θs the angle of incidence of the signal wave measured from media normal, and N.A. the numerical aperture of the reference beam.

Using a reference comprised of a finite number of plane waves separated by their Bragg spacing also results in shift selectivity [58].

Peristrophic multiplexing involves rotating the media or beams relative to each other to multiplex holograms. Peristrophic multiplexing can be momentum based or Bragg based, depending on the material thickness and geometry. For most media thicknesses, peristrophic multiplexing is dominated by Bragg selectivity. For transmission geometry, the Bragg selectivity can be expressed as the required rotation about the hologram center is given by **equation 8**:

$$\Delta\theta = \left\{\frac{2\pi}{L}\left[\frac{\cos\theta_r}{\sin\theta_r(\sin\theta_s + \sin\theta_r)}\right]\right\}^{1/2} \qquad (8)$$

where the reference plane wave is incident at θ_r measured from media normal, the center pixel of the image is incident at θ_s measured from media normal, λ is the wavelength of light, and L is the material thickness.

8.2.4.2 Momentum-based techniques

Diffraction of a hologram can be detected as long as the deviation of the angle of incidence or the position of the reference beam is within the Bragg selectivity function. However, accompanying any deviation is a change in the emergence angle of the diffracted hologram due to conservation of momentum. At a Fourier or image plane, this change in the reconstruction direction leads to a shift off the optical axis of the data path. This shift allows multiplexing of holograms at spacings less than the inherent Bragg selectivity. Momentum conservation is used to intentionally filter out the unwanted reconstructions and prevent them from reaching the detecting camera. This approach is fundamentally different from Bragg multiplexing, where, in the readout of one hologram, the diffraction from other holograms is minimized. Here, the unwanted reconstructions can have similar diffraction efficiently as the desired hologram but are blocked from detection. The most important momentum-based method is polytopic multiplexing [22]. The other methods that can be momentum based are peristrophic [57], fractal [23], and aperture multiplexing [24].

Polytopic ('many places') multiplexing is shown in **Figure 6**. In traditional angle multiplexing, the books of holograms must be spatially separated (**Figure 6(a)**). (The area of the book is much larger than the signal beam waist as the beam expands in the media and the reference beam sweeps in angle.) **Figure 6(b)** shows the polytopic-angle approach where the books (recorded by the signal and reference beams shown) overlap in the media.

Because the waists do not overlap, the reconstructions can be filtered out by placing an aperture at the beam waist. As the data beam is focused through the

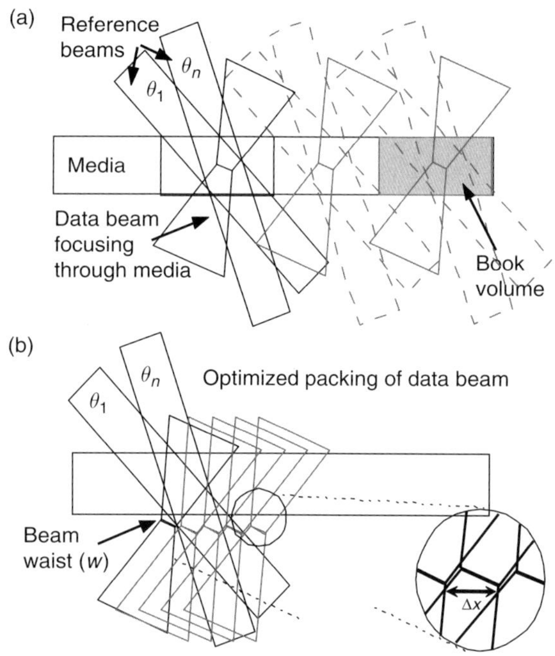

Figure 6 (a) Traditional angle multiplexing without book overlap; (b) polytopic overlap of books but without overlapping the beam waists.

media, a collimated reference beam at a chosen angle interferes with the data beam inside the media causing the interference pattern to be recorded. Changing the angle of the reference allows for a book to be recorded at that single location. Neighboring books of holograms are subsequently recorded by simply moving the media by an amount Δx given by **equation 9**:

$$\Delta x = d\cos(\theta) + d\sin(\theta)\tan(\theta + \alpha)(2.14) \qquad (9)$$

where d is the Fourier plane aperture size, θ the data beam center angle, and $\alpha = \sin^{-1}$ (N.A. of the FT lens). Moving by Δx is required to ensure that the Fourier planes of neighboring holograms do not overlap. Upon readout with the reference beam, the desired page and several of its neighbors are simultaneously reconstructed. However, the physical aperture blocks the holograms from neighboring undesired books as they are reconstructed, and only the page from the desired book is allowed to propagate to the camera and be detected. This allows the books to be separated by the Fourier plane width of the data beam, as opposed to the maximum width of the beam at the surface of the media. This can similarly be done for image plane holograms as well with the aperture located at an image plane.

The advantage of polytopic multiplexing over angle multiplexing is that it effectively uses thick media. In angle multiplexing, the spacing of the books was determined by the media thickness, set far apart enough to avoid the overlap of adjacent books, thereby limiting the achievable density. With polytopic multiplexing, the book pitch can be independent of the media thickness, thus enabling high densities.

Other examples of momentum multiplexing are fractal and shift multiplexing. In the fractal approach, different combinations of the components of the data page and the reference beam are exploited to superimpose the holograms. For shift multiplexing, if the media is moved in a plane normal to that governed by Bragg selectivity, the reconstruction is offset and can be filtered as in the polytopic case. This approach is called aperture multiplexing.

8.2.4.3 Correlation-based techniques

Correlation multiplexing uses a complex reference beam that varies in phase and amplitude across its wave front. By optimizing the structure of the wave

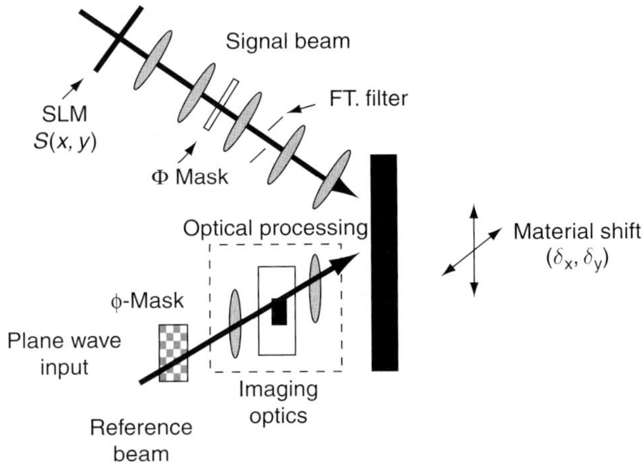

Figure 7 Configuration to introduce complex reference structure.

front, translations of the media relative to the plane of the recording beams can result in a shift selectivity that is higher than that using Bragg selectivity alone. A reference beam with the appropriate phase and amplitude structure can be achieved by passing the light through lenslet arrays, multimode fibers, or more controlled references such as a filtered phase mask.

An implementation of correlation multiplexing is shown in **Figure 7**. The geometry shown is Fourier plane recording and the data have been imaged onto a phase mask to make the Fourier transform of the data more spatially homogeneous in intensity for better recording. The reference beam originates as a plane wave and traverses through a phase mask. This phase mask can have random 2π phase or discrete values. Masks with a change (or ramp) in pixel (feature) size typically have good correlation properties. This reference beam in this Fourier transformed and the zeroth diffraction order is blocked. This high-pass filter version is then Fourier transformed back into the media as the reference. This reference beam has very sharp amplitude and phase features that provide the correlation function. In this case, a very small change in r, a few micrometer, will dramatically reduce the diffracted signal.

The selectivity is determined by the nature of the reference beam and is independent of the thickness of the media. Importantly, the detailed nature of the reference field dramatically affects the diffractive properties of the recorded image. **Figure 8** illustrates the thickness independence mentioned above. It shows the hologram shift selectivity curve (diffracted energy versus position of the media relative to the beams) in a 4 mm sample of $LiNbO_3$ and a 0.7 μm sample of holographic thermoplastic. Although the thickness difference is greater than 1000:1, the selectivity observed is essentially identical. While the selectivity is independent of the media thickness, the noise buildup is not. A thicker media will have higher capacity because the Bragg selectivity suppresses the cross-talk noise (noise from other holograms). The noise will increase as the number of superimposed holograms in the volume increases.

8.2.4.4 Combinations of multiplexing methods

Since many of the methods use reference beams of plane waves, they can be combined to increase the density. Optimum combinations include angle and polytopic multiplexing, although fractal or peristrophic multiplexing methods could also be added.[25] Shift and aperture multiplexing with a spherical reference beam are very easy to combine as well. Two or more multiplexing techniques are often used in combination to access all of the available angular bandwidth of a recording medium, which yields the maximum achievable densities.

8.3 Implementation of Holographic Storage Systems

Holographic storage was extensively researched in the 1960s and 1970s [26–34]. However, significant technical challenges, such as the development of a robust optical architecture, the availability of appropriate optical components, and the lack of suitable recording

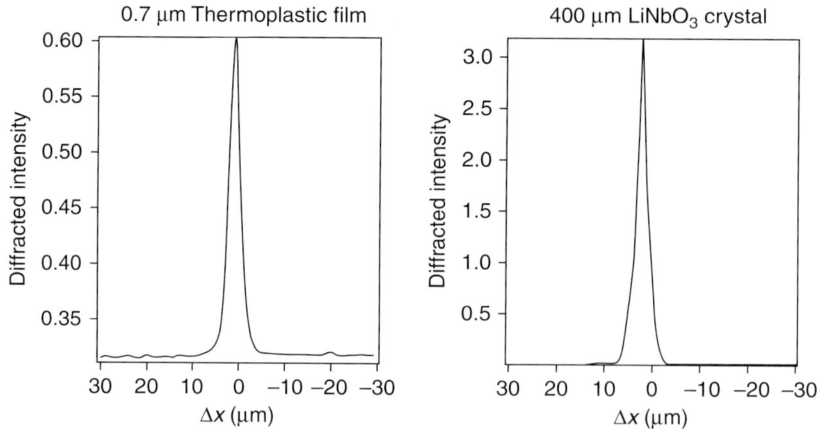

Figure 8 Hologram shift selectivity for two different thickness media. Structure drives selectivity not media thickness.

materials, hampered the progress of the technology. Recent years have seen a resurgence of activity in holographic storage with significant progress.

In the mid-1990s, the Defense Advanced Research Program Agency (DARPA), formed a consortium of companies and universities in the United States, led by IBM and Stanford University, to develop high-performance holographic storage systems [35–38]. The goal of the consortium was to demonstrated high density and transfer rates by developing the needed technology and components such as custom high-speed cameras and SLMs. In addition, development on data channel modulation and detection schemes were undertaken. Two types of storage systems, one based on a large crystal without mechanical motion as a recording medium and the other system using a spinning disk, were developed. Recording materials primarily based on photorefractive materials and on the then-available photopolymer films used for display holograms were initially used [39, 40]. These materials allowed basic demonstrations of capabilities but did not fulfill the requirements for a commercial product. The consortium grew to include Polaroid (and later Aprilis, a company spun out of Polaroid), who started developing photopolymers designed for HDS [41, 42]. This, together with the efforts of the other members, led to several significant achievements. Stanford demonstrated high data transfers from a spinning disk – up to 1Gbyte s^{-1} [36], whereas IBM demonstrated storage densities of 250 Gbytes per square inch in very thick $LiNbO_3$ crystals [43].

Also in the mid-1990s, an effort in holographic storage began at Bell Laboratories and Lucent Technologies. Aimed at developing suitable recording media in conjunction with a practically implementable drive, the program targeted systems that would lead to commercially feasible products. By designing and developing both the media and drive in concert, several important technical milestones were reached: a process (Zerowave®) allowing for optically flat recording media to be fabricated using standard optical media manufacturing methods [44]; the invention of a new type of photopolymer recording material for holography (Tapestry®, two-chemistry materials) enabling both high performance and robust lifetime characteristics; and drive designs that improved SNR and simplified the servo techniques over previous systems. By 1998, data densities of 49 Gbytes per square inch were successfully achieved in the two-chemistry materials [45]. With these technology breakthroughs, Lucent Technologies spun out an independent company, InPhase Technologies®, in 2000 to commercialize holographic storage systems.

In Phase primarily has focused on the development of a storage system suited for archival applications in the professional market. The drive system was designed for ease of implementation and operation, minimizing the use of custom-developed components and insuring environmental robustness. With this strategy, InPhase has demonstrated the highest storage density to date (663 Gbytes per square inch) of any removable storage technology, media interchange between drives for the first time, and operation over a temperature range of 40°C. In addition, InPhase has partnered with some of the leading companies and organizations in the world of optical storage to productize its system – Bayer Material Science, Hitachi Maxell Corporation, Nichia, Sanyo, Hitachi, Lite-On,

Displaytech, Cypress, University of California at San Diego, and Carnegie Mellon University.

Also in the 2000s, companies in Japan and Korea restarted research into holographic storage drives and media. In addition, several consortiums, sponsored by the Japanese government, were formed. Companies such as Sony and the start-up Optware focused their efforts on an architecture called coaxial or collinear holography to leverage CD and DVD technologies. Recently, Sony demonstrated a storage density of 415 Gbytes per square inch [46] using collinear geometries. In addition, Sony has placed some of their efforts into bitwise holographic storage, developing methods to replicate media for read-only memories (ROMs). These ROM replication efforts will be covered in Chapter 15. Most recently, Lucky Goldstar has used the InPhase architecture to design and build a miniature optical head [25] and Daewoo has used the same InPhase architecture to achieve high-speed video recording [47] and playback.

The next sections will summarize the advances on several important fronts that have enabled the implementation of practical, commercially feasible holographic storage systems. The focus will be on the applications of the advances to the InPhase architecture.

8.3.1 Robust Optical Architectures

Of the various multiplexing methods, angle multiplexing is one of the simplest to implement, requiring only changes in the incident angle of the reference beam. In addition, angle multiplexing enables high-speed recording and readout rates, allows routine interchange of media, and exhibits low sensitivity to environmental changes. However, geometrical factors ultimately limit the storage densities achievable through this approach alone. For example, angle multiplexing allows only ~140 Gbyte of storage capacity in a 130-mm-diameter disk recorded with 407 nm laser light [48]. **Figure 9** illustrates the saturation of storage capacity as a function of media thickness. The storage capacity saturates because, as the media thickness increases, sets of angle-multiplexed holograms must be spaced further apart to avoid overlap within the media. (If the sets overlap, a reference beam targeted to read a hologram within one book will also diffract constructively with a hologram in the second book recorded with the same angle.)

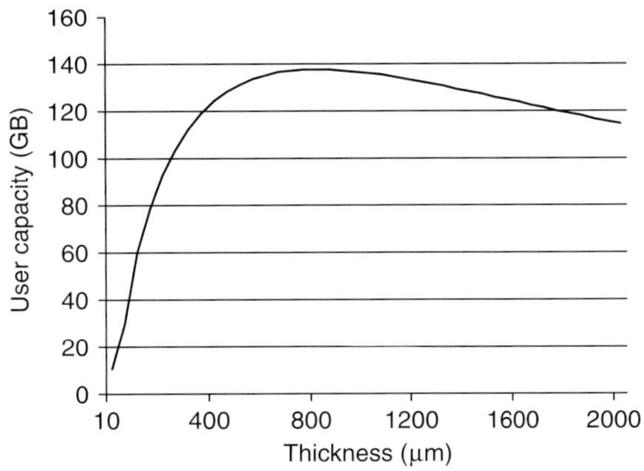

Figure 9 130 mm disk using a 405 nm blue laser, an object beam N.A. of ~0.6, a 1000 × 1000 pixel data page, and a combined data rate of 0.488 to allow for ECC and modulation coding.

Combining polytopic multiplexing with angle multiplexing eliminates these undesired reconstructions. If the books are spaced by at least one signal beam waist, a filter can be used to block the interfering reconstructions. Polytopic multiplexing breaks the dependence of book pitch and media thickness and adds a tenfold increase in the achievable storage capacity to strict angle multiplexing.

Angle-polytopic multiplexed is best implemented in a quasi-phase conjugate geometry. Unlike a traditional transmission geometry, this layout allows the optics to be placed on one side of the media and eliminates the need for complicated lenses to correct for aberrations. The basic phase conjugate architecture of InPhase's system[49,50] is shown in **Figures 10** and **11**.

8.3.2 Components for Holographic Storage

The practical implementation of holographic storage also requires the availability of optical components with appropriate characteristics, such as a tunable, coherent laser sources, large-pixel-count SLMs, and high-speed cameras.

8.3.2.1 Laser sources

Holographic storage uses the entire volume of a medium and, therefore, is sensitive to its dimensional changes. Methods to counter the effects of temperature changes are required for operation of a holographic drive. The thermal-induced dimensional changes can be compensated by changing the angle and wavelength of the reference beam used to readout the hologram. Therefore, tunable laser sources are highly advantageous for holographic storage. To date, the best candidate is a coated gallium nitride laser diode that

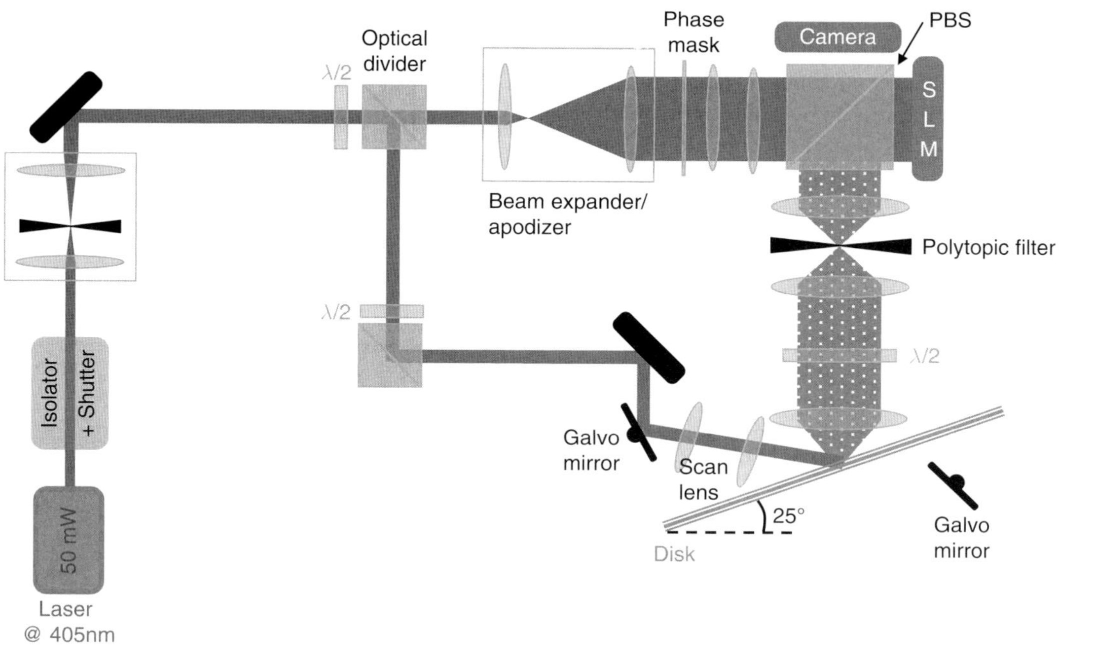

Figure 10 Diagram of light path during write operation.

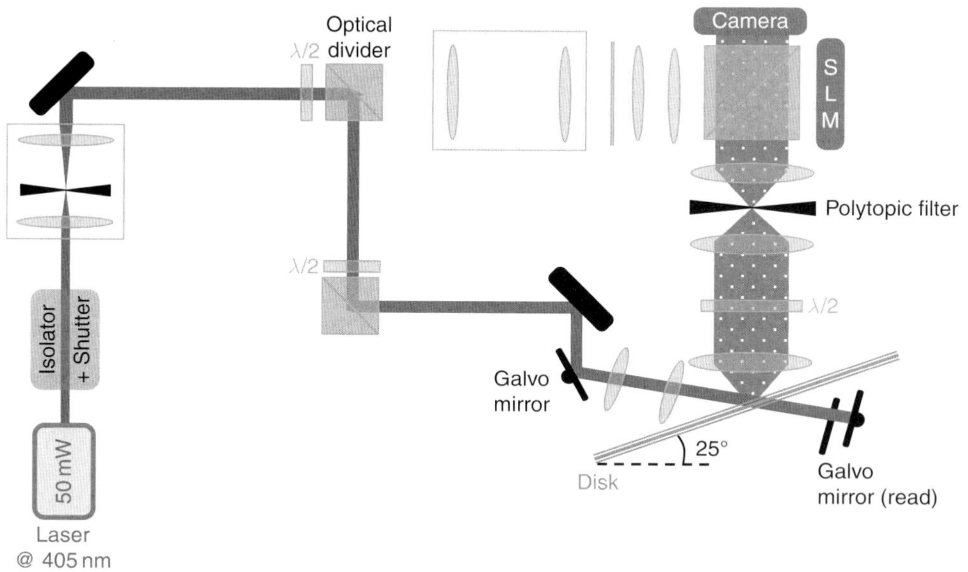

Figure 11 Diagram of light path during read operation.

forms the heart of a small, simple, stable, and relatively high-power, tunable external cavity. Using a transmissive diffractive grating as the output coupler of the laser cavity, the wavelength of the diode is stabilized for holographic recording and can be changed by rotating the grating for readout. In addition, because the grating is transmissive (rather than the typically used reflective grating) the pointing of the laser does not change making the laser compact and robust.

The tunable external cavity laser diode (ECLD) (**Figure 12**) consists of a blue laser diode [51], optics, a stepper motors, sensors, and electronics on a small board. The laser cavity consists of an antireflection coated laser diode, a lens to collimate the laser light,

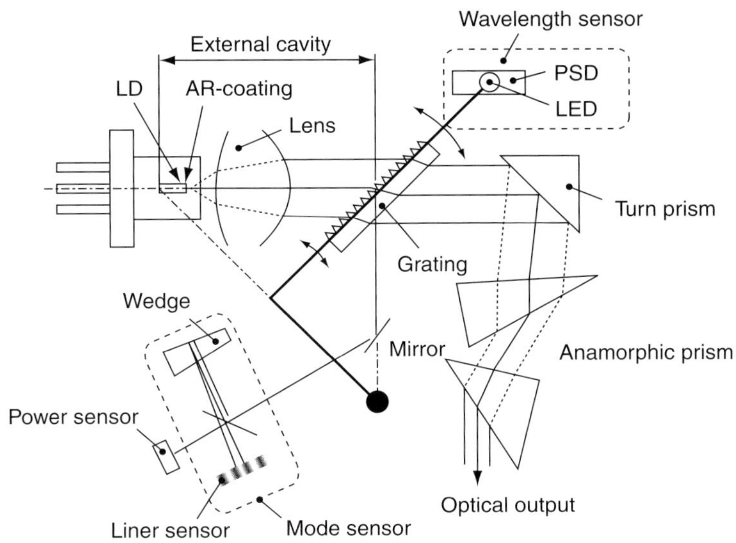

Figure 12 Diagram of ECLD laser system.

and a transmission grating that creates the external cavity and acts as the output coupler. A stepper motor adjusts the ECLD wavelength by rotating the grating. After the grating, anamorphic prisms convert the elliptical beam from the cavity into a symmetrical beam.

Figure 13 shows the performance of the laser system. The maximum output power is 53 mW at 405.7 nm. The minimum wavelength, where the ECLD remains in a single longitudinal mode, is 400.3 nm and the maximum wavelength is 410 nm. The tuning range of the ECLD is 9.7 nm. If we require the minimum output power to be 35 mW, then the tuning range is 8 nm. Because the central wavelength of the ECLDs may vary (from unit to unit) by ±1 nm, the tuning range is estimated to be 6 nm to have a reasonable yield.

8.3.2.2 Spatial light modulators

SLM is a critical interface between the data to be stored and the optical system, encoding the recording light with the data to be stored [52]. High-performance optical modulators have been developed for large-volume use in the projection display market and can be applied for holography. For holographic storage, SLM's must be capable of both amplitude and phase modulation, and rapid modulation and data transfer rates characterized by kilohertz frame rates. (Phase modulation is required of an SLM to minimize the DC component in the Fourier

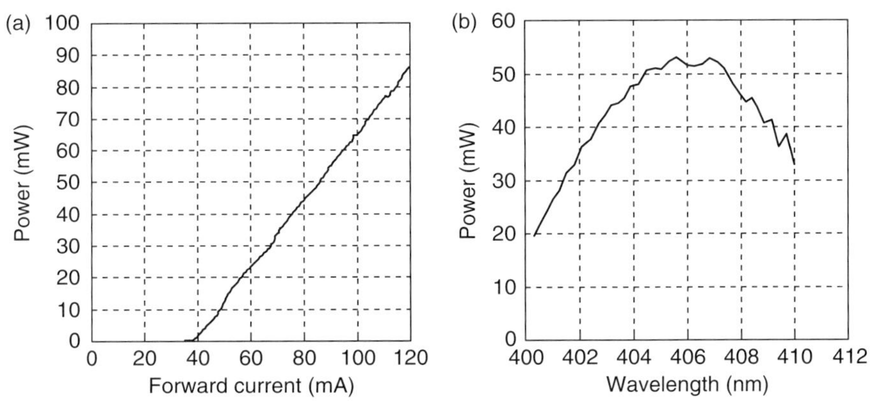

Figure 13 (a) Forward current versus output power of the ECLD; (b) wavelength versus output power of the ECLD at constant current.

Figure 14 A FLC modulator from Displaytech Corporation.

transform of the data page that results when the recording light travels through the modulator and then through a lens to the media.)

The most commonly used SLM for holographic storage are ferroelectric liquid crystal- (FLC-) based systems such as the one shown in **Figure 14**. (While nematic liquid crystals are most commonly used in SLMs for projectors, these materials are not fast enough for use in HDS systems.)

Another type of modulator is magnetooptical modulators [53]. These modulators operate on the same concept as liquid crystal device by rotating the polarization of the light. They have the potential to be very fast as the material response times can be in the ~10 ns range. However, currently these devices have substantial absorption in blue wavelengths and are in very early stages of development.

Micro electromechanical (MEM) technology for SLMs for data storage is another promising and well-developed approach [54,55]. Amplitude modulation is achieved by a tilting pixel structure or by forming a switchable diffraction grating and using a low-pass filter later in the system to block the unwanted light. Phase modulation can be achieved by moving the grating or pixel structure along the optical axis of the optical system. Amplitude and phase modulation are achieved by combining these motions. While MEM devices allow for very high speed, the requirement for hermetic packaging often leads to unwanted birefringence. In addition, some MEM devices require awkward illumination that introduces complications to the design of the drive.

8.3.2.3 Cameras

CMOS sensor arrays are the most popular choice for cameras for HDS systems. With their relatively low manufacturing and packaging costs, low power consumption, low heat generation, and ability to yield readout of sub-arrays, CMOS sensors are considered the best choice for practical holographic storage systems.

A six-transistor cell configuration for the CMOS cell is used in HDS. This pixel architecture[56] allows the readout of one data page while simultaneously capturing another data page. The sensitivity of these systems is often diminished by the fabrication process that installs an absorbing protective layer. Micro-lenses fabricated on silicon can improve the fill factor of the device.

8.3.2.4 Recording media

One of the critical challenges in the development of holographic storage systems has been the availability of suitable recording media. Parameters such as high dynamic range, high photosensitivity, dimensional stability, high optical quality, environmental robustness, readout durability, and manufacturability are required to generate a fully functional and reliable holographic recording medium.

Photopolymer recording materials are the leading candidates for holographic applications. In general, photopolymers consist of a photoreactive system dispersed within a host. During holographic recording, the optical interference pattern initiates a pattern of polymerization in the photoreactive system – polymerization is induced in the light intensity maxima of the interference pattern, whereas no polymerization occurs in the nulls. This patterned polymerization sets up a concentration gradient in the unpolymerized species. These species diffuse from the nulls to the maxima of the interference pattern to equalize their concentration in the recording area, creating a refractive index modulation proportional to the difference between the refractive indices of the photoreactive component and the host material. **Figure 15** shows the steps in the formation of a hologram in a photopolymer system.

The central mechanism of hologram formation in photopolymers, namely mass transport, imparts one of the most important advantages over many other types of recording material – nonvolatility. Instead of relying on 'local' events such as absorption changes or orientational rearrangements upon recording to provide a refractive index modulation, photopolymers exploit the diffusion of species with contrasting refractive index to produce a material that is compositionally different than that before

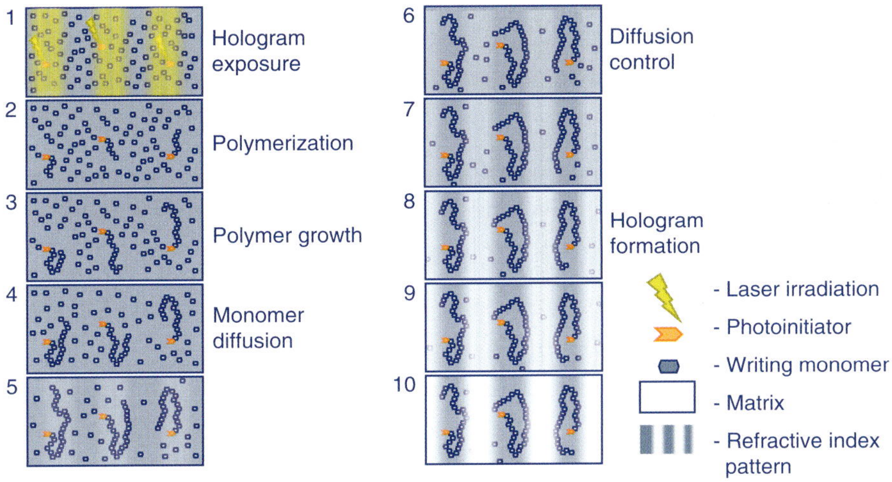

Figure 15 Hologram formation in photopolymer materials.

recording. This phenomenon is fundamentally different from those occurring in other materials where subsequent recording or readout can reverse the absorption or orientational changes and erase the recorded holograms. The mass transport of photopolymers insures the permanence of the holograms.

Another distinguishing feature of photopolymers is the amplification of the recording process. The recording light initiates a polymerization reaction in the material as the first step toward establishing a refractive index modulation. Because polymerization reactions are chain reactions – each photon creates a reactive system that propagates through the material system until a termination step – photopolymers can exhibit significantly higher sensitivities than other materials that do not benefit from this amplification of the recording light.

The great flexibility in the design of photopolymer systems also contributes to its superior performance. Because photopolymers are multicomponent systems, their performance can be tuned through judicious choices of the photoreactive system and host matrix. For example, choosing a photoreactive system with a significantly different refractive index than that of the matrix optimizes the dynamic range of the material. Choosing a host matrix with the appropriate glass transition temperature and a high-reactivity photoreactive system further increases the sensitivity of the medium. The three-dimensional network of the host matrix can be adjusted to preserve the archival life of the material. Addition of inhibitors to the photoreactive system can enhance the shelf (unrecorded) life of the material. Because of the great latitude in their design, photopolymers are an extremely powerful class of holographic recording materials.

There have been two main approaches to designing photopolymer systems for HDS systems. One approach, pioneered by Polaroid and later Aprilis, Inc., uses a low-shrinkage chemistry known as a cationic ring-opening polymerization (CROP) to control the dimensional changes that occur during holographic recording [57]. These materials consist of a photoacid generator, sensitizing dye, CROP monomers and binder [58]. The CROP monomers typically have a functionality of greater than 2 and are based on cyclohexeneoxides. In this strategy, a pre-imaging exposure is used to increase the viscosity of the material to prepare for holographic recording. The CROP imaging chemistry is also tuned for high photosensitivity.

The other approach, introduced by Bell Laboratories, Lucent Technologies, and InPhase Technologies is known as a two-chemistry approach [59]. The material is composed of two independently polymerizable systems – one system reacts to form a three-dimensional cross-linked polymer matrix in the presence of the second photopolymerizable monomer system. The photopolymerizable monomer is the imaging component, as it reacts during holographic recording. This strategy produces high-performance recording media as a result of several important attributes. The matrix is formed *in situ*, which allows thick and optically flat formats to be formed. The three-dimensional cross-linked nature of the polymer matrix creates a mechanically robust and stable medium. The matrix and photopolymerizable monomer system are chosen to be compatible in order to yield media with low levels of light scatter. The independence of the matrix and monomer systems avoids cross-reactions between the two, which can dilute the refractive index contrast due to premature polymerization of the imaging component.

Both types of materials have been used to demonstrate digital HDS [60,61]. Recent work in the CROP materials has been focused on fingerprint imaging and correlators rather than data storage applications [62]. Most recently, the InPhase two-chemistry media have demonstrated full-disk recording at a capacity of over 300 GB shown in **Figure 16**.

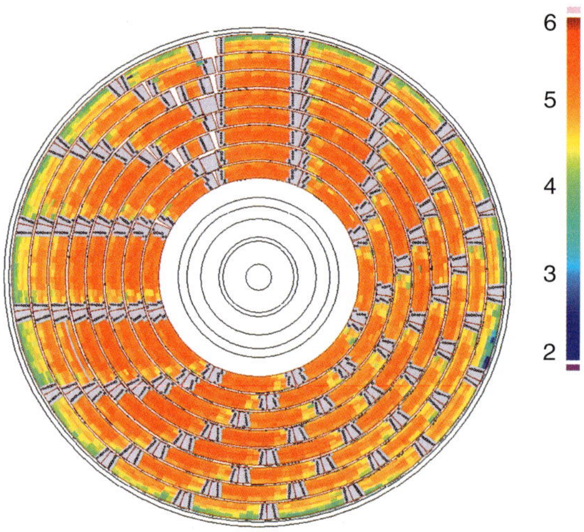

Figure 16 Map of the SNR of a fully recorded 130 mm diameter InPhase Technologies disk. The average SNR is ~4.9. (An SNR greater than 2 insures error-free recovery.)

8.4 Current Status

InPhase Technologies is introducing a family of archival storage systems delivering high storage capacities and rapid data transfer rates. The first system of the family, the 300r drive/write-once–read-many media system stores 300 Gbytes per 130-mm-diameter disk and has 160 Mbit s^{-1} data transfer rate (**Figure 17**). Later generations will have capacities up to 1.6 Tbytes per disk, with data rates to 960 Mbit s^{-1}. The storage drive fully integrates the subsystems of a holographic read/write head, auto load/unload mechanics, servo system, data channel, and electronics. Evaluation units of the product have already been tested by several customers, including Turner Network Television, which, in 2005, became the world's first network to broadcast content originating from holographic storage.

8.5 Future Developments

The applicability of holographic storage can be across a wide range of markets. With its first expected commercial introduction in the professional archival market, the technology will become further refined and find broad use. For example, improvements in performance, size, and cost will lead to consumer applications such as content distribution and home archiving. Other possibilities include areas such as content addressable memories and high-performance, low-energy fixed-

Figure 17 A test version of InPhase's holographic drive system.

disk applications. Holographic storage, with its immense performance capabilities, stands poised to play an important role in the data storage landscape.

References

1. Shimada K, Ishii T, and Ide-Hitache T (2009) High density recording using monocular architecture for 500GB consumer system. *Optical Data Storage Conference*, Florida.
2. Gabor D (1948) A new microscope principle. *Nature* 161: 777.
3. Klein WR and Cook BD (1967) Unified approach to ultrasonic light diffraction. *IEEE Transactions on Sonics and Ultrasonics* SU-14: 123.
4. Moharam MG and Young L (1978) Criterion for Bragg and Raman-Nath diffraction regimes. *Applied Optics* 17: 1757.
5. Kogelnik H (1969) Coupled wave theory for thick hologram gratings. *Bell System Technical Journal* 48: 2909.
6. van Heerden PJ (1963) Theory of optical information storage in solids. *Applied Optics* 2: 393.
7. Mok FH (1993) Angle-multiplexed storage of 5000 holograms in lithium niobate. *Optics Letters* 18: 915.
8. Staebler DL, Burke WJ, Phillips W, and Amodei JJ (1975) Multiple storage and erasure of fixed holograms in Fe-doped $LiNbO_3$. *Applied Physics Letters* 26: 182.
9. Heanue JF, Bashaw MC, and Hesselink L (1994) Volume holographic storage and retrieval of digital data. *Science* 265: 749.
10. Rakuljic GA, Levya V, and Yariv A (1992) Optical data storage using orthogonal wavelength multiplexed volume holograms. *Optics Letters* 17: 1471.
11. Campbell S, Yi X, and Yeh P (1994) Hybrid sparse-wavelength angle-multiplexed optical data storage system. *Optics Letters* 19: 2161.
12. Morozov VN (1977) Theory of holograms formed using coded reference beams. *Soviet Journal of Quantum Electronics* 7: 961.
13. Denz C, Pauliat G, Roosen G, and Tschudi T (1991) Volume hologram multiplexing using a deterministic phase encoding method. *Optics Communications* 85: 171.
14. Heanue JF, Bashaw MC, and Hesselink L (1994) Recall of linear combinations of stored data pages based on phase-code multiplexing in volume holography. *Optics Letters* 19: 1079.
15. Denz C, Pauliat G, Roosen G, and Tschudi T (1992) Potentialities and limitations of hologram multiplexing by using the phase-encoding technique. *Applied Optics* 31: 5700.
16. Curtis K, Pu A, and Psaltis D (1994) Method for holographic storage using peristrophic multiplexing. *Optics Letters* 19: 993.
17. Psaltis D, Pu A, Levene M, Curtis K, and Barbastathis G (1995) Holographic storage using shift multiplexing. *Optics Letters* 20: 782.
18. Yakimovich AP (1979) Selective properties of 3-D holographic gratings using spherical wave fronts. *Optics and Spectroscopy* 47: 530.
19. Curtis K and Wilson WL (1998) Phase Correlation Multiplex Holography. US Patent 5,719,691.
20. Curtis K (2007) Method and Apparatus for a Phase Correlated Holographic Drive. US Patent 6,909,529.
21. Curtis K, Wilson WL, Tackitt M, Hill AJ, and Campbellt S (2000) High density, high performance data storage via volume holography: The Lucent technologies hardware platform. In: Coufal HJ, Psaltis D, and Sincerbox G (eds.) *Holographic Data Storage*, pp. 359–368. New York: Springer.
22. Anderson K and Curtis K (2004) Polytopic multiplexing. *Optics Letters* 29: 1402.
23. Lee H, Gu XG, and Psaltis D (1989) Volume holographic interconnections with maximal capacity and minimal cross-talk. *Journal of Applied Physics* 65: 2191.
24. Curtis K and Wilson WL (1999) Multiplex Holography. US Patent 5,892,601.
25. Jang JS, Shin DH, and Park YS (2000) Holographic data storage by combined use of peristrophic, angular, and spatial multiplexing. *Optical Engineering* 31: 2975.
26. van Heerden PJ (1963) Theory of optical information storage in solids. *Applied Optics* 2: 393.
27. Staebler DL, Burke WJ, Phillips W, and Amodei JJ (1975) Multiple storage and erasure of fixed holograms in Fe-doped $LiNbO_3$. *Applied Physics Letters* 26: 182.
28. Fleisher H, Pengelly P, Reynolds J, Schools R, and Sincerbox G (1965) An optically accessed memory using Lippmann process for information storage. In: Tippett, J (ed.) *Optical and Electro-Optical Information Processing*, pp. 1–30. Cambridge, MA: MIT Press.
29. Leith EN, Kozma A, Upatnieks J, Marks J, and Massey N (1966) Holographic data storage in three dimensional media. *Applied Optics* 5: 1303.
30. Anderson LK (1968) Holographic optical memory for bulk data storage. *Bell Laboratories Record* 45: 319.
31. d'Auria L, Huignard JP, Slezak C, and Spitz E (1974) Experimental holographic read–write holographic storage system. *Applied Optics* 13: 808.
32. Nishida N, Sakaguchi M, and Saito F (1973) Holographic coding plate: A new application of holographic memory. *Applied Optics* 12: 1663.
33. Tsunoda Y, Tatsuno K, Kataoka K, and Takeda Y (1976) Holographic videodisc: An alternative approach to optical videodisks. *Applied Optics* 15: 1398.
34. Kubota K, Ono Y, Kondo M, Sugama S, Nishida N, and Sakaguchi M (1980). Holographic disk with high data transfer rate: Its application to an audio response memory. *Applied Optics* 19: 944.
35. Heanue JF, Bashaw MC, and Hesselink L (1994) Volume holographic storage and retrieval of digital data. *Science* 265: 749.
36. Orlov SS, Phillips W, Bjornson E, and Hesselink L (2000) High transfer rate (1 Gbit/sec) high-capacity holographic disk digital data storage system. *Conference on Lasers and Electro-Optics (CLEO 2000).* Technical Digest. Post-Conference Edition. TOPS 39:7, San Francisco, CA, USA, May.
37. Hesselink L (2000) Digital holographic demonstration systems by Stanford University and Siros technologies. In: Coufal HJ, Psaltis D, and Sincerbox G (eds.) *Holographic Data Storage*, pp. 383–397. New York: Springer.
38. Jefferson CM, Burr GW, and Hoffnagle JA (2000) IBM holographic digital data storage test platforms. In: Coufal HJ, Psaltis D, and Sincerbox G (eds.) *Holographic Data Storage*, pp. 369–382. New York: Springer.
39. Redfield S (2000) Tamarack optical head holographic storage. In: Coufal HJ, Psaltis D, and Sincerbox G (eds.) *Holographic Data Storage*, pp. 348–358. New York: Springer.
40. Ma J, Chang T, Choi S, and Hong J (2000) Digital holographic data storage with fast access. In: Coufal HJ, Psaltis D, and Sincerbox G (eds.) *Holographic Data Storage*, pp. 409–418. New York: Springer.
41. Waldman DA, Ingwall RT, Dhal PK, *et al.* (1996) Cationic ring opening photo-polymerization methods for holography. *Proceedings of the SPIE* 2689: 127.

42. Ingwall RT and Waldman D (2000) Photopolymer systems. In: Coufal HJ, Psaltis D, and Sincerbox G (eds.) *Holographic Data Storage*, pp. 171–198. New York: Springer.
43. Burr G, Jefferson M, Coufal H, *et al.* (2001) Volume holographic data storage at an areal density of 250 gigapixels/in.^{2} *Optics Letters* 26: 444.
44. Campbell S, Harris AL, and Levinos NJ (1999) Method for Fabricating a Multilayer Optical Article. US Patent 5,932,045.
45. Wilson WL, Curtis K, Tackitt M, *et al.* (1998) High density, high performance data storage via volume holography. *International Photonics Conference '98*, Taiwan.
46. Tanaka K, Hara M, Tokuyama K, *et al.* (2009) 415 Gbit/in^2 recording in coaxial holographic storage using low-density parity-check codes. *Optical Data Storage Conference*, Florida.
47. Hwang E, Pilsang Y, Nakyoung K, *et al.* (2006) Real-time video demonstration of holographic disk data storage system. *Proceedings of the SPIE* 6282: 205.
48. Li HYS and Psaltis D (1994) Three-dimensional holographic disks. *Applied Optics* 33: 3764.
49. Curtis K and Wilson WL (2006) Architecture and function of inphase's holographic drive. *Invited Talk, Asia-Pacific Data Storage Conference.* Lakeshore Hotel, Hsinchu, Taiwan, August 28–30.
50. Anderson K, Fotheringham E, Hill A, Sissom B, and Curtis K (2006) High speed holographic data storage at 500Gb/in^2. *SMPTE Motion Imaging Journal* May/June: 200–203.
51. Nakayama N, Itoh S, Ohata T, *et al.* (1993) Room-temperature continuous operation of blue-green laser-diodes. *Electronics Letters* 29: 1488.
52. Zhou G, Mok F, and Psaltis D (2000) Beam deflectors and spatial light modulators for holographic storage application. In: Coufal HJ, Psaltis D, and Sincerbox G (eds.) *Holographic Data Storage*, pp. 241–258. New York: Springer.
53. Inoue M (2007) Magneto-optic spatial light modulators for collinear holography. *International Symposium on Optical Memory.* Oct 21–25, Paper Tu-F-1, Singapore.
54. Collins DR, Sampsell JB, Hornbeck LJ, Florence JM, Penz PA, and Gately MY (1989) Deformable mirror device spatial light modulators and their applicability to optical neural networks. *Applied Optics* 28: 4900.
55. Senturia SD, Day DR, Butler MA, and Smith MC (2005) Programmable diffraction gratings and their uses in displays, spectroscopy, and communications. *Journal of Microlithography, Microfabrication, and Microsystems* 4: 041401.
56. Willems P "High Speed CMOS sensors", published in Planet Analog (http://www.planetanalog.com) and Cypress publications (http://www.cypress.com).
57. Waldman DA, Li HYS, and Horner MG (1997) Volume shrinkage in slant fringe gratings of a cationic ring-opening holographic recording material. *Journal of Imaging Science* 41: 497.
58. Waldman DA, Butler CJ, and Raguin DH (2003) CROP holographic storage media for optical data storage at greater than 100 bits/μm^2. *Proceedings of SPIE* 5216: 10.
59. Dhar L, Hale A, Katz H, Schilling M, Schnoes M, and Schilling F (1999) Recording media that exhibit high dynamic range for holographic data storage. *Optics Letters* 24: 487.
60. Shelby R, Waldman D, and Ingwall R (2000) Distortions in pixel-matched holographic data storage due to lateral dimensional change of photopolymer storage media. *Optics Letters* 25: 713.
61. Dhar L, Curtis K, Tackitt M, *et al.* (1998) Holographic storage of multiple high capacity digital data pages in thick photopolymer systems. *Optics Letters* 23: 1710.
62. Aprilis Co. Imager Products Home Page, http://www.aprilisinc.com/biometrics_imager.htm

9 Nanostructures and Surface-Enhanced Raman Spectroscopy

K M Kosuda, J M Bingham, K L Wustholz, and R P Van Duyne, Northwestern University, Evanston, IL, USA

9.1 Introduction

A major focus in the field of nanoscience is the optical properties of metallic nanostructures. Excitation of the localized surface plasmon resonance (LSPR) of metallic surfaces having nanoscale features results in local electromagnetic (EM) field enhancement. This field enhancement has dramatic effects on molecules adsorbed to the surface of plasmonic nanostructures. More than 30 years ago, it was recognized that the Raman scattering intensity of pyridine on electrochemically roughened Ag was enhanced by $\sim 10^6$ [1, 2]. This remarkable discovery led to the emergence of surface-enhanced Raman spectroscopy (SERS), a technique that proved to be a powerful tool for fundamental studies of surface adsorbates. However, the analytical utility of SERS was limited early on, owing to problems with irreproducibility and uncertainty regarding the origin of large enhancements. In the years following the discovery of SERS, intensive research focused on achieving a mechanistic understanding of SERS, which is now understood to be dominated by an EM field enhancement mechanism. Advances in nanofabrication methods that produce substrates with controlled and well-defined plasmonic properties, along with improvements in laser sources, detection schemes, and theoretical modeling provide the basis for our current understanding of SERS. In addition, the discovery of single-molecule SERS (SMSERS) in 1997 [3, 4] has generated heightened interest in both the fundamentals of the technique as well as its application to chemical analysis. SERS has developed into a sensitive analytical tool for the detection of molecules adsorbed on nanostructures, and surface modification techniques have also allowed for improved analyte selectivity. To date, SERS has been used to probe chemical problems in areas as diverse as chemistry, physics, materials science, surface science, catalysis, electrochemistry, and life sciences.

This chapter is aimed at providing a sound background for understanding the relationship between nanostructures and SERS. We first introduce the physical basis of the LSPR, with particular attention to the dielectric function of plasmonic materials as it relates to nanoparticle extinction. Experimental and computational methods of evaluating nanoparticle extinction are briefly presented. We then review various methods for the fabrication of nanostructures that support LSPRs, including chemical syntheses, laser ablation, and lithographic techniques. Factors that affect nanoparticle extinction, including nanoparticle size and shape, refractive index of the nanoparticle environment, distance dependence of the LSPR, as well as coupling among nanoparticles, are summarized with experimental examples. We then discuss a new approach to the functionalization and stabilization of nanostructures using atomic layer deposition (ALD). In particular, ultrathin Al_2O_3 films are shown to significantly protect nanoparticles from thermal annealing, laser damage, and solvent-induced structural changes.

We next focus on the fundamentals and applications of SERS. The origin of SERS enhancement is reviewed with particular emphasis on the EM mechanism. A brief discussion of enhancement factor (EF) calculation is also presented. We then describe two important consequences of the EM mechanism, the distance dependence (probed via ALD experiments) and excitation-wavelength dependence of SERS. The results of surface-enhanced Raman excitation spectroscopy experiments for resonant and nonresonant molecules reveal the important relationship between LSPR and SERS enhancement. Next, we highlight advances in the field of SMSERS. The isotopologue approach to proving SMSERS detection is presented, followed by studies aimed at elucidating the structure of SMSERS hot spots as well as evaluating the relative contributions of EM and resonance Raman enhancement to single-molecule sensitivity. We then discuss progress in the application of SERS to chemical problems, focusing specifically on *in vivo* glucose sensing, art conservation, and detection of chemical and biological warfare agents.

Finally, we turn our attention to the future of SERS. The investigation of new plasmonic materials as well as novel nanostructures promises to broaden the scope of SERS. In particular, understanding the fundamental relationships between nanostructures and their resulting optical properties is an important goal, both from a fundamental and practical standpoint. Correlated investigations of nanoparticle structure – LSPR and SERS – will ultimately enable the rational design of high-enhancing SERS substrates. We also discuss an important variation of SERS, tip-enhanced Raman spectroscopy (TERS), which combines the sensitivity and rich chemical information of SERS with the excellent spatial resolution of scanning probe microscopy. Given the advantages of complete surface generality, spatial resolution of ~10 nm, and near single-molecule sensitivity [5, 6], it is anticipated that TERS will be a powerful tool for probing chemical problems in many areas. The progress of research highlighted in this chapter points to a promising future in which SERS will have a significant impact on the field of nanoscience and technology.

9.2 Localized Surface Plasmon Resonance Spectroscopy

The LSPR is the phenomenon responsible for the brilliant colors exhibited by Ag and Au nanoparticles under illumination. These brightly colored particles have been of interest to artisans since antiquity and are responsible for the color in the Roman Lycurgus Cup [7, 8] as well as antique stained-glass windows. Scientific interest in metallic particles dates back to Michael Faraday in the nineteenth century when he examined size and color changes of Au particles [9]. To understand SERS and its capabilities, it is important to review the underlying mechanism for surface enhancement, namely the LSPR. This section provides a theoretical and experimental background of LSPR. We hope to highlight the significance of LSPR in relation to SERS and provide a sound foundation for the following SERS section in this chapter.

Surface plasmon resonance (SPR) is found in materials that have a negative real and small positive imaginary dielectric constant. In most instances, SPR occurs in metals like Ag and Au. When EM radiation is incident on these metals, it induces a collective oscillation of surface conduction electrons – a resonance. Significant attention has been focused on surface plasmons and their use for biological and chemical sensing, as well as surface-enhanced spectroscopies. In SPR spectroscopy, thin metal films (~50 nm thick) support surface-propagating plasmons, also referred to as surface plasmon polaritons (SPPs). The SPPs propagate ten to hundreds of microns in the x- and y-directions along the metal–dielectric interface and decay evanescently in the z-direction on the order of 200 nm. A variation of SPR is LSPR, which has been of particular interest since the advent of nanofabrication techniques. When surface plasmons are confined to a nanostructure much smaller than the wavelength of light, they are localized around the nanostructure with a specific frequency known as the LSPR. **Figure 1** depicts the difference between SPR and LSPR [10].

Both SPR and LSPR are sensitive to the surrounding dielectric environment and, consequently, both SPR and LSPR are used for sensing experiments. A more thorough theoretical description of these sensitivities is presented in Section 9.2.1.3. Although at present SPR spectroscopy leads LSPR in commercial development, LSPR has comparable attributes [10, 11] and is superior to SPR in many cases [10, 12, 13]. For example, SPR measurements are more sensitive to bulk changes in refractive index due to the significantly shorter EM field decay length

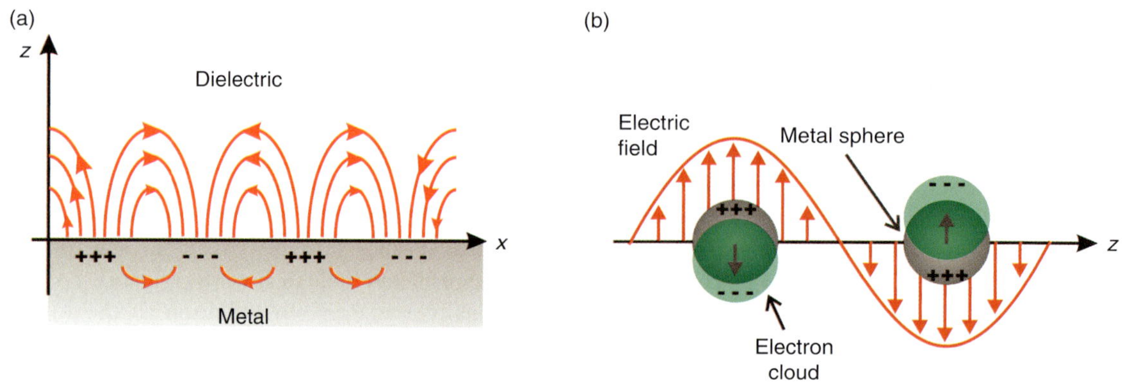

Figure 1 Schematic diagrams for (a) SPR and (b) LSPR.

of LSPR, but LSPR surpasses SPR on the absolute number of molecules detected if metal surface areas are taken into consideration [10]. The advancement of nanofabrication techniques has enabled numerous new LSPR applications and is discussed in more detail in Section 9.2.2. Moreover, LSPR is a main contributor to surface enhancement and is described more thoroughly in the following sections to provide a detailed understanding of SERS.

9.2.1 Theory

Brightly colored nanoparticles have been of interest to both the artistic and scientific communities. Therefore, it was quite significant when Gustav Mie presented a solution to Maxwell's equations that described the extinction spectra (where extinction = scattering + absorption) of spherical particles of arbitrary size, providing a classical explanation for the brilliant colors of metallic nanoparticles [14]. Mie's solutions have proved to be even more valuable over time and are a sound starting point for analyses of nonspherical particles [15], as Richard Gans extended Mie's solutions to Au spheroidal particles [16] in 1912 and to Ag spheroidal particles [17] in 1915. It is important to note that Mie's solutions to Maxwell's equations are often referred to as 'Mie theory', although the solutions do not address an independent physical phenomenon. Additionally, Mie theory is sometimes referred to as Lorenz–Mie theory or Lorenz–Mie–Debye theory, but for consistency with existing literature, the term Mie theory is used hereafter.

9.2.1.1 Solution to Maxwell's equations: Mie theory

When metallic nanoparticles are irradiated by light, the oscillating electric field causes a collective oscillation of the conduction electrons in the metal. The Coulombic attraction between the negatively charged electrons and positively charged nuclei (and repulsion among electrons) causes a restoring force to arise when the electron cloud is displaced relative to the nuclei, resulting in an oscillation of the electron cloud [15, 18, 19]. The coherent oscillation is called the dipole plasmon resonance or the LSPR and the plasmon resonant frequency is related to the metal dielectric constant that is measured as a function of wavelength for bulk metal [15]. To understand the relationship between the plasmon frequency and metal dielectric constant, the electric field surrounding the particle is considered. Here, the conditions used in Mie theory to solve Maxwell's equations for the EM field outside the particle are described, although a complete annotated derivation is described elsewhere [10]. The equation to describe the EM field outside a particle can be derived using the following limits. Briefly, the particle under consideration is spherical, with radius a, and is irradiated by z-polarized light with a corresponding wavelength, λ. In this case, a is much smaller than the wavelength of light (i.e., $a/\lambda < 0.1$), such that the electric field is approximated as constant (or static) around the nanoparticle (**Figure 1**) [10, 15]. In this 'quasi-static' approximation, the interaction is dictated by electrostatics instead of electrodynamics, but the dielectric constant of the metallic particle changes with wavelength [15]. In this approximation, the electric and magnetic components are decoupled, allowing for independent analysis of each, but the magnetic component contributes negligibly compared to the electric component [20]. Therefore, only the electric component of the EM field needs to be considered and not the magnetic component, and consequently, all future references to the EM field refer to the electric field component [10]. The electric field of the incident EM wave propagating in the z-direction is denoted by E_0 such that $E_0 = E_0\hat{\mathbf{z}}$, where $\hat{\mathbf{z}}$ is a unit vector. Solving LaPlace's equation, a fundamental equation of electrostatics, and applying boundary conditions for inside and outside the sphere, the following equation is obtained for the electric field outside a spherical particle [10, 15]:

$$E_{\text{out}}(x, y, x) = E_0\hat{\mathbf{z}} - \alpha E_0\left[\frac{\hat{\mathbf{z}}}{r^3} - \frac{3z}{r^5}(x\hat{\mathbf{x}} + y\hat{\mathbf{y}} + z\hat{\mathbf{z}})\right] \quad (1)$$

In **equation 1**, r is the distance from the center of the particle, α is the polarizabiltiy, and $\hat{\mathbf{x}}$, $\hat{\mathbf{y}}$, and $\hat{\mathbf{z}}$ are unit vectors [15]. The first term in **equation 1** represents the incident field and the second term corresponds to the induced dipole field. The induced dipole moment is $\mu_{\text{ind}} = \alpha E_0$, where $\alpha = g_d a^3$ and

$$g_d = \frac{\varepsilon_{\text{in}} - \varepsilon_{\text{out}}}{\varepsilon_{\text{in}} + 2\varepsilon_{\text{out}}} \quad (2)$$

In **equation 2**, g_d describes the polarizability of the metal particle's dielectric function (ε_{in}) and the external dielectric constant (ε_{out}) [15]. Substituting the induced dipole moment equation into **equation 1**, the solution to Maxwell's equations for the electric field outside a particle using the quasi-static approximation is [10, 15]:

$$E_{out}(x,y,x) = E_0\hat{\mathbf{z}} - \left[\frac{\varepsilon_{in} - \varepsilon_{out}}{\varepsilon_{in} + 2\varepsilon_{out}}\right] \times \alpha^3 E_0 \left[\frac{\hat{\mathbf{z}}}{r^3} - \frac{3z}{r^5}(x\hat{\mathbf{x}} + y\hat{\mathbf{y}} + z\hat{\mathbf{z}})\right] \quad (3)$$

Notably, ε_{in} is highly dependent on wavelength. Therefore, the determining factor for the dipole resonance condition of the particle is g_d. That is, when ε_{in} is approximately equal to $-2\varepsilon_{out}$, the electric field is enhanced compared to the incident field. For Ag and Au, this resonance condition occurs for wavelengths in the visible spectrum. Since this resonance condition is the basis for LSPR and the EM field enhancement mechanism for SERS, it is important to describe the dielectric constants that contribute to the plasmon resonance.

Thus far, we have considered the dipole resonance of spherical metal nanoparticles. Although the dipole resonance is most commonly described, it is important to note that higher modes of plasmon excitation (e.g., the quadrupole resonance) are observed. The quadrupole mode arises when half of the electron cloud moves parallel to the applied field and the remainder oscillates antiparallel, an effect that becomes more significant for larger particles [15]. A similar derivation is performed when determining the EM field enhancement of the quadrupole resonance, where the quadrupole polarizability (β) is $\beta = g_q a^5$ and

$$g_q = \frac{\varepsilon_{in} - \varepsilon_{out}}{\varepsilon_{in} + 3/2\varepsilon_{out}} \quad (4)$$

The quadrupole factor (g_q) in **equation 4** demonstrates an extinction dependence on the external dielectric constant, but is less sensitive than the dipole resonance [15]. Although most LSPR sensing experiments detect the changes in dipole resonance, work has demonstrated the dielectric-environment sensitivity of the quadrupole resonance [21]. In this chapter, we restrict the discussion of LSPR to dipole resonances.

9.2.1.2 Relationship between the dielectric function and nanoparticle extinction

To fully understand the LSPR resonance condition, it is important to discuss the physical meaning of the dielectric constant. The dielectric constant (ε) or relative permittivity is a dimensionless constant that embodies the electrical behavior of the material in a particular environment compared to vacuum (ε_0). In particular, the dielectric constant describes the behavior of a material when it is immersed in an electric field and is therefore also related to the polarity of the material [22]. The complex dielectric function demonstrates how a material's optical properties are dependent on the incident wavelength of light and can be described by $\varepsilon_{in} = \varepsilon' + i\varepsilon''$, where ε' and ε'' are the real and imaginary dielectric functions, respectively. The real and imaginary components of the dielectric function of a harmonic oscillator can be described by the Lorentz model as (see Ref. [23]):

$$\varepsilon' = 1 + \frac{\omega_p^2(\omega_o^2 - \omega^2)}{(\omega_o^2 - \omega^2)^2 + \gamma^2\omega^2} \quad (5)$$

$$\varepsilon'' = \frac{\omega_p^2 \gamma\omega}{(\omega_o^2 - \omega^2)^2 + \gamma^2\omega^2} \quad (6)$$

In these equations, ω_p is the plasma frequency, ω is the frequency of the incident electric field, ω_o is the natural frequency of the oscillator, taking into account its spring constant and mass, and γ is a damping factor [23]. The plasma frequency is the natural frequency of free electrons and a plasmon is a quantized plasma wave. **Equations 5 and 6** demonstrate the dependence of the dielectric function on the frequency of the incident electric field.

A refinement of the Lorentz model of the dielectric function is the Drude model, valid for conductors that have free electrons. The Drude model is similar to the Lorentz model, but the springs are 'clipped', so that the electrons move about freely [23]. Due to free electron movement, there is no restoring force in this model such that ω_o is zero. In this framework, the equations for the real and imaginary dielectric functions reduce to:

$$\varepsilon' = 1 - \frac{\omega_p^2}{\omega^2 + \gamma^2} \quad (7)$$

$$\varepsilon'' = \frac{\omega_p^2 \gamma}{\omega(\omega^2 + \gamma^2)} \quad (8)$$

Here, the plasma frequency (ω_p) is described by [23]

$$\omega_p^2 = \frac{Ne^2}{m\varepsilon_0} \quad (9)$$

where m is the electron mass, e is the electron charge, and N is the free electron density. If $\omega >> \gamma$, as is common for metals at room temperature, then **equations 7 and 8** are approximated as

$$\varepsilon' \cong 1 - \frac{\omega_p^2}{\omega^2} \quad (10)$$

$$\varepsilon'' \cong \frac{\omega_p^2 \gamma}{\omega^3} \tag{11}$$

and are identical to the high-frequency limits of the Lorentz model in **equations 5 and 6** [23]. **Equation 9** shows that the square of the plasmon frequency is proportional to the electron density. Therefore, since metals have a particularly high free electron density, the plasmon frequency is likewise high – an observation that has significant consequences for the metal dielectric function [19]. The real part of the dielectric function is approximated by **equation 10** and illustrates that when the plasmon frequency is higher than the frequency of incident light, the ratio of ω_p/ω is greater than 1, leading to a negative real dielectric constant. Therefore, metals have a negative real dielectric constant because they have a high density of free electrons leading to a high plasmon frequency. The negative real dielectric constant resulting from **equation 10** is responsible for resonant Rayleigh scattering and the imaginary dielectric constant is responsible for absorption. Ultimately, these equations demonstrate why the LSPR occurs when an electric field is incident on metals possessing negative real and positive imaginary dielectric constants. As the frequency of the incident electric field changes, so does the complex dielectric function, demonstrating the frequency dependence of the LSPR.

The dielectric function explains the physical basis for the resonance condition, but it is also important to understand how ε_{in} is related to light absorption and scattering of an incident EM field. Here, three possibilities are considered: transmission, scattering, and absorption. When light is transmitted, it has no effect on the particle (and the particle has no effect on the incident field). Scattering results from the acceleration of electrons in the particle that causes them to radiate at the same frequency as the incident wavelength and is often referred to as resonant Rayleigh scattering. Absorption originates from the imaginary component of the dielectric function and occurs when the particle absorbs the incident energy and converts it to other forms of energy (e.g., heat). As mentioned previously, extinction is the sum of absorption and scattering and the corresponding cross-sections are the areas in which these interactions occur. For example, the absorption cross-section describes the area around the particle where incident light is absorbed, often differing from the geometrical cross-section of the particle. The absorption and scattering cross-sections for a small sphere (σ_{abs} and σ_{sca}, respectively) are [15]:

$$\sigma_{abs} = \pi a^2 4x \,\mathrm{Im}(g_d) \tag{12}$$

$$\sigma_{sca} = \pi a^2 \frac{8}{3} x^4 |g_d|^2 \tag{13}$$

where $x = (2\pi a \varepsilon_{out}^{1/2})/\lambda$. [15] The cross-sections, as described by **equations 12 and 13** are highly dependent on g_d, further demonstrating its key role in determining the wavelength dependence of scattering and absorption. If the absorption cross-section for a particle is zero, then $\varepsilon'' = 0$. Under this condition, light incident on the particle is solely scattered and/or transmitted.

Incident light on a small spherical particle is transmitted when the imaginary part of the dielectric function is zero and the real part of the dielectric function is positive and equal to the external dielectric. From **equation 10,** this occurs when ω is much higher than ω_p. From the definition of g_d (**equation 2**), when ε' is positive ($\varepsilon'' = 0$), and approaches ε_{out}, g_d and therefore the polarizability become zero, such that there is no extinction and all incident light is transmitted. Alternatively, when ε' is negative ($\varepsilon'' = 0$) and twice the value of ε_{out} in magnitude (i.e., $\varepsilon_{in} = -2\varepsilon_{out}$), g_d approaches infinity and maximum scattering by the spherical particle is observed. Absorption occurs when ε'' is nonzero; that is, the imaginary dielectric function is responsible for light absorption. Because ε_{in} is highly frequency-dependent, the extinction maximum exhibits a corresponding frequency dependence [15]. Examining both **equations 2 and 10** and assuming that the imaginary component of the metal dielectric constant is small, as ε_{out} increases, ε_{in} becomes more negative, which corresponds to a lower incident light frequency that is absorbed or scattered. **Figure 2** shows the real and imaginary dielectric functions of bulk Ag and Al [24]. For Ag, the real dielectric is increasingly negative for longer wavelengths and the imaginary dielectric function is small but positive. Similarly, Al also exhibits a positive imaginary and a negative real dielectric function but in different regions of the spectrum than Ag, suggesting it has different plasmonics properties, which is discussed in section 9.4.1. The physical interpretation is the increase in the external dielectric constant damps the frequency of the oscillating surface electrons so they oscillate at lower frequencies; therefore the particle absorbs and scatters longer wavelengths to satisfy the resonance condition.

When the dielectric conditions outlined above are met such that g_d is maximized, a maximum (i.e., referred to as λ_{max}) is observed in the extinction

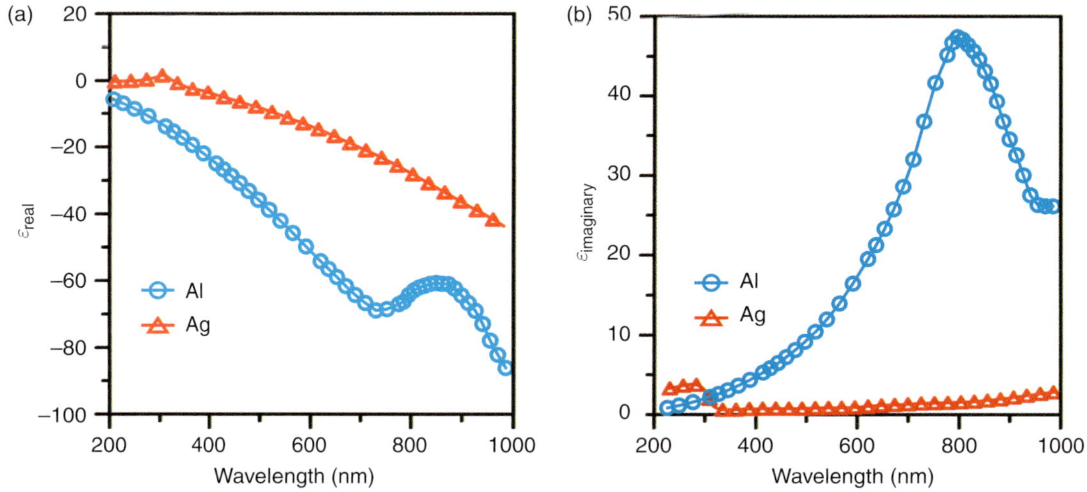

Figure 2 (a) Real and (b) imaginary components of the complex dielectric functions for Ag and Al.

spectrum of the particles. The equation describing the extinction spectrum is derived from Mie's solutions to Maxwell's equations and is as follows [10, 25]:

$$E(\lambda) = \frac{24\pi^2 N a^3 \varepsilon_{out}^{3/2}}{\lambda \ln(10)} \left[\frac{\varepsilon_{in}''(\lambda)}{(\varepsilon_{in}'(\lambda) + \chi\varepsilon_{out})^2 + \varepsilon_{in}''(\lambda)^2} \right] \tag{14}$$

In **equation 14,** $E(\lambda)$ is the wavelength-dependent extinction for spherical metallic nanoparticles, where a is the sphere radius, χ is a shape factor ($\chi = 2$ for a sphere), and N is the number of particles. The shape factor χ is a function of the nanoparticle's aspect ratio and increases proportionally. **Equation 14** also exhibits the impact of the complex dielectric function on the extinction spectrum. When $\varepsilon_{in}' = -2\varepsilon_{out}$ and ε_{in}'' is small and positive, the extinction is maximized. For nonspherical particles, the extinction maximum occurs when $\varepsilon_{in}' = -\chi\varepsilon_{out}$. Thus, for both spherical and nonspherical particles, as ε_{out} changes, a shift in λ_{max} is observed. Therefore, measurements of λ_{max} provide information about the external dielectric environment. The sensitivity of the LSPR λ_{max} to changes in dielectric environment has significant implications for sensing experiments.

9.2.1.3 Electric-field decay

The enhanced EM field surrounding metallic nanoparticles is a result of the LSPR. The LSPR of nanoparticles is sensitive to the external dielectric environment in close proximity to the surface. Mie theory is used to examine the distance dependence of the enhanced field, which is approximated to decay exponentially away from the nanoparticle surface as: $E(z) = E_0 \exp(-z/l_d)$ where l_d is a characteristic decay length [10]. For bulk changes in dielectric environment, such as a solvent or gas, the change in λ_{max} is proportional to change in refractive index (n) of the environment, $\Delta\lambda_{max} = m\Delta n$ [10, 26]. Note that the dielectric constant is the square of the complex refractive index. This equation is valid in situations where the bulk environment extends past the electric field enhancement, such that the electric-field decay lengths are irrelevant. However, for changes in the dielectric environment due to an adsorbate layer, the electric field decay length is important since the adsorbate layer thickness is less than the electric field decay length (~20 nm). Establishing boundary conditions for the thickness of the adsorbate layer and normalizing the refractive index by the relative field intensity, the following equation is obtained [10, 26]:

$$\Delta\lambda_{max} = m(n_A - n_E)[1 - \exp(-2d/l_d)] \tag{15}$$

where n_A is the refractive index of the adsorbate layer, n_E is the refractive index of the bulk environment, d is the adsorbate layer thickness, and l_d is the decay length of the electric field. A detailed derivation is found elsewhere [10, 26]. When more than one adsorbate layer is present, as is common in many biosensing experiments, additional exponential terms are added to the equation to account for the adsorbate layers occupying different distances in the electric-field decay length. Here it is demonstrated that the effect of the electric-field decay on the sensing capability of the LSPR provides the ability to not only predict λ_{max} shifts with certain adsorbates, but also to quantify adsorbate layer thicknesses with an observed λ_{max} shift.

9.2.1.4 Discrete dipole approximation for nonspherical particles

All discussion thus far has used Mie's exact solutions to Maxwell's equations for spherical particles. Although nonspherical particles are also utilized in LSPR experiments, Mie's solutions still provide an interpretation for nanoparticles that are not necessarily spherical. Because it is usually not possible to solve Maxwell's equations for nonspherical particles analytically, developing numerical methods is important [27]. The discrete dipole approximation (DDA) is a numerically exact method used to describe nonspherical metal nanoparticles and was originally proposed in 1973 [28]. Since DDA is discussed at length in computational chemistry reviews [27], only a brief description is included here. The DDA method divides the particle of interest into a large number of polarizable cubes that interact only through dipole–dipole interactions. The interaction of each dipole with a local electric field induces a polarization that is used to determine the extinction cross-section. Most commonly, the DDA method is used to predict LSPR spectra that provide a comparison with the experimental results from nonspherical nanoparticle extinction experiments. In general, good agreement between computational and experimental results is observed, demonstrating that DDA is an effective numerical method [27].

9.2.1.5 Experimental methods

As discussed previously, the extinction of metal nanoparticles is dependent on the wavelength-dependent dielectric function of the metal, the external dielectric environment, as well as the shape and size of the nanoparticles. Typically, nanoparticle extinction is measured using UV–vis spectroscopy in the transmission or reflection geometry. In the transmission geometry, white light is passed through the nanoparticle sample and the light that is not scattered or absorbed is collected by the spectrometer. Again, the maximum of the extinction spectrum is at the wavelength in which g_d is maximized. The reflectance mode is used for samples that have low transmission probability (i.e., reflective surfaces) such that incident light that is not absorbed or scattered is reflected to the spectrometer. In the case of reflectance experiments, the wavelength that satisfies the resonance term is at a minimum in the spectrum. That is, the reflectance minimum is the extinction maximum. As the external dielectric is changed, the extinction spectrum exhibits measurable changes that demonstrate the sensing capabilities of LSPR spectroscopy.

UV–vis spectroscopy is a relatively simple method of measuring the extinction spectra of metallic nanoparticles. However, this technique is optimal for ensemble-averaged measurements and not single-nanoparticle studies because the shot-noise-governed detection limit is similar to the extinction of a single nanoparticle [29]. Instead, resonant Rayleigh scattering spectroscopy is the most straightforward method for measuring the LSPR of single nanoparticles because the scattering signal is detected above a low background. One example of resonant Rayleigh scattering spectroscopy utilizes dark-field microscopy with a high numerical aperture (NA) dark-field condenser to illuminate the particle and a high-magnification oil-immersion objective with a variable NA to collect only the Rayleigh scattered light [29]. More recently, spatial modulation spectroscopy has enabled measurements of single nanoparticle extinction (i.e., scattering + absorption) [30]. Using these methods, the LSPR of single nanoparticles is measured, enabling careful determination of the structure–property relationships for single nanoparticles.

9.2.2 Nanofabrication

9.2.2.1 Chemical syntheses

Significant advances in nanofabrication techniques have enabled progress in LSPR spectroscopy. In particular, precise control over the size, shape, heterogeneity, and material composition are critical due to the dependence of LSPR on these properties. Chemical synthetic methods are among the most widely used methods for fabricating large amounts of nanoparticles in solution. For example, Ag colloids formed from the citrate reduction of $AgNO_3$ solution [31] are widely used for LSPR and SERS experiments. Indeed, citrate reduction of a metal is a common method for synthesizing nanoparticles like Ag and Au. Although this method provides a straightforward and quick route to synthesize a large quantity of nanoparticles that support localized surface plasmons, the solution of Ag colloids is quite heterogeneous. The heterogeneity of the Ag colloids is advantageous when a variety of extinction maxima are desired and colloids are easily aggregated using an electrolyte solution (e.g., NaCl). Ag colloid aggregation is an important technique used for SMSERS

[32–34], which is discussed later in the chapter. Although colloidal nanoparticles have a wide variety of applications, a more monodisperse sample is preferred when the LSPR maxima are required to be within a specific range. To circumvent the problem of nanoparticle heterogeneity, researchers have developed techniques to fabricate large amounts of metal nanoparticles in solution that are monodisperse and shape specific (e.g., nanoshells [35], triangular nanoprisms [36], cubes [37, 38], bipyramids [39], and stars [40]). These shape-specific, monodisperse particle syntheses are important for LSPR because they provide a solution of particles with a more homogeneous distribution of LSPR maxima. They also allow for the particle structure to be correlated to the corresponding LSPR spectrum more readily as compared to randomly shaped colloids.

The applications for chemically synthesized nanoparticles range from bulk LSPR measurements to single-nanoparticle LSPR spectroscopy as well as SERS experiments. Single-nanoparticle spectroscopy has proved to be a powerful method to correlate a single particle's structure and LSPR because it removes all of the inherent heterogeneities present in the bulk spectrum. The ability to determine the exact structure that gives rise to a specific LSPR spectrum is quite powerful and particularly important for determining structures that exhibit an optimal SERS enhancement.

9.2.2.2 Laser ablation

Chemical syntheses of nanostructures have been widely used for nanoparticles dispersed in solution. However, these methods often require nanoparticle surface stabilizing or functional ligands that are not always necessary or beneficial [41]. Laser ablation synthesis of nanoparticles in solution, pioneered in 1993 [42], is an effective technique that produces nanoparticles without ligands on the nanoparticle surface. Briefly, laser ablation occurs by striking a bulk metal in liquid with a laser, forming a plasma plume that condenses into nanoparticles in solution. Nanoparticles of Au, Ag, and Cu, to name a few, have been synthesized by laser ablation in both aqueous and organic solvents [41]. Although at present, laser ablation does not allow the predetermination of particle size and size distribution, it does allow for the synthesized nanoparticles to be reduced or increased in size by varying laser irradiation. As a relatively new synthetic method for nanoparticles, a complete understanding of the underlying mechanism is still fragmented; however, laser ablation is a promising approach for the fabrication of nanoparticles in solution without surface ligands. Further research, such as the formation of SERS-active colloids, should increase the utility of nanoparticles formed by laser ablation.

9.2.2.3 Nanostructured films

Nanoparticles in solution have a variety of uses and can be drop cast onto a substrate for several experiments. However, some experimental techniques are more straightforward when the nanofabrication occurs on the substrate itself. For example, Ag island films (AgIFs) are created when a thin layer of Ag (6–8 nm) is thermally or electron-beam (e-beam) deposited onto a substrate. The thin layer of Ag is not thick enough to create a continuous film across the substrate and instead forms Ag islands. Because the AgIFs support a surface plasmon and enhance the EM field on the surface, AgIFs have proved to be a simple and fast method to fabricate substrates for SERS [43]. Another nanostructured surface that supports surface plasmons is electrochemically roughened Ag and Au surfaces. Indeed, the substrate in which SERS was first observed in 1974 by Fleischman and coworkers [44] and later described independently by Jeanmaire and Van Duyne [1] as well as Albrecht and Creighton [2] in 1977 was a Ag surface that was roughened electrochemically, illustrating a natural connection between the fields of SERS and electrochemistry.

9.2.2.4 Lithographic techniques

Lithographic techniques have provided novel methods for the fabrication of periodic arrays with specific shapes on the nanoscale. These techniques include e-beam lithography, photolithography, and nanosphere lithography (NSL). E-beam lithography is a highly precise method of creating nanoparticles of arbitrary size, shape, and spacing. In this method, an electron beam is used to write a deposition mask directly into a polymer film and metal is then deposited through the mask. The film is removed to leave the remaining nanoparticle pattern on the substrate [10]. E-beam lithography has been used to systematically study the coupling between nanoparticles [45–49] (Section 9.2.3.4). Although quite powerful, e-beam lithography is time consuming, expensive, and inefficient for large-scale substrates [10]. Photolithography uses a photo-reactive mask to form arrays of micro- and nanostructures. Specifically, soft lithography utilizes a polymer, such as poly(dimethylsiloxane) (PDMS), as a stamp to transfer patterns fabricated by

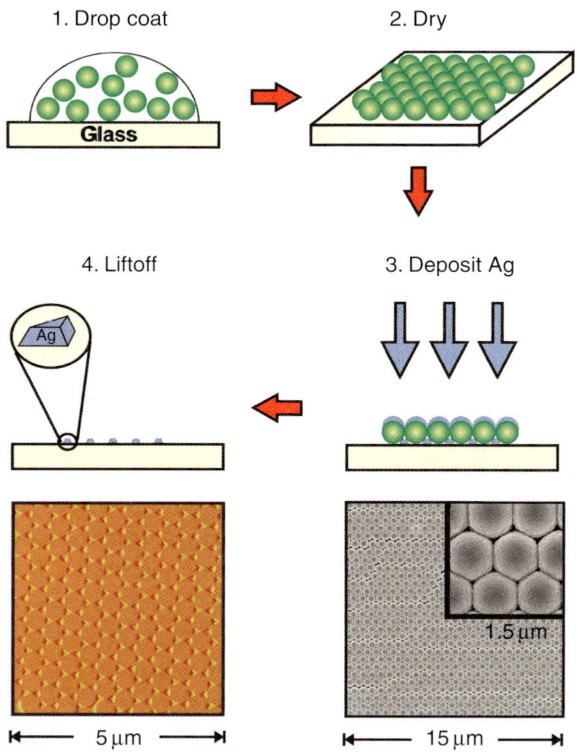

Figure 3 Schematic of NSL for the formation of periodic particle arrays and metal film over nanospheres.

photolithography [50]. These photolithographic methods are diverse and effective for creating patterns of nanostructure arrays of varying complexity and additional information is found elsewhere [50, 51]. A third lithographic technique highlighted here is NSL and in the following paragraphs, the simplicity and versatility of NSL is demonstrated.

NSL, depicted in **Figure 3**, is an extremely valuable and cost-effective technique for generating large two-dimensional (2D) arrays of nanoparticles of a specific shape, size, material, and orientation [52]. The NSL procedure is as follows: polymer or silica nanospheres are drop cast from solution onto a cleaned glass or mica substrate and allowed to dry, self-assembling into a hexagonally close-packed array. From this nanosphere mask, three different routes may be taken. First, between 10 and 100 nm of metal may be deposited onto the nanospheres mask, followed by removal of the nanospheres by sonication. This method produces a periodic particle array (PPA) of metallic nanostructures shaped as truncated tetrahedrons, covering ~7% of the glass substrate [53, 54]. Another route for NSL fabrication is to instead deposit a thicker layer of metal (~200 nm) onto the nanosphere mask to create a film over nanosphere (FON), as the nanosphere mask is not removed. Metal FONs are widely used for SERS sensing due to the high surface area of the substrate. A third approach is to subject the nanosphere mask to reactive-ion etching through the nanosphere mask generating wells between the spheres [55]. The metal is then deposited over the nanospheres and into the wells. The spherical mask is removed, and nanoparticles remain anchored to the glass substrate. Reactive-ion etching provides a unique NSL-fabricated structure because the nanoparticles are rod shaped.

In addition to FONs and PPAs, NSL is used to fabricate alternative nanoparticle structures. Typically, the nanosphere mask is a monolayer and produces truncated tetrahedron shaped PPAs; however, when a double nanosphere mask layer is used, hexagonally shaped particles are formed after metal deposition [54]. Additionally, by changing the metal deposition angle, new shapes may be fabricated [53, 54]. The variety of sizes and shapes of nanoparticles fabricated by NSL is important as it demonstrates the wide versatility of the technique. Moreover, NSL-fabricated nanoparticles may also be released into solution. To this end, NSL was used to fabricate particles on a substrate that were functionalized with a self-assembled monolayer (SAM) to prevent nanoparticle aggregation and then released into solution by sonication [56, 57]. The SAM was present on all sides of the nanoparticle except for the base that was attached to the glass substrate. Next, a dithiol was added to the solution to link two nanoparticles together. The monomers and dimers were characterized using UV–vis spectroscopy, atomic force microscopy (AFM), transmission electron microscopy (TEM), and computational methods, and the particles exhibited distinct LSPR spectra on a substrate, in solution, and as dimers.

Overall, NSL is a cost-effective and high-throughput method to fabricate nanoparticle arrays. The ability to easily vary the size and shape of the nanoparticles produces a highly tunable LSPR extinction maximum that is discussed in the following section. NSL is also indifferent to the type of metal and by using a variety of metals like Ag, Au, Al, and Cu its applications can be further expanded.

9.2.3 Characteristics of the LSPR

9.2.3.1 Dependence on nanoparticle size and shape

According to Mie theory, the LSPR of a nanoparticle is highly dependent on its size and shape. Therefore, by fabricating nanostructures with varying sizes and shapes, the LSPR can be tuned to a distinct wavelength. For example, Ag triangles, spheres, rods, cubes, and stars exhibit unique LSPR spectra because of their differing size and shape. NSL-fabricated nanoparticles are particularly useful for demonstrating the relationships between nanoparticle size and shape and LSPR. That is, the LSPR is sensitive to shape changes in the in-plane and out-of-plane directions, which can be varied independently using NSL. **Figure 4** displays the tunability of the LSPR extinction maximum across the visible range of the spectrum for a Ag PPA [53]. Increasing the in-plane nanoparticle width, *a* (by increasing the size of the nanospheres used in the mask), results in a shift towards lower-energy wavelengths. Increasing the out-of-plane height, *b*, of the nanoparticles shifts the LSPR toward higher-energy wavelengths. Therefore, varying these size parameters alone can produce a distribution of LSPR maxima. However, the LSPR is also dependent on the shape of the nanoparticles. Without any further modification, the shapes of the nanoparticles fabricated by NSL are truncated tetrahedrons. However, by thermal annealing, the truncated tetrahedrons are transformed into hemispheres. Generally, these hemispheres have LSPR extinction spectra at shorter frequencies than a truncated tetrahedron fabricated with similar parameters (**Figure 4**) [53].

As described above, the size and shape dependence of NSL-fabricated Ag nanoparticles can be systematically studied by varying the nanosphere size and metal thickness. Another method to vary nanoparticle size and shape is achieved by electrochemical modification. NSL particle arrays fabricated on indium tin oxide, a transparent conducting surface, were used for simultaneous optical and chronocoulometry measurements [58]. The controlled particle shape, size, and interparticle spacing of these arrays make them ideal candidates to monitor (1) the electrochemical modification of the nanoparticle structure and (2) the LSPR shifts resulting from the changing structure. The results are

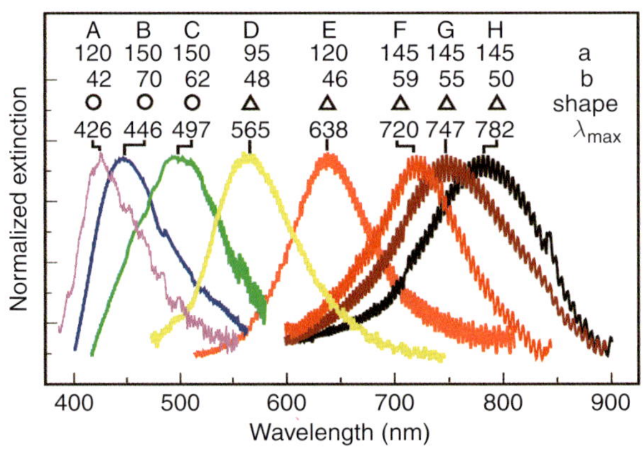

Figure 4 Modification to the LSPR maximum of periodic particle arrays with changing nanoparticle (a) in-plane width and (b) out-of-plane height.

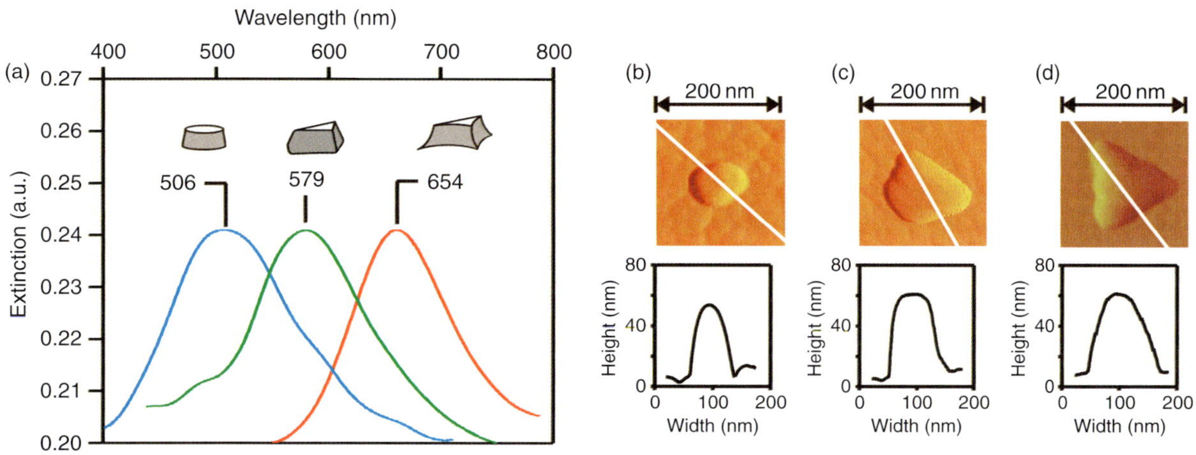

Figure 5 Electrochemical modification of Ag nanoparticles on indium tin oxide measured by (a) LSPR and (b–d) AFM.

displayed in **Figure 5**. As the Ag particles are oxidized, the shape is changed, as exhibited by scanning electron microscopy (SEM) and AFM, and the LSPR spectrum is shifted. These experiments illustrated that the particles were preferentially oxidized first at the bottom edges, next at the triangular tips, and finally the out-of-plane height. In these correlated studies, the relationship between the nanoparticle in-plane width and the LSPR maximum could be studied systematically, because the out-of-plane height does not change during the first several chronocoulometric runs. This spectroelectrochemical study demonstrated a quantitative manipulation of Ag nanoparticle geometries along with spectral monitoring [58]. The Mie theory predicts the nanoparticle size and shape dependence of the LSPR and NSL has proved to be a cost-effective, controlled experimental demonstration of this dependence.

9.2.3.2 Sensitivity to external environment

The LSPR is extremely sensitive to the environment surrounding the particle. That is, the frequency of the conduction electrons' oscillation is highly dependent on the external dielectric constant, which is related to the refractive index. The dependence of LSPR on ε_{out} can be separated into two general regimes, the bulk environment and the local environment. The bulk environment typically refers to the entire area surrounding the nanoparticle, such as air or solvent. Shifts in the LSPR extinction maximum are proportional to the change in refractive index when switching from one environment to another, demonstrated by the expression $\Delta\lambda = m\Delta n$. The refractive-index sensitivity of the nanoparticles' LSPR was demonstrated by exposing the NSL-fabricated Ag nanoparticles to N_2 and different solvents, which have different refractive indices. A plot of the change in λ_{max} versus refractive index, n, for a variety of solvents is displayed in **Figure 6(a)**, demonstrating a linear relationship between $\Delta\lambda_{max}$ and n, where the slope of this line is the refractive-index sensitivity, m [59]. **Figure 6(a)** also shows the variation of refractive-index sensitivity with different sizes of NSL-fabricated nanoparticles, where the particle arrays with the shortest out-of-plane height have the highest sensitivity. Overall, the typical refractive-index unit (RIU) sensitivity for NSL-fabricated particles is 200 nm RIU^{-1}. In other words, a change of about 0.005 RIU produces an LSPR λ_{max} shift of 1 nm. Ultimately, experiments aimed at elucidating the refractive-index sensitivity of LSPR play a critical role for the development of biosensing applications.

Refractive-index studies were repeated on SAM-functionalized Ag PPAs [60]. Thiolated SAMs are ideal since they are available with a variety of terminal functional groups, and thiols chemisorb easily to both Ag and Au. **Figure 6(b)** displays the LSPR peak shift as a function of the surrounding medium's refractive index for both unmodified (line A) and hexadecanethiol (HDT)-modified (line B) Ag nanoparticles. It can be seen that the unmodified particles have a higher refractive-index sensitivity (191 nm RIU^{-1}) compared to the HDT-modified particles (150 nm RIU^{-1}). This observation is not surprising, considering that the EM field decays away from the particle surface. The presence of the HDT monolayer diminishes the sensitivity to external environments by ~20%, as the strongest sensing capabilities occur closest to the nanoparticle surface

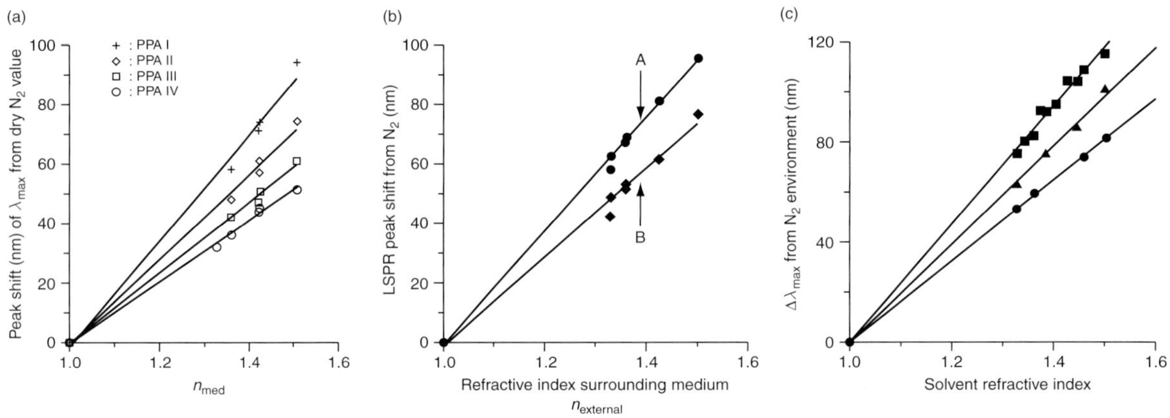

Figure 6 Refractive-index sensitivity of the LSPR spectra for Ag nanoparticles with (a) different sizes, (b) an adsorbate layer, or (c) for individual nanoparticles of various shapes.

where the HDT monolayer resides. Although the overall sensitivity of the SAM-modified PPA is decreased relative to the bare PPA, it is important to note that bulk solvent-induced shifts are readily observed and detected, indicating that the EM field extends far enough away from the particle surface to be useful for sensing experiments. Another significant result from this experiment is that an HDT monolayer produces a 40-nm LSPR shift relative to a bare PPA. An observation with important implications for sensing experiments is that this 40-nm shift corresponds to just 100 zmol of HDT. When an adsorbate layer is present, **equation 15** is used for determining the LSPR shift and is also used to estimate the surface coverage of adsorbates or the adsorbate thickness from the LSPR shift. Ultimately, the sensing experiments that examine bulk and adsorbate refractive-index changes using NSL-fabricated particle arrays provide fundamental information for future work, including single-nanoparticle sensing.

Measurements of single nanoparticles have several benefits to ensemble-averaged measurements. First, single nanoparticles are readily implemented in multiplexed detection schemes, since each unique nanoparticle possesses a distinguishable LSPR maximum. An additional advantage is that the absolute detection limit (number of analyte molecules per nanoparticle) is significantly reduced. A third advantage of single-nanoparticle spectroscopy is the ability to measure the LSPR spectrum of individual particles with a high signal-to-noise ratio [29]. Overall, the advantages of single nanoparticle spectroscopy provided motivation to examine the environmental dependence of single nanoparticle LSPR [29, 61]. The following experiments utilized resonant Rayleigh scattering spectroscopy to demonstrate the LSPR λ_{max} response of individual Ag nanoparticles to varying solvent environments and the formation of a monolayer of small-molecule adsorbates [29].

The refractive-index sensitivity of individual Ag nanoparticles was measured by exposing the nanoparticles to various solvent environments and single nanoparticles with three different shapes were used to illustrate the effect of nanoparticle shape on refractive-index sensitivity [29]. The nanoparticle shapes were inferred from TEM imaging [62], LSPR line shape, LSPR λ_{max}, and LSPR polarization dependence. Similar to the results from ensemble-averaged measurements, the LSPR λ_{max} is shifted to longer wavelengths as the refractive index of the solvent is increased, as displayed in **Figure 6(c)**. The particle with the largest aspect ratio (a rod-shaped particle) exhibited the highest refractive-index sensitivity, consistent with predictions from Mie theory. The refractive-index sensitivity decreased with decreasing particle aspect ratio. These single-nanoparticle studies exhibited the adsorption of fewer than 60 000 HDT molecules on a single Ag nanoparticle resulting in an LSPR shift of 40.7 nm, thus demonstrating low-zeptomole sensitivity [29]. Overall, current research on single-nanoparticle LSPR suggests a promising future for single-nanoparticle sensing experiments.

Significant work has focused on developing the adsorbate sensing capabilities of LSPR biosensors using both nanoparticle arrays and single nanoparticles. For example, researchers have utilized nanoparticle arrays to successfully detect and sense concanavalin A binding to a monosaccharide [12], electrostatic binding of poly-L-lysine to carboxylic acid-terminated SAM-modified Ag nanoparticles [60], avidin binding to biotinylated Ag particles [63], antibiotin binding to biotinylated Ag particles [64], Alzheimer's disease biomarkers [65, 66], and most recently, the calcium-induced conformational changes of calmodulin [13]. Single-nanoparticle LSPR spectroscopy has demonstrated biomarker detection on cancer cells *in vitro* [67]. Readers are directed to recent reviews for a survey of plasmonic biosensors [11, 68–70]. The sensing capabilities of nanoparticles have been examined by bulk and adsorbate refractive-index studies on both particle arrays and single nanoparticles.

9.2.3.3 Distance dependence

Two important consequences of the LSPR are: (1) the enhanced EM field at the surface of the nanoparticle and (2) an environmentally sensitive extinction spectrum. These consequences are understood by using a model for the refractive-index response, depicted in **equation 15,** demonstrating the change in extinction maximum is a function of refractive-index, adsorbate layer thickness, and decay length of the EM field. Although this model assumes a single exponential decay of the EM field (accurate for propagating plasmons) [10, 26, 71], which is an oversimplification for localized surface plasmons, it accurately describes the sensor's distance dependence by the adsorbate thickness, d, and EM-field decay length, l_d. Unfortunately, the equation alone does not provide nanoparticle structural information that is directly related to the refractive-index sensitivity.

In order to characterize the relationship between structure and sensitivity as well as characterize the EM-field decay, long- [72] and short-term [73] distance-dependence experiments were performed using SAMs and multilayers (layers of carboxylic terminated thiols alternating with Cu^{2+} ions). However, these studies were limited by the unknown variability in the refractive index of an SAM and a low resolution layer thickness. To better characterize the distance and structural sensitivity dependence of LSPR, a high-resolution distance-dependence study was performed using ALD, which provides atomic resolution and an extremely uniform refractive-index layer. Using ALD, up to 600 single layers (1.1 Å each) of Al_2O_3 were deposited onto Ag nanoparticle arrays and the LSPR was observed [71]. The LSPR shift with increasing Al_2O_3 film thickness is shown in **Figure 7**. Notably, an LSPR shift is apparent after a single 1.1 Å layer of Al_2O_3, demonstrating

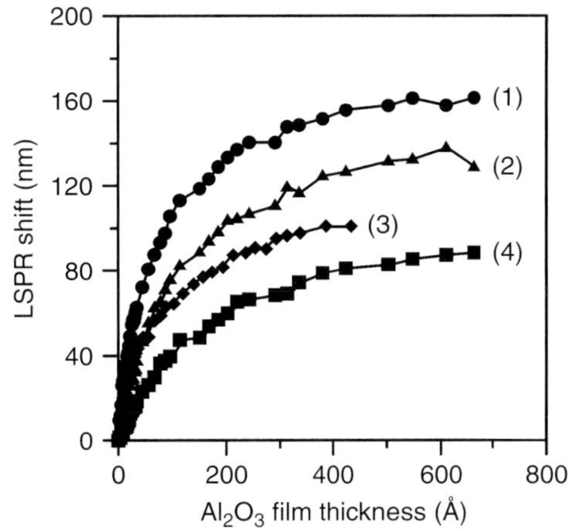

Figure 7 Observed LSPR shift from Ag nanoparticles of various shapes and sizes (1–4) as a function of Al_2O_3 film thickness.

excellent sensitivity to the local environment. Additionally, **Figure 7** shows the variability of sensitivity as the nanoparticle structure is changed: lines 1, 2, and 4 represent truncated tetrahedral-shaped particles with in-plane widths of 90 nm and out-of-plane heights of 30, 40, and 51 nm, respectively. Line 3 represents hemispherically shaped particles with in-plane width of 104 nm and out-of-plane height of 52 nm. As the out-of-plane height of the nanoparticles is increased, the refractive-index sensitivity of the particles is decreased, demonstrating a relatively modest LSPR shift. Although the different nanoparticles exhibit varying refractive-index sensitivities, for all cases (lines 1–4) the slope of the LSPR shift versus Al_2O_3 thickness is decreased as additional Al_2O_3 layers are deposited, as expected due to the EM-field decay away from the nanoparticle surface [71]. These distance-dependence studies are important for understanding the EM-field decay from the nanoparticle, an important consideration for SERS sensing experiments.

9.2.3.4 Coupling among nanoparticles

The LSPR studies presented thus far have examined bulk solutions of nanoparticles, nanoparticle arrays, or single nanoparticles. However, it is well known that the LSPR of metallic nanoparticles is dependent on other surrounding nanoparticles and, therefore, in this section the coupling between nanoparticles is examined. Nanoparticle coupling is a significant area of research because nanoscale devices comprised of assemblies or arrays of nanoparticles are greatly affected by the interactions between nanoparticles [45]. The extent of coupling can be examined by measuring changes to the LSPR maximum that occur when the interparticle spacing is varied [45, 74, 75].

The dependence of the LSPR on interparticle coupling was demonstrated by examining the LSPR shift of Ag and Au nanoparticles in varying array arrangements [45]. For Au and Ag nanoparticle arrays fabricated by e-beam lithography, the LSPR shift to shorter wavelengths when the lattice spacing of the array is decreased. Slight differences in LSPR shifting behavior were observed between nanoparticles in square or hexagonal arrays, consistent with differences in lattice density, such that denser lattices have a relatively high-energy LSPR. Blue shifts are observed when relatively large lattice spacing (i.e., approximately $\lambda_{max}/2$) is decreased due to radiative dipole interactions between nanoparticles. When the lattice spacing is small (<100 nm), the LSPR is shifted to longer wavelengths for Au nanoparticles [76]. Overall, this work provides an example of the LSPR dependence on particle spacing in an array and highlights the sensitivity of the LSPR to neighboring nanoparticles.

9.2.4 Functionalization and Stabilization of Nanoparticles

Functionalization of plasmonic nanostructures is sometimes necessary for LSPR and SERS sensing, specifically for analytes with low or no affinity to noble metals. In addition to the use of SAMs, surface modification can be achieved with ALD. ALD is a thin-film growth technique based on the alternating use of self-terminating reactions between gaseous precursor molecules and a substrate [77]. One advantage of ALD over other vapor deposition methods is the precise thickness control afforded by the self-limiting nature of the surface reactions. Film growth proceeds in a layer-by-layer fashion and coats even high-aspect-ratio and porous structures with extreme uniformity. The reaction sequence for the ALD of Al_2O_3 from trimethylaluminum (TMA) and water, one of the most commonly used ALD processes, is as follows [78]:

(A) $Al\text{-}OH^* + Al(CH_3)_3 \rightarrow Al\text{-}O\text{-}Al(CH_3)_2^* + CH_4$
(B) $Al\text{-}CH_3^* + H_2O \rightarrow Al\text{-}OH^* + CH_4$

In schemes A and B the asterisks represent surface species and the reactions are simplified to show only a single functional group. One AB sequence is referred to as a cycle, and thickness is simply controlled by the number of cycles. The growth of Al_2O_3 has been shown to proceed at a rate of ~1 Å per cycle [71, 79]. It is important to note that, because SERS intensity falls off with increasing distance from the surface, the ability to grow films with subnanometer thickness is highly desirable to minimize signal losses. Moreover, the wide variety of inorganic materials that can be grown by ALD presents new possibilities for versatile surface functionalization.

Metal oxides such as Al_2O_3 also serve to protect Ag surfaces from oxidation, thus increasing the robustness of SERS substrates. In fact, coating Ag nanoparticles with very thin Al_2O_3 films by ALD results in substrates that are also vastly more stable when exposed to high temperatures, laser excitation, and solvent environments [80, 81]. The structural integrity of SERS nanostructures can be probed by imaging techniques such as AFM and SEM. Alternatively, changes to nanoparticle structure can be interrogated by monitoring the LSPR extinction spectrum. The sensitivity of the LSPR to size and shape highlighted earlier in this chapter enables the detection of even slight shape changes that are not observable by AFM or SEM [80].

9.2.4.1 Thermal stability

At high temperatures, the tips of triangular Ag nanoparticles become rounded, eventually annealing into hemispheres, an energetically more favorable geometry. Previous work has indicated that hemispherically shaped nanoparticles exhibit smaller EM enhancement relative to triangular Ag nanoparticles [82]. These substrates, therefore, have been limited to SERS studies conducted at ambient temperatures. For certain applications, such as the study of catalytic reactions that often occur at high temperatures, the development of thermally robust SERS substrates is a necessary requirement. It has been shown that very thin Al_2O_3 layers deposited by ALD over Ag SERS substrates fabricated by NSL provide significant protection against thermal annealing [81]. LSPR spectroscopy and SEM were used to monitor the stability of Ag nanoparticles having a 0.2–1.0 nm layer of Al_2O_3 at various temperatures. SEM images revealed that after 1 h at 300°C in N_2, bare triangular Ag nanoparticles turned into hemispheres, whereas nanoparticles coated with just 0.2 nm Al_2O_3 maintained their original shape (**Figures 8(a)** and **8(b)**). The plasmon resonance of bare Ag nanoparticles at 500°C under N_2 for 1 h were blue-shifted by 192 nm, while the plasmon resonance of Ag nanoparticles having 1.0 nm of Al_2O_3 shifted just 16 nm. In addition, Al_2O_3-coated nanoparticles

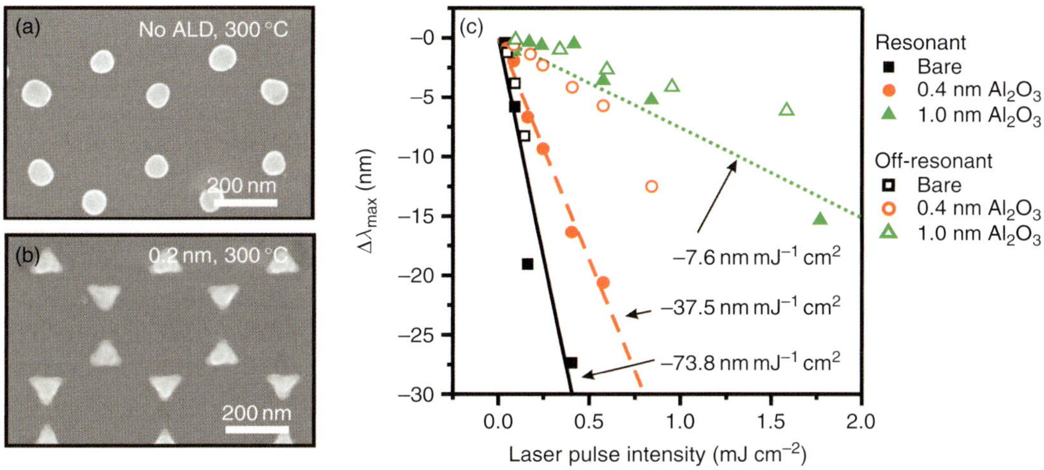

Figure 8 Modifications to Ag nanoparticle shape for bare and Al_2O_3-coated structures upon (a, b) thermal annealing and (c) exposure to laser pulses of different energies monitored by SEM and LSPR spectroscopy.

were found to be stable at 500 °C over the course of 6 hours. These results indicate that ultrathin layers of Al_2O_3 are an effective means of stabilizing SERS substrates for use at high temperatures.

9.2.4.2 Stability to laser excitation

For surface-enhanced nonlinear spectroscopy, high laser intensity is often required to achieve sufficient signal. However, metal nanoparticles are susceptible to melting or deformation caused by the high-power ultrashort laser pulses that are commonly used in nonlinear spectroscopy. Modifications to nanoparticle shape in bare and Al_2O_3-coated Ag nanoparticles upon exposure to laser pulses of varying intensity in both resonant and off-resonant conditions were monitored using LSPR spectroscopy [80]. **Figure 8(c)** shows the shift in LSPR λ_{max} for bare Ag nanoparticles and nanoparticles with 0.4 and 1.0 nm of Al_2O_3, for resonant and off-resonant conditions. For each laser pulse intensity, bare nanoparticles exhibited the largest LSPR shift and nanoparticles with 1.0 nm Al_2O_3 exhibited the smallest LSPR shift. For all cases, the shift in LSPR λ_{max} increased with increasing laser pulse intensity, showing an approximately linear relationship. The slopes of these linear fits reveal that a 1.0 nm layer of Al_2O_3 provides a factor-of-10 improvement in the stability of Ag nanoparticles. This enhanced stability is attributed to the increased surface melting temperature resulting from the decreased mean square displacement of Ag atoms located on the nanoparticle surface. Consequently, Al_2O_3-coated silver nanoparticles can serve as a stable platform for both linear and nonlinear ultrafast surface-enhanced laser spectroscopy.

9.2.4.3 Solvent stability

Exposure to organic solvents or aqueous solutions results in structural changes to unmodified triangular Ag nanoparticles fabricated by NSL which in turn affect their optical properties [60]. Specifically, the tips of the nanoparticles become rounded, resulting in a blue shift in the LSPR spectrum. Similar solvent-induced morphology changes have also been observed for other SERS substrates, such as thin Ag films [83]. The stability of Al_2O_3-coated SERS substrates has been explored in various solvent environments, which is critical for liquid-phase studies. For Ag nanoparticle arrays fabricated by NSL, the LSPR was monitored as a function of time for 2 h in each of eight different solvents having a range of refractive indices and polarities. The extent of structural changes induced by solvent exposure depended on the particular solvent. In each case, however, NSL-fabricated Ag nanoparticles coated with ~2 nm of Al_2O_3 were significantly more protected against these solvent-induced structural changes. In the case of Ag film over nanospheres (AgFONs), Al_2O_3 coatings as thin as 4 Å prevent surface oxidation and delamination of the Ag film upon exposure to water. For catalytic applications, the use of ALD to coat Ag SERS substrates with ultrathin Al_2O_3 layers thus enables the study of surface reactions in catalytically relevant liquids like water, ethanol, and methanol at elevated reaction temperatures using SERS.

9.3 Surface-Enhanced Raman Spectroscopy

9.3.1 Background

In 1928, Raman and Krishnan observed that monochromatic light incident on molecules resulted in normal Rayleigh scattering as well as modified scattered radiation of different frequencies [84]. The energies of this modified radiation corresponded to the characteristic frequencies of molecular vibrations. Therefore, Raman spectroscopy, as it was later termed, provides the unique vibrational signature of molecules. In the classical picture of Raman scattering, the incident EM field induces a dipole moment in the molecule, which in turn radiates light at its oscillation frequencies. The induced dipole moment, μ_{ind}, is described by $\mu_{\text{ind}} = \alpha E_0$, where E_0 is the applied field and α is the polarizability of molecule. The induced dipole moment may radiate at several frequencies: (1) ω, the frequency of the incident light (Rayleigh scattering), (2) ω-ω_{vib}, where ω_{vib} is the frequency of a normal mode vibration (Stokes Raman scattering), and (3) $\omega+\omega_{\text{vib}}$ (anti-Stokes Raman scattering). The intensity of Raman scattering scales as ω^4 and is proportional to the square of the polarizability derivative, $(\delta\alpha/\delta Q)^2$, where Q is the normal nuclear mode of the molecule [85]. Therefore, Raman scattering occurs only if the applied field induces a polarizability change along the nuclear mode. Although Raman spectroscopy is a powerful technique for probing the vibrational modes of molecules, it suffers from poor sensitivity due to the inherent weakness of the scattering process. Raman cross-sections are small ($\sim10^{-29}$–10^{-3} cm^2 molecule^{-1}) in comparison to typical absorption cross-sections ($\sim10^{-15}$–10^{-16} cm^2 molecule^{-1}). Therefore, for 10^{10} incident photons, only about 1 photon will be Raman scattered.

Given the weakness of the Raman effect, detecting Raman scattering from a monolayer of adsorbed molecules on a surface would be extremely difficult. For this reason, the original observation of surface Raman signal from pyridine on electrochemically roughened Ag [44] was quite remarkable. It was soon realized that the Raman signal of pyridine was enhanced by a factor of $\sim10^6$ [1, 2], implying huge enhancement due to a surface effect. Because the induced dipole involved in Raman scattering is the product of the molecular polarizability and the applied electric field, it follows that enhancement of Raman intensity would arise from an increase in one of these factors. Thus, chemical (CHEM) and EM mechanisms were proposed to account for surface-enhanced Raman scattering. Since the discovery of SERS, researchers have devoted much effort to understanding the relative contributions of the CHEM and EM mechanisms to the overall SERS EF. In general, it is now understood that the dominant effect is EM enhancement ($\sim10^4$–10^8, depending on the nanostructure) resulting from excitation of the LSPR, while CHEM enhancement may account for a factor $\leq10^2$. Strong experimental proof of the existence and magnitude of the CHEM mechanism, however, has not been well established.

9.3.1.1 Chemical enhancement mechanism of SERS

CHEM enhancement is postulated to arise from direct interaction between the adsorbed molecule and the metal surface, resulting in an increase in the Raman cross-section for the adsorbed molecule relative to the solution phase [86, 87]. Chemisorption, or bond formation between the molecule and the surface, perturbs the electronic structure and therefore the optical properties of the molecule compared to its unbound state. One way in which the Raman cross-section may be affected by electronic effects is through charge transfer between the adsorbed molecule and the surface [88–90]. Specifically, interaction between the molecular orbitals of the adsorbate and the electronic band structure of the metal allows for transfer of electrons between the molecule and the metal surface. CHEM enhancement, therefore, involves new or altered electronic states that become accessible as a result of chemisorption. If the absorbance maximum of an adsorbed molecule is altered such that it is closer to the laser excitation frequency, the cross-section of the molecule is increased. Thus, the CHEM mechanism can be understood as a type of resonance Raman effect. Evidence for the CHEM mechanism was observed in early electrochemical experiments in which enhancement was observed to vary, depending on the applied potential [91–93]. Since the electrode potential affects the Fermi level of the metal, charge transfer between the adsorbate and the electrode surface changes with applied potential. Further evidence for the CHEM mechanism was provided by the observation that SERS intensities of CO and N_2 differ by a factor of 200 under the same experimental conditions [86, 94]. This result cannot be explained using the EM mechanism in which enhancement is not molecule specific, but rather a property of the substrate.

Ultimately, the requirement for direct adsorbate–surface interaction dictates that the CHEM enhancement mechanism is a short-range effect, limited to only the first layer of adsorbed molecules.

9.3.1.2 Electromagnetic enhancement mechanism of SERS

To describe the EM mechanism of SERS, we return to the simple model of EM radiation incident on a spherical metal particle that is smaller than the wavelength of light. Recall from the earlier discussion of LSPR theory that the electric field surrounding this particle is given by **equation 1**. Since the complex dielectric function of the metal is strongly wavelength dependent, the magnitude of enhancement is likewise wavelength dependent. Note that the field enhancement decays as r^{-3}, suggesting a finite sensing volume surrounding the nanoparticle. The nature of the excitation wavelength and distance dependence of SERS will be explored in more detail in Section 9.3.2.

Since the intensity of Raman scattered light is proportional to $E_0{}^2$, which is greatly enhanced at the nanoparticle surface, the Raman intensity therefore depends on the absolute square of E_{out}, evaluated at the nanoparticle surface ($r = a$). From **equation 1**, where $\alpha = g_d\, a^3$, this is given for a small metal sphere by:

$$|E_{\text{out}}|^2 = E_0^2\left[|1-g|^2 + 3\cos^2\theta\left(2\text{Re}(g) + |g|^2\right)\right] \qquad (16)$$

where θ is the angle between the incident field vector and the vector which describes the position of the molecule on the surface. The largest field enhancements are obtained for $\theta = 0°$ or $180°$, where the molecule is aligned along the polarization direction. When g is large, the maximum enhancement approaches $|E_{\text{out}}|^2 = 4E_0{}^2|g|^2$. The field intensity as a function of position for a spherical Ag nanoparticle is shown in **Figure 9**. The maximum field intensity on a metal sphere differs from the minimum field intensity by a factor of 4, and the radially averaged intensity is $|E_{\text{out}}|^2 = 2E_0{}^2|g|^2$.

The overall physical picture of EM enhancement in SERS that emerges from this simple model is as follows. When a molecule adsorbed to the surface of a metal nanoparticle is irradiated with light, an oscillating dipole in the molecule is created which is enhanced by the metal substrate. The resulting radiation is decreased in energy by an amount equal to the vibrational frequencies of the molecule (i.e., Stokes Raman). In addition to enhancement of the incident field expressed in **equation 16**, enhancement of the emitted light from the induced dipole is also possible. Because the emitted field is not constant over the volume of the sphere as is the incident field, this enhancement is more complicated to evaluate. A rigorous treatment of enhanced emission is given by Kerker and coworkers [95, 96]. To a first-order approximation, the enhanced emission can also be modeled using **equation 16** simply by replacing the excitation frequency with the Raman Stokes-shifted frequency. Therefore, the overall EF from both incident and scattered fields can be described as

$$EF = \frac{|E_{\text{out}}|^2|E'_{\text{out}}|^2}{|E_0|^4} - 4|g|^2|g'|^2 \qquad (17)$$

where the primed symbols refer to fields evaluated at the scattered frequency. This expression describes the theoretical EF resulting from the EM mechanism of SERS. If the Stokes shift is small, then g and g' are evaluated at approximately the same wavelength such that the EF scales as g^4. This relationship is the origin of the 'E^4 enhancement' commonly referred to in the SERS literature. The derived EF expression matches exactly the results of Kerker's more rigorous approach, supporting the validity of the approximations used for this model. For a small sphere with $|g| \sim 10$, the resulting EM EF is $\sim 10^4$–10^5. EM enhancements are actually rather small for spherical nanoparticles because their plasmon resonance condition is met near the UV region

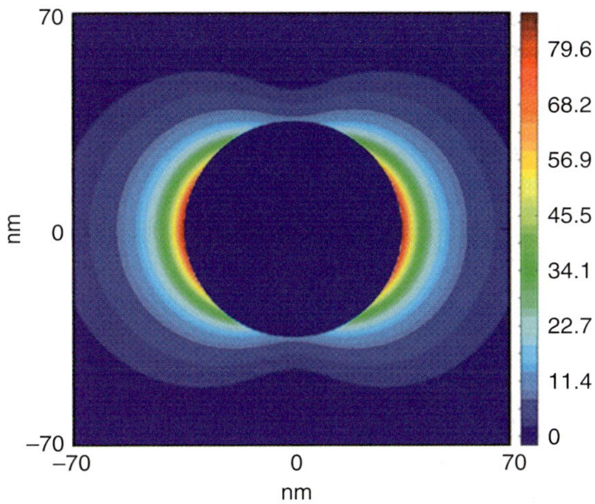

Figure 9 Calculated electric field intensity $\left(|E_{\text{out}}|^2\right)$ around a spherical Ag nanoparticle.

of the spectrum where damping due to metal interband transitions is more prevalent. Therefore, the magnitude of $|g|$ can be much larger for prolate spheroidal objects and nanoparticles of higher-order structure (e.g., aggregates).

9.3.1.3 Calculating SERS enhancement factors

While the EM EF can be predicted theoretically, the overall SERS EF (i.e., enhancement due to both CHEM and EM mechanisms) is calculated experimentally by measuring both the SERS and normal Raman spectroscopy (NRS) signals of a given analyte. The scattering intensity in NRS is given by the following expression:

$$I_{NRS} = K(v) \times N_{NRS} \times \frac{\partial \sigma}{\partial \Omega} \times I_{0,NRS} \times v^4 \times t_{NRS} \quad (18)$$

where $K(\nu)$ is the instrument response function, N_{NRS} is the number of molecules within the probe volume, $\frac{\partial \sigma}{\partial \Omega}$ is the differential Raman cross-section of the molecule, $I_{0,NRS}$ is the laser intensity, ν^4 is the frequency dependence of the scattering process, and t_{NRS} is the acquisition time. SER scattering intensity is analogous to that measured in NRS but is multiplied by a factor due to surface enhancement, expressed as EF_{SERS}:

$$I_{SERS} = K(v) \times N_{SERS} \times \frac{\partial \sigma}{\partial \Omega} \times I_{0,SERS} \times v^4 \times t_{SERS} \times EF_{SERS} \quad (19)$$

The comparison of I_{SERS} and I_{NRS} is greatly simplified if the collection geometry is similar and the same excitation frequency is used, such that $K(\nu)$ and ν^4 are constant. When the intensities are normalized for laser power and acquisition time, the ratio of the two intensities can be rearranged as

$$EF = \frac{I_{SERS}/N_{SERS}}{I_{NRS}/N_{NRS}} \quad (20)$$

This expression describes the overall enhancement in Raman scattering that is observed for molecules in close proximity to the surface of a plasmonic nanostructure. The SERS EF can therefore be calculated from the normalized SER and NR intensities of a given analyte and the number of molecules interrogated in each measurement. In practice, however, determining the number of molecules that give rise to Raman and SER signal is quite challenging.

For example, the SERS EF for benzenethiol on a Ag nanoparticle array fabricated by NSL was calculated by comparison of the intensity of the 1575 cm^{-1} mode with the corresponding peak measured from solution [97]. The SER spectrum of benzenethiol ($\lambda_{ex} = 670$ nm) on a Ag nanoparticle array is shown in **Figure 10(a)**. The surface-enhanced 1575 cm^{-1} peak intensity is 22.9 ADU s^{-1} mW^{-1} (I_{SERS}). The NR spectrum of benzenethiol in a 37 μm-thick cell at the same excitation wavelength is shown in **Figure 10(b)**. The 1575 cm^{-1} peak intensity from the NR measurement is 0.792 ADU s^{-1} mW^{-1} (I_{NRS}). To estimate the number of molecules probed in a SERS measurement, the laser spot size, surface area of the substrate, and surface coverage of the molecules must be known. In this experiment, the beam waist of the objective used was determined to be 4.0 μm by a scanning silicon knife edge test. Assuming a monolayer of benzenethiol on the surface of a Ag nanoparticle array with a packing density of 6.8×10^{14} molecules cm^{-2} and ~7 % surface coverage of the nanoparticles, approximately 6.3×10^6 molecules are probed in the SERS measurement (N_{SERS}). For the NR experiment, the number of molecules within the probe volume, approximated as a cylinder with a radius of 2.0 μm and a height of 37 μm, was 2.7×10^{12} molecules (N_{NRS}). Substitution of the N_{SERS} and N_{NRS} values along with intensities from the normalized SER and NR spectra into the above equation give an SERS EF of 1.24×10^7.

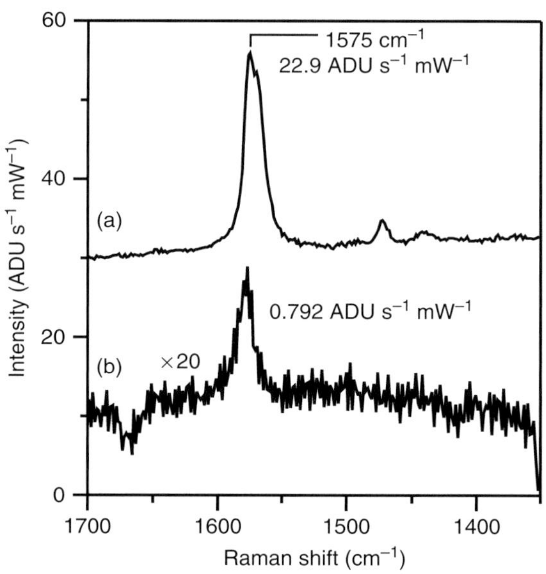

Figure 10 Calculating the enhancement factor for benzenethiol on a Ag nanoparticle array: comparison of (a) SER and (b) normal Raman spectra of benzenethiol ($\lambda_{ex} = 670$ nm).

9.3.2 Experimental Consequences of the EM Mechansim

9.3.2.1 *Distance dependence of SERS*

An important consequence of the EM mechanism is that the adsorbate is not required to be in direct contact with the surface. Enhanced EM fields generated by excitation of the LSPR extend beyond the surface of the substrate, thereby creating a volume within a few nanometers of the surface in which molecules are detected. The long-range effect predicted by the EM mechanism differs from the CHEM enhancement mechanism, which requires that molecules be directly chemisorbed to the surface. For the practical application of SERS, the ability to detect molecules near the surface, but not necessarily bound to it, is crucial since many analytes have low or no affinity to Ag or Au. In such cases, specificity for the analyte can be improved by modifying the surface with overlayers. The field enhancement around a small metal sphere decays with r^{-3}, and given the E^4 enhancement, SERS intensity should therefore scale with r^{-12}. However, at increased distances from the nanoparticle, the surface area is also increased, scaling as r^2, which leads to an overall SERS distance dependence of r^{-10}:

$$I_{SERS} = \left(\frac{a+r}{a}\right)^{-10} \qquad (21)$$

where I_{SERS} is the intensity of a particular Raman mode, a is the average size of the field-enhancing features on the surface, and r is the distance from the surface to the adsorbate [98].

To probe the distance dependence of SERS, the SER spectrum of pyridine was measured at increasing distances from an AgFON surface modified with ultrathin Al_2O_3 layers of various thicknesses deposited by ALD (**Figure 11(a)**) [99]. The relative intensity of the 1594 cm^{-1} mode as a function of Al_2O_3 thickness is shown in **Figure 11(b)**. By fitting the experimental data to **equation 21,** the average size of the enhancing feature, a, was found to be 12.0 nm. The surface-to-molecule distance at which the SERS intensity is decreased by a factor of 10, d_{10}, for this substrate is 2.8 nm. However, this value was derived assuming a complete monolayer of Al_2O_3 was formed for each ALD cycle, but quartz crystal microbalance measurements indicate the average growth rate for Al_2O_3 on Ag is ~1.1 Å per cycle [71]. Thus, the value of d_{10} could be as low as 1.0 nm. Overall, these results clearly demonstrate that SERS is a long-range effect, in agreement with the predictions of the EM mechanism.

9.3.2.2 *Excitation-wavelength dependence of SERS*

According to the EM mechanism, the maximum SERS signal is predicted to occur when both the

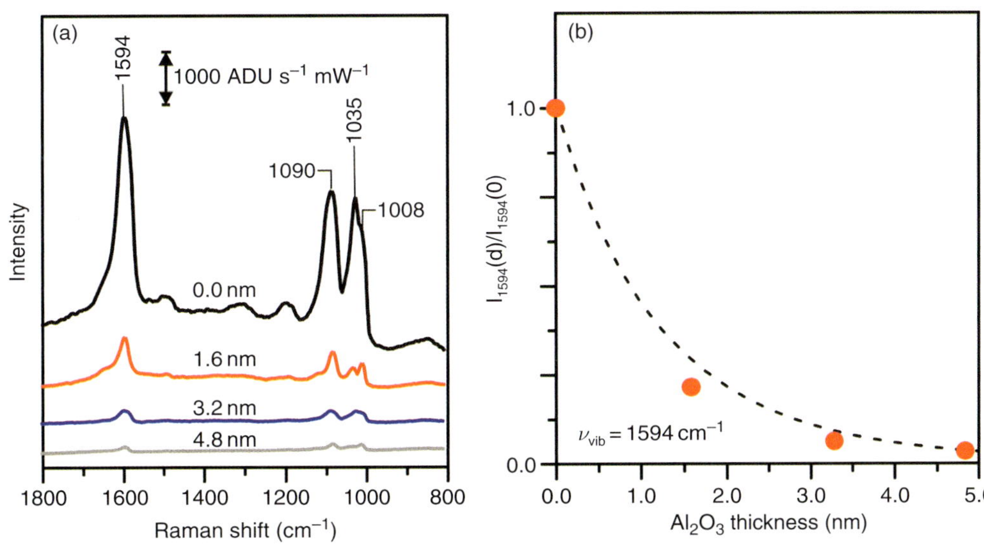

Figure 11 Distance dependence of SERS: (a) SER spectra of pyridine measured at increasing distances from an AgFON surface modified with ultrathin Al_2O_3 layers of various thickness and (b) relative intensity of the 1594 cm^{-1} mode as a function of Al_2O_3 thickness.

incident and scattered photons are strongly enhanced at the metal surface. It was recognized early on that wavelength-scanned surface-enhanced Raman excitation spectroscopy (WS SERES), in which SERS enhancement is measured at several excitation wavelengths, could be used to probe this effect. Unfortunately, early experiments were limited by fixed-frequency laser sources resulting in low data point density or limited spectral coverage [100–102]. The use of a broadly tunable laser system with a versatile detection system enabled the first detailed WS SERES experiments [97]. In particular, WS SERES profiles were measured for benzenethiol adsorbed to Ag nanoparticles fabricated by NSL. The narrow size distribution of these Ag nanoparticle arrays results in a well-defined LSPR that can be precisely tuned throughout the visible and NIR regions.

WS SERES profiles were measured for excitation wavelengths in the range of 420–500, 532–690, 532–740, and 630–800 nm for Ag nanoparticle arrays having an LSPR λ_{max} at 489, 663, 699, and 810 nm, respectively, and plotted for the 1575 cm^{-1} mode of benzenethiol. An important detail is that the relevant LSPR extinction spectra correspond to the LSPR measured after adsorption of the analyte, which is red-shifted relative to the LSPR of the bare substrate. To account for variations in intensity unrelated to enhancement by the substrate, SER spectra were corrected using the NR signal from neat cyclohexane ($\nu = 1444\ cm^{-1}$) as an intensity standard. The WS SERES profiles exhibited line shapes similar to the extinction spectrum of the substrate, and were fit to Gaussian functions. In each case, the maximum SERS enhancement was observed for excitation wavelengths slightly higher in energy than the LSPR λ_{max}, with the difference in energy corresponding to roughly half the vibrational frequency of the Raman mode. To generalize the magnitude of this energy difference, the WS SERES profiles for three different vibrational modes of benzenethiol (1575, 1081, and 1009 cm^{-1}) were compared for a single substrate (**Figure 12**). The difference in energy between the LSPR λ_{max} and the peak excitation

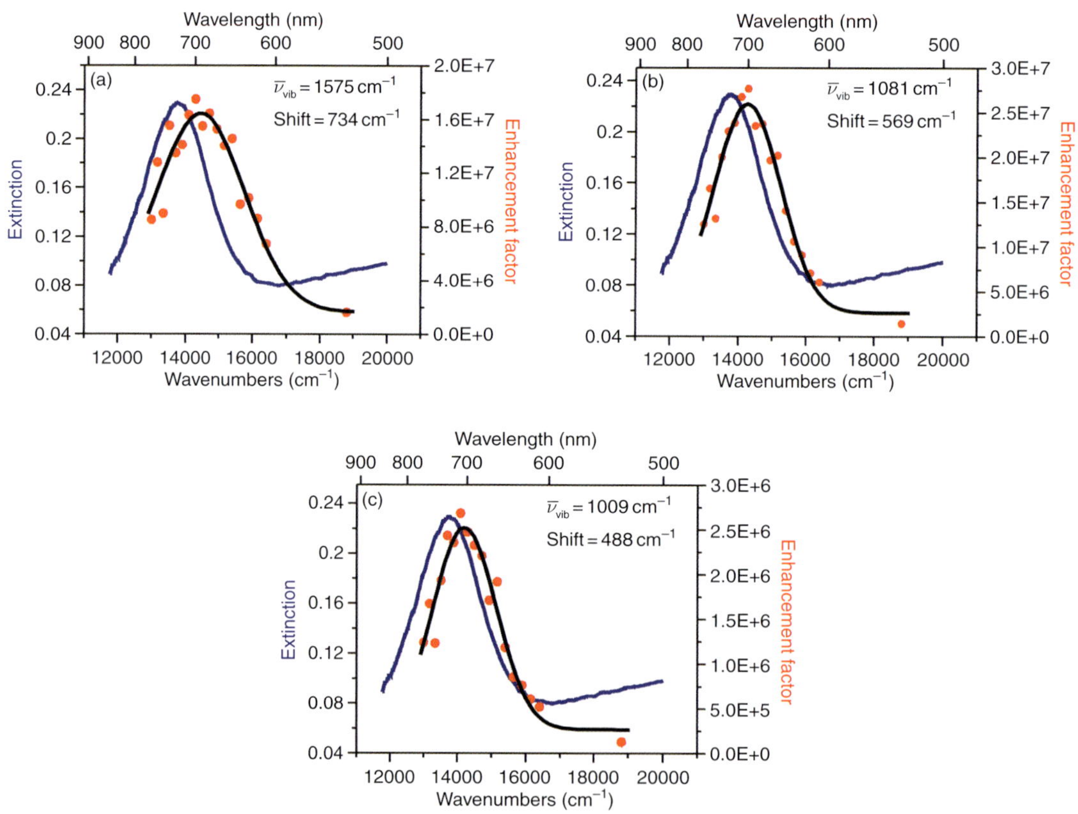

Figure 12 Wavelength-scanned SER excitation profiles and LSPR spectra measured for benzenethiol adsorbed to Ag nanoparticles fabricated by NSL.

frequency was found to increase with increasing vibrational mode frequency. For all three modes, the magnitude of the separation was approximately half the Raman frequency. Therefore, maximum SERS enhancement occurs when the LSPR λ_{max} is located between the peak excitation frequency and the vibrational frequency, wherein both incident and scattered photons are enhanced by the LSPR. In addition to supporting the fundamental EM mechanism of SERS, these experiments also highlight the practical importance of optimizing both the plasmon and excitation frequencies in order to achieve the maximum SERS enhancements for any application.

9.3.2.3 Excitation-wavelength dependence of SERRS

When a resonant Raman scatterer is adsorbed to a SERS-active surface, the relationship between the LSPR of the substrate and the frequency dependence of surface-enhanced resonant Raman spectroscopy (SERRS) is more complex owing to coupling between the molecular resonance and the plasmon resonance. Previous studies have demonstrated that when resonant molecules are adsorbed to nanoparticles, the magnitude of the observed LSPR shift depends strongly on the spectral overlap between the plasmon resonance of the nanoparticles and the molecular resonance of the adsorbed species [103–106]. In order to explore the impact of this coupling on SERRS signal as a function of excitation frequency, wavelength-scanned surface-enhanced resonance Raman excitation spectroscopy (WS SERRES) profiles were measured on Ag nanoparticle arrays fabricated by NSL [107]. The probe molecule for this study, Tris(2,2′-bipyridine)-ruthenium (II) ($Ru(bpy)_3{}^{2+}$), has an absorbance maximum at 452 nm and a shoulder at 425 nm corresponding to a metal-to-ligand charge-transfer transition. In order to ensure monolayer coverage on the Ag nanoparticles, the surface coverage of $Ru(bpy)_3{}^{2+}$ on a Ag electrode was determined using double potential step chronocoulometry. It was found that an incubation time of 30 min in 0.1 mM $Ru(bpy)_3{}^{2+}$ yielded monolayer coverage on the Ag surface ($\sim 0.65 \times 10^{14}$ molecules cm^{-2}). The LSPR shift induced by adsorption of a monolayer of $Ru(bpy)_3{}^{2+}$ on Ag was then measured for Ag nanoparticles with various LSPR λ_{max}. For Ag nanoparticles with λ_{max} blue of the $Ru(bpy)_3{}^{2+}$ resonance, the LSPR shifted by ~20 nm. However, the LSPR shift was dramatically decreased to just 5 nm when the λ_{max} was very close to the $Ru(bpy)_3{}^{2+}$ resonance. As λ_{max} was tuned to be slightly red of the $Ru(bpy)_3{}^{2+}$ resonance (453 nm), an amplified LSPR shift of 50 nm was observed. Finally, as the λ_{max} was moved further red of the $Ru(bpy)_3{}^{2+}$ resonance, the LSPR shift was decreased to ~15 nm. Overall, the sharp increase in LSPR shift from 5 to 50 nm that occurs within 10 nm of the molecular resonance is consistent with predictions by quasi-static electrodynamics modeling and previous observations for other resonant molecules [103, 106, 108].

WS SERRES profiles of $Ru(bpy)_3{}^{2+}$ on three different Ag nanoparticle arrays were measured for excitation wavelengths in the range of 400–500 nm for substrates with an LSPR λ_{max} at 434.7, 465.2, and 480.1 nm. Excitation profiles for both the 1487 and 1602 cm^{-1} modes of $Ru(bpy)_3{}^{2+}$ were fit to Gaussian functions (**Figure 13**). The observed trends in surface enhancement versus excitation wavelength differ from the nonresonant case, in which the greatest enhancement consistently occurred when the excitation wavelength was slightly blue-shifted from the LSPR λ_{max} of the adsorbate-covered substrate. This deviation is attributed to coupling of the molecular electronic resonance with both the laser excitation frequency (resonance Raman effect) and the LSPR frequency. In particular, the WS SERRES profiles for the 1487 cm^{-1} mode appear to be biased toward the $Ru(bpy)_3{}^{2+}$ electronic absorption, whereas data for the 1602 cm^{-1} mode more closely resemble nonresonant behavior. An examination of SERR spectra taken at different excitation energies reveals that when excitation energy near the molecular resonance was used (457.9 nm), the 1487 to 1602 cm^{-1} mode intensity ratio is 2.5. However, when the excitation frequency was moved away from the molecular resonance to higher energy (420 nm), this ratio changed to 1.2. Thus, the 1487 cm^{-1} mode is more resonantly enhanced relative to the 1602 cm^{-1} mode. To better understand the relationship between the LSPR, molecular resonance, and the WS SERRES profile, a theoretical model based on multiplication of EM and resonance enhancement mechanisms was proposed. The resonance enhancement profile of $Ru(bpy)_3{}^{2+}$ was approximated to have the same shape as the molecular absorbance, with a maximum resonance enhancement of 40 (based on the known value for $Fe(bpy)_3{}^{2+}$, an analog of $Ru(bpy)_3{}^{2+}$). When the resonant EF was multiplied by the theoretical EM enhancement for $Ru(bpy)_3{}^{2+}$, the predicted profiles were in good agreement with the observed experimental trends. Ultimately, the results of the WS SERRES experiment with $Ru(bpy)_3{}^{2+}$ are consistent with an overall enhancement that involves multiplicative EM and resonance Raman contributions.

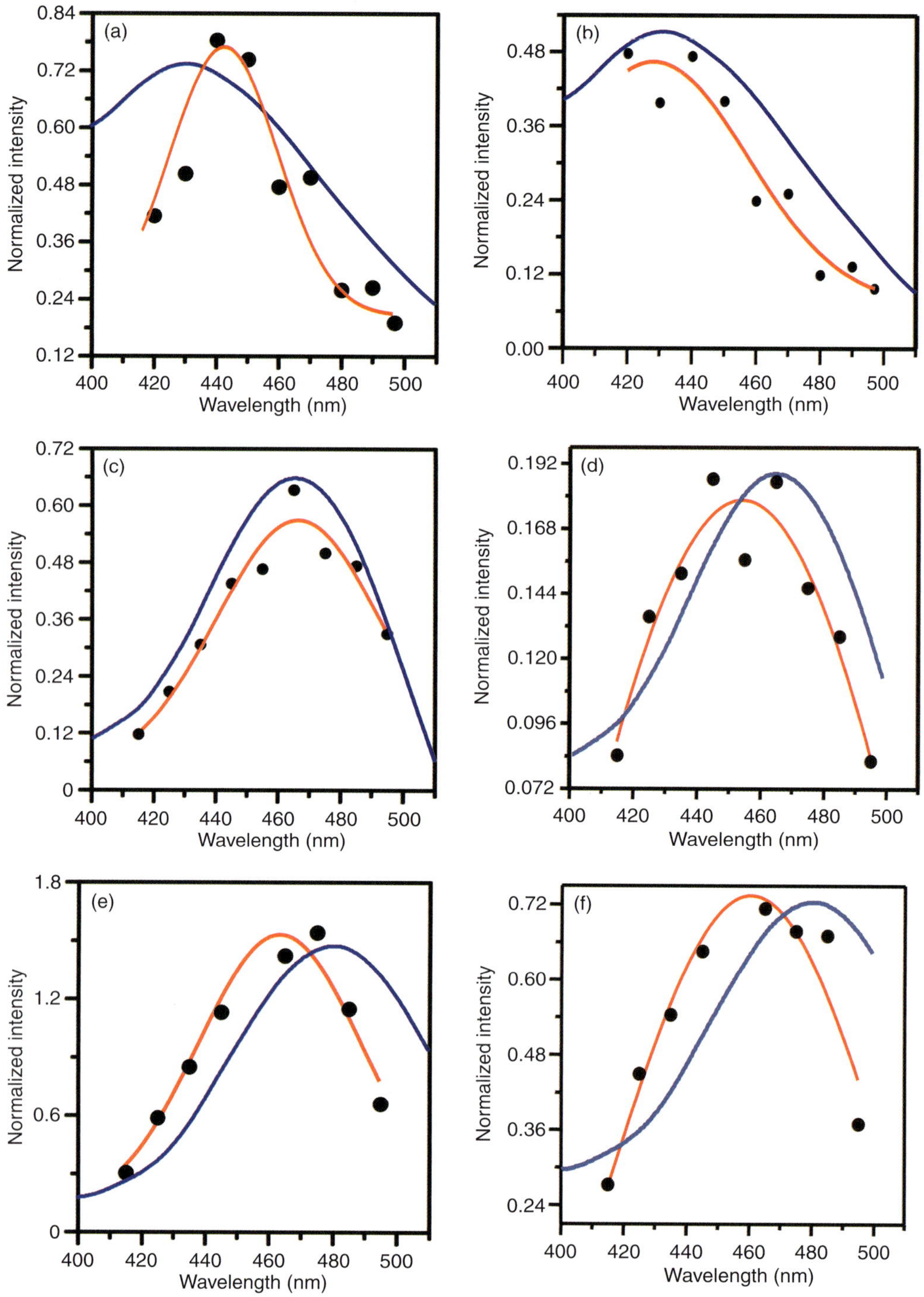

Figure 13 LSPR spectra and wavelength-scanned SERR excitation profiles measured for (a), (c), and (e) the 1487 cm^{-1} peak and (b), (d), and (f) 1602 cm^{-1} peak of $Ru(bpy)_3^{2+}$ on three different Ag nanoparticle arrays.

9.3.3 Single-Molecule SERS

The observation of surface-enhanced Raman spectra from individual molecules was independently reported in 1997 by the groups of Nie [4] and Kneipp [3]. Nie and Emory observed Raman scattering from single rhodamine 6G (R6G) molecules adsorbed on Ag nanoparticles at concentrations corresponding to zero or one analyte molecule per nanoparticle ($<10^{-10}$ M) [4]. The observations of polarization sensitivity and changes in the position and intensity of vibrational bands as a function of time suggested single-molecule behavior. Kneipp *et al.* observed SMSERS of crystal violet on Ag nanoparticles in solution. The signal intensity was quantized and corresponded to 0, 1, 2, and 3 molecules in the probe volume, and the probability of observing *n* molecules followed the Poisson distribution [3]. Given the low number of photons scattered in a typical Raman experiment ($\sim 10^{-4}$ photons s^{-1} molecule^{-1}), the detection of inelastic scattering from a single molecule is remarkable, and the initial reports of SMSERS immediately generated substantial interest. Subsequently, researchers sought a definitive proof that Raman signal originates from just one molecule. Evaluation of SMSERS based on temporal fluctuation in signal intensity, or blinking, was shown to be an unreliable proof of single-molecule detection [109]. Problems associated with the use of Poisson statistics corresponding to quantized intensities have also been discussed [110, 111].

In addition to SMSERS focused on fluctuation in the intensity domain, the observation of competing analytes in SMSER spectra provided further evidence for single-molecule behavior. In particular, Doering and Nie [112] noted spectral interchange between R6G and the citrate background (present from the synthesis of Ag nanoparticles), suggesting that blinking in SMSERS is due to surface diffusion in the proximity of a hot spot, or region of very high EM enhancement. Competitive adsorption between two analytes, R6G and benzotriazole, was then used by Etchegoin and coworkers to verify single-molecule activity [113, 114]. However, interpretation of results based on the bi-analyte approach is complicated due to differences in Raman cross-section, absorption spectra, and surface binding affinity of the two analytes [110].

9.3.3.1 A frequency domain existence proof of SMSERS

Building upon the concept of the bi-analyte approach, an isotopologue approach was developed in which two isotopologues of R6G, R6G-d_0 (non-deuterated) and R6G-d_4 (deuterated), were used to prove single-molecule activity [32]. Isotope editing circumvents the problems associated with the bi-analyte technique since isotopic substitution does not significantly alter the electronic absorbance, surface chemistry, or Raman cross-section of the molecule. However, the vibrational bands associated with C-D stretching and bending will shift in frequency relative to C-H bands, resulting in a unique vibrational signature for each molecule. Single-molecule behavior can therefore be verified by tracking the spectral signature of analyte molecules adsorbed at sufficiently low concentrations. When two isotopologues are introduced to a solution of Ag nanoparticles such that, on average, only one type of molecule is adsorbed to each nanoparticle aggregate, most SMSER spectra display the spectral features of a single isotopologue.

The SER spectra from 50 individual nanoparticle aggregates were acquired under a dry N_2 environment from randomly chosen active sites identified by dark-field scattering. **Figure 14(a)** displays representative spectra from two of the spots measured, which clearly indicate the presence of only one isotopologue within each probe volume. Analyte assignment was based on the sole presence of the 622 or 610 cm^{-1} mode, characteristic of only R6G-d_0 or R6G-d_4, respectively. Of the 50 spots sampled, 24 had solely R6G-d_4 character, 22 had solelyR6G-d_0 character, and four showed character from both isotopologues, as shown by the histogram in **Figure 14(b)**. The overall probability of measuring either isotopologue follows a combined Poisson–binomial distribution, providing strong evidence of single-molecule detection. Histograms of the peak location in the 600 cm^{-1} region for the single-molecule events fit to Gaussian functions with a full width at half-maximum (FWHM) of 3.0 and 4.8 cm^{-1} for R6G-d_4 and R6G-d_0, respectively, illustrate the degree of spectral wandering observed. Spectral wandering in SMSERS is significantly greater than spectral fluctuation due to instrument instability, measured to be 0.6 cm^{-1} for the 1028.3 cm^{-1} NR mode of cyclohexane. For the ensemble-averaged spectra of R6G-d_4 and R6G-d_0, the corresponding FWHM were 5.3 and 6.9 cm^{-1}, respectively. Therefore, the distribution of vibrational frequencies hidden by the ensemble average can be elucidated by SMSERS. This work clearly demonstrates that the ensemble-averaged spectrum is a superposition of single-molecule states. The isotopologue and bi-analyte approaches have emerged as the most definitive proofs for single-molecule detection in SMSERS to date. Indeed, it is anticipated that the isotopologue approach in particular will have substantial impact on future SMSERS experiments.

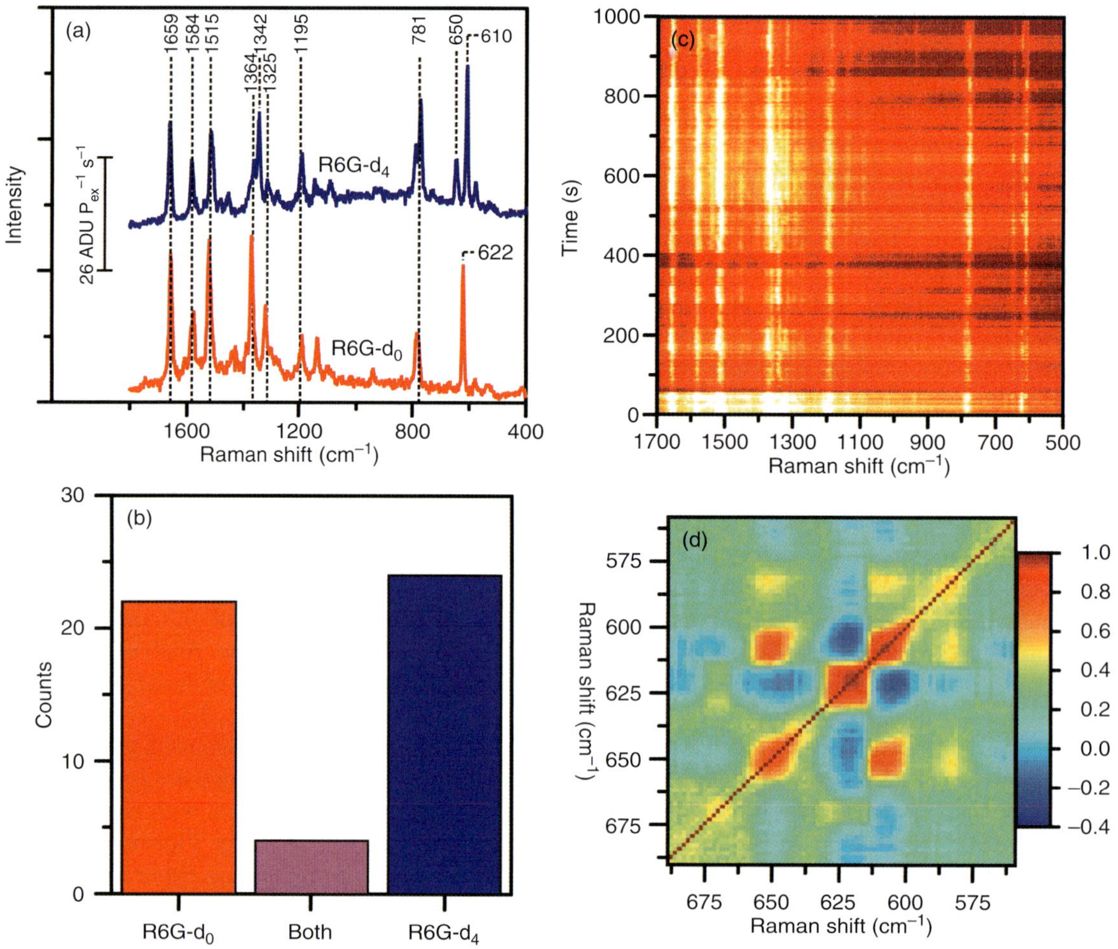

Figure 14 (a) Representative single-molecule SER spectra for R6G isotopologues and (b) corresponding histogram of deuterated, nondeuterated, or both-molecule events. (c) Waterfall plot of SER intensity versus time under ambient conditions showing surface dynamics and (d) corresponding cross-correlation of dynamic switching demonstrating the analytes compete for the same hot spot.

9.3.3.2 *Surface dynamics in SMSERS*

The effect of the nanoparticle environment on SMSERS was also investigated for the R6G/Ag system [32]. In a dry N_2 environment no blinking was observed, but under ambient conditions, and for higher coverages of R6G on Ag (statistically, ~50 R6G-d_0 and 50 R6G-d_4 molecules adsorbed per nanoparticle aggregate), surface dynamics were observed. A waterfall plot of Raman scattering as a function of time is displayed in **Figure 14(c)**. At $t = 0$ s, only the characteristic vibrational signature of R6G-d_0 is observed, which is unexpected given the higher concentration of molecules on the surface. Furthermore, the integrated intensity is similar to that seen in the single-molecule experiment described earlier, indicating that the signal is originating from only one molecule rather than the 100 molecules in the ensemble. This suggests that the EF in the hot spot is actually three to four orders of magnitude greater than the average EF for the nanoparticle, in agreement with previous results [3, 4]. Over the course of 1000 s, the spectrum switches completely from R6G-d_0 to R6G-d_4 and back to R6G-d_0, displaying dynamic behavior.

Given that SMSERS nanoparticle aggregates may contain more than one hot spot, two explanations for the observed switching behavior are possible: (1) two or more molecules are competing for the same hot spot on the surface, or (2) two or more molecules are acting independently at two different hot spots within the same probe volume. By analyzing the data with 2D cross-correlation [115] it is possible to differentiate between these cases. For two different hot spots in close proximity, the dynamic behavior of the molecules over time should be noncorrelated ($\chi = 0$). If, however, the two molecules compete for the same hot spot, their behavior should be anticorrelated ($\chi = -1$). The 2D cross-correlation of the time evolution data (**Figure 14(d)**) demonstrates strong correlation ($\chi = 0.8$) between the two vibrational bands characteristic of R6G-d_4 (650 and 610 cm^{-1}). In addition, the vibrational bands associated with the cross-peaks between R6G-d_0 and R6G-d_4 display anticorrelation ($\chi = -0.3$). In 2D cross-correlation analysis, a strong correlation of the Stokes band to the continuous SERS background is observed, such that even in the absence of a peak there is some time-correlated background contribution [116]. The correlation ($\chi = 0.2$–0.4) that remains between the background and the characteristic frequencies of the molecule is consistent with this observation. As a result, the magnitude of the anticorrelation is in fact even larger than the measured value. The results of 2D cross-correlation analysis therefore point to the existence of only one active hot spot in the probe volume, with molecules moving in close proximity to it.

9.3.3.3 Structure and enhancement factors of SMSERS hot spots

Most SMSERS studies to date have been conducted on colloidal Ag nanoparticles. It is well established that these Ag nanoparticles are extremely heterogeneous, with only a small fraction that exhibit SMSERS activity [112]. The lack of substrates having well-defined structures and reproducible single-molecule activity is a current limitation to the broad application of SMSERS. Moreover, characterizing the nature of SMSERS-active sites is also necessary to understand the underlying mechanisms. Structural determinations of SMSERS-active sites by AFM and polarization-modulation measurements suggested that 'hot' particles result from the aggregation of two or more Ag nanoparticles [4, 117, 118]. These studies led to the notion that molecules were in fact adsorbed to 'hot spots', or regions of high EM enhancement located between adjacent particles.

High-resolution TEM (HRTEM) images with correlated optical spectra of SMSERS-active nanoparticle structures and electrodynamics calculations provide further evidence for EM hot spots in nanoparticle aggregates [34]. The TEM images of nanostructures with confirmed single-molecule activity consisted of nanoparticle dimers as well as higher-order aggregates. No SMSERS active structures consisting of a single Ag nanoparticle were observed, suggesting that two or more particles are necessary for single-molecule activity. The Ag nanoparticle dimer in **Figure 15(a)**, the simplest single-molecule structure observed to date, can be approximated as a hemispherically capped rod and a sphere arranged in a T-shaped geometry. Electrodynamics calculations using the DDA with this structure as input confirm that SMSERS hot spots occur near the particle–particle junction with enhancements on the order of $\sim 10^8$ (**Figure 15(b)**), consistent with the combination of molecular resonance and EM field enhancement. Importantly, this detailed study of the specific nanostructures that give rise to SMSERS suggests that simple nanoparticle dimers can be used for the rational design of SMSERS substrates.

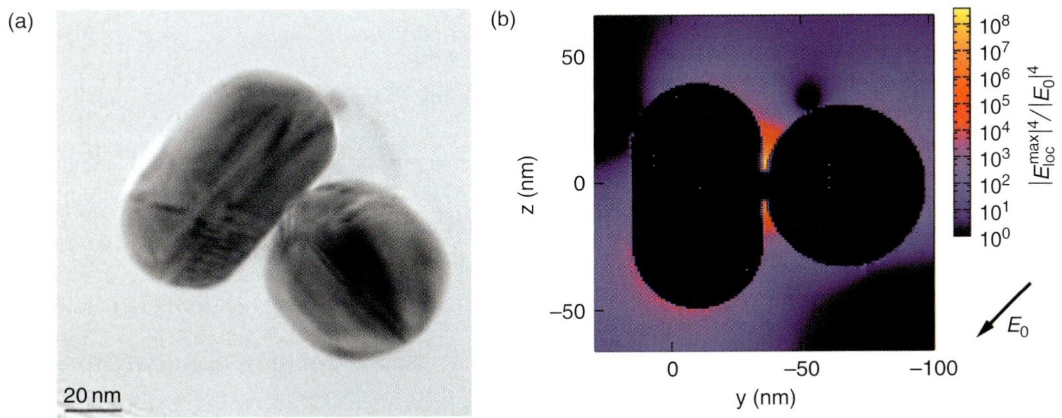

Figure 15 (a) High-resolution transmission electron micrograph of an SMSERS-active Ag nanoparticle aggregate and (b) corresponding EM enhancement map calculated with DDA.

In addition to structural characterization, substantial attention has been devoted to understanding the large enhancements that arise from SMSERS hot spots. Original estimates of the enhancement from SMSERS-active sites were $\sim 10^{14-15}$, based on a typical Raman scattering cross-section ($\sim 10^{-30}$ cm^2) [3, 4]. However, the (resonance) Raman cross-section of R6G was determined to be $\sim 10^{-23}$ cm^2, based on femtosecond stimulated Raman spectroscopy measurements [119]. This implies that resonance effects account for as much as 10^7 of the reported 10^{14-15} EFs for R6G on Ag. The remaining $\sim 10^8$ enhancement can be attributed to surface effects, which is in agreement with the DDA calculations for the EM EF of the SMSERS-active structure discussed above. Both experimental and theoretical data, therefore, support the view that SMSERS sensitivity results from multiplication of the surface plasmon excitation and resonance Raman enhancement.

9.3.3.4 Excitation-wavelength dependence of SMSERS

While progress has been made in characterizing the structure of SMSERS hot spots and it is now known that EFs of $\sim 10^8$ can support single-molecule signal, many fundamental mechanistic questions remain. The relative contributions of the EM, CHEM, and resonance Raman enhancements to single-molecule sensitivity are of particular interest. Specifically, the contributions of resonance Raman versus surface effects when excitation occurs far from the molecular resonance are not well understood. Previous SMSERS studies of R6G on Ag have employed various excitation wavelengths. The SER spectra of R6G at five different excitation wavelengths (λ_{ex} = 632.8, 532.0, 514.4, 488.0, and 457.9 nm) were measured, where maximum sensitivity ($\sim 10^{-13}$ M) was achieved with excitation at 532 nm [120]. It was also shown that SMSERS signal was observed from R6G with a SER cross-section of $\sim 10^{-17}$, corresponding to an EF of $\sim 10^9$ at $\lambda_{ex} = 633$ nm [121]. Although it is clear that SMSERS intensity is maximized at the molecular resonance of R6G and that single molecules are observable with excitation frequencies off resonance, these studies were limited to fixed-frequency laser sources and were not performed on the same molecule.

To characterize the resonance Raman contributions to SMSERS intensity about the molecular resonance, surface-enhanced Raman excitation profiles (REPs) of a single R6G molecule on a Ag nanoparticle aggregate were measured using an optical parametric oscillator to finely tune the excitation wavelength [33]. SMSERS-active sites were identified using the isotopologue approach described earlier, and only those sites having the characteristic vibrational signature of nondeuterated R6G were used in the analysis. The single-molecule surface-enhanced REPs for three vibrational modes of R6G under dry N_2 were fit to single Lorentzian functions with FWHM = 365, 451, and 463 cm^{-1} for the 610, 1504, and 1647 cm^{-1} modes, respectively. As expected, a reduction in inhomogeneous broadening results in narrower linewidths for single molecules compared with those for the ensemble average. The single-molecule surface-enhanced REP representing all vibrational modes from $\sim$500–1800 cm^{-1} and the ensemble-averaged surface absorbance of R6G are displayed in **Figure 16**. The single-molecule data are best fit to a Lorentzian function with $E_{max} = 18563$ cm^{-1} (538.7 nm) and FWHM = 449 cm^{-1}. In addition to being narrower than the surface absorbance data, the single-molecule data show a significant deviation from the ensemble average ($E_{max} = 18409$ cm^{-1}), which offers additional evidence for single-molecule detection. These experiments demonstrated that the structure of single-molecule surface-enhanced REPs is dominated by resonance Raman enhancement, while the magnitude is dominated by EM enhancement. In fact, DDA calculations show that changes in the EM EF as a function of excitation energy are relatively small, and the EM EF is $>10^7$ over a wide range of frequencies.

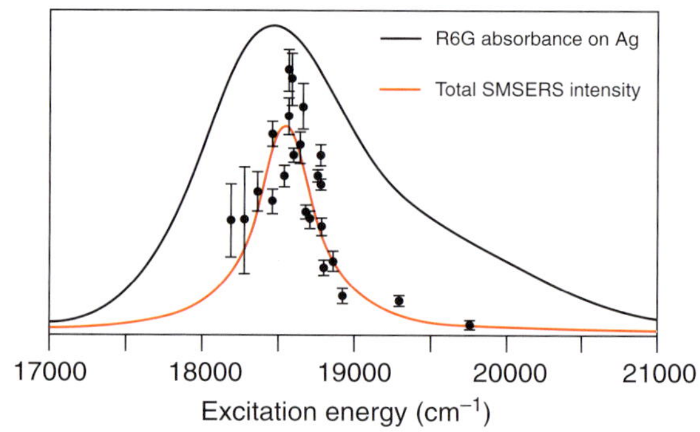

Figure 16 Surface-enhanced Raman excitation profile for a single R6G molecule fit by a Lorentzian function (red) and the ensemble-averaged surface absorbance of R6G on Ag (black).

9.3.4 SERS Sensing Applications

Vibrational spectroscopy provides chemical specificity derived from the unique vibrational modes of different molecules, and is widely used in biological and chemical sensing applications. However, infrared spectroscopy is greatly impeded by the strong absorbance of water, which is ubiquitous in most biological samples. Raman spectroscopy is limited by the inherent weakness of Raman scattering, resulting in insufficient sensitivity or the need for long integration times and high laser powers that can cause sample damage. SERS overcomes these limitations, and has emerged as a powerful sensing platform. The distance dependence of EM enhancement in SERS dictates that the analyte of interest must be within a few nanometers of the sensor surface. For analytes with no affinity for noble metals (i.e., Ag or Au), this can be achieved by surface modification with SAMs or thin metal oxide films, as discussed earlier.

9.3.4.1 In vivo *glucose sensing by SERS*

The development of *in vivo* biological sensors is particularly challenging due to the inherent complexity of the sample environment. Significant research effort has been focused on the quantitative, real-time detection of glucose, a crucial element in the management of diabetes. The feasibility of a SERS-based glucose sensor has been demonstrated using AgFON substrates functionalized with a mixed SAM consisting of decanethiol (DT) and mercaptohexanol (MH) [122]. The mixed DT/MH SAM, with dual hydrophobic/hydrophilic character, brings glucose closer to the Ag surface while excluding the interfering analytes. It also allows glucose to departition such that a quantitative equilibrium is established, analogous to partitioning in and out of the stationary phase in high-performance liquid chromatography. A continuous *in vivo* sensor must respond rapidly and be reversible in order to accurately monitor varying glucose concentrations throughout the day, and must be stable for at least several days. To evaluate stability, SER spectra of the DT/MH-functionalized AgFON surface in bovine plasma were taken over time, and intensities were found to be stable for 10 days. To demonstrate reversibility, the DT/MH-functionalized AgFON was exposed to cycles of water and 100 mM aqueous glucose (**Figure 17(a)**). The difference spectra demonstrate partitioning of glucose into the DT/MH SAM (F, G) and complete departitioning (H, I). Analysis of real-time partitioning and departitioning data reveal a time constant of less than 1 min, indicating rapid interaction between glucose and the SAM. In addition the DT/MH-functionalized AgFON surface was shown to be stable in bovine plasma for 10 days.

An exciting advance in the development of the SERS-based glucose sensor is its successful transition from *in vitro* models to *in vivo* testing using a real animal model, the Sprague–Dawley rat. Using partial least-squares chemometric analysis, quantitative glucose detection was demonstrated *in vivo* [123]. **Figure 17(b)** shows calibration and validation data for one rat plotted on a Clarke error grid, the standard for evaluating the reliability of glucose sensors in the clinically relevant concentration range. Most

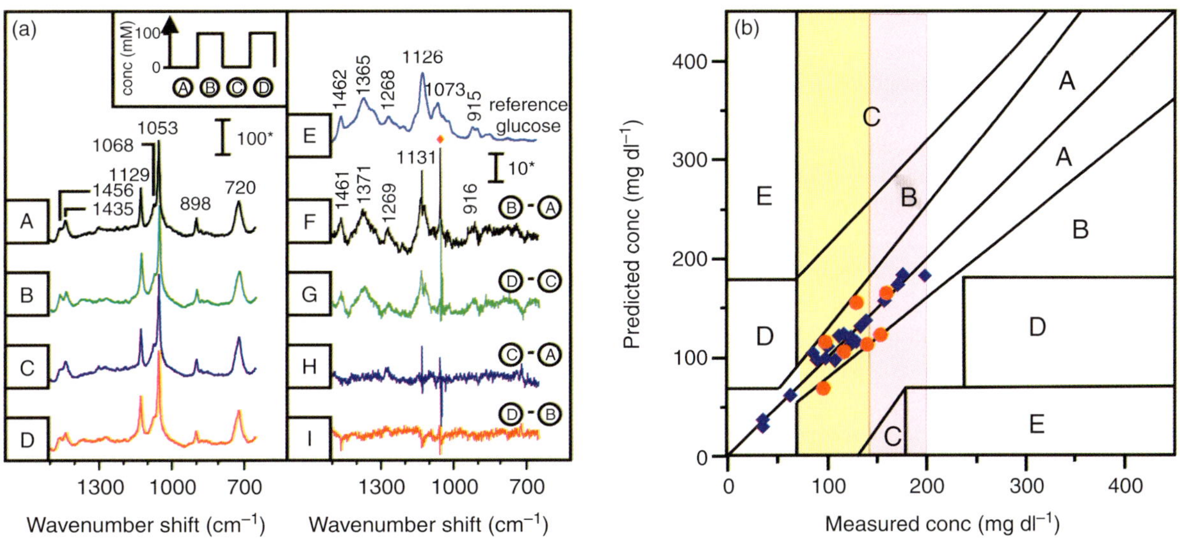

Figure 17 (a) *In vitro* and (b) *in vivo* glucose sensing using an AgFON substrate functionalized with a mixed self-assembled monolayer of decanethiol and mercaptohexanol.

of the data fall within the A and B ranges, defined as areas in which the predicted values are considered to be acceptable. Data points that fall outside of these ranges lead to erroneous diagnosis or failure to identify glucose levels outside the target range. In addition to evaluating quantitative accuracy, real-time *in vivo* glucose concentrations predicted by the SERS sensor were compared with those measured simultaneously by a commercially available electrochemical glucose sensor (Ascensia ELITE). Both sensors were able to monitor fluctuations in glucose concentration, and although the values differ, the same trends were observed.

Lastly, it was demonstrated that the DT/MH-functionalized AgFON sensor can be extended to the detection of other important biological analytes, such as lactate [124]. The potential for SERS to serve as a multi-analyte sensing platform was also shown by sequential glucose/lactate dosing with phosphate-buffered saline rinses between each step. The difference spectra demonstrate that the two analytes successfully partition and departition from the DT/MH-functionalized surface, thus demonstrating the reversibility of the sensor for multiple analytes.

9.3.4.2 Application of SERS to art conservation

The ability to detect and identify organic dyes and pigments used in historical works of art is of particular importance to the fields of art history and art conservation and restoration. For example, the identification of colorants used by an artist is useful for the authentication and dating of artwork, as well as for its long-term preservation. Thus far, high-performance liquid chromatography (HPLC) has been used to identify the largest number of colorants, but the large sample size required (~5 mm of fiber) poses a significant challenge. An alternate method for the identification of pigments is UV–vis absorbance spectroscopy, but the broad spectra typically observed can make identification difficult, especially when a mixture is present. Raman spectroscopy has also been employed, but strong fluorescence from dye molecules is often problematic. SERS overcomes some of these limitations owing to: (1) increased sensitivity compared to NR, which reduces the amount of sample necessary for analysis; (2) molecular quenching of fluorescence near metal surfaces; and (3) identification based on unique vibrational fingerprints. In fact, the potential of SERS in applications related to art conservation has been demonstrated in recent years by several groups [125].

The potential of SERS for problems in art conservation was first demonstrated by the successful identification of artists' dyestuffs in reference materials [126]. However, samples from real artworks are substantially more complex given the mixture of dyes present, matrix components, and the presence of contaminants. In addition, the harsh chemical extractions needed to remove the colorants from the artist's media can result in degradation of the dyes themselves. *In situ* SERS techniques, in which the surface-enhancing substrate (i.e., Ag nanoparticles) is brought to the dyed fiber have been successfully demonstrated [127, 128]. Recently, an *in situ* on-the-fiber extractionless nonhydrolysis SERS method using citrate-reduced colloids was developed and applied to authentic historical textiles [129]. As an example, this technique was performed on a single red weft fiber (0.5 mm × 2 μm) taken from a pre-Columbian Peruvian textile dated between 800 and 1350 AD (**Figure 18(a)**) [130]. Silver colloids were applied directly to the fiber and after drying it was imaged under a microscope (**Figure 18(b)**). The resulting SER spectra of the historical textile fiber and a reference wool fiber dyed with alizarin, a major dyestuff component of madder (*Rubia tinctoria L.*), are shown in **Figure 18(c)**. Despite significant aging and degradation of the fiber, the dye used to produce its red color is still easily identified as alizarin. *Galium corymbosum L.*, a Peruvian flower that contains large amounts of alizarin, was likely used to create this red color, a significant finding given the rarity of Peruvian textiles dyed with *Galium* species. Our understanding of the chronology and geographical distribution of this important colorant was therefore augmented by this work. In situations where more than one dye may be present in the sample, the coupling of thin-layer chromatography with SERS has also been shown to be a simple, effective technique capable of distinguishing dye molecules with very similar chemical structure [129].

9.3.4.3 SERS for chemical and biological warfare agent detection

The rapid and accurate detection of chemical and biological warfare agents is an analytical challenge of increasing significance. In addition to having high sensitivity, sensors designed for this purpose must take into account many other factors such as speed, accuracy, specificity, stability, and size. To this end, SERS-based sensors employing a portable Raman spectrometer have been developed. Initial work focused on the detection of calcium dipicolinate

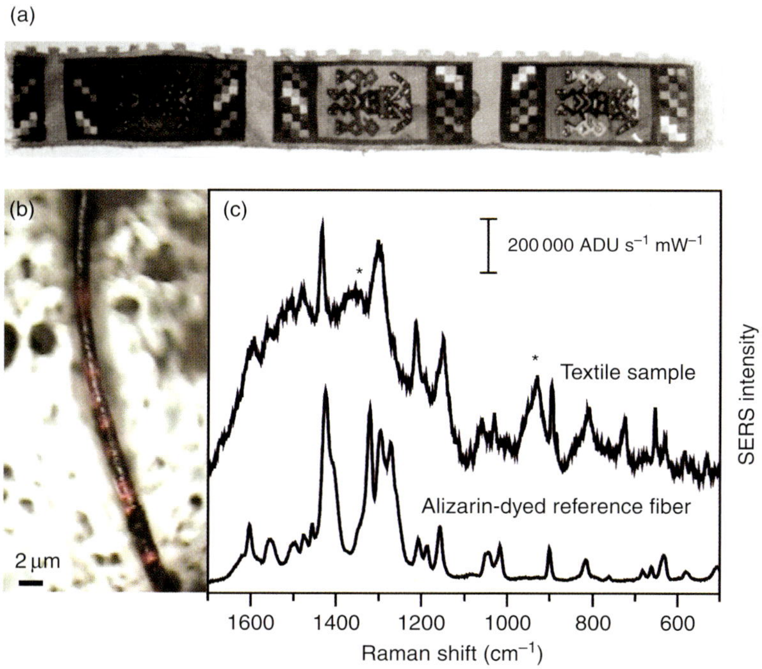

Figure 18 *In situ* on-the-fiber extractionless nonhydrolysis SERS using Ag colloidal paste. (a) Art sample (0.5 mm by 2 μm fiber) was obtained from a Peruvian textile dating to between 800 and 1350 AD, (b) coated with Ag colloidal paste, and (c) SER spectrum compared to a dyed reference fiber.

(CaDPA), a biomarker for anthrax [131]. CaDPA was extracted from *Bacillus subtilis* spores, a harmless simulant of *Bacillus anthracis* spores, and then drop coated on bare AgFON substrates. The SER spectrum of CaDPA displays a characteristic peak at 1020 cm^{-1}. The average signal intensity of this peak from varying concentrations of CaDPA was used to construct an adsorption isotherm and determine the limit of detection (LOD). At first, peak intensity is observed to increase linearly with concentration, and then saturates at higher spore concentrations as the adsorption sites on the AgFON become fully occupied. The LOD, defined as the concentration of spores for which the SERS signal of CaDPA at 1020 cm^{-1} is equal to 3 times the background SERS signal with a 1 min data acquisition, was found to be 2600 spores, well below the anthrax infectious dose of 10^4 spores. Moreover, modification of the AgFON substrate with a thin layer of Al_2O_3 deposited by ALD improved the LOD to 1400 spores [132]. The highest signal was achieved with two ALD cycles, corresponding to a thickness of ~2 Å Al_2O_3. The optimum thickness demonstrates the trade-off between increased binding affinity of CaDPA to the Al_2O_3 compared to silver, and decreased enhancement caused by decay of the EM field away from the surface. In addition to increasing sensitivity, the Al_2O_3 overlayer serves to protect the Ag surface from oxidation, thus improving the temporal and environmental stability of the sensor as well. SERS measurements of Al_2O_3-coated AgFONs suggest they are stable for >9 months.

The previous example highlights the ability of SERS to detect molecules introduced to the sensor from a solution containing the analyte of interest. However, detection of molecules from the gas phase is also desirable and particularly relevant for chemical warfare agents that may be released into the atmosphere. Initial experiments showed that 2-chloroethyl ethyl sulfide (CEES), a mustard gas simulant, could be detected at concentrations lower than the harmful limit, using an AgFON and a portable Raman spectrometer [133]. Recently, a detailed study has explored the potential of SERS for real-time gas phase detection of chemical warfare agents. The current leading technology for chemical sensing is ion-mobility spectrometry, which offers rapid analysis and compact instrumentation. However, there is no potential for remote sensing, and complex mixtures can be very challenging to identify. As a first step toward a quantitative evaluation of SERS gas detection, the adsorption kinetics of benzenethiol on

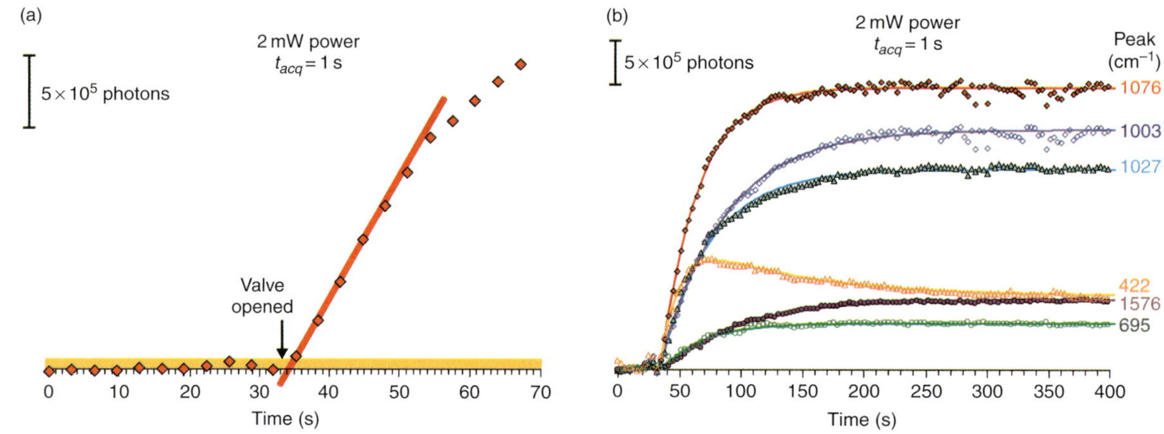

Figure 19 Real-time gas-phase sensing of benzenethiol (BT) on an AgFON. (a) SERS intensity of 8 ppm BT gas in N_2 at 358 K monitored with a portable Raman spectrometer and (b) time evolution of BT adsorption using the SERS intensity of several vibrational modes.

an AgFON were studied [134]. Benzenethiol was chosen as the test analyte because of its ability to form well-ordered SAMs from the gas phase. The Ag–S bond formation is irreversible under ambient conditions, and is the same mechanism responsible for irreversible binding of CEES to the AgFON previously described. For gas-phase dosing, ultrahigh purity N_2 was passed through a bubbler containing neat benzenethiol. The concentration was adjusted by dilution with N_2 downstream, and monitored by UV–vis spectroscopy before dosing to the flow cell. In order to establish how quickly molecules can be detected by SERS a detection limit time (DLt), dependent on the kinetic rate of analyte adsorption, defined as the minimum concentration that can be detected in one second of exposure for an acquisition time of 1 s and a spot size of 1.4×10^{-5} cm^2 was determined. The SERS signal detected from 8 ppm benzenthiol gas in N_2 dosed to an AgFON at 358 K was continuously monitored with a portable Raman spectrometer (**Figure 19**).

In **Figure 19(a)**, the intensity of the 1076 cm^{-1} mode of benzenethiol is plotted as a function of time. The initial rate of adsorption is linear and the detection threshold signal (three standard deviations (3σ) above the mean background signal), corresponding to 1.7×10^8 molecules or 1.8% of a monolayer, is reached in less than 1 s. **Figure 19(b)** shows the time evolution of benzenethiol adsorption using the SERS intensity of several benzenethiol modes. With the exception of the 422 cm^{-1} mode, all the peaks grow in according to first-order Langmuir kinetics. The 422 cm^{-1} mode displays a double-exponential behavior that is thought to result from the strong dependence of this particular mode's EM enhancement on surface orientation. The DLt was calculated to be 6 ppm-s for benzenethiol gas adsorption on an AgFON at 358 K with an acquisition time of 1 s. The DLt is fundamentally limited by the low sticking probability of benzenethiol, which was found to be 1.5×10^{-5}. The sticking probability increases with increasing temperature and the slope of the Arrhenius plot yields an activation barrier of ~13 kJ mol^{-1}. Therefore, the sensitivity of this sensor may be improved by increasing the sticking probability of the analyte, highlighting the importance of surface functionalization to optimize analyte affinity. These quantitative experiments present significant progress toward the goal of real-time gas detection of chemical warfare agents based on SERS.

9.4 Future Directions

9.4.1 New Plasmonic Materials for SERS

Despite the extraordinary surface sensitivity of SERS and the wealth of chemical information it provides, it is largely limited to studies on noble-metal nanostructured surfaces. Although Au and Ag are the mainstays of current LSPR and SERS technology, they are expensive materials and do not support plasmons in the UV region, representing an obstacle to next-generation biological investigations. Yet, in theory, any material that exhibits a negative real and small positive imaginary dielectric function may support surface plasmons. A number of other metals (i.e.,

Li, Na, Al, In, Ga, and Cu) meet this criterion but have not been widely implemented in plasmonic applications, in part due to problems associated with instability or oxidation. Surface oxidation is especially problematic as it may significantly alter the optical properties of the metal nanostructure. In some cases, it is possible to overcome this problem by removing the oxide layer with an appropriate etchant. For example, Cu is oxidized upon exposure to ambient atmosphere to form a surface layer of Cu oxides (Cu_2O and CuO) [135]. However, glacial acetic acid was found to be effective for removing the layer of Cu oxides on Cu nanoparticles without modifying the Cu metal surface [136]. Whereas untreated Cu nanoparticles fabricated by NSL exhibit broad and weak LSPR spectra, removal of the oxidized layer results in comparatively narrower and more intense extinction spectra. Electrodynamics calculations on a Cu nanoparticle coated by a Cu oxide shell of varying thickness confirmed the oxidation-induced changes to the LSPR. According to AFM and SEM images of the Cu nanoparticles, the oxide layer is of subnanometer thickness and was formed just after exposure to ambient laboratory conditions. Importantly, acid treatment did not significantly alter the shape or height of the Cu nanoparticles. Using NSL fabrication, the LSPR of Cu nanoparticles was tuned throughout the visible region, and exhibited comparable peak intensities and linewidths to Ag and Au arrays when the LSPR λ_{max} is > ~650 nm. The results of these investigations suggest that Cu, a relatively inexpensive metal, may be useful to replace Ag and Au for certain LSPR and SERS applications.

The plasmon resonance condition for Ag and Au nanoparticles is met in the visible and NIR regions of the EM spectrum. However, the development of nanostructured materials capable of supporting surface plasmons in the UV region is interesting for the potential application of UV SERS. Ultraviolet excitation would be advantageous for interrogating samples with electronic resonances in the UV, as is the case with many biological molecules, in order to benefit from combined resonance Raman and EM enhancement effects. From the dielectric functions of Ag and Al (**Figure 2**), it is evident that Ag cannot support surface plasmons below ~350 nm, where the real part of its dielectric function is positive, whereas Al should support plasmons from 200 to ~800 nm. Indeed, several recent studies of the plasmonic properties of Al have shown that Al nanoparticle arrays support localized surface plasmons throughout the UV–vis NIR spectrum [24, 137, 138]. Aluminum,

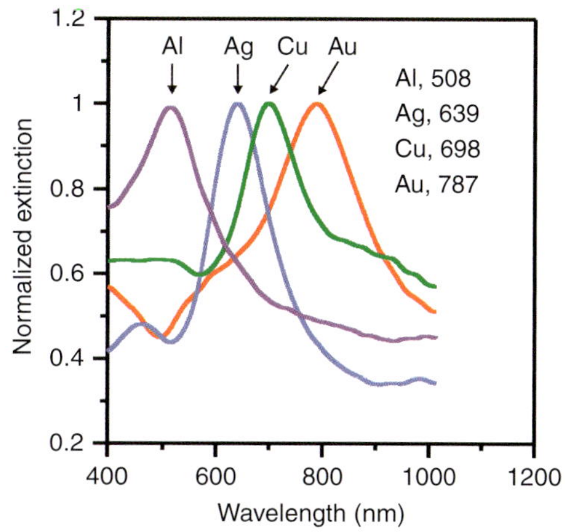

Figure 20 LSPR spectra of nanoparticle arrays with identical geometries but varying metal (Al, Ag, Cu, or Au).

however, is extremely reactive to oxygen and upon exposure to ambient conditions immediately forms a thin (~2–3 nm) Al_2O_3 layer that prevents further attack by oxygen. Interestingly, for Al nanoparticles fabricated by NSL, SEM images reveal a sharp contrast between the tips and the core of the triangular nanoparticles, suggesting concentrated oxide formation at the nanoparticle tips [24]. The native oxide shell results in a red-shifted LSPR, owing to the dielectric sensitivity of the LSPR, but does not exhibit the dampening effect of Cu oxides. However, both experimental and theoretical data show that when the LSPR of the nanoparticles is close to the Al interband transition (~800 nm), the LSPR is significantly broadened. A comparison of the normalized LSPR extinction spectra for Al, Ag, Cu, and Au nanoparticles of the same geometry, all fabricated by NSL, is shown in **Figure 20**. In general, Al nanoparticles exhibited a bluer, broader, and less intense LSPR compared to the other metals in the visible region. In the UV region, however, Al is a superior plasmonic material, and the potential utility of NSL Al triangular arrays as a substrate for UV SERS is currently under investigation. Undoubtedly, future advances in LSPR and SERS applications will rely on broadening the plasmonic periodic table to new materials as well as novel nanostructures.

9.4.2 Novel Nanostructures

Although more than 20 years have passed since the discovery of SERS, some of the fundamental questions have remained unanswered. For example, what nanoparticle structure or array provides the highest

EM enhancement? This question is significant since the development of next-generation LSPR- and SERS-based sensors requires that the most enhancing nanoparticles and nanoparticle arrays are developed. Therefore, an understanding of the fundamental relationships among the SER scattering intensity, LSPR, and nanoparticle structure must be established. Toward this end, recent interest has focused on correlated investigations of structure (i.e., with AFM or TEM imaging) and LSPR for individual nanoparticles as well as the corresponding SER spectra for adsorbed analytes, which provide the ability to measure the fundamental relationships between the structural and optical properties at the single-particle level, unobscured by ensemble averaging.

Recent advances in nanofabrication techniques have enabled the development of new nanostructures for LSPR and SER spectroscopy. For example, novel chemical synthetic methods have provided a plethora of novel nanoparticle shapes, sizes, and functionalities including Ag nanocubes [38, 139], Ag and Au nanoprisms [36], and stabilized spherical aggregates [140]. In order to determine the optimal nanostructures for LSPR and SERS sensing as well as how such nanoparticles can be incorporated in next-generation devices, the detailed relationship between the optical response and structural properties of single nanoparticles is under investigation. Cube-shaped nanoparticles fabricated using polyol synthesis [38] exhibit two LSPR maxima owing to the asymmetry induced by surface immobilization on glass [141]. Both LSPR peaks are sensitive to changes in dielectric environment, but the blue-shifted resonance has superior LSPR sensing capabilities due to its sharpness [142]. In addition, correlated LSPR and HRTEM measurements of individual Ag nanocubes revealed the dependence of the LSPR on nanocube size and shape. Corresponding electrodynamics calculations demonstrated the strong sensitivity between the nanocube optical response and the face-to-face width, corner and side rounding, and substrate of each nanocube. In a related study, correlated high-resolution SEM and polarization-dependent SERS measurements were performed on individual Ag nanocubes [143]. Although the Ag nanocube samples appeared monodisperse, various SER scattering intensities were observed from individual nanocubes as a result of nanoparticle orientation relative to the exciting laser field. Ensemble-averaged SERS experiments on the nanocubes reveal that the overall average EF is $\sim 10^5$, although detailed studies at the single-nanoparticle level are needed to elucidate the underlying details of the EM enhancement [144].

The most desirable single-particle experiment is to measure the structure, LSPR, and SER scattering intensity of the same nanoparticle. At present, SMSERS experiments have provided the best example of this correlated study. In particular, SMSERS investigations (Section 9.3.3.3) of individual R6G molecules on aggregated Ag colloids demonstrated that hot spots located between nanoparticles are responsible for the enhanced fields that enable single-molecule detection [32, 145]. However, the Ag colloids used in the majority of SMSERS experiments are randomly aggregated using electrolyte solutions and do not provide controlled, homogeneous structures. Toward this end, novel nanostructures composed of a stabilizing shell around a well-defined metallic core have been developed. For example, silica-encapsulated SERS nanotags containing Au cores, a layer of a SERS 'reporter' molecule, and silica shell have demonstrated structural uniformity, optical robustness, and the capability to use bio-relevant reporter molecules [140]. Moreover, these SERS nanotags provide an excellent system in which to study the structure–property relationships for well-defined and easily controlled nanoparticle geometries (e.g., monomers, dimers) [146]. Ultimately, correlated studies performed at the single-nanoparticle level on novel nanostructures such as nanocubes and silica-coated nanospheres hold the promise for the development of robust and high-enhancing SERS substrates and these studies are underway.

9.4.3 Tip-Enhanced Raman Spectroscopy

In addition to exploring new plasmonic materials and nanostructures, other successful approaches have been used to broaden the applicability and generality of SERS. One significant challenge to achieving a truly general SERS substrate is that surface functionalization is often required to facilitate the binding of specific analytes to specific surfaces. For example, thin catalytically active transition metal overlayers were electrodeposited on roughened Au in order to study catalytic reactions with SERS [147]. This chapter highlighted other types of overlayers such as SAMs and thin metal oxide films that are used to functionalize the surface of a SERS substrate. However, perhaps the most promising technique for achieving complete substrate generality is TERS. In TERS, enhancement arises from a noble metal

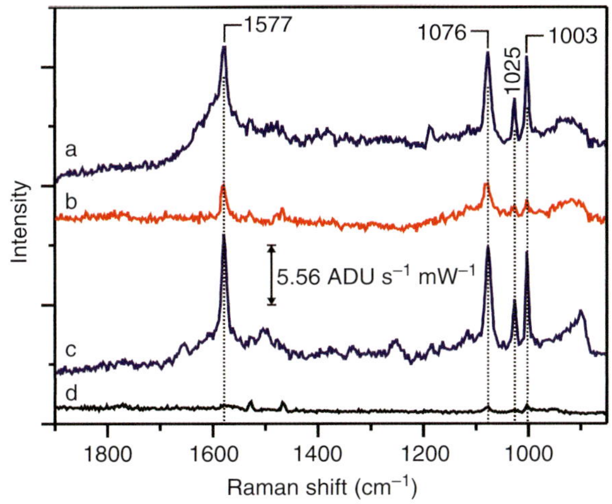

Figure 21 Tip-enhanced Raman spectra for BT adsorbed to an 8-nm Au island film. (a) AFM tip engaged, (b) tip withdrawn by 100 μm, (c) tip re-engaged, and (d) sample removed with laser focused on the tip.

scanning probe microscopy (SPM) tip rather than from the substrate. When the tip is irradiated with visible light, there is a corresponding excitation of its LSPR. The resulting EM enhancement near the tip apex is used to increase the Raman scattering from molecules in the vicinity of the tip, even without direct adsorption. The Raman scattering intensities of molecules located in the near-field region of Ag or Au SPM tips are enhanced by three to six orders of magnitude [148]. Moreover, the use of an SPM tip provides nanometer-scale resolution, a significant improvement to the diffraction-limited spatial resolution of SERS, which is approximately $\lambda_{ex}/2$ or ~250 nm for visible excitation.

Figure 21 presents representative TER spectra for benzenethiol adsorbed to an 8-nm Au island film (AuIF) that is weakly SERS active at $\lambda_{ex} = 532$ nm. In spectrum 'a', the AFM tip that is coated with 40-nm Ag is engaged and makes contact with the sample yielding a high signal-to-noise ratio TER spectrum. In 'b', the tip is withdrawn from the sample by 100 μm, while the laser spot remains focused on the sample. When the AFM tip is re-engaged in contact mode, the same high signal-to-noise TER spectrum is observed (c). After several engage–disengage cycles, the sample was removed and the spectrum (d) was taken with the laser focused on the AFM tip. These results demonstrate that benzenethiol molecules were not transferred from the AuIF to the tip, and thus the observed TERS signal originated from the molecules adsorbed on the Au surface. By taking the difference in surface area between the laser spot (~1 μm^2) and the contact area of the AFM tip (~300 nm^2) into account, the average EF is estimated to be $>10^4$, demonstrating the sensitivity resulting from tip enhancement.

TERS combines the high sensitivity and rich chemical information content of SERS with the excellent spatial resolution provided by scanning probe microscopy. In addition, ultrahigh vacuum (UHV) TERS provides integration of laser excitation and scanning tunneling microscopy (STM) capabilities and permits atomic resolution and structural elucidation of adsorbate molecules on clean, single-crystal or nanostructured surfaces in a controlled environment. In a recent UHV-TERS study, TER imaging of a single brilliant cresyl blue molecule with a spatial resolution of 15 nm was demonstrated [6]. Although TERS is still in the early stages of development, the results of these experiments demonstrate that TERS will be a powerful tool for probing chemical problems in catalysis, surface science, tribology, biological science, and nanomaterials.

9.5 Conclusion

In this chapter we aimed to present an overall description of nanostructures, LSPR, and SERS through fundamental expressions as well as experimental investigations. The research highlighted here demonstrates a promising future for SERS and significant implications for the field of nanoscience and technology.

Acknowledgments

The authors gratefully acknowledge support from the NSF (EEC-0118025, CHE-0414554, BES-0507036), the AFOSR DURIP (FA-9550-07-1-0526), DTRA JSTO (FA9550-06-1-0558), AFOSR/DARPA Project BAA07-61 (FA9550-08-1-0221), the NSF NSEC program (EEC-0647560), and the NSF MRSEC (DMR-0520513) at the Materials Research Center of Northwestern University. Conservation science research at the Art Institute of Chicago is made possible by a generous grant from the Andrew W. Mellon Foundation.

References

1. Jeanmaire DL and Van Duyne RP (1977) Surface Raman spectroelectrochemistry. Part I. Heterocyclic, aromatic, and aliphatic amines adsorbed on the anodized silver electrode. *Journal of Electroanalytical Chemistry and Interfacial Electrochemistry* 84: 1–20.
2. Albrecht MG and Creighton JA (1977) Anomalously intense Raman spectra of pyridine at a silver electrode. *Journal of the American Chemical Society* 99: 5215–5217.
3. Kneipp K, Wang Y, Kneipp H, *et al.* (1997) Single molecule detection using surface-enhanced Raman scattering (SERS). *Physical Review Letters* 78: 1667–1670.
4. Nie SM and Emory SR (1997) Probing single molecules and single nanoparticles by surface-enhanced Raman scattering. *Science* 275: 1102–1106.
5. Neacsu CC, Dreyer J, Behr N, and Raschke MB (2006) Scanning-probe Raman spectroscopy with single-molecule sensitivity. *Physical Review B* 73: 4.
6. Steidtner J and Pettinger B (2008) Tip-enhanced Raman spectroscopy and microscopy on single dye molecules with 15 nm resolution. *Physical Review Letters* 100: 236101/236101-236101/236104.
7. Brill RH (1965) The chemistry of the Lycurgus cup. In: *Proceedings of the 7th International Congress on Glass, comptes rendus 2*, pp. 1–13, Bruxelles.
8. Barber DJ and Freestone IC (1990) An investigation of the origin of the colour of the Lycurgus cup by analytical transmission electron microscopy. *Archaeometry* 32: 33–45.
9. Faraday M (1857) The Bakerian lecture: Experimental relations of gold (and other metals) to light. *Philosophical Transactions of the Royal Society of London* 147: 145–181.
10. Willets KA and Van Duyne RP (2007) Localized surface plasmon resonance spectroscopy and sensing. *Annual Review of Physical Chemistry* 58: 267–297.
11. Haes AJ and Van Duyne RP (2004) A unified view of propagating and localized surface plasmon resonance biosensors. *Analytical and Bioanalytical Chemistry* 379: 920–930.
12. Yonzon CR, Jeoung E, Zou S, *et al.* (2004) A comparative analysis of localized and propagating surface plasmon resonance sensors: The binding of concanavalin A to a monosaccharide functionalized self-assembled monolayer. *Journal of the American Chemical Society* 126: 12669–12676.
13. Hall WP, Anker JN, Lin Y, *et al.* (2008) A calcium-modulated plasmonic switch. *Journal of the American Chemical Society* 130: 5836–5837.
14. Mie G (1908) Contributions to the optics of turbid media, especially colloidal metal solutions. *Annalen der Physik (Weinheim, Germany)* 25: 377–445.
15. Kelly KL, Coronado E, Zhao LL, and Schatz GC (2003) The optical properties of metal nanoparticles: The influence of size, shape, and dielectric environment. *Journal of Physical Chemistry B* 107: 668–677.
16. Gans R (1912) The form of ultramicroscopic gold particles. *Annalen der Physik (Weinheim, Germany)* 37: 881–900.
17. Gans R (1915) Form of ultramicroscopic particles of silver. *Annalen der Physik (Berlin, Germany)* 47: 270–284.
18. Kelly KL, Jensen TR, Lazarides AA, and Schatz GC (2002) Modeling metal nanoparticles optical properties. *Metal Nanoparticles: Synthesis, Characterization, and Applications*. pp 88–118. New York: Marcel Dekker, Inc.
19. Feynman RP, Leighton RB, and Sands M (1964) *The Feynman Lectures on Physics*. Reading, MA: Addison-Wesley.
20. Doyle WT (1989) Optical properties of a suspension of metal spheres. *Physical Review B* 39: 9852.
21. Sherry LJ, Jin R, Mirkin CA, Schatz GC, and Van Duyne RP (2006) Localized surface plasmon resonance spectroscopy of single silver triangular nanoprisms. *Nano Letters* 6: 2060–2065.
22. Hecht E (1998) *Optics*. New York: Addison-Wesley.
23. Bohren CF and Huffman DR (1983) *Absorption and Scattering of Light by Small Particles*. New York: Wiley.
24. Chan GH, Zhao J, Schatz GC, and Van Duyne RP (2008) Localized surface plasmon resonance spectroscopy of triangular aluminum nanoparticles. *Journal of Physical Chemistry C* 112: 13958–13963.
25. Stiles PL, Dieringer JA, Shah NC, and Van Duyne RP (2008) Surface-enhanced Raman spectroscopy. *Annual Review of Analytical Chemistry* 1: 601–626.
26. Jung LS, Campbell CT, Chinowsky TM, Mar MN, and Yee SS (1998) Quantitative interpretation of the response of surface plasmon resonance sensors to adsorbed films. *Langmuir* 14: 5636–5648.
27. Schatz GC (2001) Electrodynamics of nonspherical noble metal nanoparticles and nanoparticle aggregates. *Journal of Molecular Structure (Theochem)* 573: 73–80.
28. Purcell EM and Pennypacker CR (1973) Scattering and absorption of light by nonspherical dielectric grains. *Astrophysical Journal* 186: 705–714.
29. McFarland AD and Van Duyne RP (2003) Single silver nanoparticles as real-time optical sensors with zeptomole sensitivity. *Nano Letters* 3: 1057–1062.
30. Billaud P, Huntzinger JR, Cottancin E, *et al.* (2007) Optical extinction spectroscopy of single silver nanoparticles. *European Physical Journal D: Atomic, Molecular, Optical and Plasma Physics* 43: 271–274.
31. Lee PC and Meisel D (1982) Adsorption and surface-enhanced Raman of dyes on silver and gold sols. *Journal of Physical Chemistry* 86: 3391–3395.
32. Dieringer JA, Lettan RB, II, Scheidt KA, and Van Duyne RP (2007) A frequency domain existence proof of single-molecule surface-enhanced Raman spectroscopy. *Journal of the American Chemical Society* 129: 16249–16256.
33. Dieringer JA, Wustholz KL, Masiello DJ, *et al.* (2009) Surface-enhanced Raman excitation spectroscopy of a single Rhodamine 6G molecule. *Journal of the American Chemical Society* 131: 849–854.
34. Camden JP, Dieringer JA, Wang Y, *et al.* (2008) Probing the structure of single-molecule surface-enhanced Raman

scattering hot spots. *Journal of the American Chemical Society* 130: 12616–12617.
35. Oldenburg SJ, Oldenburg SJ, Averitt RD, Westcott SL, and Halas NJ (1998) Nanoengineering of optical resonances. *Chemical Physics Letters* 288: 243–247.
36. Jin RC, Cao YW, Mirkin CA, *et al.* (2001) Photo induced conversion of silver nanospheres to nanoprisms. *Science* 294: 1901–1903.
37. Sun Y and Xia Y (2002) Shape-controlled synthesis of gold and silver nanoparticles. *Science (New York)* 298: 2176–2179.
38. Siekkinen AR, McLellan JM, Chen JY, and Xia YN (2006) Rapid synthesis of small silver nanocubes by mediating polyol reduction with a trace amount of sodium sulfide or sodium hydrosulfide. *Chemical Physics Letters* 432: 491–496.
39. Wiley BJ, Xiong Y, Li Z-Y, Yin Y, and Xia Y (2006) Right bipyramids of silver: A new shape derived from single twinned seeds. *Nano Letters* 6: 765–768.
40. Nehl CL, Liao H, and Hafner JH (2006) Optical properties of star-shaped gold nanoparticles. *Nano Letters* 6: 683–688.
41. Amendola V and Meneghetti M (2009) Laser ablation synthesis in solution and size manipulation of noble metal nanoparticles. *Physical Chemistry – Chemical Physics* 11: 3805–3821.
42. Fojtik A, Giersig M, and Henglein A (1993) Formation of nanometer-size silicon particles in a laser induced plasma in SiH_4. *Berichte der Bunsen-Gesellschaft* 97: 1493–1496.
43. Van Duyne RP, Hulteen JC, and Treichel DA (1993) Atomic force microscopy and surface-enhanced Raman spectroscopy. I. Silver island films and silver film over polymer nanosphere surfaces supported on glass. *Journal of Chemical Physics* 99: 2101–2115.
44. Fleischmann M, Hendra PJ, and McQuillan AJ (1974) Raman spectra of pyridine adsorbed at a silver electrode. *Chemical Physics Letters* 26: 163–166.
45. Haynes CL, McFarland AD, Zhao L, *et al.* (2003) Nanoparticle optics: The importance of radiative dipole coupling in two-dimensional nanoparticle arrays. *Journal of Physical Chemistry B* 107: 7337–7342.
46. Hicks EM, Gunnarrsson L, Rindevicius T, *et al.* (2005) Investigating narrow plasmons in nanoparticle arrays fabricated using electron beam lithography. *Materials Research Society Symposium Proceedings* 872: 317–322.
47. McLellan E, Gunnarsson L, Rindzevicius T, *et al.* (2007) Plasmonic and diffractive coupling in two-dimensional arrays of nanoparticles produced by electron beam lithography. In: *Materials Research Society Symposium Proceedings*, 951E, 0951-E0909-0920.
48. Sung J, Hicks EM, Van Duyne RP, and Spears KG (2007) Nanoparticle spectroscopy: Dipole coupling in two-dimensional arrays of L-shaped silver nanoparticles. *Journal of Physical Chemistry C* 111: 10368–10376.
49. Sung J, Hicks EM, Van Duyne RP, and Spears KG (2008) Nanoparticle spectroscopy: Plasmon coupling in finite-sized two-dimensional arrays of cylindrical silver nanoparticles. *Journal of Physical Chemistry C* 112: 4091–4096.
50. Henzie J, Lee J, Lee MH, Hasan W, and Odom TW (2009) Nanofabrication of plasmonic structures. *Annual Review of Physical Chemistry* 60: 147–165.
51. Sundaramurthy A, Schuck PJ, Conley NR, *et al.* (2006) Toward nanometer-scale optical photolithography: Utilizing the near-field of bowtie optical nanoantennas. *Nano Letters* 6: 355–360.
52. Hulteen JC and Van Duyne RP (1995) Nanosphere lithography: A materials general fabrication process for periodic particle array surfaces. *Journal of Vacuum Science and Technology A: Vacuum, Surfaces, and Films* 13: 1553–1558.
53. Haynes CL and Van Duyne RP (2001) Nanosphere lithography: A versatile nanofabrication tool for studies of size-dependent nanoparticle optics. *Journal of Physical Chemistry B* 105: 5599–5611.
54. Haynes CL, McFarland AD, Smith MT, Hulteen JC, and Van Duyne RP (2002) Angle-resolved nanosphere lithography: Manipulation of nanoparticle size, shape, and interparticle spacing. *Journal of Physical Chemistry B* 106: 1898–1902.
55. Hicks EM, Lyandres O, Hall WP, *et al.* (2007) Plasmonic properties of anchored nanoparticles fabricated by reactive ion etching and nanosphere lithography. *Journal of Physical Chemistry C* 111: 4116–4124.
56. Zhao J, Haes AJ, Zhang X, *et al.* (2006) Alkanethiol mediated release of surface bound nanoparticles fabricated by nanosphere lithography. In: *Materials Research Society Symposium Proceedings*, 900E, 0900-O0913-0908.
57. Haes AJ, Zhao J, Zou S, *et al.* (2005) Solution-phase, triangular Ag nanotriangles fabricated by nanosphere lithography. *Journal of Physical Chemistry B* 109: 11158–11162.
58. Zhang X, Hicks EM, Zhao J, Schatz GC, and Van Duyne RP (2005) Electrochemical tuning of silver nanoparticles fabricated by nanosphere lithography. *Nano Letters* 5: 1503–1507.
59. Jensen TR, Duval ML, Kelly KL, *et al.* (1999) Nanosphere lithography: Effect of the external dielectric medium on the surface plasmon resonance spectrum of a periodic array of silver nanoparticles. *Journal of Physical Chemistry B* 103: 9846–9853.
60. Malinsky MD, Kelly KL, Schatz GC, and Van Duyne RP (2001) Chain length dependence and sensing capabilities of the localized surface plasmon resonance of silver nanoparticles chemically modified with alkanethiol self-assembled monolayers. *Journal of the American Chemical Society* 123: 1471–1482.
61. Raschke G, Kowarik S, Franzl T, *et al.* (2003) Biomolecular recognition based on single gold nanoparticle light scattering. *Nano Letters* 3: 935–938.
62. Mock JJ, Barbic M, Smith DR, Schultz DA, and Schultz S (2002) Shape effects in plasmon resonance of individual colloidal silver nanoparticles. *Journal of Chemical Physics* 116: 6755–6759.
63. Haes AJ and Van Duyne RP (2002) A nanoscale optical biosensor: Sensitivity and selectivity of an approach based on the localized surface plasmon resonance spectroscopy of triangular silver nanoparticles. *Journal of the American Chemical Society* 124: 10596–10604.
64. Riboh JC, Haes AJ, McFarland AD, Yonzon CR, and Van Duyne RP (2003) A nanoscale optical biosensor: Real-time immunoassay in physiological buffer enabled by improved nanoparticle adhesion. *Journal of Physical Chemistry B* 107: 1772–1780.
65. Haes AJ, Hall WP, Chang L, Klein WL, and Van Duyne RP (2004) A localized surface plasmon resonance biosensor: First steps toward an assay for Alzheimer's disease. *Nano Letters* 4: 1029–1034.
66. Haes AJ, Chang L, Klein WL, and Van Duyne RP (2005) Detection of a biomarker for Alzheimer's disease from synthetic and clinical samples using a nanoscale optical biosensor. *Journal of the American Chemical Society* 127: 2264–2271.
67. El-Sayed IH, Huang X, and El-Sayed MA (2005) Surface plasmon resonance scattering and absorption of anti-

EGFR antibody conjugated gold nanoparticles in cancer diagnostics: Applications in oral cancer. *Nano Letters* 5: 829–834.
68. Anker JN, Hall WP, Lyandres O, *et al.* (2008) Biosensing with plasmonic nanosensors. *Nature Materials* 7: 442–453.
69. Yonzon CR, Stuart DA, Zhang X, *et al.* (2005) Towards advanced chemical and biological nanosensors – an overview. *Talanta* 67: 438–448.
70. Stuart DA, Haes AJ, Yonzon CR, Hicks EM, and Van Duyne RP (2005) Biological applications of localized surface plasmonic phenomena. *IEE Proceedings: Nanobiotechnology* 152: 13–32.
71. Whitney AV, Elam JW, Zou S, *et al.* (2005) Localized surface plasmon resonance nanosensor: A high-resolution distance-dependence study using atomic layer deposition. *Journal of Physical Chemistry B* 109: 20522–20528.
72. Haes AJ, Zou S, Schatz GC, and Van Duyne RP (2004) A nanoscale optical biosensor: The long range distance dependence of the localized surface plasmon resonance of noble metal nanoparticles. *Journal of Physical Chemistry B* 108: 109–116.
73. Haes AJ, Zou S, Schatz GC, and Van Duyne RP (2004) Nanoscale optical biosensor: Short range distance dependence of the localized surface plasmon resonance of noble metal nanoparticles. *Journal of Physical Chemistry B* 108: 6961–6968.
74. Reinhard BM, Siu M, Agarwal H, Alivisatos AP, and Liphardt J (2005) Calibration of dynamic molecular rulers based on plasmon coupling between gold nanoparticles. *Nano Letters* 5: 2246–2252.
75. Soennichsen C, Reinhard BM, Liphardt J, and Alivisatos AP (2005) A molecular ruler based on plasmon coupling of single gold and silver nanoparticles. *Nature Biotechnology* 23: 741–745.
76. Storhoff JJ, Lazarides AA, Mucic RC, *et al.* (2000) What controls the optical properties of DNA-linked gold nanoparticle assemblies? *Journal of the American Chemical Society* 122: 4640–4650.
77. Ritala M and Leskela M (2002) Atomic layer deposition. In: Naiwa HS (ed.) *Handbook of Thin Film Materials*, vol. 1, pp. 103–159. San Diego, CA: Academic Press.
78. Puurunen RL (2005) Surface chemistry of atomic layer deposition: A case study for the trimethylaluminum/water process. *Journal of Applied Physics* 97: 52.
79. Ott AW, Klaus JW, Johnson JM, and George SM (1997) Al3O3 thin film growth on Si(100) using binary reaction sequence chemistry. *Thin Solid Films* 292: 135–144.
80. Sung J, Kosuda KM, Zhao J, *et al.* (2008) Stability of silver nanoparticles fabricated by nanosphere lithography and atomic layer deposition to femtosecond laser excitation. *Journal of Physical Chemistry C* 112: 5707–5714.
81. Whitney AV, Elam JW, Stair PC, and Van Duyne RP (2007) Toward a thermally robust operando surface-enhanced Raman spectroscopy substrate. *Journal of Physical Chemistry C* 111: 16827–16832.
82. Haes AJ, Haynes CL, McFarland AD, *et al.* (2005) Plasmonic materials for surface-enhanced sensing and spectroscopy. *MRS Bulletin* 30: 368–375.
83. Roark SE, Semin DJ, Lo A, Skodje RT, and Rowlen KL (1995) Solvent-induced morphology changes in thin silver films. *Analytica Chimica Acta* 307: 341–353.
84. Raman CV and Krishnan KS (1928) A new type of secondary radiation. *Nature* 121: 501–502.
85. McCreery RL (2000) *Raman Spectroscopy for Chemical Analysis*. New York: Wiley.
86. Campion A and Kambhampati P (1998) Surface-enhanced Raman scattering. *Chemical Society Reviews* 27: 241–250.
87. Otto A, Mrozek I, Grabhorn H, and Akemann W (1992) Surface-enhanced Raman-scattering. *Journal of Physics: Condensed Matter* 4: 1143–1212.
88. Burstein E, Chen YJ, Chen CY, Lundquist S, and Tosatti E (1979) "Giant" Raman scattering by adsorbed molecules on metal surfaces. *Solid State Communications* 29: 567–570.
89. Gersten JI, Birke RL, and Lombardi JR (1979) Theory of enhance I light scattering from molecules adsorbed at the metal–solution interface. *Physical Review Letters* 43: 147.
90. Lombardi JR, Birke RL, Lu T, and Xu J (1986) Charge-transfer theory of surface enhanced Raman spectroscopy: Herzberg–Teller contributions. *Journal of Chemical Physics* 84: 4174.
91. Billmann J and Otto A (1982) Electronic surface state contribution to surface enhanced Raman scattering. *Solid State Communications* 44: 105–107.
92. Furtak TE and Macomber SH (1983) Voltage-induced shifting of charge-transfer excitations and their role in surface-enhanced Raman scattering. *Chemical Physics Letters* 95: 328–332.
93. Lombardi JR, Birke RL, Sanchez LA, Bernard I, and Sun SC (1984) The effect of molecular structure on voltage induced shifts of charge transfer excitation in surface enhanced Raman scattering. *Chemical Physics Letters* 104: 240–247.
94. Moskovits M (1985) Surface-enhanced spectroscopy. *Reviews of Modern Physics* 57: 783.
95. Kerker M, Wang D-S, and Chew H (1980) Surface enhanced Raman scattering (SERS) by molecules adsorbed at spherical particles: Errata. *Applied Optics* 19: 4159–4174.
96. Wang DS and Kerker M (1981) Enhanced Raman-scattering by molecules adsorbed at the surface of colloidal spheroids. *Physical Review B* 24: 1777–1790.
97. McFarland AD, Young MA, Dieringer JA, and Van Duyne RP (2005) Wavelength-scanned surface-enhanced Raman excitation spectroscopy. *Journal of Physical Chemistry B* 109: 11279–11285.
98. Kennedy BJ, Spaeth S, Dickey M, and Carron KT (1999) Determination of the distance dependence and experimental effects for modified SERS substrates based on self-assembled monolayers formed using alkanethiols. *Journal of Physical Chemistry B* 103: 3640–3646.
99. Dieringer JA, McFarland AD, Shah NC, *et al.* (2006) Surface enhanced Raman spectroscopy: New materials, concepts, characterization tools, and applications. *Faraday Discussions* 132: 9–26.
100. Blatchford CG, Campbell JR, and Creighton JA (1982) Plasma resonance enhanced Raman-scattering by adsorbates on gold colloids – the effects of aggregation. *Surface Science* 120: 435–455.
101. Felidj N, Aubard J, Levi G, *et al.* (2003) Optimized surface-enhanced Raman scattering on gold nanoparticle arrays. *Applied Physics Letters* 82: 3095–3097.
102. Vlckova B, Gu XJ, and Moskovits M (1997) SERS excitation profiles of phthalazine adsorbed on single colloidal silver aggregates as a function of cluster size. *Journal of Physical Chemistry B* 101: 1588–1593.
103. Haes AJ, Zou SL, Zhao J, Schatz GC, and Van Duyne RP (2006) Localized surface plasmon resonance spectroscopy near molecular resonances. *Journal of the American Chemical Society* 128: 10905–10914.
104. Wiederrecht GP, Wurtz GA, and Hranisavljevic J (2004) Coherent coupling of molecular excitons to electronic polarizations of noble metal nanoparticles. *Nano Letters* 4: 2121–2125.
105. Wurtz GA, Evans PR, Hendren W, *et al.* (2007) Molecular plasmonics with tunable exciton–plasmon coupling

strength in J-aggregate hybridized Au nanorod assemblies. *Nano Letters* 7: 1297–1303.
106. Zhao J, Jensen L, Sung JH, *et al.* (2007) Interaction of plasmon and molecular resonances for Rhodamine 6G adsorbed on silver nanoparticles. *Journal of the American Chemical Society* 129: 7647–7656.
107. Zhao J, Dieringer JA, Zhang X, Schatz GC, and Van Duyne RP (2008) Wavelength-scanned surface-enhanced resonance Raman excitation spectroscopy. *Journal of Physical Chemistry C* 112: 19302–19310.
108. Zhao J, Das A, Zhang XY, *et al.* (2006) Resonance surface plasmon spectroscopy: Low molecular weight substrate binding to cytochrome P450. *Journal of the American Chemical Society* 128: 11004–11005.
109. Andersen PC, Jacobson ML, and Rowlen KL (2004) Flashy silver nanoparticles. *Journal of Physical Chemistry B* 108: 2148–2153.
110. Domke KF, Zhang D, and Pettinger B (2007) Enhanced Raman spectroscopy: Single molecules or carbon? *Journal of Physical Chemistry C* 111: 8611–8616.
111. Etchegoin PG, Meyer M, and Le Ru EC (2007) Statistics of single molecule SERS signals: Is there a Poisson distribution of intensities? *Physical Chemistry Chemical Physics* 9: 3006–3010.
112. Doering WE and Nie SM (2002) Single-molecule and single-nanoparticle SERS: Examining the roles of surface active sites and chemical enhancement. *Journal of Physical Chemistry B* 106: 311–317.
113. Etchegoin PG, Meyer M, Blackie E, and Le Ru EC (2007) Statistics of single-molecule surface enhanced Raman scattering signals: Fluctuation analysis with multiple analyte techniques. *Analytical Chemistry* 79: 8411–8415.
114. Le Ru EC, Meyer M, and Etchegoin PG (2006) Proof of single-mokeule sensitivity in surface enhanced Raman scattering (SERS) by means of a two-analyte technique. *Journal of Physical Chemistry B* 110: 1944–1948.
115. Noda I and Ozaki Y (2004) *Two-Dimensional Correlation Spectroscopy: Applications in Vibrational and Optical Spectroscopy*. Chichester: Wiley.
116. Moore AA, Jacobson ML, Belabas N, Rowlen KL, and Jonas DM (2005) 2D correlation analysis of the continuum in single molecule surface enhanced Raman spectroscopy. *Journal of the American Chemical Society* 127: 7292–7293.
117. Jiang J, Bosnick K, Maillard M, and Brus L (2003) Single molecule Raman spectroscopy at the junctions of large Ag nanocrystals. *Journal of Physical Chemistry B* 107: 9964–9972.
118. Michaels AM, Jiang J, and Brus L (2000) Ag nanocrystal junctions as the site for surface-enhanced Raman scattering of single Rhodamine 6G molecules. *Journal of Physical Chemistry B* 104: 11965–11971.
119. Shim S, Stuart CM, and Mathies RA (2008) Resonance Raman cross-sections and vibronic analysis of Rhodamine 6G from broadband stimulated Raman spectroscopy. *ChemPhysChem* 9: 697–699.
120. Vosgrone T and Meixner AJ (2005) Surface- and resonance-enhanced micro-Raman spectroscopy of xanthene dyes: From the ensemble to single molecules. *ChemPhysChem* 6: 154–163.
121. Le Ru EC, Blackie E, Meyer M, and Etchegoin PG (2007) Surface enhanced Raman scattering enhancement factors: A comprehensive study. *Journal of Physical Chemistry C* 111: 13794–13803.
122. Lyandres O, Shah NC, Yonzon CR, *et al.* (2005) Real-time glucose sensing by surface-enhanced Raman spectroscopy in bovine plasma facilitated by a mixed decanethiol/mercaptohexanol partition layer. *Analytical Chemistry* 77: 6134–6139.
123. Lyandres O, Yuen JM, Shah NC, *et al.* (2008) Progress toward an *in vivo* surface-enhanced Raman spectroscopy glucose sensor. *Diabetes Technology and Therapeutics* 10: 257–265.
124. Shah NC, Lyandres O, Walsh JT, Glucksberg MR, and Van Duyne RP (2007) Lactate and sequential lactate-glucose sensing using surface-enhanced Raman spectroscopy. *Analytical Chemistry* 79: 6927–6932.
125. Chen K, Leona M, and Vo-Dinh T (2007) Surface-enhanced Raman scattering for identification of organic pigments and dyes in works of art and cultural heritage material. *Sensor Review* 27: 109–120.
126. Whitney AV, Casadio F, and Van Duyne RP (2007) Identification and characterization of artists' red dyes and their mixtures by surface-enhanced Raman spectroscopy. *Applied Spectroscopy* 61: 994–1000.
127. Chen K, Vo-Dinh KC, Yan F, Wabuyele MB, and Vo-Dinh T (2006) Direct identification of alizarin and lac dye on painting fragments using surface-enhanced Raman scattering. *Analytica Chimica Acta* 569: 234–237.
128. Jurasekova Z, Domingo C, Garcia-Ramos JV, and Sanchez-Cortes S (2008) *In situ* detection of flavonoids in weld-dyed wool and silk textiles by surface-enhanced Raman scattering. *Journal of Raman Spectroscopy* 39: 1309–1312.
129. Brosseau CL, Gambardella A, Casadio F, *et al.* (2009) Ad-hoc surface-enhanced Raman spectroscopy methodologies for the detection of artist dyestuffs: Thin layer chromatography-surface enhanced Raman spectroscopy and *in situ* on the fiber analysis. *Analytical Chemistry* 81: 3056–3062.
130. Wustholz KL, Brosseau CL, Casadio F, and Van Duyne RP (2009) Surface-enhanced Raman spectroscopy of dyes: from single molecules to the artists' canvas. *Physical Chemistry - Chemical Physics* 11: 7350–7359.
131. Zhang XY, Young MA, Lyandres O, and Van Duyne RP (2005) Rapid detection of an anthrax biomarker by surface-enhanced Raman spectroscopy. *Journal of the American Chemical Society* 127: 4484–4489.
132. Zhang XY, Zhao J, Whitney AV, Elam JW, and Van Duyne RP (2006) Ultrastable substrates for surface-enhanced Raman spectroscopy: Al2O3 overlayers fabricated by atomic layer deposition yield improved anthrax biomarker detection. *Journal of the American Chemical Society* 128: 10304–10309.
133. Stuart DA, Biggs KB, and Van Duyne RP (2006) Surface-enhanced Raman spectroscopy of half-mustard agent. *Analyst* 131: 568–572.
134. Biggs KB, Camden JP, Anker JN, and Van Duyne RP (2009) Surface-enhanced Raman spectroscopy of benzenethiol adsorbed from the gas phase onto silver film over nanosphere surfaces: Determination of the sticking probability and detection limit time. *Journal of Physical Chemistry A* 113: 4581–4586.
135. Kim JH, Ehrman SH, and Germer TA (2004) Influence of particle oxide coating on light scattering by submicron metal particles on silicon wafers. *Applied Physics Letters* 84: 1278–1280.
136. Chan GH, Zhao J, Hicks EM, Schatz GC, and Van Duyne RP (2007) Plasmonic properties of copper nanoparticles fabricated by nanosphere lithography. *Nano Letters* 7: 1947–1952.
137. Ekinci Y, Solak HH, and Loffler JF (2008) Plasmon resonances of aluminum nanoparticles and nanorods. *Journal of Applied Physics* 104: 083107/083101–083107/083106.
138. Langhammer C, Schwind M, Kasemo B, and Zoric I (2008) Localized surface plasmon resonances in aluminum nanodisks. *Nano Letters* 8: 1461–1471.

139. Lu XM, Au L, McLellan J, *et al.* (2007) Fabrication of cubic nanocages and nanoframes by dealloying Au/Ag alloy nanoboxes with an aqueous etchant based on $Fe(NO_3)(3)$ or NH_4OH. *Nano Letters* 7: 1764–1769.
140. Doering WE, Piotti ME, Natan MJ, and Freeman RG (2007) SERS as a foundation for nanoscale, optically detected biological labels. *Advanced Materials* 19: 3100–3108.
141. McMahon JA, Wang YM, Sherry LJ, *et al.* (2009) Correlating the structure, optical spectra, and electrodynamics of single silver nanocubes. *Journal of Physical Chemistry C* 113: 2731–2735.
142. Sherry LJ, Chang SH, Schatz GC, *et al.* (2005) Localized surface plasmon resonance spectroscopy of single silver nanocubes. *Nano Letters* 5: 2034–2038.
143. McLellan JM, Li ZY, Siekkinen AR, and Xia YN (2007) The SERS activity of a supported Ag nanocube strongly depends on its orientation relative to laser polarization. *Nano Letters* 7: 1013–1017.
144. McLellan JM, Siekkinen A, Chen JY, and Xia YN (2006) Comparison of the surface-enhanced Raman scattering on sharp and truncated silver nanocubes. *Chemical Physics Letters* 427: 122–126.
145. Camden JP, Dieringer JA, Zhao J, and Van Duyne RP (2008) Controlled plasmonic nanostructures for surface-enhanced spectroscopy and sensing. *Accounts of Chemical Research* 41: 1653–1661.
146. McMahon J, Henry A-I, Wustholz KL, *et al.* (2009) Gold nanoparticle dimer plasmonics: Finite element method calculations of the EM enhancement to surface-enhanced Raman spectroscopy. *Analytical and Bioanalytical Chemistry* 394, doi 10.1007/s00216-009-2738-4.
147. Zou SZ and Weaver MJ (1998) Surface-enhanced Raman scattering on uniform transition metal films: Toward a versatile adsorbate vibrational strategy for solid-nonvacuum interfaces? *Analytical Chemistry* 70: 2387–2395.
148. Pettinger B (2006) Tip-enhanced Raman spectroscopy (TERS). In: *Surface-Enhanced Raman Scattering: Physics and Applications*, vol. 103, pp. 217–240. Berlin: Springer.

10 Colloidal Semiconductor Nanocrystal-Enabled Organic/Inorganic Hybrid Light Emitting Devices

K R Choudhury and F So, University of Florida, Gainesville, FL, USA

Z Kafafi, National Science Foundation, Arlington, VA, USA

10.1 Light-Emitting Semiconductor Nanocrystals

10.1.1 Introduction and Background

As information processing approaches the bandwidth limits of silicon-based electronic devices, radically new technologies are emerging to offer low-cost, low-power, and high-efficiency platforms. In this context, nanostructured materials research has rapidly grown into a field of science and technology bringing together the efforts of chemists, physicists, material scientists, and engineers. Nanostructured materials generally exhibit geometries that reflect the atomistic bonding analogous to the bulk structure. When atoms or molecules organize into condensed systems, new collective phenomena develop. Cooperative interactions produce the physical properties recognized as characteristic of bulk materials. Many of these physical phenomena in both organic and inorganic materials have natural length scales between 1 and 100 nm. Hence, by controlling the physical size of materials, a wide variety of their properties can be tuned effectively. In nanometer-scale structures, such finite-size effects give rise to novel mesoscopic electronic, magnetic, optical, and structural properties, found neither in bulk nor in molecular systems. Also, due to the high surface-to-volume ratio compared to the bulk counterparts, a large fraction of atoms is present on the surface, which enables them to possess significantly different thermodynamic properties. Since an enormous range of fundamental properties are accessible simply by varying the size, nanostructured materials have been extensively studied in the past two decades. Nanomaterials effectively bridge the gap between the bulk and molecular levels leading to entirely new avenues in application [1].

An attractive new class of such building blocks on the nanometer-scale that offer considerable promise for the realization of these novel approaches are semiconductor nanocrystals, variously referred to as nanoclusters or nanoparticles, produced by colloidal chemical routes. They can exist as individual particles or clusters of different shapes and sizes. On the one hand, a nanocrystal may be regarded as a lump of bulk material. Its properties may be derived by considering the spatial confinement imposed on electrons and excitations as a result of the particle's limited size. On the other hand, it can be thought of as a big molecule that can be functionalized (capped) and can interact with its surroundings in a way similar to conventional molecules. An enormous range of fundamental and collective optoelectronic properties can be realized in these semiconductor nanocrystals by varying the size at a constant composition [2]. Specifically, the size-tunable electronic and optical properties (**Figure 1**) arise from the phenomenon of quantum confinement [3], which means electronic charges are confined to length scales (defined by the smallest physical dimension of the crystal, r) smaller than the Bohr exciton radius (a_B). It gives rise to atomic-like electronic energy levels and opens up the possibility of tailoring the properties not only through varying the chemical composition but also by utilizing the perturbation due to size [4]. A striking example is the strong dependence of the band-gap energy (E_g), the energy gap between the conduction and valence levels, on the physical size as the particle dimension approaches the de Broglie wavelength of an excited electron within the particle, manifesting shift in the excitonic band edge progressively toward higher energies and to a significant increase in the molar absorption coefficient with decreasing size [5]. The emission behavior of such nanometer-sized crystals, although much more complicated, also exhibits similar size dependence. At room temperature, homogeneous line broadening is observed in fluorescence emission corresponding to the recombination of the excited hole and electron in a single nanocrystal. For ensembles, a further broadening due to the statistical distribution of particle sizes will also be present.

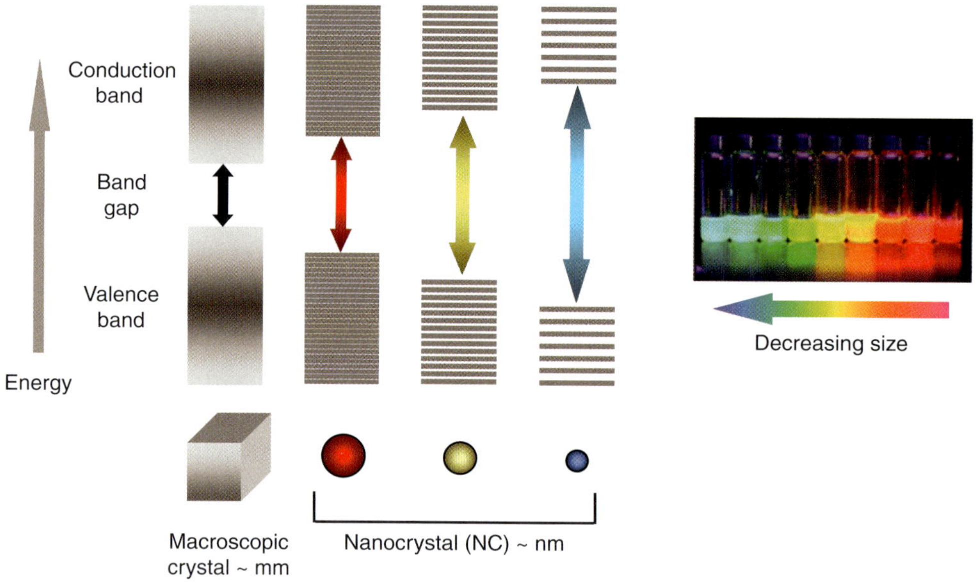

Figure 1 Diagram depicting the effect of quantum confinement on band-gap energy and nature of energy levels in semiconductor NCs. Photograph on right exhibits size-dependent emission from CdSe NCs.

Depending on the number of restricted physical dimensions, semiconductor nanocrystals can be categorized as:

- one-dimensional (1D) (nanowires, quantum wires, or hierarchical structures); and
- zero-dimensional (0D) (quantum dots or Q-dots).

Although each of these categories shows interesting optical and electronic properties, our discussion in this chapter will focus on nanocrystals (NCs) that are physically confined in all three dimensions, that is, Q-dots.

A typical 0D semiconductor NC or a Q-dot, produced by colloidal chemical routes, consists of an inorganic core, typically surrounded by an organic shell of ligands. Control over the properties of NCs can be exerted by modifying their composition (e.g., in the case of semiconductors, one can often change their stoichiometry and tune the band gap), the size and shape of NCs of a given composition, and their capping moiety. The availability of reliable colloidal syntheses, performed in organic [6, 7] or aqueous [8] solutions, provides a veritable palette of II–IV, III–V, and IV–VI compound semiconductor NCs with narrow size distribution (leading to narrow emission spectra) and variable sizes (offering tunability of the absorption onset and emission maximum from the ultraviolet (UV) to the near-infrared (NIR) spectral region) (**Figure 2**). We do not include elementary semiconductors such as silicon (Si) in this chapter, as the research on these materials mainly concentrates on producing nanostructured porous materials through electrochemical [9] rather than chemical syntheses.

Proper surface passivation of the semiconductor NCs leads to enhanced chemical stability and robustness as well as high photoluminescence quantum efficiencies (PLQEs) well in excess of ~50%. A variety of available surface manipulation techniques enable processing of NCs from different solvents and facilitate their integration with various soft matrices. The tunability of the properties of NCs achieved by controlling their composition, size, and surface offers a versatile tool in formulating new composite materials with optimized properties for various applications. The sustained interest in using unique optoelectronic properties of NCs in devices led to their successful use in optical switches [10], solar cells [11–13], light-emitting diodes (LEDs) [14,15a,b], photodetectors [16, 17], and biological labeling [18]. In particular, by virtue of their large-band-gap tunability, high luminance efficiency, narrow spectral emission, and high photostability, semiconductor NCs are an attractive choice as lumophores in next-generation LEDs with highly saturated emission and enhanced luminous power efficiency. Moreover, their compatibility with facile low-temperature solution processing opens up the possibility of fabricating organic or inorganic hybrid

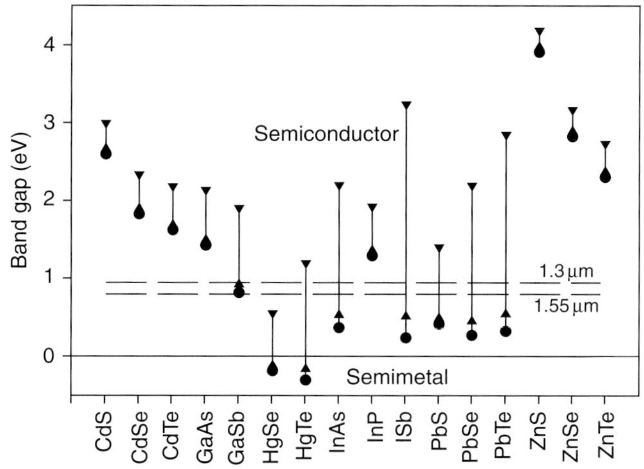

Figure 2 Variation of calculated band-gap excitation energies with varying particle size for a range of compound semiconductors. Band gaps are shown for the bulk materials (•) and for NC diameter of 10 nm (▲) and 3 nm (▼). Reproduced with permission from Harrison MT, Kershaw SV, Burt MG, *et al.* (2000) *Pure and Applied Chemistry* 72: 295.

devices at low cost with device performance similar to their inorganic counterparts.

10.1.1.1 Nanocrystals: A note on quantum size effects

The most fascinating change of properties in semiconductor NCs, typically in the 1–30-nm size range, is the drastic differences in optical absorption, excitonic energies, and electron–hole pair recombination. The optical response of a semiconductor is critically controlled by its band-gap energy (E_g) which gives the threshold energy for an electronic transition from the valence band to the conduction band [1,3,4,7]. In molecular terms, such a transition would be analogous to the lowest-energy electronic transition wherein an electron in the highest occupied molecular orbital (HOMO) is promoted to the lowest unoccupied molecular orbital (LUMO). As the reduction in size of a semiconductor NC surpasses a certain critical limit, the band gap, or difference in energies between HOMO and LUMO will begin to increase. In a typical molecular solid, it is possible to express the properties of the bulk as the sum of individual molecular contributions due to the relatively weak van der Waals forces associated with the intermolecular interactions. As a result of their comparatively diminutive nature, these intermolecular interactions are liable for no more than slight perturbations that rarely extend beyond the nearest neighbor. Under such circumstances, the size of the crystal will have no consequence over the electronic properties of the crystal. In contrast, inorganic semiconductors have no discrete molecular unit; they consist of a network of ordered atoms that interact throughout the entire lattice through strong chemical bonds. In a bulk semiconductor crystal, the absorption of an optical photon or thermal energy results in the production of a Mott–Wannier exciton [19]. This electronic excitation is composed of a loosely bound electron–hole pair which is delocalized over a certain length, typically larger than the lattice constant, known as the Bohr radius (a_B) [20]. Quantitatively, the Bohr radius, a_B, can be expressed as

$$a_B = \frac{\hbar^2 \in}{e^2}\left(\frac{1}{m_e^*} + \frac{1}{m_h^*}\right) \quad (1)$$

where $\hbar = h/2\pi$, h is Planck's constant, e is the fundamental electronic charge, $\in$ is the bulk dielectric constant, and m_e^* and m_h^* are the effective masses of the electron and hole, respectively. In materials with a higher dielectric constant, Coulombic attraction between the electron and hole is more effectively suppressed, and delocalization is enhanced [21].

Two fundamental factors, both related to the size of the individual NCs, distinguish their behavior from the corresponding macrocrystalline material:

- The high dispersity (large surface-to-volume ratio) associated with the particles, with both the physical and chemical properties of the semiconductor being particularly sensitive to the surface structure.
- The actual size of the particle, which can determine the electronic and physical properties of the material.

Charge carriers in semiconductor NCs are confined within the spatial dimensions of the crystallite. As the reduction in size of the NC approaches the Bohr radius, the motion of the electrons and holes are increasingly constrained, leading to what is known as quantum confinement [3], which modifies the density of states near the band edge. The 3D confinement effects collapse the continuous density of states of the bulk solid into the discrete electronic states of the NC. The finite size of the NC quantizes the allowed k values. Decreasing NC diameter shifts the first state to larger k values and increases the separation between states [7]. The excitonic Bohr radius is a threshold value, and the confinement effects become important when the NC diameter gets smaller than that. For small NCs, the excitonic and biexcitonic binding energies are much larger than those for bulk materials [22].

Among the many properties that exhibit a dependence upon size in NCs, two are of particular importance for their application as lumophores. The first is a shift of the absorption band edge and the maximum of emission toward shorter wavelengths as the size of the particle decreases below a particular value [23], which depends on the type of semiconductor. The effect allows a facile tuning of the energy gap, and hence light absorbed and emitted, through changes in NC size (**Figure 3**). The second important property borne out of quantum confinement is the observation of discrete, well-separated energy states in a NC compared to the continuum of states in the bulk. As a result, the electronic states of each energy level exhibit wave functions that are atomic-like with sharp emission peaks corresponding to discrete transitions. Typical intraband energy level spacings for NCs are in the range of 10–100 meV.

Several models have been proposed to explain the dependence of exciton energy on cluster size [3, 24, 25]. One of them, which is frequently used, is the effective mass approximation model, first proposed by Efros and Efros [3] and later modified by Brus [25]. It assumes a spherical particle in a potential well with an infinite potential barrier at the boundaries. For nanometer-sized particles the electron and hole are significantly closer together than in the bulk material, and as such the Coulombic interaction between electron and hole cannot be neglected; they have higher kinetic energy than in the bulk material. For strongly confined systems, Brus showed [25–28] that the size dependence on the energy of the first electronic transition of the exciton can be approximately calculated by using

$$\Delta E = \frac{\hbar^2 \pi^2}{2R^2}\left(\frac{1}{m_e^*} + \frac{1}{m_h^*}\right) - \frac{1.786e^2}{\in R} \qquad (2)$$

where R is the radius of the particle. The first term in **Equation 2** contains the familiar $1/R^2$ dependence associated with the particle in a box treatment and shows the energy shift to higher values due to quantum localization. The second term in **Equation 2** shows a $1/R$ dependence and is representative of the Coulomb interaction energy. Based on

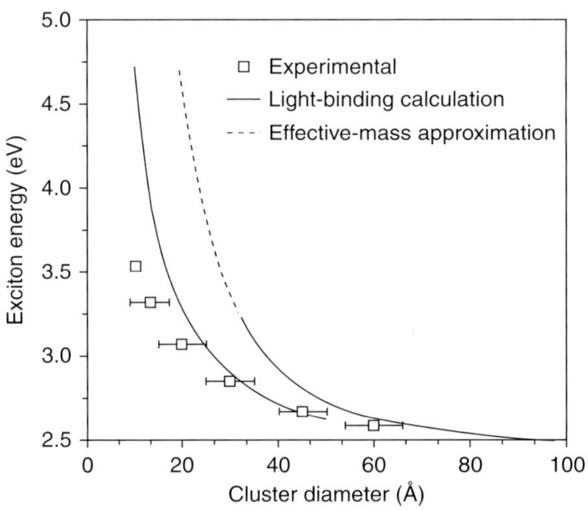

Figure 3 Experimental exciton transition energy of CdS nanoclusters as a function of cluster diameter, along with fits to the data based on the light-binding model and the effective-mass approximation. Reproduced with permission from Wang Y and Herron N (1990) *Physical Review B* 42: 7253.

Equation 2, the first excitonic transition (i.e., the band gap) increases as the particle diameter decreases. The predicted behavior of **Equation 2** has been confirmed experimentally for a wide range of semiconductor nanocrystallites [26, 27, 29, 30], with a blue shift in the onset of absorption of incident light being observed with decreasing particle diameter. Moreover, due to the quantized nature of electronic transitions allowed, as shown in **Equation 2**, the valence and conduction bands in the NCs consist of discrete sets of electronic levels corresponding to a state of matter between molecular and bulk material. Though such equations are often used to predict the quantum modulated size effects in semiconductor NCs, they lack in two important aspects. First, the energy bands can significantly deviate from the assumed parabolic character, and typically more so as the size of the particle is reduced [6,31]. The deviation is especially evident in material possessing a relatively small band gap. It implies that the effective mass approximation is no longer valid except in the region where $k = 0$. Second, while the structures found in NCs are often quite similar to those observed in the corresponding bulk materials, the differences could significantly change the effective masses of the electrons and holes. **Equation 2**, for instance, does not account for the effects arising from coupling of electronic states and those attributable to surface structure. Nevertheless, this model is particularly useful from a practical point of view in predicting size-dependent energy shifts in a large number of nanocrystalline semiconductors. Detailed theoretical studies on the size-dependent electronic properties can be found in several reviews [1, 4].

10.1.2 Homostructure and Core–Shell Nanocrystals

A typical semiconductor NC consists of an inorganic core, which is comparable to or smaller in size than the excitonic Bohr radius of the corresponding bulk material, surrounded by a layer of an organic passivating ligand. The synthesis of monodisperse NCs of desired sizes is essential to successfully study their size-dependent properties and for their use in practice. Despite advances made in colloidal synthesis approaches yielding NCs with narrow size distributions, high crystallinity, and controllable surface properties, challenges do still exist in characterizing the NC structures (surface passivation, core-only and core–shell) and their associated electronic and optical properties. For example, the selection of a suitable capping ligand that will bond to the surface of the NCs is a key issue. Capping ligands regulate the growth rate and size of the NCs, stabilize them, and at times may prevent them from photo-oxidation. They also provide a dielectric barrier at the surface, partially eliminating surface traps and rendering greater stability [7]. Additionally, typical capping ligands, being organic molecules with free functional groups, stabilize the NCs as colloidal dispersions in different solvents enabling their subsequent processing. In general, for homostructure core-only NCs, phosphene oxides (e.g., tri-*n*-octyl phosphine oxide), mercaptans (e.g., dodecanethiol), amines (e.g., hexadecylamine), and carboxylic acids (e.g., oleic acid) are the most widely used organic capping ligands [7]. However, most organic capping molecules are distorted long chains. As a result, total coverage of the surface atoms may be sterically hindered. Another crucial issue is the simultaneous passivation of both anionic and cationic surface sites using the same capping agent, which is extremely difficult. Therefore, some dangling bonds are always present when the surface is passivated by organic molecules, making them susceptible to chemical, thermal, and/or photodegradation.

In addition to organic capping groups, inorganic layers are also used to passivate NC surfaces, particularly with materials having a larger band gap, in analogy with the well-developed techniques for the growth of type I quantum wells. The passivating shell is grown either epitaxially or as a nonepitaxial crystalline or amorphous layer on the core. When the shell material adapts to the lattice parameters of the core during epitaxial growth, the resulting strain can affect the absorption and emission spectra of the core–shell NCs significantly [32]. The maximum luminance efficiency of the core–shell structures depends on the thickness of the shell layer, which is typically within a few monolayers for optimum optoelectronic properties. Thicker capping layers can lead to the formation of dislocations, which are nonradiative recombination sites and hence can decrease the emission quantum yield.

Depending on the offset of the energy bands of the core material with respect to those of the shell, core–shell NCs can exhibit both type I and type II band heterostructures (**Figure 4**). In type I core–shell NCs, the larger-band-gap semiconductor forms a closed outer shell and the band edges of the core material lie inside the band gap of the outer material.

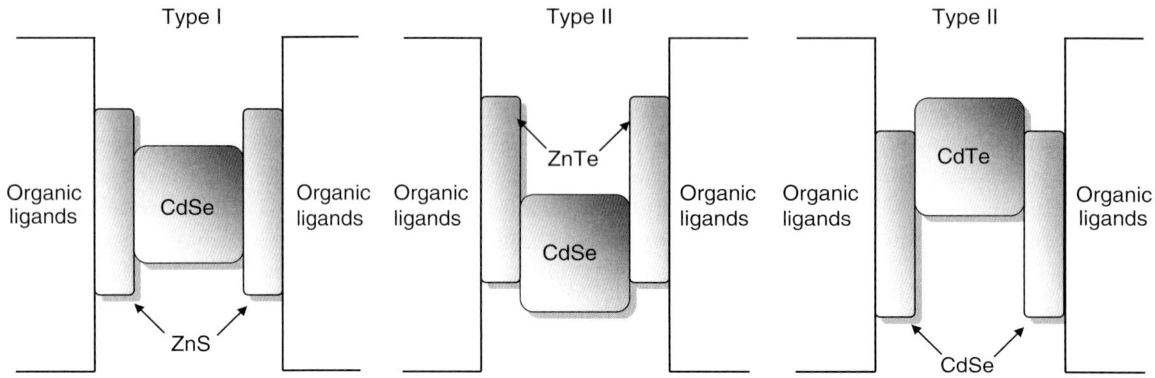

Figure 4 Schematic representation of the energy-level alignment in Type I and Type II core/shell NC systems with possible choice of semiconductors. The upper and lower edges of the rectangles correspond to the positions of the conduction- and valence-band edges (not drawn to scale) of the core and shell materials.

The outer inorganic shell efficiently passivates the core surface states and confines the charge carriers into the core region, thereby improving chemical robustness and photostability and enhancing the photoluminescence quantum yield. For example, using a thin shell of ZnS on CdSe core NCs of various sizes, narrow-band emission tunable from blue to red with room temperature photoluminescence quantum yield as high as 50–80% has been achieved [33,34].

In type II colloidal heterostructures, both the valence and conduction band offsets are in the same direction [35]. As a result, these NCs can provide spatially indirect states, with one charge carrier mostly confined to the core, and the other to the shell. For example, in CdTe/CdSe NCs the hole wave function is predominantly confined to the CdTe core, while the electron wave function is located in the CdSe shell. The situation is opposite in CdSe/ZnTe core–shell structures. Emission from such type II NCs originates from the radiative recombination of electron–hole pairs across the core–shell interface. The emission energy is smaller than the band gap of either the core or the shell due to interfacial energy offsets. Because of the reduced electron–hole wave function overlap, these structures also show extended exciton lifetimes.

One of the greatest attributes of semiconductor NCs is the tunability of the spectral response afforded by a choice of materials and through variation of the physical dimensions. The availability of reliable colloidal synthetic routes has resulted in a wide gamut of luminescent compound semiconductor NCs covering the entire visible and a large portion of the infrared (IR) spectral region. Some of the most studied colloidal semiconductor NCs listed below are grouped according to their photoluminescence spectral region of activity.

- 400–700 nm: CdS, CdSe, CdTe, ZnS, ZnSe, CdSe/ZnS. $Cd_{1-x}Zn_xSe$/ZnS, CdSe/CdS, CdSe/ZnSe.
- 700–2500 nm: InP, InAs, PbS, PbSe, PbTe, HgTe, $Cd_xHg_{1-x}Te$, InAs/ZnSe, PbSe/PbS.

10.1.3 Synthetic Routes

A process for synthesizing PbS NCs was developed more than 2000 years ago using PbO, $Ca(OH)_2$ and water, and the Romans and Greeks used these nanocrystalline materials to dye their hair [36a,b]. Control of the size of NCs in silicate glasses was one of the oldest and most frequently used techniques to manipulate the color of glass. More recently, Matijevic, among others, was the first to develop reproducible synthetic methods [37] for a wide range of near monodispersed nanosized colloidal particles.

Uncovering and mapping size-dependent nanocrystalline material properties require developing new synthetic routes to prepare Q-dots with uniform composition, size, shape, and well-defined and controlled surface chemistry. The synthesis of nanocrystalline semiconducting materials follows the conventional arrested precipitation methods first introduced by Matijevic [37]. This process requires careful control of experimental conditions and material concentration where the preparation takes place in a single solution or a mixture of solutions. The precipitation process involves several stages: nucleation in solution, growth of preformed nuclei, arresting the growth after particles reach the desired size, isolation of particles, postpreparative size fractionation, etc. To grow monodispersed NCs with narrow size distribution (standard deviation $\sigma \leq 5\%$ in diameter), it is necessary to control and temporally separate the processes of nucleation and subsequent growth [7, 38]. Nucleation can be categorized as homogeneous, heterogeneous, or secondary [39]. Homogeneous nucleation, the process favored for a tight size distribution, occurs when solute atoms or molecules combine and reach a critical size without the assistance of a preexisting solid interface. The large surface area and high surface energy lead to unstable semiconductor NCs. These bare Q-dots are stabilized by dispersing in a trapping host, thus encapsulating them. Otherwise, they agglomerate and coalesce very quickly. Various chemical methods have been developed to control the size and distribution of NCs during synthesis. It is common to grow the nuclei of a desired material in the presence of host materials such as reverse micelles or highly coordinating solvents and terminate the NC surface with organic ligands. By varying factors such as temperature, electrostatic double layer thickness, stabilizers or micelles formation, concentration of precursors, and ratios of anionic to cationic species and solvents, NCs of desired size, shape, and composition can be achieved. Among the different colloidal synthetic routes, the discussion below will be limited to two common procedures that are easily reproducible and lead to high-quality semiconductor NCs or Q-dots.

1. *Microemulsion technique.* Microemulsion processes [40–42] are popular methods for synthesizing semiconductor NCs at room temperature. The process uses two immiscible liquids and can be categorized as either normal microemulsions (e.g., oil in water), or reverse microemulsions (e.g., water in oil). In some cases, polar solvents other than water (e.g., alcohols) may be used. In practice [41,42], the polar water phase having a high dielectric constant is dispersed under agitation as very small water droplets encapsulated by a monolayer film of surfactant molecules into a continuous nonpolar oil phase (typically a water-immiscible hydrocarbon) having a low dielectric constant. Surfactants such as sodium bis(2-ethylhexyl)sulfosuccinate (AOT), cetyl trimethyl-ammonium bromide (CTAB), sodium dodecyl sulfate (SDS), or octyl phenol ethoxylate (Triton-X-100) are generally used. Since the surfactants are terminated by hydrophobic and hydrophilic groups on opposite ends, numerous tiny droplets called micelles are formed in the continuous oil medium. A minimum critical micelle concentration (CMC) is needed for the surfactant molecules to self-aggregate, giving rise to reverse micelles in organic solvents. These micelles are thermodynamically stable and can act as nanoreactors. In addition to water, aqueous solutions containing a variety of dissolved salts, including cationic and anionic species can be solubilized within the reverse micelles. Continuous exchange of the micellar contents through dynamic collisions enables the reaction to proceed inside the nanoreactors; however, since the reaction is confined within the cavity of the micelle, growth of the NC beyond the dimensions of the cavity is inhibited, and is controlled by the size of the micelle. The size of the micelle, and subsequently the volume of the aqueous pool contained within the micelle, is governed by the molar ratio of water to surfactant, W_0, where $W_0 = [H_2O]/[\text{surfactant}]$ [41]. The relation between W_0 and the radius (r) of the micelle has been reported [41] as:

$$\left(\frac{r+15}{r}\right)^3 - 1 = \frac{27.5}{W_0} \tag{3}$$

In the final stage of the synthesis, a passivating reagent is added to the continuous oil phase. This species is then able to enter the aqueous phase and bond to the surface of the contained NC, which renders the surface of the NC hydrophobic and induces precipitation (**Figure 5**).

The reverse micelle technique has been used to prepare II–VI core-only and core–shell NCs, such as CdS, CdSe, CdSe/ZnS, CdSe/ZnSe, and CdS:Mn/ZnS. This facile low-temperature method offers the advantages of easy control of the final NC size by changing the molar ratio of water to surfactant, narrow size distribution compared to some other colloidal methods like the sol-gel process, and facile dispersion of the NCs. However, this approach has a low chemical yield and is not free from the possibility of introducing impurities and crystal defects in the targeted products [41,42].

2. *Hot-solution decomposition process.* A general scheme for preparing monodisperse nanocrystalline materials requires, after the classic studies by La Mer and Dinegar [43], a single temporally discrete nucleation event followed by slower controlled growth on the existing nuclei [6,7]. The temperature of the solution should be sufficient to decompose the reagents. Rapid addition of more reagents to the reaction vessel raises the precursor concentration beyond nucleation threshold and forms a supersaturation of species in solution. A short nucleation burst partially relieves the supersaturation. Upon

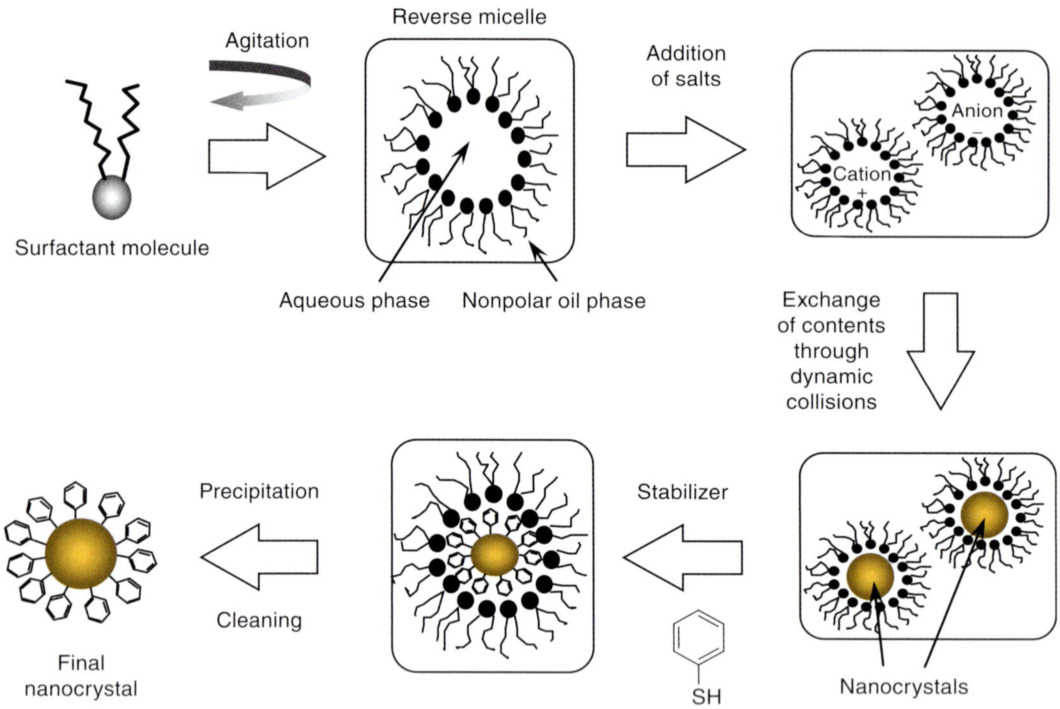

Figure 5 Schematic representation of the microemulsion technique of nanocrystal synthesis in a reverse micellar system.

nucleation, the concentration of the reactive species in solution drops below the critical concentration for further nucleation. As long as the consumption of feedstock by the growing colloidal NCs is not exceeded by the rate of precursor addition to solution, no new nuclei form. Since the rate of growth for all of the NCs is similar, the initial size distribution is largely determined by the time over which the nuclei are formed and begin to grow. The systematic adjustment of reaction parameters (e.g., reaction time, temperature, and concentration), in addition to the selection of the proper reagents and surfactants–can be applied to effectively control NC size (**Figure 6**).

In the growth of compound semiconductor NCs, the requisite supersaturation and subsequent nucleation can be triggered by rapid injection of metal-organic precursors into a vigorously stirred flask containing a hot (~150–350 °C) coordinating solvent. Such high-temperature (~300 °C) pyrolysis of organometallic compounds and/or reduction of metal salts is a well-established and highly favored route for the synthesis of semiconductor NCs [6,7,38]. In this method, the initial fast injection of the metal–organic precursors gives rise to a homogeneous nucleation, which is followed by a slow growth and subsequent annealing, resulting in uniform surface derivatization. The role of the coordinating solvents and surfactants is to terminate the particle surface with organic ligands. The stabilizing agents bind to the surface of the growing NCs preventing their growth to the bulk phase. They modify and stabilize the NC surface, increase the solubility, and prevent agglomeration [44]. Moreover, the stabilizers control the density of surface states and provide electronic passivation of surface traps to mitigate surface carrier recombination.

The solvents typically used in these syntheses are long-chain alkylphosphines, alkylphosphine oxides, alkylamines, etc. [7,45]. This method has been extensively used in the synthesis of ME type II–VI (where M = Zn, Cd, Hg; E = S, Se, Te), III–V (where M = In; E = P, As) and IV–VI (where M = Pb; E = S, Se, Te) compound semiconductor NCs by rapidly mixing and heating precursors from the respective groups in high boiling, coordinating solvents. The method also allows for the successful synthesis of core–shell architectures, both in one-pot synthesis and the more traditional two-step protocols. Also, the use of mixed precursors, such as a combination of S and Se precursors, can lead to the direct production of alloys on the nanometer scale.

The hot-solution synthesis route produces semiconductor NCs of superior quality compared to any

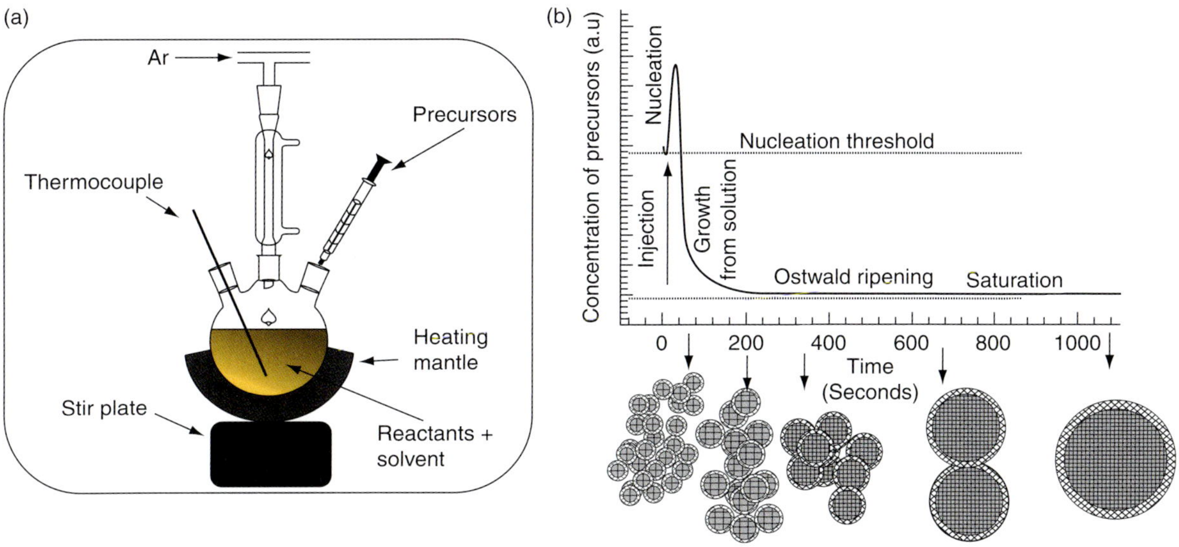

Figure 6 (a) Sketch depicting the synthetic apparatus employed in hot-solution decomposition process. (b) Schematic depicting the different stages of nucleation and growth in the preparation of monodisperse NCs after the classic studies by La Mer and Dinegar [43]. Reproduced with permission from Murray CB, Kagan CR, and Bawendi MG (2000) *Annual Review of Materials Science* 30: 545.

other colloidal synthetic process. This is because the growth of the particles is relatively slow and can be precisely controlled by modulating the temperature and time. It can easily provide a series of NC sizes from a single reaction cycle. Another significant advantage of this method is that it provides sufficient thermal energy to anneal a large number of defects during crystallization resulting in NCs of higher purity and crystallinity.

10.1.3.1 Nanocrystal synthesis: A note on focusing, defocusing, and Ostwald ripening

Bearing in mind the use of semiconductor NCs as luminescent centers in light-emitting devices (LEDs), monodispersity of any particular batch of synthesized NCs is strongly required to achieve emission color purity. In hot-colloidal synthesis, if the percentage of growth during the initial nucleation period is small compared with subsequent growth, the NCs can become more uniform over time [46]. This phenomenon has been referred to as focusing of the size distribution, which is due to the strong size dependence of the crystal surface energy influencing the kinetics of crystal growth [47]. It occurs when, in the early stages of synthesis, a considerable amount of precursor is still present in the colloidal solution in the form of a monomer, as only a small amount of the precursor was converted to NCs during nucleation. These NCs are slightly larger than the equilibrium critical size. Under these conditions, the smaller NCs grow very quickly consuming the monomer from the solution. This fast increase in particle size is accompanied by a strong narrowing or focusing of the size distribution. During the focusing stage the number of particles remains almost constant. Direct observations have confirmed more than threefold (from an initial standard deviation $\sigma \sim 20\%$ down to about $\sim 6\%$) narrowing of the size distribution in CdSe NC growth [48].

Many systems, however, exhibit a second growth phase, in which the focusing stage of growth is followed by a slow broadening or defocusing of the size distribution. When the concentration of the monomer in solution is depleted due to growth, the critical size becomes larger, and the distribution broadens as the smaller NCs shrink and eventually are totally consumed. In this process, the high surface energy of the smaller NCs promotes their dissolution, while material is redeposited on the larger NCs. The net result is a slow increase in the average size and size distribution of the NCs with a corresponding decrease in their number. This growth phase, called Ostwald ripening [7,49–51], results in rather broad size distributions with standard deviation $\sigma \sim$ 10–20% (**Figure 7**).

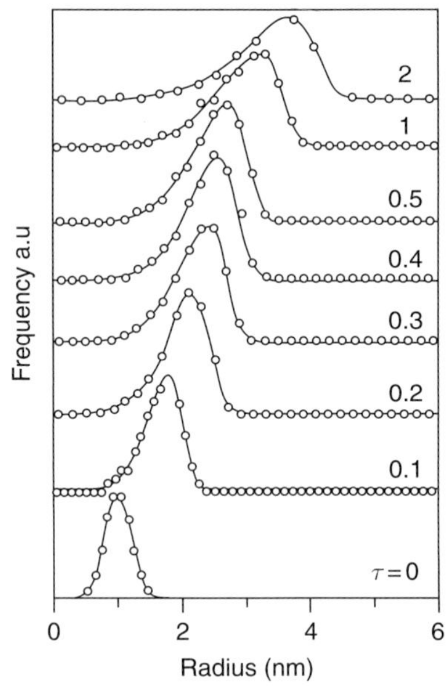

Figure 7 Temporal evolution of size distribution of an ensemble of growing NCs during Ostwald ripening under diffusion control, showing an increase in size and a broadening of size distribution. Reproduced with permission from Rogach AL, Talapin DV, Chevchenko EV, Kornowski A, Hasse M, and Weller H (2002) *Advanced Functional Materials* 12: 653.

Knowledge of these processes occurring during NC growth is important because it can be used to achieve extremely narrow size distributions, essential in many optoelectronic applications, by allowing growth of particles in the initial focusing regime and terminating the process before equilibrium between the particles and the monomer is reached, to prevent Ostwald ripening from setting in.

10.2 Structure–Property Relationships in Semiconductor Nanocrystals

10.2.1 Optical Properties: Absorption

The optical properties of semiconductor NCs are determined by factors such as size, shape, surface and bulk defects, impurities, and crystallinity. The dependence on the Q-dot size, the parameter that exerts the greatest influence, arises from changes to the surface-to-volume ratio with size, and quantum confinement, which modifies the density of states near the band edges. The effect of Q-dots size on their optical properties has been demonstrated for a variety of semiconductor nanocrystallites [1,7,26,29].

In a bulk semiconductor, the absorption of a photon leads to excitation of an electron from the valence band to the conduction band. The charge carriers are separated by distances that encompass several molecules or ions. This separation, along with the high dielectric constant of the material, makes their binding energy relatively small, leading to the formation of a Wannier-type exciton that is relatively free to move through the bulk crystal until it is trapped, annihilated in collision with another exciton, or relaxed by radiative recombination of the electron and hole. The minimum amount of energy required to form the charge carriers is known as the band-gap energy (E_g) of the semiconductor. The absorption of photons with energy similar to that of the band gap, $h\nu \geq E_g$, leads to an optical transition producing an electron in the valence band and a hole in the conduction band. Absorption of photons with energy greater than E_g produces excitations above the conduction band edge. The resulting electrons can lose the excess energy through nonradiative processes.

When the size of the semiconductor particles decreases beyond their Bohr radius, the spatial confinement of the motion of electrons leads to an increase in band-gap energy. Equally important, the actual energy of band-gap absorption becomes sensitive to the size and shape of the particles. Experimentally, this quantum size effect shows up as a shift toward higher energy values from the band edge, as compared to the typical value for the corresponding bulk material.

The absorption, A, of light by a material with thickness l can be expressed using **Equation 4** assuming that it is governed by the Beer–Lambert law

$$A = \alpha l \tag{4}$$

where α represents the linear absorption coefficient of the solid and is a function of the absorption wavelength. In the case of semiconductors, the process of optical absorption is subject to selection rules: the requirement is preservation of the wave vector. For direct band-gap semiconductors in which the wave vector for optical transitions is conserved, a useful expression for the linear absorption coefficient is given by the square root of the difference between the band-gap energy and the photon energy of a direct transition near the threshold [52]:

$$\alpha(h\nu) \propto (E_g - h\nu)^{1/2} \tag{5}$$

Using **Equation 5**, the band-gap energy of a nanocrystalline semiconductor can be experimentally estimated from its optical absorption spectrum. Interestingly, often times, NC ensembles also exhibit several peaks in the optical absorption spectrum at room temperature, revealing a large number of energy states. The lowest energy of an excited state is shown by the first observable peak, known as the quantum-confinement peak. Excitation at shorter wavelengths is possible because multiple electronic states are present at higher energy levels. In fact, the linear absorption coefficient gradually increases toward shorter wavelengths. The preparation of monodisperse NC samples makes it possible to observe, assign, and monitor the size evolution of a series of discrete and excited electronic states. For example, in samples of highly crystalline, nearly monodisperse NCs of CdSe, as many as 10 such excitonic features have been resolved [53] in a single spectrum (**Figure 8**). This results from an enhanced oscillator strength (directly proportional to the linear absorption coefficient), which increases as particle size decreases, due to the strong overlapping of the wave functions of confined charge carriers. Such effects have also been observed in CdS and PbSe NCs with different and well-defined size distributions. In these studies, decreasing the size of the particles not only leads to a shift in the primary excitonic peak to higher energies or correspondingly shorter wavelengths, but also results in an increase in the molar absorption coefficient [54a]. Bawendi and coworkers [54b] have estimated that the molar absorption coefficients of CdSe NCs are about 10^5–$10^6\,M^{-1}\,cm^{-1}$ depending on the particle size and the absorption wavelength. Interestingly, these values are 10–100 times larger than those of organic dyes active in the same spectral region. Chan and Nie [55] estimated that single ZnS-capped CdSe NCs are 20 times brighter and 100–200 times more stable than single rhodamine 6G (a widely used organic laser dye) molecules.

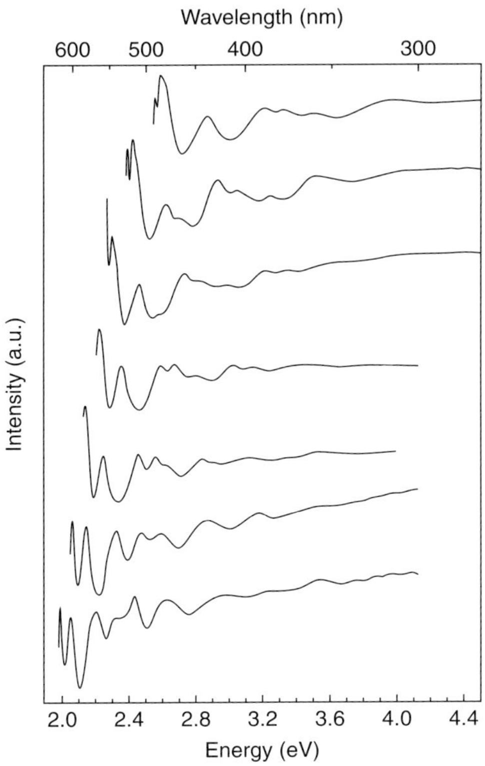

Figure 8 Normalized photoluminescence excitation (PLE) scans for seven different-sized CdSe nanocrystals, showing as many as 10 excitonic features. Nanocrystal size increases from top to bottom of plot and ranges from ~1.5 to ~4.3 nm in radius. Reproduced with permission from Norris DJ and Bawendi MG (1996) *Physical Review B* 53: 16338.

In addition to the usual band-gap excitation, also known as interband excitation, semiconductor NCs exhibit excitation and relaxation of charges within the electron or hole levels as well, referred to as intraband transitions [4,56a,b]. To understand this novel phenomenon, we need to consider the fact that the electronic excitation process in a neutral semiconductor NC can be described by molecular orbitals (MOs), just like in conjugated organic molecules [57]. In conjugated molecules, the occupied and unoccupied MOs are made of linear combinations of the same type of atomic orbitals. Furthermore, the electron and hole density in an MO is present on the same type of atom and thus are strongly correlated. However, in a compound semiconducting NC (e.g., CdSe, HgTe), the occupied MOs are made of linear combination of atomic orbitals on the negatively charged anions (e.g., Se^{2-}, Te^{2-}), while the unoccupied MOs are made of the atomic orbitals of the metallic cations (e.g., Cd^{2+}, Hg^{2+}). The band-gap excitation involves the transfer of an electron from the HOMO in the anionic electronic system to the vacant LUMO of the cationic system. Thus, the holes always occupy the anion MOs, while the excited electrons occupy the cation MOs. Due to this peculiar phenomenon and the high dielectric constant of the semiconductor material, charge carriers in semiconductor NCs are weakly coupled. As a consequence of this weak correlation, holes or electrons formed in band-gap absorption can be excited to higher-energy anion or cation MOs. This type of excitation of either charge carrier formed in the band-gap absorption is called intraband excitation [4,56a,b].

As mentioned earlier, high-quality semiconductor NCs have been successfully synthesized with tunable optical absorption from the UV to mid-IR region of the electromagnetic spectrum by varying the composition and size of ensembles. In **Figure 9**, we present examples of four different compositions (CdS, CdSe, InP, and PbSe) of semiconductor NCs that collectively address absorption over this broad band of wavelengths.

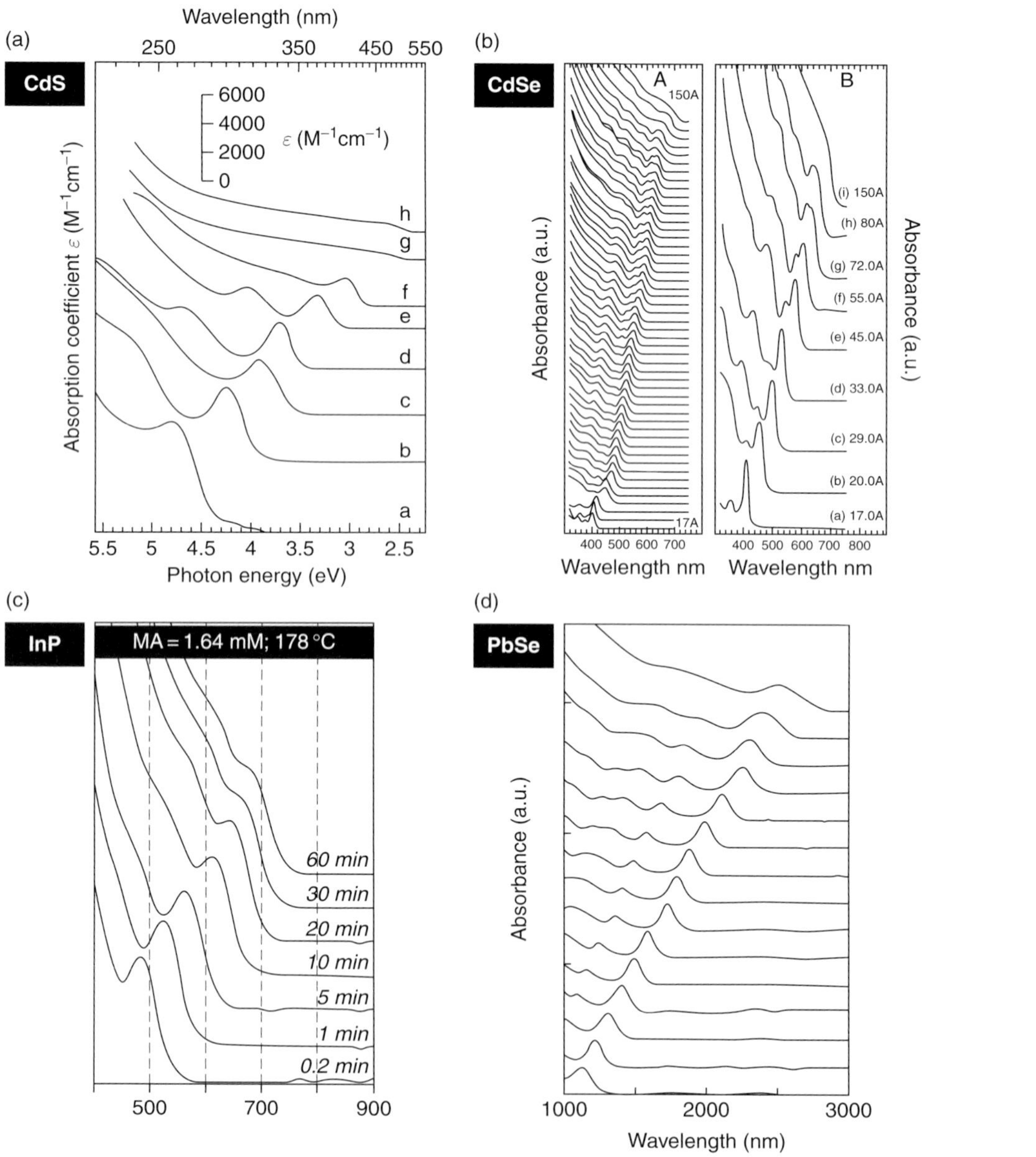

Figure 9 Discrete electronic transitions in optical absorption in NCs of (a) CdS, (b) CdSe, (c) InP, and (d) PbSe. (a) Reproduced with permission from Vossmeyer T, Katsikas L, Gienig TM, *et al*. (1994) *Journal of Physical Chemistry* 98: 7665. (b) Reproduced with permission from Murray CB, Kagan CR, and Bawendi MG (2000) *Annual Review of Materials Science* 30: 545. (c) Reproduced with permission from Xie R, Battaglia D, and Peng X (2007) *Journal of the American Chemical Society* 129: 15432. (d) Reproduced with permission from Yu WW, Falkner JC, Shih BS, and Colvin VL (2004) *Chemistry of Materials* 16: 3318.

10.2.2 Optical Properties: Emission

Photoexcitation of the NC creates an electron–hole pair. The excited electron and hole may also form an exciton, as discussed earlier. The excited electron may recombine with a hole and relax to a lower energy state, ultimately reaching the ground state. The excess energy resulting from recombination and relaxation may either be radiative (emitting photons) or nonradiative (involving phonons or Auger electrons). Radiative relaxation to the ground state of a NC results in spontaneous luminescence. Such a luminescence is called photoluminescence if it is due to photoexcitation. It can result from band edge or near band-edge transitions, or from defect and/or dopants.

10.2.2.1 Band-edge emission

The recombination of an excited electron in the conduction band with a hole in the valence band leads to band-edge emission, the most common radiative relaxation process in semiconductors. As noted before, an electron and a hole typically form an exciton with binding energy of about a few meV. Therefore, the radiative recombination of an exciton leads to near band-edge emission at energies slightly lower than the band gap. This spectral shift between the emission and absorption band edge is known as the Stokes shift [58], with the excess energy being delivered into the material through a phonon excitation. In a typical process, an electron in a nanocrystalline material is photoexcited from its ground to an excited state. Through a fast nonradiative process, the excited electron relaxes to its lowest excited electronic state, and then may radiatively recombine with a hole. For electronic relaxation in NCs, as in bulk solids or molecules, the emitted photon is red-shifted relative to the excitation photon energy due to the presence of vibrational levels in the excited as well as lower energy (including ground) states (**Figure 10**).

The emission from optically excited semiconductor NCs is called excitonic fluorescence and is observed as a sharp peak. Since the NC photophysics is ultimately determined by its electronic structure, the position of the peak emission is also governed by quantum-confinement effects, displaying spectral tunability with size. By varying the size and composition of NCs, the emission wavelength can be tuned from the blue to the NIR. For example, CdS and ZnSe NCs emit blue to near-UV light, different-sized CdSe NCs emit light across the visible spectrum. InP, InAs, PbSe, and PbTe QDs emit in various region of the IR, from the near- to mid-IR.

The full-width of the room-temperature band emission peak at half the maximum intensity (FWHM) typically lies in the range of 15–50 nm for different NCs. FWHM is affected by the size distribution of NCs in an ensemble; a broader size distribution leads to broadening of FWHM. With the advancement of synthetic techniques, NC batches with very narrow size distribution have been synthesized giving rise to extremely narrow-band emissions. For example, ZnS-capped CdSe NCs with emission spectra as narrow as

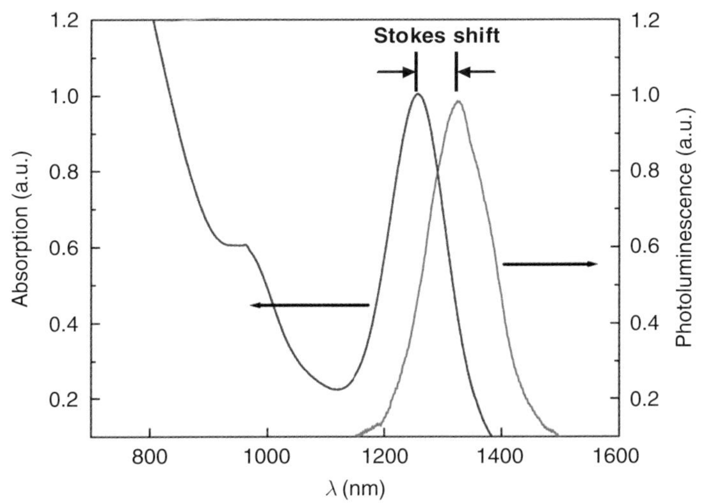

Figure 10 Typical spectral shift between the emission peak and absorption band edge of a semiconductor NC (in this case PbSe), known as Stokes shift.

FWHM = 13 nm at room temperature [59] have been achieved.

Excellent synthesis techniques have also led to the realization of NCs with high quantum yields in band-edge emission. While band-edge emission in NCs is associated with recombination of carriers in quantized states in the interior, the emission does suffer from the presence of unsaturated bonds (sometimes referred to as dangling bonds) at the surface. These surface defects can act as traps for charge carriers. Recombination of trapped carriers reduces band-edge emission and leads to a characteristic deep-trap emission band, evident as a broad tail or hump on the low-energy side of the sharper band-edge emission spectral signal. Coordinating ligands help to passivate surface trap sites and thus enhance the relative intensity of band-edge emission. High photoluminescence (PL) quantum yields are particularly indicative of a well-passivated surface. Passivation with typical organic capping ligands routinely yields 5–15% PL quantum yields. Employing specific precursor ratios and a primary amine ligand, PL efficiencies as high as 70–80% have been achieved in CdSe NCs [60], emitting in the visible spectral region. In addition, UV to blue-emitting ZnSe NCs have also been demonstrated with PL quantum yields ranging from 20% to 50% [45]. Among III–V semiconductor NCs, relatively intense band-edge emission from InP has been achieved by etching the particles with a dilute alcoholic solution of hydrofluoric acid [61a]. Using this technique, the PL quantum yield of band-edge emission is enhanced by an order of magnitude, to about 30%, while inhibiting and sometimes totally eliminating deep-trap emission that arises from radiative surface traps. High-quality NCs emitting in the IR have also been prepared by surfactant-stabilized pyrolysis reactions. Very high PL quantum yields >85% [56a,b], 180 times that of the most efficient and photostable IR dye presently available commercially (IR26), have resulted from PbSe NCs' emission over a wide IR spectral range (1–2 μm). Emission at even longer wavelengths, up to 4 μm, has also been achieved, albeit with low PL efficiency (~0.5%) (**Figure 11**) [62].

Due to the increasing surface-to-volume ratio with diminishing particle size, surface trap states exert an enhanced influence over the Q-dot photoluminescence properties, including PL quantum yield, spectral shape, position, etc. To passivate the surface, overcoating monodispersed NCs with epitaxial layers of another inorganic semiconductor, particularly a material with a larger band gap, has been proven to be quite effective. This core–shell architecture has been extensively explored, and has resulted in strongly emitting and robust NCs. In some cases, an order of magnitude enhancement in PLQE (30–70% efficiency in core–shells starting with ~5% efficiency from cores) has been realized using this approach. Luminescence enhancement has been demonstrated in core–shell configurations of

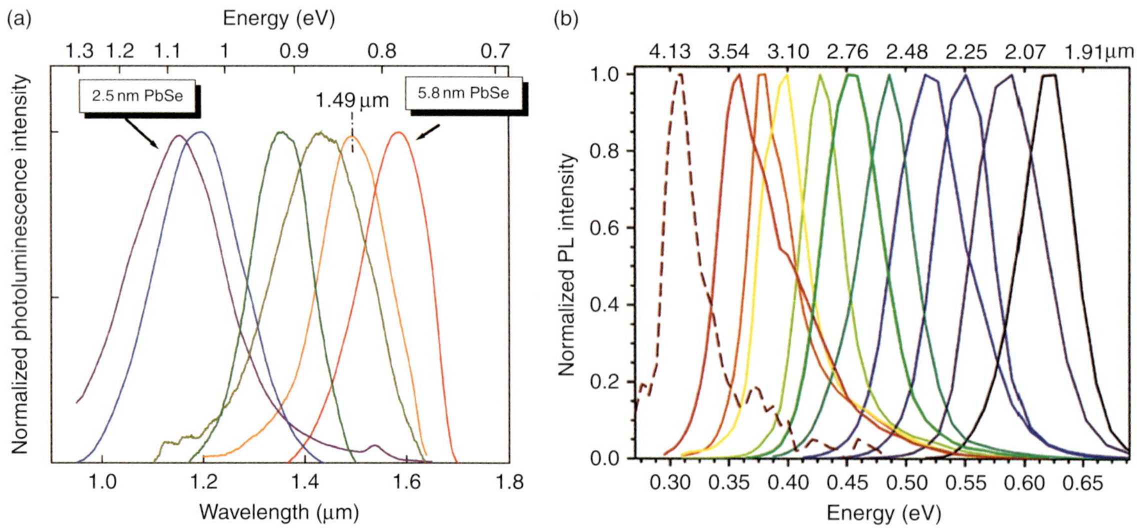

Figure 11 Example of size tunable emission: Photoluminescence in the NIR (a) and mid-IR (b) from colloidal PbSe NCs. (b) Reproduced with permission from Pietryga JM, Schaller RD, Werder D, Stewart MH, Klimov VI, and Hollingsworth JA (2004) *Journal of the American Chemical Society* 126: 11752.

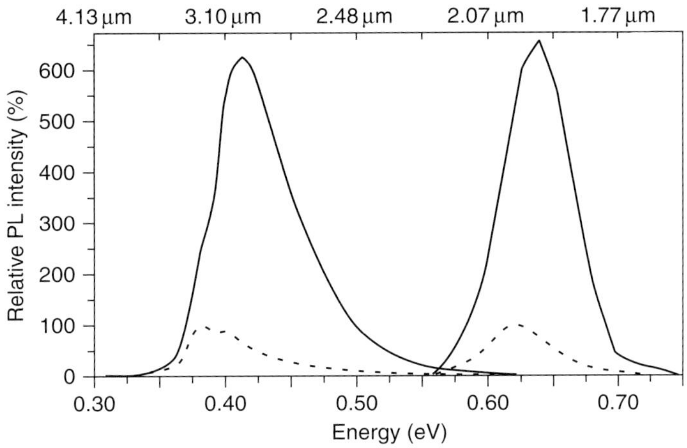

Figure 12 The effectiveness of a thin shell of a high-band-gap semiconductor in enhancing the PL intensity of a low-band-gap core material. The technique is especially useful for enhancing the emission of cores with low intrinsic PL. The figure shows relative PL intensity before (dotted line) and after (solid line) applying a thin shell of CdSe on large PbSe NC cores. Reproduced with permission from Pietryga JM, Schaller RD, Werder D, Stewart MH, Klimov VI, and Hollingsworth JA (2004) *Journal of the American Chemical Society* 126: 11752.

many Q-dots including CdSe/ZnS, CdSe/CdS, InP/CdSe, InAs/ZnSe, PbSe/CdSe, and PbS/CdS (**Figure 12**).

10.2.2.2 Defect and activator emission

Radiative emission from NCs can also arise from localized impurity and/or activator quantum states in the band gap. Defect states are called dark states when they lie inside the bands themselves [63]. Depending on the type of defect or impurity, the state can act as a donor or an acceptor. Electrons or holes are pulled to these sites of deficient or excess local charge due to Coulombic attraction. These defects can be classified as either shallow or deep-level states, where the former states have energies near the conduction or valence band edge.

Defect states are expected at the surface of NCs despite the use of various passivation methods. The concentration of surface states on the NCs depends on the synthesis and passivation processes. These surface states act as traps for charge carriers and excitons, which generally degrade the optical properties of the Q-dot by increasing the rate of its electron–hole nonradiative recombination. However, in some cases, the surface states can also lead to radiative transitions, such as in the case of green emission in ZnO [64] or long wavelength NIR emission in InP [61a,b].

Luminescence from intentionally incorporated impurities is called extrinsic luminescence. These impurities are called activators and they perturb the band structure by creating local quantum states that lie within the band gap. The predominant radiative mechanism for extrinsic luminescence is electron–hole recombination, which may occur via transitions from conduction band to acceptor state, donor state to valence band or donor state to acceptor state.

10.2.2.3 Blinking effect

NCs can often exhibit intermittent fluorescence, often referred to as blinking. In this phenomenon, a NC emits light for a period of time (on state) followed by a dark period (off state). In 1996, Nirmal and coworkers observed [65] this pattern of switching between an emitting and a nonemitting state from a single CdSe NC at room temperature. The postulated mechanism for blinking is a photo-induced ionization process [66], which leads to a charged NC resulting in separation of localized electrons and holes. Based on this model, the NCs remain dark for the lifetime of their ionized state. A nonradiative Auger recombination process would be expected to dominate the quenching of ionized NCs. However, experimental results do not completely support this model. Recently Kuno *et al.* [67] found that room-temperature intermittent fluorescence exhibits power-law statistics indicative of long-range order. The distribution of bright and dark periods follows the inverse power law given by

$$P(t) = At^{-\alpha} \tag{6}$$

where α, the power law exponent, has a value $\sim 1.5 \pm 0.1$.

10.2.3 Thin Film Properties: Absorption and Photoluminescence

Owing to their low-temperature solution-based synthesis and compatibility of processing techniques with those of organic polymers, semiconductor NCs have often been incorporated into polymeric host matrices, potentially allowing the design of novel photonic materials for optoelectronic applications. Here, we provide a brief review of the optical absorption and emission of semiconductor NCs dispersed in polymeric matrices. Special emphasis is on the modification of spectral emission properties with varying NC concentration. The modification of the Q-dot optical properties in an ensemble of dense NCs, in the form of a solid film, due to collective effects will be also discussed.

Although great progress has been made in understanding the basic electronic and optical properties of isolated noninteracting dilute ensemble of semiconductor NCs, not much work has been done in exploring the evolution of these properties as these NCs aggregate toward a dense ensemble. Artemyev *et al.* [68–70] studied the systematic evolution of the absorption spectrum of a NC ensemble, a nanocomposite with a polymer, toward collective electronic states with increasing NC concentration (**Figure 13(a)**). The overall absorption spectrum is a linear combination of the absorption due to the NCs and that from the polymer. With increasing concentration of highly monodispersed CdSe NCs in a polyethylene glycol matrix, the optical absorption due to the NC component exhibits a successive evolution from a set of a finite number of discrete sub-bands, inherent in isolated NCs, to a smooth, nearly structureless band-edge absorption spectrum similar to that of bulk semiconductors. The absorption onset also exhibits a monotonous shift toward lower energies with increasing NC concentration. The effect occurs both at room and liquid helium temperatures. These results were interpreted in terms of a systematic transition from localized individual (in the dilute ensemble) to collective (in the close-packed ensemble) electron states delocalized over a finite number of NCs. The observed behavior was considered [68] as a possible precursor to the Anderson transition in a NC solid. It must be noted here that this modification of absorption spectrum is most poignant in close-packed arrangements of NCs of relatively smaller size, in which noticeable delocalization occurs even at concentrations far from close-packing. Considering the presence of an organic capping shell, even close-packed ensembles of larger nanocrystallites may not satisfy the delocalization requirement [71].

Along with the modification of absorption features, embedding semiconductor NCs in a polymer matrix can drastically modify the photoluminescence

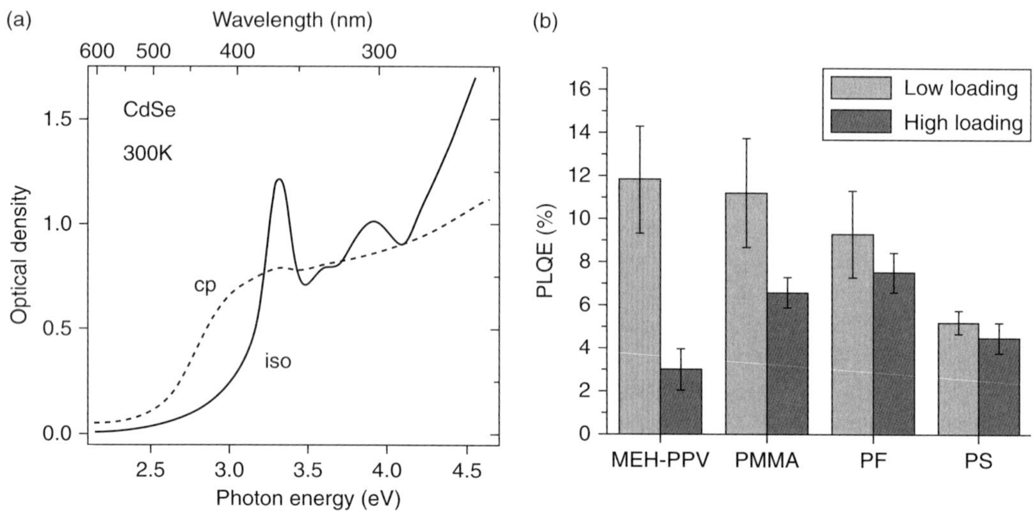

Figure 13 Absorption and photoluminescence of NCs in thin films. (a) Optical absorption spectra of isolated (solid line) and close-packed (dotted line) ensembles of CdSe NCs exhibiting a shift to structureless, red-shifted features in the latter from the cluster-like features of the former. (b) PLQE of PbS NCs in different polymer films with high (mass ratio of 1:1) and low (mass ratio 1:5) relative content of NCs. (a) Reproduced with permission from Artemyev MV, Bibik AI, Gurinovich LI, Gaponenko SV, Jaschinski H, and Woggon U (2001) *Physica Status Solid (B)* 224: 393. (b) Reproduced with permission from Chang TF, Maria A, Cyr PW, Sukhovatkin V, Levina L, and Sargent EH (2005) *Synthetic Metals* 148: 237.

properties of the nanocomposite compared to the pure polymer. Quantum size effect enables tuning of the fluorescence spectra of the composite. Effects of the NC concentration on the luminescence spectrum and emission intensity also become significant.

Semiconductor NCs dispersed in polymer matrices are used for tuning the luminescence spectra in polymer nanocomposites. Using NCs of different sizes or by using different composite semiconductors, bright emission with high PL quantum yields has been observed for these polymer composites effectively covering the entire visible spectrum [72] and different regions of the IR . Experimental results have demonstrated that an appropriate choice of polymer (keeping in mind the transparency of the polymer in the spectral region of emission from the luminescent NCs) and concentration of the Q-dot lead to high PL quantum yield (**Figure 13(b)**). In polymer composites that are effective in preventing phase segregation of the NCs in the polymer matrix, that is, in dilute ensembles, the PL quantum yield of CdSe/ZnS NCs have been shown to be as high (~40%) similar to that in solution [72]. Similar results were obtained for polymer composites where IR-active PbS NCs are dispersed in several host polymeric matrices [73]. As the size of the ordered close-packed NC domains decreases, the PL quantum yield increases. When fewer NC domains form, the distance between the particles increases. This impedes energy transfer between the NCs and increases the PL quantum yield, as it is less likely that energy is transferred to nonluminescent NCs. Moreover, the further apart the NCs are, the less likely charge separation, a photoluminescence quenching mechanism, occurs between NCs. These effects become particularly dominant in dense close-packed films of NCs, as will be discussed later. In a similar study of a nanocomposite of IR-active PbSe NCs [74], where a blend of poly (methylmethacrylate) (PMMA) and MEH-PPV was employed as the polymer, not only the intensity of the photoluminescence, but also the wavelength of emission from the NCs could be tuned by their concentration and excitation of the polymer matrix. When pumped with a broadband incandescent source, the fundamental emission from the NCs, governed by their size, could be tuned by adjusting their concentration. Further, excitation of the polymer host in an absorption band that is resonant with the second absorption transition of the NCs results in an additional emission peak indicating energy transfer from the polymer. Finally, Kagan *et al.* [75] showed that in dense close-packed NC solids, significant electronic energy transfer arising out of dipole–dipole interaction between nearest-neighbor particles modifies the emission intensity and spectrum of these films. In an ensemble of different-sized NCs, the luminescence from the smallest particles get quenched while that of the larger ones is enhanced due to this pattern of energy transfer, the net result being a red shift in the emission wavelength and a reduction of PL quantum yield compared to the values obtained in solutions.

10.2.4 Electrical Transport Properties

NCs are discrete particles, which can be physically separated from one another either by the surrounding host medium or by a shell of insulating capping ligands. The electronic states in a semiconductor NC are typically confined within the physical boundaries of the particle by significantly large potential barriers. The discrete nature of the NCs' density of states in a confined geometry gives rise to novel electronic properties. For example, transport between NCs can be considered a hopping problem between localized states similarly to that in molecular solids. For semiconductor NCs in the strong quantum confinement regime, charge transport is limited by interparticle electron transfer rather than transport within individual nanocrystallites. The physics governing charge carrier transport in semiconductor NCs is therefore similar to that in molecular systems rather than that of the traditional semiconductor band transport. This is consistent with the temperature dependence of conductivity in films of semiconductor NCs which has been observed to be thermally activated, and thus suggests an activated hopping transport mechanism similar to the hopping model described for molecular organic semiconductors. In this case, energetic disorder arises from the size distribution of the particles and geometric disorder from the separation of particles, spatially or by ligands. Unlike most conjugated polymers, NCs can transport both electrons and holes with comparable mobilities. Moreover, varying the NC size allows tuning of the electron affinity and ionization potential, which can influence charge transfer. More importantly, due to the large surface-to-volume ratio in NCs, surface traps and defects can significantly affect charge transport properties [4].

10.2.4.1 Nanocrystal–polymer hybrid thin films

Incorporation of semiconductor NCs into conducting or semiconducting polymeric matrices has led to the development of a novel class of electronic and optoelectronic media for a number of potential technological applications. The tunability of NC properties by controlling their composition, size, and interface with the matrix provide an advantage in formulating new composite materials with optimized properties for functional optoelectronic devices such as LEDs, photovoltaic cells, and photorefractive devices. Semiconductor NCs can, in principle, be used in multiple roles in such composites as photocharge generators, luminescent recombination centers, or transport modifiers. Knowledge of the influence of NC concentration on charge carrier transport properties in such nanocomposites is important in optimizing device properties.

Greenham *et al.* [76] showed, through photoluminescence quenching and photoconductivity measurements, that in a photoconductive nanocomposite of CdSe NCs in MEH–PPV matrix, the photoconductive efficiency improves with increasing concentration of NCs. At concentrations high enough to exceed the percolation threshold, the NCs form a connected network in the plane of the film. Thus, both electrons and holes can be transported through continuous pathways to the electrodes provided by the NC and polymer components of the composite material. Moreover, when the surface of the NCs were treated by removing the long insulating capping groups or replacing them with short-chain ligands, charge separation and transport were significantly enhanced. These observations were supported later by results from hybrid photodiodes consisting of NCs and/or nanorods blended in a semiconducting polymer matrix [77], confirming the role of NCs as a mediating medium of charge transport in such hybrid devices.

In another surprising finding, Roychoudhury *et al.* [78] observed that NCs present in a composite with a semiconducting polymer influenced the hole mobility in the polymer matrix. Carrier transport was described in terms of the stochastic continuous-time random-walk model by Scher and Montroll [79] in a dispersive (non-Markoffian) process. Significant enhancement of the effective carrier mobility up to an order of magnitude was noticed with the increase of NC concentration which was still well below the percolation limit.

10.2.4.2 Close-packed nanocrystal solid thin films

The ability to successfully and efficiently incorporate NCs into optoelectronic devices requires a fundamental understanding of the conduction of charge carriers in the system. Solids of chemically synthesized monodispersed NCs provide an opportunity for developing materials with novel optical and electronic properties. Not only can the properties of the constituent particles be tuned through quantum-size effects, but even collective properties of the thin film can be adjusted by controlling interparticle coupling and order. This is because charge transport in an array of NCs separated by insulating capping ligands depends on the energy levels of the neighboring NCs, the exchange coupling energy between the NCs, and the Coulomb charging energy of the array. Since thermally generated carriers will be highly suppressed in semiconductor NCs, apart from playing a role in the narrow-band-gap IR-active entities, conductivity will be dominated by carriers injected from the electrodes [80], generated by photoexcitation [81], or generated through chemical doping [82].

In highly monodispersed NCs, quantum confinement leads to discrete energy states with narrow homogeneous line widths, which then leads to extraordinarily high density of states. Furthermore, the carrier wave functions are confined to the physical extent of the NC, and strong electronic overlap is unlikely to occur. It can be expected, in these situations, that long-range Coulomb interaction and space-charge effects will play a dominant role in transport [80]. Additional considerations peculiar to these systems further complicate the matter. In nanocrystalline systems, trap states exist at the surface of the particles and, the number and depth of traps are highly sensitive to surface passivation. Studies have revealed low electronic conductivity in semiconductor NC arrays because of large concentrations of surface dangling bonds that trap carriers in mid-gap states [80,83,84]. Additionally, electron transport in such a thin film is a local process sensitive to the immediate environment with individual NCs. Thus, disorder, in the form of distribution of particle sizes and distribution of interparticle separations, can play an important role in determining transport.

Several experimental methods and device geometries have been used in exploring charge transport in solid thin films of semiconductor NCs, which include temperature- and field-dependent dark dc-conduction in a planar or vertical geometry, electrochemical gating, photoconduction, and thin-film field-effect transistor architectures [4]. Results of electron injection efficiency into close-packed CdSe NC films by Ginger and Greenham [80] showed that low-work-function metals provide good electron injection capabilities in these films and the currents in the devices are limited by bulk transport properties. In the case of photogeneration of charges by an assembly of NCs of CdSe, Leatherdale *et al.* [81] have shown that a resonant tunneling model describes the probability of electron–hole pair separation as a function of electric field. The photocurrent increases with illumination intensity, suggesting a single-photon mechanism for carrier generation with first-order recombination kinetics. This indicates that the recombination of the exciton is the rate-limiting step dominated by a high density of traps [4]. An interesting feature of the current–voltage characteristics of these films is the exhibition of history-dependent memory effects in conductivity, where the device current (both in the dark and under illumination) is found to decay with stretched exponential kinetics [80,85]. This was explained with a space-charge-limited current model [80] with a fixed number of deep-trap sites.

Temperature-dependent conductivity measurements have also provided valuable insights into the microscopic physics of carrier transport in these materials. Interestingly, Arrhenius-type temperature dependence (in a CdSe system) [86], typical of a simple activated hopping process, as well as non-Arrhenius behavior (in a PbSe system) [87] have been observed. The non-Arrhenius behavior was explained by conduction following a variable range hopping model proposed by Mott [88]. This model describes low-temperature conduction in amorphous semiconductors, which was expanded by Efros and Shklovskii [89] to include Coulombic interactions between electrons. According to the variable range hopping model, the temperature dependence of the conductance, *G*, follows a relation

$$G(T) \propto \exp(-(\beta/T)^{\nu}) \tag{7}$$

For a 3D system, the exponent $\nu = 1/4$ at moderate to high temperatures and $\nu = 1/2$ at very low temperatures (**Figure 14**).

In thin-film field-effect transistors of close-packed PbSe NCs, Talapin and Murray [90] achieved chemical activation by treatment with hydrazine in acetonitrile to desorb the insulting capping ligands and enhance the conductance of the films by ~10 orders of magnitude. NC field-effect transistors allowed reversible switching between n-type and p-type transport through hydrazine treatment, yielding high field-effect electron and hole mobilities of 0.9 and 0.2 $cm^2 V s^{-1}$, respectively. In a similar study, Law *et al.* [91] performed different amine-based chemical treatments on close-packed solid films of PbSe. The

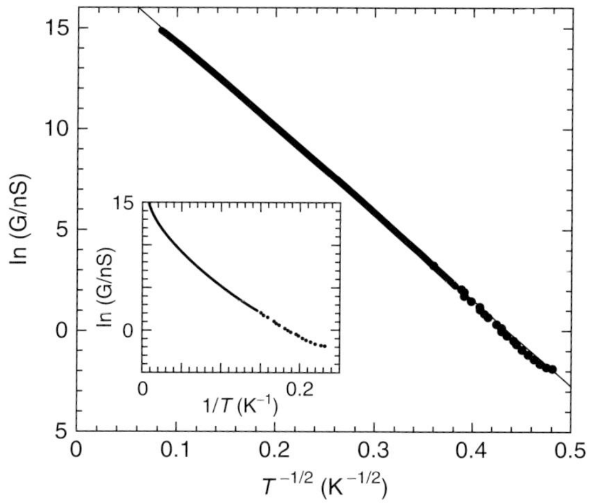

Figure 14 Temperature dependence of the conductance at low applied bias for a film of PbSe NCs exhibiting variable range hopping transport (**Equation 7**) with an exponent $\nu = 1/2$, in the temperature range of 4.3–135 K. The inset shows deviation from Arrhenius behavior. Reproduced with permission from Wehrenberg BL, Yu D, Ma J, and Guyot-Sionnest P (2005) *Journal of Physical Chemistry B* 109: 20192.

original oleate ligands were successfully removed to a great degree using ethanol-based hydrazine treatments. Depending on the particular chemical treatment employed, the process yielded n-type or p-type transistors with high mobility of charge carriers.

10.3 Organic/Inorganic Hybrid LEDs

Size-tunable band gaps, high PLQEs, narrow spectral emission line-widths, and high photostability make colloidal semiconductor NCs an attractive choice as lumophores as well as sensitizers in a variety of applications (e.g., optoelectronic devices). Their compatibility with facile low-temperature solution processing opens up the possibility of fabricating organic/inorganic hybrid devices at low cost with device performance similar to or better their inorganic or organic counter-parts. For electroluminescent device applications, combining the advantages of the rapidly developing organic light-emitting diode (OLED) technology with the attractive properties of semiconductor NCs has led to remarkable advances: the external quantum efficiency has improved by over two orders of magnitude, highly saturated, spectrally pure color emission across the visible spectrum has been achieved, and better color rendering than other existing technologies without power consumption trade-offs has been realized. If developed properly, these new hybrid LEDs have the potential of outperforming existing counterparts in flat panel displays and solid-state lighting in terms of brightness, color purity, power conservation, and design flexibility.

In the following sections, an overview will be given on hybrid organic/inorganic LEDs based on semiconductor NCs operating over a broad range of wavelengths, from the visible blue to the NIR. The attraction behind NC-based hybrid LEDs is to the generation of efficient light emission over the entire visible spectrum using the same nanocrystalline material. Continuous, narrow band, and tunable color emission can be achieved by varying, for instance, the size of the NC. The advantage of using an hybrid LED is the separate control, optimization of charge carrier transport, and emission properties. The organic component nonconjugated semiconducting polymer, such as polyvinylcarbazole (PVK), or organic molecules, such as aluminum-tris-(8-hydroxyquinoline) (Alq_3) or *N,N′*-bis(4-butylphenyl)-*N,N′*-bis(phenyl)benzidine (α-NPD). In the following sections, recent advances in device architectures, processing techniques, and material development, including progress made toward achieving white hybrid LEDs will be reviewed.

10.3.1 Hybrid LEDs Emitting in the Visible

Core/shell colloidal NCs exhibit exceptional luminescent properties such as narrow emission spectra tunable throughout the entire visible spectrum and high PL quantum yields. Their sharp and tunable PL spectra lead to highly saturated emission colors, that is, superior color purity and a very attractive color gamut. These desirable properties for device applications such as high-definition flat-panel displays and solid-state lighting make them a serious competitor with established (LCD for displays) and emerging (LEDs for solid-state lighting and OLEDs for both) technologies.

10.3.1.1 Nanocrystal/polymer hybrid LEDs

NC/polymer-based hybrid OLEDs have been fabricated in a variety of configurations, including bilayer heterojunctions [92–94], NC polymer intermixed composites [95–97], close-packed NC films [98], and even self-assembled stacks of NC and organic monolayers [99,100] (**Figure 15**). Electroluminescence arises in these devices from

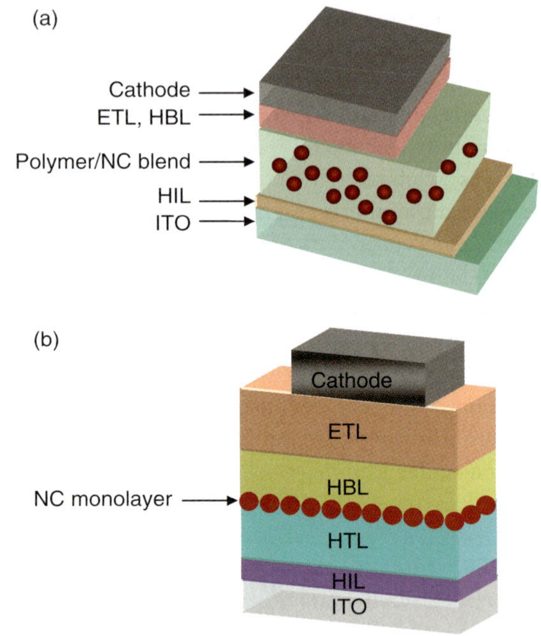

Figure 15 Two basic device architectures employed in hybrid NC-based OLEDs: (a) NC/polymer blend active layer; (b) multilayer NC/organic materials. Abbreviations: HIL: hole-injection layer, HTL: hole-transport layer, HBL: hole-blocking layer, ETL: electron transport layer.

the radiative recombination of electrons and holes by either injection of electrons and holes into the NC or from exciton energy transfer from the electronically excited organic host to the NC. The electroluminescence (EL) spectrum arising from these LEDs closely resembles the photoluminescence solution of the NCs. Blue or red shifts in the EL spectrum toward higher [92] or lower [93] energies relative to the PL spectrum have been observed. Typically, the EL spectrum is broader than that of the corresponding solution PL, which may be due to the influence of local electric fields, variation in the local dielectric environment of the emitting sites, size-dependent charge injection, and exciton transfer rates. Optical interference effects due to the photonic cavity structure of the device may be another cause of these spectral shifts.

The first published work on hybrid OLEDs incorporating colloidally grown semiconductor NCs, by Alivisatos and coworkers [92], appeared in 1994. They reported a bilayer device employing a 100 nm-thick layer of soluble PPV derivative as the hole transporting layer, and a 15–25nm-thick layer of CdSe NCs deposited with the help of a bifunctional dithiol linker as the electron transporting layer. The structure was sandwiched between an ITO-coated glass substrate anode and an Mg/Ag cathode. Under forward bias, holes are injected into the PPV layer and electrons into the CdSe layer. The device turned on around 4 V when emission characteristic of CdSe NCs was exhibited. By varying the size of the NCs, the electroluminescence band could be tuned from yellow to red. The current–voltage (*I–V*) characteristics of these devices were limited by electron injection due to the poor electrical contact at the Mg/CdSe interface and poor charge transport in the NC film (with a thickness of several monolayers). The band offsets in the device favor hole transfer from PPV to CdSe; hence the EL spectrum is dominated by NCs at low drive voltages. Due to the large imbalance in carrier concentration and carrier mobility, the recombination zone which lies within the NC layer close to the CdSe/PPV interface at low bias, spreads to the polymer layer at higher operating voltages. This gives rise to emission from both PPV and the CdSe NCs, and lead to voltage-dependent emission color depending on the spectral contribution of the two emitting components, namely the NCs and the organic material.

A subsequent paper by Dabbousi *et al.* [95] reported a single-layer NC OLED structure incorporating CdSe NCs homogeneously dispersed into a composite organic layer consisting of the hole transporting polymer PVK and the electron transporting *t*-Bu-PBD [2-(4-biphenylyl)-5-(4-*tert*-butylphenyl)-1,3,4-oxadiazole]. In this case, the NCs, having a volume fraction (5–10%) well below the percolation limit to rule out any role in charge transport, are solely responsible for charge trapping and recombination. The narrow EL spectrum (FWHM < 40 nm) could be tuned by varying the size of NCs. The devices were nonrectifying and the *I–V* characteristics exhibited near-inversion symmetry. The electroluminescence in reverse bias showed no change in spectral line shape compared to that in forward bias, indicating that the NCs are not directly involved in carrier transport and appear to serve only as trap sites. The injection of electrons and holes by tunneling through the organic capping ligands was considered the mechanism of excitation of the NC lumophores, although the possibility of Förster energy transfer from excitations formed in the organic host was also not ruled out. While the external EL quantum efficiency was low, 0.001–0.01%, these early studies demonstrated the possibility of generating electroluminescence from NC-based hybrid OLED structures, opening up the technological potential of such devices. The low EL quantum efficiency of these devices can be attributed to the energetics of the polymer–CdSe interface. Although electron transfer from PBD to CdSe (having a much higher electron affinity) is favorable, hole transfer from PVK is not leading to most of the injected electrons and holes passing through the film without recombining on the NCs.

Subsequent work explored the use of core–shell NCs along with bilayer structures, to enhance the intrinsic PL quantum yield of the NCs and reduce charge separation at the NC/polymer interface. In a bilayer device employing a spin-coated PPV hole transporting layer followed by a subsequent layer of CdSe/CdS core–shell NCs spin-coated from toluene, a 20-fold enhancement in EL quantum efficiency over that of a device based only on CdSe was achieved [93]. These hybrid OLEDs exhibit tunable emission from the green to the red with a peak external EL quantum efficiency of ~0.22% at a current density of 1 A cm^{-2} A brightness of 600 cd m^{-2} was obtained at an operating voltage of ~4 V. The stability of these devices was greatly improved with a half-life time of several hundred hours under continuous operation, a nearly 100-fold improvement over the earlier devices based on the core CdSe NCs. Interestingly, these devices had a slow response time (several seconds) to applied electric fields which can be explained in terms of a charge build-up of charges within the device. This is due to the charges being trapped either within the NC or at the hetero-interface with the polymer. In a similar device embodiment utilizing CdSe NCs overcoated with a ZnS shell, very similar enhanced device characteristics were reported [94]. **Figure 16** shows typical size-dependent electroluminescence spectra of hybrid devices employing CdSe NCs as the light emitter.

Several different hybrid NC/polymer OLEDs employing different II–VI compound semiconductor NCs with emission in the visible spectrum have been developed more recently, including single-layer devices consisting of ZnS NCs synthesized directly in a polymethylacrylic acid (PMA)-polystyrene (PS) matrix doped with tetraphenylbenzidine as a hole transporting material [96]. CdSe/ZnS-based OLEDs using different thicknesses of the emitting layer sandwiched between an ITO substrate coated with PEDOT:PSS and different metal electrodes (Ba, Al, and Au) have also been reported [101]. The performance of these devices was upon the introduction of a hole-transporting layer of PVK and an electron-conducting layer of (TPBI) 1,3,4-tris(2-*N*-

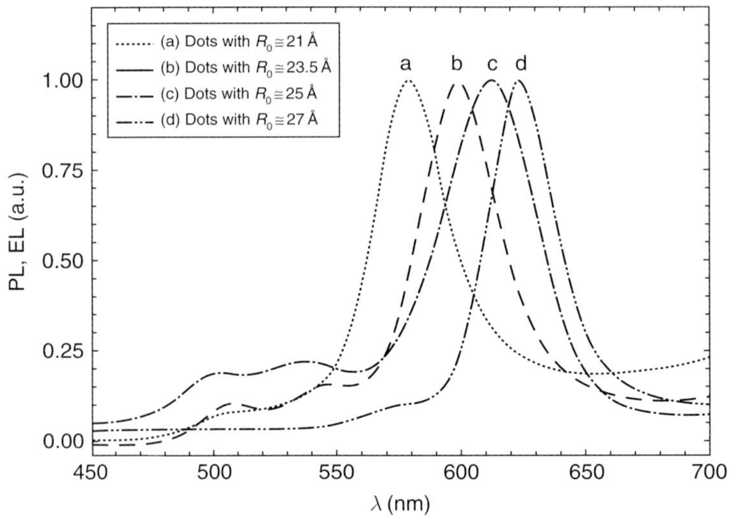

Figure 16 Size-dependent electroluminescence from CdSe NCs in a NC/PPV hybrid LED demonstrating color tunability. Reproduced with permission from Mattoussi H, Radzilowski LH, Dabbousi BO, Thomas EL, Bawendi MG, and Rubner MF (1998) *Journal of Applied Physics* 83: 7965–7974.

phenylbenzimidazol)-benzene [102]. Using 1D elongated core–shell nanorods of CdSe/CdS sandwiched between the hole and electron transporting layers in an oriented manner, emission of polarized light was also realized [102]. Multilayer hybrid OLEDs containing manganese-doped ZnS [103] or isolated CdSe/ZnS NCs capped by different organic ligands have also been demonstrated [104]. However, their EL quantum efficiencies have been rather low in comparison to OLEDs without NCs. In a recent development, Jen and coworkers employed a thermally cross-linked solvent-resistant hole transporting layer [105]. Colloidal CdSe/CdS NCs were deposited on the polymerized underlayers by spin-coating. Precise control of film thicknesses was easily achieved. This approach facilitated generation of hybrid OLEDs with nearly pure electroluminescence from the NCs and virtually no emission from the organic layers. External EL quantum efficiencies as high as 0.8% at a brightness of $100\,\mathrm{cd\,m^{-2}}$ and a maximum brightness exceeding $1000\,\mathrm{cd\,m^{-2}}$ were achieved.

10.3.1.2 Multilayer nanocrystal/organic small-molecule hybrid OLEDs

A fundamentally different device architecture of hybrid OLEDs consisting of a single monolayer of semiconductor NCs sandwiched between thin films of organic semiconducting materials was introduced by Coe *et al.* [106]. In these trilayer hybrid devices, the organic layers transport charge carriers to the vicinity of the NC monolayer from which the narrow band electroluminescence originates. This design differs from most of the earlier device concepts with thicker NC layers, where the NCs performed the dual function of transporting electrons and serving as the layer for exciton recombination. In this case, the NCs only function as luminescent centers. Several technical and operational shortcomings can be avoided in this device configuration. In bilayer polymer/NC devices, poor charge conduction in the NC multilayers leads to an imbalance of carriers in the recombination zone, resulting in low luminance power efficiency. Since the NC layer no longer participates in charge conduction, this bottleneck is eliminated. The design also does away with the possibility of pinhole defects that occur more frequently in thicker NC layers, and hence improves device production yields.

This approach utilized an elegant phase-segregation technique [107] in which the NC monolayer and the hole transporting layer are deposited in a single spin-casting step. Self-segregation of trioctylphosphine oxide (TOPO) capped CdSe/ZnS core–shell NCs from the aromatic molecules of the hole transporting material *N,N″*-diphenyl-*N,N″*-bis(3-methylphenyl)-(1,1′-biphenyl)-4,4′-diamine (TPD), when spin-coated from a mixture in chloroform having an appropriate concentration of NCs, led to the formation of a complete single NC monolayer on top of the 35 nm-thick TPD film. Subsequent thermal evaporation of 10 nm of 3-(4-biphenyl)-4-phenyl-5-

tert-butylphenyl-1,2,4-triazole (TAZ), followed by 40 nm of Alq_3 and Mg/Al electrode completed the device structure. The function of TAZ, having a deep HOMO level and a relatively high triplet energy level, was to block holes and confine excitons within the NC monolayer, resulting in narrower emission bands. The external EL quantum efficiency of the best devices exceeded 0.4% for a broad range of current density, peaking at 0.52%. The highest brightness achieved in the device was about 2000 cd m^{-2}, a 25-fold improvement over previously reported NC-based OLEDs. Additionally, these devices exhibited a much shorter response time than the earlier device systems (**Figure 17**).

The mechanism of exciton generation in the NC monolayer was attributed to two processes occurring in tandem: direct injection of charge carriers and exciton energy transfer from the organic molecules in the vicinity. A subsequent study by the same group [108] demonstrated that among these two possible routes of generating excitations, Förster energy transfer of excitons created in the organic molecules to the NCs is the dominant for electroluminescence. However, it was also shown that at higher currents the width of the exciton generation region and hence the recombination region broadens, exceeding the organic-NC Förster energy transfer radius, R_F, and resulting in an increased contribution of EL from the organic molecules. Another interesting feature of hybrid OLEDs fabricated with this multilayered device configuration is the dependence of device performance on the NC size. For the most efficient operation of NC-based hybrid OLEDs, NCs with high intrinsic photoluminescence quantum yield, η_{PL}, and large absorption cross-section, maximizing R_F is highly desirable. Since the absorption cross-section and η_{PL} generally increase with NC size, there is a need for larger-size particles. However, NC sample-size distributions are more difficult to control for larger-size particles. In this case, it is possible to lose color purity and form disordered monolayer films. Thus, high efficiency LEDs with no compromise in color purity have been best demonstrated only with intermediate-sized NCs (e.g., 3–4-nm diameter NCs for CdSe). Another interesting point in this monolayer NC device architecture is the possibility of generating very high excitation densities at precisely defined positions, opening up the possibility of achieving electrically pumped lasing in organic-based LEDs (**Figure 18**).

With further improvements in device processing techniques and more progress in achieving ultrahigh material purity, multilayer hybrid OLEDs containing CdSe/ZnS core–shell NCs with highly monochromatic emission in the red (peak at 615 nm, FWHM of 27 nm) were reported with an external EL quantum efficiency in excess of 2% [107], a luminous power efficiency >1 lm W^{-1}, and a maximum brightness of over 7000 cd m^{-2}. Hybrid OLEDs with a monolayer of NCs with blue emission (peak at 468 nm, FWHM of 30 nm) optimized for display applications have also been realized in a multilayered device structure utilizing CdS/ZnS core–shell NCs [109]. To extend the applicability of this material platform in

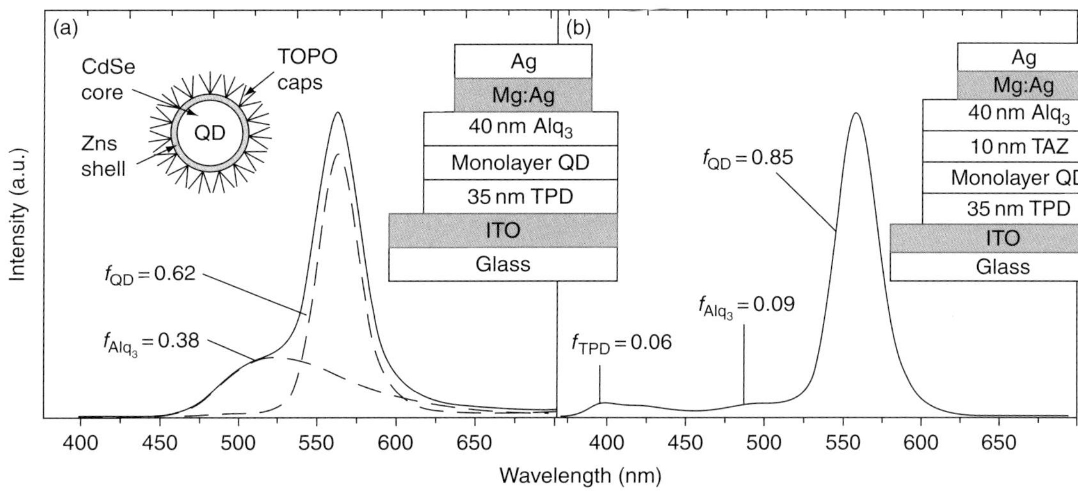

Figure 17 Electroluminescence spectra of multilayer NC/organic light-emitting devices with two different architectures (with and without the hole-blocking layer of TAZ) emitting in the visible. Reproduced with permission from Coe S, Woo W-K, Bawendi MG, and Bulovic V (2002) *Nature* 420: 800.

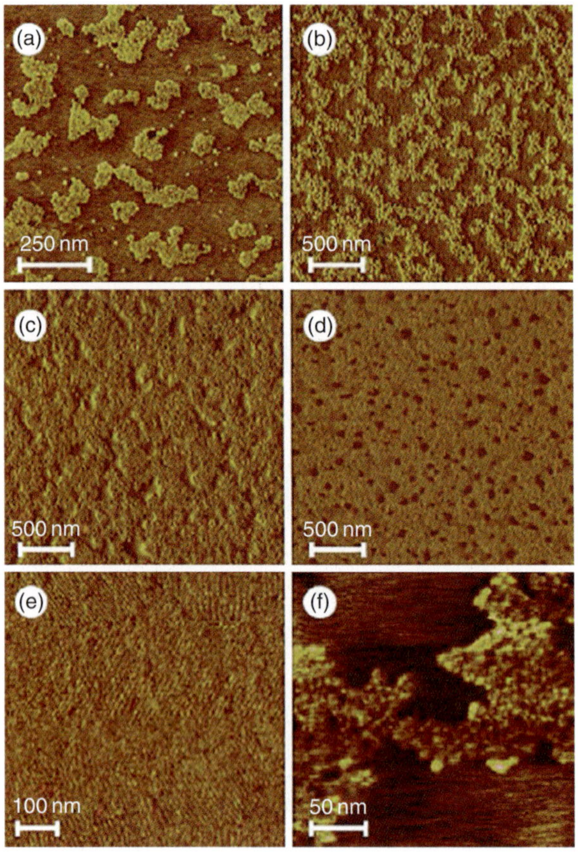

Figure 18 Demonstration of the technique of forming a monolayer of nanocrystals via phase segregation. Atomic force microscopy (AFM) images show how the coverage fractions of a nanocrystal monolayer on a smooth TPD film can be tailored from (a) 20%, (b) 50%, (c) 80%, (d) 90%, to (e) 100%. (f) Demonstrates the need for highly monodispersed nanocrystals. Disordered clusters are formed when the nanocrystals are spin coated from a sample with large-size distribution. Reproduced with permission from Coe-Sullivan S, Woo W-K, Steckel JS, Bawendi MG, and Bulovic V (2003) *Organic Electronics* 4: 123.

achieving hybrid OLEDs emitting across the visible spectrum, highly color saturated green-emitting hybrid OLEDs were also built using the same device concept with core–shell alloyed $Cd_xZn_{1-x}Se$/$Cd_yZn_{1-y}S$ NC as the active emissive monolayer [110]. The devices have an emission peak at 527 nm with a FWHM of 35 nm. External EL quantum efficiencies of 0.5% with operating voltages lower than 10 V were achieved. Additionally, the color coordinates of this device on the Commission Internationale de l'Eclairage (CIE) chromaticity diagram lie far outside the standard NTSC color triangle, indicating its potential usefulness toward display applications.

Interestingly, after these successful developments, trilayer hybrid NC/polymer devices, comprising a film of CdSe/ZnS NCs a few monolayers thick sandwiched between films of PVK and butyl-PBD [111] were also reported. All the layers in this device were deposited by spin-casting from dissimilar solvents. The devices showed 20 times improvement in external EL quantum efficiency and less than half the threshold voltage of a single-layer device based on a blend of polymers and NCs. The improvements from a blend to a layered structure were attributed to better balance of charge carriers and enhanced recombination in the NC-emitting layer.

10.3.2 Hybrid LEDs Emitting in the Near-Infrared

Large area NIR emitters can have a number of applications including night vision and biomedical sensing. Further, electroluminescent sources with activity in the 1.3–1.55-μm window can specifically find applications in wave-guided, free-space optical circuits, and wireless communications. Owing to the paucity of molecular and polymeric organic semiconductors with efficient tunable emission beyond the wavelength of 1 μm, even when modified through ligation of transition metals or when doped with inorganic complexes, the use of semiconductor NCs emitting in the NIR region of the electromagnetic spectrum offers a particularly attractive alternative. Rapid advances in the colloidal synthesis of high-quality inorganic semiconductor NCs with band gaps germane to this spectral region have opened a wide range of possibilities for tunable NIR emitters to be used in NC-based hybrid OLEDs.

10.3.2.1 Nanocrystal/polymer hybrid NIR OLEDs

The first report on NIR OLEDs based on colloidal NCs appeared in 2002 [112]. These single-layer composite blend devices employed InAs/ZnSe core–shell NC emitters in a semiconducting polymer –either MEH-PPV or poly(9,9-dehexyl-fluorenyl-2,7-diyl)-co-(1,4-{benzo-(2,1′,3)thiadiazole}) (F6BT) – deposited by spin-casting from toluene solutions. The devices exhibited broad NIR emission tunable from 1 to 1.3 μm, and a rather high turn-on voltage of 15 V. The width of the emission peak was attributed to inhomogeneous broadening caused by size distribution of the ensemble of NCs. The authors noticed a systematic increase of the turn-on voltage in a series of OLEDs with increasing volume fraction of NCs in the polymer matrix consistent with a picture of charge carriers trapped at the NC sites [114] leading to trap-limited transport. However, PL and EL quenching of the host polymer was also observed, and suggests a possible role of energy transfer from excitons on the polymer to the NCs. Despite the high turn-on voltage, these devices exhibited a reasonably good EL external quantum efficiency reaching 0.5% at high operating voltages (**Figure 19**).

In a subsequent report, based on blends of PbS NCs with MEH-PPV or poly(2-(6-cyano-6′methylheptyloxy)1,4-phenylene) (CN-PPP), Bakueva *et al.* [114] demonstrated tunable NIR electroluminescence from 1 to 1.6 μm in single-layered

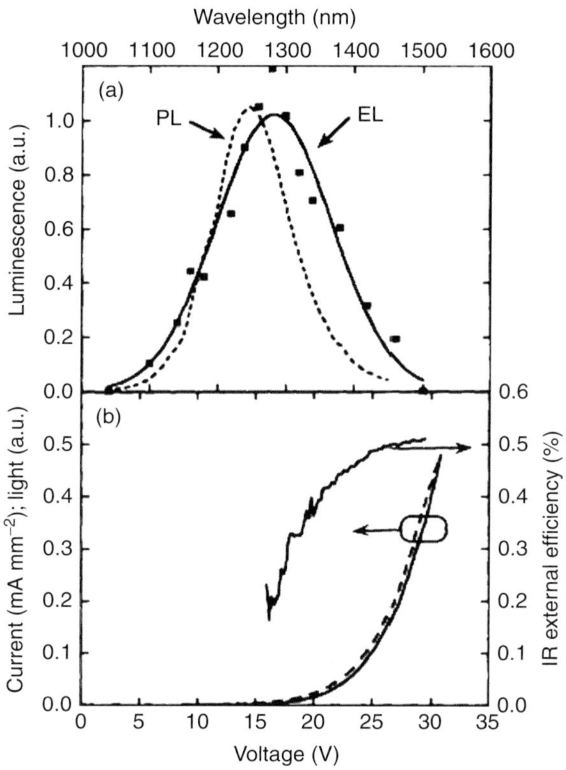

Figure 19 (a) The PL spectrum of InAs/ZnS NCs in solution (dashed line) and the EL spectrum of an InAs/ZnS NCs/MEH-PPV NIR hybrid OLED (■ and solid line). (b) Current (dashed line), light output (solid line), and external EL quantum efficiency (solid line, right axis) of the NIR hybrid OLED as a function of applied voltage. Reproduced with permission from Tessler N, Medvedev V, Kazes M, Kan SH, and Banin U (2002) *Science* 295: 1506–1508.

devices. However, the external EL quantum efficiency was low (0.008%). The authors noticed a dependence of the EL quantum efficiency on the length of the organic capping ligands used to passivate the NCs, which they attributed it to the larger impedance caused by the longer-chain organic ligand for transfer of excitons as a result of both direct carrier injection or transfer of energy from the polymer. This finding was further explored in a later work by the same group [115]. Normalized photoluminescence excitation spectra were used to corroborate their earlier findings. A threefold increase in excitation transfer efficiency was estimated in blends with the longer-chain (~2 nm) ligands replaced by shorter ones (~1 nm). Subsequently [116], optimization of structural and material components led to an improvement of the external EL quantum efficiency which reached a value of 0.27%, with an emission peaked at 1160 nm, and FWHM of ~170 nm.

In his recent work, Roychoudhury *et al.* [117] demonstrated solution-processed hybrid NIR tunable OLEDs with enhanced EL quantum efficiency using PbSe NCs as the lumophores dispersed in the conjugated polymer MEH-PPV. The significant suppression of device current and increase in turn-on voltage with increasing NC content in the composite led to the conclusion that PbSe NCs act as carrier traps and radiative recombination centers. With an optimized concentration of NC emitters and active layer thickness, the devices emitting at 1280 nm exhibited an external EL quantum efficiency of 0.83% (**Figure 20**), which is an improvement by a factor of 3 compared to those of earlier reported

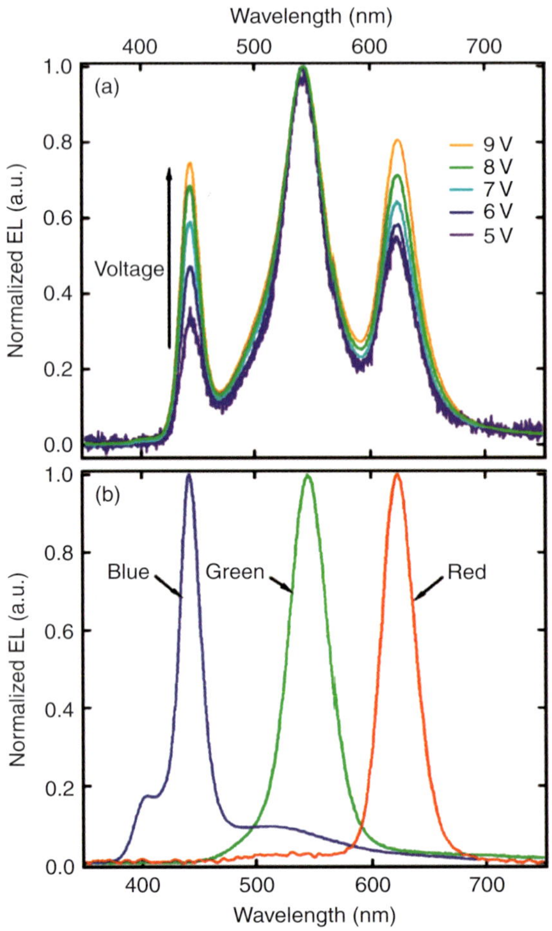

Figure 20 (a) Normalized EL spectra of a white hybrid OLED at increasing applied voltages. (b) Normalized EL spectra of red, green, and blue monochrome hybrid OLEDs of each constituent NC of the white OLED. Reproduced with permission from Anikeeva PO, Halpert JE, Bawendi MG, and Bulovic V (2007) *Nano Letters* 7: 2196.

single-layer devices. Single-layer hybrid NIR OLEDs have also been developed from blends of HgTe NCs and methyl-substituted poly(*para*-phenylene) (MeLPPP) spin-coated from a toluene solution and sandwiched between ITO and Al electrodes [118]. The emission wavelength was tunable between 900 nm and 2 μm. Infrared emission was turned on at 10 V and the devices attained a maximum external EL quantum efficiency of ~0.001%.

10.3.2.2 Multilayer nanocrystal/organic small-molecule hybrid infrared OLEDs

To extend the device concept of visible multilayer hybrid NC-based OLEDs, Steckel *et al.* [119] fabricated trilayer NIR OLEDs comprising a single monolayer of PbSe NCs sandwiched between two layers of organic semiconductors. The fabrication technique employed was similar to the visible CdSe/ZnS-based OLEDs, relying on the phase-segregation process of the PbSe NCs capped with an aliphatic chain (oleic acid) from the hole transporting aromatic organic molecules of TPD or NPD in chloroform. The narrow size distribution yielded monolayers of hexagonally closed-packed NCs after spin-casting. An electron transporting layer of Alq_3 and/or BCP was thermally subsequently deposited. The NIR electroluminescence spectra were tunable from 1.33 to 1.56 μm by varying the size of the PbSe NCs. The external EL quantum efficiency was measured to be 0.001%, limited by the reduced photoluminescence quantum yield of closely packed PbSe NCs in the solid state. The device design was not optimized for emission in this spectral region, resulting in a poor spectral overlap between the organic and the semiconductor nanocrystalline emitters. Additionally, given the short-range dependence of Förster and to a lesser extent of Dexter energy transfer, this crucial step must compete with radiative and nonradiative recombination of excitons prior to energy transfer, and may limit the EL quantum efficiency of NIR emission originating from the NCs.

To eliminate the requirement for exciton generation on organic molecules and their subsequent transfer to NCs which is prone to the loss mechanisms mentioned above, Bourdakos *et al.* [120] recently fabricated a hybrid device in which excitons are directly created on the NCs. Although the device configuration included a layered structure, it deviates considerably from the earlier design of a single monolayer of NCs sandwiched between the organic charge transporting layers. The reported devices employed a significantly thick layer of PbS NCs (110 nm) deposited on a hole-injecting layer of pentacene on ITO. A thin (10 nm) electron transporting and hole-blocking layer of BCP was deposited on top. NIR light emission with a peak at 1.2 μm and a device turn-on voltage of ~1 V were observed. The hybrid NIR OLEDs exhibited a maximum external EL quantum efficiency of 1.15% at 1.6 V, representing significant improvement over devices previously reported. Contrary to the approach used for previous multilayer device configuration pioneered by the MIT group [119], which attempted to isolate the functions of carrier transport and exciton recombination, this device actually relies on the thick NC layer to inject carriers injected from the organic layers as well as acts as a carrier recombination layer for exciton formation.

10.3.3 Multicolor and White Hybrid OLEDs

In addition to NC-based hybrid OLEDs with high monochromatic color saturation, devices with broad spectral emission of white light and a high color rendering index have also attracted a lot of interest due to their potential use in large area displays and solid-state lighting. For white color generation, NC lumophores have been utilized as replacements for red or green color components in hybrid white OLEDs employing organic emitters. For example, white hybrid OLEDs have been fabricated using this approach [121] employing stable red-emitting CdSe/ZnS core–shell NCs which were doped into a blue-emitting polymer (PFH-MEH) along with green-emitting Alq_3. A fairly pure white color with CIE coordinates of (0.30, 0.33) was realized by controlling energy and charge transfer between the different components. Maximum external EL quantum efficiencies of up to 0.24% were reported. In a more recent publication [122], CdSe/ZnS core–shell NCs were used to fabricate white-light-emitting OLEDs by combining the green to red emission of the NCs with blue emission from the organic molecule 2,7-bis[2-(4-diphenylaminophenyl)-1,3,4-oxadiazole-5-yl]-9, 9-dihexylfluorene (BADF). Maximum external EL quantum efficiency of 0.41% was achieved for white light emission with CIE color coordinates of (0.34, 0.32) at an applied bias of 13 V.

Using just NCs of the three primary colors as emitters in an electrically driven structure, white-light emission was also reported by two groups. Li *et al.* [123] first demonstrated white-light emission from a guest–host ternary NC composite using three different sizes of CdSe/ZnS NCs blended in the organic matrix of 4,4′, *N,N″*–diphenylcarbazole (CBP). PEDOT-PSS was used as the hole-injection layer, while Alq_3 served as the electron injection and transporting layer. Efficient emission of the three primary colors from the NCs gave rise to a balanced near-white overall emission with CIE coordinates of (0.32, 0.45) at an operating voltage of 13 V. Exciton generation on the NCs occurred via the parallel processes of charge trapping and energy transfer from CBP and Alq_3. To obtain white electroluminescence, the different color emissions from the NCs have to be accurately balanced. This was done by varying the concentration of the NCs to control the energy transfer and charge trapping mechanisms, while optimizing the photoluminescence quantum yields of the three types of NCs. A turn-on voltage of 6 V and maximum brightness of $1050\,cd\,m^{-2}$ at $58\,mA\,cm^{-2}$ operating current density was achieved which corresponded to a current efficiency of $1.8\,cd\,A^{-1}$.

In a more recent publication, the MIT group, using the multilayered device architecture earlier introduced by them, reported efficient broad spectral emission in a hybrid organic/inorganic structure [124]. Spectrally broad electroluminescence, tunable across the CIE color space, was generated from a single monolayer of NCs emitting multiple colors. Independent processing of the organic charge transporting layers and the NC luminescent monolayer allowed for this precise tuning of the emission spectrum without changing the device structure. By adjusting the relative concentration of NCs emitting different colors in the active layer, white-light-emitting hybrid OLEDs were demonstrated. The device structures were very similar to those used earlier for monochromatic hybrid OLEDs, which consist of a hole transporting layer, an emissive monolayer, a hole and exciton blocking layer, and an electron transporting layer. Three types of semiconductor NCs were used to generate the red, green, and blue components of the white light: CdSe/ZnS core–shells for red, ZnSe/CdSe alloyed cores with a ZnS shell for green, and ZnCdS alloyed NCs for blue. The external EL quantum efficiency of the corresponding monochrome hybrid OLEDs were found to vary from 1.6% for red-emitting devices to 0.65% and 0.35% for green- and blue-emitting OLEDs, respectively. This difference in performance, due to an increasing difficulty of resonant energy transfer from the organic transport materials to the wider-band-gap blue NCs makes the operation of the blue-emitting OLEDs more reliant on direct charge injection which is easier at higher voltages. As a result, the CIE color coordinates of the white device shifted to the blue with increased operating voltages. Nevertheless, white OLEDs with external EL quantum efficiencies of 0.36% with CIE coordinates of (0.35, 0.41) at video brightness $\sim 100\,cd\,m^{-2}$ and a high color rendering index of 86 were demonstrated (**Figure 20**).

10.3.4 Hybrid OLEDs Processed from Aqueous Solutions

An interesting variation of the hybrid NC-enabled OLEDs, borne out of the difference in the synthetic approach for obtaining colloidal NCs includes light-emitting devices processed partially or wholly from water as a solvent. State-of-the-art aqueous-based synthesis protocols yield NCs with high PL quantum yields from the short-wavelength blue (ZnSe), through the visible (CdTe), and extend into the NIR (HgTe, CdHgTe). NCs synthesized using this approach can easily be charged at specific pH values due to the free functional groups on the organic ligand molecules. This feature, in turn, allows their processing using the layer-by-layer (LBL) assembly technique, based on alternating adsorption of oppositely charged species. The method enables production of large-area high-quality films with nanometer scale control over thickness and composition.

The first hybrid OLED fabricated by the LBL assembly [99], formed by stacking 20 alternating double layers of a PPV precursor and CdSe NCs capped by thiolactic acid, emitted white light originating from recombination at NC trap sites with an external EL quantum efficiency of 0.0015%. Green to red electroluminescence was obtained [100] from LBL-assembled hybrid OLEDs based on CdTe NCs of different sizes capped by thioglycolic acid and poly(diallyldimethylammonium chloride) (PDDA). These LEDs exhibited a significantly low turn-on voltage of 2.5–3.5 V with peak external EL quantum efficiencies of 0.1%. In these LBL-assembled devices, the quality and uniformity of the emissive multilayer are crucial for achieving better efficiency and reliability. It was effectively demonstrated [125] in a red-emitting hybrid OLED based on thiol-capped CdTe NCs in combination with PDDA. The results indicated that devices with higher homogeneity in the NC layer and uniform thickness of each bilayer stabilized the device against localized high-field areas prone to electrical breakdown. An external EL quantum efficiency of 0.51% with a peak brightness of $1.42\,cd\,m^{-2}$ was reported in these hybrid OLEDs.

10.3.5 Recent Developments in Processing of Hybrid OLEDs

Since the first demonstration of using NCs in LEDs, significant strides have been made toward the development of a novel material platform that holds the promise of being at par with existing LED technologies in terms of performance while opening up the possibility of new form factors. However, this goal has not been achieved yet since the EL quantum efficiency of this new class of hybrid devices is still low and falls short of that required for commercial applications in many areas such as flat panel displays and solid-state lighting. Here, we outline a few recent developments addressing issues with processing these hybrid devices, to both enhance their performance and explore processes for commercial manufacturing.

In fabricating hybrid NC-based OLEDs, one of the synthetic designs favored by researchers involves the spin coating of a monolayer or multilayer of NCs onto an underlying solution-processed organic hole transporting layer. A critical issue in this case is the use of solvents that are mutually incompatible for the NC layer and the organic layer. Most colloidal synthetic designs producing high-quality semiconductor NCs render the surface of the NC hydrophobic, thus making them soluble in typical hydrophobic organic solvents. At the same time, most organic hole transporting layers are also soluble in common hydrophobic organic solvents. This raises an important processing issue with these devices and, demands the development and use of solvent-resistant organic hole transporting layers. The problem was addressed by Zhao *et al.* [105] who employed a thermally cross-linked PS-TPD-PFCB that is resistant to common organic solvents after polymerization. The researchers fabricated hybrid OLEDs by spin-casting highly uniform layers of CdSe/CdS core–shell NCs of controlled thickness on top of the underlying hole transporting layers. The resulting hybrid OLEDs exhibit narrow (FWHM $\sim$ 30 nm) electroluminescence, with virtually no emission from the organic materials at any applied bias. The use of multiple spin-on organic hole transporting layers, made possible by thermal cross-linking, resulted in an improved external EL quantum efficiency of 0.8% at a brightness of 100 cd m^{-2}. Use of such organic hole transporting layers in these hybrid devices, which are able to resist interfacial erosion by organic solvents during the deposition of emissive layers, makes them promising candidates for processing robust NC-based OLEDs.

Another recent work explored the possibility of replacing the organic transporting layers with high-band-gap inorganic hole transporting layers that are chemically and electrically stable. In a particular device embodiment with CdSe/ZnS core–shell emitters, wide-band-gap NiO was used [126] to replace TPD as the hole transporting layer. By controlling the resistivity of the sputtered NiO layer to balance the density of charge carriers at NC recombination sites, a maximum external electroluminescence quantum efficiency of 0.18% was achieved.

To achieve full-color pixilated structures in hybrid LEDs for full-color display applications, there is a need for a reliable processing method to form close-packed NC monolayers allowing for lateral patterning. Existing solution-based methods of fabricating large-area films like Langmuir–Blodgett techniques or spin coating suffer from the limitations of low throughput, incompatibility with patterning requirements, and issues with solvent compatibility requirements. A recent work addressed this processing issue by demonstrating a novel facile, low-cost contact-printing method to deposit patterned, solvent-free NC monolayers [127]. The researchers employed a poly(dimethylsiloxane) (PDMS) elastomer stamp coated with the aromatic polymer parylene-C to transfer monolayers of NCs onto the substrate by contact printing. The method allows for controlled deposition of NC films with low roughness (rms $\sim$ 0.5 nm). The method was applied to fabricate both patterned and unpatterned NC LEDs emitting red, green, and blue with external EL quantum efficiencies of 1%, 0.5%, and 0.2%, respectively, at video brightness. This solvent-free deposition technique enables the insertion of NC layers without exposing the constituent films to solvents. It also enables the formation of arbitrarily complex multilayer structures in devices containing NCs.

10.4 Summary

Superior optoelectronic properties combined with easy processing techniques and novel design options have set up colloidal semiconductor NCs as a viable material platform for next-generation optoelectronic applications. There is great progress on the research on NC-based hybrid LEDs in the last decade. More than two orders of magnitude improvement in external EL quantum efficiency, tunable highly saturated emission colors across a broad spectral range, and decent device lifetimes have already been realized.

Despite this progress the device figures of merit are considerably lower than their all-organic counterparts, OLEDs, and fall short of the expectations for commercialization. Extensive research efforts are underway to overcome the technological drawbacks. In review of the progress that has been made since the early demonstration of an organic-based LED (OLED) and its subsequent technological applications, it can be expected that semiconductor NC-based hybrid OLED will fulfill a lot of the scientific and technological promises that are yet to be shown.

References

1. Trindade T, O'Brien P, and Pickett NL (2001) *Chemistry of Materials* 13: 3843.
2. Nirmal M and Brus L (1999) *Accounts of Chemical Research* 32: 407.
3. Efros AL and Efros AL (1982) *Soviet Physics Semiconductors* 16: 772.
4. Klimov VI (2004) *Semiconductor and Metal Nanocrystals: Synthesis and Electronic and Optical Properties*. New York: Marcel Dekker.
5. Efros AL and Rosen M (2000) *Annual Review of Materials Science* 30: 475.
6. Murray CB, Norris DJ, and Bawendi MG (1993) *Journal of the American Chemical Society* 115: 8706.
7. Murray CB, Kagan CR, and Bawendi MG (2000) *Annual Review of Materials Science* 30: 545.
8. Gaponik N, Talapin DV, Rogach AL, *et al.* (2002) *Journal of Physical Chemistry B* 106: 7177.
9. Hodes G (2001) *Electrochemistry of Nanomaterials*. Weinheim: Wiley-VCH.
10. Guyot-Sionnest P (2003) Infrared and visible electrochromic semiconductor colloid quantum dots. In: Liz-Marzán LM and Kamat PV (eds.) *Nanoscale Materials*, pp. 181–190. Berlin: Springer.
11. Huynh WU, Dittmer JJ, and Alivisatos AP (2002) *Science* 295: 2425.
12. Gur I, Fromer NA, Geier ML, and Alivisatos AP (2005) *Science* 310: 462.
13. Sun B and Greenham NC (2006) *Physical Chemistry – Chemical Physics* 8: 3557.
14. Rogach AN, Gaponik N, Lupton JM, *et al.* (2008) *Angewandte Chemie International Edition* 47: 6538.
15a. Rogach AL, Eychmuller A, Hickey SG, and Kershaw SV (2007) *Small* 3: 536.
15b. Harrison MT, Kershaw SV, Burt MG, *et al.* (2000) *Pure and Applied Chemistry* 72: 295.
16. Konstantatos G, Howard I, Fischer A, *et al.* (2006) *Nature* 442: 180.
17. Cho N, Roy Choudhury K, Thapa RB, *et al.* (2007) *Advanced Materials* 19: 232.
18. Bruchez MP, Morrone M, Gin P, Weiss S, and Alivisatos AP (1998) *Science* 281: 2013.
19. Kayanuma Y (1988) *Physical Review B* 38: 9797.
20. Wang Y (1996) *Pure and Applied Chemistry* 68: 1475.
21. Guzelian AA, Katari JEB, Kadavanich AV, *et al.* (1996) *Journal of Physical Chemistry* 100: 7212.
22. Klimov VI (2006) *Journal of Physical Chemistry B* 110: 16827.
23. Yoffe AD (2001) *Advances in Physics* 50: 1.
24. Wang Y, Suna A, Mahler W, and Kasowski R (1987) *Journal of Chemical Physics* 87: 7315.
25. Brus LE (1983) *Journal of Chemical Physics* 79: 5566.
26. Steigerwald ML and Brus LE (1990) *Accounts of Chemical Research* 23: 183.
27. Bawendi MG, Steigerwald ML, and Brus LE (1990) *Annual Review of Physical Chemistry* 41: 477.
28. Brus LE (1984) *Journal of Chemical Physics* 80: 4403.
29. Weller H (1993) *Advanced Materials* 5: 88.
30. Alivisatos AP (1996) *Journal of Physical Chemistry* 100: 13226.
31. Wang Y and Herron N (1990) *Physical Review B* 42: 7253.
32. Chen XB, *et al.* (2003) *Nano Letters* 3: 799.
33. Hines MA and Guyot-Sionnest P (1996) *Journal of Physical Chemistry* 100: 468.
34. Holder E, Tessler N, and Rogach AL (2008) *Journal of Materials Chemistry* 18: 1064.
35. Kim S, Fisher B, Eisler HJ, and Bawendi M (2003) *Journal of the American Chemical Society* 125: 11466.
36a. Walter P, Martinetto1 P, Tsoucaris G, *et al.* (1999) *Nature* 397: 483.
36b. Walter P, Welcomme E, Hallégot P, *et al.* (2006) *Nano Letters* 6: 2215.
37. Matijevic E (1985) *Annual Review of Materials Science* 15: 483.
38. Sugimoto T (2001) *Monodisperse Particles*. Amsterdam: Elsevier.
39. Burda C, *et al.* (2005) *Chemical Reviews* 105: 1025.
40. Chatterjee M and Patra A (2001) *Journal of the American Ceramic Society* 84: 1439.
41. Masui T, Fujiwara K, Machida K, *et al.* (1997) *Chemistry of Materials* 9: 2197.
42. Lisiecki I, Bjorling M, Motte L, Ninham B, and Pileni MP (1995) *Langmuir* 11: 2385.
43. La Mer VK and Dinegar RH (1950) *Journal of the American Chemical Society* 72: 4847.
44. Murray CB, Sun S, Gaschler W, Doyle H, Betley TA, and Kagan CR (2001) *IBM Journal of Research and Development* 45: 47.
45. Hines MA and Guyot-Sionnest P (1998) *Journal of Physical Chemistry B* 102: 3655.
46. Reiss H (1951) *Journal of Chemical Physics* 19: 482.
47. Lifshitz IM and Slesov VV (1958) *Soviet Physics, JTEP* 35: 479.
48. Peng X, Wickham J, and Alivisatos AP (1998) *Journal of the American Chemical Society* 120: 5343.
49. De Smet Y, Deriemaeker L, and Finsy R (1997) *Langmuir* 13: 6884–6888.
50. Oswald W (1901) *Zeitschrift für Physikalische Chemie* 37: 385.
51. Rogach AL, Talapin DV, Chevchenko EV, Kornowski A, Hasse M, and Weller H (2002) *Advanced Functional Materials* 12: 653.
52. Pankove JI (1970) *Optical Processes in Semiconductors*. New York: Dover Publications.
53. Norris DJ and Bawendi MG (1996) *Physical Review B* 53: 16338.
54a. Vossmeyer T, Katsikas L, Gienig TM, *et al.* (1994) *Journal of Physical Chemistry* 98: 7665.
54b. Leatherdale CA, Woo W-K, Mikulec FV, and Bawendi MG (2002) *Journal of Physical Chemistry B* 106: 7619.
55. Chan WCW and Nie S (1998) *Science* 281: 2016–2018.
56a. Wehrenberg BL, Wang C, and Guyot-Sionnest P (2002) *Journal of Physical Chemistry B* 106: 10634.
56b. Yu WW, Falkner JC, Shih BS, and Colvin VL (2004) *Chemistry of Materials* 16: 3318.
57. El-Sayed MA (2004) *Accounts of Chemical Research* 37: 326.
58. Lakowicz JR (1983) *Principles of Fluorescence Spectroscopy*. New York: Plenum Press.

59. Dabbousi BO, Rodriguez-Viejo J, Mikulec FV, *et al.* (1997) *Journal of Physical Chemistry B* 101: 9463–9475.
60. Qu L and Peng XG (2002) *Journal of the American Chemical Society* 124: 2049.
61a. Micic OI, Sprague JR, Lu Z, and Nozik AJ (1996) *Journal of Applied Physics* 68: 3150.
61b. Xie R, Battaglia D, and Peng X (2007) *Journal of the American Chemical Society* 129: 15432.
62. Pietryga JM, Schaller RD, Werder D, Stewart MH, Klimov VI, and Hollingsworth JA (2004) *Journal of the American Chemical Society* 126: 11752.
63. IssacA, von Borczyskowski C, and Cichos F (2005) *Physical Review B* 71: 16.
64. Alexandra BD and Leung YH (2006) *Small* 2: 944.
65. Nirmal M, *et al.* (1996) *Nature* 383: 802.
66. Efros AL and Rosen M (1997) *Physical Review Letters* 78: 1110.
67. Kuno M, Fromm DP, and Hamann HF (2000) *Journal of Chemical Physics* 112: 3117.
68. Artemyev MV, Bibik AI, Gurinovich LI, Gaponenko SV, and Woggon U (1999) *Physical Review B* 60: 1504.
69. Artemyev MV, Woggon U, Jaschinski H, Gurinovich LI, and Gaponenko SV (2000) *Journal of Physical Chemistry B* 104: 11617.
70. Artemyev MV, Bibik AI, Gurinovich LI, Gaponenko SV, Jaschinski H, and Woggon U (2001) *Physica Status Solid (B)* 224: 393.
71. Murray CB, Kagan CR, and Bawendi MG (1995) *Science* 270: 1335.
72. Lee J, Sundar VC, Heine JR, Bawendi MG, and Jensen KF (2000) *Advanced Materials* 12: 1102.
73. Chang TF, Maria A, Cyr PW, Sukhovatkin V, Levina L, and Sargent EH (2005) *Synthetic Metals* 148: 237.
74. Voitenko I, Muth JF, Gerhold M, Cui D, and Xu J (2007) *Materials Science and Engineering* 27: 1078.
75. Kagan CR, Murray CB, and Bawendi MG (1996) *Physical Review B* 54: 8633.
76. Greenham NC, Peng X, and Alivisatos AP (1996) *Physical Review B* 54: 17628.
77. Huyunh WU, Peng X, and Alivisatos AP (1999) *Advanced Materials* 11: 923.
78. Roychoudhury K, Samoc M, Patra A, and Prasad PN (2004) *Journal of Physical Chemistry B* 108: 1556.
79. Scher H and Montroll EW (1975) *Physical Review B* 12: 2455.
80. Ginger DS and Greenham NC (2000) *Journal of Applied Physics* 87: 1361.
81. Leatherdale CA, Kagan CR, Morgan NY, Empedocles SA, Kastner MA, and Bawendi MG (2000) *Physical Review B* 62: 2669.
82. Shim M and Guyot-Sionnest P (2000) *Nature* 407: 981.
83. Jarosz MB, Porter VJ, Fisher BR, Kastner MA, and Bawendi MG (2004) *Physical Review B* 70: 195327.
84. Vanmaekelberg D and Lijeroth P (2005) *Chemical Society Reviews* 34: 299.
85. Kohlrausch R (1847) *Annalen der Physik (Leipzig)* 12: 393.
86. Yu D, Wang C, Wehrenberg BL, and Guyot-Sionnest P (2004) *Physical Review Letters* 92: 216802-1.
87. Wehrenberg BL, Yu D, Ma J, and Guyot-Sionnest P (2005) *Journal of Physical Chemistry B* 109: 20192.
88. Mott NF (1993) *Cconduction in Non-Crystalline Materials*. Oxford, UK: Clarendon Press.
89. Efros A and Shkloviskii B (1975) *Journal of Physics C* 8: L49.
90. Talapin DV and Murray CB (2005) *Science* 310: 86.
91. Law M, Luther JM, Song Q, Hughes BK, Perkins CL, and Nozik AJ (2008) *Journal of the American Chemical Society* 130: 5974.
92. Colvin VL, Schlamp MC, and Alivisatos AP (1994) *Nature* 370: 354.
93. Schlamp MC, Peng XG, and Alivisatos AP (1997) *Journal of Applied Physics* 82: 5837–5842.
94. Mattoussi H, Radzilowski LH, Dabbousi BO, Thomas EL, Bawendi MG, and Rubner MF (1998) *Journal of Applied Physics* 83: 7965–7974.
95. Dabbousi BO, Bawendi MG, Onitsuka O, and Rubner MF (1995) *Applied Physics Letters* 66: 1316–1318.
96. Yang Y, Xue S, Liu S, Huang J, and Shen J (1996) *Applied Physics Letters* 69: 377.
97. Mattoussi H, Radzilowski LH, Dabbousi BO, *et al.* (1999) *Journal of Applied Physics* 86: 4390.
98. Artemyev MV, Sperling V, and Woggon U (1997) *Journal of Applied Physics* 81: 6975.
99. Gao MY, Richter B, and Kirstein S (1997) *Advanced Materials* 9: 802–805.
100. Gao MY, Lesser C, Kirstein S, Mohwald H, Rogach AL, and Weller H (2000) *Journal of Applied Physics* 87: 2297–2302.
101. Hikmet RAM, Talapin DV, and Weller H (2003) *Journal of Applied Physics* 93: 3509.
102. Hikmet RAM, Chin PTK, Talapin DV, and Weller H (2005) *Advanced Materials* 17: 1436.
103. Yang H, Holloway PH, and Ratna BB (2003) *Journal of Applied Physics* 93: 586.
104. Zhao J, Zhang J, Jiang C, Bohnenberger J, Basche T, and Mews A (2004) *Journal of Applied Physics* 96: 3206.
105. Zhao J, Bardecker JA, Munro AM, *et al.* (2006) *Nano Letters* 6: 463.
106. Coe S, Woo W-K, Bawendi MG, and Bulovic V (2002) *Nature* 420: 800.
107. Coe-Sullivan S, Steckel JS, Woo W-K, Bawendi MG, and Bulovic V (2005) *Advanced Functional Materials* 15: 1117.
108. Coe-Sullivan S, Woo W-K, Steckel JS, Bawendi MG, and Bulovic V (2003) *Organic Electronics* 4: 123.
109. Steckel JS, Zimmer JP, Coe-Sullivan S, Stott NE, Bulovic V, and Bawendi MG (2004) *Angewandte Chemie International Edition* 43: 2154.
110. Steckel JS, Snee P, Coe-Sullivan S, *et al.* (2006) *Angewandte Chemie International Edition* 45: 5796.
111. Chaudhary S, Ozkan M, and Chan WCW (2004) *Applied Physics Letters* 84: 2925.
112. Tessler N, Medvedev V, Kazes M, Kan SH, and Banin U (2002) *Science* 295: 1506–1508.
113. Kao KC and Hwang W (1981) *Electrical Transport in Solids, Volume 14: International Series in the Science of the Solid State*. New York: Pergamon.
114. Bakueva L, Musikhin S, Hines MA, *et al.* (2003) *Applied Physics Letters* 82: 2895–2897.
115. Chang T-WF, Musikhin S, Bakueva L, *et al.* (2004) *Applied Physics Letters* 84: 4295–4297.
116. Konstantatos G, Huang C, Levina L, Lu Z, and Sargent EH (2005). *Advanced Functional Materials* 15: 1865–1869.
117. Roychoudhury K, Song DW, and So F (submitted).
118. Koktysh DS, Gaponik N, Reufer M, *et al.* (2004) *ChemPhysChem* 5: 1435.
119. Steckel JS, Coe-Sullivan S, Bulovic V, and Bawendi MG (2003) *Advanced Materials* 15: 1862.
120. Bourdakos KN, Dissanayake DMNM, Lutz T, Silva SRP, and Curry RJ (2008) *Applied Physics Letters* 92: 153311.
121. Li Y, Rizzo A, Mazzeo M, *et al.* (2005) *Journal of Applied Physics* 97: 113501.
122. Ahn JH, Bertoni C, Dunn S, *et al.* (2007) *Nanotechnology* 18: 335202/1.
123. Li Y, Rizzo A, Cingolani R, and Gigli G (2006) *Advanced Materials* 18: 2545.

124. Anikeeva PO, Halpert JE, Bawendi MG, and Bulovic V (2007) *Nano Letters* 7: 2196.
125. Bertoni C, Gallardo D, Dunn S, Gaponik N, and Eychmuller A (2007) *Applied Physics Letters* 90: 034107.
126. Caruge J-M, Halpert JE, Bulovic V, and Bawendi MG (2006) *Nano Letters* 6: 6991.
127. Kim L, Anikeeva PO, Coe-Sullivan S, Steckel JS, Bawendi MG, and Bulovic V (2008) *Nano Letters* 8: 4513.

INDEX

NOTES:

Cross-reference terms in italics are general cross-references, or refer to subentry terms within the main entry (the main entry is not repeated to save space). Readers are also advised to refer to the end of each article for additional cross-references - not all of these cross-references have been included in the index cross-references.

The index is arranged in set-out style with a maximum of three levels of heading. Major discussion of a subject is indicated by bold page numbers. Page numbers suffixed by *t* and *f* refer to Tables and Figures respectively. *vs.* indicates a comparison.

This index is in letter-by -letter order, whereby hyphens and spaces within index headings are ignored in the alphabetization. Prefixes and terms in parentheses are excluded from the initial alphabetization.

A

B

C

D

E

F

G

H

I

L

INT/ABP/APP/08/10
$1,215.00

O

P

R

S